"Bagemihl's monumental *Biological Exuberance* embraces paradox and ce brates seemingly incompatible phenomena while forging a compel' argument about the very essence of existence.... It is a landmark in the lit erature of science."
—*Chicago Tribune*

"A brilliant and important exercise in exposing the limitations of received opinion, this book presents to the lay reader and specialist alike an exhaus tively argued case that animals have multiple shades of sexual orientation.... What might so easily have turned into a tub-thumping activist tract hitched to the need for acceptance of homosexuality in humans is instead elevated to a hugely inclusive, celebratory biological interpretation of the world."
—*Publishers Weekly* (starred review)

"*Biological Exuberance* is a welcome antidote to the deluge of zoological research which equates sex with reproduction. It contains a wealth of infor mation and is truly impressive in terms of its scope and its depth. Bagemihl avoids pat explanations in favor of rich analyses that do real justice to the complexities of his subject matter. His treatment of homophobia and hete rocentrism in zoology is both brave and honest. *Biological Exuberance* will surely be considered the definitive source on the subject of nonreproductive sexual behavior in animals for many years to come and could well prove to be a watershed in terms of future research in this area."
—Dr. Paul L. Vasey, zoologist, Concordia University

"Now and then a work comes along and firebombs a set of passionately held convictions.... *Biological Exuberance,* while not exactly a scientific rev olution, is at least fodder for a stunning paradigm shift, this time in the realm of animal sexuality, and ultimately, human sexuality. It does, in fact, challenge our whole notion of the word 'natural.'"
—*Gear* magazine

"In an encyclopedic tour de force, Bruce Bagemihl demonstrates just how natural homosexuality actually is.... As interesting as this catalog is, the task Bagemihl sets out for himself is far more fascinating. In the first half of his book, he takes as close a look at the scientists who have studied animal behavior as he does at the behavior they have studied.... We see how even well-meaning scientists bring culturally determined, preconceived biases to their research.... In the face of such homophobia in the scientific arena and in light of the amount of data he has amassed, it would not have been sur prising if Bagemihl had turned his text into a piece of potent political writ ing, arguing that his data demand acceptance of homosexuality in humans. Instead, he allows the scientific record to speak for itself, and it certainly speaks more powerfully than could any political tract."
—Dr. Michael Zimmerman, biologist, author of *Science, Nonscience, and Nonsense*

"Bagemihl's work is tinged with comedy as he describes how biologists and zoologists have for years stifled or skirted the fact that animals under their observation are up to all sorts of naughtiness...his book is more than a polemic of sexual politics or a queering of zoology.... Instead Bagemihl is more or less taking the recent revolution in attitudes to human sexuality into the 'natural' world."
—*The Times Higher Education Supplement* (London)

"There are certain books that seem, as soon as they have appeared, to have been nothing less than predestined. *Biological Exuberance* is just such a work. It's a masterpiece of unconscious humor...alternately indignant about the way the truth about animal homosexuality has been suppressed in the past and rhapsodic about its riches around the world...[With] page after page of irresistible entertainment...this book has a lot to add to the gaiety of nations."
—*Evening Standard* (London)

"Biologist Bruce Bagemihl has 'outed' nature and brought it into a post-Darwinian, gay-friendly world.... General readers will find the text elegantly written, convincing, and extremely engaging. The photographs and illustrations documenting sexual diversity also make it an unusual coffee table book."
—*Washington Blade*

"Much, much more fun than reading E. O. Wilson. A thoroughly researched refutation of the attitudes of sociobiology...It pops the Victorianism out of Darwinism and muscles the uptight, mechanical model of evolution toward a more accurate understanding of Nature's version as sloppy and exuberant.... I haven't read a more stimulating book in biology in a decade."
—Dr. Peter Warshall, biologist and editor, *Whole Earth* magazine

"The topic is hot stuff, and it could have easily generated some superficial and even sensationalistic pop biology. Instead, Bruce Bagemihl has produced a lengthy, scholarly, and multifaceted work.... In an era when increasing numbers of people grow up in urban environments with little contact with nature, it is important that biologists transmit to each new generation the capacity to marvel at the extravagance of nature. The author's contribution is to show that animal sexual behavior may well be another manifestation of this same richness and therefore equally deserving of appreciation and wonder."
—Dr. Elizabeth Adkins-Regan, zoologist, Cornell University

"Bagemihl [has] produced one phenomenal book on gay, lesbian, bisexual, and transgendered animal life.... Fascinating, page-turning in its own way, and full of pictures of homosexual matings and sexual congress among our furry and feathered friends, *Biological Exuberance* is one of the most readable scientifically based books of the year. Get this one. It is amazing."
—barnesandnoble.com (official review)

"In this astounding book, Bruce Bagemihl shows that homosexuality is little short of ubiquitous in nature.... Bagemihl draws on, and persuasively interprets, a vast amount of data, going back many decades...[and] is eloquent about the wrongheadedness of the dominance argument.... It's a small criticism of *Biological Exuberance* to say that it comprises two or three books—only one of them a world-changing piece of work."
—*The Observer* (London), Literary Editor's selection

"Bagemihl amasses a wealth of information on these topics [homosexuality and nonreproductive heterosexuality], which he manages to present in an accessible yet scholarly manner.... I recommend this book to anyone interested in animal sexual behavior."
—*Animal Behaviour*

"What humans share with so many other animals, it now appears, is free-wheeling homosexuality.... According to Bagemihl, the animal kingdom is a more sexually complex place than most people know.... Bagemihl's ideas have caused a stir in the higher, human community.... For a love that long dared not speak its name, animal homosexuality is astonishingly common."

—*Time*

"In his new book *Biological Exuberance,* author Bruce Bagemihl portrays an animal kingdom that embraces a whole spectrum of sexual orientation. From female grizzly bears coparenting their cubs to trysting male lions fending off other envious males, the book paints a complex mosaic that resembles humanity....*Biological Exuberance* brings the dusty facts to light as Bagemihl deconstructs the all-heterosexual Noah's Ark we've been sold."

—*The Advocate*

"Bruce Bagemihl...throws you straight into the gay underworld of animals...He's a driven man, this author, and it soon becomes clear that...he is taking revenge on centuries of homophobia among scientists, who have chosen to marginalize or ignore homosexual behavior among animals. And as I read the book, I felt myself being caught up in his sense of injustice.... This book does leave you feeling enriched, as well as slightly shocked at the true raunchiness of the animal kingdom."

—*Mail on Sunday* (London)

"It seems to me that this book should have been unnecessary. But there has been such a lengthy and deafening silence on this enormous subject, the sexual behavior of animals, that the first revolutionary survey of the scientific literature had to be over 700 pages long! And it is a splendid job, massive enough to fill the gap of the centuries, comprehensive enough to address every question which comes to mind after the initial awakening triggered by the front cover alone."

—Dr. Ralph Abraham, chaos scientist

"For anyone who has ever doubted the 'naturalness' of homosexual, bisexual, and transgendered behaviors, this remarkable book—which demonstrates and celebrates the sexual diversity of life on earth—will surely lay those doubts to rest. The massive evidence of the wondrous complexity of sexuality in the natural world that Bagemihl has marshaled will inform, entertain, and persuade academic and lay readers alike. *Biological Exuberance* is a revolutionary work."

—Dr. Lillian Faderman, historian, author of
Odd Girls and Twilight Lovers

"A scholarly, exhaustive, and utterly convincing refutation of the notion that human homosexuality is an aberration in nature.... Bagemihl does realize that some among us will never be convinced that homosexuality occurs freely and frequently in nature. But his meticulously gathered, cogently delivered evidence will quash any arguments to the contrary."

—*Kirkus Reviews*

Biological Exuberance

Animal Homosexuality and Natural Diversity

BRUCE BAGEMIHL, Ph.D.

Illustrated by John Megahan

ST. MARTIN'S PRESS · New York

DESIGN BY HELENE WALD BERINSKY

Library of Congress Cataloging-in-Publication Data

Bagemihl, Bruce.
    Biological exuberance : animal homosexuality and natural diversity /
by Bruce Bagemihl ; illustrated by John Megahan.
        p.   cm.
    Includes bibliographical references and index.
    ISBN 0-312-19239-8 (hc)
    ISBN 0-312-25377-X (pbk)
    1. Homosexuality in animals.   I. Title.
QL761.B24   1999
591.56'2—dc21
                                                            98-28528
                                                            CIP
                                                            AC

10 9 8 7 6 5 4 3

Snow

The room was suddenly rich and the great bay-window was
Spawning snow and pink roses against it
Soundlessly collateral and incompatible:
World is suddener than we fancy it.

World is crazier and more of it than we think,
Incorrigibly plural. I peel and portion
A tangerine and spit the pips and feel
The drunkenness of things being various.

And the fire flames with a bubbling sound for world
Is more spiteful and gay than one supposes—
On the tongue on the eyes on the ears in the palms of one's hands—
There is more than glass between the snow and the huge roses.

—LOUIS MacNEICE

. . . hugest whole creation may be less
incalculable than a single kiss

—E. E. CUMMINGS

Contents

Acknowledgments

This project has been a labor of love, one that would not have been possible without the participation and contributions of numerous individuals and organizations. I owe a tremendous debt of gratitude to the many people who helped bring this book to life.

I am especially grateful to all the zoologists, wildlife biologists, natural history photographers, and zoo biologists who generously provided (often previously unpublished) information, data, and/or original photographs in response to my inquiries regarding various species (any errors in fact or interpretation, however, remain solely my responsibility): Arthur A. Allen/David G. Allen (Bird Photographs, New York)—Canada Geese; John J. Craighead/John W. Craighead (Craighead Wildlife-Wildlands Institute, Montana)—Grizzly Bears; James D. Darling (West Coast Whale Research Foundation, Canada)—Gray Whales; Bruno J. Ens (Institute of Forestry and Nature Research [IBN-DLO], Netherlands)—Oystercatchers; Ron Entius (Artis Zoo, Netherlands)—Flamingos; J. Bristol Foster (Sierra Club of British Columbia)—Giraffes; Clifford B. Frith (Frith & Frith Books, Australia)—Birds of Paradise; Masahiro Fujioka (Applied Ornithology Laboratory, National Agricultural Research Center, Japan)—Egrets; Michio Fukuda (Tokyo Sea Life Park)—Great Cormorants; Valerius Geist (University of Calgary)—Bighorn Sheep; Jeremy Hatch (University of Massachusetts)—Roseate Terns; Dik Heg (University of Groningen, Netherlands)—Oystercatchers; Denise L. Herzing (Wild Dolphin Project/Florida Atlantic University)—Bottlenose/Atlantic Spotted Dolphins; Katherine A. Houpt (New York State College of Veterinary Medicine, Cornell University)—Przewalski's Horses; George L. Hunt Jr. (University of California–Irvine)—Western Gulls; Alan R. Johnson (Station Biologique de la Tour du Valat, France)—Flamingos; Catherine E. King (Rotterdam Zoo, Netherlands)—Flamingos; Tamaki Kitagawa (Ichikawa High School, Japan)—Black-winged Stilts; Walter D. Koenig (Hastings Natural History Reservation/UC–Berkeley)—Acorn Woodpeckers; Adriaan Kortlandt (United Kingdom)—Great Cormorants; James N. Layne (Archbold Biological Station, Florida)—Botos; Michael P. Lombardo (Grand Valley State University, Michigan)—Tree Swallows; Dale F. Lott (University of California–Davis)—American Bison; Stephen G. Maka (Wildlife/ Environmental Photography, Massachusetts)—Giraffes; Michael Martys (Alpenzoo Innsbruck)—Greylag Geese; Donald B. Miles (Ohio University/University of Washington)—Whiptail Lizard identification; Gus Mills (Hyena Specialist Group, Kruger National Park, South Africa)—Spotted Hyenas; Daniel K. Niven (Smithsonian Environmental Research Unit/Illinois Natural History Survey)—Hooded Warblers; Jenny Norman (Macquarie University, Australia)—Eastern Gray Kangaroos; Yoshiaki Obara (Tokyo University of Agriculture and Technology)—Cabbage White Butterflies, UV perception; David Powell (University of Maryland)— Flamingos; Mitch Reardon (Photo Researchers, New York/Okapia Bild-Archiv, Germany)—African Elephants; Juan C. Reboreda (Universidad de Buenos Aires)— Greater Rheas; Caitlin Reed (University of North Carolina/Cambridge University)—Crested Black Macaques; H. D. Rijksen (Institute of Forestry and Nature Research [IBN-DLO]/Golden Ark Foundation, Netherlands)—Orang-utans; Leonard Lee Rue III/Len Rue Jr. (Leonard Rue Enterprises, New Jersey)—White-tailed Deer, Bighorn Sheep; Susan Savage-Rumbaugh (Language Research Center, Georgia State University)—Bonobos; Carolien J. Scholten (Emmen Zoo, Netherlands)—Humboldt Penguins; John W. Scott/John P. Scott (Bowling Green State University, Ohio)—Sage Grouse; Paul E. Simonds (University of Oregon)—Bonnet Macaques; L. H. Smith (Australia)—Superb Lyrebirds; Judie Steenberg (Woodland Park Zoological Gardens, Washington)—Tree Kangaroos; Elizabeth Stevens (Zoo Atlanta/Disney World Animal Programs)—Flamingos; Yukimaru Sugiyama (Primate Research Institute, Kyoto University)—Bonnet Macaques; Angelika Tipler-Schlager (Austria)—Greylag Geese; Pepper W. Trail (Oregon)—Guianan Cock-of-the-Rock; Paul L. Vasey (Université de Montréal)—Japanese Macaques, other species; Frans B. M. de Waal (Emory University/Yerkes Regional Primate Research Center, Georgia)—Bonobos; Juichi Yamagiwa (Kyoto University)—Gorillas. Thanks also to the photo archives of Yellowstone National Park and the American Museum of Natural History for supplying images, and to the many publishers, journal editors, and scientists for permission to reprint previously published photographs (see photo credits on pp. 733–735).

I am profoundly grateful to Michael Denneny, not only for his editorial acumen and invaluable insights, but also for championing this project with unflagging enthusiasm and personal devotion. This book would simply not have come into being without his guiding hand at its helm. I would also like to acknowledge the many other people at St. Martin's Press who worked on this project, including: Robert Cloud, Helene Berinsky, Steven Boldt, and Sarah Rutigliano. A heartfelt thanks as well to Natasha Kern and Oriana Green, who believed in this book from the very, very beginning and helped steer it through the (sometimes treacherous) waters of the publishing industry. I am also grateful to Robert Jones and Eric Steel for their early support of this project.

A special note of appreciation goes to John Megahan (Museum of Zoology, University of Michigan), whose superb drawings turned my vision of this book into a reality. John brought nearly two hundred animals to life with accuracy, aplomb, and an exuberant visual style, all the while weathering the numerous pressures of this project with grace and good cheer. John would like to thank his wife, Anne, for her invaluable support and feedback during this project. Many thanks also to the other individuals and organizations who contributed illustrations or assisted in the preparation of visual materials: Stuart Kenter and Tom McCarthy (Stuart Kenter Associates)—photo research and permissions; Gary Antonetti and Jon Daugherity (Ortelius Design)—cartography; Phyllis Wood (Phyllis Wood Associates Scientific Illustration)—icon design; Turid Owren—legal services; and the Guild of Natural Science Illustrators.

I would also like to thank the librarians, information specialists, and other staff at the following libraries, who guided me through numerous arcane bibliographic and electronic searches and helped me navigate their collections: University of Washington libraries, University of British Columbia libraries, Simon Fraser University Library, University of California–Los Angeles libraries (including the Rare Books collection of the biomedical library), University of Hawaii libraries, National Marine Mammal Laboratory Library, Woodland Park Zoological Gardens Library, Vancouver Public Aquarium Library, Seattle Public Library, Vancouver Public Library. Many rare journals, books, monographs, dissertations, and technical reports were not available at these institutions and were obtained primarily through the interlibrary loan division of the Seattle Public Library. Thanks to the hardworking staff in this division, and to the following institutions and organizations for loaning or otherwise making available from their library collections rare or hard-to-find items: California Academy of Sciences, Evergreen State College, Humboldt State University, Idaho State University, Montana State University, Oregon Regional Primate Research Center, Portland State University, The Royal Australasian Ornithologists' Union, University of Alaska, University of Alberta, University of California–Davis, University of California–Irvine, University of California–Santa Cruz, University of Kansas, University of Minnesota, University of Montana, University of Oregon, University of Texas–Austin, University of Utah, University of Victoria, University of Wisconsin–Milwaukee, University of Wyoming, U.S. Fish and Wildlife Service Northern Prairie Wildlife Research Center, Vancouver Public Library, Washington State University. For help with translations of scientific articles, thanks to Sergei V. Mihailov (Russian), John R. Van Son (Dutch), and Courtney Searles-Ridge (German).

Finally, I am truly indebted to the many other people who provided personal support, encouragement, feedback, commiseration, and inspiration over the many years that this book was in gestation (and labor!), including Dawn Bates, Nicola Bessell, Thom Feild, Neal Graves, Ed Kaplan, Clara Ma, Nathan Ohren, Jackal Plumb, Michael Rochemont, and Liza White. Thank you most deeply, Nicola, for holding my hand as I leapt into the void. . . .

Introduction

The most beautiful thing we can experience is the mysterious. It is the source of all true art and science. He to whom this emotion is a stranger, who can no longer pause to wonder and stand rapt in awe, is as good as dead: his eyes are closed.
—ALBERT EINSTEIN[1]

*A*ny book on homosexuality and transgender in animals is necessarily unfinished, a work in progress. The subject is so vast, the types of behaviors so varied, and the number of species involved so large, as to defy any attempt at comprehensiveness. And the scientific research in this area is only in its infancy: new developments and discoveries are continually being made, and the extent of uncharted and as yet unknowable terrain is so great as to render any attempt at completeness hopelessly premature.

Notwithstanding such formidable challenges, this book endeavors to present a reasonably extensive and up-to-date account of the subject. To help narrow the field, certain parameters have been chosen: only examples of homosexual behavior or transgender that have been scientifically documented, for example, are covered in this book (such documentation includes published reports in scientific journals and monographs, and/or firsthand observations by zoologists, wildlife biologists, and other trained animal observers, corroborated by multiple sources whenever possible). Not only does this limit the number of species to be included (many more cases undoubtedly occur but have not been so documented), it establishes a uniform and verifiable platform of data on which to base further discussion. In addition, the book focuses primarily on mammals and birds—not because other types of animals are somehow less interesting or "important," but simply because space and time limitations necessitate that not all species can be covered. These two groups are considered to be sufficiently representative and to have a broad enough appeal to warrant their inclusion, however arbitrary the exclusion of others may be.

Even with these parameters in place, however, an enormous amount of ground must still be covered. In addition to discussing an extensive array of species (nearly 300 mammals and birds), the book draws upon more than two centuries of scientific research. Some of the findings reported here in a few sentences represent literally lifetimes of work on the part of biologists, who often devote their entire careers to studying one very specific and complex aspect of one type of behavior, in one particular population of one particular species. With this in mind, the book should be seen not as a final, definitive pronouncement on the subject, but rather as a beginning or overture, an invitation to further research and discussion.

Any account of homosexuality and transgender in animals is also necessarily an account of human interpretations of these phenomena. Because animals cannot speak directly for themselves the way people can, we must rely on human observations of their behavior. This presents both special challenges and unique advantages to the study of the subject. On the one hand, certain behaviors such as sexual acts can be observed directly (and even quantified), which is often extremely difficult, impossible, or unethical to do in studies of sexuality among people (especially stigmatized or alternative forms of sexuality). On the other hand, we are in the dark about the internal experiences of the animal participants: as a result, the biases and limitations of the human observer—in both the gathering and interpretation of data—come to the forefront in this situation. In many ways this is the reverse of what occurs in some studies of homosexuality among people (including well-informed historical or anthropological studies of different cultures or time periods). With people, we can often speak directly to individuals (or read written accounts) about what their sexuality and associated phenomena mean—and so get a sense of their emotional and motivational states—without necessarily being able to verify their actual sexual behaviors. With animals, in contrast, we can often directly observe their sexual (and allied) behaviors, but can only infer or interpret their meanings and motivations. As a result, many contentious assertions, theories, interpretations, and explanations have been put forward (and continue to be made) within the field of zoology about the function(s) and meaning(s) of homosexuality and transgender. This book seeks to address this historical and very human dimension of the subject, while still maintaining a focus on the animals, their behaviors and lives.

The unique historical moment we find ourselves in also necessitates the book being geared as much as possible toward specialist and nonspecialist alike, and informs the organization and two-part structure of the book. Because of the current inaccessibility of a large body of scientific information, a primary aim is to present the technical material to a general (nonacademic) readership, without sacrificing accuracy or sensationalizing what is often a controversial and difficult subject matter. However, because no comprehensive survey (and synthesis) of this material is yet available *within* the scientific literature—indeed, many zoologists are themselves unaware of much of this material—and because a considerable amount of misinformation and misunderstanding surrounds the subject even among trained biologists, the volume will also be of interest to the scientific community. Consequently, every effort has been made to provide full documentation in the form of notes and references, and to include relatively exhaustive and detailed coverage of a wide range of species. However, this more technical material is positioned in such a way that it can easily be skipped by readers who do not wish to delve into such matters.

In a book such as this which is intended for both an academic and a nonacademic readership, the question of terminology poses special challenges. I have attempted to steer a course between more accessible but overly anthropomorphic or loaded vernacular, on the one hand, and more "neutral" but highly technical jargon or awkward circumlocutions, on the other. In particular, *homosexual(ity)* and *same-sex* are utilized as the labels of choice. Since the words *gay* and *lesbian* are burdened with human connotations (cultural, psychological, historical, and/or political) and may not be regarded as appropriate designations for animals, I have been careful to avoid using these terms throughout most of the book (as pointed out in chapter 1). When referring specifically to animals and their behaviors, for example, *gay* is never employed, while *lesbian* is used only sparingly (it occurs in less than 3 percent of the more than 3,000 instances in the text where animal homosexuality is named). Even then, *lesbian* is usually reserved only for cases of linguistic expedience, when alternate phrasings such as "female homosexual(ity)" or "same-sex . . . among/between females" would become repetitive, cumbersome, or otherwise infelicitous.

Nevertheless, it is important to recognize that a precedent has been established within the zoological discourse for using the less "neutral" (or more culture-bound) designations. The words *gay* and *lesbian* are applied by scientists to animals and their behaviors in a number of scholarly publications spanning the past quarter century, including three separate instantiations in the prestigious journal *Nature*. As in *Biological Exuberance*, *lesbian* is more widely used than *gay*, e.g., "lesbian females" in Fruit Flies (Cook 1975), a "lesbian pair" in Black-billed Magpies (Baeyens 1979), a Common Chimpanzee behaving "in a lesbian fashion" (de Waal 1982), "gay" Snow Geese (Diamond 1989), "gay courtship" in Long-legged Flies (Dyte 1989), "lesbian behavior" in Bonobos (Kano 1992), "lesbian pairs" in Black Stilts (Reed 1993), "lesbian females" in Lesser Flamingos (Alraun and Hewston 1997), "lesbian copulations" in Oystercatchers (Heg and van Treuren 1998); see chapter 3. As for other terms such as *transvestism* and *transsexual(ity)*, these are also used in the zoological literature with meanings largely divorced from their human connotations (though other labels are employed as well, such as *male/female mimicry* or *sequential hermaphroditism*).

It should also be pointed out that the term *homosexual*—which many people feel is preferable to *gay* or *lesbian* when referring to animals—is not devoid of anthropomorphism. It too is a culture-specific, historical construction with very particular human connotations (the same for other putatively "neutral" designations such as *hermaphroditism, mimicry*, etc.). In fact, a wide variety of terms used routinely in the zoological literature—e.g., *courtship, parent(ing), monogamy, adoption, consort(ship)* or, for that matter, *heterosexual, male*, etc.—carry the same baggage of human referents. In addition, the range of variation between (and within) animal species in behaviors that are labeled with the same terms is sometimes as great as—if not greater than—the variation in corresponding behaviors between animals and people. In other words, the differences between "mothers" (or "homosexual copulations") in flies and chimpanzees probably equal if not exceed the differences between "mothers" (or "homosexual copulations") in chimpanzees and humans. Yet such terms are applied to a wide range of animals with the understanding that a given word can have variant meanings in different contexts, and that the human connotations are specifically not implied when such vocabulary is used in a zoological context. This issue is discussed more fully in chapter 3, where I offer a careful rationale for the continued use of such terms—specifically with reference to the supposedly anthropomorphic/centric label *homosexual* and the historical reluctance of zoologists to utilize even this designation.

Furthermore, within this book such terminology is not used in a vacuum: it is accompanied by explicit discussion of the meanings of all such labels when applied to animals—including overt disavowal of their human connotations and extensive consideration of the inappropriateness of making unwarranted human-animal comparisons (see chapter 2). In order to contextualize the discussion, I also address a number of related issues such as the precedent for employing these words within the zoological literature; the problems inherent in any choice of terminology; and the widespread use within scientific discourse of anthropomorphic labels and descriptions for *heterosexual* animals and behaviors. Finally and perhaps most importantly, I point out in *Biological Exuberance* that terminological debates themselves are not ahistorical—they reflect and embody very specific cultural and historical streams both within the scientific community and in society at large; they recapitulate (and lag behind) debates regarding "appropriate" terminology for homosexuality in humans; and the effect of such debates within the scientific discourse has often been to distract from the phenomena designated by such terms rather than to clarify them.

Virtually no terminology for animal behavior—particularly sexual behavior—is entirely free of human (cultural, historical, etc.) associations. When confronted with this situation, we have two options: construct an alternative vocabulary of relatively opaque labels and unwieldy circumlocutions that attempts to avoid such bias (but inevitably falls short of this ideal); or use the already available terms with careful qualification of their meanings and an understanding of their historical context, such that they become uncoupled from their anthropomorphic connotations. In *Biological Exuberance,* I opt for the latter.

The book is organized into two complementary sections. Part 1, *A Polysexual, Polygendered World,* offers a wide-ranging exploration of all aspects of animal homosexuality and transgender: their diversity, history, and meanings. Part 2, *A Wondrous Bestiary,* presents a series of profiles of individual homosexual, bisexual, and transgendered animals. Where the first part of the book follows a linear, narrative progression, part 2 is organized in a nonlinear, reference format. The two halves of the book are linked via the animals themselves: throughout part 1, the reader is referred to specific animals that are profiled in part 2 and may at any point consult those profiles to supplement the narrative (names of profiled species or groups of related species are capitalized to indicate their inclusion in part 2 and the appendix). Alternatively, those readers more interested in a general cross-species survey or the interpretive/historical aspect can focus almost exclusively on part 1, while those who wish to gain a more in-depth understanding of particular animals can focus primarily on part 2. This dual structure allows the reader to access information on animal homosexuality/transgender in a variety of ways, suited to his or her own reading style.

Chapter 1, "The Birds and the Bees," presents a broad overview of animal homosexuality and transgender, exploring the full range of behaviors and phenomena covered by these terms. Comparisons between animal and human homosexuality are the focus of chapter 2, "Humanistic Animals, Animalistic Humans," including a discussion of the advisability and implications of making such comparisons in the first place. This chapter also exposes the false dichotomy of the "nature versus nurture" debate, by examining the sociocultural dimensions of homosexuality within animal communities. Next, the history of the scientific study of animal homosexuality is chronicled in chapter 3, "Two Hundred Years of Looking at Homosexual Wildlife." This includes documentation of systematic prejudices within the field of zoology in dealing with this subject, which have often hampered our understanding of the phenomenon. Chapter 4, "Explaining (Away) Animal Homosexuality," continues the historical perspective by examining the many attempts to interpret and determine the "function" or "cause" of animal homosexuality and transgender. Most such efforts to find an "explanation" have failed outright or are fundamentally misguided—particularly when they try to show how homosexuality might contribute to heterosexual reproduction. In the next chapter, "Not for Breeding Only," animal life and sexuality are shown *not* to be organized exclusively around reproduction. A wide range of nonprocreative heterosexual activities are described and exemplified, as are the diverse ways that homosexual, bisexual, heterosexual, and transgendered animals structure their relationship to breeding.

The final chapter of part 1, "A New Paradigm: Biological Exuberance," calls for a radical rethinking of the way we view the natural world. This revisioning begins with an exploration of another, alternative set of human interpretations: traditional beliefs about animal homosexuality/transgender in indigenous cultures. Particular attention is paid to the ways in which these ideas are relevant to contemporary scientific inquiry. As it turns out, Western science has a lot to learn from aboriginal cultures about systems of gender and sexuality. In the remainder of the chapter, a synthesis of a number of "new" sciences is suggested, including chaos theory, post-Darwinian evolutionary theorizing, biodiversity studies, and the theory of General Economy. The approach taken throughout this chapter is exploratory rather than explanatory. Ultimately, this synthesis leads to a worldview in which animal homosexuality and other nonreproductive behaviors suddenly "make sense," while still remaining, paradoxically, "inexplicable"—a worldview that is also remarkably consistent with indigenous perspectives on gender and sexuality.

In the second half of the book, *A Wondrous Bestiary,* the reader is treated to a series of individual profiles of homosexual, bisexual, and transgendered animals, from Antbirds to Zebras. Each profile is a verbal and visual "snapshot" of one (or several closely related) species, allowing the reader to "meet" the animal and "get to know" it in detail. Part 2 is divided into two major sections, one for mammals and one for birds, each of which in turn is organized around the formal subgroupings of animals in that category. The section on mammals, for example, includes separate groupings for primates, marine mammals, hoofed mammals, and so on. Each profile within these groupings contains a wealth of information—everything from detailed descriptions of courtship displays to statistics on frequency of homosexual behaviors, to background information on the animal's social organization.

Although its focus is primarily on animal homosexuality and transgender, the book actually moves far beyond these subjects to consider much broader patterns in nature and human society. Sexual and gender variance in animals offer a key to a new way of looking at the world, symbolic of the larger paradigm shifts currently underway in a number of natural and social sciences. The discussion is rooted in the basic facts about animal homosexuality and nonreproductive heterosexuality, information that is presented most fully in the individual animal profiles. Using these to expose the hidden assumptions behind the way biology looks at natural systems, a fresh perspective is developed, based on the melding of contemporary scientific insights with traditional knowledge from indigenous cultures. Taking a broad interdisciplinary perspective, the narrative builds upon a solid foundation of scientific and cultural research to arrive at some conclusions that have the potential to fundamentally alter the way we think about the world and our position in it. *Biological Exuberance* is, ultimately, a meditation on the nature of life itself, and a celebration of its paradoxes and pluralities.

As such, the book seeks not only to convey "the facts" about animal behavior but, perhaps as importantly, to capture something of their "poetry" as well. The beauty and mystery of nature can be found in many forms. And one particular form of natural beauty is the diversity of sexuality and gender expression throughout the animal world. In addition to being interesting from a purely scientific standpoint, these phenomena are also capable of inspiring our deepest feelings of wonder, and our most profound sense of awe.

Part I

A Polysexual, Polygendered World

Chapter **1**

The Birds and the Bees

The universe is not only queerer than we suppose, it is queerer
than we can suppose.

—evolutionary biologist J. B. S. HALDANE[1]

*I*n the dimly lit undergrowth of a Central American rain
forest, jewel-like male hummingbirds flit through the
vegetation, pausing briefly to mate now with a male, now with a female. A whale
glides through the dark and icy waters of the Arctic, then surges toward the surface
in a playful frenzy of churning water and splashing, her fins and tail caressing another female. Drifting off to sleep, two male monkeys lie gently in each other's
arms, cradled by one of the ancient jungles of Asia. A herd of deer picks its way cautiously through a semidesert scrub of Texas, each animal simultaneously male but
not-quite-male, with half-developed, velvety antlers and diminutive, fine-boned
proportions. In a protected New Zealand inlet, a pair of female gulls—mated for
life—tend their chicks together. Tiny midges swarm above a bleak tundra of northern Europe, a whirlwind of mating activity as males couple with each other in
midair. Circling and prancing around her partner, a female antelope courts another
female in an ageless, elegant ritual staged on the African savanna.

Although biologist J. B. S. Haldane was not (necessarily) referring to homosexuality when he spoke of the "queerness" of the natural world, little did he know how
accurate his statement would turn out to be. The world is, indeed, teeming with homosexual, bisexual, and transgendered creatures of every stripe and feather. From
the Southeastern Blueberry Bee of the United States to more than 130 different bird
species worldwide, the "birds and the bees," literally, are queer.[2]

On every continent, animals of the same sex seek each other out and have probably been doing so for millions of years.[3] They court each other, using intricate and
beautiful mating dances that are the result of eons of evolution. Males caress and
kiss each other, showing tenderness and affection toward one another rather than
just hostility and aggression. Females form long-lasting pair-bonds—or maybe just

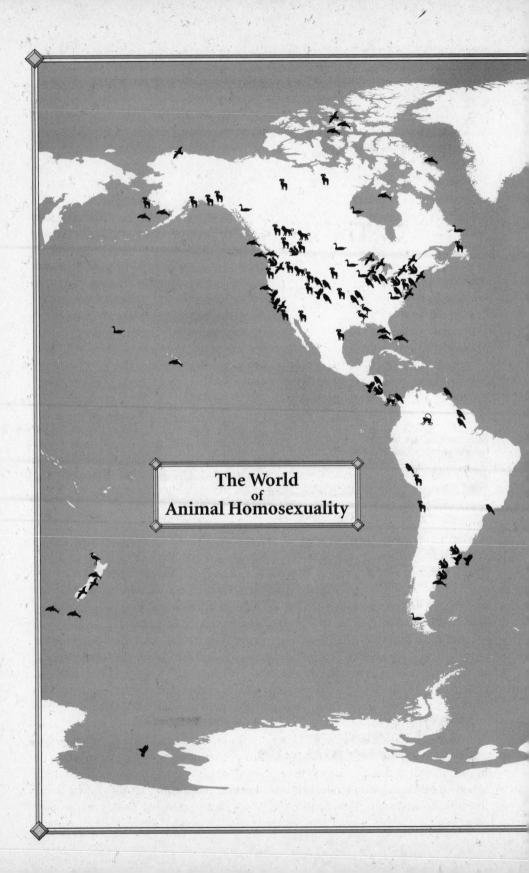

The World
of
Animal Homosexuality

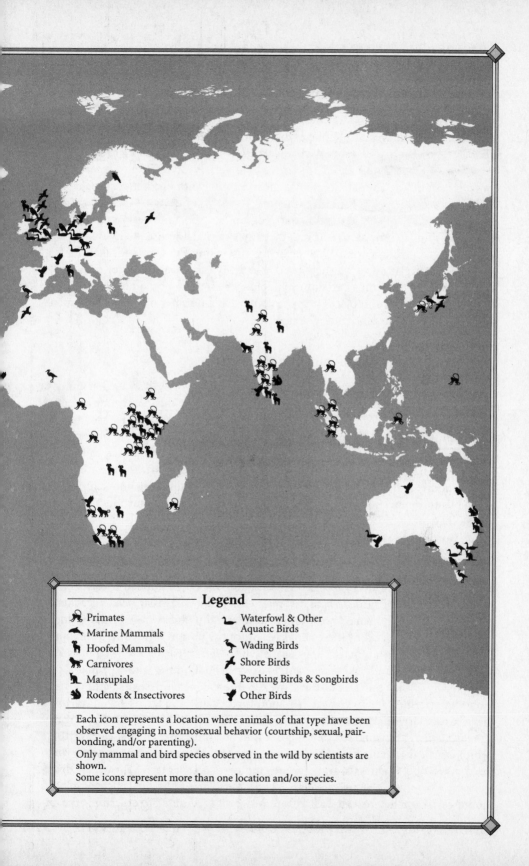

Legend

- Primates
- Marine Mammals
- Hoofed Mammals
- Carnivores
- Marsupials
- Rodents & Insectivores
- Waterfowl & Other Aquatic Birds
- Wading Birds
- Shore Birds
- Perching Birds & Songbirds
- Other Birds

Each icon represents a location where animals of that type have been observed engaging in homosexual behavior (courtship, sexual, pair-bonding, and/or parenting).

Only mammal and bird species observed in the wild by scientists are shown.

Some icons represent more than one location and/or species.

meet briefly for sex, rolling in passionate embraces or mounting one another. Animals of the same sex build nests and homes together, and many homosexual pairs raise young without members of the opposite sex. Other animals regularly have partners of both sexes, and some even live in communal groups where sexual activity is common among all members, male and female. Many creatures are "transgendered," crossing or combining characteristics of both males and females in their appearance or behavior. Amid this incredible variety of different patterns, one thing is certain: the animal kingdom is most definitely not just heterosexual.

Homosexual behavior occurs in more than 450 different kinds of animals worldwide, and is found in every major geographic region and every major animal group.[4] It should come as no surprise, then, that animal homosexuality is not a single, uniform phenomenon. Whether one is discussing the forms it takes, its frequency, or its relationship to heterosexual activity, same-sex behavior in animals exhibits every conceivable variation. This chapter presents a broad overview of animal homosexuality and places it in the context of a number of other phenomena involving alternative genders and sexualities.

The Many Forms of Animal Homosexuality

For most people, "homosexuality" means one thing: sex. While it's true that animals of the same gender often interact sexually with each other, this is only one aspect of same-sex expression. Animal homosexuality represents a vast and diverse range of activities: it is neither a monolithic nor an exclusively sexual phenomenon. This section offers a survey of the full range of homosexual activity found in the animal world, organized around five major behavioral categories: courtship, affection, sex, pair-bonding, and parenting. While these categories are not mutually exclusive and often blend imperceptibly into one another, they offer a useful introduction to the multiplicity of homosexual expression in the animal kingdom.

A word on terminology is in order. In this book, *heterosexuality* is defined as courtship, affectionate, sexual, pair-bonding, and/or parenting behaviors between animals of the opposite sex, while *homosexuality* is defined as these same activities when they occur between animals of the same sex. When applied to people, the terms *homosexual, gay,* or *lesbian* can refer either to a particular behavior when it occurs between two men or two women, or to an individual whose primary "identity" involves any or all of these activities. Since the notion of identity is inappropriate to ascribe to animals, these terms will be reserved for the behaviors that animals engage in and, where relevant, to describe individuals whose primary "orientation" is toward animals of the same sex where courtship, sexual, and/or pair-bonding activities are concerned. In addition, because the terms *gay* and *lesbian* have particularly human connotations, these will generally be avoided in favor of *homosexual(ity)* or *same-sex* (although it must be remembered that each of these words can have specific meanings independent of their human connotations when used in relation to animals, and they are employed as cover terms for widely divergent activities even among humans). When a particular individual engages in both homosexual and heterosexual activity, these words are limited to describing the animal's specific behaviors (depending on the gender of the animal's partner), while the animal itself is described as *bisexual.*[5]

Pirouette Dances, Ecstatic Displays, and Triumph Ceremonies: Courtship Patterns

To attract the attention and interest of a potential partner, animals often perform a series of stylized movements and behaviors prior to mating, sometimes in the form of a complex visual or vocal display. This is known as *courtship behavior,* and it usually indicates that one animal is advertising his or her presence to prospective mates or is sexually interested in another individual. If the interest is mutual, this may lead to mating or other sexual activity and possibly pair-bonding (for example, in birds). Some animals also use special courtship behaviors to conclude, as well as initiate, sexual activity, or to reinforce their pair-bonds. Courtship behavior is a common feature of homosexual interactions, occurring in nearly 40 percent of the mammals and birds in which same-sex activity has been observed.

Same-sex courtship assumes a dizzying array of forms, and zoologists often use evocative or colorful names as the technical terms to designate these most striking of animal behaviors (which are usually part of heterosexual interactions as well). Many species perform elaborate dances or kinetic displays, such as the "strutting" of female Sage Grouse, who spread their fanlike tails; or the spectacular acrobatics and plumage displays of Birds of Paradise and Superb Lyrebirds; or the courtship encounters of Cavies, who "rumba," "rumble," "rump," and "rear" each other in an alliterative panoply of choreographed behaviors. In other cases, subtler poses, stylized postures, or movements are used, such as the foreleg kicking found in the courtship displays of many hoofed mammals; "rear-end flirtation" in male Nilgiri Langurs and Crested Black Macaques; ritual preening and bowing during courtship

A male Superb Lyrebird (foreground) *courting a younger male in the forests of Australia. He is singing and performing the dramatic "full-face display," in which he fans his elaborate tail plumes forward over his head.* ▼

interactions in Penguins; "tilting" and "begging" postures in Black-billed Magpies; "jerking" by female Koalas; and "courtship feeding"—a ritual exchange of food gifts seen in same-sex (and opposite-sex) interactions among Antbirds, Black-headed and Laughing Gulls, Pukeko, and Eastern Bluebirds. Sometimes two courting individuals perform mutual or synchronized displays, such as the "triumph ceremonies" of male Greylag Geese and Black Swans; the "mutual ecstatic" and "dabbling" displays of Humboldt and King Penguins, respectively; synchronous aquatic spiraling in male Harbor Seals and Orcas; the elaborate "leapfrogging" and "Catherine wheel" courtship displays by groups of Manakins; and synchronized wing-stretching and head-bobbing in homosexual pairs of Galahs. Many birds have breathtaking aerial displays, including tandem flying in Griffon Vultures, shuttle displays and "dive-bombing" in Anna's Hummingbirds, "hover-flying" in Black-billed Magpies, "song-dancing" in Greenshanks, and the "bumblebee flight" of Red Bishop Birds.

Animals sometimes exploit specific spatial and environmental elements in their courtship activities as well. Special display courts are used in same-sex (and opposite-sex) interactions in many species, including the "drumming logs" of male Ruffed Grouse, the elaborate architectural creations of Regent Bowerbirds, and the traditional group or communal display areas known as leks found in animals as diverse as Kob antelopes, Long-tailed Hermit Hummingbirds, and Ruffs. In other species, dramatic chases that may cover great distances are part of same-sex interactions: aerial pursuits occur in Greenshanks, Golden Plovers, Bank Swallows, and Chaffinches; ground chases take place during courtships in Mule Deer, Cheetahs, Whiptail Wallabies, and Redshanks; aquatic pursuits occur in Australian Shelducks; while Black-billed Magpies combine both ground *and* aerial pursuit in their courtship behavior known as chase-hopping. Perhaps most amazing of all are the light-related displays of a number of bird species, which are designed to utilize specific properties of sunlight or other luminosity in the bird's environment. Guianan Cock-of-the-Rock, for example, position their leks and courtship displays in special "light environments" that maximize the visibility of the birds through a sophisticated interaction of the ambient light, the reflectance and coloration of the bird's own (brilliant orange) plumage, and the forest geometry in which they are located. Anna's Hummingbirds precisely orient the trajectory of their stunning aerial climbs and dives to face into the sun, thereby showing off their iridescent plumage to its best. As a male swoops toward the object of his attentions (either male or female), he resembles a brilliant glowing ember that grows in intensity as he gets closer. To advertise their presence on the lek, male Buff-breasted Sandpipers perform a wing-raising display that exploits the midnight sun of their arctic habitat. Seen from a distance, the brilliant white underwings of each bird flash momentarily against the dull tundra background, reflecting the weak late-night sunlight and thereby creating a luminous semaphore that attracts other birds, both male and female, to their territories.[6]

In addition to spectacular visual displays, homosexual courtship—like the corresponding heterosexual behaviors—can involve a veritable cacophony of different sounds. Female Kob antelopes whistle, male Gorillas pant, female Rufous Rat Kangaroos growl, male Blackbuck antelopes bark, female Koalas bellow, male Ocellated Antbirds carol, female Squirrel Monkeys purr, and male Lions moan and hum. The "snap-hiss" ceremonial calls of Black-crowned Night Herons, the croaking of male Moose, "geckering" and "snirking" of female Red Foxes, the chirp-squeaks of male West Indian Manatees, "yip-purr" calls of Hammerheads, the yelping and babble-singing of Black-billed Magpies, "lip-smacking" in several Macaque species, the humming call of Pukeko, "stutters" and "chirps" of male Cheetahs, the "vacuum-slurping" of male Caribou, and pulsive scream-calls in Bowhead Whales are just some of the vocalizations heard during same-sex courtship and related interactions. Sometimes pairs of birds execute synchronized vocal displays, as in the duets of rolling calls performed by Greylag gander pairs, or the precisely syncopated "moo" calling of pairs of male Calfbirds. In a few cases, courtship activities involve nonvocal sounds or sounds produced in unusual ways. Male Guianan Cock-of-the-Rock, Ruffed Grouse, Victoria's Riflebirds, and Red Bishop Birds, for example, make distinctive whistling, drumming, or clapping sounds by beating or fanning their wings (which in some cases have specially modified, sound-producing feathers), while male Anna's Hummingbirds produce a shrill popping sound as a result of air passing through their tail feathers during display flights. Some of the most extraordinary sounds during same-sex courtship are made by aquatic animals: Walruses generate eerily metallic "bell" sounds by striking special throat pouches with their flippers and castanet-like "knocks" by chattering their teeth, while Musk Ducks have an entire repertoire of courtship splashing sounds made by kicking their feet during displays variously named the paddle-kick, plonk-kick, and whistle-kick. Finally, some Dolphins appear to engage in a sort of sonic "foreplay": male Atlantic Spotted Dolphins have been observed stimulating their partner's genitals with pulsed sound waves, using a type of vocalization known as a genital buzz.

In most species the same courtship behaviors are used in both homosexual and heterosexual interactions. Sometimes, however, same-sex courtship involves only a subset of the movements and behaviors found in opposite-sex displays. For example, when Canada Geese court each other homosexually, they perform a neck-dipping ritual also found in heterosexual courtships, but do not adopt the special posture that males and females use after mating. In animals like the Western Gull or Kob antelope, individuals vary as to how many courtship behaviors they use in same-sex interactions. Some exhibit only one or two of the typical courtship postures and movements, while others go through the entire elaborate courtship sequence. Perhaps most interesting are those creatures that have a special courtship pattern found only in homosexual interactions. Male Ostriches, for example, perform a unique "pirouette dance" only when courting other males, while female Rhesus Macaques engage in courtship games such as "hide-and-seek" that are unique to lesbian interactions.

Kisses, Wuzzles, and Necking: Affectionate Behaviors

Many animals of the same sex touch each other in ways that are not overtly sexual (they do not involve direct contact of the genitals) but that do nevertheless have clear sexual or erotic overtones. These are referred to as *affectionate* activities and are found in nearly a quarter of the animals in which some form of homosexual activity occurs. Although many of these behaviors (grooming, embracing, play-fighting) can occur in other contexts, their erotic nature in a same-sex context is usually obvious: the two animals may be visibly sexually aroused, the behavior may directly precede or follow homosexual copulation or courtship, or the affectionate activity may occur in a same-sex pair-bond.

One type of affectionate activity is simple grooming or rubbing. Male Lions "head-rub" and roll around with each other before having sex together; Bats such as Gray-headed Flying Foxes and Vampire Bats engage in erotic same-sex grooming and licking; male Mountain Sheep rub their horns and faces on other males, sometimes becoming sexually aroused; Whales and Dolphins stroke and rub each other with their flippers or tail flukes, as well as rub bodies together; while numerous primates such as Apes, Macaques, and Baboons frequently caress and groom each other in both sexual and nonsexual contexts. A few birds such as Humboldt Penguins, Pukeko, Black-billed Magpies, and Parrots also indulge in preening—the avian equivalent of grooming—in their homosexual interactions or pair-bonds.

Some animals also "kiss" each other: male African Elephants, female Rhesus Macaques, male West Indian Manatees and Walruses, female Hoary Marmots, and male Mountain Zebras (among others) all touch mouths, noses, or muzzles during their homosexual encounters. Even some birds, such as Black-billed Magpies, engage in mutual beak-nibbling or "billing" as part of same-sex courtship. In primates, kissing (in both homosexual and heterosexual contexts) can bear a startling resemblance to the corresponding human activity: a number of species such as Squirrel Monkeys and Common Chimpanzees engage in full mouth-to-mouth contact, while male Bonobos kiss each other with "passionate" open-mouthed kisses with considerable mutual tongue stimulation.

Numerous species of Monkeys and Apes also "hug" or embrace same-sex partners in homosexual contexts (usually face-to-face, although male Bonobos and Vervets also embrace while standing in a front-to-back position). Among non-primates, female Bottlenose Dolphins clasp each other during homosexual activity, male West Indian Manatees embrace one another underwater, while Gray-headed Flying

(PHOTO BY FRANS DE WAAL)

◀ *Two younger male Bonobos kissing*

Foxes wrap their wing-membranes around same-sex partners while stimulating each other. A striking form of same-sex embracing is the "sleeping huddle" found in Stumptail and Bonnet Macaques: a pair of males often sleep together in a front-to-back position, one male wrapping his arms around the other and sometimes even holding on to his partner's penis. A similar sleeping arrangement occurs, surprisingly, among male Walruses, who often sleep in same-sex pairs or extended "chains" of males, all clasping each other in a front-to-back position as they float at the water's surface.

A number of mammals also engage in mock battles or "play-fights" that have erotic overtones. Although they superficially resemble aggressive behavior, these "battles" or "contests" do not involve any physical violence and are clearly distinguished from actual cases of aggressive or territorial behavior in these species. Male African Elephants, for example, frequently become sexually aroused and develop erections when they perform ritualized erotic jousting matches, while numerous hoofed mammals such as male Giraffes, Bison, Blackbuck antelopes, and Mule Deer mount each other during play-fights or ritualistic jousting. Among primates such as Orang-utans, Gibbons, and Proboscis Monkeys, males sometimes engage in playful wrestling matches that can develop into sexual encounters, while male Australian and New Zealand Sea Lions also indulge in play-fighting combined with same-sex mounting. Although play-fighting is most common among male mammals, female Cheetahs sometimes engage in "mock fighting" with each other as part of same-sex courtship sequences, while female (and male) Galahs and Orange-fronted Parakeets in same-sex pairs have playful "fencing bouts" with their bills.[7]

Many other types of affectionate and contactual behaviors occur between animals of the same sex. Sometimes animals gently bite, nibble, or chew on each other's ears (female Hoary Marmots), or wings and chests (Gray-headed Flying Foxes), or rumps (male Dwarf Cavies), or necks (male Savanna Baboons). Male African Elephants intertwine their trunks, while female Japanese Macaques sometimes suck each other's nipples, and male Crested Black Macaques and Savanna Baboons affectionately pat or grab other males' rear ends. Pairs of animals may sit, huddle, or lie together in close proximity, sometimes touching hands or putting an arm around the shoulder (female Gorillas, Squirrel Monkeys, and Japanese Macaques, male Siamangs), while male Hanuman Langurs "cuddle" together by sitting back-to-front, one male between the other's legs with his partner's hands resting on his loins. Male Lions and female Long-eared Hedgehogs slide the lengths of their bodies along their partner's, while male Bowhead Whales, Killer Whales, and Gray Seals roll their bodies over each other, and same-sex companions in Gray Whales and Botos swim side by side while gently touching each other with their fins.

Some animals have developed unique forms of touching that combine several different types of affectionate activities along with courtship and sexual behaviors. Male Giraffes engage in "necking", a multifaceted activity that incorporates elements of play-fighting, courtship, and sexuality, in which they rub their necks along each other's body while also licking, sniffing, and becoming sexually aroused by one another. In Giraffes and other species, these types of activities sometimes involve multiple animals interacting simultaneously in near "orgies" of bodily contact. Spinner Dolphins, for example, participate in "wuzzles"—group sessions of mutual caressing and sexual activity (both same-sex and opposite-sex)—while West Indian Manatees have a similar sort of "free-for-all" group activity known as cavorting, which can involve rubbing, chasing, and sexual interactions, among many other activities. Among birds, Hammerheads, Acorn Woodpeckers, and Blue-bellied Rollers have ritualized bouts of courtship and mounting activity that may involve groups of individuals and both same-sex and opposite-sex partners. The distinctive and, in many cases, unabashedly sensual and playful aspects of some of these activities are aptly reflected in the descriptive names given to them by zoologists. In fact, the term *wuzzle*—though used as a technical designation for this behavior in the scientific literature—is actually a nonsense word coined by a marine biologist, whose whimsical "etymology" for the name could be right out of Lewis Carroll: "The term comes from W. E. Schevill of Woods Hole Oceanographic Institution, who, when asked what the behavior was, replied without hesitation, 'Why, it looks like a wuzzle to me.'"[8]

Mounting, Diddling, and Bump-Rumping: Sexual Techniques

Affectionate activity often leads to, or is inseparable from, overtly sexual behavior—defined here as any contact between two or more animals involving genital stimulation. Stumptail and Crab-eating Macaques, for example, kiss their same-sex partners during sexual mounting. In fact, mounting is the most common type of sexual behavior found in homosexual contexts: one animal climbs on top of the other in a position similar to heterosexual intercourse, usually from behind in a front-to-back position (that is, one animal mounted on the back of the other). More than 95 percent of mammal and bird species use this position, for both male and female homosexual interactions. On the other hand, some animals—particularly primates such as Gorillas, Bonobos, and White-handed Gibbons—use a face-to-face position (in addition to, or instead of), and in some cases this is more common in homosexual encounters than in heterosexual ones. Belly-to-belly copulation is also the norm for both homosexual and heterosexual interactions in Dolphins. Occasionally more unusual or "creative" mounting positions are used, particularly by female animals. In Bonobos, Stumptail Macaques, and Japanese Macaques, for instance, females sometimes interact in a supine or semirecumbent position, one individual behind the other with her partner between her legs or sitting "in her lap" (which may also be done in a face-to-face position). Occasionally female Warthogs, Rhesus and Japanese Macaques, Koalas, and Takhi mount their female partner from the side rather than from behind; lateral mounts also sometimes occur during heterosexual interactions in these (and other) species. And in some animals a "backward," head-to-tail mounting position is occasionally used, e.g., in Botos, Hammerheads, Ruffs, and Western Gulls. Most same-sex interactions involve only two individuals at a time, but group sexual (and courtship) activity—involving anywhere from three or four (Giraffes, Lions) to six or more (Bowhead Whales, Mountain Sheep) partners—occurs in over 25 different species.

The actual type of genital contact varies widely. Full penetration in male anal intercourse occurs in some species (for example, Orang-utans, Rhesus Macaques, Bison, and Bighorn rams), while female penetration of various types occurs during lesbian interactions in Orang-utans (insertion of the finger into the vagina), Bonobos (insertion of the erect clitoris into the vulva), and Bottlenose and Spinner Dolphins (insertion of a fin or tail fluke into the female's genital slit). Simple pelvic thrusting and rubbing of the genitals on the rump of the other animal is widespread in both male and female homosexual mounts (occurring in the Northern Fur Seal, Lion, and Proboscis Monkey, among others), and simple genital-to-genital touching is the form of homosexual (and heterosexual) contact in species where males do

▲ *A male Giraffe mounting another male*

not have a penis (as in most birds, such as the Pukeko and Tree Swallow). A more unusual type of male homosexual contact involves various forms of *non*-anal penetration. In Whales and Dolphins, both males and females have a genital slit or opening; when not aroused, the male's penis is contained in the cavity leading to this slit. Homosexual activity in Bowhead Whales, Bottlenose Dolphins, and Botos sometimes involves insertion of the penis of one male into the genital slit of the other. Other more unusual forms of penetration have also been documented: male Botos occasionally insert the penis into a male partner's blowhole (on the top of his head!), while male Orang-utans have even been observed retracting their penis to form a sort of "hollow" or concavity that another male can penetrate. Clitoral rubbing or other types of genital tribadism are found in female Bonobos, Gorillas, and Rhesus Macaques (among others), while males in several species (e.g., White-handed Gibbons, West Indian Manatees, and Gray Whales) rub their penises together or on each other's body. In male Bonobos, mutual genital rubbing sometimes takes the form of an activity with the colorful name of "penis fencing," in which the males hang suspended by their arms and rub their erect organs against each other.

Oral sex of various kinds also occurs in a number of species. This may involve actual sucking of genitals (fellatio between males in Bonobos, Orang-utans,

Siamangs, and Stumptail Macaques); licking of genitals (cunnilingus in Common Chimpanzees, Long-eared Hedgehogs, and Kob antelopes; penis-licking in Thin-horn Sheep and Vampire Bats; genital licking in female Spotted Hyenas and male Cheetahs); mouthing, nuzzling, or "kissing" of genitals (female Gorillas, male Savanna Baboons, Crab-eating Macaques, and West Indian Manatees); and genital sniffing in female Pronghorns and Marmots as well as scrotal sniffing in Whiptail and Red-necked Wallabies. Male Stumptail Macaques even perform mutual fellatio in a sixty-nine position, while males of a number of primate species (including Gibbons, Bonnet and Crested Black Macaques, and Nilgiri Langurs) sometimes actually eat or swallow their partner's (or their own) semen—though usually after mutual genital rubbing or manual stimulation rather than oral sex.[9] Dwarf Cavies and Rufous Bettongs occasionally indulge in anal licking, nuzzling, and sniffing with same-sex (and opposite-sex) partners. Another sort of "oral" sexual activity is called beak-genital propulsion and occurs among both male and female Bottlenose and Spinner Dolphins: one animal inserts its snout or "beak" into the genital slit of another, simultaneously stimulating and propelling its partner forward while swimming (a similar behavior in Orcas, involving simple nuzzling or touching of the genitals with the snout, is known as beak-genital orientation).

Another type of activity found during homosexual interactions is masturbation, in which one animal stimulates its own or its partner's genitals with a finger, hand, foot, flipper, or some other appendage. For example, male Savanna Baboons often touch, grab, or fondle the genitals of another male—this behavior is known aptly as diddling—while male Bottlenose Dolphins and West Indian Manatees sometimes rub another male's penis with their flippers. Male Rhesus and Crested Black Macaques, female Gorillas, male Vampire Bats, female Proboscis Monkeys, and male Walruses sometimes masturbate themselves when mounting, courting, or interacting sexually with another animal of the same sex. Mutual masturbation in a side-by-side sixty-nine position occurs in female Crested Black Macaques, while male Bonnet and Stumptail Macaques masturbate each other and even fondle one another's scrotums. Another form of mutual masturbation in these species involves two males backing up toward each other and fondling each other's genitals between their legs. In Bonobos and Common Chimpanzees, individuals often rub their anal and genital regions together while in this rump-to-rump position, prompting zoologists to give these behaviors names like "rump-rubbing" and "bump-rump." Other more unusual forms of "manual" stimulation include mutual genital stimulation using trunks in female Elephants, and anal stimulation and penetration with fingers by male Common Chimpanzees, Siamangs, and Crab-eating Macaques.

Consorts, Satellites, and Triumvirates: Same-Sex Mates and Pair-Bonding

Wild animals often form significant pair-bonds with animals of the same sex. Homosexual pair-bonding takes many different forms, but two broad categories can be recognized: "partners," who engage in sexual or courtship activities with each other, and "companions," who are bonded to each other but do not necessarily engage in overt sexual activity with one another. More than a third of the mam-

▲ *A mated pair of female Canada Geese*

mals and birds in which homosexual activity occurs have at least one of these types of same-sex bonding. The archetypal example of a "partnership" is the mated pair: two individuals who are strongly bonded to one another in a way that is equivalent to heterosexually paired animals of the same species. Partners engage directly in courtship, sexual, and/or parenting behaviors; they usually spend a significant amount of time with each other; and they do similar activities together. This is found primarily in birds (more than 70 different species)—not surprisingly, since heterosexual pairing is typical of feathered creatures (but generally rare in other animal groups). Examples of homosexual mates are found in male Black Swans and Black-headed Gulls, and female Black-winged Stilts and Silver Gulls (among many others). In mammals, partnerships take many different forms, including "consortships" in female Rhesus and Japanese Macaques, "sexual friendships" in Stumptail and Crab-eating Macaques, "tending bonds" between male Bison, and "coalitions" between male Bonnet Macaques, Savanna Baboons, and Cheetahs. Some animals, while not necessarily forming same-sex bonds, do have "preferred" or "favorite" sexual and affectionate partners with whom they tend to interact more often than with others: this is true for Bonobos, Gorillas, Killer Whales, and Dwarf Cavies, among others.

Many forms of same-sex partnership are exclusive or monogamous, and partners may even actively defend their pair-bond against the intrusion of outside individuals (for instance in male Gorillas, female Japanese Macaques, and male Lions). Animals of the same sex sometimes also compete with each other for the attentions of homosexual partners, as in male Gorillas and Blue-winged Teals; female Orangutans, Japanese Macaques, and Orange-fronted Parakeets may even compete with *males* for "preferred" female partners. Some partnerships, however, are "open" or nonmonogamous: female Bonobos and Rhesus Macaques, for instance, may have sexual relations with several different "favorite" partners or consorts (of both sexes). Males in homosexual pairs of Greylag Geese, Laughing Gulls, Humboldt Penguins, and Flamingos sometimes engage in "promiscuous" copulations with birds (male or female) other than their mate (heterosexual pairs in these species are also sometimes nonmonogamous). Another form of nonmonogamy occurs among lesbian pairs in a number of Gulls and other birds: one or both females sometimes mate with a male (while still maintaining their same-sex bond) and are thereby able to fertilize their eggs and become parents.

The second main type of homosexual pairing is the "companionship." Two animals of the same sex may bond with each other, often spending most of their time together exclusive of the opposite sex, but they do not necessarily engage in recognizable courtship or sexual activities with each other. For example, older African Elephant bulls sometimes form long-lasting associations with a younger "attendant" male: these animals are loners, spending all their time with each other rather than with other Elephants, helping each other, and never engaging in heterosexual activity. Male Calfbird companions display and travel together and also sometimes share a "home" with one another (a special perch known as a retreat where they spend time away from the display court). Similar same-sex associations are found in many other species, including Orang-utans, Gray Whales, Grizzly Bears, Vampire Bats, and Superb Lyrebirds. Younger same-sex attendants are known as satellites in male Moose and shadows in male Walruses, while companions are called duos in male Hanuman Langurs and spinsters in female Warthogs—the latter is something of a misnomer, though, since Warthog companions do occasionally participate in sexual activity with males or females, but not necessarily with their companions.

Sometimes more than two animals bond together, forming a "trio" (in either partnership or companionship form). This arrangement can consist of three animals all of the same sex who are bonded with each other, as occasionally happens among female Ring-billed Gulls and male African Elephants, White-tailed Deer, and Black-headed Gulls. Trios can also be bisexual, consisting of two females and one male (e.g., Canada Geese, Common Gulls, and Jackdaws) or two males and one female (Greylag Geese, Black Swans, Sociable Weavers); in Oystercatchers, both types occur. In either form of a bisexual trio, there is significant bonding, courtship, and/or sexual behavior between the two animals of the same sex. This distinguishes such associations from heterosexual trios, in which two animals of the same sex are bonded with an opposite-sexed individual but not to each other. Same-sex trios of closely bonded male Greylag Geese or female Grizzly Bears are also sometimes known as triumvirates, while bisexual (and heterosexual) trios in Flamingos are called triads. In a few species, "quartets" involving simultaneous homosexual and heterosexual bonds between four individuals sometimes occur: in Greylag Geese and Black-headed Gulls, for instance, three males and a female sometimes bond with each other, while in Galahs, two males and two females may associate in a quartet with various bonding arrangements between them.

Homosexual pair-bonds vary not only in their type, but also in their duration. Same-sex bonding often follows the species-typical pattern for heterosexual pairing in terms of how long it lasts. In species such as the Greylag Goose, for example, which remain mated for life (or else for many consecutive years), male pairs are also generally long-lasting or lifelong, while in Bison, tending bonds usually last only a few days or hours in both heterosexual and homosexual situations. In some cases, long-term pair bonding involves continuous association throughout the year, as among male Ocellated Antbirds. This contrasts with seasonal association, for example among several species of Gulls, in which females re-pair with the same fe-

male only during the mating season. Homosexual pairs may also be of shorter duration than heterosexual ones in some species: Black-headed Gull male couples, for
example, appear to be more prone to divorce than heterosexual ones. However, in
many cases homosexual pairings, particularly companionships, actually exceed
heterosexual ones in their stability and duration. Among Lions and Elephants, for
example, the bond between male companions is closer and longer-lasting than any
heterosexual bonds (which, in these and many other species, are virtually nonexistent beyond mating), while mated gander pairs in Greylag Geese are often more
strongly bonded than heterosexual pairs. Consortships between Japanese Macaque
females sometimes develop into yearlong friendships, unlike the majority of heterosexual associations in this species. In fact, in a number of animals the only pair-
bonds that occur are homosexual, not heterosexual. Male Bottlenose Dolphins, for
example, form lifelong partnerships with each other, while males and females in
this species do not generally pair-bond with one another at all. Other animals with
same-sex but not opposite-sex pairings (often in the form of companionships) include Musk-oxen, Wapiti, White-tailed Deer, Warthogs, Cheetahs, Eastern Gray
Kangaroos, Red Squirrels, and Calfbirds.

Formidable Fathers and Supernormal Mothers: Homosexual Parenting

Same-sex pairs in many species (especially birds) raise young together. Not only
are they competent parents, homosexual pairs sometimes actually exceed heterosexual ones in the number of eggs they lay, the size of their nests, or the skill and extent of their parenting. How are such animals able to have offspring in the first
place if they are in homosexual associations? Many different strategies are used, including several in which one or both partners are the biological parent(s) of the
young they raise together. The most common parenting arrangement of this type is
found in lesbian pairs of several Gull, Tern, and Goose species: one or both female
partners copulate with a male to fertilize her eggs. No bonding or long-term association develops between the female and the male (who is essentially a "sperm
donor" to the homosexual pair), and the youngsters are then jointly raised by both
females without any assistance from a male parent. Because female birds can lay
eggs regardless of whether they are fertilized, however, each partner in a lesbian pair
usually contributes a full clutch of eggs to their nest even if she hasn't mated with a
male. As a result, female homosexual pairs often lay what are called supernormal
clutches, that is, double the number of eggs usually found in nests of heterosexual
pairs.[10]

Sometimes two female animals who already have offspring join forces, bonding
together and raising their young as a same-sex family unit (among mammals, female coparents may even suckle each other's young): this occurs in Grizzly Bears,
Red Foxes, Warthogs, Dwarf Cavies, Lesser Scaup Ducks, and Sage Grouse. Notably,
heterosexual pairs do not occur in these species, and most offspring are otherwise
raised by single females.[11] In some species, a nonbreeding animal bonds with a (single) breeding animal and helps parent its young: this occurs in Squirrel Monkeys,
Northern Elephant Seals, Jackdaws (where a widowed female with young may pair

with a single female), and Greater Rheas (where one male may help another incubate his eggs and then raise the young together). In most such joint parenting arrangements (as opposed to homosexual mated pairs), there is not necessarily any overt courtship or sexual activity between the bonded coparents, although in some species (e.g., Squirrel Monkeys, Northern Elephant Seals, Emus, Sage Grouse), homosexual activity does occur in contexts *other* than between coparents. Still other birds (e.g., Greylag Geese, Common Gulls, Oystercatchers) may form bisexual parenting trios, mating with the opposite-sexed partner(s) in their association while maintaining homosexual and heterosexual bonds simultaneously, with all three birds then raising the resulting offspring together. A variation on this arrangement in Black Swans involves a sort of "surrogate motherhood": established male homosexual pairs sometimes associate temporarily with a female, mating with her to father their own offspring. Once the eggs are laid, however, they chase her away and raise the cygnets on their own as a homosexual couple.

In a number of cases, homosexual pairs raise young without being the biological parents of the offspring they care for. Some same-sex pairs adopt young: two female Northern Elephant Seals occasionally adopt and coparent an orphaned pup, while male Hooded Warblers and Black-headed Gulls may adopt eggs or entire nests that have been abandoned by females, and pairs of male Cheetahs occasionally look after lost cubs. Sometimes female birds "donate" eggs to homosexual couples through a process known as parasitism: in many birds, females lay eggs in nests other than their own, leaving the parenting duties to the "host" couple. This occurs both within the same species, and (more commonly) across species, and usually involves heterosexual hosts. Male pairs of Hooded Warblers, however, sometimes receive eggs from Brown-headed Cowbirds (and possibly also from females of their own species) in this way; within-species parasitism may also provide eggs for male

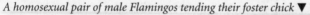

A homosexual pair of male Flamingos tending their foster chick ▼

pairs of Black-headed Gulls and female pairs in Roseate and Caspian Terns. The opposite situation is thought to occur in Ring-billed Gulls: researchers believe that some homosexually paired females actually lay eggs in nests belonging to heterosexual pairs. Finally, some birds in same-sex pairs take over or "kidnap" nests from heterosexual pairs (e.g., in Black Swans, Flamingos) or occasionally "steal" individual eggs (e.g., in Caspian and Roseate Terns, Black-headed Gulls); homosexual pairs in captivity also raise foster young provided to them.

In a detailed study of parental behavior by female pairs of Ring-billed Gulls, scientists found no significant differences in quality of care provided by homosexual as opposed to heterosexual parents. They concluded that there was not anything that male Ring-billed Gull parents provided that two females could not offer equally well.[12] This case is not exceptional: homosexual parents are generally as good at parenting as heterosexual ones. Examples of same-sex pairs successfully raising young have been documented in at least 20 species, and in a few cases, homosexual couples actually appear to have an advantage over heterosexual ones.[13] Pairs of male Black Swans, for example, are often able to acquire the largest and best-quality territories for raising young because of their combined strength. Such fathers—dubbed "formidable" adversaries by one scientist—consequently tend to be more successful at raising offspring than most heterosexual pairs.[14] And in many species in which single parenting is the rule (because there is no heterosexual pair-bonding), same-sex pairs provide a unique opportunity for young to be raised by two parents (e.g., Squirrel Monkeys, Grizzly Bears, Lesser Scaup Ducks). Moreover, in some Gulls, female pairs are consigned (for a variety of reasons) to less than optimal territories, yet they still successfully raise young: in many cases they compensate by investing more parental effort—and are more dutiful in caring for their chicks—than male-female pairs.[15] There are exceptions, of course: some female pairs of Gulls, for instance, tend to lay smaller eggs and raise fewer chicks (although this is also true of heterosexual trios attending supernormal clutches), while same-sex parents in Jackdaws, Canada Geese, and Oystercatchers may experience parenting difficulties such as egg breakage or nonsynchronization of incubation duties. By and large, though, same-sex couples are competent and occasionally even superior parents.

Birds in homosexual pairs often build a nest together. Usually they construct a single nest the way most heterosexual pairs do, but other variations also occur: female Common Gulls and Jackdaws sometimes make "twin" or "joint" nests containing two cups in the same bowl, while male Greater Rheas and female Canada Geese may use "double" nests consisting of two adjacent or touching nests. Female Mute Swans occasionally construct two separate nests in which both birds lay eggs. Nests belonging to male couples in some species (e.g., Flamingos and Great Cormorants) are often impressive structures, exceeding the size of heterosexual nests because both males contribute equally to their construction (in heterosexual pairs of these species, usually only one sex builds the nest, or males and females make unequal contributions). Many same-sex pairs construct nests regardless of whether they lay fertile eggs. Male pairs of Mute Swans, Flamingos, Black-crowned Night Herons, and Great Cormorants, for example, usually build nests even though they never acquire eggs, and the male "parents" may even sit on the nests as if they contained eggs, while female pairs frequently build nests in which they lay supernormal clutches that are entirely infertile. Same-sex parents often share incubation duties, either taking turns sitting on their nest (the most common arrangement), or else incubating simultaneously on a single nest (female Red-backed Shrikes, male Emus) or side by side on a twin or double nest (female Jackdaws, male Greater Rheas).

In addition to parenting by homosexual couples, some animals raise young in alternative family arrangements, usually a group of several males or females living together. Gorilla babies, for example, grow up in mixed-sex, polygamous groups where their mothers may have lesbian interactions with each other, while Pukeko and Acorn Woodpeckers live and raise their young in communal breeding groups where many, if not all, group members engage in courtship and sexual activities with one another (both same-sex and opposite-sex). In such situations, individuals that engage in homosexual courtship or copulation activities may either reproduce directly because they also mate heterosexually (Pukeko), or they may assist members of their group in raising young without reproducing themselves (Acorn Woodpeckers).[16] Other alternative family constellations include bisexual trios (mentioned above), homosexual trios (as in Grizzly Bears, Dwarf Cavies, Lesser Scaup Ducks, and Ring-billed Gulls) where three mothers jointly parent their offspring, and even quartets, in which four animals of the same (Grizzlies) or both sexes (Greylag Geese) are bonded to each other and all raise their young together.[17]

Finally, some animals that have homosexual interactions are "single parents." Many female mammals, for example, that court or mate with other females also mate heterosexually and raise the resulting young on their own or in female-only groups (as is typical for exclusively heterosexual females in the same species as well). This is especially prevalent among mammals with polygamous or promiscuous heterosexual mating systems, such as Kob and Pronghorn antelopes and Northern Fur Seals (where males, and sometimes females, usually mate with more than one partner). Males in many polygamous species are often bisexual as well, fathering offspring in addition to courting or mating with other males; typically, however, they do not actively parent their offspring regardless of whether they are bisexual or exclusively heterosexual.[18]

What's Good for the Goose . . . :
Comparisons of Male and Female Homosexuality

Is homosexuality more characteristic of male animals or female animals? And does it assume different forms in the two sexes—or, to paraphrase a popular saying, is the behavior of the "goose" essentially similar to that of the "gander"? As it so happens, homosexuality in three species of Geese—Canada, Snow, and Greylag—exemplifies some of the major patterns of male and female homosexuality and the range of variation found throughout the rest of the animal world. In Canada Geese, both males and females participate in the same basic type of homosexual activity, forming same-sex pairs and engaging in some courtship activities. Within these same-sex bonds, however, there are gender differences in some less common behaviors: sexual activity is more characteristic of females (especially if they are part of a bisexual trio), as is nest-building and parenting activity. There are also differences in the frequency of participation of the two sexes: although same-sex pairs are relatively common, accounting for more than 10 percent of pairs in some populations, a greater proportion of the male population participates in same-sex pairing. In contrast, homosexual activity in Snow Geese is vastly different in males than in females, although it is relatively infrequent in both sexes. Females form long-lasting pair-bonds with other females in which sexual activity is not necessarily very prominent, although parenting activity is: both partners lay eggs in a joint nest and raise their young together (they fertilize their eggs by mating with males). Ganders, on the other hand, limit their homosexual activity to same-sex mounting of other males during heterosexual group rape attempts and do not form same-sex pairs (although interspecies gander pairs with Canada Geese sometimes do occur). Finally, in Greylag Geese homosexual activity is found exclusively in males, who form gander pairs that engage in a variety of courtship, sexual, pair-bonding, and parenting activities.

When we look at the full range of species and behaviors, we find that male homosexuality is slightly more prevalent, overall, than female homosexuality, although the two are fairly close. Same-sex activity (of all forms) occurs in male mammals and birds in about 80 percent of the species in which homosexuality has been observed, and between females in just over 55 percent of these (the figures add up to more than 100 percent because *both* male and female homosexuality are found in some species). It must also be kept in mind that the prevalence of female homosexuality may actually be greater than these figures indicate, but has simply not been documented as systematically owing to the general male bias of many biological studies.[19] There is also variation between different animal subgroupings: in carnivores, marsupials, waterfowl, and shorebirds, for example, male and female homosexuality are almost equally common (in terms of the number of species in which each is found), while in marine mammals and perching birds male homosexuality is more prevalent. And in many species same-sex activity occurs only among males (e.g., Boto, a freshwater dolphin) or only among females (e.g., Puku, an African antelope).

The frequency of same-sex behavior in males versus females can also be assessed within a given species, and once again, many different patterns are found: in Rhesus Macaques, Hamadryas and Gelada Baboons, and Tasmanian Native Hens, for example, 80–90 percent of all same-sex mounting is between males, while homosexual activity is also more prevalent among male Gray-headed Flying Foxes.[20] In other species, female homosexual activity assumes prominence: more than 70 percent of same-sex copulations in Pukeko are between females, and 70–80 percent of homosexual activity in Bonobos is lesbian. Females account for almost two-thirds of same-sex behaviors in Stumptail Macaques and Red Deer, while homosexual activity is also more typical of females in Red-necked Wallabies and Northern Quolls.[21] In some species, however, male homosexuality is so predominant that same-sex activity in females is often missed by scientific observers or rarely mentioned (e.g., Giraffes, Blackbuck, Bighorn Sheep), while the reverse is true in other species (e.g., Hanuman Langurs, Herring Gulls, Silver

Gulls). In contrast, Pig-tailed Macaque same-sex mounting, Galah pair-bonds, and Pronghorn homosexual interactions are fairly equally distributed between the two sexes (although actual same-sex mounting is more common in male Pronghorns).[22]

As with the species of Geese mentioned above, gender differences are also apparent in various behavior types. Of those mammal and bird species in which some form of homosexual behavior occurs, each of the activities of courtship, affectionate, sexual, or pair-bonding are generally more prevalent in male animals. They occur among males in 75–95 percent of the species in which they are found, while among females these activities occur in 50–70 percent of the species (again, however, the possible gender bias of the studies these figures are based on must be kept in mind). The one exception is same-sex parenting, which is performed by females in more than 80 percent of the species where this behavior occurs, but by males in just over half of the species that have some form of such parenting. Of course, not all these forms of same-sex interaction always co-occur in the same species, and animals sometimes differ as to which activities males as opposed to females of the same species tend to participate in (as in the Geese). In Silver and Herring Gulls, for example, females form same-sex pairs that undertake parenting duties while males engage in homosexual mounting; in Cheetahs and Lions, both sexes engage in sexual activity, but males in each species also develop same-sex pair-bonds while female Cheetahs participate in same-sex courtship activities. In Ruffs, males engage in sexual, courtship, and (occasional) pairing activity with each other, while Reeves (the name for females of this sandpiper species) participate primarily in sexual activity with one another.

Within each of the categories of courtship, sexual, pairing, and parenting behaviors, further gender distinctions can be drawn. Consider various types of sexual behavior. Mounting as a same-sex activity is ubiquitous and occurs fairly regularly in both males and females (although there are exceptions—in African Elephants, for example, sexual activity between males assumes the form of mounting while female same-sex interactions consist of mutual masturbation). Oral sex (which includes activities as diverse as fellatio, cunnilingus, genital nuzzling and sniffing, and beak-genital propulsion) is about equally prevalent in both sexes. Group sexual activity is more common in males (only occurring among females in 6 species, including Bonobos and Sage Grouse), as are interactions between adults and adolescents (only occurring among females in 9 species, including Hanuman Langurs, Japanese Macaques, Ring-billed Gulls, and Jackdaws, but among males in more than 70 species). Although penetration is also more typical of male homosexual interactions, there are notable exceptions (e.g., Bonobos, Orang-utans, and Dolphins, as mentioned previously). Gender differences sometimes also manifest themselves in the minutiae of various sexual acts. Same-sex mounting in Gorillas, for instance, is performed in both face-to-face and front-to-back positions, but the two sexes differ in the frequency with which these two positions are used: females prefer the face-to-face position, adopting it in the majority of their sexual interactions, while males use it less often, in only about 17 percent of their homosexual mounting episodes.[23] In contrast, the frequency of full genital contact during homosexual copulations is nearly identical for both sexes of Pukeko: females achieve cloacal contact in about 23 percent of their same-sex mounts while males do so in about 25 percent of theirs (in comparison, genital contact occurs in a third to half of all heterosexual mounts). Among Flamingos, though, genital contact is more characteristic of copulations between females than between males.[24]

Or consider pair-bonding and parenting. Stable, long-lasting pair-bonds are generally not more characteristic of females (contrary to what one might initially expect); mated pairs or partnerships are almost equally common in both sexes (in terms of number of species in which homosexual pair-bonding occurs), while same-sex companionships are more prevalent between males. Likewise, long-term pair-bonds are just as likely to be found between males as females, while non-monogamy and divorce occur in male and female couples in roughly equal numbers of species. Nor are male couples less successful parents: male coparents or partners are not overrepresented among the few species in which same-sex parents occasionally experience parenting difficulties. One area where a gender difference in same-sex parenting does manifest itself is in the way that homosexual pairs "acquire" offspring. Particularly among birds, female couples can raise their own offspring by simply having one or both partners mate with males without interrupting their homosexual pair-bond. This option is not as widely available for male couples, who usually father their own offspring by forming a longer-lasting (prior or simultaneous) association with a female (as in Black Swans, Greylag Geese, and Greater Rheas).

Japanese Macaques offer a particularly compelling example of the multiple ways that homosexual activity can differ between males and females. Although homosexual mounting occurs in both sexes in this species, males and females differ in the specific details of their sexual interactions. Homosexual mounts are usually initiated by the mounter between males but by the mountee between females, make use of a wider variety of mounting positions between females, are accompanied by a unique vocalization only between males, and involve pelvic thrusting and multiple mounts more often between females than between males.[25] The two sexes also differ in their partner selection and pair-bonding activities: females generally form strongly bonded consortships and have fewer partners than males, while the latter tend to interact sexually with more individuals and develop less intense bonds (although some do have "preferred" male partners). Finally, there is a seasonal difference in male as opposed to female same-sex interactions: homosexual mounting is more common outside the breeding season in males but during the breeding season in females, while same-sex bonds in females, but not males, may extend into yearlong associations that transcend the breeding season.

Whether we're talking about ganders and geese or Ruffs and Reeves, whether it's Botos and Bonobos or Pukus and Pukeko, male and female homosexuality can be either surprisingly similar to each other or decidedly distinctive from one another. In any case, a complex intersection of factors is involved in the expression of homosexuality in each gender. As with other aspects of animal homosexuality, preconceived ideas about how males and females act must be reassessed and refined when considering the full range of animal behaviors. In some species such as Silver

Gulls, male and female homosexualities conform to stereotypes commonly held about similar human behaviors: females form stable, long-lasting lesbian pair-bonds and raise families while males participate in promiscuous homosexual activity. In other species these gender stereotypes are turned completely on their heads, as in Black Swans, where only males form long-term same-sex couples and raise off-spring, and Sage Grouse, where only females engage in group "orgies" of homosexual activity.[26] And in the majority of cases, male and female homosexualities present their own unique blends of behaviors and characteristics that defy any simplistic categorization—such as Bonobos, where sexual penetration occurs in female rather than male same-sex activity, where sexual interactions between adults and adolescents are a prominent feature of female interactions, and where males do not form strongly bonded relationships with each other the way females do, but engage in less homosexual activity overall and more affectionate activity such as openmouthed kissing. Once again, the diversity of animal homosexuality reveals itself down to the very last detail of expression.

A Hundred and One Lesbian Acts:
Calculating the Frequency of Homosexual Behavior

$$\sqrt{a} = \frac{1}{2\sqrt{E}}\left(C - \frac{D^2}{4E}\right) \qquad b = \frac{D}{2E}\left(C - \frac{D^2}{4E}\right)$$

where a, b, C, D, and E represent the number of nests with 2–6 eggs respectively
—formulas used in estimating the number of
female homosexual pairs in Gull populations[27]

While studying Kob antelopes in Uganda, scientists recorded exactly 101 homosexual mounts between females. In Costa Rica, 2 copulations between males were observed during a study of Long-tailed Hermit Hummingbirds. In which species is homosexuality more frequent? The answer would appear to be obvious: Kob. However, simply knowing the total number of homosexual acts observed in each species is not sufficient to evaluate the prevalence of homosexuality. For example, it could be that the Kob were observed for a much longer period of time than the Hummingbirds, in which case the greater number of same-sex mounts would not necessarily reflect any actual difference between the two species. What we really need is a measure of the *rate* of homosexual activity—that is, the number of homosexual "acts" performed during a given period of time. To determine this, we have to know the duration of the study period for each species and how many animals were being observed. In this case, 8 female Kob antelopes were studied for a total of 67 hours, whereas 36 male Long-tailed Hermit Hummingbirds were observed over several hundred hours—so indeed the Kob have a much higher rate of same-sex activity (both in general and per individual), on the order of many hundreds of times higher than the Hummingbirds.

Rate of occurrence is only one measure of frequency, however. It could be that sexual activity in general is much rarer in Hummingbirds than in Kobs, in which case comparing absolute numbers or rates gives a distorted or incomplete picture. A more meaningful comparison would be to look at how many heterosexual acts are performed during the same time period and express the frequency of homosexual activity as a *proportion* of all sexual activity. In fact, sexual activity *is* incredibly common among Kob and much rarer in Long-tailed Hermits: during the same study period, 1,032 heterosexual mounts among Kob were tabulated while only 6 heterosexual matings in Hummingbirds were observed. Thus, homosexual mounts constitute only 9 percent of all sexual activity among the Kob, whereas one-fourth of all copulations in Long-tailed Hermits are between males. This is diametrically opposed to the frequency rate or absolute count of homosexual activity in the same species.[28]

These two cases offer a good example of the many complications that arise when attempting to answer the question "How common or frequent is homosexuality in animals?" The most valid answer—clichés aside—is, "It depends." It depends not only on the measure of frequency being used, but also on the species, the behaviors being tabulated, the observation techniques that are employed, and many other factors. In this section we'll explore some of these factors and try to arrive at some meaningful generalizations about the prevalence of homosexuality in the animal kingdom.

One broad measure of frequency is the total number of species in which homosexuality occurs. Same-sex behavior (comprising courtship, sexual, pair-bonding, and parental activities) has been documented in over 450 species of animals worldwide.[29] While this may seem like a lot of animals, it is in fact only a tiny fraction of the more than 1 million species that are known to exist.[30] Even considering the two animal groups that are the focus of this book—mammals and birds—homosexual behavior is known to occur in roughly 300 out of a total of about 13,000 species, or just over 2 percent. However, comparing the number of species that exhibit homosexuality against all known species is probably an inaccurate measure, since only a fraction of existing species have been studied in any depth—and detailed study is usually required to uncover behaviors such as homosexuality. Scientists have estimated that at least a thousand hours of field observation are required before more unusual but important activities will become apparent in a species' behavior, and relatively few animals have received this level of scrutiny.[31] Unfortunately, it is not known exactly how many species have been studied to this depth, although it has been estimated that perhaps only 1,000–2,000 have begun to be adequately described. Using these figures, the proportion of animal species exhibiting homosexual behavior comes in at 15–30 percent—a significant chunk.[32]

In fact, the percentage is probably even higher than this, when we consider how easy it is for common behaviors to be missed during even the most detailed of study. A caveat of any scientific endeavor, particularly biology, is that much remains to be learned and observed, and many secrets await discovery—and this is especially true where sexual behavior is concerned. Nocturnal or tree-dwelling habits, elusiveness, habitat inaccessibility, small size, and problems in identifying individual animals are just some of the factors that make field observations of sexuality in many species exceedingly difficult.[33] Consider heterosexual mating, a behavior that is known to occur in all mammals and birds (and most other animals), usually with great regularity.[34] Yet in many species this activity has *never* been seen: "Despite literally thousands of hours of observations made by biologists over many years in the West Indies, Hawaii, and elsewhere, actual copulation in humpback whales has yet to be observed."[35] Lucifer hummingbirds, northern rough-winged swallows, black-and-white warblers, red-tailed tropic birds, and several species of cranes (such as wattled and Siberian cranes) are just a handful of the birds in which heterosexual mating has never been recorded. In some cases, opposite-sex mating has been observed, but only a handful of times at most: in magnificent hummingbirds and black-headed grosbeaks, for example—the latter a common North American bird—copulation between males and females has only been seen once during the entire history of the scientific study of these species. Heterosexual copulation in Victoria's Riflebirds was not documented until the mid-1990s (and then only several times), even though the species has been known to Western science for nearly a century and a half. During a ten-year study of Cheetahs, no opposite-sex matings were seen over the course of 5,000 hours of observation, and copulation has only been observed a total of five times in the wild during the entire scientific study of this animal. Similar patterns are characteristic of other species: in the akepa (a Hawaiian finch), only five copulations were witnessed during five years of study, only five heterosexual matings were seen in a four-year study of Spotted Hyenas, and only three matings in a three-year study of Agile Wallabies. Nests and eggs of many birds such as swallows and birds of paradise have yet to be discovered, while the first nest of the marbled murrelet was found in 1959, more than 170 years after discovery of the species by Western science.

And of course new revelations about heterosexual behavior are being made all the time: female initiation of mating activity in Orang-utans, for example, was not documented until 1980 in spite of nearly 22,000 hours of observation over the preceding 20 years (and prior extensive field studies often failed to report *any* heterosexual copulations). As recently as 1996, the existence of polygamous trios in the tanga'eo or Mangaia kingfisher (of the Cook Islands near New Zealand) were uncovered for the first time, and the full extent of heterosexual mating by Common Chimpanzees with animals outside their group was not understood until 1997. Multiple heterosexual matings by female Harbor Seals were not verified until 1998; even then, the behavior was never directly observed during three years of study (including continuous, 24-hour videotape surveillance of captive animals over an entire breeding season), and had to be verified indirectly through DNA testing.[36] If direct observation by scientists were used as the sole criterion for the existence of a behavior, we would have to conclude that many species never engage in heterosexuality (or in certain forms of heterosexuality)—yet we know this cannot be true. So the fact that homosexuality has not been seen in many animals does not necessarily mean that it is absent in those species—only that it has yet to be observed.

Ironically, many species in which heterosexuality has rarely or never been observed are ones in which homosexual activity *has* been recorded. No information on the heterosexual mating system of wild Emus was available prior to 1995, for example, although homosexual copulation in the same species had been observed in captivity more than 70 years earlier. Heterosexual mating has never been observed in Black-rumped Flameback Woodpeckers—although homosexual copulation has—while some studies of Nilgiri Langurs, Harbor Seals, Northern Quolls, and Gray-capped Social Weavers failed to record any instances of opposite-sex mounting, although same-sex mounting did occur. Similarly, documentation of sexual activity between male Walruses—including photographs—preceded by almost a decade comparable descriptions and photographic evidence of sexual activity between males and females. In Acorn Woodpeckers—a species that regularly engages in same-sex mounting—only 26 heterosexual copulations were recorded in over 1,400 hours of observation devoted specifically to recording opposite-sex mating. Likewise, heterosexual copulations in Australian Shelducks (a species in which females sometimes form homosexual pairs) were observed only nine times during nearly a decade of study, and on only three of these occasions was a complete behavioral sequence involved. Because of the difficulty of observing heterosexual copulation, the mating system of Killer Whales is still poorly understood and, according to one scientist, "may never be known with certainty." Homosexual activity in the same species has already been documented, although its study is also still in its infancy.[37] Obviously, then, an activity can be part of the regular behavior of a species and still be completely missed by observers or documented only rarely, in spite of conscientious and in some cases exhaustive observational regimens (both in the wild and in captivity).

Scientists have often characterized homosexuality in animals as "extremely rare" or "quite common," for example, or as occuring "regularly" or "infrequently"—often without any numerical or contextual information. Yet such statements are virtually meaningless without a common standard of measurement and an agreed-upon point of reference. In an attempt to standardize the evaluation of homosexual behavior, therefore, many scientists have collected quantitative information—usually tallies of particular behaviors (sexual, courtship, pairing, etc.). In a few cases, the difficulty of field observations has precluded the direct observation of both heterosexual and homosexual activity, and several innovative techniques have been developed to calculate the frequency of same-sex activity based on indirect measures. The sex of Gulls, for example, is often difficult to determine under field conditions, and in colonies that may contain tens of thousands of breeding pairs, the task of determining which couples are homosexual and which are heterosexual is a daunting one. However, once researchers discovered that lesbian pairs typically lay supernormal clutches, the frequency of same-sex pairs could be much more easily tallied by counting the number of nests with double the usual number of eggs. One ornithologist even developed a mathematical formula (see the beginning of this section) for estimating the total number of lesbian pairs in a population based on a sample of clutch sizes, taking into account same-sex pairs that lay smaller than supernormal clutches (or heterosexual pairs that lay larger than average clutches).[38] Likewise, determining the sex of mating Dragonflies can be extremely challenging while the animals are still alive, since they copulate in flight. Scientists discovered, though, that insects (both male and female) usually suffer head injuries from being clasped by a male during mating. These injuries can easily be recognized and counted once individual Dragonflies are collected—revealing that an average of nearly 20 percent of males in 11 different species (and more than 80 percent of males in some species) experience homosexual copulations.[39]

Even when quantitative information is available, assessments of frequency are often subjective and contradictory. For example, one scientist observed 24 homosexual copulations between Pukeko (7 percent of all sexual activity) and classified the behavior as "common," while another zoologist observed nearly identical numbers and proportions of same-sex mounts in Pronghorns (23 mounts, 10 percent of all mounting activity) yet classified the behavior as "rare."[40] The problem is that there are many different ways of measuring and interpreting the frequency of same-sex behaviors. Besides tallying absolute numbers of particular activities or determining the proportion of all sexual activity that is homosexual, frequency rates and activity budgets can also be calculated, along with the percentage of the population that engages in same-sex activity. Frequency *rate* refers to the number of homosexual acts performed during a given time period, either by each individual or within a group of animals as a whole. For example, among Hanuman Langurs each female participates in a homosexual mounting, on average, once every five days, while in some populations of Ostriches courtship between males occurs at the rate of two to four times a day. An *activity* or *time budget,* on the other hand, refers to what proportion of an individual animal's activity or time is devoted to homosexual interactions. For instance, a male Killer Whale spends more than 10 percent of his time interacting socially and sexually with other males; about 15 percent of courtship display time in male Regent Bowerbirds involves displaying to other males; more than a third of some male White-handed Gibbons' time is devoted to homosexual interactions; while 10 percent of a male Crab-eating Macaque's interaction with other males involves mounting activity.[41] The percentage of the population that engages in homosexual activity varies widely, from only one or two individuals in a flock of several thousand Gulls, to virtually the entire population of male Bighorn Sheep, and everything in between. Of course, some of these individuals also engage in heterosexual activity (exhibiting various degrees of bisexuality), while others are more or less exclusive in their homosexual relations; this will be discussed in more detail in chapter 2, where the question of sexual orientation is explored.

Because of the diversity and complexity of homosexual expression across a wide range of species, it is not always a straightforward matter to calculate various measures of frequency such as these. Three different species exemplify some of the issues that are involved in just one measure of frequency, the proportion of sexual activity that is homosexual. Observed versus actually occurring behaviors in Giraffes, seasonal variations in sexual activity in Mountain Sheep, and alternative standards of reference in Gray Herons complicate the calculation of homosexual frequency in each of these species. During an exhaustive study of Giraffes in the Arusha and Tarangire National Parks of Tanzania, researchers recorded 17 homosexual mounts and 1 heterosexual mount in more than a year (and 3,200 hours) of observation. Thus, 94 percent of all observed mounting activity was same-sex. Does this reflect the actual proportion of homosexual activity in Giraffes? Certainly more than one heterosexual mating occurred during that time, since over 20 calves were born that year in one population alone. However, these populations did have relatively low birth rates, and heterosexual mating appeared to be genuinely rare. In addition, if heterosexual matings were being missed by the observers, probably homosexual ones were as well (unless same-sex mountings consistently took place in a more visible setting than opposite-sex ones). This means that the same proportion of homosexual activity could have been involved regardless of whether some mountings were missed.[42] In Mountain Sheep, there are sharp seasonal differences in the proportion of same-sex activity. During the rut (about two months out of the year), approximately a quarter of all mounts are between males; during the rest of the year, virtually all mounts are homosexual, although only a small fraction of rams' interactions with each other involves mounting.[43] Thus, homosexual activity is more common during the rut if frequency rates, time budgets, or absolute numbers are tallied, but more prevalent outside the rut if proportions of sexual activity are calculated. Gray Herons, like many other birds, often engage in promiscuous copulations: males try to mate with birds, both male and female, other than their partner. One study revealed that about 8 percent of such promiscuous copulation attempts were homosexual. However, if the *total* number of copulations—both promiscuous and between bonded partners—is taken as the point of reference, the proportion of homosexual mounting drops to 1 percent.[44] As these three examples show, differences between populations, seasons, and behaviors (among other factors) must be taken into consideration when assessing frequency.

Recognizing that measures of frequency often obscure important behavioral distinctions between (and within) species and may reflect widely divergent observational methodologies, it is nevertheless still useful to compare the prevalence of various homosexual activities across a wide variety of animals. The following summaries focus on the proportion of three behavioral categories—courtship, sexual, and pair-bonding—that occur between animals of the same sex, as well as the percentage of the population that engages in homosexual activity. These measures are the most widely available for a large number of animals, and they lend themselves fairly well to cross-species comparisons. Interestingly, similar average proportions are obtained for a number of these measures, even though they represent many different species and behaviors.[45]

In those animals where homosexual activity occurs, an average of about a quarter of individuals in the population (or of a given sex) engage in same-sex activity—ranging from 2–3 percent of male Ostriches and female Sage Grouse, to nearly half of all male Giraffes and Killer Whales, to entire troops of Bonobos. Concerning specific behaviors, an average of about 25 percent of courtship activity occurs between animals of the same sex in those species that exhibit homosexual courtship interactions—ranging from less than 5 percent in Herring Gulls and Calfbirds, to more than 50 percent in Dwarf Cavies and Giraffes. A nearly identical proportion of sexual activity, just over one-quarter, occurs between animals of the same sex in those species where homosexuality has been observed—from as little as .3 percent in Silvery Grebes and 1–2 percent in Dusky Moorhens and Tasmanian Native Hens, to more than half of all sexual activity in Bison and Bonnet Macaques, and a whopping 94 percent of observed mountings in some populations of Giraffes (as mentioned above). Finally, an average of 14 percent of all pairs are homosexual in those species that have some form of same-sex (and opposite-sex) pair-bonding: ranging from as few as .3–.5 percent in Herring Gulls and Snow Geese, to more than a quarter of all consortships (on average) in Japanese Macaques, to more than half of all pair-bonds among young Galahs.

Combining these three behavioral categories yields a figure of just over 20 percent: roughly one-fifth of all interactions, on average, are homosexual in mammal and bird species that have at least some form of same-sex courtship, sexual, and/or pair-bonding activities. If one figure could be said to represent the overall frequency of homosexual activity in animals, this would probably be the one—but it is virtually impossible to come up with a truly "representative" number. A figure such as this collapses a multiplicity of behaviors (both between and within species), it represents only a fraction of the animals in which homosexuality has been documented, it glosses over many observational and theoretical uncertainties (not the least of which is widely differing sample sizes), and it misleadingly equates often radically unlike phenomena (different forms of heterosexual and homosexual behaviors, disparate social contexts, and so on). A less satisfying, but ultimately more meaningful, "formula" for understanding frequency is to recognize that there is no one overall frequency, no single formula. As in all aspects of animal homosexuality, different species exhibit an extraordinary range of rates, quantities, periodicities, and proportions of same-sex behavior—a diversity that is equal to the variation in the behaviors themselves. We can make tallies for particular species, develop formulas for certain populations or behaviors, and calculate percentages, time budgets, and so on—thereby trying to gain some overall impressions regarding the prevalence of homosexuality in animals. In the end, though, we must acknowledge that our measures are at best imperfect—and what we are attempting to quantify is, in many senses, incalculable.

Within Genders, Without Genders, Across Genders

The traditional view of the animal kingdom—what one might call the Noah's ark view—is that biology revolves around two sexes, male and female, with one of each to a pair. The range of genders and sexualities actually found in the animal world, however, is considerably richer than this. Animals with females that become males, animals with no males at all, animals that are both male and female simultaneously, animals where males resemble females, animals where females court other females and males court other males—Noah's ark was never quite like this! Homosexuality represents but one of a wide variety of alternative sexualities and genders. Many people are familiar with transvestism or transsexuality only in humans, yet similar phenomena are also found in the animal kingdom. Although this book focuses primarily on homosexuality, it is helpful to compare this with related phenomena that are often confused with homosexuality, and to discuss some specific examples of each.

Many animals live without two distinct genders, or with multiple genders. In hermaphrodite species, for instance, all individuals are both male and female simultaneously, and hence there are not really two separate sexes; in parthenogenetic species, all individuals are female and they reproduce by virgin birth. A number of other phenomena in the animal kingdom—for which we will use the cover term *transgender*—involve the crossing or traversing of existing gender categories: for example, transvestism (imitating the opposite sex, either behaviorally, visually, or chemically), transsexuality (physically becoming the opposite sex), and intersexuality (combining physical characteristics of both sexes).[46]

Early descriptions of animal homosexuality often mistakenly called the animals "hermaphrodites," since any "transgression" of gender categories (such as sexual behavior) was usually equated with physical gender-mixing. True hermaphroditism, however, involves animals that have both male and female reproductive organs at the same time. This phenomenon is found in many invertebrate organisms, such as slugs and worms, as well as in a number of fish species (for example, lantern fishes and some species of hamlets and deep-sea fishes). Some hermaphrodites can fertilize themselves, but mating in many hermaphroditic species involves two individuals having sex with each other in order to mutually exchange both eggs and sperm.[47] Since both such individuals have identical biologies, i.e., are of the same (dual) sex, technically such behavior could be classified as homosexual. However, such activity differs from actual homosexuality because it occurs in a species that does not have two separate sexes, and because it typically *does* result in procreation. In species that do have two distinct sexes, there are other types of hermaphroditism or intersexuality, in which individuals combine various physical features of both sexes. These differ from species-wide, true hermaphroditism because such animals are not able to reproduce as both males and females simultaneously, and they usually comprise only a fraction of the otherwise nonhermaphroditic population. Further examples of this type of transgender will be discussed in chapter 6.

Virgin birth, or parthenogenesis, is not just the stuff of religions: it is actually found in over a thousand species worldwide and is a "natural" form of cloning. Each member of a parthenogenetic species is biologically female (that is, capable of producing eggs). Rather than requiring sperm to fertilize these eggs, however, she simply makes an exact copy of her own genetic code. Virgin birth is found in a number of fishes, lizards, insects, and other invertebrates. In most parthenogenetic species, individuals do not have sex with each other, but in some species, such as the Amazon Molly and Whiptail Lizards, females actually court and mate with one another, even though no eggs (or sperm) are exchanged in such encounters.

Whereas homosexuality and bisexuality involve activity *within* the same gender, hermaphroditism and parthenogenesis involve courtship and sexual behavior *without* genders (at least, without one class of individuals that are male and another class that are female). In contrast, transvestism and transsexuality are a kind of "crossing over" from one gender or sexual category to another, or the combining of elements from each category. In transvestism, individuals of one biological sex take on the characteristics of the other sex, either behaviorally or physically, without actually changing their own sex. In transsexuality, individuals actually become the opposite sex, so that a male turns into a female or vice versa (where *male* and *female* are used strictly in the reproductive sense to refer to animals that produce sperm or eggs, respectively).

Transvestism is widespread in the animal kingdom and takes a variety of forms.[48] Both male-to-female and female-to-male transvestism occur: some female African swallowtail butterflies, for example, resemble males in their wing coloration and patterning, while in some species of squid, males imitate female arm postures during aggressive encounters.[49] Physical transvestism can involve almost total physical resemblance between males and females, or mimicry of only certain primary or secondary sexual characteristics. For instance, in several species of North American perching birds, young males resemble adult females in their plumage—making them distinct from both adult males and juvenile females. In some birds, such as the painted bunting, the resemblance between adult females and juvenile males is nearly total, while in others, younger males are more intermediate between adult males and females in appearance.[50] Several species of hoofed mammals show a different type of physical transvestism: female mimicry of the horns or tusks found in males.[51] Female Chinese water deer, for example, grow special tufts of hairs on their jaws that resemble the tusks of the male, while female Musk-oxen have a patch of hair on their foreheads that mimics the males' horn shield. Physical transvestism can also be chemical or scent-based: some male Common Garter Snakes, for example, produce a scent that resembles the female pheromone, causing males to mistake them for females and attempt to court and mate with them.

In behavioral transvestism, an animal of one sex acts in a way that is characteristic of members of the opposite sex of that species—often fooling other members of their own species. For example, males of several species of terns imitate female food-begging gestures to steal food from other males. Behavioral transvestism does *not* mean animals behaving in ways that are thought to be "typically" male or female in other species. In sea horses and pipefishes, for instance, the male bears and gives birth to the young. Even though these are activities usually thought to be "female," this is not a genuine case of transvestism because it is part of the regular behavior patterns and biology of the species (i.e., it is true for all males and no females). Female sea horses never bear young, nor are they fooled into thinking that males aren't male because they do bear young. The same goes for initiation of courtship: in some species the female is more aggressive in initiating courtship and copulation (e.g., in greater painted-snipes), yet this could only be considered "transvestism" with reference to other species in which females do not initiate such activity.[52]

The question of transvestism is an important one for animal homosexuality because these two phenomena have often been confused. Many scientists have labeled all examples of animal homosexuality male or female "mimicry" since they consider any same-sex behavior to be nothing more than imitation of the opposite sex. True, many animals, when courting and mating homosexually, employ behavior patterns that the opposite sex also employs. In most cases, however, this simply involves making use of the available behavioral repertoire of the species rather than being an attempt to mimic the opposite sex. Moreover, the resemblance to "heterosexual" patterns is often partial at best, while in some species entirely distinct courtship and copulation patterns are used for homosexual activity.[53]

A good example of the difference between behavioral transvestism and homosexuality is in the Bighorn Sheep. In this species, males and females lead almost entirely separate lives: they live in sex-segregated herds for

▲ Homosexuality as a "masculine" activity: a male Bighorn ram mounts another ram. Males who mimic females in this species (behavioral transvestites) do not generally permit other males to mount them, unlike nontransvestite rams.

most of the year and come together for only a few short months during the breeding season. Among males, homosexual mounting is common, while females do not generally permit themselves to be mounted by males except when they are in heat (estrus). A small percentage of males, however, are behavioral transvestites: they remain in the female herds year-round and also mimic female behavior patterns. Significantly, such males also generally refuse to allow other males to mount them, just the way females do. Thus, among Bighorn Sheep, being mounted by a male is a typically "masculine" activity, while refusal of such mounting is a typically "feminine" behavior. Males who mimic females specifically *avoid* homosexuality. This is the exact opposite of the stereotypical view of male homosexuality, which is often considered to be a case of males "imitating" females. It is also a striking reminder of how important it is not to be misled by our preconceptions about human homosexuality when looking at animals.[54]

Transsexuality or sex change is a routine aspect of many animals' lives, especially in invertebrates: shrimp, oysters, and sow bugs, for example, all undergo complete reversals of their sex at some stage in their lives.[55] It is among coral-reef fish, however, that the most remarkable examples of transsexuality are found. More than 50 species of parrot fishes, wrasses, groupers, angelfishes, and other species are transsexual. In all such cases, the reproductive organs of the fish undergo a complete reversal. What were once fully functional ovaries, for example, become fully functional testes, and the formerly female fish is able to mate and reproduce as a male.[56]

The types of sex change that are found, the number and fluidity of genders, and the overall social organization of these species are so complex that a detailed terminology has been developed by scientists to describe all the variations. In some species, females turn into males (this is called *protogynous* sex change), while in others males become females (called *protandrous* sex change). In some fish, sex change is *maturational;* that is, it happens automatically to all individuals when they reach a certain age or size or else occurs spontaneously at different times for each individual. In other species, sex change appears to be triggered by factors in the social environment of the fish, such as the size, sex, or number of neighboring fish. In female-to-male fish species, many different gender profiles are found. In some cases, all fish are born female, and males result only from sex change (such a system is called *monandric*). In other cases, both genetic males (born male) and sex-changed males are found (this arrangement is called *diandric*). In these fishes, genetic males are sometimes referred to as *primary males* while sex-changed males are called *secondary males*. Often these two types of males differ in their coloration, behavior, and social organization so that transsexual males form a distinct and clearly visible "gender" in the population.

Things get even more complicated in some species. Among secondary males, some change sex before they mature as females (*premagurational* secondary males), while others change sex only after they live part of their adult life as females (*postmaturational* secondary males). Many species also have two distinct color phases: fish often begin life with a dull color and drab patterning, then change into the more brilliant hues typically associated with tropical fish as they get older. Which individuals change color, when they change, and their gender at the time of the color change can yield further variations. Many species of parrot fishes, for example, have multiple "genders" or categories of individuals based on these distinctions. In fact, in some families of fishes, transsexuality is so much the norm that biologists have coined a term to refer to those "unusual" species that *don't* change sex—*gonochoristic* animals are those with two distinct sexes in which males always remain male and females always remain female.

As an example of how elaborate transsexuality can become in coral-reef fish, consider the striped parrot fish, a medium-sized species native to Caribbean and Atlantic waters from Bermuda to Brazil (the name refers to the fact that its teeth are fused together like a parrot's beak).[57] Striped parrot fish, like many sex-changing fishes, have both males that were born as males and males that were born as females. In fact, more than half of all males in this species used to be females. Moreover, *all* female striped parrot fish eventually change their sex, becoming male once they reach a certain size; the sex change can take as little as ten days to be completed. Sex-changed males have fully functional testes that used to be fully functional ovaries when the fish was female; they are able to mate and fertilize eggs the same way that genetic males do. Striped parrot fish have one of the most complex polygendered societies in the animal kingdom. There are five distinct genders, distinguished by biological sex, genetic origin, and color phase. Biological sex refers to whether the fish has ovaries (= female) or testes (= male). Genetic origin refers to whether the fish was born that sex or has changed from another sex (= transsexual). Color phase refers to the two types of coloration that striped parrot fish exhibit: *initial-phase* fish (so named because all fish start out with this coloration) are a drabber brown or bluish gray, while *terminal-phase* fish are a brilliant blue-green and orange. These three categories intersect to create the following five genders (percentages refer to what proportion of the total population, at any given time, belongs to each gender): (1) genetic females: born female, each of these initial-phase fish will eventually become a male and change color (45 percent); (2) initial-phase transsexual males: born female, these fish become male before they change into their bright colors and are fairly rare (1 percent); (3) terminal-phase transsexual males: born female, these fish become male at the same time they changed color, usually at a later age than genetic males (27 percent); (4) initial-phase genetic males: born male, most of these will change color as they get older (but won't change sex) (14 percent); and (5) terminal-phase genetic males: born male, these fish start out as initial-phase males and change color (but not sex) at a younger age than transsexual males (13 percent).

Along with its numerous genders and fluid changes between them, striped parrot fish society is characterized by a number of intricate systems of social organization and mating patterns, each found in a particular geographic area. One system, known as group spawning or explosive breeding assemblages, is common in Jamaican striped parrot fish. Large groups of up to 20 initial-phase males and females gather to spawn together, swimming in dramatic formations that rapidly change direction. Often, terminal-phase males try to disrupt this mating activity. Another system is found in the waters off Panama and is known as haremic because the basic breeding group consists of one terminal-phase male and several females. These individuals are known as territorials since they live in permanent locations that they defend against intruders. Other fish in the same area, however, associate with each other in different kinds of groups: "stationaries" are celibate (nonbreeding) fish in both initial and terminal phases, while "foragers" gather together to feed in large groups of up to 500 fish. Some of these foraging groups are composed of females and initial-phase genetic males, while others are made up only of terminal-phase males; half of all the females, and all the males, in such groups are nonbreeders. Finally, striped parrot fish in the waters off Puerto Rico and the Virgin Islands associate together in "leks," clusters of small, temporary territories that both initial-phase and terminal-phase males defend and use to attract females for spawning.

Further variations in transsexuality are found in other species. The paketi or spotty, a New Zealand fish, combines transsexuality with transvestism (some females become males before changing color, thus "masquerading" as females), while the humbug damselfish combines transsexuality with same-sex pairings and associations. An even more complex gender system, involving hermaphroditism, transsexuality, transvestism, and apparent homosexual activities, exists in the lantern bass and other fishes. In addition to nontranssexual males and females, some individuals are hermaphrodites (both male and female at the same time) and others are secondary (transsexual) males, while in a few cases individuals exhibit courtship and mating patterns typical of the opposite sex (directed toward individuals of the same sex). All female Red Sea anemonefish start out as males; once they change sex, however, they become dominant to males and tend "harems" of up to nine males, all but one of whom are nonbreeders. Finally, although most transsexual fishes are one-way sex changers, in a few species sex change actually occurs in both directions. In the coral goby, for instance, some individuals go from male to female, others from female to male, and some even undergo multiple sequential changes, "back and forth" from male to female to male, or female to male to female.[58]

As these examples show, not only are transgendered and genderless biologies a fact of life for many animals, they have developed into incredibly sophisticated and complex systems of social organization and behavioral patterning in many species. For those of us used to thinking in terms of two unchanging and wholly separate sexes, this is extraordinary news indeed. Likewise, animal homosexuality itself is a rich and multifaceted phenomenon that is at least as complex and varied as heterosexuality. Animals of the same sex court each other with an assortment of special— and in some cases, unique—behavior patterns. They are both affectionate and sexual toward one another, utilizing multiple forms of touch and sexual technique, ranging from kissing and grooming to cunnilingus and anal intercourse. And they form pair-bonds of several different types and durations and even raise young in an assortment of same-sex family configurations. If, as scientist J. B. S. Haldane stated, the natural world is queerer than we can ever know, then it is also true that the lives of "queer" animals are far more diverse than we could ever have imagined. In the next chapter, we'll take a look at how these different forms of sexual and gender expression in animals compare to similar phenomena in people.

Chapter 2

Humanistic Animals, Animalistic Humans

*T*itus and Ahab—male Gorillas—often courted and had sex with one another in the mountains of Rwanda, while Marchessa sought out her own sex during her pregnancy. In Florida, Bottlenose Dolphins Frank, Floyd, and Algie participated in homosexual activity with each other, as did Gabe and Moe-Miller, West Indian Manatees. Les and Sam (Siamangs) were doing the same in Milwaukee, while Kiku, a female Bonobo living in Congo (Zaire), had sex with her female "mentor" Halu more often than with anyone else in the new troop she joined. Cato and Mola (male Crested Black Macaques), Depp and Nice (male Rhesus Macaques), as well as Saruta and Oro (male Japanese Macaques) and Daddy and Jimmy (Crab-eating Macaques), also mounted one another. On the isle of Corsica, Le Baron and Le Valet (Asiatic Mouflons) were inseparable, as were Marian and her female Grizzly companion in the high mountains of Wyoming. Apolli and Arima—Long-eared Hedgehogs in Vienna—each refused to mate with males after they were separated from one another. In Austria, Greylag gander Pepino had a brief liaison with Florian but was later courted by Serge, while Max, Odysseus, and Kopfschlitz formed a threesome and went on to raise a family with Martina. A White-handed Gibbon named Floyd became sexually involved with George (his father) in Thailand, while Sibujong and Bobo, male Orang-utans, had sexual interactions with one another in Indonesia.[1]

As these examples show, zoologists sometimes bestow names upon the animals they study, lending an unintentionally—and eerily—human quality to their reports of homosexual activity. Although most scientists are careful to avoid anthropomorphizing their subjects, their use of human names such as these reminds us at once of each creature's individuality as well as the dangers of projecting human qualities onto animals. Such naming also demonstrates the nearly universal human preoccupation with seeking connections between ourselves and other species. Where animal behavior—especially sexual behavior—is concerned, it seems that comparisons will inevitably be made between animals and people (even by scientists).

There are a number of genuine connections and points of correspondence between animal and human homosexuality, as well as significant differences. There are also numerous pitfalls in attempting to extrapolate from animal to human behavior, or vice versa. This chapter explores a number of specific animal-human comparisons and the issues surrounding them. For example, we'll address claims that certain aspects of homosexuality are uniquely human, such as various types of sexual orientation, or the treatment of homosexual and transgendered individuals in the larger society. Also to be discussed are the special insights into human behavior offered by primate homosexuality (especially cultural behavior), and the rationale and motivations behind making cross-species comparisons in the first place (especially where the dubious concept of "naturalness" is concerned). Overall, a cautionary note must be sounded: while it is tempting to jump to broad conclusions about human homosexuality based on animal behavior (or vice versa), the full complexity and richness of homosexual expression in both animals and people must be considered. Only then will we begin to understand both the uniqueness and the commonalities of each.

From Pederasty to Butch-Femme: Uniquely Human?

One of the most significant results to emerge from the study of human homosexuality over the past few decades is the enormous variety of forms that this activity takes. From pederasty or "boy love" in ancient Greece, to ritualized homosexual initiation in New Guinea, to butch-femme lesbian relationships, to situational homosexuality in prisons, to contemporary North American gay couples—homosexuality has assumed many guises across history, cultures, and social situations. Thus, while homosexual desires and activities are probably ubiquitous, the specific forms that they assume are intimately shaped by particular sociohistorical contexts. Instead of talking about homosexuality, we should really speak in terms of *homosexualities,* plural, for there are many variations on the theme of same-sex relations.[2]

Animal homosexuality puts a new twist on this observation, since nearly every type of same-sex activity found among humans has its counterpart in the animal kingdom. Comparisons between animal and human homosexuality are inevitably muddled, however, by the lack of an adequate understanding and classification of different types of homosexuality. The confusion surrounding this subject is readily apparent: activities as different as pair-bonding between female Kangaroos and same-sex mounting in male Bighorn Sheep and Bottlenose Dolphins, for example, have all been compared to the sort of homosexual activity that occurs among human beings in prisons.[3] The problem with analogies like this—which are inevitably imperfect and inaccurate—is that something like "prison homosexuality" is itself actually a conflation of many different behavioral variables and diverse patterns of same-sex activity, as are the "corresponding" animal behaviors.[4] In addition to the actual form of the homosexual activity involved (pair-bonding, sexual behavior, etc.), many other factors must be considered, such as consensuality, age, gender presentation of partners, and so on. Thus, a particular example of homosexual activity—whether animal or human—is in reality a unique amalgamation or "blend" of multiple factors, any one of which may be shared with other forms of homosexual activity without necessarily conferring identity between the overall patterns they represent. Comparisons of homosexuality in animals and humans that fail to recognize such complexities are simply misleading.

It is helpful in this regard to think of homosexuality in terms of a number of independent axes, each of which is a continuum joining two "opposite" ends of a particular category (as suggested by researchers Stephen Donaldson and Wayne Dynes, who have developed a typology for human homosexuality based on this framework).[5] For example, one axis might represent the degree to which the homosexual interaction is gendered or role-based (ranging from the heavily role-oriented homosexuality of Native American two-spirits or Euro-American butch-femme lesbians, to the nongendered homosexuality of the South African San peoples or some gay couples in contemporary Euro-American culture). Another axis would represent the age relationship of the partners involved (ranging from no age difference to a clearly age-differentiated interaction); another represents sexual orientation of participants (homosexual ↔ bisexual ↔ heterosexual); another consensuality (forced or nonconsensual ↔ freely chosen or consensual); another genetic relatedness of partners (incestuous ↔ unrelated); another social status or position of same-sex activity (socially sanctioned ↔ socially condemned); and so on.

The utility of such a system is that homosexuality in any given context (or species) can be seen as the intersection at various points on a number of such axes, thereby allowing comparisons to be made across multiple factors. In this chapter a number of these typological axes will be explored in greater detail to show that both animal and human homosexuality exhibit a comparable variability when examined from virtually every angle.[6] Ultimately, we will see that the plurality of homosexualities in both animals and people suggests a blurring of the seemingly opposite categories of nature and culture, or biology and society. On the one hand, it is no longer possible to attribute the diversity of human (homo)sexual expression solely to the influence of culture or history, since such diversity may in fact be part of our biological endowment, an inherent capacity for "sexual plasticity" that is shared with many other species. On the other hand, it is equally meaningful to speak of the "culture" of homosexuality in animals, since the extent and range of variation that is found (between individuals or populations or species) exceeds that provided by genetic programming and begins to enter the realm of individual habits, learned behaviors, and even community-wide "traditions."

Comparisons between animals and people almost inevitably focus on behaviors that are supposed to be uniquely human. As biologist James Weinrich points out, nearly every behavior that was at one time believed to be practiced only by people has been found to have an analogue among animals—including homosexuality:

> There is a long and sordid history of statements of human uniqueness. Over the years, I have read that humans are the only creatures that laugh, that kill

other members of their own species, that kill without need for food, that have continuous female receptivity, that lie, that exhibit female orgasm, or that kill their own young. Every one of these never-never-land statements is now known to be false. To this list must now be added the statement that humans are the only species that exhibit "true" homosexuality. Does anyone ever state that we alone exhibit true heterosexuality?[7]

While many scientists now accept that animals engage in homosexuality, claims about human uniqueness continue to be made regarding the *specifics* of homosexual interactions: people, but not animals, engage in exclusive homosexuality, for example; people, but not animals, exhibit greater variety or "genuine" sexual motivation in their homosexuality; people, but not animals, react with hostility toward homosexuality and live in groups segregated by sexual orientation; and so on.

As we come to understand more and more about animal behavior, premature blanket statements like these have generally proven to be naive, if not incorrect—and this is especially true where homosexuality is concerned, since so much still remains to be learned regarding such activities in animals. In this section we'll address a number of these claims and explore some broader issues surrounding each (this theme will also be taken up in subsequent chapters with regard to the other typological "axes"). While there is some truth to these statements of human uniqueness, none is an absolute line of demarcation between human and nonhuman animals. As always, animal sexuality and social life are far more complex and nuanced than previously imagined: perhaps the only true difference in behavior between the species is that people, but not animals, are prone to make simplistic generalizations.

Exclusively Homosexual, Simultaneously Bisexual: Sexual Orientation

> *... preferential or obligatory adult homosexuality is not found naturally in any mammalian species other than* Homo sapiens.
>
> —W. J. GADPAILLE, 1980

> *Homosexual human couples who remain together throughout their adult life have few, if any, counterparts in wild mammals as far as is known at present.*
>
> —ANNE INNIS DAGG, 1984

> *Exclusive homosexual behavior appears to be absent among nonhuman primates...*
>
> —PAUL L. VASEY, 1995[8]

An oft-repeated claim about homosexuality is that exclusive, lifetime, or "preferential" homosexual activity is unique to human beings, or at least rare among animals (especially among primates and other mammals). This is really a question of sexual orientation—that is, to what extent do animals engage in sexual and related activities with members of the same sex without also engaging in such activities with members of the opposite sex? In fact, exclusive homosexuality of various types occurs in more than 60 species of nondomesticated mammals and birds, including at least 10 kinds of primates and more than 20 other species of mammals.[9] In this section we'll consider these various forms of homosexual orientation and compare them to the wide variety of bisexualities that are also found throughout the animal world.

When discussing the question of exclusive homosexuality, several factors need to be distinguished: the length of time that exclusivity is maintained (short-term versus long-term, including lifetime), the social context and type of same-sex activity involved (pair-bonding versus promiscuity in nonbreeding animals, for example), the type of animal involved (e.g., mammal versus bird), and the degree of exclusivity (e.g., absolute absence of opposite-sex activity versus primary homosexual associations with occasional heterosexual ones, and vice versa). These factors combine in various ways and interact with each other to produce a number of different patterns. To begin with, we will consider long-term or extended exclusivity, since this pattern appears to be the most contested as to its existence among animals. Because species vary widely as to their life expectancy, onset of sexual maturity, and period of adulthood, it is difficult to come up with an absolute definition of *long-term* that has wide applicability. For the purposes of this discussion, though, we will somewhat arbitrarily consider homosexual activity that continues for less than two consecutive years (or breeding seasons) to be short-term, while anything continuing longer is considered extended or long-term, with the understanding that the latter category includes a wide spectrum of possibilities, anywhere from 3 years to a life span of over 40 years.

The only way to absolutely verify lifetime exclusive homosexuality is to track a large number of individuals from birth to death and record all the various homosexual or heterosexual involvements they have. Needless to say, this is a difficult task to accomplish (especially in the wild) and has been achieved for only a few species—indeed, in many cases the comparable evidence for lifetime exclusive *heterosexuality* is not available either, for precisely the same reasons. Nevertheless, in at least three species of birds—Silver Gulls, Greylag Geese, and Humboldt Penguins—fairly extensive tracking regimes have been conducted, and individuals who form only homosexual pair-bonds throughout their entire lives have been documented. In some cases these are continuous pair-bonds that last upward of 15 years in Greylag Geese and 6 years in Humboldt Penguins (until the death of the individuals involved), while in other cases (e.g., Silver Gulls) individuals may also have several same-sex partnerships during their lives (either because of "divorce" or death of the partners).[10]

While absolute verification of lifetime homosexuality is not directly available for other species, extended periods of same-sex activity, perhaps even lifelong, are strongly suggested. In Galahs, Common Gulls, Black-headed Gulls, Great Cormorants, and Bicolored Antbirds, for example, specific homosexual partnerships have been documented as lasting for as long as six years (or individuals having several consecutive homosexual associations for that length of time); in most of these cases the absence of heterosexual activity for at least one partner has been docu-mented or is highly likely. In many other bird species, same-sex partnerships that last anywhere from several years to life probably also occur: Black Swans, Ring-billed Gulls, Western Gulls, and Hooded Warblers, for instance. Although these du-rations have not been confirmed in specific individuals, homosexual pairs that continue for at least two years or birds who consistently form same-sex pairs for that time have been verified.[11] In still other cases, long-term same-sex bonds un-doubtedly occur because homosexual pairs in these species typically follow the pat-tern of heterosexual pairs, which are usually lifelong (or of many years duration): Black-winged Stilts, Herring Gulls, Kittiwakes, Blue Tits, and Red-backed Shrikes, among others. Finally, it must also be remembered that in many animals (e.g., Pied Kingfishers), same-sex (and opposite-sex) pair-bonds that last two to three years can still be lifelong, owing to the relatively short life span of the species.

In mammals, cases of long-term, exclusively homosexual pairing are indeed rare. One example is male Bottlenose Dolphins: the majority of males in some pop-ulations form lifelong homosexual pairs, specific examples of which have been ver-ified as lasting for more than ten years and continuing until death. Although the sexual involvements (both same- and opposite-sex) of such individuals have not in all cases been exhaustively tracked, it is quite likely that at least some of these ani-mals have little or no sexual contact with females (since breeding rates tend to be low in Bottlenose communities, with many individuals not participating in repro-duction each year and, by extension, possibly throughout their lives).[12] Absolute verification in this species, however, may not be forthcoming, since it is virtually impossible to continuously monitor the sexual behavior of all individuals within a given population of an oceangoing species. Bottlenose Dolphins are exceptional, however, in that the homosexual pattern in this species is distinct from the hetero-sexual one: opposite-sex pair-bonding does not occur among Bottlenose Dolphins. In most other species, homosexual and heterosexual activities tend to follow the same basic patterns, whether this means pair-bonding, polygamy, promiscuity, or some other arrangement.[13] Lifetime homosexual couples are not prevalent among mammals, therefore, for the same reason that lifetime heterosexual couples are not: monogamous pair-bonding is simply not a common type of mating system in mammals (it is found in only about 5 percent of all mammalian species).[14]

Nevertheless, long periods of exclusive homosexuality among mammals have been documented for other social contexts besides pair-bonding. In many species, significant portions of the population do not engage in breeding or heterosexual pursuits for at least a part of their lives. Because some of these animals continue to engage in same-sex interactions, however, they are exclusively homosexual for at least that time, which can be considerable. Among Gorillas, for example, males of-ten live in sex-segregated groups where homosexual activity takes place. The aver-age length of stay in a male-only group is more than six years, although some males remain in such exclusively homosexual environments for much longer. One indi-vidual lived in an all-male group for ten years, staying until his death, and nearly a third of the males who joined the group over a thirteen-year study period were still with the group at the end of that time. Likewise, Hanuman Langur males may spend upward of five years in male-only bands in which homosexual activity takes place, and some individuals live their entire adult lives in such groups.[15] In a number of hoofed mammals, a similar form of exclusivity based on sex segregation occurs: only a few individual males participate in heterosexual mating, while the remainder live in "bachelor herds" where homosexual activity often takes place.[16] Among Mountain Zebras, for example, males stay an average of three years in such groups before joining breeding groups, and some remain their whole lives without ever mating heterosexually. Analogous patterns occur in a number of other species where only a relatively small percentage of males ever breed: antelopes and gazelles, including Blackbuck, Pronghorn, and Grant's and Thomson's Gazelles; Giraffe; Red Deer; Mountain Sheep; seals such as Northern Elephant Seals and Australian and New Zealand Sea Lions; and birds such as Ruffed Grouse, Long-tailed Hermit Hummingbirds, and Guianan Cock-of-the-Rock. In some hoofed mammals such as American Bison, a related age-based pattern is found. Males generally do not participate in heterosexuality until they are five to six years old; prior to that time, many engage in homosexual activities, entailing a period of exclusively same-sex activity of up to five years for some individuals.[17]

Other patterns of exclusivity occur as well. In Nilgiri Langurs and Hamadryas Baboons, for instance, generally only the highest-ranking male in a group mates with females; remaining males, if they engage in sexual activity at all, are sometimes involved only in homosexual pursuits. In Nilgiri Langurs, cases of nonbreeding males having only same-sex interactions for at least four years have been documented. In Ruffs, there are several different categories of males, many of whom rarely, if ever, mate heterosexually; some of these individuals participate in homosexual activities and may do so over an extended period, perhaps even for life. Finally, in some species same-sex activity may be exclusive because it is incestuous, involving a parent and its nonbreeding offspring. In male White-handed Gibbons, for instance, father-son sexual relations may continue for several years; the son is not involved in concurrent heterosexual activity, and sometimes even his father may have little or no opposite-sex mating during this time. Red Fox daughters can remain with their family group for many years—sometimes they never leave—during which time they may be involved in occasional same-sex mounting with their mothers (or each other) but no heterosexual activity.[18]

Thus, while in many species documentation of exclusive long-term homosexuality (or heterosexuality, for that matter) is not directly available, exclusivity can be inferred from the general patterns of social organization in the species. For example, a system that involves large numbers of nonbreeders (including individuals who never mate heterosexually during their entire lives), combined with homosexual activities among at least a portion of these nonbreeding animals (sometimes in sex-segregated groups), will invariably entail some individuals whose only sexual contacts are with animals of the same sex. For some animals this period of exclusive homosexuality lasts no more than a few years; for others, it may extend considerably longer, even for the duration of their lives.

Shorter periods of exclusive or "preferential" homosexuality also occur. Sexual "friendships" in Stumptail Macaques and Rhesus Macaques, for example, and homosexual consortships in Japanese Macaques, last anywhere from a few days to several months, during which time there are no heterosexual involvements. During the seasonal aggregations of male Walruses and Gray Seals, same-sex activity usually occurs to the exclusion of opposite-sex behavior. Female Marmots forgo breeding for a couple of years but may still have sexual contact with other females. Same-sex pair bonds in King Penguins and homosexual associations in female Orang-utans are also exclusive for their duration. Of course, many of these animals are actually bisexual because they also engage in heterosexual pursuits at other times during their lives, but while they are involved in same-sex activity, they do not simultaneously engage in opposite-sex behavior. Thus, when considering various forms of exclusive homosexuality it is also necessary to understand the different types of *nonexclusive* homosexuality—that is, bisexuality.

The participation of an individual in both homosexual and heterosexual activities is widespread among animals: bisexuality occurs in more than half of the mammal and bird species in which same-sex activity is found. Nevertheless, there are many different forms and degrees of bisexuality, and these must be carefully distinguished when discussing sexual orientation in animals. A useful differentiation to start with is *sequential* as opposed to *simultaneous* bisexuality, a distinction that hinges on the temporal or chronological separation between homosexual and heterosexual pursuits. In sequential or serial bisexuality, periods of exclusively same-sex activity alternate with periods of exclusively opposite-sex activity. In simultaneous bisexuality, homosexual and heterosexual activities co-occur or are interspersed within a relatively short period (say, within the same mating season). Thus, many of the "shorter" periods of exclusive homosexuality that we have been considering actually fall into a larger pattern of sequential bisexuality, which itself forms a continuum in which same-sex activity may occupy anywhere from several months to several decades of an animal's life. Moreover, the "sequentiality" of bisexual experience assumes many different forms: a seasonal pattern (for example, in Walruses, who engage in homosexuality primarily outside of the breeding season, or in

◄ *A group of male Walruses off the coast of Round Island (Alaska). Pairs of males are participating in courtship and other activities with each other while floating in the water. Male Walruses are often seasonally bisexual, engaging in homosexual pursuits outside the breeding season.*

Gray Whales, during migration and summering); an age-based pattern (e.g., in Bison or Giraffe, where same-sex activity is more characteristic of younger animals, or in which the earlier years of an animal's life are occupied largely with homosexual pursuits, to be followed by heterosexual activity in later years—or the reverse, as in some African Elephants); onetime "switches," in which individuals change over from heterosexual to homosexual activity at a specific point in time (e.g., Herring Gulls, Humboldt Penguins), or from homosexual to heterosexual (e.g., Great Cormorants); as well as less structured sequencing, in which several periods of same- and opposite-sex activity of varying lengths may alternate with each other (e.g., Gorillas, Silver Gulls, King Penguins, Bicolored Antbirds).[19]

Simultaneous bisexuality also assumes many guises. At one extreme, sexual activity with same-sex and opposite-sex partners takes place at *literally* the same time: "pile-up" copulations, for example, in which a male mounts another male who is mounting a female (e.g., Wolves, Laughing Gulls, Little Blue Herons), or group sexual activity in which some or all participants are interacting with both males and females (e.g., Bonobos, West Indian Manatees, Common Murres, Sage Grouse). At the other extreme, individuals court or mate with both sexes separately, over short but relatively distinct spans of time, as in Crab-eating Macaques, Mountain Goats, Redshanks, and Anna's Hummingbirds. In between these extremes are other patterns, such as ongoing bisexual trios and quartets, in which both same-sex and opposite-sex partners are bonded to one another concurrently (e.g., Greylag Geese, Oystercatchers, Jackdaws). Another form of simultaneity involves an animal in a pair-bond with a member of the opposite sex who has occasional courtship and/or sexual encounters with a member of the same sex (or vice versa). For example, male Herring and Laughing Gulls, Herons, Swallows, and Common Murres who have female partners, and female Mallard Ducks who have male partners, sometimes mount birds of the same sex. Conversely, female Snow Geese, Western Gulls, and Caspian Terns and male Humboldt Penguins and Laughing Gulls who have same-sex partners sometimes mate with opposite-sex partners. Still another variation is found in Lesser Flamingos: males in homosexual pairs sometimes try to mate with females who are themselves in homosexual pairs. And in some animals such as Bottlenose Dolphins, Black-headed Gulls, and Galahs, the combinations are even more varied: different forms of sequential and simultaneous bisexuality, as well as exclusive homosexuality (and heterosexuality) are found in different individuals within the same species and may even combine in the same individual at different points in time.

Even within a given category of bisexuality—say, simultaneous bisexuality involving interspersed homosexual and heterosexual activity—each individual within a population generally exhibits a unique sexual orientation profile, consisting of his or her own particular combination of same- and opposite-sex activity. The concept of a scale or continuum as developed by Alfred Kinsey for describing human sexual orientation is useful here: within each species, individuals generally fall along a range from those exhibiting predominantly or exclusively heterosexual behavior, to those exhibiting a balance of both, to those exhibiting predominantly or exclusively homosexual behavior, and every variation in between.[20] Species as a whole also differ as to where the majority of individuals fall along this continuum, and how many engage in more exclusive homosexuality or heterosexuality as opposed to more equal bisexuality. Thus, among Bonobos every female participates in both homosexual and heterosexual activity, but the proportion of same-sex behavior exhibited by each of the females in one particular troop varied between 33 percent and 88 percent (averaging 64 percent); in female Red Deer, from 0–100 percent (averaging 49 percent); among Bonnet Macaque males, between 12 percent and 59 percent (averaging 28 percent); in male Pig-tailed Macaques, from 6–22 percent (averaging 18 percent); and among Kob females, from 1–58 percent (averaging 11 percent).[21] In other words, within an overall pattern of bisexuality, individual animals exhibit varying "degrees" of bisexuality—different "preferences," as it were, for homosexual as opposed to heterosexual activity.

These findings are particularly relevant since the concept of a scale or continuum of (homo)sexual behavior and orientation is yet another example of something still thought to be "uniquely human." The Bonobo data (as well as that for the other species) directly refute one primatologist's recent claim that "all wild primates we have seen within a particular species are equally homosexual. . . . If you lined up ten female bonobos, it's not like one would be a 6 on the Kinsey scale and another a 2. They would all be the same number. It's only humans who adopt identities."[22] Of course, the Kinsey scale is specifically a measure of behaviors and *not* identities (it was designed expressly to bypass the often problematic issue of people's "self-identification"), and certainly no animal studies purport to assess anything as subjective as sexual "identity." In its intended usage, though, the Kinsey scale (or a comparable measure of sexual gradations) in fact appears to be particularly apt for Bonobos. The figures cited above are based on the work of Dr. Gen'ichi Idani in Congo (Zaire), who studied a troop consisting of (coincidentally) exactly ten female Bonobos and tabulated all their homosexual genital rubbings versus heterosexual copulations over a three-month period. The percentages of homosexual activity in these individuals were 33, 36, 47, 68, 68, 70, 75, 75, 82, and 88 percent. Idani also tabulated the number of different male and female partners of each female (another possible measure of degree of bisexuality or behavioral "preference"). Again, the percentage of partners that were same-sex exhibits a range across all females: 36, 50, 50, 54, 67, 67, 67, 69, 71, and 80 percent. Clearly these individuals fall into a spectrum in terms of their sexual behavior and thus exhibit different degrees of bisexuality in terms of their sexual orientation (although none are in fact exclusively heterosexual or homosexual).[23]

"Preference" for same-sex activity is, admittedly, a rather elusive concept to measure when dealing with nonhumans (though not nearly as slippery as "identity"). Although we cannot access their internal motivations or "desires," animals do offer a number of other clues as to their individual "preferences" in addition to the proportion of their behaviors or partners that are same-sex. These include homosexual activity being performed in (spite of) the presence of members of the opposite sex, individuals actively competing for the attentions of same-sex partners (rather than "resorting" to such activity), advances of opposite-sex partners being ignored and/or refused, and "widowed" or "divorced" individuals continuing to pair with same-sex partners after the loss of a homosexual mate (even when opposite-sex partners are available). These types of behaviors have in fact been re-ported in more than 50 mammals and birds (see the profiles for some examples), indicating that for at least some individuals in these species, same-sex activity has "priority" over opposite-sex activity in some contexts. The converse is also true for species such as Canada Geese, Silver Gulls, Bicolored Antbirds, Jackdaws, and Galahs: in situations where opposite-sex partners are not available, only a fraction of the population engages in same-sex activity, indicating more of a heterosexual "preference" in the remainder of the population.[24] Animals who do participate in same-sex activity in such a situational context could perhaps be said to exhibit a "latent" bisexuality; i.e., a predominantly heterosexual orientation with the poten-tial to relate homosexually under certain circumstances. Another factor to be con-sidered when evaluating individual "preferences" or degrees of bisexuality is the consensuality of the sexual interaction. Female Canada Geese and Silver Gulls in homosexual pairs, for example, may engage in occasional heterosexual copulations under duress; i.e., they are sometimes forcibly mated or raped by males. Likewise, heterosexually paired males in Common Murres, Laysan Albatrosses, Cliff Swal-lows, and several Gull species may be forcibly mounted by other males. Technically, all such individuals are "bisexual" because they engage in both homosexual and heterosexual activity, but the sort of bisexuality they exhibit is far different from that of a female Bonobo or a male Walrus, for instance, who willingly mates with animals of both sexes.

Broad patterns of sexual orientation across individuals show almost as much variation as that within individuals. In some species, the majority of animals are ex-clusively heterosexual, but a small proportion engage in bisexual activities (e.g., Mule Deer) or exclusively homosexual activities (e.g., male Ostriches). In others, the vast majority of individuals are bisexual and few if any are exclusively hetero-sexual or homosexual (e.g., Bonobos). Other species combine a pattern of nearly universal bisexuality with some exclusive homosexuality (e.g., male Mountain Sheep). In still other cases, the proportions are more equally distributed, but still vary considerably. In Silver Gulls, for instance, 10 percent of females are exclusively homosexual during their lives, 11 percent are bisexual, and 79 percent are hetero-sexual. Homosexual-bisexual-heterosexual splits for specific populations of other species include: 22-15-63 percent for Black-headed Gulls; 9-56-35 percent for Japanese Macaques; and 44-11-44 percent for Galahs.[25]

Thus, sexual orientation has multiple dimensions—social, behavioral, chrono-logical, and individual—which must all be taken into account when assessing pat-terns of heterosexual and homosexual involvement. It is true that exclusive homosexuality in animals is less common than bisexuality—but it is not a uniquely human phenomenon, for it occurs in many more species than previously supposed. Moreover, because of the wide prevalence of bisexuality—both within and across species—exclusive *heterosexuality* is also certainly less than ubiquitous. Animals, like people, have complex life histories that involve a wide spectrum of sexual ori-entations, with many different degrees of participation in both same- and opposite-sex activities. To the question "Do animals engage in bisexuality or exclusive homosexuality?" we must therefore answer "both and neither." There is no such thing as a single type of "bisexuality" nor a uniform pattern of "exclusive homosexuality." Multiple shades of sexual orientation are found throughout the animal world—sometimes coexisting in the same species or even the same individual—forming part of a much larger spectrum of sexual variance.

Nonchalant Onlookers and Gay Ghettos: Social and Spatial Responses

> While homosexual behavior is widespread among our primate relatives, aggression specifically directed toward individuals that engage in it appears to be a uniquely human invention.
>
> —PAUL L. VASEY[26]

An aspect of animal homosexuality that has received little attention in both popular and scientific discussions is the position or "status" of homosexual, bisexual, and transgendered individuals in the larger society. What kind of social response do they evoke from the animals around them? What is their spatial relation to the rest of the population—are they segregated, fully integrated, or somewhere in between? Primatologist Paul L. Vasey suggests that homosexual behavior in primates is characterized by a noticeable lack of hostility and segregation from the animals around them, and by and large this does appear to be true—not only for primates, but also for the vast majority of other species in which homosexual activity occurs. Almost without exception, animals with "different" sexualities and/or genders are completely integrated into the social fabric of the species, eliciting little of the attention, hostility, segregation, or secrecy that we are accustomed to associating with homosexuality in our society. Observer after scientific observer has commented on how homosexual behavior in animals is greeted with nonchalance from nearby animals. Individuals move effortlessly between their homosexual activities and other social interactions or behaviors without eliciting so much as a second glance from the animals around them.[27]

Where individuals engaging in homosexual activity do attract attention, it is usually out of simple curiosity (e.g., African Buffalo, Musk-oxen), or else because other animals want to participate.[28] In a number of species such as Bonobos, Killer Whales, West Indian Manatees, Giraffes, Pronghorns, Common Murres, and Sage Grouse, homosexual interactions between two animals often develop into group sessions as more and more animals are drawn to the activity and join in. This is also true for heterosexual interactions in many of these species, and sometimes homosexual and heterosexual activity are part of the same group interaction. This illustrates an important point concerning the integration of homosexual activity within the larger social framework: when bisexuality is prevalent in a species, or when a large proportion of the population engages in homosexual activity (as is often the case), the distinction between "homosexual" and "heterosexual" animals melts away, as does the potential for aggressive responses based on those categories. An "observer" of homosexual activity could just as easily be a participant at some other time, and any separation between animals that engage in same-sex activity and those that don't becomes essentially arbitrary.

Even in species where homosexuality, bisexuality, or transgender are not wide-spread, animals that participate in same-sex behaviors (or transgendered individuals) are not generally treated to adverse reactions from the majority around them. Rather, homosexual activity is regarded as routinely as heterosexual activity is. In fact, in many species it is heterosexual, not homosexual, behaviors that draw a negative response. In numerous primates and other animals, for example, male-female copulations are regularly harassed and interrupted by surrounding animals. Same-sex activity in these species is either disregarded altogether (e.g., Stumptail Macaques), or else is subject to a much lower rate of harassment and interruption than opposite-sex matings (e.g., Hanuman Langurs, Japanese Macaques).[29] Adult male Bonobos interfere with the heterosexual pursuits of younger males while ignoring (or even participating in) their homosexual activities, while heterosexual breeding pairs of Jackdaws, rather than same-sex pairs, are sometimes terrorized by nonbreeding heterosexual pairs (who may even kill their young). And in Guianan Cock-of-the-Rock, heterosexual courtship interactions are routinely interrupted and harassed by other males while homosexual activities are not. In fact, females defer to males engaging in same-sex courtship or copulation (by leaving or avoiding the display grounds while this activity is going on), and males may actually interrupt heterosexual interactions by initiating homosexual ones.[30]

Not only are homosexuality and transgender largely devoid of negative responses from other individuals, in some cases they actually appear to confer a positive status on the animals involved. In species that have a ranked form of social organization, for instance, homosexual activities are often found among the highest-ranking individuals (e.g., Gorillas, Bighorn Sheep, Takhi, Gray-capped Social Weavers). Likewise, transgendered animals sometimes have high status in a population (e.g., Savanna Baboons) or are more successful than other animals at obtaining sexual partners (e.g., Red Deer, Common Garter Snakes).[31] While the benefits experienced by these individuals are not necessarily a direct result of their transgender or homosexuality, in a few cases individuals actually do appear to rise in status or obtain other positive results specifically because of their homosexual activities. Black Swans and Greylag Geese who form homosexual partnerships, for example, often become powerful, high-ranking forces in their flocks, in part because the combined strength of the paired males gives them an advantage that single males and heterosexual pairs do not have. In fact, Black Swan male pairs sometimes acquire the largest and most desirable territories in their domain, relegating other birds to a distinctly disadvantaged status.[32]

Same-sex couples in many animals routinely defend their home territories against intruders or assist their partners in conflicts with other individuals (as do heterosexual pairs).[33] However, some homosexual and transgendered individuals in a number of species take this a step further, not merely defending themselves but actually going on the offensive. Gander or cob pairs often become so powerful that they are able to "terrorize" an entire flock, attacking individuals (as in Greylag Geese) or even forcing heterosexual pairs to give up their nests and eggs (as in Black Swans), which they take over to raise as their own. Male pairs of Flamingos have also been known to steal nests from other birds, while single males occasionally pursue and harass heterosexual pairs out of interest in the male (rather than the female) partner. Female pairs of Orange-fronted Parakeets often behave aggressively toward heterosexual pairs and may actively "dominate" them through attacks and threats, even successfully competing against them for possession of nesting sites. Laughing Gull homosexual pairs sometimes intrude on territories belonging to neighboring heterosexual pairs and harry the owners, as do pairs of male Rose-ringed Parakeets. A similar pattern has also been reported for Nilgiri Langurs, in which two males who live in the same troop and sometimes participate in same-sex mounting with each other (without necessarily being bonded to one another) may cooperate in attacking males from neighboring troops. Male Lions who are involved in homosexual courtship and sexual activities may attack other males who get too close to them, leading to intense fights in which the courting pair is often assisted by other group members who are not themselves directly involved in the homosexual activity. Homosexual consortships between female Rhesus Macaques can develop into powerful and highly aggressive alliances when the partners take the initiative in attacking other individuals and even driving them from the troop; female Japanese Macaques often vigorously compete with males (and other females) for access to female sexual partners. One female Common Chimpanzee that had sexual relationships with other females was consistently aggressive toward other individuals and came to be feared by Chimps of both sexes. Sometimes the aggression is directed at rival heterosexual partners: a female Livingstone's Fruit Bat who sexually pursued her own mother, for example, successfully fought off males that were also interested in mating with her mother. Finally, a transgendered Savanna (Chacma) Baboon was one of the strongest and highest-ranking members of her troop; described as exhibiting "courage and determination," she routinely interfered in heterosexual matings by threatening, "capturing," and then "carrying off" the male partner so she could mate with him.[34]

Ironically, then, some of the most aggressive interactions surrounding variant sexualities/genders in animals involve heterosexual individuals being attacked, harassed, or invaded by homosexual, bisexual, and transgendered individuals. Nevertheless, the converse situation is not unknown: there are a number of examples of homosexual animals being targeted by heterosexual ones. All such cases involve a male interfering with homosexual activity between two females, often in an attempt to gain sexual access to one of the females. Male Brown Capuchins, Rufous Rat Kangaroos, and Sage Grouse occasionally try to break up mating activity between females, while male Gorillas have been known to attack two females that are having sex together. A male Bonobo repeatedly tried to interfere in the sexual activity between females by screaming, jumping, and sometimes even hitting them; rather than preventing their sexual interactions, however, this simply caused the females to have sex with each other furtively until he gave up harassing them, after which they could do so openly. Male Canada Geese and Wapiti sometimes try to separate female pairs and mate with one member by driving the other away or isolating her from her companion (the females usually do manage to get back together), while female Japanese and Rhesus Macaques in homosexual consortships are occasionally threatened and charged by males. Jackdaw females who are bonded to each other as part of a bisexual trio may be hindered in their joint parenting efforts by their male partner, who sometimes prevents one of the females from having access to their nest. In some cases, this may lead to a loss of eggs or young.[35] Notably, these reaction patterns are not typical for most of these species, since on other occasions animals usually have no adverse response to same-sex activity (e.g., in Bonobos, Gorillas, Rufous Rat Kangaroos, and Sage Grouse). And rarely do these attempts at interference (even when violent) force individuals to permanently cease homosexual activities: rather, they simply alter their patterns of relating or resume their activities once the interference has stopped.

In contrast, there is a consistent pattern among White-tailed Deer of highly aggressive attacks against the transgendered "velvet-horns" (individuals who combine both male and female characteristics). These animals are continually hounded by nontransgendered Deer of all ages and sexes, who drive them away and prevent them from approaching feeding grounds. Sometimes a "gang" of up to half a dozen bucks will attack a velvet-horn, charging, chasing, and severely wounding it with their antlers. Possibly as a result of this social ostracism, velvet-horns tend to associate only with other velvet-horns, forming their own groups and generally avoiding other Deer.[36] Other than this example, though, it is rare to find animals with different sexualities or genders living separately because of persecution from members of their own species.

In many cases, animals that are involved in homosexual interactions do live in segregated groups, but their spatial and social separation from other individuals is based on factors other than their sexuality (since such groups typically also contain individuals who do not engage in homosexuality). Among such factors are age, sex, breeding status, social rank, activity patterns, and various combinations of these. For example, homosexual activity is characteristic of groups of younger, nonbreeding, and/or lower-ranking Northern Elephant Seals; of nonbreeding males in the sex-segregated "bachelor herds" characteristic of many hoofed and marine mammals; of groups of Cliff Swallows engaged in mud-gathering activities away from the nesting colonies; of older, solitary African Elephant males; of nonbreeding Pied Kingfishers who are not involved in helping heterosexual pairs; and of groups of male Gray Seals who gather together during the molting season. Physical disabilities can also isolate individuals into their own groups: in Greenshanks, for instance, flocks of one-legged birds have been observed socializing and migrating separately from other individuals. This is probably because they are unable to keep up with other birds rather than because of social ostracism, since two-legged birds are also sometimes found in such flocks.[37] In contrast, although some Greenshanks participate in homosexual activity, no corresponding "flocks" of clearly homosexual or bisexual birds are known in this species.

Still other factors besides hostility from other animals may be involved in the occasional segregation of individuals that participate in homosexual activity. Among Ring-billed Gulls, for instance, female pairs are sometimes relegated to poorer-quality nesting sites or smaller territories, or they end up congregated together in the spaces between territories belonging to heterosexual pairs. Although this could be due to active hostility from neighboring birds, it is just as likely due to the fact that female pairs are generally not as aggressive as male-female pairs and consequently are unable to defend their nest sites from the encroachments that *all* pairs must endure in crowded colonies. In addition, heterosexual pairs of Ring-billed Gulls that are younger or less experienced also tend to end up in suboptimal locations, and in some colonies female pairs are fully integrated or randomly distributed rather than being peripheralized or clustered. This indicates that hostility toward female pairs is not ubiquitous and, if present at all, is not directed exclusively toward female pairs. In some species segregation is actively initiated by the individuals who are involved in homosexual activity. Female Japanese Macaques in homosexual consortships, for instance, isolate themselves physically and socially from other troop members, including their relatives, to spend time together. Likewise, Black Swan male pairs may end up physically separated from other individuals, but this is because their territories are the most expansive and also because they are aggressive toward other birds that approach them. Greylag gander pairs tend to occupy a peripheral position in their flocks, but it is unlikely that they have been "forced" to the edges, since male pairs are typically more domineering than any other birds in the flock. Some scientists have suggested that such homosexual pairs may actually be performing the role of "sentinels" or guards for the group as a whole, hence their position at the flock's "border." There is also evidence that homosexual Mallard Ducks prefer each other's company and tend to congregate together: when large numbers of male pairs were brought together in captivity, for instance, they tended to form their own flocks and socialize with each other rather than with heterosexual birds. The reason such flocks are not often seen in the wild, then, may simply be a matter of numbers. Because same-sex pairs tend to make up a minority of the population in this species, it is unlikely that enough homosexual individuals would ever be present together in a wild flock to form their own large groups.[38]

Nevertheless, the virtual absence of segregated subgroups of homosexual or bisexual individuals in the animal world is probably related at least in part to the general lack of overt hostility toward homosexuality among animals. Of course,

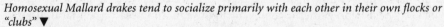

Homosexual Mallard drakes tend to socialize primarily with each other in their own flocks or "clubs" ▼

multiple factors are undoubtedly involved, as is true for the formation of segregated groups of homosexuals among people. The emergence of "gay ghettos" or subcultures in some human societies is a complex process related to many things besides gaining refuge from persecution, such as the need to find and associate with one's own kind, the formation of a homosexual "identity," the development of economic independence, and so on. Nor are such groups merely a defensive response to a hostile society: as with many other minorities, such "ghettos" may begin as a necessary survival tactic but then develop into vital and enriching subcultures of their own. For animal societies we have already seen that many other factors—widespread bisexuality, for instance, or small numbers of animals participating in same-sex activities—can mitigate against the formation of separate groups. Conversely, segregated social units in which homosexual activity takes place often form for reasons that are (initially) unrelated to the sexuality of the animals involved. However, it is striking that both active hostility toward individuals involved in homosexuality and segregation of such individuals are rare occurrences in the animal kingdom. While neither of these social responses to homosexuality is uniquely human (as has been claimed), they are generally uncharacteristic of animal societies. Homosexuality, bisexuality, and transgender are usually as much a part of animal social life as heterosexuality, regardless of their prevalence or frequency of occurrence. In this respect, the vast majority of other creatures have an approach to sexual and gender variance that is decidedly *humane*, rather than human—and they might even offer us models of how our societies could integrate differently oriented or ambiguously gendered individuals into the fabric of social life.

Sexual Virtuosos: Heterosexuality and Homosexuality Compared

The complete *pattern of human heterosexuality is not found in any other species (social-class differences in sexual behavior, pair-bonding, face-to-face copulation, hidden menstrual/estrous cycles, oral and anal intercourse, etc.), although any* single *aspect of human heterosexuality can be found in some animal species. The same statement can be made about human homosexuality.*

—JAMES D. WEINRICH[39]

It is ironic that this last assertion about how human beings are unique in their sexual behavior should have been made by the same scientist who commented on how rarely such statements of human uniqueness prove to be true. Indeed, now that more detailed and comprehensive information is available about animal homosexuality, it appears that at least three species rival, if not equal, human beings in the variability and "completeness" of their sexual expression: Bonobos, Orangutans, and Bottlenose Dolphins. For each of the features mentioned above, an identical or equivalent aspect of behavior can be found, at least in a same-sex context. While none of these species has rigidly stratified "social classes," there are discernible differences in sexual behavior between animals of different ages and social statuses. Homosexual activity is often more frequent in younger, lower-ranking female Bonobos who have recently joined a new troop, for instance, while younger individuals are often "on the top" during female homosexual interactions and "on the bottom" during male homosexual interactions. There is also some evidence that sexual activity between females occurs more often when they belong to distant rank classes. Adolescent or younger-adult Orang-utans participate in same-sex activities more than older, higher-ranking individuals and also exhibit distinctive heterosexual patterns. Younger male Bottlenose Dolphins tend to form their own groups in which same-sex activity is more common than among pair-bonded individuals— while adult males in this species generally do form lifelong bonds with each other.

Although Bonobos do not have exclusive pair-bonding per se, females do form long-lasting bonds with each other that include sexual interactions; adolescent females also typically pair up with an older "mentor" female when they arrive in a new group and engage in sexual activity most frequently with her. Orang-utans often form pairlike consortships in heterosexual contexts, and similar sorts of associations also occur in homosexual contexts. Sexual interactions between female Bonobos usually occur in a face-to-face position, as do heterosexual (and some homosexual) interactions in Orang-utans, as well as most copulations in Bottlenose Dolphins (for the latter, a "belly-to-belly" position is perhaps a more apt characterization). "Hidden estrous cycles" refers to the fact that no overt physical changes signal the various phases of a woman's sexual cycle. A female Bonobo's sexual skin swells with her cycle, although it is present for the majority of the cycle and is not associated specifically with ovulation. Bottlenose Dolphins do not generally give any visual cues as to their sexual cycles or timing of ovulation, nor do female Orang-utans.[40] In any case, all three species engage in sexual activity throughout the female's cycle the way human beings do, which is one important consequence of concealed sexual cycles in humans. Finally, Bonobos, Orang-utans, and Bottlenose Dolphins all engage in forms of anal and oral sex (the latter including fellatio in Bonobos and Orangs, cunnilingus in Orang-utans, and beak-genital propulsion in Dolphins).[41]

One area relating to sexual variability where animal and human homosexuality have been claimed to be comparable, rather than different, concerns the variety of sexual acts or positions used in homosexual as opposed to heterosexual contexts. Masters and Johnson found that gays and lesbians in long-term relationships often had better sexual technique and more variety in their sex lives than married heterosexuals. James Weinrich has suggested a parallel to this observation among animals, claiming that animals that engage in same-sex activity are, in a sense, sexual "virtuosos," employing a wider range of sexual behaviors, positions, or techniques than do their heterosexual counterparts.[42] Although the validity of this claim with respect to human beings cannot be directly addressed here, its accuracy with respect to animals can be assessed—and it appears that in this case the situation is considerably more complex than previously supposed.

It is certainly true that homosexual repertoires have a wider range of sexual acts than heterosexual repertoires in a number of species. In Stumptail and Crab-eating Macaques, for instance, oral sex and mutual masturbation occur primarily, if not exclusively, between same-sex partners. Male Bonobos have a form of

mutual genital rubbing known as penis fencing that is unique to same-sex interactions. Male West Indian Manatees employ a wider variety of positions and
forms of genital stimulation during sex with each other than do opposite-sex
partners. Various types of anal or rump stimulation (besides mounting or intercourse) occur in homosexual but not heterosexual contexts in several Macaque
species, Siamangs, and Savanna Baboons. Only same-sex partners participate in
reciprocal mounting in at least 15 species, including Bonnet Macaques, Mountain Zebras, Koalas, and Pukeko.[43]

This does not, however, appear to be part of an overall pattern, especially where
sexual activities other than mounting or intercourse are concerned. For example, of
the 36 species (exhibiting homosexual behavior) in which some form of oral sex is
practiced, in only 10 (28 percent) of these is oral-genital stimulation limited to homosexual contexts, and in some cases (e.g., Rhesus Macaques, Caribou, Walruses,
Lions) genital licking is a uniquely heterosexual act. Similarly, manual stimulation
of the genitals or masturbation between partners is limited to same-sex interactions in 15 of the 27 species (55 percent) where this behavior occurs; in the remaining animals, both heterosexual and homosexual (or, in some cases, only
heterosexual) partners are involved. Even anal or rump stimulation (besides intercourse) is found in heterosexual contexts in half of the species (6 of 12) that engage
in such activities. Combining these observations, we find that a variety of sexual
acts are part of both heterosexual and homosexual repertoires in the majority of
cases, with behaviors unique to same-sex interactions occurring only in about 40
percent of the cases.

As a matter of fact, in most instances both heterosexual and homosexual acts
are equally "uninspired," involving nothing more exotic than mounting behavior in
the front-to-back position typical of mating in most animals. Even considering animals where other mounting positions are used, however, it is overly simplistic to
claim that same-sex activity involves more "versatility." In many species a variety of
positions are employed in *both* heterosexual and homosexual situations. Furthermore, even though their frequency of use differs depending on the context, the major distinctions in mounting positions often lie along lines of gender rather than
sexual orientation. The differentiating factor is not whether sexual activity involves
partners of the same or opposite sex, but whether it involves males (in either heterosexual or homosexual contexts). For example, a face-to-face position is used for
roughly 99 percent of sexual interactions between female Bonobos, but rarely in
male-female interactions. However, a face-to-face position is almost equally rare in
male homosexual interactions, occurring in only about 2 percent of activity between males. Thus, male homosexuality is more similar to heterosexuality than either is to female homosexuality in terms of the frequency of use of these two basic
positions. A similar pattern occurs in Gorillas: although the face-to-face position is
virtually absent in heterosexual encounters and much more common in homosexual ones, the two sexes have almost opposite preferences for this position. Almost
three-quarters of female homosexual encounters involve the face-to-face position,
while more than 80 percent of male homosexual mountings involve the front-to-
back position (also preferred for heterosexual encounters). In Hanuman Langurs,

male homosexuality is also more similar to heterosexuality than is female homosexuality in terms of the way that interactions are initiated: both males and females typically invite males to mount them by performing a special "head-shaking" display, which is much less characteristic of mounting between females.[44]

In Japanese Macaques, female homosexuality is more similar to heterosexuality than to male homosexuality in terms of the *variety* of positions used. In contrast, male homosexuality is more similar to heterosexuality than female homosexuality in terms of the *frequency* that various positions are used. In this species, fully seven different mounting positions can be identified, including four varieties of the front-to-back posture (with the mounting animal sitting, lying, or standing—with or without clasping its partner's legs—behind the mountee), two types of face-to-face positioning (sitting or lying down), and sideways mounts. All seven of these positions are found in both heterosexual and female homosexual encounters, while male homosexual mounting employs only five of the seven (sitting or lying on the partner in a front-to-back position are not used). However, sexual encounters between females differ from both heterosexual *and* male homosexual mounts in using the face-to-face position more often, and in using the double-foot-clasp posture less than 20 percent of the time (compared to 75–85 percent of the time for sexual encounters where males are involved, either heterosexual or homosexual).[45]

Other patterns based on the intersection of gender and sexual orientation also occur. In Stumptail Macaques, for example, female homosexual encounters use three basic positions (standing or sitting front-to-back and sitting face-to-face), heterosexual activity uses two of these (standing or sitting front-to-back), while male homosexual activity uses only one of these (standing front-to-back). (Males do, however, employ a wider range of oral and manual forms of genital stimulation in their encounters with each other.) Copulations between female Flamingos generally resemble heterosexual matings more than they do mountings between male Flamingos, but same-sex copulations in birds of either sex differ from heterosexual activity in their lack of a particular "hooking" posture.[46] Male White-handed Gibbons interact sexually with other males only in a face-to-face position and with females only in a front-to-back position—thus, homosexual and heterosexual interactions are equally "flexible" or "inflexible" in this species, but differ in which position is preferred in each context. Even reciprocal or reverse mounting—in which partners take turns mounting each other—is part of the heterosexual repertoire in more than three-quarters of the species that engage in this activity (in either same- or opposite-sex contexts); it is unique

◀ *Female Japanese Macaques in a sexual embrace. This face-to-face position is less common during heterosexual and male homosexual interactions.*

to heterosexual relations in many of these (including Western Gulls and Silvery Grebes) and present in many animals that do not engage in homosexual behavior at all.

In fact, it is sometimes the case that opposite-sex partners show *more* variability or flexibility in their sexual activity. Heterosexual copulation in Botos occurs in three main positions (all belly-to-belly, either head-to-head, or head-to-tail, or at right angles) while homosexual copulation usually uses only one of these (head-to-head).[47] Both heterosexual and homosexual encounters in this species can involve two different forms of penetration (genital slit or blowhole), although same-sex activity also includes a third option, anal penetration. Among birds, the overwhelming majority of species mate in the standard position of one individual mounted on the other's back, in both heterosexual and homosexual contexts. The only examples of other positions being used with any regularity involve male-female mounts: a facing position (extremely unusual for birds) is used by stitchbirds, for instance, who mate with the female lying on her back and the male on top of her, and in purple-throated Carib hummingbirds, who mate belly-to-belly while perched on a branch. Copulation in red-capped plovers is achieved by the male first throwing himself on the ground on his back, then pulling the female on top of him in a facing position. Vasa parrots have an elaborate and (for birds) unusual form of genital contact in which the male inserts his genital protrusion (a bulbous swelling surrounding his genital orifice) inside the female's cloaca, which extends and envelops his organ while the two birds transfer from a regular mounting position to a side-by-side position (full penetration does not usually occur in bird matings). Vervain hummingbirds actually mate in midflight while traversing an 80-foot trajectory low above the ground. Finally, several species of woodpeckers are true heterosexual virtuosos: in an acrobatic sequence the male first performs a standard front-to-back mount and then drops to one side of the female, making genital contact with his tail underneath hers and sometimes ending up on his back or with his entire body in a perpendicular or even upside-down position.[48]

"Virtuosity" in other areas of behavior is not generally exclusive to homosexual encounters either. The vast majority of courtship interactions, for example, involve the same set of behaviors typical for the species regardless of whether they are being performed between partners of the same or the opposite sex. There are notable exceptions, of course: the courtship "games" of female Rhesus Macaques and solicitations of female Japanese Macaques; "necking" interactions between male Giraffes; pirouette dances in male Ostriches; the vocal duets of Greylag gander pairs; aspects of courtship feeding in Laughing Gulls, Antbirds, Superb Lyrebirds, and Orange-fronted Parakeets; alternative bower displays in Regent Bowerbirds; and unique vocalizations during homosexual but not heterosexual interactions in male Emus and Japanese Macaques. Occasionally courtship activities are also performed at different rates or with different intensities: in same-sex pairs of Black-winged Stilts and Black-headed Gulls, for instance, certain courtship behaviors occur more frequently in same-sex pairs, others more commonly in opposite-sex pairs. All of these represent behavioral innovations in same-sex contexts, but they are atypical. Usually both homosexual and heterosexual courtships draw upon the same repertoire of behaviors, and in many cases same-sex interactions actually involve only a subset of the full behavioral suite that is characteristic of the species.

Thus, while homosexuality among animals is sometimes characterized by innovative or exceptional behaviors not found in heterosexual interactions, the opposite situation is equally, if not more, prevalent. It seems, then, that neither virtuosity nor mundanity of sexual expression are exclusive to either homosexual or heterosexual contexts. This is really not surprising: as we have already seen, a hallmark of sexual (and related) behaviors in animals is the tremendous range of variation found between species as well as among different individuals. For just about any pattern or trend that can be discerned, one that is contradictory or equivocal can be found. It stands to reason, then, that something like "sexual technique" would exhibit a similar range of diversity. And although Masters and Johnson may have found a greater level of technical proficiency in sex among some homosexual couples, this is probably an overly simplistic generalization even among people. A wider study sample that includes extensive cross-cultural information, as well as closer attention to age, gender and class differences, social contexts, and other factors, would likely reveal that (once again) human beings are much more like other species in this regard.

Primate (Homo)Sexuality and the Origins of Culture

Homosexuality is part of our evolutionary heritage as primates: anyone looking at the prevalence and elaboration of homosexual behavior among our closest relatives in the animal kingdom will be led, eventually, to this conclusion. In fact, primatologist Paul L. Vasey traces the occurrence of homosexuality in primates back to at least the Oligocene epoch, 24–37 million years ago (based on its distribution and characteristics among contemporary primates).[49] Some of the most organized and developed forms of homosexuality among animals can be found in the more than 30 species of monkeys and apes where this behavior occurs. Bonobos, for instance, engage in both male and female homosexual interactions with disarming frequency and enthusiasm, and they have also developed many unique forms of sexual expression, including a type of lesbian tribadism known as genito-genital rubbing. Similar elaborations of homosexual patterns are found among Stumptail Macaques, Gorillas, Hanuman Langurs, and many other monkey and ape species. In addition to highly developed systems of same-sex interaction and diverse sexual techniques, a number of other aspects of homosexual activity in primates are particularly salient. Among these are various forms of pair-bonding such as consortships, "favorite" partners, or sexual friendships; evidence for exclusive or preferential homosexual activity in some individuals (as discussed in the preceding section); female orgasm in monkeys and apes, in both homosexual and heterosexual contexts; female-centered or matrifocal societies, as well as male alliances and other groups of cooperating males in some species; and the wide range of nonreproductive heterosexual activities found in many primates.[50]

In addition to being part of our evolutionary heritage, homosexuality is also part of our *cultural* heritage as primates—for same-sex activity in monkeys and apes offers us some startling examples of cultural traditions among animals. Although "culture" is something that we typically associate with human beings, many animals innovate behaviors and then pass them on from generation to generation through learning. Zoologists speak of this as "cultural" behavior in animals—or, if the activity is less well developed, as "precultural" or "protocultural" behavior. Animal cultural traditions are widespread and often highly complex, occurring in many different kinds of species and involving behaviors as diverse as foraging and hunting techniques, communication patterns and song dialects, forms of social organization, response to predators, characteristics and locations of home sites and shelters, and migration patterns.[51] Perhaps the most famous example of animal cultural behavior concerns food-gathering techniques in wild Japanese Macaques: in the mid-1950s, one female invented several ways of accessing novel food items (introduced by investigators), including sweet-potato washing, peanut digging, and "placer mining" for wheat. Within ten years 90 percent of all troop members had acquired these habits, which were being learned spontaneously by younger animals and passed on to subsequent generations.

"Culture" can also involve social behaviors: male caretaking of infants in Japanese Macaques, for example, is characteristic of only certain populations and appears to be a learned behavior, acquired by some individuals or troops and not others. Sexuality—including homosexual activity—can also bear the hallmarks of cultural activity. Scientists studying mounting behavior by females—once again in Japanese Macaques—suggest that whether and how females mount male or other female partners may represent a form of protocultural behavior. Certain mounting positions, for example, seem to become more "popular" in some troops over time, only to wane and be replaced by others. Likewise, masturbation among females appears to be learned through observation or other social channels. Although a *capacity* for homosexual activity (along with reverse mounting and masturbation) is probably an innate characteristic of the species (as evidenced by at least some level of these activities in most populations), its occurrence between different troops and individuals is highly variable. Key aspects of such activities are apparently being learned and passed on through space and time. This indicates that "traditions" or patterns of sexual activity may be innovated and then transmitted via a web of social interactions, moving between and within population groups, geographic areas, and generations. Sexuality, including aspects of both same-sex and opposite-sex interactions, is also considered to exhibit aspects of cultural traditions in at least two other primates, Stumptail Macaques and Savanna Baboons.[52]

Not only is sexuality itself a form of cultural behavior, it can also impact and intersect with other sorts of cultural innovations in primates, often in surprising ways. In fact, nonreproductive sexual activities, including homosexual behaviors, may have contributed to the development of a number of significant cultural "milestones": hallmarks of evolutionary and cultural change that are considered to be defining characteristics of "humanness," yet which also exist in prototypical form in some of our primate relatives (and presumably also in our protohuman ancestors). In this section we'll briefly consider the role that sexuality might have played in the development of primate communication systems and the origin of language, in the manufacture and use of tools, and in the creation of social taboos and rituals. Caution must always be exercised in making direct comparisons between animals and people, and most of these areas are only beginning to be studied in any detail. Nevertheless, primate (homo)sexuality and the "traditions" associated with it offer us a striking mirror of some of our most "human" characteristics—and perhaps even a window into our evolutionary past and cultural history.

Language

Bonobos (also sometimes known as Pygmy Chimpanzees) have one of the most varied sexual repertoires of any species, with a wide variety of behaviors and positions used in sexual interactions, both heterosexual and homosexual. As a result, some Bonobos have developed an extraordinary system of gestural communication that is used specifically during sex. First discovered by pioneering ape-language researcher Susan Savage-Rumbaugh and her colleagues in the mid 1970s, this gestural system has far-reaching implications for our understanding of primate communication systems and the development of human language.[53]

A "lexicon" of about a dozen hand and arm gestures—each with a specific meaning—is utilized by Bonobos to initiate sexual activity and negotiate various body positions with a partner (of the same or opposite sex). For example, one gesture involves flicking the hand back-and-forth sideways from the wrist, meaning (approximately), "Move your genitals around"; this is used to get one's partner to position his or her genitals so as to facilitate a sexual interaction. Another gesture, lifting the arm with the palm downward, is employed when a Bonobo wants a sexual partner to move into a facing position for copulation. A catalog of some of the other gestures is given in the accompanying illustration. Hand signals may even be strung together in short sequences, and there is some evidence that the order of the gestures is significant.[54] These manual signs are used during both heterosexual and homosexual activity, although it appears that they may be more prevalent in opposite-sex interactions, and males and females may also employ some of the gestures with different frequencies.

Most of the hand signals are *iconic*, which means that they bear a physical resemblance to the meanings (i.e., body movements) they represent, although some are less transparently representational than others. Thus, the meaning "Turn your body around" can be related to the circular movement of the corresponding gesture, but this meaning is also conventionalized, since the intended action conveyed by the gesture must be agreed upon and understood by both participants for the gesture to be effective. An individual who hasn't learned the gestural system, for instance, wouldn't necessarily know what sort of "turning around" is intended by such a gesture. In conjunction with these hand gestures, a variety of positioning movements involving direct touching and placement of a partner's body or limbs are also used to facilitate sexual interactions. Together with the more abstract manual gestures, a total of at least 25 signals used during sex have been identified. In addition, patterns of eye contact and gaze also appear to have significant communicative value.

A "Lexicon" of Ten Hand Gestures Used during Bonobo Sexual Interactions

▲ Description: *Arm is partially extended, hand flicks back and forth with sideways movement from wrist*
Meaning: *"Move your genitals around"—used to get partner to position his or her genitals to facilitate a sexual interaction*

▲ Description: *Arm is extended, hand bends at wrist and makes rapid, vigorous circular motions*
Meaning: *"Turn your body around"—also used as an invitation for a sexual interaction if other methods of initiation have not worked*

▲ Description: *Hand is held outstretched toward partner (arm extends upward and outward with palm of hand angled toward the other individual)*
Meaning: *"Approach"—used to get partner to move closer; also used simply as an invitation for a sexual interaction*

▲ Description: *Arm is extended, hand bends at wrist and flips toward self*
Meaning: *"Come here"—used to get partner to move closer during a sexual interaction*

▲ Description: *Arm is partially extended and raised while hand flips upward at wrist*
Meaning: *"Stand up"—used to get partner to stand on hind legs before engaging in a face-to-face sexual interaction*

▲ Description: *Arm is extended and raised till about head level with palm facing downward and placed lightly on another individual's back or shoulder*
Meaning: *An invitation for a sexual interaction*

▲ Description: *Knuckles are rested on partner's arm or back and arm moves back toward self*
Meaning: *"Move closer"—used to get partner to assume a stance compatible with front-to-back sexual interaction*

▲ Description: *Arm is lifted up with palm down*
Meaning: *"Position yourself"—used to get partner to move into a face-to-face position*

▲ Description: *Hand and forearm move across body in a sweeping motion*
Meaning: *"Turn around"—used to get partner to rotate his or her entire body*

Description: *Both arms wave or open out from body*
Meaning: *"Spread your legs and/or arms"—used to get partner to open limbs to facilitate face-to-face positioning* ▶

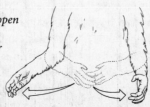

Researchers studying the Bonobo gestural system have suggested that the more abstract hand signals may have developed out of the simpler positioning movements.[55] Communication during sex may initially have involved only fairly crude attempts to move one's partner, from which more ritualized touching and directive gestures may have evolved, which then gradually became more and more stylized until, in some cases, fairly opaque hand gestures resulted. This sequence is significant, because it represents the sort of progression from purely representational gestures to highly codified manual signals that has been identified in the development of human sign languages.[56] More broadly, it shows the beginnings of abstraction or arbitrariness—that is, the creation of *symbols*—which is a hallmark of human language in general. The Bonobo gestural repertoire is certainly not a "language" in the true sense of a complete human linguistic system, and by no means does it have the complexity or subtlety of even the simplest human gestural systems, let alone of a fully developed signed, spoken, or written language. Nevertheless, it is a formalized communication system that exhibits a level of sophistication unparalleled in any other nonhuman primate—perhaps even with a rudimentary "syntax" (in terms of the ordering of gestures)—and it may indeed represent a precursor to human language.

Even more significantly, Bonobos devised this system spontaneously: they invented the hand signals on their own and were not taught to use them by people. Attempts to teach apes various forms of human sign language or other communication systems have demonstrated that our primate relatives in many instances possess formidable linguistic capacities, but in all cases human prompting and intervention (at least initially) are involved. What is unique about the sexual gestures is that Bonobos themselves developed the hand signals and taught them to each other (or learned them from one another), in response to a communicative need that arose naturally within their own social interactions.[57] Moreover, the specific social context that prompted this development is also unique: it was sexual behavior, or rather, the highly variable and plastic nature of Bonobo sexuality, that led to this development. Because of the wide variety of heterosexual and homosexual activities that characterize Bonobo sexual interactions, a supplementary communication system arose to help negotiate sexual interactions. In response to an unsurpassed sexual capacity in this species—including prominent homosexual activities—an unsurpassed animal communication system was created.

Not only is the Bonobo gestural system an outstanding example of the spontaneous development of cultural traditions in animals, it offers some clues into the origin of the human linguistic capacity. A number of theorists have suggested that the first human communication system may indeed have been a gestural language, that is, a system of hand signals.[58] Why language should have evolved in the first place among human beings, however, is a subject shrouded in mystery and controversy. Of the many theories that have been put forward, a number suggest that language developed in response to social factors, such as the need to coordinate complex group activities such as hunting or farming. The Bonobo system demonstrates that another factor may also have been involved—one that is rarely, if ever, considered in discussions of the origin of language: sexuality. In particular, as sexual interactions became more variable over the course of evolution, gestural systems of greater complexity may have developed to facilitate sexual encounters.

Primate evolution has been characterized by an ever-increasing separation of sexuality from its reproductive "functions," including the development of numerous types of homosexual and nonprocreative heterosexual activities—most prominent in human beings and Bonobos (considered by some to be the primate most similar to humans), less so in Common Chimpanzees, Gorillas, and other apes. Scientists have also identified a corresponding increase in complexity of communication systems used during sexual interactions among apes, proceeding from Gorillas to Common Chimpanzees to Bonobos (and onward, of course, to human beings).[59] The progression is probably not quite as orderly as this sequence suggests, and multiple factors are surely involved in the genesis of each species' particular communication systems. The general trend, however, is clear: as sexual interactions become more variable, sexual communication systems become more sophisticated. It is possible, therefore, that sexuality—particularly the fluidity associated with nonreproductive sexual practices—played a significant role in the origin and development of human language.

Tools

A hallmark of human cultural evolution was the development of tools, later elaborated into the full array of material technologies that we know today. But many animals, especially primates, also use inanimate objects to manipulate or affect things in their environment in ways that can be seen as precursors to similar activities in human beings. Over 20 different types of tool use have been identified in primates and other species—Common Chimpanzees, for example, employ objects as weapons, as levers, and as drinking and feeding implements of various sorts (such as the well-known example of sticks being used to capture and eat termites or ants). Tools can also be used to affect an animal's own (or another's) body, for example as part of a "hygiene" or grooming regimen. Chimpanzees and other primates, for example, often use leaves, twigs, straw, rags, or other objects to clean themselves and wipe away bodily secretions (such as saliva, blood, semen, feces, and urine). Chimps and Savanna Baboons have also been observed using sticks, twigs, and stones to clean their own or each other's teeth (and even perform dental extractions). Chimps also sometimes tickle themselves with various items such as stones or sticks, and Japanese Macaques occasionally use similar items to groom one another.[60]

Less well known, however, is the use of objects for purposes of sexual stimulation. A number of primates employ various implements as masturbatory aids (in both the wild and captivity), and this aspect of tool culture has not received widespread attention in discussions of the development of animal and human object manipulation. Female Orang-utans, for instance, sometimes masturbate by rubbing objects on the clitoris or inserting them into the vagina; tools used for this purpose include pieces of liana bitten off to an appropriate size or (in captivity) pieces of wire. Male Orangs also use objects to stimulate their genitals, including

▲ *A female Orang-utan in the forests of Sumatra masturbating with a tool she made from a piece of liana*

one individual who ingeniously fashioned an implement by pushing a hole through a leaf with his finger. He inserted his erect penis into this "orifice," then rubbed the leaf up and down the shaft to stimulate himself. Males also sometimes hold a piece of fruit (such as an orange peel) in their hand and masturbate against it.[61]

Common Chimpanzees have also developed several innovative masturbation techniques using a variety of different tools. One female gathered a small collection of sticks, pebbles, and leaves, from which she would carefully select a particular item to stimulate herself with. By placing a leaf underneath her vulva, for example, and flicking the stem with her knuckle, she made the leaf vibrate and thereby externally stimulate her genitals. She also repeatedly inserted the stem into her vagina, often lubricating it with saliva and manipulating it with her hand so as to stimulate herself internally. In one instance, she rocked back and forth with the stem inserted, rubbing the leaf against a vertical surface so that the stem actually vibrated inside her. On other occasions, she repeatedly inserted and withdrew a pebble from her vagina or used a small stick to stimulate her genitals. Other female Chimps have also been observed rubbing or tickling their external genitals with items or inserting them into the vagina, including pieces of mango, twigs, and leaves, as well as man-made objects such as small boxes or balls. Similarly, several young males assembled collections of stones, fruits, or even pieces of dried dung, which they would thrust against to stimulate their genitals. Male and female Bonobos occasionally employ inanimate objects for masturbation as well, stimulating themselves with (or thrusting against) branches, wood shavings, and other items.[62]

Like Orang-utans, one female Bonnet Macaque invented some relatively sophisticated techniques of tool manufacture, regularly employing five specific methods to create or modify natural objects for insertion into her vagina. For example, she stripped dry eucalyptus leaves of their foliage with her fingers or teeth and then broke the midrib into a piece less than an inch long. She also slit dry acacia leaves in half lengthwise (using only a single half) and fashioned short sticks by breaking longer ones into several pieces or detaching portions of a branch. Implements were also sometimes vigorously rubbed with her fingers or between her palms prior to being inserted into her vagina, and twigs, leaves, or grass blades were occasionally used unmodified.[63]

The use and manufacture of tools by primates is considered an important example of cultural behavior in animals, and a forerunner of the activities that are so widespread among human beings. Although many different forms and functions are evident in animal tool use, these examples show that nonreproductive sexual activities are part of the overall behavioral pattern: the primate capacity for object manipulation extends seamlessly into the sexual sphere. Apes and monkeys use a variety of objects to masturbate with and even deliberately create implements for sexual stimulation by cutting or forming materials such as leaves or twigs (often in highly creative ways). Similar types of activities occur among people, of course, and sexual implements of various sorts have a long and distinguished history in human culture. Dildos or phalli made of stone, terra-cotta, wood, or leather, for example, were used in ceremonial "deflowering" and fertility rites—as well as for masturbation and inducing sexual pleasure in a partner—in ancient Egypt, Rome, Greece, India, Japan, and Europe. Examples have been recorded from at least as far back as the Paleolithic through medieval times—including some biblical references—as well as in the ongoing traditions of many indigenous peoples throughout the world.[64] However, few (if any) anthropologists have ever considered the possibility that sexual stimulation may have been a component of tool use among early humans or even played a part in the origin and elaboration of material culture. Of course, technological complexity is not the only measure of cultural development—some of the most complex linguistic and oral history traditions, for example, are to be found among the South African San peoples and the Australian Aborigines, whose material culture is relatively simple. And certainly many more "utilitarian" functions can be identified in the development of tool use among our human, protohuman, and primate ancestors. Nevertheless, the pursuit of sexual pleasure may have contributed, in some measure, to our own heritage as creatures whose tool-using practices are among the most polymorphous of any primate.

Taboo

The vast majority of human cultures prohibit sexual relations between people who are related. There is still ongoing debate among scientists as to whether this prohibition—commonly known as the incest taboo—is instinctual or learned. Regardless of the extent to which biological factors are involved, there are clearly strong social and cultural components to incest avoidance. Different human cultures and societies vary widely in how they define incestuous relations and to what extent such activities are both stigmatized and practiced. For example, although parental incest (father-daughter, mother-son) is prohibited in virtually all societies (yet still occurs, despite such prohibition, with varying frequencies), there is wider latitude regarding other blood relations. Cousin marriage is considered acceptable in some cases, unacceptable in others, while some societies make a further distinction between relations with cross cousins as opposed to parallel cousins—a biologically arbitrary distinction, since there is no evidence of any greater genetic "harm" in one form of cousin marriage than another. Brother-sister marriage was widely practiced in ancient Roman Egypt, and among the royal families of some central African and Balinese societies, ancient Incans, Hawaiians, Iranians, and Egyptians—in fact, Cleopatra is thought to have been the product of 11 generations of incestuous marriages within the Ptolemaic dynasty.[65]

Further evidence of a learned or cultural component to incest prohibitions relates to the role played by social familiarity as opposed to genetic relatedness in choice of partners. In our culture, sexual relations between adoptive or stepfamily members are generally frowned upon even though the individuals involved are not related by blood. Conversely, people who are genetically related but, because of social circumstance (e.g., separation at birth), are unaware of their biological connection may develop a relationship (at least until they learn of their relatedness). Other societies vary considerably in this regard: in the Israeli kibbutzim, for instance, unrelated individuals who are brought up together hardly ever marry one another. In contrast, a traditional form of Taiwanese marriage involved girls being adopted into families as children and then, on reaching adulthood, marrying their stepbrothers, although such marriages were considered less preferable than other arrangements. Among the Arapesh people of New Guinea, a similar practice of stepsister marriage was widely accepted and preferred.

The fact that homosexual relations are usually prohibited between related individuals also points to the importance of nonbiological factors in the incest taboo. In most human cultures that "permit" some form of same-sex eroticism, from contemporary America to indigenous tribes of New Guinea, the choice of homosexual partners is subject to distinctions of "kin" and "nonkin." This is in spite of the fact that no children, and hence no potentially harmful genetic effects, can result from such unions. Typically the same restrictions are applied to homosexual as to heterosexual relations. In a number of New Guinean societies, however, slightly different kinship constraints regulate the choice of same-sex and opposite-sex partners. In fact, homosexual partners in some tribes are actually required to be *more distantly related* than heterosexual ones—the exact opposite of what would be expected if incest taboos were based solely on biological factors.[66]

This is significant, because most theories about the biological basis of the incest taboo focus on the potential for increased rates of birth defects and lower genetic variability as a result of inbreeding. Even for heterosexual relations, though, the evidence is not nearly as unequivocal as one would suppose: numerous studies of small populations that have practiced inbreeding for many generations reveal no deleterious effects, owing to the rapid elimination of genetic defects and subsequent stabilization of the gene pool.[67] To adduce further evidence for a biological basis to the prohibition, scientists often point to the existence of "incest taboos" in animals. Ironically, though, many animal species actually show evidence of a "cultural" or "social" dimension to their avoidance of sexual activity between relatives that parallels the human examples—most notably among primates, and most notably involving homosexuality.

There is a great diversity of incestuous activity among animals, not only in the frequency and types of relations that occur but also in the degree to which such activity is avoided or pursued, in both heterosexual and homosexual contexts. Even among primates, many different scenarios and versions of "taboos" are found. In Rhesus Macaques, for instance, incest of any sort is not common although mother-son, brother-sister, and brother-brother relations do occur (some males actually appear to prefer mating with their mothers). In Gibbons, heterosexual incest (both parental and sibling) is sometimes practiced and homosexual relations are almost always incestuous, while both heterosexual and homosexual activity between siblings (or half-siblings) occurs in Gorillas. Most strikingly, several species appear to have developed systematic homosexual "incest taboos," each with its own socially defined set of "acceptable" and "unacceptable" partners. In some cases, these restrictions differ significantly from those governing the corresponding heterosexual relations (as in some human populations).

Homosexual consortships (pair-bonding with sexual activity) among Japanese Macaques, for example, virtually never occur between mothers and daughters or sisters. In contrast, heterosexual brother-sister or mother-son relations, while not common, are much more prevalent than homosexual incest in this species. Interestingly, aunts and nieces among Japanese Macaques do not generally recognize each other as kin—when intervening on behalf of individuals during aggressive encounters, for example, aunts do not assist their nieces any more often than do non-relatives, and significantly less often than do mothers, grandmothers, and sisters. Consequently, some blood relatives *are* able to form consortships with one another: about a third of all aunt-niece dyads interact homosexually. In other words, Japanese Macaques have an overall pattern of incest avoidance unique to homosexual relations, within which apparently incestuous aunt-niece pairs are "permitted" because such partners do not count as kin in the larger social framework.

In Hanuman Langurs, both heterosexual and homosexual incest taboos are in effect, but with slightly different restrictions. Heterosexual incestuous relations of any kind are generally avoided; sexual activity between mothers and daughters is also "prohibited" (accounting for only about 1 percent of all homosexual mounting). In contrast, half sisters (females with the same mother but different fathers) are "allowed" to have sexual relations with one another—and in fact, more than a quarter of all mounts between females occur between half sisters. In Bonobos, incestuous relations between females also generally appear to be avoided: when females immigrate into new troops as adolescents, they are usually unrelated to most of the other troop members, but sexual activity is not practiced with those females to whom they are related.[68]

These examples demonstrate that, at least in some nonhuman primates, homosexual (as well as heterosexual) relations are subject to various social prohibitions regarding choice of appropriate partners, especially where relatives are involved. These choices are not due to instinct (i.e., avoidance of an activity that would yield harmful genetic effects), because no offspring result from such relations, and because not all incest taboos are identical. Nor are they simply a "carryover" from heterosexual taboos, because same-sex and opposite-sex relations often have different

prohibitions. Crucial distinctions exist between species, populations, and even between heterosexual and homosexual activity in the same species, concerning "al-
lowable" incestuous relations—differences that cannot be attributed solely to
biological (genetic) factors. Only some relatives actually "count" as related for the
purpose of incest taboos, and which individuals are "tabooed" is, to a large extent,
arbitrary. In other words, primates must *learn* what sort of kinship system(s), if any,
govern sexual associations—both homosexual and heterosexual—in their social
group. While the occurrence and expression of same-sex activity in these animals
very likely has an instinctual or genetic component as well, homosexual relations
exhibit important "cultural" characteristics that probably involve a high degree of
social learning. "Taboos" exist in animals, and homosexuality is one area where
such prohibitions manifest themselves in particularly compelling ways.

Ritual

Where do human rituals such as taking an oath come from? In a fascinating
study of Savanna Baboon social systems, primatologists Barbara Smuts and John
Watanabe offer a startling answer: they suggest that such symbolic gestures might
be traced to the ritualized homosexual activities that take place between male Baboons.[69] As part of their social interactions, male Savanna Baboons perform a variety of formalized sexual and affectionate behaviors with each other, most notably
"diddling," that is, fondling of the penis and scrotum. Other ritualized homosexual
activities include mounting; grabbing, fingering, and nuzzling of the rump; kissing
and nuzzling of the genitals; and embracing and kissing on the head or mouth

*Two male Bonnet Macaques embracing each other in a "greeting" gesture. The male on the left is
fondling the scrotum of the other male with his right hand, an activity also found in male
Savanna Baboons, where it is known as "diddling."* ▼

(similar activities are also found in a number of other primates, including Common Chimpanzees, Bonnet Macaques, and Crested Black Macaques). Although these behaviors undoubtedly have an affectionate or "pleasurable" tactile component as well as a sexual dimension, they have also been characterized by some scientists as "greetings" interactions, and it is thought that they may serve to negotiate and solidify cooperation between males as well. Indeed, two males sometimes pair up and form a stable "coalition" with each other in which their mutual defense and aid is symbolized by the reciprocity of their ritual sexual exchanges with one another. Smuts and Watanabe suggest that sexual gestures such as diddling, which involve one male placing his most vulnerable and intimate organs literally in the hands of another, are in a sense a prototypical form of oath-swearing: one male, by his actions, is indicating his trust and commitment to cooperate with the other.

But what does this have to do with human rituals of oath-taking? In our society at least, oaths usually involve gestures such as raising of the right hand, crossing the heart, or perhaps even placing the hand on a Bible, but certainly nothing so forward as fondling of the genitals. Surprising as it may seem, though, Smuts and Watanabe present some intriguing clues that gestures similar to the ritual homosexual activities of Baboons (and other primates) may in fact have been a part of human oath-taking at one time and are even still used in some contemporary cultures. In a number of Australian Aboriginal tribes, for example, holding of the penis is traditionally used as a gesture to express male allegiance and cooperation, as well as a ritual part of resolving disputes between "accused" and "defending" parties. Among the Walbiri and Aranda people, when different communities get together or when grievances need to be settled in formal "trials," men participate in what is known variously as touch-penis, penis-offering, or the penis-holding rite. Each man presents his semi-erect organ to all the others in turn, pressing it into each man's palm and drawing it along the length of the upturned hand (held with the fingers toward the testicles). By offering and grasping each other's penis—said to represent "paying with one's life"—the men make an avowal of mutual support and goodwill between them, or symbolize and solidify the agreement they have reached during the settling of a dispute. A similar gesture involving stroking of the genitals and/or scrotum is used as a greeting in some New Guinea tribes such as the Eipo and Bedamini.[70]

Closer to home, there is historical—even biblical—evidence that similar rituals may have been a part of the Judeo-Christian and Euro-American heritage. Ironically, the book that is today used in so many of our own oath-swearing ceremonies contains within it an allusion to these earlier practices. In Genesis 24:9 there is a reference to the servant of Abraham swearing an oath by placing his hand under his master's "loins." Moreover, according to the *Oxford English Dictionary* the words *testify, testimony,* and *testicle* are probably all related, sharing the common root *testis,* which originally meant "witness." Although these connections are somewhat speculative, they suggest a line of continuity between ritualized homosexual behavior in primates and human social rituals such as oath-taking. As Smuts and Watanabe point out, notions of truth and sanctity as expressed by the human ceremonies are vastly different from those of Baboons (if present at all in the nonhuman context). Nevertheless, the *forms* of these rituals—and their social outcomes—are strikingly similar.

Language, tools, taboo, ritual—each of these is part of a larger puzzle or *matrix* of cultural development that is traditionally seen as distinct and distant from sexuality. Nevertheless, the occurrence of a number of remarkable primate behaviors revolving around homosexuality and nonreproductive heterosexuality suggests that these domains are much more intimately associated than previously imagined. Sexual gesture systems, masturbatory tools, homosexual incest taboos, and ritualized same-sex "oath-taking" offer extraordinary juxtapositions of culture, biology, society, and evolution. Primate (homo)sexual behaviors such as these exemplify both cultural traditions and evolutionary inheritances. In turn, they may have contributed to the development of some of the most hallowed and cherished landmarks of human cultural history as well.[71]

Unnatural Nature

> *Animals don't do it, so why should we? Can you even imagine a queer grizzly bear? Or a lesbian owl or salmon?*
>
> —from a letter written to Dean Hamer, coauthor of
> *The Science of Desire: The Search for the Gay Gene and the*
> *Biology of Behavior*[72]

Many people, such as the man quoted above, believe that homosexuality does not occur in nature and use this belief to justify their opinions about human homosexuality. In fact, rarely is homosexuality in animals discussed on its own: inevitably, cross-species comparisons are drawn to ascribe moral value to the behavior—both positive and negative. Nowhere is this more apparent than in the notion of "naturalness" and the entire complex of animal-human comparisons that this problematic term evokes. The prevailing view is an overly simplistic one: if homosexuality is believed to occur in animals, it is considered to be "natural" and therefore acceptable in humans; if it is thought not to occur in animals, it is considered "unnatural" and therefore unacceptable in humans. The debate seems clear and the lines of distinction inviolable.

Any careful consideration of the logic behind the equation *occurs in animals = natural = acceptable in humans* will show, however, that this line of reasoning is flawed. As many people have pointed out, humans engage in a wide variety of behaviors that do not occur in nature, from cooking to writing letters to wearing clothes, and yet we do not condemn these activities as "unnatural" because they are not found among animals. As author Jon Ward explains, with regard to a friend who asserted "You can't argue with biology" (believing that homosexuality was "unnatural"):

> Has he never fried an egg? The whole of human history is an "argument with biology." The very civilization which the most homophobic ideologues are eager to defend is the *antithesis* of nature: law and art.[73]

We also use our biology and anatomy in ways that "nature did not intend for them to be used" without ascribing a moral value to such activities. As James Weinrich observes, the tongue's primary biological purpose is for the act of eating, yet its use in acts of speech, bubble-gum-blowing, or kissing is not therefore considered "unnatural." In addition, many things that do occur spontaneously in animals—diseases, birth defects, rape, and cannibalism, for example—are not considered to be "natural" or desirable conditions or behaviors in most humans. Weinrich aptly remarks, "When animals do something that we like, we call it natural. When they do something that we don't like, we call it animalistic."[74]

The Natural History of Homosexuality

The historical record also shows that attitudes toward homosexuality have little to do with whether people believe it occurs in animals or not, and consequently, in its "naturalness." True, throughout much of recorded history, the charge of "unnaturalness"—including the claim that homosexuality did not occur in animals—was used to justify every imaginable form of sanction, control, and repression against homosexuality. But many other interpretations of "naturalness" were also prevalent at various times. Indeed, the very fact that homosexuality was thought to be "unnatural"—that is, not found in nature—was sometimes used to justify its *superiority* to heterosexuality. In ancient Greece, for example, same-sex love was thought to be purer than opposite-sex love because it did not involve procreation or "animal-like" passions. On the other hand, homosexuality was sometimes condemned precisely because it was considered *closer* to "nature," reflecting the base, uncontrolled sexual instincts of the animal world. The Nazis used this reasoning (in part) to target homosexuals and other "subhumans" for the concentration camps (where homosexual men subjected to medical experiments were referred to as "test animals"), while sexual relations between women were disparagingly characterized as "animal love" in late eighteenth-century New England. The irrationality of such beliefs is highlighted in cases where charges of "unnaturalness" were combined, paradoxically, with accusations of animalistic behavior. Some early Latin texts, for instance, simultaneously condemned homosexuals for exhibiting behavior unknown in animals while also denouncing them for imitating particular species (such as the hyena or hare) that were believed to indulge in homosexuality.[75]

In our own time, the fact that a given characteristic of a minority human population is biologically determined has little to do with whether that population should be—or is—discriminated against. Racial minorities, for example, can claim a biological basis for their difference, yet this has done little to eliminate racial prejudice. Religious groups, on the other hand, can claim no such biological prerogative, and yet this does not invalidate the entitlement of such groups to freedom from discrimination. It should be clear, then, that whether homosexuality is biologically determined or not, whether one chooses to be gay or is born that way, or whether homosexuality occurs in nature or not—none of these things guarantees the acceptance or rejection of homosexuality or in itself renders homosexuality "valid" or "illegitimate."

The debate about the "nature" and origin of homosexuality often invokes seemingly opposite categories: genetics versus environment, biology versus culture, nature versus nurture, essentialism versus constructionism. Indeed, the very categories "homosexual" and "heterosexual" are themselves examples of such a dichotomy. By using these categories, biologists and social scientists hope to discover what aspects of homosexuality, if any, are biologically determined. Yet by framing the debate in terms of such categories, it is easy to forget that more complex interactions between factors must be considered. For example, most research shows that both environment *and* biology are relevant in determining sexual orientation in people (and probably also animals). Some individuals may have an innate predisposition for homosexuality, but the right combination of environmental (including social) factors is required for this to be realized. And how meaningful is it to talk about a culture-nature distinction when, as we have just seen, some animal species have themselves developed forms of cultural behavior? Similarly, by focusing attention on the "causes" of homosexuality, the determinants of heterosexuality are considered irrelevant—or, alternatively, heterosexuality is assumed to be inevitable unless something "goes wrong." And not all sexuality fits neatly into the categories of exclusive homosexuality or exclusive heterosexuality—the large realm of experience that involves bisexuality is easily glossed over in discussions about the origins of homosexuality/heterosexuality. So, too, with the question of whether homosexuality is "natural" and what its occurrence in animals can tell us about this: things are considerably more complicated than they initially appear.

What is remarkable about the entire debate about the naturalness of homosexuality is the frequent absence of any reference to concrete facts or accurate, comprehensive information about animal homosexuality. Those who argue against the naturalness of homosexuality assert with impunity that same-sex behavior does not occur in nature (like the man quoted above) and is therefore self-evidently abnormal. Those who argue in favor of a biological origin for homosexuality often ignore the complexities of animal behavior arising from social, protocultural, or individual life-history factors (relying on the behavior of laboratory animals injected with hormones, for example, instead of long-term studies of animals interacting in their own social groups or communities).[76] This is because naturalness is more a matter of interpretation than facts. Now that the widespread occurrence of animal homosexuality is beginning to be documented, little if anything is likely to change in this discussion. More information about same-sex activity in animals simply means more possible interpretations: the information can be used to support or refute a variety of positions on the naturalness or acceptability of homosexuality, depending (as before) on the particular outlook of whoever is drawing the conclusions.

As James Weinrich points out, the only claim about naturalness that is actually consistent with the facts is the following: homosexual behavior is as natural as heterosexual behavior.[77] What this means is that homosexuality is found in virtually all animal groups, in virtually all geographic areas and time periods, and in a wide variety of forms—as are heterosexuality, divorce, monogamy, and infanticide, among other things. Conversely, heterosexuality is as "unnatural" as homosexuality is, since it often exhibits social elaboration or cultural "embellishment," as well as many of the "unacceptable" features stereotypically associated with same-sex relations, such as promiscuity, nonreproduction, pursuit of sexual pleasure, and interactions marked by instability, ineptitude, and even hostility.[78] But whether this means that homosexuality is "biologically determined" and/or "socially conditioned"—and by extension, (un)acceptable in humans—is largely a question of interpretation. Of course, from a scientific perspective, the sheer extent and variety of homosexual expression in the animal world reveals an aspect of nonhuman biology and social organization that is unexpected—one with far-reaching (perhaps even revolutionary) implications. It demands careful consideration and suggests a rethinking of some of our most fundamental notions of environment, culture, genetics, and evolutionary and social development. But to automatically conclude that because homosexuality occurs in animals, it must be biologically determined oversimplifies the debate and does an injustice to the facts.

For most people, animals are symbolic: their significance lies not in what they are, but in what we think they are. We ascribe meanings and values to their existence and behaviors in ways that usually have little to do with their biological and social realities, treating them as emblems of nature's purity or bestiality in order to justify, ultimately, our views of other human beings. The animals themselves remain enigmatic, mute in the face of this seemingly endless onslaught of human interpretations of their lives. If this were merely a matter of debate among people, it could perhaps be put in its proper perspective as simply yet another human folly. Unfortunately, the interpretations applied to animal (sexual) behaviors by people are far from innocuous: they can have grave consequences, or even be a matter of life or death—for both humans and animals alike. When a gay man or lesbian is assaulted or murdered because the attacker thinks that homosexuality is "unnatural," for example, or when politicians' legislative and judicial decisions concerning homosexuality are coded in such terms as "crimes against nature," much more is at stake than the scientific interpretation of animal behavior.[79]

The moral value ascribed to animal sexuality can also impact directly on the welfare of the creatures themselves. In 1995 a biologist with the U.S. Fish and Wildlife Service briefed Senator Jesse Helms's staff about the value of saving an endangered bird, the red-cockaded woodpecker, which lives in the southeastern United States.[80] His presentation stressed the supposed "family values" of the species, referring to the birds' monogamous and relatively long-lasting heterosexual pair-bonds. In other words, the right of this species to exist—as determined by legislators voting on the Endangered Species Act—was predicated not on its intrinsic value, but on how closely its behavior could be made to resemble what is currently considered acceptable conduct for humans. And this is most definitely a case of presenting an idealized "image" of the species: the red-cockaded woodpecker's "family values" are in reality far more complex, messy, and "questionable" than what the politicians were told.

True, this species usually breeds in long-term, monogamous pairs, but its social life is replete with variations on this theme, some of which Senator Helms would have found downright horrifying.[81] Many family groups in this species are unstable:

one six-year investigation found that only six out of thirteen breeding pairs remained together, while studies of the species in Helms's home state of North Carolina revealed that nearly 20 percent of females in this population desert their mates and switch family groups. Males sometimes leave their partners as well, and the overall (species-wide) divorce rate is about 5 percent; nonmonogamous copulations also occasionally occur, with slightly more than 1 percent of nestlings being fathered by a male other than their mother's mate. Red-cockaded woodpeckers also frequently live in "stepfamilies" or "blended families": more than a quarter of the younger birds who live in breeding groups and help with parenting duties may be related to only one parent, and 5–11 percent are related to neither. Some of these "helper" birds engage in decidedly un-family-like activities, such as ousting a parent from its group or even committing "stepfamily incest" by mating with the remaining parent. Incest involving full or half siblings, though rare, also occurs. Other helpers forgo reproduction entirely (continuing to live with their parents as adults for several years), and there are also solitary nonbreeding birds in the population, as well as all-male groups. Some red-cockaded woodpecker groups may also be polygamous or "plural" breeding units, with two females both breeding (or trying to breed) at the same time.

Would the red-cockaded woodpecker be considered less "deserving" of protection if Senator Helms and his staff learned that these birds participate in nonreproductive sexual activity (mating during incubation, or long before egg-laying), or siblicide and starving of offspring, or infanticide and chick-tossing from the nest? All of these behaviors have been documented in this species, yet none were included in the scientific presentation to the politicians in whose hands this bird's fate rests, for they would shatter the illusion of its "family values." Homosexual activity has not (yet) been observed in red-cockaded woodpeckers, although it does occur in related species such as Acorn Woodpeckers and Black-rumped Flamebacks. Should such behavior come to light, one can only dread the consequences for this, or any other, endangered species whose survival depends on human assessment of its "moral conduct."

Homosexuality has a "natural history" in every sense of the term: that is, it has both biological ("natural") and social or cultural ("historical") dimensions that are interconnected and inseparable. It is not a uniform phenomenon in either animals or people: it takes many forms, and it exhibits numerous variations and idiosyncrasies. The interplay of biology and environment in shaping these features—and indeed, the very definitions of what is "cultural" as opposed to "biological"—is far more complex than polarized debates would have us believe. Because the discussion is often framed in terms of misleading dichotomies such as "nature versus nurture" or "genetics versus environment," the possibility that both are relevant (and can influence one another) is repeatedly overlooked, as is the possibility that sexual behavior in some animals has a significant sociocultural component. Yes, homosexuality occurs in nature and apparently always has. But does this make it "natural" or simply "animalistic"? The answer to this question is entirely in the eye of the beholder, rather than in any inherent quality or context of the phenomenon itself.

Homosexuality in the Home and on the Farm: Pets and Domesticated Animals

Nowhere are questions of interpretation more vividly illustrated than with the animals we often consider to be the most "human"—our pets. Here it is not genetic similarity (as with primates) but our emotional and physical proximity that establishes the animal-human connection, whether it is with the creatures we keep as companions or domesticated animals on farms and ranches. Same-sex activity is readily familiar to pet owners and animal trainers/handlers, many of whom can give examples of homosexual mounting, pair-bonding, or other same-sex (or bisexual) activities in their own animals or those of friends. These anecdotal reports have been confirmed by scientific studies of domesticated animals.[82] Same-sex pair-bonding as well as homosexual mounting (including ejaculation during interactions between males) has been documented in Dogs, including breeds such as Beagles, Basenjis, Cocker Spaniels, and Weimaraners. Examples of both female and male homosexual behavior in Cats have also been confirmed, including mutual genital stimulation and mounting between females, and mounting leading to orgasm among males. Homosexuality has also been verified in other animals kept as pets, including same-sex courtship and mounting in Guinea Pigs; homosexual mounting in female domesticated Rabbits and Hamsters; and same-sex pair-bonding, courtship, and mounting among caged birds such as Zebra Finches, Bengalese Finches, and Budgerigars. Many common aquarium fish also exhibit homosexuality or transsexuality.

Homosexual behavior has been studied in a wide variety of domesticated and farm animals as well: Cattle, Sheep, Goats, Pigs, and Horses of both sexes participate in homosexual mounting, while same-sex pair-bonds have also been reported for Pigs, Sheep, and Goats. In fact, homosexual activity is so routine among domesticated hoofed mammals that farmers and animal breeders have coined special terms for such behavior: mounting among male Cattle is referred to as the buller syndrome (steers who are mounted are called bullers, the males who mount them are riders), female sows who mount each other are described as "going boaring," mares who do so are said to "horse," while cows are said to "bull." Same-sex activity is often utilized, paradoxically, in breeding programs. In some species homosexual mounting among females is used as a reliable indicator of when they are in heat, while young bulls or steers (known as teasers) are often presented to mature bulls to arouse them and allow their semen to be collected (for later use in artificial insemination).[83]

For each domesticated species that exhibits homosexual behavior, there are one or more wild "relatives" in which homosexuality has also been observed: Lions and other wild Cats, Wolves and other wild Dogs, Cavies (the wild ancestors of Guinea Pigs), American Bison and other wild Buffalo, Bighorn Sheep and other wild Sheep and Goat species, Zebra and other wild Horses, and so on. In some cases, the same-sex activities observed in domesticated species and their wild ancestors exhibit striking similarities: the group sexual interactions or "huddles" of Goats and Sheep (both wild and domesticated), the frequency of same-sex mounting in domestic

Cattle and wild Buffalo, female Cats and Lions placing themselves underneath a female partner to invite mounting, or the same-sex courtship displays by female Turkeys and wild Sage Grouse. In other cases, there are equally striking dissimilarities: pair-bonding and mounting among domesticated boars, for instance, contrasted with the virtual absence of same-sex activity in male wild Pigs; or fairly extensive homosexual courtship activities in female Cheetahs, contrasted with little, if any, such courtship activities in domestic Cats.

Although scientifically verified, homosexuality in pets and other domesticated animals continues to evoke many meanings for the people who simply live or work with such creatures, independent of the "facts." As with all human observations of the animal world, people tend to see only what they are prepared to accept. This is illustrated quite clearly in two contrasting views of homosexual behavior in farm animals, symbolizing the contradictory interpretations of same-sex activity that are applied to both people and animals. Anita Bryant, in a particularly brilliant turn of logic, once asserted that "even barnyard animals don't do what homosexuals do." When informed that barnyard animals and many wild species actually do "stoop" to the level of human homosexuals, she retorted, "That still doesn't make it right."[84] Not surprisingly, noted lesbian author and historian Lillian Faderman offers a markedly different view:

> It's ridiculous for people not to recognize it [homosexuality] in nature. My partner once had a ranch, and I was just fascinated with the way the female animals would often mount other female animals as well as be mounted. . . . Mammals are simply sexual.[85]

Each of these women has strong opinions about animal homosexuality, and each woman's viewpoint is informed by her feelings about homosexuality in people. Of the two, however, Lillian Faderman's perspective is closer to the scientific reality of homosexuality in the animal kingdom. The next chapter explores in more detail the way that people have interpreted animal homosexuality throughout the history of science. Unfortunately, biologists themselves have often espoused views that have more in common with Anita Bryant's than Lillian Faderman's.

Chapter 3

Two Hundred Years
of Looking
at Homosexual Wildlife

1764: . . . *three or four of the young [Bantam] cocks remaining where they could have no communication with hens . . . each endeavoured to tread his fellow, though none of them seemed willing to be trodden. Reflection on this odd circumstance hinted to me, why the natural appetites, in some of our own species, are diverted into wrong channels.*

—GEORGE EDWARDS,
Gleanings of Natural History

1964: *Another example of an irreversible sexual abnormality concerns an orang-utan. This ape, a young male, was kept with another young male and they spent a great deal of time playing together. This included some sex play and anal intercourse was observed on a number of occasions.*

—DESMOND MORRIS, "The Response of
Animals to a Restricted Environment"

1994: *There are several explanations for homosexual behavior in non-human animals. First, it is possible that the pursuers misidentified male 42 as a female because the plumage of after-second-year female Tree Swallows resembles that of males . . .*

—MICHAEL LOMBARDO et al., "Homosexual
Copulations by Male Tree Swallows"[1]

*A*nimal homosexuality is by no means a "new" discovery by modern science. Some of the earliest statements regarding homosexual behavior in animals date back to ancient Greece, while the first detailed scientific studies of same-sex behavior were made in the 1700s and 1800s. From the very beginning, descriptions of homosexuality in animals were accompanied by attempts to interpret or explain its occurrence, and observers who witnessed the behavior were almost invariably puzzled, astonished, and even upset by the simple fact of its existence. As the quotes above illustrate, many of these same attitudes have continued to this day. With more than 200 years of scientific attention devoted to the subject, how is it that so many people today—many scientists included—are unaware of the full extent and characteristics of animal homosexuality, and/or continue to be puzzled by its occurrence? This chapter seeks to answer this question, first by chronicling the history of the study of homosexuality in animals, and then by documenting the systematic omissions and negative attitudes of many zoologists in dealing with this phenomenon. As we will see, a history of the scientific study of animal homosexuality is necessarily also a history of human attitudes toward homosexuality.

A Brief History of the Study of Animal Homosexuality

The history of animal homosexuality in Western scientific thought begins with the early speculations of Aristotle and the Egyptian scholar Horapollo on "hermaphroditism" in hyenas, homosexuality in partridges, and variant genders and sexualities in several other species.[2] Although much of their thinking was infused with mythology and anthropomorphism, and there are notable inaccuracies in their observations (the Spotted Hyena, for example, is not hermaphroditic), the discussions of these scholars represent the first recorded thoughts on homosexuality and transgender in animals. The earliest scientific observations of animal homosexuality are those of the noted French naturalist (and count) Georges-Louis Leclerc de Buffon, whose monumental fifteen-volume *Histoire naturelle générale et particulière* (1749–67) includes observations of same-sex behavior in birds. Additional observations on homosexuality in birds were made in the eighteenth century by the British biologist George Edwards, and (as indicated above) they also include some of the first pronouncements about the supposed "causes" and "abnormality" of such behavior.[3]

The beginning of the modern study of animal homosexuality was heralded by a number of early descriptions of same-sex behavior in insects (e.g., by Alexandre Laboulmène in 1859 and Henri Gadeau de Kerville in 1896), small mammals (e.g., by R. Rollinat and E. Trouessart on Bats in 1895), and birds (e.g., by J. Whitaker on Swans in 1885 and Edward Selous on Ruffs in 1906), while the German scientist Ferdinand Karsch offered, in the year 1900, one of the first general surveys of the phenomenon.[4] Since then, the scientific study of animal homosexuality has expanded enormously to include a wide variety of investigations, reported in close to 600 scientific articles, monographs, dissertations, technical reports, and other publications in over ten different languages. These range from field observations of animals that only anecdotally mention homosexual behavior, to more extensive descriptions of homosexuality in a wide range of species studied in the wild, to observations of captive animals (including at many zoos and aquariums throughout the world), to experiments on laboratory animals, to more recent studies devoted to examining all aspects of homosexual behavior in a particular species (often in the wild), to

▲ *The earliest photographic record of animal homosexuality: a pair of male Mute Swans photographed in 1923 on the nest they built together in Scotland. A female pair in the same species was first observed in 1885.*

more comprehensive general surveys of the phenomenon. Some reports have received wide attention, such as the discovery of female pairing in various Gull and Tern species that initiated a flurry of scientific and media interest in the late seventies and early eighties. On the other hand, many reports of animal homosexuality have gone unnoticed even by other zoologists, languishing in small specialty or regional journals such as *The Bombay Journal of Natural History, Ornis Fennica* (the journal of the Finnish Ornithological Society), *Revista Brasileira de Entomologia* (the Brazilian Journal of Entomology), or the *Newsletter of the Papua New Guinea Bird Society*. In a few cases, well-known scientists have published descriptions of animal homosexuality, including Desmond Morris on Orang-utans, Zebra Finches, and Sticklebacks, Dian Fossey on Gorillas, and Konrad Lorenz on Greylag Geese, Ravens, and Jackdaws.[5] Aristocracy has even been involved: in addition to Count Buffon's observations in the eighteenth century, in the 1930s the Marquess of Tavistock in England coauthored a report on bird behavior with scientist G. C. Low that included descriptions of same-sex pairs in captive waterfowl. Like Desmond Morris's account of same-sex activity in Orang-utans quoted above, however, his report was somewhat less than "objective," containing as it did a statement about how "ludicrous" were a pair of male Mute Swans that remained together and built a nest each year.[6]

While most scientific studies of homosexuality in animals have simply involved careful and systematic observation and recording of behavioral patterns (occasion-

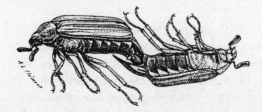

◄ *A drawing from 1896 showing two male Scarab Beetles copulating with each other. This is one of the first scientific illustrations of animal homosexuality to be published.*

ally supplemented by photographic documentation), in some cases more elaborate measures have been employed. The study of animal behavior has now become extremely sophisticated and even "high-tech," and many of these techniques have been applied with great effect to the recording, analysis, and interpretation of same-sex activities and their social context. DNA testing, for example, has been employed to ascertain the parentage of eggs belonging to lesbian pairs of Snow Geese, to determine the genetic relatedness of female Oystercatchers and Bonobos who engage in same-sex activity, to verify the sex of Roseate Terns (some of whom form homosexual pairs), and to investigate the genetic determinants of mating behavior in different categories of male Ruffs. The extent and characteristics of homosexual pair-bonding in Silver Gulls and Bottlenose Dolphins have been revealed by long-term demographic studies that identified and marked large numbers of individuals, who were then monitored over extended periods. Because most sexual activity in Red Foxes takes place at night, investigators only discovered same-sex mounting in this species by setting up infrared, remote-control video cameras that automatically recorded the animals' nocturnal activities (night photography was also required to document similar activity in wild Spotted Hyenas). Radio tracking (biotelemetry) of individual Grizzlies revealed the activities of bonded female pairs, while similar techniques applied to Red Foxes yielded information about their dispersal patterns and overall social organization that relate to the occurrence of same-sex mounting. Videography, including "frame-by-frame" analysis of taped behavioral sequences, has been utilized in the study of courtship interactions in Griffon Vultures and Victoria's Riflebirds, as well as of communicative interactions during Bonobo sexual encounters (both same-sex and opposite-sex). One ornithologist even x-rayed the eggs belonging to a homosexual pair of Black-winged Stilts to see if they were fertile (they weren't).[7]

Unfortunately, in a few cases scientists have subjected animals to more extreme experimental treatments, procedures, or "interventions." During several studies of captive animals, same-sex partners in Rhesus Macaques, Bottlenose Dolphins, Cheetahs, Long-eared Hedgehogs, and Black-headed Gulls (among others) were forcibly separated, either because their activities were considered "unhealthy," or in order to study their reaction and subsequent behavior on being reunited, or to try to coerce the animals to mate heterosexually. A female pair of Orange-fronted Parakeets was forcibly removed from their nest—which they had successfully defended against a heterosexual pair—in order to "allow" the opposite-sex pair to breed in their stead (based in part on the mistaken assumption that female pairs are unable to be parents). Female Stumptail Macaques had electrodes implanted in their uteri in order to monitor their orgasmic responses during homosexual encounters, while female Squirrel Monkeys were deafened to monitor the effect on vocalizations made during homosexual activities.

Although intended ostensibly to reveal important behavioral and developmental effects, the "treatments" applied to animals have in some cases been disturbingly similar to those administered to homosexual people in an attempt to "cure" them (separation or removal of partners, hormone therapy, castration, lobotomy, and electroshock, among others). Numerous primates, rodents, and hoofed mammals, for example, have been subjected to hormone injections to see how this might affect their homosexual behavior or intersexuality. Macaques were castrated as part of behavioral studies that included investigations of homosexual activity, as were White-tailed Deer to determine the "cause" of transgender in this species. Cats have even been lobotomized in order to study the effect on their (homo)sexuality. In some cases, biologists have gone so far as to kill individuals participating in same-sex activities (e.g., Common Garter Snakes, Hooded Warblers, Gentoo Penguins) in order to take samples of their internal reproductive organs.[8] The reasons for this—usually to verify their sex or to determine the condition of their reproductive systems, including the presence of any "abnormalities"—reveal the incredulity as well as the often distorted preconceptions that many scientists harbor about homosexuality. As we will see in the next sections, these attitudes often carry over into the "interpretation" or "explanation" of homosexuality/transgender as well.

"A Lowering of Moral Standards Among Butterflies": Homophobia in Zoology

> . . . I have talked with several (anonymous at their request) primatologists who have told me that they have observed both male and female homosexual behavior during field studies. They seemed reluctant to publish their data, however, either because they feared homophobic reactions ("my colleagues might think that I am gay") or because they lacked a framework for analysis ("I don't know what it means"). If anthropologists and primatologists are to gain a complete understanding of primate sexuality, they must cease allowing the folk model (with its accompanying homophobia) to guide what they see and report.
>
> —primatologist LINDA WOLFE, 1991[9]

There is an astounding amount and variety of scientific information on animal homosexuality—yet most of it is inaccessible even to biologists, much less to the general public. What has managed to appear in print is often hidden away in obscure journals and unpublished dissertations, or buried even further under outdated value judgments and cryptic terminology. Most of this information, however, simply remains unpublished, the result of a general climate of ignorance, disinterest, and even fear and hostility surrounding discussion of homosexuality that exists to this day—not only in primatology (as Linda Wolfe describes), but throughout the field of zoology. Equally disconcerting, popular works on animals routinely omit any mention of homosexuality, even when the authors are clearly aware that such information is available in the original scientific material. As a result, most people don't realize the full extent to which homosexuality permeates the natural world.

Scientists are human beings with human flaws, living in a particular culture at a particular time. Although the profession demands standards of "objectivity" and nonjudgmental attitudes, a survey of the history of science shows that this has not always been the case. For example, the sexism of much biological thinking has been exposed by a number of feminist biologists over the past two decades.[10] They have shown that not only are scientists fallible human beings, but most are men—and their scientific theorizing has often been (and in many cases continues to be) detrimentally colored by their own and their culture's (often negative) attitudes toward women. This observation can be taken a step further: scientists (who are often heterosexual) frequently project, consciously or unconsciously, society's negative attitudes toward homosexuality onto their subject matter. As a result, both scientific and popular understanding of the subject have suffered.[11]

There are notable exceptions, of course. A number of scientists have presented relatively value-neutral descriptions of same-sex activity in various species without feeling a need to overlay their own commentary on the behavior, and several authors have recognized that homosexual activity is a "natural" or routine component of the behavioral repertoire in certain animals. Zoologist Anne Innis Dagg, for example, offered a groundbreaking survey of the phenomenon among mammals in 1984 that was light-years ahead of her contemporaries, while the more recent work of primatologist Paul L. Vasey is beginning to directly address some of the inadequacies and biases of previous studies.[12] Aside from these few examples, though, the history of the scientific study of animal homosexuality has been—and continues to be—a nearly unending stream of preconceived ideas, negative "interpretations" or rationalizations, inadequate representations and omissions, and even overt distaste or revulsion toward homosexuality—in short, homophobia.[13] Moreover, not until the 1990s did zoologists begin to address such biased attitudes: Paul Vasey and Linda Wolfe are, so far, the only scientists to acknowledge in print that there may be a problem in their profession (and Wolfe the only one to name this specifically as homophobia). The full extent, history, and ramifications of the problem, however, have not been previously discussed or documented.

The Perversion of Scientific Discourse

> From a distance this might be mistaken for fighting, but perverted sexuality is the real keynote. . . . In fact, the birds seem sometimes hardly to understand themselves, or to know where their feelings are leading them. . . . My principal observation during the earlier part of the time . . . was a repetition of what I have before noted in regard to the sexual perversion, as one calls it—a term which serves to save one the trouble of thinking. . . .
> —from a scientific description of Ruffs in 1906

> Three unnatural tending bonds were observed: . . . On July 16 a two-year-old bull closely tended a yearling bull for at least four hours in the Wichita Refuge and attempted mounting with penis unsheathed. . . .
> —from a scientific description of American Bison in 1958

> Among aberrant sexual behaviors, anoestrous does were very occasionally seen to mount one another. . . .
> —from a scientific description of Waterbuck in 1982[14]

In many ways, the treatment of animal homosexuality in the scientific discourse has closely paralleled the discussion of human homosexuality in society at large. Homosexuality in both animals and people has been considered, at various times, to be a pathological condition; a social aberration; an "immoral," "sinful," or "criminal" perversion; an artificial product of confinement or the unavailability of the opposite sex; a reversal or "inversion" of heterosexual "roles"; a "phase" that younger animals go through on the path to heterosexuality; an imperfect imitation of heterosexuality; an exceptional but unimportant activity; a useless and puzzling curiosity; and a functional behavior that "stimulates" or "contributes to" heterosexuality. In many other respects, however, the outright hostility toward animal homosexuality has transcended all historical trends. One need only look at the litany of derogatory terms, which have remained essentially constant from the late 1800s to the present day, used to describe this behavior: words such as *strange, bizarre, perverse, aberrant, deviant, abnormal, anomalous,* and *unnatural* have all been used routinely in "objective" scientific descriptions of the phenomenon and continue to be used (one of the most recent examples is from 1997). In addition, heterosexual behavior is consistently defined in numerous scientific accounts as "normal" in contrast to homosexual activity.[15]

The entire history of ideas about, and attitudes toward, homosexuality is encapsulated in the titles of zoological articles (or book chapters) on the subject through the ages: "Sexual Perversion in Male Beetles" (1896), "Sexual Inversion in Animals" (1908), "Disturbances of the Sexual Sense [in Baboons]" (1922), "Pseudomale Behavior in a Female Bengalee [a domesticated finch]" (1957), "Aberrant Sexual Behavior in the South African Ostrich" (1972), "Abnormal Sexual Behavior of Confined Female *Hemichienus auritus syriacus* [Long-eared Hedgehogs]" (1981), "Pseudocopulation in Nature in a Unisexual Whiptail Lizard" (1991).[16] The prize, though, surely has to go to W. J. Tennent, who in 1987 published an article entitled "A Note on the Apparent Lowering of Moral Standards in the Lepidoptera." In this unintentionally revealing report, the author describes the homosexual mating of Mazarine Blue butterflies in the Atlas Mountains of Morocco. The entomologist's behavioral observations, however, are prefaced with a lament: "It is a sad sign of our times that the National newspapers are all too often packed with the lurid details of declining moral standards and of horrific sexual offences committed by our fellow *Homo sapiens;* perhaps it is also a sign of the times that the entomological literature appears of late to be heading in a similar direction."[17] Declining moral standards—in butterflies?! Remember, these are descriptions by *scientists* in respected *scholarly publications* of phenomena occurring in *nature!*

In addition to such labels as *unnatural, abnormal,* and *perverse,* a variety of other negative (or less than impartial) designations have also been employed in the scientific literature. Once again, these span the decades. Mounting among Domestic Bulls is characterized as a "male homosexual vice" (1983), echoing a description from nearly a century earlier in which same-sex activities between male Elephants are classified as "vices" and "crimes of sexuality" that are "prohibited by the rules of at least one Christian denomination" (1892). Courtship and mounting between male Lions is called an "atypical sexual fixation" (1942); same-sex relations in

Buff-breasted Sandpipers are described in an article on "sexual nonsense" in this species (1989); while courtship and mounting between female Domestic Turkeys are referred to as "defects in sexual behavior" (1955). Homosexual activities in Spinner Dolphins (1984), Killer Whales (1992), Caribou (1974), and Adélie Penguins (1998) are characterized as "inappropriate" (or as being directed toward "inappropriate" partners), and same-sex courtship among Black-billed Magpies (1979) and Guianan Cock-of-the-Rock (1985) is called "misdirected." In what is perhaps the most oblique designation, one scientist uses the term *heteroclite* (meaning "irregular" or "deviant") to refer to Sage Grouse engaging in homosexual courtship or copulations (1942).[18]

Besides labeling same-sex behavior with derogatory or biased terms, many scientists have felt the need to embellish their descriptions of homosexuality with other sorts of value judgments. Repeatedly referring to same-sex activity in female Long-eared Hedgehogs as "abnormal," for example, one zoologist matter-of-factly reported that he separated the two females he was studying for fear that they might actually "suffer damage" from continuing to engage in this behavior. Similarly, in describing pairs of female Eastern Gray Kangaroos, another scientist suggested that only in cases where there was no (overt) homosexual behavior between the females could bonding be considered to represent a "positive relationship between the two animals." In the 1930s, homosexual pairing in Black-crowned Night Herons was labeled a "real danger," while one biologist (upon learning the true sex of the birds) referred to his discovery and reporting of same-sex activities in King Penguins as "regrettable disclosures" and "damaging admissions" about "disturbing" activities. More than 50 years later, a scientist suggested that homosexual behavior between male Gorillas in zoos would be "disturbing to the public" were it not for the fact that people would be unable to distinguish it from "normal heterosexual mating behavior." Same-sex pairing in Lorikeets has been described as an "unfortunate" occurrence, while mounting activity between female Red Foxes has been characterized as being part of a "Rabelaisian mood." Finally, in describing the behavior of Greenshanks, an ornithologist used unabashedly florid and sympathetic language to characterize an episode of heterosexual copulation, referring to it as a "lovely act of mating" and concluding, "The grace, movement, and passion of this mating had created a poem of ecstasy and delight." In contrast, homosexual copulations in the same species were given only cursory descriptions, and one episode was even characterized as a "bizarre affair."[19]

In a direct carryover from attitudes toward human homosexuality, same-sex activity is routinely described as being "forced" on other animals when there is no evidence that it is, and a whole range of "distressful" emotions are projected onto the individual who experiences such "unwanted advances."[20] One scientist surmises that Mountain Sheep rams "deem it an insult to be treated as a female" (including being mounted by another male), while Rhesus Macaques and Laughing Gulls are described as "submitting" to homosexual mounts even when there is clear evidence that they are willing participants (for example, by initiating the activity). Cattle Egrets who are mounted during homosexual copulations are characterized as "suffering males," while female Sage Grouse mounted by other females are their "victims." Orang-utan males who participate in homosexuality are said to be "forced into nonconformist sexual behavior" by their partners even though they display none of the obvious signs of distress (such as screaming and struggling violently) that are characteristic of female Orangs during heterosexual rapes. Scientists describing same-sex courtship in Kob antelopes imply that females try to "avoid" homosexual attentions by circling around the other female or butting her on the shoulder. In fact, these actions are a formally recognized ritual behavior called mating-circling that is a routine part of heterosexual courtships, and not indicative of disinterest or "unwillingness" on the part of the courted female. Females who do not want to be mounted (by male partners) actually drop their hindquarters to the ground (a behavior not observed in homosexual contexts). Same-sex courtship in Ostriches is deemed to be a "nuisance" that goes "on and on" and is perpetrated by "sexually aberrant" males. The calm stance of a courted male (referred to as the "normal" partner) in the face of such homosexual advances is described as "astonishing," while the recipient's occasional acknowledgment of the activity is downplayed in favor of those times when he makes no visible response (interpreted as disinterest). Yearling male Guianan Cock-of-the-Rock are consistently described as "taking advantage of" or "victimizing" adult males that they mount, while their partners are said to "tolerate" such homosexual activity. This is at odds with the descriptions, by the same scientists, of the adult partners as willing participants who actively facilitate genital contact during homosexual mounts and allow the yearlings to remain on their territories (unlike unwanted adult intruders who are chased away or attacked). Finally, male Mallard Ducks that switch from heterosexual to homosexual pairings are described as being "seduced" by other males, while Rhesus Macaques are characterized as reacting with a sort of "homosexual panic" to same-sex advances—both echoing widely held misconceptions about human homosexuality.[21]

In other cases, zoologists have problematized homosexual activity or imputed an inherent inadequacy, instability, or incompetence to same-sex relations, when the supporting evidence for this is scanty or questionable at best and nonexistent at worst. For example, the fact that male homosexual pairs in Greylag Geese engage in higher rates of pair-bonding and courtship behavior is ascribed to an (unsubstantiated) "instability" of same-sex pair-bonds. In fact gander pairs in this species have been documented as lasting for 15 or more years and are described as being, in many cases, more strongly bonded than heterosexual mates.[22] Similarly, even though pair-bonds between male Ocellated Antbirds can last for years, one ornithologist insisted on portraying them as "fragile" and liable to dissolve at the mere appearance of a "nubile female." Antbird same-sex pairs do sometimes divorce, but so do heterosexual ones, and any generalizations about the comparative stability of each cannot be made without comprehensive, long-term studies of pair-bonding—which have yet to be undertaken for this species.[23] The fact that sexual activity between female Gorillas generally takes longer than heterosexual copulations is speculatively attributed to "mechanical difficulties" involved in sex between two females—it is apparently inconceivable to the investigator that females might be experiencing closer bonding or greater enjoyment with each other (as reflected by their face-to-face position and other features that also distinguish homosexual from heterosexual activity in this species). In the same vein, accounts of same-sex mounting in Western Gulls, Guianan Cock-of-the-Rock, and Red Foxes refer to the "disoriented," "bumbling," or "fumbling" actions of some individuals—terms that are rarely used to describe nonstandard mounting attempts in heterosexual contexts (even when they are equally "incompetent"). Conversely, one primatologist is willing to concede that affiliative gestures (such as mutual touching, grooming, or preening) between animals of the opposite sex may be "tender" and even "an expression of love and affection," yet similar or identical activities between same-sex participants are never characterized this way.[24]

This double standard is particularly apparent where descriptions of same-sex pairs in Gulls are concerned. When a male Laughing Gull in a homosexual pair courted and mounted a female, for example, this was taken by one investigator to mean that his pair-bond was unstable and that he was "dissatisfied" with his homosexual partnership (rather than as simply an instance of bisexual behavior). In contrast, homosexual activity by birds in heterosexual pairs is never interpreted as "dissatisfaction" with heterosexuality or as reflecting the tenuousness of opposite-sex bonds. In a study on pair-bonding in Black-headed Gulls, the term "monogamous" (implying stability) was reserved for heterosexual pairs, even though homosexual pairs in this species can also be stable and monogamous, and heterosexual pairs are sometimes nonmonogamous. Likewise, the stability of female pairs of Herring Gulls was claimed to be lower than heterosexual pairs. Yet in making this assessment, researchers were considering females to have broken their pair-bond if they were simply not seen at the nesting colony the following year—when in fact they or their partner could have died, relocated, or been missed by observers. Among those females that *were* subsequently observed at the colony (a more accurate measure, and the standard way of calculating mate fidelity for heterosexual pairs), the rate of pair stability was in fact nearly identical to that of opposite-sex pairs.

Similarly, the parenting abilities of female pairs in many Gull species are often implied to be substandard because such couples usually hatch fewer chicks than heterosexual pairs. However, calculations of the hatching success of homosexual pairs typically include infertile eggs in the overall count; since many females in same-sex pairs do not mate with males, large numbers of their eggs are infertile and so of course a larger proportion of their clutches do not hatch. In addition, all of the traits taken to indicate poor quality of parenting in some female pairs—e.g., smaller eggs, slower embryonic development, lower hatching rate of fertile eggs, reduced weight and greater mortality of chicks, higher rates of loss or abandonment—are also characteristic of supernormal clutches attended by heterosexual parents (usually polygamous trios). In other words, they are related to the larger-than-average clutch size rather than the sex of the parents per se. In fact, most studies of Gulls have shown that the parenting abilities of homosexual pairs are at least as good as those of heterosexual pairs. Moreover, heterosexual parents in many Gull species can be severely neglectful or overtly violent toward their chicks, causing youngsters to "run away" from their own families and be adopted by others (or even perish). Needless to say, this behavior is never interpreted as being representative of all heterosexual pairs or as impugning heterosexuality in general (even though it is usually far more widespread than homosexual inadequacies).[25] Thus, many zoological studies evidence the same inconsistency often found in discussions of human homosexuality: any difficulties or irregularities in same-sex relations are generalized to *all* homosexual interactions (or else focused on to the exclusion of other examples), whereas comparable problems in opposite-sex relations are seen in the proper perspective, simply for what they are—individual (or idiosyncratic) occurrences that, while noteworthy, do not reflect the entirety of heterosexuality nor warrant disproportionate attention.

Homophobia in the field of zoology is not always this overt or virulent; nevertheless, ignorance or negative attitudes that are not directly expressed usually have identifiable consequences and important ramifications for the way the subject is handled. Discussion of animal homosexuality has in fact been compromised and stifled in the scientific discourse in four principal ways: presumption of heterosexuality, terminological denials of homosexual activity, inadequate or inconsistent coverage, and omission or suppression of information.

Heterosexual Until Proven Guilty

> . . . after about twenty minutes I realized that what I was watching was three whales involved in most erotic activities! . . . Then one, two, and eventually three penes appeared as the three whales rolled at the same time. Obviously, all three were males! It was almost two hours after the first sighting . . . and up to that point I was convinced I was watching mating behavior. A discovery—and a stern reminder that first impressions are deceiving.
> —JAMES DARLING, "The Vancouver Island Gray Whales"[26]

Many behavioral studies of animals operate under a presumption of heterosexuality: a widespread—if not universal—assumption among field biologists is that all courtship and mating activity is heterosexual unless proven otherwise. This is particularly prevalent in studies of animals in which males and females are not visually distinguishable at a distance. The scientific literature is filled with examples of biologists who were convinced that the sexual, courtship, or pair-bonding activity they had been observing was between a male and female—until confronted with

Two male Gray Whales participating in homosexual activity off the coast of Vancouver Island. Only the erect penises of the whales are visible above the surface of the water, but this enabled scientists to verify the sex of the animals. Without this confirmation, observers would probably have mistaken this for heterosexual mating activity. ▶

clear evidence of homosexuality, such as a glimpse of *two* sets of male genitalia, or a nest containing more eggs than just one female could have laid.[27]

Moreover, many zoologists still routinely determine the sex of animals in the field based on their behavior during sexual activity—with the (often unstated) assumption that there must be both a male (the one doing the mounting) and a female (the one being mounted) in any such interaction. Of course, this automatically eliminates any "chance" of observing homosexual activity in the first place. A field study of Laughing Gulls, for example, utilized the following assumptions in determining the sex of birds: "(1) any bird copulating more than twice on top was presumed a male, and (2) the mate of a male was presumed to be a female." Yet other studies of this species in both the wild and captivity have revealed that male homosexual mounting and pairing do in fact occur in Laughing Gulls. Scientists studying sexual behavior in Common Murres admitted that they probably underestimated the frequency of homosexual mounting because they assumed that sexual activity involved opposite-sex partners unless they had direct evidence to the contrary. Amazingly, this practice is even used in species where homosexual behavior is known to occur from previous studies of either captive or wild animals, such as Kittiwakes and Griffon Vultures.[28] True, some biologists have critiqued this method of sex determination—but only on the grounds that it can miss examples of reverse *heterosexual* mounting (where females mount males).[29] And in spite of its obvious shortcomings, behavioral sex determination continues to be employed in recent studies, some of which constitute the first and only documentation of little-known species. One can only guess at how many examples of homosexual activity have been and will continue to be overlooked because of this.[30]

Even in captivity, the sex of animals is often mistaken, and the consequent "amending" of mating or courtship activity from heterosexual to homosexual sometimes results in elaborate retractions, revisions, and reinterpretations. Renowned German ornithologist Oskar Heinroth, for example, published one of the first descriptions of heterosexual mating in Emus—only to discover that the two birds he had been observing in captivity were in fact both males, prompting him to publish a "correction" to his earlier description three years later. In reviewing the earliest descriptions of courtship behavior in captive Regent Bowerbirds, scientists realized that what had previously been described as heterosexual activity was in fact display behavior performed between two males. This resulted in rather confusing citations of the earlier material such as the following, in which the true sex of the birds is indicated by the later author's bracketed insertions (prefaced by the assertion "I make no apology for revising in brackets his text to make it meaningful"): "'These love-parlours, each one built by a female [immature male] for her [his] sole use . . . were of the shape of a horseshoe. . . . The female would enter and squat in her [in the immature male's] love-parlour, the tail remaining towards the entrance . . . the rejected females [immature males in adult female dress] . . . built or partly built three love-parlours in different spots.'" The very first description of "heterosexual" courtship and mating in Dugongs (a marine mammal) was published, ironically, in a scientific article prefaced with lines of romantic verse about the "heaving bosoms" of mermaids and sea nymphs (creatures that the animal has historically been mistaken for). Ironic, because nearly a decade later biologists confirmed that both animals involved in this sexual activity were actually males.[31]

Perhaps the most convoluted—and humorous—mix-up of this sort involves a set of King Penguins that were studied at the Edinburgh Zoo from 1915 to 1930. The various permutations and shufflings of mistaken gender identities (on the part of human observers, not the birds) reached truly Shakespearean complexity. The sex of the penguins was initially determined on the basis of what was thought to be heterosexual behavior, and the birds were given (human) names accordingly. Following this, however, some "puzzling" observations of apparently homosexual activity were made. Subsequent re-pairings and breeding activity eventually revealed—more than seven years later!—that in fact the sex of all but one of the birds had been misidentified by the scientists. At this point a comprehensive "sex change" in the names of the birds was hastily instituted to reflect their true genders: "Andrew" was renamed Ann, "Bertha" turned into Bertrand, "Caroline" became Charles, and "Eric" metamorphosed into Erica ("Dora" had correctly been identified as a female). Ironically, although some previous "homosexual" interactions could be reclassified as heterosexual once the true sex of the birds was known, other less straightforward revisions were also required. Two penguins that had initially been seen engaging in "heterosexual" activity—"Eric" and Dora—later turned out to be same-sexed, while premature observations of lesbian mating between "Bertha" and "Caroline" were confirmed as homosexual—but actually involved the *males* Bertrand and Charles![32]

Sometimes the presumption of heterosexuality concerns not the sex of animals but the context in which courtship or pairing activity occurs. This can be characterized as a "heterocentric" view of animal behavior, i.e., one that tends to see all forms of social interaction as revolving around heterosexual activity (see chapter 5). For example, female homosexual pairs in a number of birds, such as Snow Geese, Ring-billed Gulls, Red-backed Shrikes, and Blue Tits, were initially thought to represent the female portion of heterosexual trios. The females were erroneously assumed to be bonded not to each other but to a third, male, bird (that had yet to be observed)—to the extent that several researchers felt compelled to provide explicit evidence and argumentation that no male was associated with such female pairs. Likewise, courtship and mounting activity between male Guianan Cock-of-the-Rock was categorized as a form of "disruption" of heterosexual courtships in one study, when in fact the majority of same-sex activity took place outside of heterosexual courtships when females weren't even around. In a similar vein, same-sex behavior in Stumptail Macaques was classified as sexual in one study only if it occurred "during or immediately after or between heterosexual copulations." In summarizing the pairing strategies adopted by widowed Jackdaws, one scientist enumerated only *heterosexual* mating patterns and failed to include the formation of female homosexual pairs, even though his own data showed that 10 percent of widowed females attracted new female mates. Likewise, one author's discussion of homosexual activity in male Cheetahs focused on a single case where males mounted each other in apparent "frustration" during heterosexual courtship activities, when in fact the majority of same-sex interactions did not occur in this type of context. Finally, sexual activity and bonding between female Bonobos has traditionally been interpreted as a derivative extension of heterosexuality and subsumed under the general patterns of male-female relations. Recent work, however, shows that female bonding and homosexuality in this species are in fact autonomous from heterosexuality, not geared toward attracting opposite-sex partners, and actually much stronger and more primary than male-female bonding.[33]

Similar assumptions have frequently guided the treatment of actual sexual behavior, most blatantly when same-sex activity is excluded entirely from the definition of what constitutes sexual activity. One researcher, for example, only considered cases involving "insertion of the penis into the vagina" to be genuine examples of sexual penetration in Savanna (Olive) Baboons, and a study of Right Whales classified behavior as sexual only if it occurred in groups containing both males and females. A recent study of Moose defined sexual mounting behavior solely as "a male mounting a female," while any mounting activity in Cattle Egrets "in which male-female cloacal [genital] contact appeared to be impossible" was classified a priori as "incomplete" or unsuccessful sexual activity.[34] Anal and oral intercourse are not the only forms of penetration excluded by these sorts of definitions. In discussing homosexual activity in female Squirrel Monkeys, one scientist bluntly asserted that clitoral penetration—the insertion of one female's clitoris into another's vagina—was anatomically impossible: "Because of the structure of the female genitalia, however, intromission between females is not possible." In fact, the clitoris in Squirrel Monkeys and many other female mammals becomes conspicuously erect during sexual arousal, and actual clitoral penetration *has* been documented during lesbian sexual activity in Bonobos, and it may also occur in Spotted Hyenas.[35] The phallocentric viewpoint expressed in comments such as these is merely the most recent manifestation of attitudes that can be traced back to some of the earliest descriptions of animal homosexuality. In 1922, for example, one scientist wrote of female homosexual interactions in Savanna (Chacma) Baboons, "The physical completion of the act was, of course, impossible and it seemed more like an impulsive action in which there was no real sexual excitement involved."[36] This perfectly epitomizes the sort of stereotypes and misinformation that have continued to engulf homosexuality to this day, in both animals and people.

Mock Courtships and Sham Matings

The attitude that homosexual activity is not "genuine" sexual, courtship, or pair-bonding behavior is also sometimes made explicit in the descriptions and terminology used by researchers. In spite of witnessing two male homosexual mounts during a morning spent observing Ruffs, for example, one ornithologist reported offhandedly that "there were no real copulations" because no heterosexual mounting took place; a similar comment was made by a scientist studying Bonnet Macaques.[37] This attitude is also encoded directly in the words used for homosexual behaviors: rarely do animals of the same sex ever simply "copulate" or "court" or "mate" with one another (as do animals of the opposite sex). Instead, male Walruses indulge in "mock courtship" with each other, male African Elephants and Gorillas have "sham matings," while female Sage Grouse and male Hanuman Langurs and Common Chimpanzees engage in "pseudo-matings." Musk-oxen participate in "mock copulations," Mallard Ducks of the same sex form "pseudo-pairs" with each other, and Blue-bellied Rollers have "fake" sexual activity. Male Lions engage in "feigned coitus" with one another, male Orang-utans and Savanna Baboons take part in "pseudo-sexual" mountings and other behaviors, while Mule Deer and Hammerheads exhibit "false mounting." Bonobos, Japanese and Rhesus Macaques, Red Foxes, and Squirrels all perform "pseudo-copulations" with animals of the same sex.[38] Amid this abundance of counterfeit sexual activity, one thing is all too real: the level of denial on the part of some zoologists in dealing with this subject.[39]

Even use of the term *homosexual* is controversial. Although the majority of scientific sources on same-sex activity classify the behavior explicitly as "homosexual"—and a handful even use the more loaded terms *gay* or *lesbian*[40]—many scientists are nevertheless loath to apply this term to any animal behavior. In fact, a whole "avoidance" vocabulary of alternate, and putatively more "neutral," words has come into use. "Male-male" or "female-female" activity is the most common appellation, although some more oblique designations have also appeared, such as "male-only social interactions" in Killer Whales or "multifemale associations" for same-sex pairs in Roseate Terns and some Gulls. Homosexual activities are also called "unisexual," "isosexual," "intrasexual," or "ambisexual" (meaning single-sex, same-sex, within-sex, and bisexual, respectively) in various species such as Gorillas, Ruffs, Stumptail Macaques, Hooded Warblers, and Rhesus Macaques. The use of "alternate" words such as *unisexual* is sometimes advocated precisely because of the homophobia evoked by the term *homosexual:* one scientist reports that an article on animal behavior containing *homosexual* in its title was widely received with a "lurid snicker" by biologists, many of whom never got beyond the "sensationalistic wording" of the title to actually read its contents.[41]

Occasionally there are directly opposing assertions regarding the suitability of the term *homosexual* for the same behavior and species. In a relatively enlightened treatment of same-sex activities in Giraffes, for example, one zoologist stated, "Such usage [of the term *homosexual*] is acceptable provided it is used without the usual human connotation of stigma and sexual abnormality.... In giraffes the erection of the penis, mounting, and even possibly orgasm leaves little doubt as to the sexual motivation behind these actions." In contrast, a decade later another zoologist objected, "Considerable significance has been attached to the fact that necking males sometimes show penis erections and that one may mount the other ... such behavior has been called 'homosexual.' However ... I ... do not feel that the use of the term *homosexual,* with its usual (human) connotation, is justified in this context."[42] Ironically, where the first scientist objected only to the stigma associated with the term as applied to people, the second objected to the connotation of genuinely sexual behavior in the term as applied to people.

When it comes to heterosexual activities, however, scientists are not at all adverse to making analogies with human behaviors. Opposite-sex courtship-feeding in birds is described as "romantic" and reminiscent of human lovers kissing, male canaries whose vocalizations attract female partners are said to sing "sexy" songs, while avian heterosexual monogamy and foster-parenting are compared to similar activities in people (in spite of the acknowledged differences in the behaviors involved). Even more flagrant anthropomorphizing sometimes occurs: male-female interactions in Savanna Baboons, for example, are likened to "May-December romances," "flirting," and other human courtship rituals in a "singles bar"; polyandry in Tasmanian Native Hens is termed "wife-sharing"; opposite-sex bonds between cranes who readily pair with one another are characterized as "magic marriages"; and heterosexually preco-cious male Bonobos are dubbed "little Don Juans." Female fireflies that lure males of other species by courting and then eating them are labeled "femmes fatales," and one scientist even uses the term *gang-bang* to describe group courtship and forced het-erosexual activity in Domestic Goats. Regardless of whether these characterizations are appropriate, among zoologists it is still more acceptable (in practice if not in the-ory) to draw human analogies where heterosexuality is concerned.[43]

Many scientists' denial that same-sex courtship, sexual, pair-bonding, and/or parenting activities should be put in the category of "homosexuality" are based on spurious or overly restrictive interpretations of the phenomenon (or the word). For example, Konrad Lorenz claims that gander pairs in Greylag Geese are not actually "homosexual" because sexual behavior is not necessarily an important component of such associations (not all members of gander pairs engage in sexual activity), and because not all such birds pair exclusively with other males over their entire lifetime. By the same criteria, however, opposite-sex pairs would fail to qualify as "heterosexual": sexual activity is not an important component of male-female pairings in this species (as Lorenz himself acknowledges), and not all such birds pair exclusively with opposite-sex partners during their lives. Yet Lorenz has no qualms about labeling such pairs "heterosexual."[44] In fact, what we have here is sim-ply an attempt to equate homosexuality with only one characteristic or type of same-sex activity (sexual versus pair-bonding, or sequential bisexuality versus ex-clusive homosexuality).

In a parallel discussion of female pairs in Western Gulls, one researcher suggests that previous descriptions of such pairs as "homosexual" or "lesbian" or "gay" is in-appropriate because they do not resemble homosexual pairings in humans.[45] But which homosexual pairings, in which humans? As discussed in chapter 2, there is no single type of same-sex pair-bonding in people: homosexual couples differ vastly in a wide range of factors such as their sexual behavior, social status, forma-tion process, sexual orientation of members, participation in parenting, duration, and so on, and they vary enormously between different cultures, historical periods, and individuals. Assuming, however, that this author is referring to Euro-American lesbian couples, it is difficult to see what specific similarities are required before the label of *homosexual* would be considered acceptable. Same-sex pairs in both Gulls and humans engage in a variety of courtship, pair-bonding, sexual, and parenting activities and exhibit parallel variability in their formation, social status, and the sexual orientation of their partners. In fact, it is fallacious to suggest that a same-sex activity should resemble some human behavior before we can label it homosexual. A more reasonable approach (the one used in this book as well as in many scientific sources) is to take comparable behaviors *in the same or closely related species* as the point of reference: any activity between two animals of the same sex that involves behaviors independently recognized (usually in heterosexual contexts) as courtship, sexual, pair-bonding, or parenting activities is classified as "homosexual." By this criterion, same-sex pairs of Gulls are "homosexual" because all of the characteristics they exhibit are well-established components of pair-bonding in heterosexual pairs of the same species—to the extent that same-sex couples have often been mistaken for heterosexual ones and unhesitatingly labeled a "mated pair" before their true sex was discovered.

More generally, a number of scientists have suggested that the term *homosexual* should be reserved for overt sexual behavior, and that it is inappropriate to apply this word to other behavior categories such as same-sex courtship, pair-bonding, or parenting arrangements. We might characterize this as a "narrow" definition of homosexuality (such as that assumed by Lorenz). On the other hand, *homosexuality*, as the term is used in this book, refers not only to overt sexual behavior between animals of the same sex, but also to related activities that are more typically associated with a heterosexual or breeding context. This usage is consistent with a number of studies in the zoological literature, in which the word is employed as a cover term for both sexual and related behaviors (e.g., courtship, pairing, parenting).[46] We might characterize this as a "broad" definition of homosexuality. Although overt sexual behavior is by far the single most common type of same-sex activity found in various species—hence the original terminology—the other behavior categories also occur in a sizable proportion of cases in which same-sex activities have been documented. In many (but not all) species, behaviors of various categories co-occur (e.g., sexual and courtship activity with pair-bonding, courtship or bonding with parenting, and so on). There are also numerous cases where only one behavior type is instantiated, or where several behavior categories co-occur in the same species but are not necessarily observed in the same individuals (e.g., sexual behavior may be seen between some animals, courtship behavior between others, etc.). In some cases this represents actual discontinuities of behaviors; in others, it represents observational gaps. When the term *homosexuality* is employed in the broad sense for these cases, it is always with the understanding that only selected behavior categories or co-occurrences may be involved (as in observations of heterosexual behavior).[47]

The difference between these two usages of the term *homosexual* can be illustrated with an example involving two different forms of same-sex activity (each widely attested in birds, sometimes both in the same species). On one hand, consider two female birds that are pair-bonded to each other for life, regularly engage in courtship activity with one another, build a nest each year in which they jointly lay eggs, and on one occasion raise chicks together (fathered via a single heterosexual copulation that season by one of the partners), yet never mount each other. On the other hand, consider a male bird who is mated to a female partner for life—with whom he regularly copulates and raises offspring—but who participates in a single copulation with another male (and never again engages in such behavior for the remainder of his life). A narrow definition of homosexuality would require us to consider the first case to be somehow less "homosexual" than the second simply because no overt sexual behavior takes place between the two females. A broad view of homosexuality, on the other hand, recognizes that both cases involve homosexual behavior—but of two distinct types that need to be carefully distinguished in terms of their social context as well as the other sexual and pairing activities of the participants (since both scenarios actually exemplify contrasting forms of *bisexuality*). Unlike the narrow definition, this usage acknowledges the complexities and variability of same-sex interactions in the animal world, while providing a useful framework for cross-species comparisons and generalizations; it also offers the possibility of more precise and nuanced characterizations of sexual orientation.

Most scientists are understandably wary of anthropomorphizing animals with terms that have wide applicability in a human context—as well they should be—and obviously not all zoologists who avoid the word *homosexual* are motivated by homophobia. Nevertheless, the lengths that are taken to circumvent terminology that can easily be clarified with a simple explanatory statement often border on the absurd.[48]

"Not Included in the Tabulated Statistics"

Even when homosexual behavior is recognized as such, detailed study of it is often omitted or passed over, or the phenomenon is marginalized and trivialized. For instance, numerous published reports on the courtship and copulation behavior of animals provide excruciatingly detailed descriptions and statistics on frequency of mounts, number of ejaculations, duration of penile erections, number of thrusts, timing of estrous cycles, total number of sexual partners, and so on and so forth— but all for *heterosexual* interactions. In contrast, homosexual activity is often mentioned only in passing, not deemed worthy of the exhaustive coverage that is afforded "real" sexual behavior.[49] In a detailed study of Spinner Dolphin sexual activity, for example, only heterosexual behavior is quantified and given a thorough statistical treatment, even though the author recognizes the prominence of homosexual activity in this species and actually states directly that its frequency exceeds that of heterosexual behavior. Another study of the same species mentions homosexual copulations without providing the total number observed, unlike heterosexual matings. In a tabulation of homosexual and heterosexual activity in Kob antelopes, the number of male partners of each female is cataloged while the number of female partners is not. Likewise, articles on Crested Black Macaque and Brown Capuchin sexual behavior acknowledge the occurrence of female homosexual activity yet offer no statistics on this behavior, even though it is said to be more common (in Crested Blacks) than male homosexual activity (which, along with heterosexual behavior, *is* quantified). Finally, graphs of the frequency of various Giraffe activities in one study fail to provide adequate information on homosexual mounts: all same-sex interactions are lumped into the category of "sparring" (a form of fighting) without distinguishing actual sparring from necking (a ritualized, nonviolent form of play-fighting and affection) or mounting activity.[50]

Sometimes certain aspects of homosexual activity are excluded or arbitrarily eliminated from an overall analysis or tabulation—often resulting in a distorted picture of same-sex interactions (regardless of whether the omission is deliberate or well-motivated). For instance, a female Western Gull who exhibited the most overt sexual activity with her female partner was "not included in the tabulated statistics" of a study comparing heterosexual and homosexual behaviors. By failing to incorporate data from this individual (intentionally or not), researchers undoubtedly helped foster the (now widely cited) impression that sexual activity is a uniformly negligible aspect of female pairing in this species. Along the same lines, scientists surveying pair formation in Black-crowned Night Herons only tabulated homosexual couples that they considered to be "caused" by the "crowded" conditions of captivity. They ignored a male pair whose formation could not be attributed to such conditions and also overlooked the fact that such "crowded" conditions regularly occur in wild colonies of the same species. And all data concerning same-sex pairs or coparents in Laughing Gulls, Canary-winged Parakeets, Greater Rheas, and Zebra Finches were excluded from general studies of pair-bonding, nesting, or other behaviors in these species.[51]

The significance of homosexual activity is sometimes also downplayed in discussions of its prevalence or frequency. Certainly many variables must be considered when trying to quantify same-sex activity, and the task is rarely straightforward (as we saw in chapter 1). Nevertheless, in some instances homosexual frequency is interpreted or calculated so as to give the impression that same-sex activity is less common than it really is or else is de-emphasized in terms of its importance relative to other species. In Gorillas, for example, homosexual activity in females is classified as "rare" because investigators observed it "only" 10 times on eight separate days. However, these figures are incomplete unless compared with the frequency of heterosexual interactions during the same period. In fact, 98 episodes of heterosexual mating were recorded during the same period, which means that 9 percent of all sexual activity was homosexual—a significant percentage when compared to other species.[52] Similarly, investigators studying lesbian pairs in Western Gulls state, "We have estimated female-female pairs make up *only* 10–15 percent of the population" (emphasis added), when in fact this is one of the higher rates recorded for homosexual pairs in any bird species (and certainly the highest rate reported at that time). Homosexual mounting in female Spotted Hyenas is claimed to be much less frequent than in other female mammals, yet no specific figures are offered; the one species that is mentioned in comparison is the Guinea Pig, a domesticated rodent that is not necessarily the best model for a wild carnivore.[53]

It is also important to consider the behavioral type and context when evaluating frequency. Homosexual copulations in Tree Swallows, for example, have been characterized as "exceedingly rare" because they have been observed only infrequently and are much less common than heterosexual matings between pair-bonded birds. However, homosexual copulations are nonmonogamous matings (i.e., they typically involve birds that are not paired to one another and may even have heterosexual mates); it is insufficient in this case to compare the frequency rates of *two different kinds* of copulation (within-pair and extra-pair). In fact, the more comparable heterosexual behavior—nonmonogamous copulations involving males and females—are also "rarely" seen. Early observers considered them to be exceedingly uncommon (or nonexistent), while a later study documented only two such matings during four years of observation, and subsequent research has yielded consistently low levels of observed promiscuous (heterosexual) copulations. Yet scientists now know that such matings must be common because of the high rates of offspring resulting from them—in some populations, more than three-quarters of all nestlings (as verified by DNA testing). Thus, it is likely that the frequency of homosexual nonmonogamous matings has been similarly underestimated.[54]

Many scientists, on first observing an episode of homosexual activity, are also quick to classify the behavior as an exceptional or isolated occurrence for that species. In contrast, a single observed instance of heterosexuality is routinely interpreted as representative of a recurrent behavior pattern, even though it may occur (or be observed) extremely rarely or exhibit wide variation in form or context. This sets up a double standard in assessing and interpreting the prevalence of each behavior type, especially since opposite-sex mating can be a less than ubiquitous or uniform feature of an animal's social life (see chapter 5). It also conflicts with the patterns established for other species. In repeated instances, homosexual activity was initially recorded in only one episode, dyad, or population (and usually interpreted—or dismissed—as an isolated example), but was then confirmed by subsequent research as a regular feature of the behavioral repertoire of the species—often spanning many decades, geographic areas, and behavioral contexts.[55] It is no longer possible to claim that homosexuality is an anomalous occurrence in a certain species simply because it has only been observed a handful of times.

In some cases, conflicting verbal assessments of the prevalence of homosexual activity are offered by the same investigators, when the actual quantitative data show a relatively high occurrence. Homosexual courtship/copulation in Pukeko, for example, is described as being both "common" and "relatively rare"—the actual rate of 7 percent of all sexual activity is in fact fairly high compared to other species (and the same-sex courtship rates are even higher). Likewise, a report on Black-headed Gulls states, "Homosexual pairs were also rare," then a few pages later counterasserts that "male-male bonds occurred rather commonly"—and at approximately 16 percent of all pairs observed, the actual rates support the latter interpretation more than the former.[56] Not only are these assessments inconsistent and unfair with regard to the observed rates of homosexuality, they also run counter to a standard cross-species measure of heterosexual frequency. Although there is no absolute or universal criterion for what is "rare" or "common," biologists do recognize a "threshold" of 5 percent as being significant where at least one heterosexual behavior is concerned—polygamy. When this mating system is exhibited by only a minority of the population (as is true in many birds, for example), it is nevertheless considered to be a "regular" feature of the species' behavioral repertoire when its incidence reaches 5 percent. This is certainly far less than the rate of homosexuality in many species where same-sex behavior is regarded as "uncommon" or "exceptional."[57]

In a vivid example of the marginalization that often surrounds discussion of animal homosexuality, scientists sometimes find their own descriptions of same-sex activity published with "amendments," "asides," or "explanations" inserted by journal or reprint editors who are uncomfortable with the content or appellation. For example, one ornithologist's description of homosexual activity in House Sparrows and Brown-headed Cowbirds was embellished with a note from the editor of the journal where it appeared, offering several implausible "reinterpretations" of the behavior that eliminated any sexual motivation. Likewise, when descriptions of homosexual activity in Baboons from the 1920s were reprinted nearly half a century later, a scientist who penned the introduction to the new edition felt compelled to annotate the offending passages with the "modern" viewpoint that such activity is not really homosexual. And editors of the journal *British Birds* scrambled to try to "explain" a case of homosexual pairing in male Kestrels as actually involving a "male-plumaged female" (i.e., a female bird that looked exactly like a male). They added in their published postscript to the article that this putative plumage variation was, in their opinion, "of much more interest than the copulation or attempted one between the two males" that was the primary focus of the author's report.[58]

In a similar vein, one scientist who observed a pair of female Chaffinches hedged his bets by saying only that "female-plumaged" birds were involved, leaving open the possibility that one might still have been a male (and consequently part of a heterosexual pair)—even though there was absolutely no evidence that either bird could have been a male. He finally had to concede that the birds "were surely females." Sometimes this strategy backfires, as in the case of an early description of courtship display in Regent Bowerbirds (mentioned previously), in which the presumed "female-plumaged" birds *both* turned out to be males—and therefore still participants in homosexual activities.[59] These cases show that scientists are sometimes reluctant even to commit to the *sex* of the animals they are observing if it seems that homosexuality might be involved—in stark contrast to the haste with which they usually judge (or assume) participants to be opposite-sexed on the scantiest of evidence.

The Love That Dare Not Bark Its Name

Although the first reports of homosexual behavior among primates were published >75 years ago, virtually every major introductory text in primatology fails to even mention its existence.

—primatologist PAUL L. VASEY, 1995[60]

In the 1890s, Oscar Wilde's lover, Lord Alfred Douglas, characterized homosexuality as "the Love that dare not speak its name," referring to the silence and stigma surrounding disclosure of homosexual interests and discussion of same-sex activities.[61] An analogue to this silencing and stigmatization exists in the pages of zoology journals, monographs, and textbooks, and in the wider scientific discourse. Discussion of homosexual activity in animals has frequently been stifled or eliminated, and a number of examples can only be considered active suppression of information on the subject. When several comprehensive reference works devoted to every conceivable aspect of an animal's biology and behavior are published, including chapters by scientists who originally observed homosexuality in the species, and yet consistently no mention is made of that homosexual behavior, one has to wonder about the "objectivity" of these scientific endeavors.

At one extreme, there are cases of apparently deliberate removal of information. In 1979, a report on Killer Whale behavior was issued by the Moclips Cetological Society, a nonprofit scientific organization devoted to whale study. Sexual activity between males—classified explicitly as "homosexuality" in the report—was discussed at some length, concluding with the statement, "Homosexual behavior has been observed in many animals including cetaceans, canids, and primates, and, in some cases, it has significance for social order." A year later, when this report was published as a government document for the U.S. Marine Mammal Commission, all mention of homosexuality was eliminated even though the remainder of the report was intact.[62] At the other extreme are cases where homosexuality is discussed but is buried in unpublished dissertations, obscure technical reports, foreign-language journals, or in articles whose titles give no clue as to their content. For example, the earliest reports of same-sex courtship and mounting in wild Musk-oxen appeared in an unpublished master's thesis at the University of Alaska and a (published) report for the Canadian Wildlife Service. Consequently, a study on homosexual activity in captive Musk-oxen conducted more than 20 years after the initial discovery fails to mention any occurrence of this behavior in the wild. Similarly, the first reports of Walrus homosexual activity, complete with photographs, were published in an article with the rather opaque title of "Walrus Ethology I: The Social Role of Tusks and Applications of Multidimensional Scaling," while all records of homosexual behavior in Harbor Seals are contained in unpublished reports and conference proceedings that are only available at a handful of libraries in the world. This perhaps explains why virtually every subsequent discussion of homosexuality in animals omits any mention of these two species.[63]

Between these extremes are numerous examples where homosexuality is "overlooked" or fails to gain mention. Describing itself as "the culmination of years of intensive research and writing by more than 70 authors"—all experts on the species—the massive book *White-tailed Deer: Ecology and Management* (1984) presents in minute detail every imaginable aspect of this animal's biology and behavior, no matter how obscure or rare. There's even room in the book's nearly 900 pages for lengthy discussion of "abnormal" and pathological phenomena (a category in which homosexual activity is often placed). Although the chapter on behavior was coauthored by the scientist who originally described homosexual mounting in White-tailed Deer, there is no mention anywhere in the book of this particular behavior. Nor is there discussion of the transgendered deer found in Texas, even though a whole chapter is devoted to this regional population. A decade later, the same scenario was repeated when another volume of the same scope and on the same species was put out by the same publishers. Similarly, a standard scientific source book, *The Gray Whale*, Eschrichtius robustus (1984), omits any reference to homosexuality in this species even though it includes a chapter by the first biologist to record same-sex activity in Gray Whales.[64] Several comprehensive reference volumes on woodpeckers fail to mention homosexual copulations in Black-rumped Flamebacks, even though no other (hetero)sexual behavior has ever been observed in this species. This omission cannot be due to the putative rarity or "insignificance" of such behavior, since one book does mention another behavior that has only ever been observed once in wild woodpeckers—bathing.[65] Other in-depth surveys of individual species follow suit, eliminating any mention of homosexuality even when they make direct use of other information from the very sources that describe same-sex activity.[66]

Because of the omission and inaccessibility of information on animal homosexuality in the scientific literature, many zoologists are themselves unaware of the full extent of the phenomenon. One of the most unfortunate consequences of this is that misinformation (and absence of information) about the subject is widely disseminated and perpetuated from one source to the next. On discovering homosexual activity in a particular species they are studying firsthand—and being unable to find more than a handful of comparable examples in a cursory literature search—many zoologists acquire the mistaken impression that their observations of this behavior are somehow unique or unusual. At that point they may issue blanket statements to the effect that homosexual activity is rare or previously unreported in the form or species they are observing. Such statements are then often repeated by other biologists and become definitive pronouncements on the subject. As recently as 1993, for example, a scientist reporting on Hooded Warblers could claim that male homosexual pairs had not previously been seen in wild birds—when, in fact, such pairs were documented more than a quarter century earlier in Antbirds, Orange-fronted Parakeets, Golden Plovers, and Mallard Ducks, and thereafter in Black Swans, Scottish Crossbills, Black-billed Magpies, and Pied Kingfishers, among others.[67] Scientists studying same-sex pairs of Black-headed Gulls in captivity asserted in 1985 that this behavior had yet to be seen in this species in the wild—apparently unaware of a description of a male homosexual pair in wild Black-headed Gulls published in a Russian zoology journal just a year earlier. And researchers who discovered same-sex matings in Adélie and Humboldt Penguins and in Kestrels stated that they did not know of any comparable phenomena in other species of penguins or birds of prey, when in fact homosexual activity in King Penguins, Gentoo Penguins, and Griffon Vultures *had* previously been reported in the literature.[68]

Sadly, omission and misinformation on the subject of animal homosexuality have ramifications far beyond the individual scientific articles in which they occur. Reference works such as those mentioned above are frequently consulted by researchers in other fields, and they are also the source of much of the information on animal behavior that is presented to the general public. As the quote at the beginning of this subsection indicates, the cycle is also perpetuated through each new generation of scientists as the textbooks they use (or the professors who instruct them) continue to offer inaccurate or incomplete information on the subject (when they aren't completely silent on the topic). It is no surprise, then, that many scientists—and, by extension, most nonscientists—continue to harbor the erroneous impression that homosexuality does not exist in animals or is at best an isolated and anomalous phenomenon. When erasure and silence surround the subject among zoologists, misinformation and prejudice readily fill in the gaps—both in the scientific community and beyond.

To conclude this examination of homophobic attitudes in the scientific establishment, one simple observation can be made: given the considerable obstacles encountered in the recording, analysis, and discussion of the subject, it is remarkable that *any* descriptions of animal homosexuality make it to the pages of scientific journals and monographs (or to a wider audience). A great deal of progress is being made, and the situation today is certainly improved over that of even a decade ago. Moreover, none of this discourse would even be possible without the invaluable work of zoologists and wildlife biologists who study animals firsthand and report their findings—however flawed that study and reporting may be at times. Nevertheless, the examples of animal homosexuality currently contained in the zoological literature represent only the tip of the iceberg. Many more remain to be discovered, recorded, and granted the scientific attention that has so repeatedly been denied them in the past.

Anything but Sex

As we have seen, one way that zoologists have tried to avoid classifying same-sex activity as "homosexuality" is by using terminology and behavioral categories that deny it is sexual activity at all. This approach also extends to the interpretations, explanations, and "functions" attributed to same-sex behavior, even when it involves the most overt and explicit of activities. Astounding as it sounds, a number of scientists have actually argued that when a female Bonobo wraps her legs around another female, rubbing her own clitoris against her partner's while emitting screams of enjoyment, this is actually "greeting" behavior, or "appeasement" behavior, or "reassurance" behavior, or "reconciliation" behavior, or "tension-regulation" behavior, or "social bonding" behavior, or "food exchange" behavior—almost anything, it seems, besides *pleasurable sexual* behavior.[69] Similar "interpretations" have been proposed for many other species (involving both males and females), allowing scientists to claim that these animals do not really engage in "genuine" (i.e., purely sexual) homosexual activity. But what heterosexual activity is ever "purely" sexual?

(PHOTO BY FRANS DE WAAL)

◀ *Two female Bonobos participating in "GG (genito-genital) rubbing"*

Most biologists are not as candid as Valerius Geist, who, in *Mountain Sheep and Man in the Northern Wilds,* readily admits to his discomfort and homophobia in trying to "explain" homosexuality in Bighorn Rams as "aggressive" or "dominance" behavior:

> I still cringe at the memory of seeing old D-ram mount S-ram repeatedly. . . . True to form, and incapable of absorbing this realization at once, I called these actions of the rams *aggressosexual* behavior, for to state that the males had evolved a homosexual society was emotionally beyond me. To conceive of those magnificent beasts as "queers"—Oh God! I argued for two years that, in [wild mountain] sheep, aggressive and sexual behavior could not be separated. . . . I never published that drivel and am glad of it. . . . Eventually I called the spade a spade and admitted that rams lived in essentially a homosexual society.[70]

This section will examine a number of nonsexual interpretations, including attempts to classify homosexuality as dominance or aggressive behavior, as a form of play, as a social interaction that relieves group tension, and as a greeting activity. In many cases, these "explanations" are not so much genuine attempts to understand the phenomenon as they are ways of denying its existence in the first place. Often these interpretations are simply incompatible with the facts, especially where "dominance" is involved. Furthermore, while in many instances animal homosexuality does have components of all these (nonsexual) activity types, this does not cancel its sexual aspects. As Paul L. Vasey observes, "Just because a behavior which is sexual in form serves some social role or function doesn't mean it cannot be simultaneously sexual."[71] Indeed, both animal and human *heterosexualities* also share aspects of these nonsexual functions without losing their classification as "sexual" activities.

The Dominant Paradigm

In many animal societies, individuals can be ranked with respect to each other on the basis of a number of factors—aggression, access to food or heterosexual mating opportunities, age and/or size, and so on. The resulting hierarchy of individuals and their interaction within this system is often subsumed under the term *dominance.* Many scientists have suggested that mounting and other sexual behaviors between animals of the same sex are not in fact sexual behavior at all, but rather express dominance relations between the two individuals. The usual interpretation is that the "dominant" partner mounts the "subordinate" one and thereby asserts or solidifies his or her ranking relative to that individual. This "explanation" of homosexual behavior is firmly entrenched within the scientific establishment: one of the earliest statements of this position is a 1914 description of same-sex mounting in Rhesus Macaques, and since then dominance factors have regularly been invoked in discussions of animal homosexuality.[72] Most scientists have appealed to dominance as an explanation for animal homosexuality only in relation to the particular species (or at most, animal subgrouping) that they are studying—and sometimes only for one sex within that species—without regard for a broader range of considerations. Once the full panoply of animal types, behaviors, and forms of social organization is taken into account, however, it becomes quite clear that dominance has little, if any, explanatory power. While dominance may be relevant in a few specific cases, it cannot account for the full range of homosexual interactions found throughout the natural world. Moreover, even in particular instances where dominance seems to be important, mitigating factors usually render its influence suspect, if not irrelevant.

At the most basic level, dominance is neither a sufficient nor a necessary condition for the occurrence of homosexual behavior in a species. Just because an animal has a dominance-based or ranked form of social organization does not mean that it exhibits homosexuality, and just because homosexual behavior occurs in a species does not mean that it has a dominance hierarchy. For example, many animals with dominance hierarchies have never been reported to engage in homosexual mounting. Dominance systems are found in "the vast majority of mammal species forming groups with any degree of social complexity"—most primates, seals, hoofed mammals, kangaroos, and rodents, for instance—yet only a fraction of these participate in same-sex mounting. Specific examples of birds with dominance hierarchies but no reported homosexuality include curlews, silvereyes, Harris's sparrows, European jays, black-capped chickadees, marabou storks, white-crowned sparrows, and Steller's jays.[73] Conversely, homosexuality is found in many animals that do not have a dominance hierarchy or in which the relative ranking of individuals plays only a minor role in their social system: for example, some populations of Gorillas, Savanna (Olive) Baboons, Bottlenose Dolphins, Mountain and Plains Zebras, Musk-oxen, Koalas, Buff-breasted Sandpipers, and Tree Swallows.[74]

Often, the relevance of dominance to homosexuality contrasts sharply in two closely related species: Pukeko have a well-defined dominance hierarchy that some scientists believe impacts on the birds' homosexual behavior, yet in the related Tasmanian Native Hen, same-sex mounting occurs in the absence of a dominance hierarchy. Male homosexual mounting has been claimed to correlate with dominance in Cattle Egrets, yet in Little Blue Herons this connection is expressly denied. And the white-browed sparrow weaver (and several other species of weaver birds) has an almost identical social organization and dominance system as the Gray-capped Social Weaver, yet mounting between males is only found in the latter species.[75] Not only cross-species but also cross-gender comparisons are relevant here. A particularly good example of the problematic relationship between dominance and same-sex activity becomes apparent when one looks at males and females within the same species. In many animals both sexes have their own dominance hierarchies, yet homosexuality occurs in only one sex—male but not female Wolves, for example, and female but not male Spotted Hyenas. A corollary to this is that in some species, only one sex exhibits a stable dominance hierarchy, yet homosexuality occurs among *both* males and females. In Squirrel Monkeys, for example, female interactions are not consistently organized around a dominance or rank system, yet same-sex mounting and genital displays are not limited to males. In Bottlenose

Dolphins, stable dominance hierarchies (if they exist at all) are more prominent among females, yet homosexual activity occurs in both sexes.[76] Finally, homosexual mounting sometimes occurs between animals of different species. Although cross-species dominance relations have been documented (e.g., in birds), in the majority of the cases involving homosexual activity there is no well-established hierarchical relationship between the participating animals of different species.[77] Clearly, then, dominance cannot be the only factor involved in the occurrence of homosexuality in a given species.

Even in animals where there is a clear dominance hierarchy, same-sex mounting is often not correlated with an individual's rank, and it rarely follows the idealized scenario of "dominant mounts subordinate, always and without exception." In many species, there is simply no correlation between rank and mounting behavior, since subordinate animals frequently mount dominant ones. In Rhesus Macaques, for example, 36 percent of mounts between males are by subordinates on dominants, while 42 percent of all female Japanese Macaque homosexual mounts go "against" the hierarchy, as do 43 percent of mounts between male Common Chimpanzees.[78] Both dominant-subordinate and subordinate-dominant mounting occur in Bonobos, Lion-tailed Macaques, Squirrel Monkeys, Gelada Baboons, and Ruffs, among others, while mounting of older, larger, and/or higher-ranking animals by younger, smaller, and/or subordinate individuals has also been reported for numerous species: Common Marmosets, Australian and New Zealand Sea Lions, Walruses, Bottlenose Dolphins, White-tailed Deer, Mule Deer, Père David's Deer, Wapiti, Moose, Mountain Goats, Red Foxes, Spotted Hyenas, Whiptail Wallabies, Rufous Rat Kangaroos, Mocó, Préa, Guianan Cock-of-the-Rock, Emus, and Acorn Woodpeckers. Oftentimes, while a large proportion of mounts may seem to follow the dominance hierarchy in a particular species, mounting by subordinates on dominants also takes place in the same species. This is true for Hanuman Langurs, Bonnet Macaques, Musk-oxen, Bighorn and Thinhorn Sheep, Cattle Egrets, and Sociable Weavers.[79]

The precise opposite of the "standard" dominance-based system of mounting is often found as well: mountings by subordinates on dominants occur more frequently than the reverse in many species. In Crested Black Macaques, for example, 60–95 percent of mounts are subordinate on dominant, while nearly two-thirds of Bison male homosexual mounts are by subordinates on dominants. To complicate things further, this is often combined with a gender difference in the relationship between mounting and dominance, with female mounts "following" the hierarchy and male mounts going "against" it. For instance, in Pig-tailed Macaques mounting between females is usually by a dominant individual on a subordinate one, but more than three-quarters of mounts between males are just the opposite. Similarly, in both Red Deer and Pukeko, females tend to mount lower-ranking animals while males tend to mount higher-ranking ones. There are often individual or geographic differences as well: in some consortships between female Japanese Macaques, all mounting is done by the lower-ranking individual on the higher-ranking partner, while in some populations of Bighorn Sheep, mounting of dominant rams by subordinates is much more prevalent than in other populations.[80] Furthermore, homosexual mounting in many species is reciprocal, which means that partners exchange positions—mounter becomes mountee, and vice versa—either in the same mounting session or in alternation over longer periods of time. This behavior, which is found in at least 30 different species, is potent evidence of the irrelevance of dominance for homosexual interactions, since mounting should only occur unidirectionally if it strictly followed the rank of the participating individuals.[81] Finally, in some species mounting can also occur between individuals of the same or close ranks—for example, in Common Chimpanzees, White-faced Capuchins, Musk-oxen, Blackbucks, Cavies, and Gray-capped Social Weavers.[82]

In a dominance-based view of homosexual mounting, it is often assumed that the animal being mounted is somehow a less willing participant in the interaction, "submitting" to the will of the more dominant individual, who thereby asserts his or her "superiority." In fact, in more than 30 species the mounted animal actually initiates the interaction, "presenting" its hindquarters to the other individual as an invitation to mount, sometimes even actively facilitating anal penetration (among males) or other aspects of the interaction. Where the presenting animal is subordinate, this could be interpreted as simply a reinforcement of the dominance system, but in a number of species it is actually the more *dominant* individual who presents and actively encourages the lower-ranking animal to mount.[83] In addition, dominance "explanations" often ignore the clear differences between consensual and nonconsensual mounts (or rapes), as well as evidence for sexual arousal and even enjoyment on the part of mounted animals.[84]

The relationship between sexuality and dominance is complex and multifaceted, differing greatly from the frequent simplistic equating of homosexual mounting with nonsexual rank-based or aggressive behavior. In many species a gradation or continuum exists between sexual mounts and dominance mounts, with one type "blending" into the other so that any distinction between the two is essentially arbitrary. Thus, same-sex mounting can have an unmistakable sexual component even when it still follows a dominance pattern. Among Hanuman Langurs, for example, usually only dominant females mount subordinate ones, yet so inextricably linked are signs of sexual excitement with this behavior that scientists have concluded, "It seems virtually impossible to separate 'sexual mounting' from 'dominance mounting.' . . . Sexual arousal and dominance are obviously not mutually exclusive in langur females, since mounting between females is related to both dominance and sexuality."[85] At the other end of the spectrum, in some species a sharp distinction does in fact exist between two types of mounting, *both* of which occur between same-sex partners: a nonsexual form associated with dominance and/or aggression, and a clearly sexual form that occurs in other contexts (often within a homosexual pair-bond or consortship). This is true for female Japanese Macaques, Rhesus Macaques, and Black-winged Stilts, and male Greylag Geese, among others.[86]

Finally, in some animals dominance and mounting are entirely separate, with social rank being expressed through obviously nonsexual activities. For example, male Walrus dominance interactions involve fighting and tusk displays that usually occur during the breeding season and often involve younger animals. Male homosexual mounting is not associated with either of these activities and usually takes place in the nonbreeding season among males of all age groups (a similar pattern is also seen in Gray Seals). Oystercatchers use a special ritualized "piping display" (neck arched, bill pointed downward, accompanied by shrill piping notes) to negotiate their dominance interactions, while same-sex mounting and courtship occur in other contexts.[87] Dominance in many other animals is expressed through fighting and aggressive encounters, access to food or feeding frequency, body size or age, physical displacement (causing another individual to move off through posture, threats, staring, or other activities), access to heterosexual mating opportunities, or a combination of these or other factors, and specifically does not involve mounting or the other homosexual interactions that occur in these species. Savanna (Yellow) Baboons, (female) Hamadryas Baboons, Bottlenose Dolphins, Killer Whales, Caribou, Blackbucks, Wolves, Bush Dogs, Spotted Hyenas, Grizzly Bears, Black Bears, Red-necked Wallabies, Canada Geese, Scottish Crossbills, Black-billed Magpies, Jackdaws, Acorn Woodpeckers, and Galahs are all species in which this is the case.[88]

Another limitation in looking at homosexual interactions from the perspective of dominance is that only mounting behavior lends itself to such an interpretation. A whole host of other homosexual activities do not fit neatly into the dominance paradigm—either because, by their very nature, they are reciprocal activities, or because neither participant can be assigned a clearly "dominant" or "subordinate" status on the basis of what "position" it assumes during the activity. For example, mutual genital rubbing—in which two animals rub their genitals on each other without any penetration—often occurs with neither participant "mounting" the other. Gibbon and Bonobo males frequently engage in this activity when hanging suspended from a branch, facing each other in a more "egalitarian" position. In aquatic animals such as Gray Whales, West Indian Manatees, Bottlenose Dolphins, and Botos, males rub their penises together or stimulate each other while rolling and clasping one another in constantly shifting, fluid body positions that defy any categorization as "mounter" or "mountee." Reciprocal rump rubbing and genital stimulation—found in Chimpanzees and some Macaques—also renders meaningless a dominance-based view of homosexual interactions. When two males or two females back toward each other and rub their anal and genital regions together, sometimes also manually stimulating each other's genitals—which one is "dominating" the other? Or when a male Vampire Bat grooms his partner, licking his genitals while simultaneously masturbating himself, which one is behaving "submissively"? By the same token, Crested Black Macaque females have a unique form of mutual masturbation in which they stand side by side facing in opposite directions and stimulate each other's clitoris—again, because of the pure reciprocity, it makes little sense to interpret this behavior as expressing some sort of hierarchical relationship between the partners.

Genital rubbing, masturbation of one's partner, oral sex, anal stimulation other than mounting, and sexual grooming occur among same-sexed individuals in more than 70 species—yet virtually all of these forms of sexual expression fall outside the realm of clear-cut dominance relationships.[89] These more mutual, reciprocal, or dominance-ambiguous sexual activities are commonly found alongside

▲ *Male Stumptail Macaques manually stimulating each other's genitals. Mutual or reciprocal sexual behaviors such as this are good examples of homosexual activity that is not "dominance" oriented.*

homosexual mounting behavior in the same species—but the former are typically ignored when a dominance analysis is advocated.[90] Ironically, another entire sphere of homosexual activity eludes a dominance interpretation—any same-sex interaction that is *not* overtly sexual. Courtship, affectionate, pair-bonding, and parenting behaviors that do not involve genital contact or direct sexual arousal— yet still occur between same-sex partners—are routinely omitted from any discussion of the relevance of dominance to the expression of homosexuality.[91] The exclusion of nonsexual behaviors such as these from dominance considerations contrasts, paradoxically, with the way that mounting behavior itself is ultimately rendered nonsexual by its *inclusion* under the category of dominance.

A final indictment of a dominance analysis is that the purported ranking of individuals based on their mounting or other sexual behavior often fails to correspond with other measures of dominance in the species. Male Giraffes, for example, have a well-defined dominance hierarchy in which the rank of an individual is determined by his age, size, and ability to displace other males with specific postures and stares. Homosexual mounting and "necking" behavior is usually claimed to be associated with dominance, yet a detailed study of the relationship between these activities and an individual's social standing according to other measures revealed no connection whatsoever. Mounting position also fails to reflect an individual's rank as measured by aggressive encounters (e.g., threat and attack behavior) and other criteria in male Crested Black Macaques, male Stumptail Macaques, and female Pig-tailed Macaques. In only about half of all male homosexual mounts among Savanna (Olive) Baboons is there a correlation between dominance status, as determined in aggressive or playful interactions, and the role of an animal as mounter or mountee. In male Squirrel Monkeys, dominance status affects an individual's access to food, heterosexual mating opportunities, and the nature of his interactions with other males, yet the rank of males as evidenced by their participation in homosexual genital displays does not correspond in any straightforward way to these other criteria. Among male Red Squirrels, there is no simple relationship between aggressiveness and same-sex mounting: the most aggressive individual in one study population indeed mounted other males the most frequently, yet he was also the recipient of mounts by other males the most often, while the least aggressive male was hardly ever mounted by any other males. This is also true for Spinifex Hopping Mice, in which males typically mount males who are *more* aggressive than themselves. Similarly, although male Bison fairly consistently express dominance through displays such as chin-raising and head-to-head pushing, these behaviors do not offer a reliable predictor of which will mount the other. Although some mounts between male Pukeko appear to be correlated with the dominance status of the participants (as determined by their feeding behavior, age, size, and other factors), there is no consistent relationship between these measures of dominance and another important indicator of rank—a male's access to heterosexual copulations (or the number of offspring he fathers). Finally, dominance relations in Sociable Weavers are not always uniform across different measures either: one male, for example, was "dominant" to another according to their mounting behavior, yet "subordinate" to him according to their pecking and threat interactions.[92]

In fact, multiple nonsexual measures of dominance often fail to correspond even among themselves, and this has led some scientists to suggest that the entire concept of dominance needs to be seriously reexamined, if not abandoned altogether. While it may have some relevance for some behaviors in some species, dominance (or rank) is not a fixed or monolithic determinant of animal behavior. Its interaction with other factors is complex and context-dependent, and it should not be accorded the status of a preeminent form of social organization that it has traditionally been granted.[93] Primatologist Linda Fedigan advocates a more sophisticated approach to the role of dominance in animal behavior, eloquently summarized in the following statement. Although her comments are specifically about primates, they are relevant for other species as well:

> We often oversimplify the phenomena categorized together as dominance, as well as overestimating the importance of physical coercion in day-to-day primate life. . . . An additional focus on *alliances* based on kinship, friendship, consortship, and roles, and on social power revealed in phenomena such as leadership, attention-structure, social facilitation, and inhibition, may help us to better understand the dynamics of primate social interaction. Also it may help us to place competition and cooperation among social primates in proper perspective as intertwined rather than opposing forces, and female as well as male primates in their proper perspective as playing major roles in primate "politics" through their participation in alliance systems.[94]

Considering the wide range of evidence against a dominance analysis of animal homosexuality—as well as a number of explicit statements by zoologists questioning or entirely discounting dominance as a factor in same-sex activities[95]—it is surprising that this "explanation" keeps reappearing in the scientific literature whenever homosexual behavior is discussed. Yet reappear it does, even in several studies published in the 1990s. As recently as 1995, in fact, dominance was invoked in a discussion of mounting between male Zebras, and this explanation still has enough currency that scientists felt compelled to refute it in a 1994 account of homosexual copulation in Tree Swallows. In looking through the many examples of the way that this "explanation" has been used, it becomes apparent that the relevance of dominance is often asserted without any supporting evidence, then cited and re-cited in subsequent studies to create a chain of misconstrual, as it were, extending across many decades of scientific investigation. Again and again, early characterizations of homosexual activity as dominance behavior—often hastily proposed on the initial (and unexpected) discovery of this behavior in a species— have been refuted by later, more careful investigations of the phenomenon.[96] Yet frequently only the earlier studies are cited by researchers, perpetuating the myth that this is a valid characterization of the behavior. For example, in a 1974 report that described same-sex mounting in Whiptail Wallabies, a zoologist referred to dominance interpretations of Rhesus Macaque homosexuality even though more recent studies had invalidated—or at the very least, called into question—such an analysis for this species.[97]

At times, the very word *dominance* itself becomes simply code for "homosexual mounting," repeated mantralike until it finally loses what little meaning it had to begin with. "Dominance" interpretations have in fact been applied to same-sex mounting regardless of how overtly sexual it is. The relatively "perfunctory" mounts between female Tree Kangaroos or male Bonnet Macaques, as well as interactions involving direct clitoral stimulation to orgasm between female Rhesus Macaques, and full anal penetration and ejaculation in Giraffes, have all been categorized as nonsexual "dominance" activities at one time or another. Even though many scientists have gone on record against a dominance interpretation—thereby challenging the stronghold of this analytic framework—information that contradicts a dominance analysis is sometimes troublesomely discounted or omitted from studies. For example, in several reports on dominance in Bighorn rams, same-sex mounting and courtship activities (as well as certain aggressive interactions) were deliberately excluded from statistical calculations because they frequently involved "subordinates" acting as "dominants," i.e., they did not conform to the dominance hierarchy. One scientist even classified some instances of same-sex mounting in Crested Black Macaques as "dysfunctional" because they failed to reflect the dominance system or exhibit any other "useful" properties.[98]

Nor is this merely a question of relevance to scientists, or simply a matter of esoteric academic interpretation. The assertions made by zoologists about the "functions" of homosexual behavior are often picked up and repeated, unsubstantiated, in popular works on animals, becoming part of our "common knowledge" of these creatures. In a detailed survey of primate homosexuality published in 1995, zoologist and anthropologist Paul L. Vasey finally and definitively put the dominance interpretation of homosexuality in its proper perspective, stating that "while dominance is probably an important component of some primate homosexual behavior, it can only partially account for these complex interactions."[99] We can only hope that his colleagues—and ultimately, those who convey the wonders of animal behavior to all of us—will take these words to heart once and for all.

The Desexing of Homosexual Behavior

> ... two males (Dinding and Durian) regularly mouthed the penis of the other on a reciprocal basis. This behavior, however, may be nutritively rather than sexually motivated.
>
> —T. L. MAPLE, Orang-utan Behavior[100]

In nearly a quarter of all animals in which homosexuality has been observed and analyzed, the behavior has been classified as some other form of nonsexual activity besides (or in addition to) dominance. Reluctant to ascribe sexual motivations to activities that occur between animals of the same gender, scientists in many cases have been forced to come up with alternative "functions." These include some rather far-fetched suggestions, such as the idea (quoted above) that fellatio between male Orang-utans is a "nutritive" behavior, or that episodes of cavorting and genital stimulation between male West Indian Manatees are "contests of stamina."[101] At various times, homosexuality has also been classified as a form of aggression (not necessarily related to dominance), appeasement or placation, play, tension reduction, greeting or social bonding, reassurance or reconciliation, coalition or alliance formation, and "barter" for food or other "favors." It is striking that virtually all of these functions are in fact reasonable and possible components of sexuality—as any reflection on the nature of sexual interactions in humans will reveal—and indeed in some species homosexual interactions do bear characteristics of some or all of these activities. However, in the vast majority of cases these functions are ascribed to a behavior *instead of*, rather than *along with*, a sexual component—and only when the behavior occurs between two males or two females. According to Paul L. Vasey, "While homosexual behavior may serve some social roles, these are often interpreted by zoologists as the primary reason for such interactions and usually seen as negating any sexual component to this behavior. By contrast, heterosexual interactions are invariably seen as being primarily sexual with some possible secondary social functions."[102]

Thus, a widespread double standard exists when it comes to classifying behavior as "sexual." Desexing is selectively applied to homosexual but not heterosexual activities, according to a number of different strategies. The first and most obvious is when scientists explicitly classify the same behavior as sexual when it takes place between members of the opposite sex and nonsexual when it involves members of the same sex. This is readily apparent in the following statement: "Mounting [in Bison] can be referred to as 'mock copulation.' It seems appropriate to classify this action as sexual behavior only when it is directed towards females. The gesture, however, was also directed to males which suggests that it also has a social function." Likewise, because a behavior often associated with courtship in Asiatic Mouflons and other Mountain Sheep (the foreleg kick) was observed more frequently between individuals of the same sex than of the opposite sex, one zoologist concluded that this activity must therefore be aggression rather than courtship. Primatologists reassigned what they had initially classified as sexual behavior in Stumptail Macaques to the category of aggressive or dominance behavior when it took place in homosexual pairs, while marine biologists reclassified courtship and mating activity in Dugongs as nonsexual play behavior once they learned both participants were actually male. Ornithologists studying the courtship display of Laysan Albatrosses also questioned whether this behavior was "truly" related to pair-bonding or mating after they discovered that some courting birds were of the same sex. Finally, because (male) Dwarf Mongooses and Bonnet Macaques are as likely to mount same-sex as opposite-sex partners, scientists decided this behavior must be nonsexual.[103] This is not to say that behaviors cannot have different meanings or "functions" in same-sex versus opposite-sex contexts, only that the erasure by zoologists of sexual interpretations from same-sex contexts has been categorical and nearly ubiquitous.

Not only do zoologists apply nonsexual interpretations to behaviors when they know that the participants are of the same sex, they also do the reverse, assuming that a superficially nonsexual behavior—especially if it involves aggression—must involve animals of the same sex. A particularly interesting example of these assumptions concerns the flip-flop in interpretation of sexual chases in Redshanks, part of the courtship repertoire of this sandpiper. Because of their somewhat aggressive nature, these chases were originally interpreted as a nonsexual, territorial interaction and assumed to involve two males—in spite of the fact that some scientists reported seeing chases between birds of the opposite sex. Subsequently, more detailed study involving banded birds (enabling individual identification) revealed that most chases did in fact involve a male and a female and occurred early in the breeding season—at which point the behavior was reclassified as a form of courtship. However, it was also discovered that in a few instances two males were actually chasing each other—and of course scientists then tried to claim that, only in these cases, the chasing was once again nonsexual (in spite of the fact that the two males often copulated with each other as well).[104]

Sometimes the arbitrary categorization of behaviors reaches absurd levels. In a few instances components of one and the same activity are given separate classifications, or the undeniably sexual character of a homosexual interaction is taken to mean only that the activity is "usually" heterosexual. For example, in one report on female Crested Black Macaques, a behavior labeled the "mutual lateral display" is classified as a "sociosexual" activity, i.e., not fully or exclusively sexual. It is described as a "distance-reducing display" or a form of "greeting" that "precedes grooming or terminates aggression between two animals." Yet the fact that females masturbate each other's clitoris during this "display"—about as definitively sexual as a behavior can get—is inexplicably omitted from the description of this activity. Instead, this detail is included separately in the "sexual behavior" section of the report under the heading "masturbation"—a contradictory recognition that, apparently, part of this behavior is not truly sexual yet part of it is! In the same species, males often become sexually aroused while grooming one another, developing erections and sometimes even masturbating themselves to ejaculation.

Amazingly, this is interpreted by another investigator not as evidence of the sexual nature of grooming between males, but rather that grooming is probably an activity "typically" performed by females to males prior to copulation. Apparently such overt sexuality could only be a case of misplaced heterosexuality, not "genuine" homosexuality.[105]

Often a behavior is automatically assumed to involve courtship or sexuality when its participants are known to be of the opposite sex—and the criteria for a "sexual" interpretation are generally far less stringent than those applied to the corresponding interactions between like-sexed individuals. In other words, heterosexual interactions are given the benefit of the doubt as to their sexual content or motivation, even when there is little or no direct evidence for this or even overt evidence to the contrary. For example, simple genital nuzzling of a female Vicuña by a male—taking place outside of the breeding season, and without any mounting or copulation to accompany it—is classified as sexual behavior, while actual same-sex mounting in the same species is considered nonsexual or "play" behavior. In Musk-oxen, foreleg-kicking in heterosexual contexts is often much more aggressive than in homosexual contexts. The male's blow to a female's spine or pelvis is sometimes so forceful that it can be heard up to 150 feet away, yet this behavior is still classified as essentially courtship-oriented. If this level of aggression were exhibited in foreleg-kicking between males, the behavior would never be considered homosexual courtship (as it is, this classification is granted only grudgingly, accompanied by the obligatory reference to its "dominance" function between males).

When a male Giraffe sniffs a female's rear end—without any mounting, erection, penetration, or ejaculation—he is described as being sexually interested in her and his behavior is classified as primarily, if not exclusively, sexual. Yet when a male Giraffe sniffs another male's genitals, mounts him with an erect penis, and ejaculates—then he is engaging in "aggressive" or "dominance" behavior, and his actions are considered to be, at most, only secondarily or superficially sexual. In one study of Bank Swallows, all chases between males and females were assumed to be sexual even though they were rarely seen to result in copulation. Indeed, the majority of bird studies label dyads composed of a male and female as "[heterosexual] pairs" in spite of the fact that overt sexual (mounting) activity is rarely verified for all such couples. In contrast, most investigators will not even consider classifying same-sex interactions in birds to be courtship, sexual, or pair-bonding activity—even when they involve the same behavior patterns used in heterosexual contexts—unless mounting is observed. Certain associations between male and female Savanna Baboons and Rhesus Macaques are described as "sexual" relationships or "pair-bonds" even though they often do not include sexual activity. In contrast, bonds between same-sex individuals in these species are characterized as nonsexual "coalitions" or "alliances" even though they may involve sexual activities (as well as the same intensity and longevity found in heterosexual bonds). Finally, the "piping display" of the Oystercatcher described earlier was initially assumed to be a courtship behavior, largely because it is a common activity between males and females. Subsequent studies have shown that this is in fact a primarily nonsexual (territorial or dominance) interaction.[106]

Another strategy adopted by scientists when confronted with an apparently sexual behavior occurring between two males or two females is to deny its sexual content in both same-sex *and* opposite-sex contexts. For example, because female Crested Black Macaques show behavioral signs of orgasm during homosexual as well as heterosexual mounts, one scientist concluded that this behavior is not reliable evidence of female orgasm in *either* situation. The fact that intercourse and other sexual interactions occur between like-sexed individuals in Bottlenose and Spinner Dolphins is often taken to be "proof" that such behaviors have become largely divorced from their sexual content and are now forms of "greeting" or "social communication," even in heterosexual contexts. Similarly, copulation in Common Murres has many nonreproductive features: in addition to occurring between males, in heterosexual pairs it frequently takes place before the female becomes fertile. Drawing an explicit analogy with "nonsexual" mounting in primates, one ornithologist suggested that in this bird heterosexual mounting must therefore serve an "appeasement" function rather than being principally a sexual behavior, i.e., females invite their male partners to mate in order to deflect aggression from them. Likewise, nonprocreative copulations in both heterosexual and homosexual contexts in Blue-bellied Rollers are categorized as a form of ritualized aggression or appeasement.[107]

A difference *in form* between homosexual and heterosexual behaviors is often interpreted as a difference in their sexual content. The reasoning is that if same-sex activity does not resemble opposite-sex activity, and only opposite-sex activity is by definition sexual, then same-sex activity cannot be sexual. For example, in Rhesus Macaques most heterosexual copulations involve a series of mounts by the male, only the last of which typically involves ejaculation. Because mounts between males are often single rather than series mounts, they are frequently classified as nonsexual, even when they include clear signs of sexual arousal such as erection, pelvic thrusting, penetration, and even ejaculation. A similar interpretation has also been suggested for mounting between male Japanese Macaques. In contrast, significant differences in form also exist between heterosexual copulation in Macaques (with series mounting) and male masturbatory patterns, yet both activities are clearly sexual and are typically classified as such.[108]

In other animals the very characteristics that are used to claim that same-sex activities are nonsexual—their briefness, "incompleteness," or absence of signs of sexual arousal, for example—are as typical, if not more typical, of opposite-sex interactions that *are* classified as sexual behavior. Nearly a third of all mammals in which same-sex mounting occurs also have "symbolic" or "incomplete" heterosexual mounts in which erection, thrusting, penetration, and/or ejaculation do not occur; "ritual" heterosexual mountings are also typical of many bird species.[109] In Kob antelopes, 52 percent of heterosexual copulations involve at least one mount by the male without an erection; in contrast, 56 percent of homosexual mountings between male Giraffes—sometimes classified as nonsexual—*do* involve erections. Likewise, only one in four to five heterosexual mounts among northern jacanas results in cloacal (genital) contact, and ejaculation probably occurs in less than three-quarters of Orang-utan heterosexual mounts.[110] Evidence for sexual arousal or "completed" copulations is often entirely lacking in heterosexual contexts, yet such male-female mounts are still considered "sexual" behavior. In Walruses, Musk-oxen, Bighorn Sheep, Asiatic Mouflons, Grizzly Bears, and Olympic Marmots, for example, penetration and ejaculation are rarely, if ever, directly observable during heterosexual mounts, while male erections are routinely not visible during White-tailed Deer copulations, ejaculation can only be "assumed" to occur in observations of Orang-utan, White-faced Capuchin, and Northern Fur Seal heterosexual mating, and genital contact is difficult to verify during Ruff male-female mounts (among many other species).[111]

In fact, actual sperm transfer during heterosexual copulations in many species is so difficult to observe that biologists have had to develop a variety of special "ejaculation-verification" techniques. In birds such as Tree Swallows, for example, tiny glass beads or "microspheres" of various colors are inserted into males' genital tracts. If the birds ejaculate during a heterosexual mating, these beads are transferred to the female's genital tract, where they can be retrieved by scientists and checked for their color coding to determine which males have actually transferred sperm. For rodents and small marsupials, biologists actually inject several different radioactive substances into males' prostate glands. During ejaculation, these are carried via semen into females, who are then monitored with a sort of "sperm Geiger counter" to determine which males, if any, have inseminated them.[112] If such elaborate lengths are required to verify a fundamental and purportedly self-evident aspect of heterosexual mating, is it any wonder that homosexual matings should sometimes appear to be "incomplete"?

Because of such difficulties in observation and interpretation, scientists have often employed similarly extreme measures in an attempt to "verify" homosexual intercourse. In the early 1970s, for example, a controversy arose concerning to what extent, if at all, mounting activity between male animals was truly "sexual." As proof of its "nonsexual" character, some scientists claimed that full anal penetration never occurred in such contexts (thus equating penetration with "genuine" sexuality). Researchers actually went to the trouble of filming captive male Rhesus Macaques mounting each other in order to record examples of anal penetration; they even anesthetized the monkeys afterward to search for the presence of semen in their rectums. Needless to say, the cinematographic proof of anal penetration they obtained did little to quell any subsequent debate about whether such mounts were "sexual"—all it did was institute a revised definition of "sexual" activity. The fact that they were able to document penetration but not ejaculation simply meant that a new "standard" of sexuality could now be applied: only mounts that culminated in ejaculation were to be considered "genuine" sexual behavior. Ironically, none of these researchers were apparently aware of an earlier field report of homosexual activity in Rhesus Macaques in which both anal penetration *and* ejaculation were observed.[113]

This near-obsessive focus on penetration and ejaculation—indeed, on "measuring" various aspects of sexual activity to begin with—reveals a profoundly phallocentric and "goal-oriented" view of sexuality on the part of most biologists. Not just homosexual activity, but noninsertive sexual acts, female sexuality and orgasmic response, oral sex and masturbation, copulation in species (such as birds) where males do not have a penis—any form of sex whatsoever that does not involve penis-vagina penetration falls off the map of such a narrow definition. The fact is that both heterosexual *and* homosexual activities exist along a continuum with regard to their degree of "sexuality" or "completeness." Male mammalian mounting behavior, for example, çan involve partial mounting, full mounting but no thrusting, thrusting but no erection, erection but no penetration, penetration but no ejaculation, ejaculation but no penetration, penetration and ejaculation without series mounting, and so on and so forth.[114] Each stage along this continuum has at one time or another been considered a defining threshold of "true" sexual behavior—often so as to exclude same-sex interactions—rather than as one possible manifestation of a broader sexual capacity that is sometimes, but not always, orgasmically (or genitally) focused.

A nonsexual component of homosexual behavior does appear to be valid in a number of species; in equally many species, there are clear arguments against various nonsexual interpretations, and some zoologists have themselves explicitly refuted nonsexual analyses.[115] Overall, though, three important points must be considered in relation to nonsexual interpretations of behaviors between animals of the same sex. First, the question of causality—or the primacy of the nonsexual aspect—must be addressed. Just because an apparently sexual behavior is associated with a nonsexual result or circumstance does not mean that the sole function or context of the behavior is nonsexual. For example, female Japanese Macaques often gain powerful allies by forming homosexual associations, since their consorts typically support them in challenging (or defending themselves against) other individuals. However, a detailed study of partner choices showed that such nonsexual benefits are of secondary importance: females choose their consorts primarily on the basis of sexual attraction *rather* than on whether they will make the best or most strategic allies.[116] Likewise, mounting (or other sexual activity) between animals of the same sex is described in many species (e.g., Bonobos) as a behavior that serves to reduce aggression or tension between the participants. Indeed, individuals who mount each other may be less aggressive to one another or may experience less tension in their mutual interaction, and homosexuality probably does serve a tension-reducing function for at least some animals in some contexts (as does heterosexuality). However, the situation is considerably more involved than this. Tension reduction is as likely to be a consequence of an affiliative or friendly relationship between individuals—a relationship that is also expressed through sexual contact—as it is to be a direct result of their sexual activity. Moreover, as some researchers have pointed out for Bonobos, the causal relationship may also be the reverse of what is usually supposed. That sexual behavior and situations involving tension often co-occur in this species can give the impression that sexuality is functioning only to reduce tension, when in fact it may also create or generate its own tension. Indeed, homosexual activity in male Gorillas often results in increased rather than decreased social tension.[117]

Second, even if behaviors are classified as nonsexual or having a nonsexual component, the behavioral categories to which they are assigned (aggression, greetings, alliance formation, etc.) are not monolithic. Many important questions remain concerning the forms and contexts of such behaviors—questions that are often overlooked once they receive their "classification." Just because we "know" that a given behavior is "nonsexual" does not mean that we then know everything about that behavior. Apparently sexual behaviors between males in both Bonnet Macaques and Savanna Baboons, for example, are classified as social "greetings" interactions. Yet there are fundamental differences between these two species, not only in the types of activities involved, but in the frequency of participation, the types of participants, the social framework and outcome of participation, and so on.[118] Ultimately, classifying such behavior as "nonsexual" is as meaningless, misleading, and unilluminating as many investigators claim a sexual categorization is, if it obscures these differences or fails to address their origin.

Finally, the relationship between sexual and nonsexual aspects of behavior is complex and multilayered and does not fit the simple equations that are usually applied, namely *same-sex participants = nonsexual, opposite-sex participants = sexual.* In many species, there is clear evidence of the genuinely sexual aspect of behaviors between animals of the same sex, using the same criteria that are applied to heterosexual interactions—for example, penile or clitoral erection, pelvic thrusting, penetration (or cloacal contact), and/or orgasm.[119] In still other species there is a gradation or cline between sexual and nonsexual behaviors that defies any rigid categorization—or else there is a sharp distinction between the two, with *both* occurring among animals of the same sex. Most importantly, the sexual and nonsexual aspects of a behavior are not mutually exclusive. An interaction involving genital stimulation between two males or two females can be a form of greeting, or a way of reducing tension or aggression, or a type of play, or a form of reassurance, or any number of different things—and *still* be a sexual interaction at the same time. Ironically, by denying the sexual component of many same-sex activities and seeking alternative "functions," scientists have inadvertently ascribed a much richer and varied palette of behavioral nuances to homosexual interactions than is often granted to heterosexual ones.[120] Because heterosexuality is linked so inextricably to reproduction, its nonsexual "functions" are often overlooked, whereas because homosexuality is typically disassociated from reproduction, its sexual aspects are often denied. By bringing these two views together—by recognizing that both same-sex and opposite-sex behaviors can be all these things and sexual, too—we will have come very close indeed to embracing a fully integrated or *whole* view of animal life and sexuality.

Chapter 4

Explaining (Away)
Animal Homosexuality

*I*n August 1995 a historic event took place: a special symposium on sexual orientation in animals was held at the 24th International Ethological Conference (ethologists are zoologists who study animal behavior). This was an unprecedented occurrence: the first time that animal homosexuality was formally recognized by a zoological organization as a legitimate subject of inquiry unto itself. As hundreds of zoologists and other scientists gathered from more than 40 countries around the world to discuss the latest findings and hypotheses, this conference held the promise of inaugurating a new era in the study of animal homosexuality—one characterized by an absence of the judgmental attitudes chronicled in the previous chapter.

Unfortunately, what actually transpired at the conference is symbolic of the pitfalls that have plagued discussions of animal homosexuality throughout the scientific study of this topic. The symposium's stated mission was to explore "behavioral correlates of sexual plasticity"; its organizer's opening remarks even invoked Paul L. Vasey's recent work on primate homosexuality—to the visual accompaniment of giant photographs of human gay couples projected on the screen.[1] Yet only a handful of papers at the symposium even mentioned homosexuality, let alone dealt with it in any depth. Most were concerned with the hormonal and neurological correlates of male and female differences in behavior and anatomy—reflecting the still widespread view that homosexuality is simply an example of gender "inversion" or "gender-atypical" behavior (e.g., males exhibiting "female" behavior patterns and vice versa). Ironically, among the conference attendees were a veritable who's who of zoologists who have observed homosexual behavior firsthand in wild animals—a treasure trove of information on the topic that went virtually untapped by the symposium's organizers and all but unnoticed by the conference-goers.[2] On the final day of the conference, after it became apparent that animal homosexuality would receive no more than a cursory discussion in any of the formal presentations, one zoologist tacked a hand-scrawled note on the public message-board: "I am looking for examples of homosexual affairs in insects, please contact . . ."—a cogent reminder of both the desire for, and lack of, information on this subject at the very locus where it should be most available.

What happened at this conference is not unusual. The scientific discourse surrounding animal homosexuality has been preoccupied with finding an *explanation* for the phenomenon, often at the expense of providing comprehensive descriptive information about, or acknowledgment of, the actual extent and diversity of same-sex activity throughout the animal kingdom. Rather than being seen as part of a spectrum of natural variation in sexual and gender expression, homosexuality and transgender are viewed as exceptions or anomalies that somehow stand outside the natural order and must therefore be "explained" or "rationalized." In most respects, by trying to answer the question "Why do some animals engage in homosexual behavior?" scientists have simply found an opportunity to continue many of the same homophobic attitudes documented in the preceding chapter (while ignoring the biases inherent in such a question in the first place). Significant numbers of zoologists are willing to concede that same-sex courtship, copulation, and pair-bonding are indeed "sexual" or "homosexual" activities. However, they commonly propose alternative explanations for these behaviors premised on the notion that this activity is still in some way "anomalous" or "aberrant." Ultimately, most such attempts to find an "explanation" have failed outright or are fundamentally misguided. In this chapter we'll explore four such "explanations" that crop up repeatedly in the scientific and popular discourse surrounding animal homosexuality—the idea that homosexuality is an imitation of heterosexuality, a "substitute" activity when the opposite sex is unavailable, a "mistake," or a pathological condition. These explanations need to be addressed not only because they are widespread within the scientific establishment, but also because they form part of the popular mythology surrounding animal homosexuality. Each of these ideas or analyses is in fact incorrect—or at the very least, only partially relevant.

Significantly, each of these explanations has also been proposed at various times as the "cause" or "reason" for human homosexuality, and equally as often shown to be false. In fact, the language and logic of many of these explanations for animals are directly out of the psychopathological analyses of human homosexuality from the 1940s and 1950s (which, in turn, are a continuation and elaboration of earlier prejudicial attitudes about "abnormal" behaviors). So similar are they to the luridly homophobic accounts of these eras that many such descriptions would be entirely interchangeable were it not for use of the word *animals* in one and *people* in the other. The nearly seamless continuity between attitudes toward human and animal homosexuality is exemplified by the following pair of "observations," each of which reduces homosexuality to a form of role-playing imitative of heterosexuality:

> . . . one woman lying on top of another and simulating in movements the act
> of intercourse . . . gratifies her masculine component. . . . Some authorities

regard [the partners of these] women . . . as pseudohomosexuals. The number of sex-starved women who yield to homosexuality . . . is much greater than one might suppose.

—F. S. CAPRIO, *female homosexuality*, 1954

female[s] . . . occasionally carry out elaborate homosexual pseudocopulatory manoeuvres. Usually one female assumes the male role and mounts another female . . . and the two animals then perform a remarkably realistic pseudo-copulation.

—from a scientific description of Northern Fur Seals, 1959[3]

Sadly, such perspectives on animal homosexuality are still prevalent today among both scientists and nonscientists alike. In many cases, people are still reapplying to animals the same outmoded views of homosexuality that were used to condemn and pathologize the behavior in humans throughout most of this century. Such "explanations" have since been shown to be untenable (if not downright laughable) for people, and they should similarly have been abandoned long ago by scientists studying animals.

"Which One Plays the Female Role?"—Homosexuality as Pseudoheterosexuality

One of the most prevalent and pernicious misconceptions about animal homosexuality is that it is simply an imitation of heterosexuality and heterosexual gender roles.[4] In numerous species, animals that participate in homosexual interactions are assigned—sometimes arbitrarily—to one of two roles: "male" or "female." *Masculine* or *feminine, malelike* or *femalelike, male-acting* or *female-acting, male mimicry* or *female mimicry, pseudo-male* or *pseudo-female* are just some of the other terms widely used to refer to the participants in homosexual interactions.[5] In other words, homosexuality is seen merely as a replica of heterosexuality—male-female patterns transposed onto same-sex partners. In perhaps the most extreme example of this viewpoint, one scientist actually treated the homosexual couples in his captive population of Orange-fronted and Aztec Parakeets as stand-ins for heterosexual pairs. Because of the rather embarrassing fact that there were more same-sex than opposite-sex pairs in his flock, he used several homosexual couples as male-female surrogates in his experiments on "heterosexual" pair-bonding behavior. For this to work, however, "it was necessary . . . to assume that in homosexual pairs one bird assumes the role of the male, the other of the female, and that behavioral events between such birds are those typical of heterosexually paired birds." This assumption entirely disregarded the fact that female pairs in this species differ in important respects from heterosexual pairs (for example by exhibiting mutual, as opposed to one-way, courtship feeding) and probably also hindered the discovery of other such differences.[6]

The idea that homosexual relations in animals are necessarily gendered along heterosexual lines has its origins in Freud's (and others') view of (human) homosexuality as sexual *inversion,* the adoption by one partner in a same-sex interaction of the behaviors or roles "typical" of the opposite sex.[7] In fact, some zoologists have used the very terms *sexual inversion* and *inverse* (or even *reverse*) *sexuality* to describe homosexual activity in animals. Desmond Morris developed this idea further with respect to animals in a series of papers in the 1950s, in which he introduced the terms *pseudo-male* and *pseudo-female* to describe animals who exhibit behavior patterns more commonly seen in the opposite sex; these terms are still used to this day in scientific publications that describe same-sex activity.[8] Also still employed is the analytical framework represented by such terms, which argues that the occurrence of homosexuality in a species can be directly attributed to, and characterized by, opposite-sex or "gender-atypical" behavior. The argument goes something like this: certain animals in a population are prone to "pseudo-female" or "pseudo-male" behavior; that is, imitation of behavioral patterns found in the opposite sex. This mimicking of heterosexuality automatically triggers "homosexual" behavior in individuals who are essentially "deceived" into thinking they are dealing with a member of the opposite sex, hence they respond with sexual or courtship behaviors.

Often, an attempt is also made to correlate sexual "role inversion" with other behavioral or physical traits that are supposedly characteristic of the opposite sex (or of certain gender roles), such as higher levels of aggression in female Takhi and Mallard Ducks that mount other females. One scientist, in describing a male Snow Goose that supposedly adopted the "male role" in a homosexual pairing, even goes so far as to comment on the bird's "much enlarged penis" in addition to his greater aggressiveness. As we shall see later in this chapter, this is reminiscent of descriptions of human "inversion" from the early sexological literature, which often focused on the appearance of a person's genitals as somehow indicative of the "abnormality" or pathology of their homosexuality.[9]

Reciprocal Homosexuality and Heterosexual "Inverts"

In spite of apparently unabated popularity in scientific circles, a "pseudo-heterosexuality" interpretation imposes a restrictive and often erroneous framework on animal homosexuality, and there are numerous arguments against it.[10] To begin with, in an overwhelming number of cases homosexual behavior cannot possibly be construed as mimicking heterosexuality. In a number of species, unique sexual or courtship behaviors occur between animals of the same sex that are not found in heterosexual interactions. For example, homosexual but not heterosexual interactions in Bonobos, Gibbons, Stumptail Macaques, Crested Black Macaques, West Indian Manatees, and Gray Whales often involve mutual genital rubbing or manual and oral stimulation of the genitals.[11] The actions of both partners are often identical or reciprocal, and therefore neither animal can be construed as adopting a stereotypically "male" or "female" role.[12] In species such as Bottlenose

Dolphins, Cheetahs, and Grizzly Bears (among others), same-sex pair-bonding occurs to the exclusion of opposite-sex pairing; thus, the "roles" of individuals in homosexual pairs cannot be modeled after male-female (heterosexual) "roles" because there simply are no such models in these species.

Even for animals where identical or similar behaviors occur in both homosexual and heterosexual interactions, the same-sex activities often do not fall neatly into the gendered patterns expected under a "pseudoheterosexual" interpretation. For example, homosexual mounting is often reciprocal, which means that the animals take turns in the mounter/mountee positions, with neither preferring exclusively "male" or "female" roles. Various forms of reciprocal mounting have been documented in at least 30 species (and probably occur in many more): simultaneous reciprocity, in which the partners exchange roles during the same mounting bout (as in Pukeko or Black-rumped Flamebacks); and sequential reciprocity, in which partners trade roles at different points in time—the latter can involve frequent alternation over an extended period as in Japanese Macaques, or perhaps a onetime switch as has been reported for some Bottlenose Dolphins. Moreover, in many species heterosexual mounting can be "reversed" or "inverted," in that the female mounts her male partner. Thus, the "mounter" and "mountee" positions cannot be absolutely equated with fixed "male" and "female" roles even in opposite-sex interactions. Most "pseudoheterosexual" interpretations of homosexuality, therefore, involve stereotyped views not only of same-sex activity but also of male-female relations.[13]

Reciprocal, mutual, or non-"inverted" homosexual activities also characterize many other behavior categories besides mounting and sexual activity. In Laughing Gulls and Antbirds, for example, courtship feeding occurs in both heterosexual and homosexual contexts; between two males, however, this activity has a number of distinctive features owing to the fact that neither male is playing a "female" role. Males often engage in reciprocal courtship feeding by passing the food gift back and forth between them, and either partner may initiate the exchange (in heterosexual courtship feeding, typically the male initiates the activity and the female does not reciprocate). In a number of other bird species, including Greylag Geese, Mallard Ducks, Greenshanks, and Humboldt Penguins, both males in homosexual pairs exhibit typically *male* sexual, courtship, and pair-bonding behaviors, i.e., neither part-

Male West Indian Manatees employ a wide variety of positions and forms of genital stimulation during their homosexual encounters, including manual, oral, and mutual genital contact. These are not characteristic of heterosexual interactions in this species and are a good example of homosexual behavior not being modeled on stereotypical "male" and "female" roles. ▼

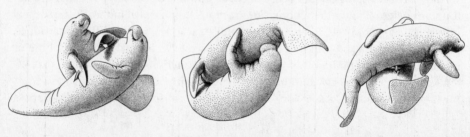

ner adopts a "feminine" role.[14] Likewise, *both* females in homosexual pairs of Snow Geese, Mute Swans, Lovebirds, Red-backed Shrikes, and Blue Tits incubate eggs—an activity typical only for females in heterosexual pairs of these species—while *both* males in Emu and Greater Rhea same-sex associations incubate the eggs and raise the chicks (an activity performed only by males in heterosexual associations).

▲ *A Greylag gander pair performing a synchronized duet of "rolling" calls. In this and other species,* both *males in a homosexual pair perform mutual or typically "male" activities rather than one bird adopting a "male" role and the other a "female" role.*

In "pseudoheterosexual" explanations of homosexuality, it is usually assumed that a same-sex interaction is initiated by the animal who is adopting behavior patterns of the opposite sex. That is, a sexual or courtship episode between males is triggered by one male performing typically female "invitations" to the other male, while an analogous interaction between females is initiated by one female making typically "male" advances toward the other. While the initiation of homosexual activities sometimes does follow this pattern, exactly the opposite is seen in as many—if not more—cases. Sexual activity between females is often initiated by the mountee making typically female solicitations or overtures toward another female—this is true for species as diverse as Lions, Squirrel Monkeys, Rhesus Macaques, Hanuman Langurs, and Sage Grouse. Conversely, it is also common for sexual interactions between males to be initiated by the mounter making a typically "male" approach to another male. Even in courtship activities, the "roles" of the participating animals often do not fall into the patterns predicted by a "pseudoheterosexual" interpretation. In Ostriches, for example, homosexual courtships are not prompted by "female" behavior on the part of a male, but rather are initiated by one male approaching another using behaviors unique to same-sex interactions. Similarly, male Musk Ducks perform their courtship displays without being "triggered" by female behaviors on the part of either males or females; rather, Ducks of both sexes are attracted to males who are already displaying.

Often only one animal in a same-sex interaction is classified by scientists as truly "homosexual"—the one engaging in the putative "gender-atypical" behaviors. Thus, a male animal that solicits and is mounted by another male is considered to be the "true" homosexual, while the male who mounts him is a "normal" heterosexual male who is reacting to "opposite-sex" mimicry. This kind of logic frequently leads to absurd and contradictory classifications of animals. We've already discussed cases of reciprocal mounting, where the exchange of "roles" between animals necessitates a corresponding switch in which one is considered to be engaging in "homosexual" behavior for the moment. Sometimes an animal is actually both mounter and mountee simultaneously: in Wolves, Laughing Gulls, Little Blue Herons, Sage Grouse, and other species, an animal mounting another individual (of the same or opposite sex) is sometimes itself mounted by an animal of the same sex. Thus, an individual can exhibit gender "typical" and "atypical" mounting behavior at the same time and can perform concurrent "homosexual" and "heterosexual" acts with same-sex partners. In other cases, these behaviors occur in the same individual but separated in time, and in ways that do not conform to a "pseudoheterosexual" interpretation. Typically, the "true" homosexual animal is thought to be limited to opposite-sex behavior patterns and hence incapable of actual heterosexual relations (e.g., a male playing the "female role" with a male partner is considered incapable of playing the "male role" with a female partner). However, bisexual animals who successfully mate and breed with opposite-sex partners often perform the "gender-atypical" role during their homosexual interactions, while strictly heterosexual animals may perform the "gender-atypical" role during their heterosexual interactions—showing that there is no necessary connection between homosexual and heterosexual "roles."[15]

In Red Deer, for example, one study revealed almost all possible combinations. Some Red Deer females who do not participate at all in homosexual activity play the "male" role in reverse heterosexual mounts, while others who are not involved in heterosexual activity play the "female" role in homosexual interactions (or assume both "roles" equally). One female who exhibited the most heterosexual behavior was only the "mounter" during homosexual interactions (i.e., she did not play the "female" role in that context), while the female who showed the most activity in the "male" role during homosexual interactions only played the "female" role in heterosexual interactions.[16] Moreover, as neuroscientist William Byne has pointed out, a "pseudoheterosexual" interpretation taken to its logical conclusion would have to regard each of the animals performing a reverse heterosexual mount as "homosexual," since each is exhibiting the mounting behavior of the opposite sex (male being mounted, female doing the mounting). We are left with the nonsensical result that same-sex mounting is a "heterosexual" act for some of its participants (those in the "gender-typical" role) while opposite-sex mounting can sometimes be a "homosexual" act for its participants (those in the "gender-atypical" role).[17]

Gendering and Transgendering

Just as most examples of homosexuality cannot be attributed to opposite-sex mimicry or "pseudoheterosexual" behavior, many examples of genuine transgender or sexual mimicry are not associated with homosexuality. In species such as northern jacanas, arctic terns, squid, and numerous reptiles and insects, animals imitate the behavior of members of the opposite sex in various contexts without inducing homosexual activity in animals of the same sex. In fact, more often than not such opposite-sex mimicry or behavioral transvestism is associated with *heterosexual* courtship, mating, or interaction. In jacanas, for example, males regularly adopt the female's copulation posture to solicit sexual behavior from females, yet this does not trigger homosexual mounting from other males; likewise for male arctic terns that utilize females' food-begging gestures.[18]

Not only is this true for species such as these where homosexuality has not been reported at all, homosexuality and "pseudoheterosexual" behavior (or transgender) often co-occur in the same species without having anything to do with each other. For example, when confronted aggressively by another male, Chaffinch males sometimes adopt the female's sexual solicitation posture to prevent an attack, yet this does not trigger homosexual mounting by the other male. Nonbreeding males in this species also sometimes behave like females when trespassing on another male's territory, but this does not cause the other male to begin courting him. Sexual chases between males, as well as female pairing, do occur in Chaffinches, but in contexts that are unrelated to such opposite-sex mimicry. Rufous-naped Tamarin males perform a "pseudo-female" behavior called upward tail-curling, typically used by females as a prelude to mating; however, males use this display during ambivalent or hostile encounters with females and not during episodes of homosexual mounting with other males. Likewise, Mountain Zebra bachelor stallions imitate the facial expressions and calls of mares in heat when they meet territorial breeding stallions, yet this opposite-sex mimicry does not incite homosexual mounting on the part of the territorial stallion. Rather, same-sex mounting in this species takes place almost exclusively between territorial stallions or between bachelors, rarely if ever between a territorial stallion and a bachelor.

Female Black-crowned Night Herons and Kittiwakes, and male Koalas, occasionally perform courtship behaviors typical of the opposite sex, but in none of these cases are such behaviors associated with the homosexual activity that does occur in these species—in fact, they are typical of animals in heterosexual interactions.[19] In Northern Elephant Seals, too, younger males imitate females specifically to gain access to *heterosexual* mating opportunities, "camouflaging" themselves from older males (who would attack them if they were discovered trespassing among females). Yet this does not specifically trigger homosexual mounting from the older male, and same-sex mounting is typical of contexts outside of female mimicry in this species. In fact, transgendered individuals in Northern Elephant Seals and a number of other species (e.g., Red Deer, Black-headed Gulls, Common Garter Snakes) are often *more* successful at heterosexual mating than many non-transgendered individuals—in other words, animals that look and/or act like the opposite sex can actually be "more heterosexual" than ones that do not.[20]

In a number of animals, some homosexual interactions have characteristics that could be interpreted as involving "pseudoheterosexuality" or transgendered behaviors, yet these constitute only a portion of same-sex activity in the species—and hence, only a partial "explanation," at best, for the occurrence of these activities. In Tasmanian Native Hens, for example, males adopt a posture following heterosexual copulation that resembles the female's invitation to mate—yet only one homosexual mounting recorded in this species was apparently triggered by this posture; the rest occurred in other contexts. Rhesus Macaque females who mount other females sometimes display typically "male" behaviors such as various head movements, the way they carry their tails, or other patterns—but just as many females, if not more, do not exhibit these behaviors as a part of their homosexual interactions.[21]

Perhaps the most compelling example of how homosexuality, transgender, and gender roles interact in unexpected ways concerns "femalelike" males in Mountain Sheep. In Bighorn and Thinhorn Sheep, being mounted by another male is a typically "male" activity. As described in chapter 1, most males participate in homosexual mounting throughout the year, while females generally refuse to allow males to mount them except for the two or so days out of the year when each of them is in heat. Consequently, transgendered males—rams who associate with females throughout the year (unlike most other males) and exhibit other female behavioral characteristics—do not typically allow other males to mount them. In other words, homosexual activity is characteristic of "masculine" males rather than "feminine" males in these species. Moreover, because same-sex mounting has such primacy in the social organization of these animals, heterosexual activity is actually patterned after homosexual interactions and not the other way around. Females in heat typically imitate the courtship patterns of male homosexual interactions in order to arouse the sexual interest of males—a remarkable example of the exact opposite of a "pseudoheterosexual" pattern.[22]

Homosexual "Role-Playing": Gender Blending and Amalgamation

In many animals gender roles of some sort do exist in homosexual interactions, but it is overly simplistic to consider these mere replicas of male and female behaviors. Gendered activities in a same-sex context are never an exact copy of heterosexual roles, and in many cases animals actually exhibit a complex *mixture* of male and female behavior patterns. This type of gender-role mixing assumes three basic forms: a continuum among individuals, role-differentiated combinations, and behavioral amalgams.[23] In some species, individuals vary along a scale or continuum in the extent to which their behaviors in homosexual interactions resemble "male" or "female" patterns. In Kob antelope, for example, some females utilize the full array of courtship patterns typically employed by males, others make use of none or few of these, while most females range somewhere in between these extremes.[24] Ruff males fall into four categories along a spectrum of most "malelike" to most "femalelike" in terms of appearance (presence and color of neck ruff, size), aggressive behavior, courtship behaviors, and other characteristics. However, these categories cut across aspects of sexual behavior, including participation in the "male" role of mounter and the "female" role of mountee in homosexual interactions. The most "malelike" males (residents) perform both roles as do the most "femalelike" males (naked-napes), while of the intermediate categories, some participate in both roles (satellite males) and some rarely engage in either (marginal males). In a number of species such as Gorillas, Hanuman Langurs, and Rhesus, Bonnet, and Pig-tailed Macaques, some individuals clearly prefer (or end up mostly participating) in the "mounter" as opposed to the "mountee" roles during same-sex activity, while for other individuals the reverse is true. Yet these patterns represent the two poles of a continuum, since many individuals in these species actually fall along the entire range in terms of their mounting activities.[25]

To specifically address the question of "pseudoheterosexual" roles, scientists studying homosexual pairing in Western Gulls made detailed observations regarding whether one partner is more "feminine" and the other more "masculine," in terms of which courtship, sexual, and territorial behaviors they exhibit. They found that most females employ a mixture of typically male and typically female patterns, although pair-bonds vary in the extent to which there is role differentiation between the partners. In some pairs, one bird performs the majority of mounting and courtship feeding (typically "male" activities) and less "head-tossing" (a typically "female" courtship behavior). In others, there is less of a distinction between the two partners, while in still others the two females participate nearly equally in gendered behaviors. Overall, however, scientists found that both partners in homosexual pairs are more similar to heterosexual females than to males in terms of the amount of time they spend on their nesting territories and their aggressive responses to intruders.[26]

Another pattern of gender mixing involves role-differentiated combinations, in which same-sex interactions are largely gendered or separated into "male" and "female" roles, yet each individual still combines elements of both to varying degrees. This is a crossing or intermixing of "masculine" and "feminine" traits—in the domains of sexual, courtship, or parenting and pair-bonding behaviors—set against an overall pattern of polarity between the two. For example, male couples in Hooded Warblers often divide up their parenting duties into typically male and female roles: one male builds the nest and incubates the eggs ("female" duties) while the other defends the territory and sings ("male" activities). Yet layered on top of this are more subtle meldings of gender roles: the more "feminine" partner may also engage in the typically male activity of singing (although with a distinctive song type), while the more "masculine" partner may also feed his mate during incubation (an activity rarely exhibited by either partner in heterosexual pairs).[27]

Other examples related to pair-bonding and parenting activities abound. The "masculine" partner in some Canada Goose (and Chiloe Wigeon) lesbian couples still carries out the quintessentially female activities of egg laying, incubation, and nest-building (the latter usually done only by females in these species). One female in Orange-fronted Parakeet homosexual pairs typically performs the "male" activity of nest-tunnel excavation, yet both partners may initiate courtship feeding (characteristic of males in opposite-sex pairs). And in Mute Swan female pairs, one partner stands guard and defends the territory (like a male), yet both females lay eggs (and both build the nest, typical also of both partners in heterosexual pairs). Some Lovebirds in same-sex pairs are role-differentiated, while others engage in combinations of "male" and "female" courtship and sexual activities. In either case, though, if two females are involved, they both perform the typically "female" roles of nest-building, egg laying, and incubation, while neither of two paired males shows any interest in nest-building (which is not characteristic of either heterosexual role). Some "feminine" partners in Chaffinch lesbian couples also exhibit characteristically male behavioral patterns such as singing, while both partners in role-differentiated Jackdaw homosexual couples (or trios) preen each other—a behavior typical only of females in heterosexual pairs. Similar patterns are to be found where sexual and courtship activities are concerned as well. In Long-eared Hedgehog lesbian interactions, for example, one female may be more "malelike" in initiating and carrying out various courtship and sexual behaviors, yet both partners may perform characteristically "female" invitation postures or typically "male" mounting attempts. Likewise, in courtship interactions between male Victoria's Riflebirds or Blue-backed Manakins, the more "femalelike" partner that is being courted often responds with his own distinctly male display patterns.[28]

A final type of gender-role mixing seen in homosexual interactions involves behavioral amalgams—more balanced combinations of "male" and "female" traits in the same individual, a sort of behavioral "androgyny." This can involve sexual activities: during homosexual interactions between male Gorillas, for instance, the mounter (i.e., the animal "playing the male role") usually also "plays the female role" of initiating the interaction (female Gorillas typically initiate sexual activity in heterosexual contexts). Mallard females who perform the "male" activity of mounting other females nevertheless display postcopulatory behaviors typical of females, while the mountee in male Black-crowned Night Heron homosexual encounters may perform typically "male" courtship behaviors. The mounter in Hanuman Langur female homosexual encounters often exhibits otherwise "female" behaviors such as initiating the sexual interaction and grooming her partner following the mount. Behavioral amalgams can also involve courtship and parenting activities. When one male Emu is courting another, for example, he stretches his neck and erects his neck feathers—a behavior characteristic of *both* females and males in heterosexual courtships[29]—yet neither male makes the booming vocalization typical of females, and each may follow the other (usually only males follow females in heterosexual courtships). Younger male Swallow-tailed Manakins that are courted by adult males exhibit a combination of male and female behavioral traits that makes them distinct from either (and also parallels their plumage, which is a mixture of adult male and female appearance). Their vocalizations and participation in some noncourtship displays are distinctly masculine, while their generally quiet and inconspicuous demeanor is unlike adult males, and in courtship interactions they may assume the role that the female usually does.[30] In Snow Goose homosexual pairs, *both* females perform typically female activities such as incubation *and* typically male activities such as defense of the goslings.[31]

A multiplicity of gender-role mixtures that defy categorization into any of these three types is the norm in species like the Black-headed Gull. Detailed comparisons of both heterosexual and homosexual pairs showed that birds in same-sex pairs exhibit neither stereotypically "male" nor "female" behaviors. Rather, the frequency with which they perform various courtship and pair-bonding activities tends to be distinct from, or intermediate between, that of both males *and* females in opposite-sex pairs. For example, the maximum rate of "ceremonial encounters" (a type of courtship interaction) in homosexual pairs exceeds that of both partners in heterosexual pairs. On the other hand, rates of "long-calling" and "head-flagging" (other forms of courtship) tend to be intermediate between those of heterosexual males and females, while the rate of courtship "begging" by males in homosexual couples is generally as low as that of males in heterosexual pairs (which itself is generally lower than that of heterosexual females).[32] In addition, both males in homosexual pairs usually build the nest (which is a typically "male" activity in heterosexual couples), although there is also variation between individuals in this regard, with only one partner contributing to the nest in some male pairs.

The "pseudoheterosexual" interpretation of animal behavior offers striking parallels to stereotypical views about human homosexuality. Scientific puzzlement over assigning animals "male" or "female" roles echoes the refrain often heard by gay and lesbian people, who are frequently asked, "Which one plays the man (or woman)?" The assumption is that homosexual relationships must be modeled after heterosexual ones—a view that is as narrow a conception of human relationships as it is of animal sexuality. Each partner in a gay or lesbian relationship (or sexual encounter) is thought to "play" one-half of a heterosexual couple. In reality, far more complex and multidimensional expressions of gender categories are involved, even (or perhaps especially) when the partners appear most "heterosexual" to outside observers. Some people do not structure their homosexual interactions along gendered lines at all; others do, but re-create typically "male" and "female" patterns in new configurations. To give just one example: butch-femme lesbian relationships have long been viewed as simplistic imitations of heterosexuality, in which the butch partner is the "man" and the femme partner is the "woman." Lesbians whose erotic lives are organized along these lines, however, describe eloquently how their actual experiences are far different from this. Neither partner is "copying" heterosexual roles; rather, each is taking elements of masculinity and femininity and alloying them in different combinations and intensities to create female-specific genders. As one lesbian has said about the kind of women she is attracted to, a masculine lesbian is not an imitation man, but a real butch.[33] If even this most superficially "heterosexual" gender presentation is more than what it appears, imagine the possibilities when homosexual interactions are gender-role-defined in other ways, or not at all. Such "possibilities" are in fact everyday realities in the lives of both humans and animals.

Over the past thirty years, a sophisticated analysis of gender categories has been emerging from within the feminist, gay and lesbian, and transgender movements, one that challenges basic notions such as "male" and "female," "masculine" and "feminine," "mannish" and "effeminate." These movements are also calling for a recombining and reimagining of categories such as these, rather than simply their denigration or abolishment. Unfortunately, zoologists for the most part are still operating under an earlier, outmoded conception of gender roles (both heterosexual and homosexual)—one that is inconsistent with the actualities of sexual and gender expression within the animal and human worlds. If any progress is to be made in the study and understanding of animal homosexuality and transgender, scientists and nonscientists alike will need to acquire the sort of multifaceted view of gender and sexuality that is now being articulated within these human liberation movements.

"The Lengths to Which Deprived Creatures Will Go"— Homosexuality as Substitute Heterosexuality

One of the most prevalent myths about animal homosexuality is that it is invariably caused by a shortage of members of the opposite sex. This is typically attributed to skewed sex ratios in the population (more males than females, or vice versa), or the unavailability of opposite-sex partners due to sex segregation, hostility or indifference on the part of potential mates, or other factors. This belief is widespread among nonscientists and is also the most common "explanation" that biologists have proposed for the occurrence of homosexual behavior in animals. In more than 65 species of mammals and birds, for example, same-sex activity is claimed by zoologists to result from individuals being "unable" to mate heterosexually. Sometimes this is attributed to a predominance of one sex over the other in wild or captive populations: the formation of lesbian pairs in Australian Shelducks and Ring-billed Gulls, for instance, is supposedly "caused" by an excess number of females (65 percent females in Shelduck populations, 55 percent females in Ring-billed Gulls).

Homosexual pairs in Mute Swans occurring in populations with unbalanced sex ratios are said to be "examples of the lengths to which deprived creatures will go to satisfy their natural urge to reproduce." In some cases, homosexual behavior is labeled a "substitute" for heterosexuality or "redirected" heterosexual behavior, resulting from a variety of factors. For example, it is claimed that individuals are "prevented" from mating with (or otherwise having access to) the opposite sex by other (often higher-ranking) animals, or by the overall social organization (e.g., in Mountain Sheep, Bottlenose Dolphins, or Killer Whales). Alternatively, it has been suggested that individuals resort to homosexuality when their heterosexual advances are met with refusal or disinterest (e.g., in White-handed Gibbons, West Indian Manatees, Asiatic Elephants). In a few cases (e.g., Hanuman Langurs, Lions, Sage Grouse) scientists have even suggested that females turn to one another because they have not been "satisfied" or received enough attention from male partners—a version of the widespread stereotype about the "cause" of lesbianism among people.[34]

The line of reasoning in "explanations" such as these is curious, since it implies that unless there is an adequate supply of the opposite sex, homosexuality will inevitably ensue. This is actually an unintentional assertion of the relative strength of the homosexual urge, or correspondingly, the relative weakness of the heterosexual imperative—for the stronghold of heterosexuality must be tenuous indeed if such factors are capable of upsetting the balance. Besides this, however, unavailability of the opposite sex—what we will call the shortage hypothesis—is simply incompatible with the facts.

Surplus Homosexuality

The shortage hypothesis cannot be a universal explanation for animal homosexuality because of the many examples of animals engaging in same-sex activity when opposite-sex partners are freely available.[35] In Orang-utans, Japanese Macaques, Stumptail Macaques, Rhesus Macaques, Common Gulls, Black-headed Gulls, King Penguins, Galahs, and more than 40 other species, scientists have documented individuals either ignoring opposite-sex partners and seeking out same-sex partners instead, or else engaging in homosexual activity more or less concurrently with heterosexual activity (i.e., even when opposite-sex partners are accessible, as already mentioned in the discussion of simultaneous bisexuality).[36] In fact, in a surprisingly large number of species, homosexual activity is *positively* correlated with heterosexual activity: same-sex interactions actually increase as animals gain access to opposite-sex partners and decrease in their absence. This is the exact reverse of what would be expected if homosexuality resulted from a lack of access to heterosexual mating opportunities.

Homosexual activity among male Bottlenose Dolphins in captivity, for instance, actually declined when females were removed from their tank, while aggressive interactions between the males increased. Conversely, female Squirrel Monkeys in one study engaged in virtually no homosexual activity when kept in same-sex groups, yet showed significant rates of homosexual mounting and other activities (along with heterosexual behaviors) when males were introduced into their group. Another study of this species found that females with the most attention from heterosexual partners also engaged in the most homosexual pursuits. In Bonobos, Stumptail Macaques, Savanna (Yellow) Baboons, and West Indian Manatees, same-sex activity is often stimulated by opposite-sex activity (and vice versa), with the result that sessions may involve both heterosexual and homosexual encounters among multiple participants. Homosexual mounting in Pukeko is most prevalent in breeding groups that have the greatest amount of heterosexual activity, while homosexual mounts in Common Murres become more common as promiscuous heterosexual mounts also increase in frequency (although the latter may, ironically, result from a decrease in available females). In some species, individuals that participate in the most heterosexual matings may also engage in the most homosexual ones, as in Sociable Weavers and Bonnet Macaques. Conversely, animals that are the least active heterosexually are often the least active homosexually. In Ruffs, for example, the class of males who do not generally mate with females (known as marginal males) also rarely participate in homosexual matings, while in one study of Japanese Macaques, the only female who did not consort with any other females also failed to consort with any males.[37] Finally, in a number of species such as Swallows, Laysan Albatrosses, and Herons, same-sex mounting occurs primarily among breeding individuals (i.e., those who already have heterosexual mates) and is largely absent among nonbreeders.

A number of species do have skewed sex ratios, but (like dominance) this is neither a sufficient nor a necessary prerequisite for the occurrence of homosexuality in a population. Male homosexuality is not reported for red-winged blackbirds or giant cowbirds, for example, even though some populations are 80–84 percent male, nor for pintail duck populations with two-thirds males, or kiwis with 58 percent males, or purple finches with 57 percent males. Likewise, female homosexuality is absent in boat-tailed grackles even though males may comprise only a third of the population, and in sparrow hawks, where there is also a "surplus" of females (less than 40 percent males). In contrast, homosexuality occurs in numerous species or populations that have equal (or nearly equal) sex ratios, including Bonobos, Bonnet Macaques, West Indian Manatees, Snow Geese, California Gulls, and Pukeko.[38] Moreover, closely related species or different populations of the same species that have identical (or similar) sex ratios and forms of social organization often exhibit strikingly different patterns of homosexuality. Many Seals and Sea Lions with polygamous mating systems, for instance, have strongly female-biased sex ratios (three to five females for every male) and social systems that often include sex segregation and/or exclusion of large numbers of males from breeding opportunities. Some of these species exhibit male homosexuality (e.g., Gray Seals, Northern Elephant Seals, Walruses), others have female homosexuality (e.g., Northern Fur Seals), some have both (e.g., Australian Sea Lions), while the majority have no homosexuality at all (e.g., California sea lions, southern fur seals). Likewise, lunulated antbirds, salvin's antbirds, Bicolored Antbirds, and Ocellated Antbirds all live in populations that have an excess of males, yet homosexual pairing is only found in the latter two species.[39]

In many animals that have skewed sex ratios, homosexuality only occurs (or is more common) in the sex that is in *shorter* supply rather than in the "surplus" sex. In some populations of Crab-eating Macaques, for example, females outnumber males by more than two to one, yet same-sex activity only occurs among males. Female homosexuality accounts for more than 80 percent of same-sex activity in Pukeko even though some populations are more than 70 percent male. The reverse is true for Rhesus Macaques: in some populations females outnumber males nearly three to one, yet the majority of same-sex activity (over 80 percent) is between males. Tree Swallow populations often have a surplus of females, but only male homosexuality occurs. Likewise, female pairs have formed in captive populations of Galahs and Scarlet Ibises that have an excess of males, while male pairs of Flamingos are reported from populations that have more females than males. In Nilgiri Langurs, there is a female-biased sex ratio in the overall population (and individuals live in groups with more females than males), yet only male same-sex activity is reported. Finally, in Little Egrets and Little Blue Herons there is a "surplus" of unpaired males, yet same-sex mounting occurs almost exclusively among paired males rather than in the population of birds that are unable to find heterosexual mates.[40]

While homosexual activity in some species may appear to be associated with an unavailability of the opposite sex, the patterns of its occurrence are often far more complex than a shortage explanation would indicate. Although lesbian pairs in Black Stilts, for example, generally do occur in populations where the sex ratio is biased toward females, in other populations of the same species with a surplus of males, no male homosexual pairs have formed. The same is true, in reverse, for captive Humboldt Penguins: male pairs form when there is a surplus of males but female pairs do not form when there is a surplus of females. Among some populations of Savanna (Yellow) Baboons, the sex ratio becomes skewed among older juveniles, where males outnumber females two to one—and indeed, 10 percent of such animals' mounting is homosexual. However, the sex ratio is equal among adults and younger juveniles, and the prevalence of homosexual mounting in these segments of the population is the exact opposite of what the shortage hypothesis would predict: 17–24 percent of their mounting is same-sex. In other words, older juvenile males actually exhibit the lowest proportion of homosexual activity and the greatest participation in heterosexual mounting of any segment of the population (accounting for more than half of all male-female mounts), even though their age group contains the greatest surplus of males. Sex ratios in wild Mallard Ducks fluctuate during the breeding season, with fewer females being present in some months than others. Although male pairs sometimes form at these times, during other months when there is also an excess of males in the population, there are no male pairs.[41]

If access to heterosexual mates were the only factor involved in the occurrence of homosexuality, both males and females in sex-segregated populations should exhibit the same degree of homosexual activity. However, in the majority of species that have some form of sex segregation, homosexual activity is found in only one sex (e.g., Walruses, Gray Seals, Warthogs, American Bison) or is much more common in one sex (usually males) than the other (e.g., Giraffes, Blackbucks, Mountain Sheep, Australian Sea Lions). Conversely, in some species that have unbalanced sex ratios (in wild or captive contexts), such as Pig-tailed Macaques, Bottlenose Dolphins, Cheetahs, Koalas, Canada Geese, and Flamingos, homosexuality occurs in *both* sexes (although it may be more common in the "surplus" sex). This indicates that more is involved than simply a "shortage" of available heterosexual partners.[42] Likewise, where populations of the same species vary in their sex ratios, homosexuality is sometimes less common in nonskewed populations, but it is still present. In Japanese Macaques, Giraffes, and Greylag Geese, for example, same-sex activity may increase in populations with an excess of one sex, but it still occurs at a fairly steady rate in other circumstances regardless of the sex ratio and may even be present in the "limiting" sex (e.g., in Giraffe populations with more than 60 percent females, male homosexuality still occurs). Even in populations of Japanese Macaques with highly skewed sex ratios, most individuals still manage to participate in both heterosexual *and* homosexual activities, indicating that they are not turning to same-sex partners as a result of being completely "deprived" of opposite-sex partners.[43]

Similarly, in a number of species where homosexuality sometimes occurs in the absence of opposite-sex partners (due to sex segregation, heterosexual refusal, captive situations, etc.), same-sex activity is not limited to these contexts, but also occurs in mixed-sex groups (e.g., Gorillas, Hanuman Langurs, Crested Black Macaques, Squirrel Monkeys, Walruses, Lions, Mallard Ducks, Black-headed Gulls) or in contexts where it is not a response to the refusal or unavailability of the opposite sex (e.g., West Indian Manatees, Cheetahs).[44] If same-sex activity were due entirely to an absence of the opposite sex, it should disappear completely once opposite-sex partners are available, yet these examples show that it does not. Conversely, homosexuality does not arise automatically or immediately when animals are deprived of opposite-sex partners, nor does heterosexuality necessarily ensue once they have access to such partners. Homosexual activity in a captive group of female Squirrel Monkeys, for instance, did not develop until fully one year after they had been sequestered away from males, while female Long-eared Hedgehogs that were homosexually involved with each other in the absence of males did not participate in heterosexual mating for more than two years after they were given access to males.[45]

Multiple Possibilities

Even if homosexuality in some species only occurs in populations where there is more of one sex than the other, this is, at the very least, evidence of a "latent" bisexual capacity among some individuals. Moreover, the skewed sex ratio is probably only a contributing factor rather than a determining "cause" of same-sex interactions in such cases. Typically only a portion of the "surplus" sex in these populations actually participates in homosexuality, and sometimes "available" opposite-sex partners are even passed over. This is most obvious in Silver Gulls, where nearly half of all females are "unable" to find a male partner each year, yet lesbian pairs constitute only about 6 percent of the population—in other words, the vast majority of "surplus" females remain single rather than forming homosexual pairs. Furthermore, about 14 percent of all males are unpaired, which means that females who form same-sex bonds do so in spite of the presence of single males in the population. Likewise, some female Mallard Ducks remain unpaired even in populations with more males than females. In one semi-wild population of Canada Geese with an excess of males, some of the unpaired males failed to form homosexual pairs; furthermore, some females also remained unpaired or formed homosexual bonds even though opposite-sex birds were "available." While approximately 10 percent of widowed Jackdaws form homosexual pair-bonds, the majority of widowed birds who do not find male partners actually remain single rather than pairing with female partners. Lesser Scaup Duck populations generally consist of 60–80 percent males, yet only a fraction of these individuals engage in homosexual mounting (and none form homosexual pair-bonds). Similarly, herds of Caribou may contain only 30–40 percent males, yet same-sex activity among females is not overwhelming.[46] Other species in which only a portion of the "surplus" individuals form same-sex bonds include Flamingos, Laughing Gulls, Humboldt Penguins, Gentoo Penguins, Pied Kingfishers, Peach-faced Lovebirds, Galahs, and Bicolored Antbirds.[47]

Although homosexual involvements in such species may be the "result" of skewed sex ratios, any "explanation" of homosexuality that relies on this factor alone needs to address why only some individuals "choose" this strategy, and why this strategy rather than another. For in addition to remaining single or forming same-sex pair-bonds, a wide variety of other behavioral responses occur among animals in populations that have a surplus of one sex, or in situations where the opposite sex is "unavailable." For example, in many otherwise monogamous species that have more females than males (or vice versa), some individuals form polygamous heterosexual trios (so-called "bigamy") or even quartets ("trigamy"). These options occur alongside homosexual pairings in Flamingos and Humboldt Penguins, and instead of same-sex pairing in Cattle Egrets, emperor penguins, and dippers (among many others). Individuals in the same population may also adopt different strategies or combine these strategies, to varying degrees: in Oystercatcher communities, for instance, with large surpluses of nonbreeding birds unable to find heterosexual mates or breeding territories of their own, only a small fraction of these birds form polygamous trios (most remain unmated); and only a portion of these in turn go on to develop homosexual bonds within their trio. Australian noisy miners (a bird species with a heavily male-biased sex ratio) have developed a complex, specieswide system of communal breeding that involves, among other arrangements, polyandry (several males associating with each female, without any same-sex bonding). The reverse situation occurs in spotted sandpipers, in which "surplus" birds actually participate in monogamy rather than polygamy. In this species, females usually mate with several males and generally leave the parenting duties to them; females unable to find polygamous mates, however, often "revert" to monogamous (heterosexual) pairing and parenting, helping one male partner with incubation and brooding.[48]

Polygamy is just the tip of the iceberg as far as alternate strategies are concerned. Surplus female Redshanks participate in promiscuous matings with already paired males rather than forming bonds involving other females; "extra" male mustached warblers help already established pairs raise their families; while surplus female Ostriches and Greater Rheas lay their eggs in other females' nests or abandon them rather than forming bonds with either males or females. Female Tree Swallows, male tropical house wrens, and male barn swallows unable to find mates of their own often invade the nests of existing heterosexual pairs and forcibly acquire a partner (either through direct attack and eviction of the other mate, or by killing their young and causing the pair's breakup). In some species of Penguins and Egrets, individuals form temporary or serial heterosexual pair-bonds or divorce their partners more often in response to a surplus of one sex. In Black Stilt populations with a surplus of one sex, birds regularly seek heterosexual partners outside their species (hybridizing with the closely related Black-winged Stilt), while female Silver and Herring Gulls in some colonies with a "shortage" of adult males often simply pair with much younger males. A common response of male African Elephants that are sexually aroused but unable to find receptive female partners is simply to roll in mud wallows or take dust baths (thereby perhaps "diffusing" their arousal), rather than engaging in any sexual behavior (with either males or females). In all of these species, homosexual activity (if it occurs at all) is simply one "option" that some individuals adopt alongside many other alternatives.[49]

A shortage explanation cannot adequately account for the occurrence of such multiple strategies, or for the choice of one over the other. By claiming that animals "resort" to homosexuality in times of need, scientists often overlook other more plausible (heterosexual) alternatives—unintentionally providing support for the idea that homosexuality may actually be the most appealing option for some individuals in these circumstances. For example, homosexual activity in White-handed Gibbons has been attributed to the unavailability (or unwillingness) of a male's female partner to have sex with him. But why don't such males seek heterosexual matings outside of their pair-bond or simply masturbate (strategies that both occur in other contexts for this species)? Likewise, homosexual courtships in Ostriches are claimed—in a particularly convoluted sequence of logic—to result from a *balanced* sex ratio in some populations. Because Ostriches often mate polygamously—one male with several females—a population that has equal numbers of males and females would supposedly be unable to support such multiple matings. One scientist suggested that in this case males turn to their own sex—but why would this occur instead of males simply remaining in monogamous heterosexual pairings (as do spotted sandpipers)? Some same-sex mountings in Buff-breasted Sandpipers are said to result from the scarcity of females late in the breeding season. But once females stop visiting the breeding territories at this time, males often move their courtship displays to where the females are (at their nests) and may even copulate with them during the egg-laying period. Why would some males "resort" to relocating their heterosexual activities, while others would "abandon" them for homosexual activities?[50] Even if same-sex activity in these species could be attributed to the unavailability of the opposite sex, important questions such as these remain unaddressed under a shortage explanation. On the other hand, if participation in homosexuality is seen as the expression of individual variability and plasticity in sexual orientation—rather than as being "caused" or "determined" by a shortage of the opposite sex—then the variety of sexual responses and capacities actually seen in such circumstances is no longer incongruous.

Deprived of Heterosexuality?

One particularly common version of the shortage explanation is that males turn to homosexuality after being prevented from mating with females, due to the active interference or inhibiting presence of higher-ranking males. This is often suggested for mammals with polygamous or promiscuous mating systems in which only relatively few males ever get to mate heterosexually. While this may contribute to same-sex activity in some species (e.g., Mountain Sheep, Northern Elephant Seals), there is considerable evidence that this explanation is overly simplistic. Homosexual mounting in American Bison, for instance, is especially common among younger males that do not breed (ages one to six), but this cannot be attributed solely to their being "denied" access to females by older, higher-ranking males. Same-sex activity is much less common among bulls four to six years old than it is in males one to three years old, even though both groups are "prevented" (or abstain) from engaging in breeding activities. Also, studies of populations in which older bulls have been removed (in order to give younger males more "access" to females) show that although heterosexual activity increases among the younger bulls in the absence of the higher-ranking males, so does homosexual activity. In fact, more than 55 percent of mounts in such groups are still between males even though there are no older bulls to "prevent" the males from mating heterosexually (and even though females outnumbered males in such populations). Likewise, as they are growing up, male Bonobos experience a sharp drop in their access to heterosexual partners that is not reflected in a corresponding increase in their homosexual activity. As infants and juveniles they are very sexually active with mature females, but once they reach adolescence, they are generally prevented by older males from interacting sexually with females. However, their participation in same-sex activity increases steadily throughout this period rather than rising sharply when female partners become "unavailable" (or dropping when they are once again available). In fact, homosexual activity reaches its maximum level during adulthood when heterosexual activity also peaks.[51]

Although younger male Musk-oxen may be excluded from heterosexual mating opportunities by older males, this does not "cause" the homosexual activity that occurs in this species. Courtship and mounting activity between males generally takes place in the breeding (harem) herds and is initiated by adult bulls toward yearlings—in other words, by the males who *do* have access to females. Most other adult males are excluded from breeding as well, yet homosexual behavior is not characteristic of the all-male herds where such bulls often spend much of their time. Additionally, in captive groups lacking older males to "prevent" heterosexual mating, homosexual mounting still occurs at a fairly high rate alongside opposite-sex mounting. Likewise, homosexual behaviors in male Asiatic Elephants occur in both breeding and nonbreeding individuals. When older or higher-ranking males (who usually monopolize female partners) are prevented from mating heterosexually in captive groups, younger or lower-ranking males are then able to copulate with females. However, same-sex activities continue in both age/rank groups regardless of whether they have access to female partners or not. In fact, males who are heterosexually active may exhibit higher levels of homosexual activity than those who do not mate with females. In New Zealand Sea Lions, most younger males are excluded from breeding, but homosexual activity also occurs among adult breeding males. Lower-ranking male Wolves are prevented from mating with the highest-ranking female in the pack, but rather than mating with the available lower-ranking females, such males often mount each other. Finally, although homosexual activity in adolescent Killer Whales is attributed to their exclusion from heterosexual mating opportunities, adult males who have access to females also sometimes participate in same-sex activity. In addition, adolescents spend roughly the same proportion of their time as do adults engaging in sexual and related social behaviors with females.[52]

Homosexual activity in birds also generally fails to be correlated with patterns of exclusion from (or nonparticipation in) heterosexual mating. In Ruffs, for instance, mountings between males are actually more common when females are present on the breeding grounds than when they are absent. Even though males sometimes do try to prevent each other from mating heterosexually, same-sex mounts are not simply "redirected" or "substitute" heterosexual copulations. Homosexual mounts occur among a variety of different classes of males even when they are not directly "prevented" from mating with females (such as individuals known as satellites and naked-nape males). Conversely, one class of males—so-called marginal males—that *is* routinely excluded from opposite-sex interactions (by direct attacks from other males) does not usually engage in homosexual activity either. In Pukeko, a significant portion of the population assists other birds in raising their young rather than breeding themselves; however, homosexual activity in this species is characteristic of breeding individuals rather than such nonbreeding "helpers." In Ocher-bellied

Flycatcher and Ruffed Grouse populations that have a "surplus" of nonbreeding birds, breeding territories often go unused, indicating that nonbreeders are not being "prevented" from mating heterosexually (or at least are choosing to remain nonbreeders until better territories become available). Moreover, only a portion of such nonbreeders ever participate in same-sex activity, which in these species generally involves at least one partner that *does* have his own breeding territory. Likewise, the incidence of homosexual bonding among Oystercatchers (in the form of bisexual trios) does not increase significantly under higher population densities, when many individuals are unable to acquire their own opposite-sex mates and breeding territories. Brown-headed Cowbird populations generally have a significant surplus of males, and higher-ranking males also actively prevent lower-ranking males from courting and pairing with females. Yet the large number of birds thereby "excluded" from heterosexual mating opportunities—half to two-thirds of all males—do not regularly court or mate with one another. In fact, the only homosexual activity observed in this species involves male Cowbirds occasionally being mounted by males of *another* species, House Sparrows. Finally, Guianan Cock-of-the-Rock exhibit a pattern of homosexual and heterosexual participation that is strikingly similar to that of American Bison. Both yearling and two-year-old males are generally "excluded" from breeding, yet extensive homosexual activity only occurs in the former age group—and almost always with adult males as partners, many of whom are *not* excluded from heterosexual mating.[53]

Homosexuality is not generally the result of animals being "deprived" of heterosexual mating opportunities—this can be seen quite clearly in the behavior of individuals toward members of the *opposite* sex in skewed or segregated populations (in both the wild and captivity). Potential heterosexual partners are often ignored or even actively refused in such situations—they are rarely inundated with attentions as would be expected if animals were being excluded from participation in opposite-sex mating. In Giraffe populations with a majority of males, for instance, females are not swamped with heterosexual attentions, and mating opportunities with females are sometimes even bypassed in favor of homosexual mounting and other activities. Female Japanese Macaques and Hanuman Langurs engaging in homosexual activities usually disregard males entirely and may actually threaten or attack them if they make sexual overtures. Gray Seal and Killer Whale homosexual activities usually take place in all-male groups; although a few females are occasionally present in such groups, they are subject to little sexual attention from the males, the majority of whom ignore them altogether. This is in stark contrast to the often violent sexual attacks by large numbers of males on females during the breeding season in Gray (and other) Seals. Female Plains and Mountain Zebras living alone or in "bachelor" (or nonbreeding) herds are rarely approached sexually (or competed for) by either "bachelor" males or herd stallions (both of whom sometimes participate in homosexual activities). Stallions may even actively prevent new females from joining their herds. "Surplus" male Great Cormorants—some of whom form homosexual pairs—never display any interest in the (heterosexually paired) females among them. Conversely, in populations of Orange-fronted Parakeets with an "excess" of females, homosexual pairing is not limited to birds

▲ *A herd of male Gray Seals hauled out on land during the molting period on Ramsay Island (England). Just before this picture was taken, the two bulls at the water's edge near the large rock were engaging in homosexual activity—a common pursuit among males of all ages at this time of year. Although a few females may be present in these spring haul-outs, they are largely ignored by the males, demonstrating that their homosexual activity is not simply a "substitute" for heterosexual mating.*

that "can't find" heterosexual mates, since females paired with males also regularly form same-sex bonds (as part of bisexual trios). Adult bull Wapiti often show no sexual interest in females they happen to encounter outside the breeding season (even if the females are in heat at that time), while younger bulls do often show such interest. Yet homosexual activity in this species occurs in both age groups outside the breeding season—and therefore in neither case can it be due to a lack of access to females. Finally, in many Duck species that have a surplus of males, females *are* swamped by males trying to rape them—but in almost all cases, such males are *already paired* to females. Unpaired males without access to heterosexual partners rarely, if ever, engage in forced copulations with females (regardless of whether homosexual behavior also occurs in the species).[54]

Even same-sex activity that *does* have its genesis in the absence of opposite-sex partners—so-called situational homosexuality—often shows remarkable longevity and durability, rarely conforming to the stereotype of being "fragile" or liable to disintegrate once heterosexual mates are available. Captive animals that bond sexually with one another when opposite-sex partners are completely unavailable often resist later attempts to "convert" them to heterosexuality. They may even exhibit a longer-term "preference" for same-sex mates that outlasts their initial "situational" introduction to homosexuality. A pair of male White-fronted Amazon Parrots, for example, vigorously refused the advances of female birds even though their homosexual bond was formed "because" no females were available, while two female Long-eared Hedgehogs who were sexually involved with each other in the absence of males refused to mate heterosexually for up to two and a half years after they were separated (as mentioned earlier). The bonding in same-sex pairs of male Steller's Sea Eagles and female Barn Owls (housed without any heterosexual mates) was strong enough to enable successful coparenting of chicks, and in some cases the birds ignored subsequent introductions of opposite-sex partners. Male Rhesus Macaques, Crab-eating Macaques, Bottlenose Dolphins, Cheetahs, and Black-headed Gulls with homosexual bonds resist the attentions of opposite-sex partners or are clearly distressed when separated from one another, and/or they promptly renew their relationship on being reunited—often showing visible signs of affection and excitement when seeing their male partner again. This is also true for male Mallard Ducks that are raised together, in whom homosexual pairing typically becomes their lifetime "orientation." They consistently seek the company of other males even when opposite-sex mates are available and maintain their homosexual bonds year after year (or re-pair with males after the death of a partner) in spite of persistent overtures from females.[55]

The Contamination of Homosexuality

Of all the scientists who have advocated a shortage explanation for homosexuality, not one has ever specified a critical sex ratio that will consistently "induce" homosexuality, or a crucial threshold of members of the opposite sex that must be present in order to unfailingly prevent individuals from "resorting" to homosexuality. Is a mere 5 percent surplus of one sex enough to tip the scales? Apparently, since a population of Ring-billed Gulls with only 55 percent females is claimed to have enough of a skew to "cause" homosexual pairing. Yet a 5 percent excess of males in other species such as Greylag Geese is apparently not sufficient to "precipitate" homosexual pairing.[56] In fact, it is highly unlikely that a single critical sex ratio could ever be specified, because the proportion that "causes" homosexuality in one species (or population) has no such effect at all in other species, even where enormous "surpluses" (of, say, 80 percent or more of one sex) are concerned. More broadly, the underlying assumption behind the shortage hypothesis—that sex ratios actually *determine* a species' mating habits and social systems—has already been shown to be false for other types of mating behaviors. Scientists now recognize that there is not a clear, one-way causal relationship between how many males or females are available in a population, and the form that their mating system takes (e.g., polygamy as opposed to monogamy). Rather, a complex interplay of many factors is at work.[57] Unfortunately, the subtlety of this interaction is generally only recognized where heterosexual mating systems are concerned.

The shortage hypothesis is not only suspect on theoretical grounds, it is often applied to particular cases in a hasty or inconsistent fashion. Skewed sex ratios in animals exhibiting same-sex activity are often presumed without adequate supporting evidence, or else questionable "explanations" are proposed for the origin of such skewed ratios.[58] This is best illustrated by the species in which the shortage explanation is most prominent: Gulls. In the late 1970s and early 1980s scientists noticed that high levels of DDT and other environmental contaminants seemed to be associated with some populations of Western and Herring Gulls where nests contained supernormal clutches (often belonging to lesbian pairs). The following chain of "causation" was proposed to explain the apparent correlation: toxins (such as DDT) cause "feminization" of male Gull embryos, which in turn leads to female-biased sex ratios, which in turn results in lesbian pairs, who then attempt to breed, ultimately laying supernormal clutches.[59] Let's set aside for the moment the fact that this explanation is only of limited applicability—homosexual pairing is not associated with environmental toxins in over 70 other bird species, including several Gulls (e.g., Ring-billed Gulls, Common Gulls, and Kittiwakes).[60] Let's also set aside the fact that it is only of limited explanatory value—even if it could be shown conclusively that same-sex pairing results from skewed sex ratios that in turn result from toxins, the fact that only some species (and only some individuals in each species) respond to such conditions with homosexuality would still need to be addressed. Even for the Gull species where this explanation is supposedly relevant, however, each link in the overall chain is weak.

First, although laboratory experiments have shown that some toxins may cause male bird embryos to develop some ovarian tissue, no "feminized" male chicks or adults have actually been found in the wild among Western Gulls (or other species) living in contaminated areas.[61] Second, it is unlikely that toxin-induced feminization of males would result in populations with more breeding females than males (since it is most definitely *not* the case that toxins actually "convert" male embryos into female birds with fully functional ovaries). It would have to act either directly on the health of males, causing more deaths, or indirectly, by resulting in behavioral changes in males that would prevent them from mating with females. But there is no direct evidence that toxins cause anything beyond some physiological alterations in the reproductive organs of Gulls.[62] A higher mortality rate among males exposed to toxins has never been demonstrated, nor have behavioral differences among such males been observed that might lead them to forgo mating or otherwise to be "unavailable" as mates.[63] It has been suggested that "chemically sterilized" males simply fail to join the breeding colonies, or that such males are "no longer interested" in copulating heterosexually. Yet this begs the question of how or why sterility (or other physiological modifications) causes such males to exempt themselves from reproductive activities, or what exactly prevents them from pairing (or even copulating) with females even if they are sterile. Just because an animal is intersexed or transgendered (e.g., "feminized") does not necessarily mean that it is asexual or that its reproductive organs or behavior are "dysfunctional." "Masculinized" female Deer, Bears, and Spotted Hyenas, for example, regularly mate with males, give birth, and raise offspring—even though such individuals often have highly modified reproductive anatomies and hormonal profiles. Even when they are sterile, transgendered and intersexual animals in other species engage in courtship, copulation, and/or pair-bonding. Thus, it is overly simplistic to equate changes in reproductive physiology with an absence of sexual, pairing, or even procreative abilities. It should also be pointed out that a skewed sex ratio is not necessarily an "unnatural" result of environmental contamination: many Gull populations are in fact "naturally" biased in favor of females independently of the effects of toxins, owing to the overall higher survival rate of females (among other factors).[64]

Regarding the third link in this proposed chain: skewed sex ratios do not in fact automatically result in homosexual pairs—some populations of Western and Herring Gulls with a disproportionate number of females, for example, have few (if any) same-sex pairs.[65] Even in populations that do have a surplus of females, only a subset of the birds actually form homosexual pairs: most unpaired female Herring Gulls remain single (lesbian pairs constitute less than 3 percent of all pairs, and sometimes as few as 1 in 350), and some males remain unpaired even in populations with more females than males (indicating that some "extra" females are bypassing heterosexual mates). Granted, scientists were able to "induce" the formation of female pairs in a population of Ring-billed Gulls by removing males. However, this simply demonstrates that many females in this species have a latent bisexual capacity that manifests itself when males are in short supply—not that *all* same-sex pairs in this (or any other species) result from a shortage of the opposite sex. Moreover, the sex ratio that was required to "trigger" homosexual pairing (77 percent females) was much higher than the proportion of females in naturally occurring populations with homosexual pairs (55 percent females).[66] Apparently not all bird species have this latent bisexual capacity either (or at least not to the same extent), since homosexual pairing does not occur in most species that have their sex ratios experimentally manipulated. Removing males (or adding females) to populations of willow ptarmigans, bufflehead ducks, Pied Flycatchers, great tits, Brown-headed Cowbirds, and song, seaside, and savanna sparrows, for example, results in females mating polygamously with males or remaining single (among other strategies) rather than forming same-sex pairs. When alternate heterosexual behaviors such as polygamy are "induced" in usually monogamous species such as these, scientists do not interpret this as evidence that the behavior is somehow "artificial" or that its occurrence is due solely to the experimentally triggered demographic changes. Rather, it is taken to indicate that the species possesses an inherent capacity for polygamy (and more broadly, a flexibility of mating behavior), perhaps expressed at relatively low levels in most populations but manifesting itself on a larger scale under the appropriate conditions.[67] Significantly, this interpretation has not generally been afforded homosexual pairing.

Fourth, the evidence for homosexual pairing as a breeding strategy is slim. According to scientists, females bond with same-sex partners in order to raise young that result from copulation (but not pairing) with males (since two-parent care is generally required in these species). However, only a relatively small proportion of females in homosexual pairs actually mate with males and lay fertile eggs: 0–15 percent of Western Gull eggs belonging to female pairs are fertile, while only 4–30 percent of Herring Gulls' are fertilized, indicating that few such females are actually breeding.[68] Most importantly, females that could potentially benefit from same-sex pairing—were it a reproductive strategy—do not generally "avail" themselves of this option. Researchers found that unpaired Herring Gull females that copulate with males do not in fact go on to form homosexual pairs in order to raise any resulting offspring, nor do they even try to form such pairs. Likewise, Ring-billed Gull, Western Gull, and Roseate Tern mothers that have lost their male partners (and are otherwise unable to find another male) do not establish same-sex pair-bonds with available females, even though they supposedly need to find a new mate to assist them with parenting. In addition, some unpaired and homosexually paired Ring-billed females may actually lay eggs in the nests of other (heterosexual) pairs. This shows that: (*a*) single females need not seek pair-bonding (with birds of either sex) in order to have their young raised by two parents, and (*b*) at least some females in homosexual pairs lay eggs that they have no intention of caring for themselves.[69]

Lastly, there is not an absolute correlation between female pairs and supernormal clutches. True, in some species most lesbian pairs lay supernormal clutches, and most supernormal clutches belong to lesbian pairs. However, in many cases female pairs lay "normal"-sized clutches (or lose eggs so they end up with regular-sized clutches), while oversized clutches also regularly result from many other factors. These include egg stealing or adoption, supernumerary clutches laid by one female, nest-sharing by two heterosexual pairs, egg laying by outside females (not paired to the nest owners), and heterosexual trios, among others. In many gulls and other species, the connection between supernormal clutches and homosexual pairs has never been established (e.g., glaucous-winged gulls) or has been refuted (e.g., black-tailed gulls, brown noddies). Hence, studies that show correlations between toxins and increases in supernormal clutches cannot reliably be extrapolated to homosexual pairing unless it has been independently established that female pairs in that species lay larger than average clutches.[70]

Scientists also frequently point out a "correlation" between the two end points of this chain—toxins and supernormal clutches—without also providing evidence for all the intervening links.[71] To show conclusively a relationship between the two phenomena, all the intermediate sequences need to be established, and they should preferably be established *for the particular species in question.* Sometimes, extrapolations are made in these links *between species:* that is, toxins are shown to be in the environment of one Gull species, feminization from toxins in another Gull species, skewed sex ratios in a third, female homosexual pairs in a fourth, and supernormal clutches in others—yet rarely (if ever) have all these conditions been shown to co-exist in the same species or geographic area.[72] Moreover, in many Gull studies this chain is collapsed entirely or rendered circular. If homosexuality occurs in a species, and there is also evidence of contamination or pollutants in the environment, the two are automatically assumed to be linked. Homosexual pairing is regarded as a self-evidently "dysfunctional" phenomenon (typically characterized as "reproductive failure"), hence investigators often feel no need to address the actual details of occurrence or causation in the supposed link to toxins. Indeed, the very existence of homosexuality is often subtly equated with environmental contamination and disease even when no actual pollutants have been discovered in the population in question. Ultimately, female pairing is seen as more than simply a behavioral response to certain demographic parameters, which may or may not be indirectly traceable to certain chemical effects. Rather, it assumes the status of a pathological "symptom" directly induced by man-made toxins, symbolizing the larger havoc that people have wreaked on the environment—nature gone awry as a result of human meddling.[73] In the end, homosexuality becomes not merely the *result* of pollution, but the very "contamination" that is itself poisoning otherwise healthy—that is, purely heterosexual—species.

In summary, then, unavailability of the opposite sex is, at best, a tenuous "explanation" for the occurrence of animal homosexuality. Aside from having questionable theoretical and methodological underpinnings, this explanation is in many cases simply incompatible with the facts. In other cases, while same-sex activity does occur in contexts where opposite-sex partners are unavailable, many additional factors are involved, and many important questions concerning its occurrence remain to be addressed. Why, for example, do only some individuals or species with sex-skewed populations exhibit homosexual activity, while others manifest a wide variety of alternative behavioral responses? And why have social systems that entail sex segregation or skewed sex ratios—and hence that supposedly "favor" homosexual activities—evolved in the first place, and in so many species? Where it is relevant, unavailability of the opposite sex should be seen as only one of many contributing factors—and the *beginning* of further study of other more complex issues surrounding the occurrence of homosexuality in animals. Unfortunately, this explanation continues to be offered as a *final* scientific pronouncement on the "cause" of same-sex activity. Not only does this do a disservice to the actual richness of animal behavior, it effectively discourages further investigation of a phenomenon whose true intricacies are just beginning to be understood.

"The Errors of Their Ways"—Homosexuality as Mistaken Sex Identification

One surprisingly common scientific "explanation" for the occurrence of animal homosexuality is that it simply results from an inability on the part of animals to "properly" differentiate males from females, or else it represents an "indiscriminate" mating urge (i.e., any perceived differences between the sexes are ignored). This explanation is common for some "lower" animals such as insects and amphibians, where there is limited evidence that mating may indeed be random between homosexual and heterosexual.[74] However, this type of "indiscriminatory" mating or mistaken sex identification has also been proposed for higher animals, including more than 55 mammals and birds—mostly species in which adult males and females superficially resemble each other (e.g., Cliff Swallows), or in which adolescent or juvenile males supposedly resemble adult females (e.g., Blackbucks, Birds of Paradise).

The gist of this explanation is that when animals engage in homosexuality they are just "making a mistake"—they intend to mate heterosexually, but simply misidentify the sex of their partner because of the physical resemblance between the sexes. Indeed, homosexual interactions are explicitly labeled as "mistakes" or "errors" in several species. Male Cock-of-the-Rock who mount other males have actually been described as "confused" and "bumbling"; the "aberrant sexual behavior" of male Giraffes who mount each other is attributed to their "muddled reflexes"; Black-billed Magpies are characterized as "confused" when they engage in "misdirected" courtship activity with birds of the same sex; and one scientist even suggested that same-sex courtship in Mountain Sheep would probably never occur if males could properly distinguish females from young males.[75] Often, the very existence of homosexuality in a species is taken to be "proof" that the animals cannot distinguish males from females: "In many waders the sexes are difficult to distinguish, not only to the observer, but on occasions to the birds themselves, as records of males attempting to copulate with other males have been recorded." The circularity in this line of reasoning is blatant, since usually no further evidence is offered to indicate that sex misrecognition is prevalent in the species.[76] Conversely, the absence of homosexuality in species such as yellow-eyed penguins and its infrequency in Silvery Grebes and Red-faced Lovebirds is offered by scientists as evidence that there are no "problems" with sex recognition in these species.[77]

Bumbling and Confused?

Quite clearly, sex misrecognition cannot be a widespread "cause" of homosexuality in animals. Same-sex courtship, copulation, and/or pair-bonding occur in numerous species in which males and females look very different from each other: many primates and hoofed mammals, for example, and birds as varied as Ostriches, Grouse, Black-rumped Flameback Woodpeckers, and Scottish Crossbills, to name just a few. Conversely, homosexuality is not found in many animals in which males and females are visually indistinguishable. For example, same-sex activities are not reported for any of the 31 species of North American perching birds in which younger males significantly resemble adult females, while homosexuality occurs in only a small fraction of the hundreds (if not thousands) of birds in which adult males and females are identical to each other.[78] Moreover, in the majority of species where homosexuality is attributed to mistaken sex identification, only *one sex* is involved in homosexual activity (usually males). If the animals truly could not tell males and females apart, we would expect both sexes to participate in homosexuality at comparable rates—unless, of course, only one sex has trouble identifying the other, which seems improbable. Furthermore, in many species where homosexual interactions between adult and adolescent males are attributed to the resemblance of the younger males to females, homosexuality also occurs between adults or older males, or between females, where sex misrecognition is not likely. This is true for Blackbucks, Mountain Goats, Elephant Seals, Bishop Birds, Swallow-tailed Manakins, and Superb Lyrebirds, among others. Adult-adolescent homosexuality also occurs in many species where younger males do not resemble females, or between females (where neither partner specifically resembles a male).

In some mammals and birds where homosexuality is attributed to the resemblance between younger males and adult females (e.g., Blackbucks, Manakins, Birds of Paradise), the two sexes are not necessarily identical. Rather, older adolescent and younger adult males exhibit physical characteristics that are actually *intermediate* between those of adult females and adult males, and they are often recognizably male.[79] Even in species where homosexuality is claimed to result from the *identical* appearance of males and females, there are often slight but noticeable physical differences between the sexes that may be discernible to individuals. These include

▲ *An adult male Blackbuck courting a younger male by "presenting the throat," a stylized courtship display. Some scientists have suggested that homosexual activity in this species is triggered by the resemblance between younger males and adult females (e.g., their lighter coat color), yet younger bucks are clearly identifiable as male because of their horns and other anatomical features.*

body and horn size in Mountain Goats, wing length in Bishop Birds (with juvenile males distinct from adult females), iris color and other aspects of eye structure in Galahs, relative size and other body measurements in Humboldt and King Penguins, patterning of tail feathers in male and female (and between adult female and juvenile male) Superb Lyrebirds, wing and tail length (and, in some populations, wing feather notching) in Ocher-bellied Flycatchers, presence of a brown forehead patch and shorter wings in female Tree Swallows, and bill structure and tail coloration between adult female and juvenile male Anna's Hummingbirds.[80]

Are these (often subtle) differences actually perceptible to the animals themselves? Implicit in many scientists' pronouncements of sex misrecognition is the assumption that just because males and females look alike to our eyes, they must be indistinguishable to the animals as well. Species differ widely in their visual acuity, color perception, and other sensory abilities, so each case needs to be evaluated individually before any conclusions can be made about animals' sex recognition abilities—and this has most definitely not been systematically investigated for cases involving animal homosexuality. Nevertheless, one thing is certain: we are only beginning to understand many aspects of animal perception, including heretofore unimagined powers of visual, acoustic, and temporal recognition. Scientists recently discovered, for example, that a number of birds such as starlings, Zebra Finches, bluethroats, and Blue Tits use ultraviolet vision in distinguishing between individuals and between sexes. Birds that appear identical in ordinary light have different patterns under UV that are recognized and used by other members of their species to choose mates. Likewise, males and females of some butterfly species that are indistinguishable to us have radically different appearances in UV light. In the acoustic and temporal realms, analysis of tape recordings of Lyrebird vocal mimicry has revealed that their perception of time may be ten times greater than that of humans, giving them the extraordinary ability to imitate the calls of five different birds simultaneously.[81] It is quite likely, then, that animals can perceive differences in appearance or other minute sensory cues that are distinguishable only to human measuring instruments and not to human eyes (or ears).

Further evidence that animals can differentiate between males and females that appear identical to us comes from the different frequencies of homosexual and heterosexual interactions in species with "indistinguishable" sexes. Animals often preferentially court, mate, or bond with individuals of one or the other sex. Male Mountain Goats. for example, court male yearlings more frequently than female yearlings even though they are supposedly "unable" to differentiate between the two. The opposite scenario occurs in Musk-oxen: although adult males court both yearling males and females, they interact with females more than with males. Likewise, Dwarf Cavy adult males court juvenile males more often than they do juvenile females (and even seek out specific male partners). In contrast, adult males in the closely related Aperea court only juvenile females and never males, even though in both of these species juvenile males and females are purportedly indistinguishable. Among Bighorn Sheep, rams are claimed to be sexually interested in other males in direct proportion to how closely the latter resemble females—yet yearling males, which resemble females the most, still receive far less sexual attention than do females, indicating that some form of sexual differentiation still occurs. Although male Common Murres are said to mount other males because they have difficulty distinguishing the sexes, females are still mounted at a much higher rate than males. Supposedly "indiscriminate" sexual chases by male Flamingos actually involve many more pursuits of females than males. Finally, adult male Pronghorns court and mount yearling and two-year-old males, both of whom superficially resemble females. However, adults actually direct more sexual behavior toward two-year-olds, who are as "femalelike" as (if not more "malelike" than) yearlings in terms of the size of their horns and black cheek patches.[82]

A related argument against sex misrecognition as a factor in precipitating homosexuality is that males and females are often *behaviorally* distinct even when they are physically identical. A male who "looks like" a female will frequently perform identifiably male behavior patterns during a homosexual interaction, seriously casting doubt on the notion that his partner has failed to recognize his actual sex. Male Antbirds "mistaken" for females actually initiate and reciprocate courtship feeding with their male partners (something females never do). Younger male Swallow-tailed Manakins and Regent Bowerbirds that participate in courtship with adult males may physically resemble females, but they exhibit distinctly "masculine" behaviors, displays, or vocalizations. Among male Greenshanks, both participants in homosexual copulations display in a typically male fashion prior to same-sex mounting, and both female partners in Jackdaw homosexual pairs preen each other (a typically female activity). There is hardly a more identifiably "male" activity than copulating with a female, yet male Laughing Gulls sometimes make sexual advances toward males who are in turn mating with their female partners (creating a three-bird "pile-up"). Nor are such homosexual mounts simply attempts by the topmost male to mate with the female, since he often remains mounted on the other male (or continues to remount him repeatedly) even after the female becomes "available" once her partner dismounts. Conversely, there is hardly a more definitively "female" activity for birds than laying eggs, yet male Black-headed Gulls have been observed bypassing females *in the very act* of laying an egg in order to try to copulate with her male partner![83] It seems highly unlikely that homosexual activity in a case such as this is due to faulty sex recognition (especially since heterosexual copulation attempts on laying or incubating females are fairly routine among Gulls), yet this is a prominent "explanation" for same-sex behavior in this species.

Deceptively Clear

In many animals where only a subset of the population resembles the opposite sex (or is transgendered), the occurrence of homosexuality is often directly counter to what would be expected if confusion between the sexes were "causing" homosexual behavior. For example, adolescent male Scottish Crossbills resemble females in their plumage coloration, yet homosexual pairs in this species form between adult males, not between adult and juvenile males. In Ruffs, some males resemble females in that they lack the elaborate neck feathers and other distinctive plumage characteristics of other males, yet homosexuality in this species is not limited to these "naked-nape" males. Males who do not resemble females also court and mount each other, while "femalelike" males often mount more "masculine" males. Tree Swallows are unusual among North American perching birds in that females retain the drab gray-brown plumage of adolescence during the first year that they breed, making them resemble adolescent males more than adult females. Thus, one would expect that either (*a*) adult males would be more apt to "mistake" brown-plumaged females for males, perhaps responding more aggressively to them (i.e., as if they were males); or (*b*) homosexuality in this species would manifest itself as an age-based system, with males pursuing only younger brown-plumaged males because they "mistake" them for first-year breeding females. Neither of these scenarios is true, however: males have no trouble recognizing the sex of brown-plumaged females (and in fact are significantly less aggressive toward them), and homosexuality in this species involves adult males interacting with each other, not adults being "confused" by brown-plumaged males.[84]

Black-headed Gull males and females are nearly identical in appearance, except that males have, on average, slightly longer heads and bills than females. However, some males are more "femalelike" in that they have shorter head and bill lengths than average. If sex misrecognition were operative in this species, one would predict that smaller males (i.e., birds who more closely resemble females in size) would be more likely to form homosexual pairs (since males would "mistake" them for females) and less likely to form heterosexual pairs (since females would "mistake" them for other females). On the contrary, scientists studying sex recognition in this species found that female-resembling males are just as likely to form heterosexual as homosexual pair-bonds. In fact, smaller males are *more successful* at maintaining long-lasting heterosexual bonds and fathering chicks than more "masculine"-appearing males—paralleling other cases of greater heterosexual prowess in some transgendered animals.[85]

Other species in which both transgender and homosexuality occur are particularly cogent examples of how ineffective sex misrecognition is in "explaining" homosexuality. Typically, the patterns of same-sex and opposite-sex interactions in these species do not follow the clear divisions that would be expected if individuals were simply "mistaking" their partners for the opposite sex. In Hooded Warblers, for example, some females have transvestite plumage, appearing almost identical to males because of their dark hoods (which are usually found only in males). Others have intermediate plumage, darker or more melanistic than most females but without the complete hood pattern of males, while others have no "malelike" head feathers at all. Males, though, are typically heavier and have longer wings than females, hooded or otherwise. It has been suggested that male homosexual pairs initially form in this species because of the visual resemblance between some females (transvestites) and males. Yet if males in homosexual pairs tended to confuse hooded females with males, one might expect them to pair with individuals whose sex is especially "blurred" or hard to decipher: darker, more malelike females and/or smaller, more femalelike males. However, at least one bisexual male chose just the opposite kinds of mates. His male partner did not have female body proportions but, on the contrary, was exceptionally "masculine" in this regard, exceeding the average weight and wing lengths of most males. Conversely, his heterosexual pairings involved "obviously" female partners, i.e., nontransvestite or only moderately melanistic individuals. Moreover, males that are supposedly mistaken for females in homosexual pairings do not develop brood patches (a distinctive bare patch of skin on the belly used for incubating eggs, characteristic only of females). So it is unlikely that such males are mistaken for hooded females.[86] There is also evidence that male Hooded Warblers do not generally confuse transvestite or melanistic females with males. First of all, males are differentially aggressive toward other males, attacking them during territorial encounters more often and ignoring them less often than they do the darkest, most malelike females. Furthermore, "masculine-appearing" (melanistic) females are generally as successful as nontransvestite females in finding male partners and are as subject to promiscuous copulation attempts by males as are nontransvestite females.[87] If males tended to confuse hooded females with males, they would probably avoid darker birds (including melanistic females) during heterosexual mating interactions (since such birds would more likely be other males), yet this does not appear to be the case.

Transgendered Hooded Warblers: females of this species usually have little or no black on their heads (far left), but some individuals are plumage transvestites, exhibiting a full malelike black hood and chin strap (far right). Other females exhibit a gradation of plumage patterns that fall between these two extremes (center). ▼

Even in species where some individuals clearly are "tricked" into same-sex relations by transgendered animals, the situation is considerably more complex than this. In Common Garter Snakes, for example, some males produce a pheromone that is similar to the scent of females. These individuals are called she-males by scientists, and they attract as many male suitors as female snakes do. Most males who court she-males are apparently "deceived" into thinking they are interacting with a genetic female. However, she-males and genetic females are not identical: chemical analysis has shown that the pheromones of she-males, rather than being indistinguishable from those of females, are actually intermediate between those of males and females. When given a choice, most nontransvestite males prefer genetic females—demonstrating that they can distinguish between the two under the appropriate circumstances. Moreover, nontransvestite males sometimes abandon their courtship of females to pursue she-males, and up to 20 percent of males may actually prefer courting she-males rather than females when given a choice—indicating that not all individuals who interact with transgendered snakes do so entirely under "false pretenses." Even though their pheromones resemble females', she-males also have no trouble finding opposite-sex partners—in fact, some studies indicate that they may be more successful in mating with females than males who are not transvestite (she-males actually have more than three times as much testosterone as do males). In addition, male Garter Snakes also occasionally court each other in situations that do not appear to involve transvestism—and therefore not all same-sex interactions can be attributed to (transgender-induced) mistaken sex identification. In many other species where a subset of the population is transgendered, homosexuality does not occur at all, and transgendered individuals again have no difficulty in attracting mates of the opposite sex. This is true for female red-winged blackbirds that have malelike epaulets, female Pied Flycatchers that have the white forehead patches characteristic of males, female lesser kestrels that have male rump and tail coloration, and younger male long-tailed manakins whose plumage resembles that of females.[88] If sex misrecognition were a "cause" of homosexual pairing, one would expect same-sex pairing, courtship, or mounting to be prevalent in these species as a result of "confusion" between transvestite individuals and members of the opposite sex. One would also expect transvestite individuals to be avoided by members of the opposite sex because they do not resemble "obviously" heterosexual partners. Once again, neither of these scenarios generally occurs.

Another problem with attributing homosexual interactions to mistaken sex identification is that it can (at most) account for the initial interest of one animal in another of the same sex. It cannot explain why the animal "mistaken" for the opposite sex often willingly participates in the homosexual interaction or may even initiate it. Even if homosexual pairs in Antbirds, for example, result from an initial failure on the part of a courting male to distinguish between the sexes (as has been claimed), such pairs could not persist for years unless both males were actively fostering the bond between them (or at the very least, not resisting the homosexual relationship). As scientists studying homosexual matings in Tree Swallows have pointed out, even if males mistake other males for females (which is not likely), the males they copulate with nevertheless do not resist their homosexual advances and even actively facilitate genital contact. Notably, they do not adopt the specific tactics used by birds in this species to deter unwanted sexual advances (typically displayed by females in heterosexual contexts). While male Black-crowned Night Herons may court males and females indiscriminately, their male partners are nevertheless sexually stimulated by the performance and may go on to form a homosexual pair-bond with them. In Regent Bowerbirds, "female-resembling" adolescent males may actually initiate courtship display toward adult males (the reverse of the usual scenario in cases of "mistaken" sex identification). Finally, male Greenshanks who visit other males' territories and are "mistaken" for females actively precipitate homosexual courtship pursuits: they depart from the territory using a special swerving flight pattern that invites the other male to follow them (also used by females during heterosexual courtships). If they did not want to spark a homosexual courtship, they could simply employ any of the several strategies used by females to deter males' advances in this species, such as leaving the territory in a direct flight path or "leapfrogging" over a pursuing male during a ground chase—yet these are not typically part of homosexual interactions.[89] Thus, even if mistaken sex recognition is responsible for bringing two animals of the same sex together, it is ultimately irrelevant in explaining why those two animals often remain together to continue their interaction and bring it to its full conclusion, be it a completed courtship or mating episode, or a pair-bond lasting many years.[90]

In summary, a whole host of considerations cast serious doubt on mistaken sex recognition or indiscriminate mating as an explanation with wide applicability (or credibility). Once again, the complexities of animal behavior elude the broad brushstrokes of human interpretation. Numerous interconnected elements must be factored in, such as the subtleties of actual physical differences between the sexes, the strength and acuity of animals' various perceptual abilities, differential behaviors between males and females, the active participation of individuals "mistaken" for the opposite sex, and the intricacies that arise when transgender is layered over homosexuality. In the end, the most significant "misrecognition" is probably not that of animals who overlook each other's sex, but that of scientists who fail to recognize the importance and interplay of these factors. Nevertheless, *even if* mating or courtship in some species is in fact random or indiscriminate between males and females, such "randomness" is actually compelling evidence (once again) for a bisexual capacity in such creatures. This in itself is a vital observation that is frequently downplayed by scientists, who all too readily discount the homosexual part of this mating equation as a necessary "error" made by animals on their path to achieving greater heterosexual output. In such a mechanistic view, animals simply mate with as many partners as they can—male or female—to maximize their reproductive success, even if it means that some of their matings will be nonreproductive. The fact remains, however, that such animals have the ability to respond sexually to individuals of their own sex—and they do so repeatedly, with apparent enthusiasm, and (one might add) noticeable disregard for the "mistakes" they are making.

"Gross Abnormalities of Behavior"— Homosexuality as Pathology

Homosexuality in animals has frequently been regarded as a pathological condition. Such terms as *abnormal* and *aberrant* are routinely applied to this phenomenon (as mentioned in chapter 3), often with no further justification or explication—homosexuality is considered sufficient in itself to warrant the label of disease, disorder, dysfunction, or deviance. A number of researchers, however, are more specific in their pathologizing of homosexuality and transgender, and in this section we'll examine two of the principal "explanations" of this sort that have been put forward: the claims that homosexuality is caused by the artificial conditions of captivity, and that homosexuality/transgender is the manifestation of a physiological abnormality.

Something Amiss at the Zoos

For a long time, scientists discounted examples of animal homosexuality because some of the earliest descriptions were based on captive animals. In many cases, biologists continue to classify this behavior as "abnormal" and attribute it to the "unnatural" circumstances of confinement or contact with humans. One scientist, for example, writes of homosexual pairs in Swans (as well as other "sexual aberrations" such as heterosexual trios and interspecies matings): "Captive swans, like many other animals, sometimes show gross abnormalities of behavior. These are due almost entirely to the artificial conditions under which the birds are kept."[91] As recently as 1991, homosexuality in Wattled Starlings was ascribed to their captivity. Other species for which similar "explanations" have been proposed (including appeals to factors such as crowding and/or stress in captivity) include Common Chimpanzees, Gorillas, Stumptail Macaques, Musk-oxen, Koalas, Long-eared Hedgehogs, Vampire Bats, and Black-crowned Night Herons.[92] Sometimes the only context where same-sex activity is discussed is to exemplify the types of "pathologies" that arise in captivity. Homosexuality in Dolphins, for instance, was offered as an illustration of the "sexual aberrancy" that can result from confinement in aquariums, while a case of female coparenting in Barn Owls was included in a report on "Abnormal and Maladaptive Behavior in Captive Raptors"—part of a monograph on (of all things) *diseases* in birds of prey. Homosexual activity in Rhesus Macaques was even presented (along with a number of other "abnormal" behaviors) as an illustration of the deleterious effects of malnutrition.[93] In a perfect example of the sort of circular reasoning that is employed in many scientific discussions, homosexuality in captive animals is often cited as the "proof" of the artificiality of their captive conditions. One zoologist proclaims, "Homosexual behavior [in Cheetahs] . . . is reported in zoos as quite frequent, which to me indicates that something is amiss at the zoos," while another states, "The very occurrence of female-female pairs [in Zebra Finches] suggests behavioral pathology."[94] This is chillingly reminiscent of the not-so-distant "medicalized" views of human homosexuality, where the mere existence of same-sex attraction or activity was sufficient to "diagnose" pathology or mental illness.

While it is true that captivity sometimes does induce unusual behaviors in animals, the bulk of the evidence does not support this as a "cause" of animal homosexuality. As primatologist Linda Fedigan observes, "Although . . . homosexual relationships in animals *can* occur as a result of stressful captive conditions, we would suggest that all such behavior should not be dismissed as pathological or dysfunctional, a practice which results in 'explaining it away' rather than explaining it."[95] On statistical grounds alone there is no substantiation for a greater incidence of homosexuality in captive animals—in fact, just the reverse is true. In more than 60 percent of the mammals and birds in which same-sex activity has been documented, this behavior occurs spontaneously in the wild. In more than two-thirds of these species, homosexuality has *only* been observed in the wild, while in the remaining cases it occurs in both wild and captive animals.[96] A number of scientists have remarked on a higher *rate* of homosexual activity in captivity compared to in the wild when the behavior occurs in both situations. In other words, there may be a quantitative, rather than qualitative, difference between wild and captive conditions, although the occurrence of homosexuality itself cannot be attributed to confinement. However, even this difference is less than clear-cut. In some species such as Orang-utans, Hamadryas Baboons, Mule Deer, and Musk-oxen, there does indeed appear to be a higher rate of homosexual courtship and/or sexual activity—as well as heterosexual activity—in captivity compared to the wild, although in some instances this is based on impressionistic observations.[97] In contrast, two species for which detailed quantitative information is available show nearly identical rates of same-sex activity in the wild versus captivity: in Bonobos, studies of wild animals have revealed that 45–46 percent of all sexual activity is homosexual, while a captive study yielded a figure of 49 percent; in Black Swans, one investigator found that 5 percent of captive pairs were homosexual while 6 percent of wild ones were.[98]

Failure to observe homosexuality in the wild is more often due to incomplete study or inadequate observational techniques rather than an actual absence of the behavior in free-ranging animals. Time and again, same-sex activity has initially been seen only in captive animals and therefore declared to be definitively not a part of the "normal" sexual repertoire of the species in the wild. Yet when detailed field studies of the same species are finally conducted—often decades later—homosexuality is inevitably discovered. In fact, so pervasive or routine is the behavior now known to be for some species in the wild, that scientists have had to completely revise prior assessments of same-sex activity as "artificial" for these animals in captivity. In Bottlenose Dolphins, for example, male pairs engaging in homosexual behavior were originally observed in aquariums and were considered to be the "aberrant" result of keeping males together without females. Detailed longitudinal and demographic studies of the species—more than forty years later—revealed that male pairs, as well as sex segregation, are a prominent feature of the social organization of this species in the wild. By 1998, zoologists were actually advocating that captive male Bottlenose Dolphins be kept (and reintroduced into the wild) as bonded pairs, recognizing that these constitute a "natural functional social unit" of the species that can assist captive individuals in adjusting to life in the wild upon their release. Another example of a complete turnaround on the part of scientists concerns Gorillas. Early studies of this species reported that homosexuality was not seen in wild Gorillas; three decades later, extensive same-sex activity had been documented in both males and females in the mountain forests of Africa. By 1996, biologists and zookeepers were (at last) openly acknowledging that homosexuality in all-male groups was not an "artificial construct of captivity," and were even encouraging the formation of such groups in zoos to approximate the species' natural social patterns.[99]

Many other examples of field studies confirming earlier captive observations of homosexuality (and disproving initial assessments of its "artificiality") can be found. In 1935, Konrad Lorenz asserted that the formation of same-sex pairs in female Jackdaws "does not appear to occur under natural conditions"; it wasn't until more than forty years later that ornithologists confirmed the occurrence of homosexual pair-bonding in wild Jackdaws. Same-sex activities between male Elephants in captivity were first reported in the scientific literature in 1892 and characterized as "aberrations" and "perversions"; almost 75 years later, similar and more extensive homosexual interactions were documented among wild Elephants. In 1997 zoologists presented the first descriptions of same-sex activities in wild Crested Black Macaques, finally confirming captive observations made more than thirty years earlier. Because no detailed field studies of this species had been conducted before the 1990s, all prior reports of homosexual activity were based on observations in captivity, leading some scientists to suggest that same-sex activity was not likely to be found in wild Crested Black Macaques—a prediction we now know was incorrect. Homosexual pairing in Parrots was long considered to be "induced or brought forth by the conditions of confinement," but in 1966 an ornithologist documented a male pair of Orange-fronted Parakeets in the forests of Nicaragua—the first confirmation of homosexuality in wild Parrots. Ironically, the sex of the birds was verified only because the scientist mistook them for a heterosexual pair copulating unusually early in the breeding season (and therefore he wanted to check the condition of their internal reproductive organs). Initial observations of homosexuality in captive female Lions, made in 1942, were confirmed in the wild in 1981, while observations of male pairs in wild Great Cormorants in 1992 corroborated early observations of this phenomenon among zoo birds in 1949. Likewise, same-sex courtship in Regent Bowerbirds was first described on the basis of aviary observations in 1905, but display between wild males was not documented until nearly a century later. And homosexual activity between different species of Dolphins, long observed in aquariums, was finally verified in a wild population in 1997.[100] Today, homosexuality in many species is still known only from captive studies, but it is likely that most, if not all, of these will follow this same pattern and eventually be confirmed by field studies. Perhaps it is finally time for scientists to acknowledge that homosexuality in captive animals is nearly always an expression of their normal behavioral repertoire, rather than a result of their captivity.

Another point to keep in mind is that the distinction *wild* versus *captive* is in some sense a false dichotomy, since in actuality there is a continuum of degrees of confinement, "artificiality," and human intervention in the living conditions of animals. At one extreme are truly "wild" animals that have experienced no, or virtually no, contact with humans—an increasingly rare phenomenon in the contemporary world. At the other extreme are domesticated animals that have been bred and raised in captivity for many generations, often to the point of being genetically distinct (as a separate species or subspecies) from their wild counterparts. In between, there is a whole spectrum of contexts and factors. Toward the more "wild" end of the continuum, there are free-ranging species that have nevertheless experienced varying degrees of human contact or interference, such as Killer Whale populations that have been heavily poached, or wild Tree Swallows that nest in colonies of human-supplied nest boxes, or Grizzly Bears living in "disturbed" habitats, or wild Atlantic Spotted Dolphins that are habituated to the presence of people.[101] There are also *semi-wild* animals, a cover term that includes a host of different situations. For instance, unconfined animals may nevertheless be tame (e.g., Greylag Geese), while animals on reserves may be wild or "free-ranging" within a confined but extensive territory, often hundreds or thousands of acres in size (e.g., Bison, Cheetahs). Transplanted populations consist of entire troops or herds that have been moved, their social organization and demographics intact, to a new (often more restricted) environment, sometimes because they are endangered in their natural habitat (e.g., Rhesus Macaque troops transplanted from India to Puerto Rico, Blackbuck herds moved from India to France). Another semi-wild situation involves animals that are recently extinct in the wild and therefore can *only* be observed in captivity (e.g., Takhi, Père David's Deer). In many cases such species are kept in conditions that approximate their "former" wild habitat and social organization as closely as possible, and in some instances they are even being slowly reintroduced to the wild from their captive populations. "Provisioned" animals are wild but supplied with food and varying amounts of human contact (e.g., Japanese Macaques), while "rehabilitated" animals are formerly captive (and possibly wild-born), but reintegrated into wild populations (e.g., Orang-utans). Finally, feral species are domesticated animals that have escaped and "gone wild," establishing their own free-ranging populations (e.g., Water Buffalo, Mute Swans, Rock Doves).

Among captive animals, a wide variety of factors—each of which represents a continuum of its own—must also be considered when assessing the "artificiality" of their confinement: Are the animals wild-born or raised in captivity? Are they tame and/or trained, or do they have little or no contact with humans? Are they free-ranging within an outdoor enclosure (e.g., at a zoo or wild-animal park) or are they kept in restrictive cages (e.g., in a laboratory)? Are they living in mixed-sex groups or sex-segregated groups—and which is typical of wild populations? How closely does their social organization in captivity approximate that of wild animals, in terms of the size and number of social groups, the sex and ages of the animals making up those groups, and the transiency or stability of such groups? With regard to the occurrence of homosexuality, virtually every situation along this continuum—and every semi-wild context—has been claimed to be "artificial" to one degree or another, or else "natural" enough not to warrant concern. Yet the fact that homosexuality has been observed in virtually every one of these contexts argues for the relative independence of this behavior from whatever conditions of "captivity" or "wildness" may prevail. Moreover, what constitutes a "natural" captive context is often directly counter to preconceived ideas. Regarding Cheetahs, for example, a number of researchers have commented that keeping males and females together in captivity is actually something of an "artificial" situation (since it seems to contribute to an inhibition of heterosexual courtship and mating); conversely, same-sex pair-bonds appear to be integral to the "psychosocial well-being" of males. Ironically then, sex segregation in captivity is actually more "natural" for this species, since it reflects the Cheetah's social organization in the wild (a situation that is also true for many other species in which males and females typically live apart from each other).[102]

It should also be pointed out that something of a double standard exists regarding what is interpreted as "natural" as opposed to "captivity-induced" behavior. It is common zoological practice, for example, to study mate choice in pair-bonding species (such as birds) by setting up captive situations where individuals are only given a "choice" of opposite-sex partners. It is also standard practice to keep zoo animals strictly in heterosexual pairs for breeding purposes. Thus, a sizable portion of reported "heterosexual" behavior is in fact based on situations that would be considered "artificial" if they were used to study homosexual behavior. In other words, if animals are kept only with members of their own sex and then subsequently exhibit homosexual activity, this is overwhelmingly interpreted as "situational" behavior that would not otherwise happen. In contrast, if they are only given access to opposite-sex partners and subsequently exhibit heterosexual behavior, this is without exception interpreted as an expression of their "natural" tendencies. Although researchers readily regard homosexuality to be the result of external or artificial factors operating on otherwise heterosexual animals, no one has dared suggest that the reverse situation might also sometimes occur—that heterosexuality could be "forced" on otherwise homosexual (or largely same-sex-oriented) animals. In fact, zoos and other captive breeding programs offer countless reports of animals "failing" to breed in captivity for no apparent reason when placed with opposite-sex partners. Even after exhausting the long list of factors that could be involved, animal breeders uniformly overlook the possibility that some of these individuals may simply have a preference for same-sex activities and/or partners.

In the majority of species where homosexuality has only been observed in captive or semi-wild conditions, researchers have confirmed that other aspects of behavior or social organization in captivity—including sexual behaviors—are comparable to those of wild animals. In some instances, behaviors once considered to be "abnormal," "artificial," or "unusual" products of captivity have also been documented in the wild. For example, Botos frequently play with man-made objects in aquariums (carrying and manipulating rings, brushes, and so on) and also interact playfully with animals of other species kept in their tanks. Wild Botos have also been observed in similar behaviors, playing with sticks, logs, fruit pods, and even fishermen's paddles, as well as with other species such as river turtles. Tool use and manufacture by Orang-utans had long been known from studies of captive and semi-wild animals, but until the behavior was documented in wild Orang-utans in 1993, it was considered typical only of "artificial" situations. One researcher, studying captive Savanna Baboons, asserted that "certain types of behavior such as copulation during pregnancy or lactation may be related to caged life, and not be the norm in natural populations," yet later studies of wild populations revealed that these behaviors do in fact occur regularly. Likewise, until it was documented in the wild, cross-species herding behavior by male Thomson's Gazelles was thought to be caused by the unavailability of same-species groupings in captivity. Parenting trios, mate-switching, promiscuous copulations, and egg stealing were all initially observed in captive King Penguins and considered to be "unusual" (if not "abnormal") behaviors. Yet detailed study of this species in the wild nearly thirty years later verified the occurrence of each one of these activities, as well as many other "unexpected" behavioral patterns. In a few cases, a more "unusual" behavior has only been documented in wild populations, or else is more prevalent in the field than in captivity: for example, reverse mounting in Black-headed Gulls and divorce in Flamingos.[103] Thus, while homosexuality has not yet been observed in many of these species in the wild, it is probably only a matter of time before it is.

Other situations involving homosexuality in captive animals also occur. Often same-sex activity in one species has only been observed in captivity (e.g., Siamangs, Mute Swans, Sociable Weavers), yet a closely related animal does exhibit similar or identical behavior in the wild (e.g., White-handed Gibbons, Black Swans, and Gray-capped Social Weavers, respectively). In other cases, one form of homosexuality is seen in captivity and another form in the wild. In Griffon Vultures, for example, homosexual pairs and sexual activity have been observed in captivity while same-sex courtship and pair-bonding display flights have been seen in the wild. In Emus, sexual activity between males has been documented in captivity and male coparenting in the wild. In Galahs, homosexual pairs have been observed extensively in captivity but not in the wild, although "supernormal clutches"—nests with double the number of eggs, typical of female pairs in other birds—have been verified in the field. And in Cheetahs, same-sex courtship and sexual activity have been seen in captivity while male pair-bonds have been observed in both wild and captive animals. This suggests that the absence of certain behaviors in studies of wild animals are probably accidental "gaps" that will be filled once more extensive field studies are conducted. This is particularly likely when one considers that the proper observational techniques for identifying homosexual activity are often not employed, even in species where same-sex activity has previously been verified in captivity. In the most recent ongoing field studies of Griffon Vultures, for example, the sex of birds is determined "behaviorally" by their position during mounting (top bird = male, bottom bird = female), or not verified at all, thereby precluding the possibility of detecting homosexual pairs. "Behavioral" sexing has also been employed in the major long-running studies of large populations of wild King Penguins, Gentoo Penguins, and Flamingos—in some cases combined with "morphological" sexing, i.e., the larger bird in a pair is assumed to be male and the smaller female, without actual verification of sex—all species in which same-sex pairs have been observed in captivity but not yet documented in the wild. And the sex of wild Dugongs participating in mating behavior has never been unequivocally determined in nearly two decades of field observations; researchers invariably assume that the interactions are heterosexual, even though same-sex activity has been observed in captivity (and in the wild in the related West Indian Manatee).[104]

It must also be remembered that it is often extremely difficult to observe some species in the wild or obtain detailed information about their behavior. Many animals in which homosexuality has only been seen in captivity present formidable challenges to field study. Some are nocturnal (active only at night) or crepuscular (active at dusk or dawn), such as Lesser Bushbabies (and other Lemurs), Wolves, Rufous Bettongs, and Black-crowned Night Herons. Others are diurnal (active in the daytime) but engage in sexual behavior mostly at night (e.g., Red Deer). This can greatly hamper efforts to observe sexual activity: homosexual mounting in Red Foxes, for example, was only discovered by setting up remote-control infrared video cameras to continuously monitor nighttime activities in a captive population—virtually impossible to do under field conditions. Other species are highly elusive: Bush Dogs, for example, have rarely even been sighted in the wild, let alone studied, and the most complete analysis of their social organization in captivity was only published in 1996. Likewise, the elusiveness of Pig-tailed Macaques precluded detailed field observations until the early 1990s, while the first in-depth behavioral studies of wild Crested Black Macaques were published in 1997. Sometimes the inaccessibility of the animal's habitat poses nearly insurmountable hurdles: Siamang Gibbons, for example, frequent the jungle canopy as much as 120 feet above the ground, and homosexuality in the closely related, and equally arboreal, White-handed Gibbon was not discovered in the wild until 1991. Whales and Dolphins spend less than 20 percent of their time at the surface of the water, and underwater observation (where sexual activity often occurs) is frequently impractical.[105] This is compounded by the fact that recognition of individual animals and determination of their sex—essential for obtaining detailed behavioral data—is also usually extremely difficult. An animal's size can also be a factor: few behavioral observations of wild Apereas have been made because they are so small and their social activities

Because sexual activity in Red Foxes usually takes place at night, scientists were only able to document homosexual activity by using remote-control video cameras, set up in an enclosure illuminated by infrared light. In these two stills taken from the videotape, a younger female Red Fox mounts her mother. ▼

are often hidden in dense grass and brush. Small size (among other factors) also hampers field observations of Squirrel Monkeys, Rufous-naped Tamarins, and Rufous Bettongs. The latter species is also largely asocial or solitary, a problem encountered as well in Bears and numerous other carnivores, where many thousands of hours of observation in the field often yield precious little information about social or sexual interactions.

Even for species that are not difficult to observe, enormous time must still be invested in observation and quantification of behaviors before a reasonably complete picture of the animals' habits can be pieced together. Zoologists have estimated that to obtain a good working understanding of a species, three field-workers would need to invest two years and 2,000 hours of observation time—yet even this may not be enough. Over a dozen scientists studying Orangutans, for example, have collectively spent more than ten times this amount—20 years and a combined total of 22,000 hours of field observation—yet they admit that many aspects of the behavior of this species are still poorly understood. Likewise, zoologists involved in a comprehensive study of Oystercatcher behavior in the wild did not observe homosexual activities until nearly a decade into their research project.[106] It's no wonder, then, that many behaviors, including homosexuality, are just beginning to be documented in the wild or have yet to be observed outside of captivity. In summarizing wild versus captive comparisons of animal behavior, Jane Goodall has remarked, "If a primate shows behavior in captivity which has not been observed in the wild, this by no means implies that it does *not* occur in the wild."[107] The history of the study of animal homosexuality has shown this to be a truism, not just for primates, but for all species.

Hormonal Imbalances and Other Monstrosities

Unable to find any other "reason" for same-sex activity in animals, many scientists have tried to argue that homosexuality is itself a physical abnormality or the manifestation of some pathological condition. The most common physiological "malfunctions" that are suggested to "explain" homosexual behavior in animals are some sort of hormonal imbalance, and an "abnormal" condition of the sex organs. Female Sage Grouse who court and mount other females are described as suffering from "hormonal or hermaphroditic irregularities," for example, while scientists speculate on the "endocrine balance" of female Rhesus Macaques that participate in homosexual activity, the possible "hormonal defects" of female Fat-tailed Dunnarts that mount other females, and the influence of "abnormal physiological particulars" on the lesbian behavior of Long-eared Hedgehogs. Scientists have even suggested that homosexual mounting by Takhi mares who are pregnant is due to the male hormones circulating in their system as a result of carrying a male fetus.[108]

Scientific studies of homosexuality often seek evidence of "irregularities" in the form or condition of an animal's sex organs. This reflects in large part the widespread misconception (stemming from early sexological discussions of humans) that homosexuality is tantamount to hermaphroditism—i.e., any gender "transgression" is mapped onto an anatomical or physiological "abnormality." In 1937, scientists carefully examined the external genitalia of a male Common Garter Snake that had engaged in sexual behavior with another male to verify that it had "normal" male sex organs (it did). They then killed and dissected the animal to see if it had female gonads, reporting "no ovarian tissue was discovered." Lest one think this merely reflects the outmoded views of the time, nearly 60 years later this scenario was repeated with uncanny parallelism. In 1993 scientists performed a laparotomy (a surgical technique to examine the internal sexual organs) on a male Hooded Warbler that repeatedly formed homosexual pairs, in order to verify its sex and determine the condition of its male organs; the bird was later killed to obtain tissue samples. They reported that his sex organs were indistinguishable from other males', adding—in words echoing those used more than half a century earlier—"No ovarian tissue was present."[109] Just as early medical descriptions of homosexuality in humans often focused attention on the supposedly abnormal development or condition of the external genitals (along with hormonal factors), so, too, have scientists studying same-sex activity in animals tried to link this behavior to genital "peculiarities." In describing male companions in African Elephants, one zoologist emphasized that animals in such partnerships may exhibit physical "defects" including "enlarged external genitalia," while an ornithologist describing a male Snow Goose in a homosexual pair felt compelled to remark, "His much enlarged penis indicated a strong endocrine stimulation."[110]

There is no evidence to support a hormonal or other physiological "explanation" of animal homosexuality, and there is considerable evidence against it. Comprehensive and rigorous endocrinological analyses, as well as gonad measurements, of homosexual Western and Ring-billed Gull females show conclusively that there are *no* significant hormonal or anatomical differences between birds in homosexual and heterosexual pairs that could account for same-sex pairing. Specifically, investigators found that females in homosexual pairs are not hormonally "masculinized," i.e., they do not have higher levels of male hormones (androgens) than do females in heterosexual pairs. If anything, homosexual females may be more hormonally "feminine" than heterosexual females: some lesbian Ring-billed Gulls actually have significantly higher levels of progesterone, a female hormone associated with nest-building and incubation behavior.[111] Likewise, studies of a variety of primate species have shown no correlation between hormone levels and homosexual activity.[112]

Conversely, researchers *have* found hormonal differences between individuals in some species that exhibit homosexual behavior, but *not* in animals that participate specifically in same-sex activity. Endocrinological studies of Pied Kingfishers, for example, reveal that some males have lowered testosterone levels (as well as smaller gonads), but these individuals constitute one class of nonbreeding "helpers" who assist their parents in raising young. Few, if any, such males are involved in the homosexual pair-bonding or mounting activity that sometimes occurs in this species. Likewise, a certain category of nonbreeding Orang-utans (those with "flanges") often have elevated estrogen levels—but homosexual activity is neither characteristic of, nor exclusive to, such individuals. In other species scientists have determined that an "unusual" hormonal profile is found in the majority of individuals, but this is not linked to homosexual activity. In Spotted Hyenas, females generally have higher levels of a particular "male" hormone (a type of androgen) than do males, yet only a fraction of them actually participate in same-sex mounting; moreover, pregnant females also have elevated levels of testosterone (regardless of the sex of their fetus) but are not more prone to same-sex mounting. Likewise, *all* female Western Gulls have high levels of androgens (including testosterone)—nearly equal to those of males—regardless of whether they are in homosexual or heterosexual pairs.[113] These examples illustrate another important point: in many species a portion of the population routinely exhibits hormonal profiles (or other physiological characteristics) that differ from the "norm"—sometimes correlated with nonbreeding—yet only when individuals display overt homosexuality or transgender is the label of *abnormality* or *dysfunction* applied.

In most instances where a physiological "explanation" is advocated, this is purely conjectural, not based on any actual hormonal studies of the animals involved, and often highly improbable on independent grounds. For example, the connection between male fetal hormones and a pregnant mother's behavior—advocated as an "explanation" for mounting among female Takhi—is entirely speculative, since endocrinological profiles were not drawn up for the specific individuals involved in same-sex activity. Moreover, even if there were a connection, it would be at most only a partial explanation for this (and other) species. One Takhi mare mounted *males* when carrying a male fetus rather than mounting other females and also failed to show similar behavior the next year when she was again pregnant with a male fetus.[114] Thus, additional factors must be involved in determining whether such mares participate in homosexual, bisexual, and/or heterosexual (reverse) mounting behavior, if any of these. More generally, this explanation does not have wide applicability to other species. For example, only a fraction of Domestic Horses (which are closely related to Takhi) ever show mounting behavior of any sort when pregnant.[115] In addition, homosexual behavior by pregnant females occurs in less than 8 percent of all mammals in which female homosexual mounting has been documented, and in none of these species is the behavior *exclusive* to pregnant females (or to pregnant females carrying male fetuses). A fetal hormonal "explanation" is irrelevant, as well, for huge segments of homosexual activity, such as same-sex behaviors in animals that do not get pregnant (males of virtually all species and females of egg-laying species, for example).

In addition to being empirically unfounded, physiological explanations are also suspect on conceptual grounds. Almost without exception, hormonal or other pathological accounts of homosexuality focus on the animal exhibiting "gender-atypical" behavior, e.g., the male being mounted or the female doing the mounting. The partners of these individuals are usually considered to be physically "normal" animals whose behavior warrants no further consideration. Yet in many cases the "gender-conforming" partners are equally active participants in homosexual activity, sometimes even initiating same-sex interactions. As we saw in the discussion of "pseudoheterosexual" explanations, this categorization of animals into gender-conforming versus nonconforming, or "truly homosexual" versus "not-quite-homosexual" individuals, is in most cases arbitrary. It reflects not so much any inherent qualities or meaningful behavioral attributes in the animals themselves, but rather the observer's biases or conceptual categories.[116]

The pathologizing of "gender-atypical" behavior is taken to its extreme in the discussion of transgendered animals. Early descriptions of intersexual animals often labeled them "monstrosities."[117] More recently, hermaphroditism, chromosomal and other forms of gender mixing, and physical and behavioral transvestism are invariably considered diseased states, birth defects, physiological abnormalities, or otherwise dysfunctional. Yet researchers have usually been as unsuccessful in determining the physical "causes" for transgender as they have for homosexuality. For example, in discussing what they call "effeminate" behavior in Bighorn rams (males who exhibit some of the behavioral and social characteristics of females), scientists have tried to appeal to hormonal factors. Yet they were forced to conclude that this is an unsatisfactory explanation, since such males are physically "normal" and differ from other rams only in their behavior. The entire discourse surrounding transgender in White-tailed Deer centers on describing this as a "pathological condition" and attempting to find its physiological source. Velvet-horns (gender-mixing male deer) in Texas were subjected to a comprehensive battery of tests, including sampling and dissection of their sex organs to look for infection or "anomalies," blood tests for possible microorganisms or contaminants, dietary profiles, hormone injections, and chromosomal studies, none of which turned up any "cause." Investigators finally concluded that this "condition" must be due to a naturally occurring toxin in the soil where the animals live, yet admitted that no specific substance that might have this effect could be pinpointed or isolated in the animals' environment. Similarly, a gender-mixing Savanna (Chacma) Baboon in South Africa was shot and dissected to "study" its reproductive organs. Another was captured and given hormone "treatments" to see if it would behave like a "normal" female (defined, in this case, as participation in heterosexual intercourse with a male). Investigators stated that this individual could have been a "successful female in the wild" if only it had "normal functioning ovaries."[118]

These cases highlight one of the primary reasons that transgendered animals are usually considered abnormal: they often cannot (or do not) reproduce. Yet this is a limited and erroneous definition of "normalcy" that overlooks crucial facts about the lives of transgendered (and nontransgendered) animals. For one thing, transgendered animals arise "spontaneously" and repeatedly in natural populations, and they *do* survive successfully in the wild. Gender-mixing Baboons similar to the one given hormonal treatments have been observed in the same area of South Africa as far back as the early 1900s and are probably a regularly occurring feature of this and other populations. Moreover, such individuals are fully integrated members of their troops and may even assume high-ranking or "leadership" positions. The truth is, the gender-mixing individual described above (and others like it) was able to survive and even prosper *without* "normal functioning ovaries." Similarly, velvet-horns have been reported in a wide range of geographic areas and at least as far back as 1910–20, again indicating a long-standing, regular feature of natural Deer populations.[119] Although such individuals are sometimes "ostracized" by other Deer, they have developed their own forms of social organization, living in distinct "communities" with unique behavior patterns.

Conversely, many nontransgendered animals fail to participate in reproduction and may in fact never successfully procreate during their entire lives (numerous examples will be discussed in the next chapter). If failure to reproduce were sufficient grounds to exclude an individual from "normalcy," the majority of animals in some populations and species would not make the roster. In contrast, many transgendered animals *do* reproduce, such as intersexual Bears and gender-mixing female White-tailed Deer, and may in fact be more heterosexually successful than nontransgendered animals (as in transvestite Northern Elephant Seals, Red Deer, Black-headed Gulls, and Common Garter Snakes[120]). The final irony is that nonbreeding animals (including transgendered individuals) are also sometimes *more* healthy than breeders, precisely *because* they do not reproduce. Velvet-horn White-tailed Deer, for instance, are generally in much better physical condition than breeding males because they do not undergo the extreme physical rigors of the rutting season, which often cause severe weight loss and may even stunt growth in young bucks. Likewise, the mortality rate of breeding Bighorn rams is nearly six times higher than that of nonbreeding males. Clearly, then, participation in reproduction can be a liability rather than an asset to an individual's survival and success.

The vehement pathologizing of transgender encapsulates the entire discussion surrounding the "cause" of alternate sexual and gender expression in animals. Phenomena such as homosexuality or gender mixing are never seen as neutral or expected variations along a sexual and gender continuum (or continua), but rather as abnormal or exceptional conditions that require explanation. At the root of this perception is the idea that homosexuality and transgender are dysfunctional behaviors or conditions because they do not lead to reproduction. In the next chapter, we'll explore in greater detail the role of procreation in the animal kingdom and its complex interrelationships with homosexuality, bisexuality, transgender, and heterosexuality. Some of our most fundamental assumptions regarding the significance of reproduction must be revised as we come to understand the often surprising ways that animals structure their breeding and nonbreeding lives.

Chapter 5

Not for Breeding Only: Reproduction on the Periphery of Life

*H*eterosexual animals that never reproduce, homosexual animals that regularly procreate—breeding and sexual orientation often combine in unexpected and paradoxical ways. In an attempt to understand the origin and function of homosexuality, many scientists have suggested that same-sex activity might actually contribute in some way to reproduction or the perpetuation of the species. In this way, they have tried to carve out a "place" for homosexuality in the scheme of things—but a place on the sidelines, with breeding and heterosexuality decidedly in the center. What many people fail to realize is that reproduction itself often occupies a peripheral position in animal life—either being a "marginal" activity among apparently heterosexual animals, or else a common activity among seemingly "marginal" animals such as those involved in homosexuality. In this chapter we'll explore some of the various attempts to find a "useful" place for homosexuality in the larger patterns of life and consider why these attempts have often been as misguided as efforts to deny such a "purpose" for homosexuality in the first place.

The Evolutionary "Value" of Homosexuality

In 1959 noted evolutionary biologist George Evelyn Hutchinson published a proposal that was radical for its time (and even now remains controversial): he advanced the first theory of the evolutionary *value* of homosexuality.[1] Hutchinson argued that since homosexuality appears to be a biological constant, appearing in generation after generation (in both humans and animals) at a rate that far exceeds that of biological "mistakes," it must perform some useful function rather than be an aberrant behavior, and moreover, it must have a genetic basis.[2] Nearly

20 years later, in 1975, renowned biologist Edward O. Wilson published his semi-nal work *Sociobiology*, in which he took up the same theme: homosexuality must be beneficial to a species if it keeps reappearing. Since then, many other "positive" explanations have been proposed for animal homosexuality: some provocative, some absurd, but all revolving around the idea that breeding, heterosexuality, or the overall reproductive profile of an individual or species may be enhanced by homosexuality.[3]

A number of these proposals have been formulated with reference to homosex-uality in human beings and have not been rigorously evaluated (in either people or animals), in part because of the difficulty of finding relevant data or situations with which to test them. Many have not been applied to the domain of animal homo-sexuality at all, in part because of the inaccessibility of information about same-sex activity in nonhumans. In this section we will explore a number of these "explana-tions," evaluating—in many cases, for the first time—whether they hold true for a variety of different species. While many of these proposals are a welcome change from the view that homosexuality is "abnormal," they still face significant prob-lems. Often such explanations are simply not consistent with the facts about ho-mosexuality across a broad spectrum of animals. In addition, the underlying assumptions of many of these proposals—especially with regard to the participa-tion (or not) of homosexual, bisexual, transgendered, and heterosexual animals in breeding—are frequently incorrect.

For the Good of the Family and the Species?

Homosexual or transgendered individuals in many human societies perform a special role, acting as shamans, teachers, or caretakers for the benefit of the tribe as a whole, or for particular families. A number of biologists have suggested that ho-mosexuality in animals may work in a similar fashion. One proposal is that homo-sexual animals, while not reproducing themselves, act as "helpers" in raising the offspring of their relatives, thereby contributing indirectly to the passing on of their own genes. Another idea is that homosexuality, because it is nonreproductive, acts as a self-regulating mechanism to control a species' population growth.[4] Both of these theories have generated considerable controversy, yet little concrete evidence to either support or refute them has been brought forward. Moreover, neither of these proposals has been evaluated with respect to animals—even though they are directly testable with data from animal species—probably because a comprehensive and detailed survey of nonhuman homosexuality has not been previously available. Once the relevant facets of behavior and social organization are considered, how-ever, it becomes quite clear that neither of these hypotheses can be correct.

Underlying each of these proposals is the assumption that animals who engage in homosexuality do not reproduce (and must therefore "contribute" in some other way)—yet this is patently false. As we saw in earlier chapters, bisexuality is wide-spread in the animal kingdom: in more than half of the mammal and bird species in which homosexuality occurs, at least some individuals engage in both same-sex and opposite-sex interactions. Moreover, actual breeding by animals who participate in homosexuality has been verified in more than 65 species. This includes animals who are heterosexually paired and raise offspring but have outside homosexual interactions (Greenshanks, Little Egrets, Tree Swallows, Gray-capped Social Weavers); animals who engage in homosexuality as single parents (Japanese Macaques, Hanuman Langurs, Northern Fur Seals); animals who raise offspring in bisexual trios or quartets (Black Swans, Greylag Geese, Oystercatchers, Jackdaws) or in same-sex pairs as a result of outside heterosexual matings (Ring-billed Gulls, Western Gulls); females who participate in homosexual activity while pregnant (Gorillas, Takhi, Vicuñas) or even while their infants are clinging to them (Bonobos); animals who breed at some point in their lives prior to or following a period of homosexuality (Orang-utans, Rufous Rat Kangaroos, Emus, Silver Gulls, Bicolored Antbirds); homosexuality among those individuals in a population who monopolize most of the breeding opportunities (Nilgiri Langurs, Mountain Zebras, Bighorn Sheep, Ruffs, Pukeko); and animals that have incestuous homosexual relations with their own offspring (White-handed Gibbons, Red Foxes, Livingstone's Fruit Bats, Ocellated Antbirds). Thus, animals use multiple strategies to combine homosexuality with breeding, and even animals who may "prefer" homosexuality or have more same-sex than opposite-sex interactions can successfully raise offspring.[5] It is simply not true that animals who participate in homosexuality are unable to reproduce and pass on their genes to future generations. Of course, some animals are exclusively homosexual and never reproduce (as discussed in chapter 2) or else are unsuccessful breeders in either a heterosexual or a homosexual context, but reproduction is most definitely not limited to animals that only have heterosexual contacts.

Setting aside the fact that the initial premise of these two hypotheses is incorrect, is there nevertheless any validity to the substance and implications of each of these proposals? As it turns out, the animal world offers us a ready-made natural "laboratory," as it were, to test the first hypothesis, that homosexual animals act as "helpers" for other members of their own species or families. Numerous animals have developed a variety of "helping systems" in which individuals contribute to the care and upbringing of youngsters that are not their own offspring (although they may be relatives). These arrangements take several different forms: communal or cooperative breeding systems (group-living arrangements in which only some animals breed while the others assist them); "day-care" systems such as crèches or nursery groups, in which youngsters from

◄ A pregnant female Takhi (Przewalski's Horse) mounting another female from the side. Breeding animals in many species participate in homosexual activity.

more than one family are pooled together and watched over by one or two caretakers; alloparenting, in which individuals assist parents in duties such as feeding, protecting, carrying, or even "baby-sitting" their offspring; and adoption, involving foster-parenting of orphaned, lost, or abandoned youngsters.[6] Yet virtually none of these helper systems is preferentially "staffed" by homosexual animals or associated in any particular way with homosexuality. True, some individuals that engage in homosexuality certainly do act as helpers in some of these systems, but there is not a privileged association between homosexuality and helping as has been hypothesized. In fact, in some instances the connection between homosexuality and helping is the exact opposite of what is predicted by this hypothesis.

Consider the example of communal breeding systems: this form of social organization is especially prevalent among birds, where it is found in at least 222 species—yet homosexuality occurs in only 8 (4 percent) of these.[7] Although caution must always be exercised when drawing conclusions based on the *absence* of homosexuality in a species, nevertheless this proportion is far less than would be expected if helpers were somehow predisposed to homosexuality, or if individuals that participated in homosexuality were somehow predisposed to helping.[8] Moreover, in each of these 8 cases the specifics of which birds participate in homosexuality and/or helping do not follow the predicted patterns. In Pukeko and Gray-capped Social Weavers, for example, only *breeding* birds, not their helpers, engage in homosexuality—directly counter to what is predicted by this hypothesis. In other cases, homosexuality is not limited to helpers, but is also found in breeders: all Acorn Woodpecker communal group members—breeder and helper alike—participate in same-sex mounting, while in Tasmanian Native Hens, Dusky Moorhens, and Mexican and San Blas Jays, homosexuality occurs in both breeders and helpers, but only in a small proportion of each (i.e., many or most helpers do not engage in same-sex relations at all). Finally, homosexuality in Pied Kingfishers is characteristic of neither breeders nor helpers: rather, a subset of the nonbreeding population that does not participate in the helper system is involved in same-sex activity.[9]

Likewise, the other forms of parental help found in animals do not support any connection between helping behavior and homosexuality. Crèches, alloparenting, and adoption occur in numerous species without homosexuality. Of those mammals and birds in which at least some individuals do engage in homosexual activity, these types of helping systems are found in less than a third of the species, and they also occur in less than half of the species in which at least some individuals are exclusively homosexual. Moreover, in none of these cases is there a specific association between homosexuality and helping. For example, in many animals helping is performed only by members of the sex in which homosexuality is absent (e.g., Nilgiri Langurs, in which females may help take care of each other's offspring, but only males participate in homosexuality) or else it is characteristic of (heterosexual) breeding animals who assist other heterosexual breeders (for example, parents who take turns watching over a crèche, or who help feed and protect other parents' youngsters).[10] In no instance is helping restricted to animals that engage exclusively, primarily, or even sporadically in homosexuality, nor is it even more prevalent in such individuals.[11] In some species we find even more confounding situations: among Hanuman Langurs, for instance, "helpers" actually enable *breeding* animals to participate in homosexual activity. Mothers in this species often engage in same-sex mounting, but only when they have been temporarily "freed" from their parental duties by other individuals who "baby-sit" their young.[12]

What about the idea that homosexuality acts as a mechanism to regulate population growth? Again, little concrete evidence supports this hypothesis, and there are also serious problems with its underlying premises.[13] Aside from the fact that many animals engaging in homosexual activity continue to reproduce (as already mentioned), it is unlikely that population growth would be seriously affected even if a large proportion of animals were exclusively homosexual. Most animal populations can and do support large numbers of nonbreeding individuals without suffering a decrease in numbers: indeed, in many species a majority of individuals do not reproduce without any adverse effects on the population as a whole. In Damaraland mole-rats, for example, 90–98 percent of all individuals never breed during their lifetime, yet the population sustains itself and even continues to grow. Scientists have also calculated that a stable Killer Whale population can include up to 30 percent nonreproducing females without experiencing any decline. A significant pool of nonbreeding individuals exists in many other species, and up to 90 percent or more of one sex may fail to mate and/or breed.[14] Thus, exclusive homosexuality on a much more massive scale than that seen in any species would have to occur before homosexuality could even begin to impact on population growth and size.

A number of animals experience periodic and often dramatic fluctuations in their numbers, sometimes undergoing regular five- or ten-year cycles of population increase and decrease—for example, snowshoe hares, lemmings, voles, and some species of finches, sandpipers, falcons, and grouse.[15] If homosexuality were correlated with population size, one might expect that it would feature prominently in such species. One might also predict that its occurrence would "shadow" or fluctuate along with the population cycles, becoming more prevalent when population size or growth rate reaches its maximum, and less prevalent or nonexistent when the population is at its ebb. In fact, homosexual behavior has not been reported for most such species, and in the few cases where it has—Scottish Crossbills, Kestrels, and Grouse, for example—it does not appear to be related to either the cyclic or the irregular population increases ("eruptions," as they are sometimes known) that occur in these species.[16]

Similarly, if homosexuality actually resulted in a significant decrease in population growth, one might expect it to be disproportionately represented among animals that are suffering a severe decline in numbers, i.e., in endangered species. However, of the 2,203 mammals and birds in the world that are currently classified as threatened (either critically endangered, endangered, or vulnerable), homosexuality has been documented in just over 2 percent of these.[17] Moreover, the distribution of homosexuality across different species clearly has nothing to do with their endangered status: there are examples of two closely related species, such as the Pukeko and the takahe—two birds of New Zealand—in which homosexuality only occurs in the *nonendangered* one (the Pukeko); or animals in which one subspecies is endangered (e.g., the Asiatic Lion) yet homosexuality is not restricted to this subspecies; or else cases in which one or more subspecies are threatened (e.g., the Baja California and Sonoran Pronghorns), yet homosexual behavior is found in the nonthreatened subspecies of the same animal (the American Pronghorn); or two closely related species, in one of which homosexuality is common yet the species is not endangered (Hanuman Langur), the other in which homosexuality is much less common but the species is threatened (Nilgiri Langur). Conversely, if homosexuality were a form of self-preservation for a species as a whole—a "safety-valve," as it were, activated in times of overpopulation—one would not expect to find it *at all* in animals suffering severe population declines. Nevertheless, same-sex activity is reported in at least 50 endangered species. Perhaps the most dramatic example is the nearly extinct Black Stilt: less than 50 of these birds are left in the wild, yet some individuals still form lesbian pairs.[18]

Animals are perfectly capable of "regulating" their population size with far more efficient and effective strategies than homosexuality. A wide variety of mechanisms for reducing density and/or growth rates have been documented, including emigration, stress-induced hormonal changes that inhibit reproduction, decreased fertility, delayed maturation or slowed development, infanticide, and cannibalism (not to mention "outside" checks on population size such as predators).[19] In summary, then, it appears that homosexuality is neither useful to the species as a way of controlling population growth, nor useful to individual families as a mechanism whereby breeding animals are supplied with nonbreeding "helpers."

Bisexual Superiority and the Genetics of Homosexuality

In attempting to argue for the evolutionary value of homosexuality, scientists are confronted with an apparent paradox: if homosexuality is a valuable trait, it should have a genetic basis—yet how can a gene that doesn't lead directly to reproduction continue to be passed on from one generation to the next? Perhaps, some have suggested, because the putative gene for homosexuality does not operate on its own, but rather is acting in tandem with another gene to promote reproduction. An often-cited analogy involves the genetics of sickle-cell anemia and malaria resistance in humans. People who receive a sickle-cell gene from one parent and a regular hemoglobin gene from the other parent are resistant to malaria; those who receive two sickle-cell genes (one from each parent) succumb to sickle-cell anemia, while those who receive two regular hemoglobin genes are more likely to succumb to malaria. Thus, genes that (on their own) can potentially decrease an individual's reproductive capacity continue to be passed on because they are beneficial when combined with each other. Scientists have suggested that this might also be the case with homosexuality, as follows: Suppose there were one gene that predisposed an individual to homosexuality, and another that predisposed an individual to heterosexuality. Those individuals who receive two homosexual genes (one from each parent) would be exclusively homosexual; others would receive two heterosexual genes and be exclusively heterosexual; while those receiving one of each would be bisexual. If individuals who have one homosexual and one heterosexual gene were somehow more successful at reproducing, then the gene for homosexuality would confer an advantage and would continue to be passed on, even though it would sometimes result in individuals who do not reproduce (those who receive a homosexual gene from each parent).[20]

At first glance, this hypothesis seems counterintuitive: regardless of the genetic mechanism involved, why should *bisexual* individuals be superior at procreating or have a reproductive advantage? On the contrary, one would expect individuals with two heterosexual genes—those who are exclusively or "doubly" heterosexual, as it were—to be more successful breeders than bisexuals. Nevertheless, this hypothesis accords surprisingly well with a number of aspects of animal homosexuality that remain puzzling under other accounts. First of all, as noted previously, bisexuality is widespread in the animal kingdom. Unlike other theories about the evolutionary value of homosexuality, this hypothesis recognizes that many individuals who participate in homosexual activity may also be involved in heterosexual behavior, and therefore capable of reproducing and passing on their genes. Additionally, the incidence of bisexuality within populations is often high: in a number of animals such as Bonobos, Japanese Macaques, Bottlenose Dolphins, Mountain Sheep, Giraffe, and Kob, for instance, virtually all members of the species (or of one sex) participate in both same-sex and opposite-sex interactions (either concurrently or at different points in their life). Again, this hypothesis predicts that such situations should exist, since it argues for the maximization of bisexuality in a population— that is, if bisexual individuals are more successful breeders, they should tend to make up the majority of a population.

Even more startling, in a few species bisexual animals actually do appear to be more successful than exclusively heterosexual individuals at reproduction, heterosexual mating, and/or attracting members of the opposite sex. As we have already discussed, pairs of male Black Swans, who can father cygnets by associating temporarily with a female and then raise the resulting offspring on their own, are generally more successful parents than heterosexual pairs. In part, this is because such same-sex pairs are more aggressive than male-female pairs and are therefore able to acquire larger and better-quality territories, which are essential for successfully raising cygnets. They may also have an advantage because both males contribute to incubating the eggs, whereas in heterosexual pairs males may take part in less of the incubation duties. Over a three-year period, 80 percent of male pairs in one study were found to be successful parents, while only about 30 percent of heterosexual pairs successfully raised offspring (unsuccessful parents either deserted their clutches, lost them to predators or other hazards, or ended up having their cygnets die). Homosexual pairs constituted up to a quarter of all successful parents even though they made up only 13 percent of all breeding pairs or associations in the study population.[21]

Animals who participate in homosexual activity are also sometimes more successful at attracting members of the opposite sex, or participate more often in heterosexual mating. For example, male Ruffs who display with and mount male partners on their courtship territories attract females for mating more often than males who display by themselves. Because of their superior strength and courage, as well as their high rank in the flock, Greylag Geese in gander pairs or other homosexual associations are also sometimes attractive to the opposite sex. Females may associate themselves with a gander pair and eventually form a bisexual trio, mating with one or both of the males and raising their goslings together. In Pukeko, breeding groups in which homosexual interactions take place between males are also the groups in which the most intense heterosexual copulatory activity occurs. Adolescent Guianan Cock-of-the-Rock males who participate in the most visits to adult males' display territories, during which homosexual courtship and mounting often occur, sometimes acquire their own territories at a younger age. With earlier access to heterosexual mating opportunities, this may give them a "head start" on breeding. Likewise, female Oystercatchers in bisexual (as well as heterosexual) trios may have an advantage in acquiring their own breeding territories and heterosexual mates in subsequent years.[22]

A number of studies have also shown that animals that are the most active heterosexually are sometimes also the most active homosexually. In specific populations of Sociable Weavers, Bonnet Macaques, and Asiatic Elephants, for example, the top two males in terms of heterosexual mountings and other behaviors also participated in the most homosexual activities. Some of the most complete male homosexual behavior in Japanese Macaques, including full copulations with ejaculation, was exhibited by "one of the most vigorously heterosexual males in the troop," while in another study the one female in a troop who failed to form any homosexual consortships also did not participate in any heterosexual consortships.[23] And as mentioned in the preceding chapter, in a number of birds such as Common Murres, Laysan Albatrosses, and Swallows, most individuals who participate in homosexual copulations are in fact breeders who have heterosexual mates, rather than nonbreeders who are heterosexually inactive.

In spite of these rather unexpected confirmations, however, the bulk of the evidence does not actually favor this hypothesis and in fact disconfirms many of its predictions. Most of the examples cited above that seem to support the idea of "bisexual superiority" are misleading because they are based on anecdotal, rather than quantitative, information, and because they only look at a few individuals at a single point in time (or, at most, over the span of a few breeding seasons). To assess whether bisexual animals are more successful at reproducing, what is actually needed is a long-term study of large numbers of individuals that tracks them over their entire lifetimes, comparing the total number of offspring produced by bisexual animals to the total number produced by heterosexual individuals. Needless to say, this would be a huge and difficult undertaking, complicated by the logistics of keeping track of hundreds or even thousands of animals over many years and potentially large geographic areas, tabulating not only the reproductive output of each individual but also his or her entire sexual history to determine which animals are bisexual and which are exclusively heterosexual. Not surprisingly, few longitudinal studies of this type have been conducted, and those that have rarely involve species in which homosexual or bisexual activity is prominent (or else they do not take into account such behavior when it is present).

However, one scientist—James A. Mills—has conducted exactly this sort of long-term, comprehensive study on the Silver (Red-billed) Gull in New Zealand, a species in which there is extensive bisexuality and homosexuality. His results show that bisexual individuals are in fact significantly less successful breeders than heterosexual ones. Over more than 30 years, Dr. Mills and his colleagues banded over 80,000 individual gulls, tabulating detailed lifetime reproductive and sexual profiles of more than 5,000 of these. Because of the enormity of this project, special computer programs had to be developed to analyze and keep track of all the data. The Silver Gull is an ideal species in which to test this hypothesis, because the sexual orientation of females (in terms of their pairing behavior) falls into three clear-cut categories: some form only homosexual pairs during their entire lifetimes and hence are exclusively lesbian, while others have both same-sex and opposite-sex partners during their lives and are therefore unequivocally bisexual, while other females only pair with male partners and thus are exclusively heterosexual.[24] Moreover, Mills and his team looked not only at how many chicks were hatched and raised by heterosexual versus bisexual (and homosexual) individuals, but also at how many of those chicks survived to adulthood and became breeders themselves—the true measure of whether an individual is actually passing on his or her genes.

Mills's final results were conclusive: "Females which were bisexual during their life produced 14 percent fewer chicks than females in exclusively male-female pairings."[25] Furthermore, fewer of those chicks went on to join the breeding population as adults: exclusively heterosexual birds raised chicks who survived to breed at a rate that was more than one and a third times higher than that of bisexual females. Nor was the lower overall reproductive output of bisexual females due to their participation in (potentially less productive) homosexual pairings at some point in their life: such females also "tended to be less successful breeders even with male partners."[26] It would be difficult to find a more definitive or better-documented refutation of the bisexual-superiority hypothesis. Not only do bisexual females hatch and raise fewer chicks than heterosexual females, they also contribute fewer offspring to the pool of breeding individuals in the population, and their decreased reproductive output appears to be independent of whether they happen to be breeding with a male or a female partner.

One criticism that has been leveled at the bisexual-superiority hypothesis is that it is so difficult to test, and a number of scientists have even remarked that they cannot imagine a relevant experiment or study that could possibly verify or falsify its claims.[27] Amazingly, although it has all of the elements needed to evaluate the bisexual-superiority hypothesis, Mills's study was not specifically designed to test this proposal, nor even to focus on the reproductive performance of bisexual animals in particular. Indeed, it is doubtful that Mills was even aware of this hypothesis—it had yet to be formulated at the time he initiated his project in 1958, and it was not widely known or discussed in the scientific community even after it had been published and revised in various forms over the next 30 some years.[28] Nevertheless, the procedures and analyses Mills used were almost tailor-made to assess the validity of this hypothesis, and it is a testament to his expertise that his results should prove useful for a line of inquiry so far removed from their original purpose.

Unfortunately, studies of a similar scale and quality have yet to be undertaken for most other relevant species. Nevertheless, although it is possible that different patterns of reproductive performance across sexual orientations may be revealed in other animals, this is unlikely. Most reports of same-sex parenting and/or breeding in other species appear to be in line with the Silver Gull results.[29] Notwithstanding the Black Swan case (to which we'll return shortly), animals in homosexual pairs who also reproduce are generally only as successful or less successful than heterosexual parents in raising offspring, not more successful. Moreover, in a number of instances homosexual activity on the part of breeding animals actually interferes with their reproductive performance: in female Jackdaws, Oystercatchers, Canada Geese, and Calfbirds, for example, homosexual associations may in fact be *detrimental* to the successful raising of offspring, often by interfering with incubation (these examples will be discussed more fully later in this chapter). Same-sex activity in Buff-breasted Sandpipers often discourages heterosexual mating and breeding opportunities, while male Cheetahs living in bonded pairs or trios often disrupt, compete with, or prevent their companions from mating heterosexually (and thereby reduce their reproductive output).[30] Although differential breeding success can be associated with sexual variance in some species, typically *transgendered* rather than bisexual (or homosexual) individuals are more reproductively successful (as in the examples of Northern Elephant Seals, Red Deer, Black-headed Gulls, and Common Garter Snakes discussed in the preceding chapter).

There are further arguments against the bisexual-superiority hypothesis. If bisexual animals were more successful breeders, one would expect them to make up the majority of the population in any given species, with much smaller proportions being exclusively heterosexual or homosexual—yet the distribution of sexual orientations does not, in fact, typically follow this pattern. In Silver Gulls, heterosexual versus bisexual percentages are in accord with what we have just seen about their relative reproductive proficiencies: 79 percent of all females are exclusively heterosexual, 11 percent are bisexual, and 10 percent are exclusively lesbian. This pattern is characteristic of many other species for which we do not have information about the lifetime reproductive output of a cross-section of individuals: bisexual animals generally make up a much smaller percentage of the population, sometimes even less than the proportion of exclusively homosexual individuals. For example, the heterosexual-bisexual-homosexual proportions for male Black-headed Gulls are 63-15-22 percent, respectively, and for Galahs, 44-11-44 percent.[31] In many other species the proportion of animals who engage in bisexual activity is even smaller.

Moreover, in some cases there do not appear to be any bisexual individuals at all in a population (i.e., same-sex activity occurs only in nonbreeding animals). For example, female homosexual pairs in Kittiwakes, Red-backed Shrikes, and Mute Swans, among others, appear to consistently lay infertile eggs (indicating that they do not mate with males); in Pied Kingfishers, homosexuality is typical of nonbreeding birds who are not likely to reproduce later in life; while male Ostriches who court other males do not appear to have heterosexual relations. Although longitudinal studies are needed in each case to verify that such individuals are not in fact sequentially bisexual, these patterns do not fit well with a bisexual-superiority hypothesis. More broadly, species in which homosexuality or bisexuality is only found in individuals of one sex—or in which all individuals are exclusively heterosexual—are extremely common and are further evidence against this hypothesis (since they are examples of bisexuality failing to be "maximized").

What about species in which the majority of individuals are bisexual, i.e., the examples of maximization of bisexuality mentioned above? In all of the animals in which this is the case (Bonobos, Dolphins, Mountain Sheep, etc.), individuals differ significantly in the *degree* to which they are bisexual. Some animals participate very little in homosexual and/or heterosexual activity while others account for the majority of (one or both) such activities, and same-sex versus opposite-sex interactions make up varying proportions of each individual's sexual encounters as well (as we saw in chapter 2). Thus, if bisexuality were related to reproductive success, one would expect animals to differ depending on "how bisexual" they are—successful breeders (i.e., those animals who are the most active heterosexually) should engage in a greater proportion of homosexual activity as well. Again, long-term studies of reproductive output are required to test this, but data on the sexual activity of individual animals in a number of species where bisexuality is widespread do not support this idea. If we take the number of heterosexual copulations that an animal participates in to be a rough measure of its reproductive prowess, then we do not generally find that there is a positive correlation between degree of bisexuality and breeding success.

For example, in Kob antelopes—in which virtually all females engage in both same-sex and opposite-sex mounting—there is generally an inverse relationship between an individual's heterosexual and homosexual activity. One study revealed that a female who had the most homosexual mounts had the fewest heterosexual ones and vice versa, while individuals who ranked in the upper quarter or third of the population in terms of heterosexual activity often ranked much lower in their homosexual participation. Furthermore, the female whose heterosexual and homosexual activities were most equal—i.e., the most "bisexual" individual—actually participated in the fewest total number of heterosexual matings. Similarly, all Bonobo females interact sexually with both males and females, but differ widely in the extent of their bisexuality. In one troop, three females participated in the most heterosexual copulations—two-thirds of all mating activity—yet these same females accounted for less than one-third of all homosexual activity, and one had among the fewest same-sex encounters of any of the females. Nor were these females necessarily "balanced" in terms of their individual proportions of same-sex and opposite-sex activity. One had fairly equal ratios of homosexual and heterosexual interactions, but the other two were less "proportional" bisexuals, with the majority (two-thirds) of their sexual encounters skewed toward opposite-sex partners. Likewise, those female Japanese Macaques who were the most involved in homosexual activity in each of four mating seasons (the top two in terms of the proportion of time they spent) were rarely as involved in heterosexual interactions and were often among the *least* heterosexually active members of their troop. Another pattern was revealed in a study of Pig-tailed Macaques. Although all the males in one troop mounted both females and other males, they had roughly the same number of homosexual encounters regardless of their participation in heterosexual activity (which varied enormously). The male who was the most heterosexually active was also the least "bisexual" individual (same-sex mounting made up only 8 percent of his sexual activity, compared to an average of 48 percent for the other males).[32]

Of course, heterosexual activity (i.e., number of opposite-sex matings) is not necessarily an accurate measure of reproductive success, and none of these studies tracked individual animals and the number of offspring they produced throughout their entire lives.[33] Nevertheless, there does not appear to be the sort of connection between homosexual and heterosexual activity that would be expected if bisexuality contributed to an animal's reproductive prowess or success. Moreover, in most of the species where bisexuality seems to be "maximized," it is usually the case that one sex participates in homosexual activity to a greater extent than the other: females in Kob, Bonobos, and Japanese Macaques, males in Mountain Sheep and Bottlenose Dolphins, for example. Even if bisexuality were somehow an advantageous reproductive strategy, it would remain to be explained why there should be a gender difference in its "efficacy" (and why it should pertain to different genders in different species).

Finally, most of the specific cases mentioned above (e.g., Black Swans, Pukeko, Ruffs) that seem to support some sort of connection between bisexuality and reproductive prowess are not as convincing as they initially appear. In each instance, closer investigation reveals that the connection is doubtful, if not completely spurious.[34] For example, although male pairs in Black Swans tend to be more successful parents, such couples are not necessarily made up of bisexual birds, nor do they always raise their own offspring. Same-sex pairs in this species often "adopt" cygnets by taking over or stealing nests from heterosexual pairs (rather than mating with a female)—thus many successful male pairs need not have been involved in any heterosexual activity at all and may be exclusively homosexual rather than bisexual. Moreover, even if such individuals prove to be bisexual over their entire lives (e.g., by subsequently pairing with females), much of their parenting success involves raising offspring that are not related to them (by virtue of having been "adopted"). This situation is inimical to the rationale behind the bisexual-superiority hypothesis, which depends on bisexual individuals being more successful at passing on their *own* genes, not other animals'.

Similar problems or qualifications are apparent in the other cases. Greylag gander pairs do sometimes attract females, it is true, but there is no evidence that they are *more* attractive to the opposite sex than single, exclusively heterosexual males. While male homosexuality in Pukeko is associated with the most heterosexually active groups, female homosexuality—which is more common, and more highly developed in terms of the courtship behaviors involved—is not. In addition, the greater levels of heterosexuality found in some groups is not necessarily a *result* of the homosexual or bisexual involvements of their participants. It is just as likely that the increased heterosexuality *and* homosexuality are both manifestations of a third factor, perhaps something akin to a generally higher sexual "state," level of activity, or arousal in such groups. This is supported by observations in a number of other species (e.g., Bonobos, Gorillas, Squirrel Monkeys, Wolves, Common Tree Shrews, Bottlenose Dolphins) where homosexual activity actually peaks or increases dramatically along with heterosexual activity (in different age/sex classes or social contexts).[35]

The matter of causality is also relevant for several of the other species discussed above. For example, although participation in heterosexuality and homosexuality appear to be linked in male Sociable Weavers, Bonnet Macaques, and Asiatic Elephants, this is primarily true for higher-ranking individuals—and such animals tend to have access to more individuals (including sexual partners) of either gender. In other words, greater heterosexual mating opportunities for such individuals are probably not a consequence of their bisexuality, but rather of their status—which also grants them greater homosexual mating opportunities. Similarly for the Guianan Cock-of-the-Rock: although adolescent males who engage in more homosexual encounters seem to have an advantage in their subsequent ability to acquire breeding territories, scientists admit that this may be due to a third factor (such as higher levels of aggression or "initiative" on the part of such males, or even physiological differences between them) rather than being a direct consequence of their same-sex activity. Furthermore, while bisexuality in this species may appear to be related to breeding success for adolescent males, it is definitely not conducive for reproduction in adult males (who nevertheless continue to participate in such activity). Homosexual courtships and sexual activity often interrupt and displace heterosexual activity, and females usually stay away from breeding territories while their owners are having homosexual encounters with adolescents. Likewise, the future reproductive advantages that may accrue to female Oystercatchers in trios are not specifically a function of whether they are bisexual. Compared to nonbreeders, such individuals are more likely to acquire heterosexual mates and breeding territories of their own in subsequent years, but this is regardless of whether their current trio is bisexual (with bonding and sexual activity between the same-sex partners) or strictly heterosexual (with no such same-sex activity). In fact, females in bisexual trios may actually be *less* likely than females in heterosexual trios to acquire their own mates subsequently, since bisexual trios tend to be more stable and longer-lasting than heterosexual trios. And as in Guianan Cock-of-the-Rock, homosexual activity does not promote reproductive output for such individuals while they remain within bisexual trios.[36]

Even though some of the most complete sequences of homosexual behavior in Japanese Macaques are seen in some of the most heterosexually active males, this pattern is not universal in either this species or others. In one study of Kob antelopes, for example, a female who exhibited the most fully developed sequence of lesbian courtship also participated in the second-fewest number of heterosexual matings of any of the study animals.[37] And while homosexual copulations (as well as promiscuous heterosexual matings) are characteristic of heterosexually paired (breeding) males in a number of bird species (e.g., Swallows, Herons), there is not necessarily a correspondence between specific amounts of same-sex and opposite-sex activity for individual birds. In Cattle Egrets, for example, males often try to mate with birds—male or female—other than their female partner. However, one study revealed that a male who completed the most promiscuous copulations with females—and therefore was probably the most reproductively "successful"—did not engage in any homosexual copulations. Other males had homosexual encounters regardless of whether they also sought nonmonogamous heterosexual activity, indicating no necessary connection between bisexuality and breeding success.[38]

Paradoxically, some of the strongest evidence against the bisexual-superiority hypothesis, as well as against genetics as the sole determinant of homosexuality, comes from the Ruff—a species in which same-sex activity between males clearly does attract females to breeding territories. To see why, we need to take a closer look at some social and biological patterns in this bird. Male Ruffs fall into four distinct classes—residents, marginals, satellites, and naked-napes—who differ from each other physically, behaviorally, and sexually.[39] While it is true that females are drawn to resident males' display territories by homosexual (and other behavioral) interactions between satellites and residents, satellites actually *interfere* with heterosexual mating by resident males once females have been attracted. Less than 3 percent of copulations occur when satellite males are on a resident's territory: not only does their presence inhibit heterosexual interactions, they sometimes directly prevent residents from mating by interposing themselves between the male and the female, or by trying to knock the resident off a female's back.[40] Moreover, not all homosexual activity is associated with attracting females: same-sex mounting and courtship also occur between males who are not involved in breeding (naked-napes), between males when females are not present, and during the nonbreeding season. In addition, not all resident males participate in homosexuality: some display on their own without a satellite "partner." If same-sex activity were vital for attracting females (and therefore breeding success) in this species, one would expect all males to engage in it. Further geographic and population differences in the occurrence of homosexual activity also argue against its being an essential component of successful reproduction.

Four classes of male Ruffs, which differ in their physical appearance, social and sexual behavior, and genetics. Clockwise from upper left: *resident, marginal, naked-nape, and satellite males.* ▼

Perhaps the most important piece of evidence concerns genetic differences between the classes of males. Scientists recently discovered that the distinctions between some categories of males are genetically determined—but the genetic differences *cut across* differences in their homosexual behavior rather than falling in line with their sexual variations. Detailed chromosome and heredity studies revealed that whether a male becomes a resident or a satellite is genetically controlled—a finding corroborated by the fact that these two categories of males are the most physically distinct from one another in their plumage, and also by the fact that category changes between the two types are virtually impossible (satellite males never become residents or vice versa).[41] Yet both residents *and* satellites engage in homosexual behavior—in fact, it is their joint participation in such activity that often attracts females. In stark contrast, residents and marginals are not genetically distinct: the two share many plumage characteristics, and a male may change his class membership from marginal to resident or vice versa. Yet it is precisely these two categories of males who are the *most* different sexually: resident males are commonly involved in both heterosexual and homosexual encounters, while marginal males are nonbreeders who rarely participate in either same-sex or opposite-sex activity.

This certainly does not mean that homosexuality lacks a genetic basis in this (or any other) species. Rather, it demonstrates the importance—the primacy, even—of nongenetic factors in the expression of homosexuality, regardless of whether it has a genetic component. A male Ruff may begin his adult life as a marginal, engaging in no sexual activity whatsoever, then change over to resident status and begin copulating with both males and females, or only females, or only males. He may even revert back to marginal status later in life, becoming asexual once again—or he may never engage in same-sex activity even as a resident or perhaps never become a resident in the first place. Other males live their entire lives as either residents or satellites, with or without homosexual activity—but in all cases, the manifestation of their sexuality is dependent on the social and behavioral contexts in which they find themselves as much as, if not more so than, on their genes. This is not to say that genetic programming or an innate predisposition for homosexuality does not exist or is unimportant—only that many other factors are involved as well.

This is in line with what else we know about the genetics of homosexuality in animals (and people). Direct evidence for a genetic component is accumulating: in several species of insects, for example, scientists have recently isolated genetic markers for homosexuality (and there are parallel findings of genetic links to homosexuality in humans).[42] Yet it is also clear that social, behavioral, and individual factors are at least as important as genetic ones, especially in "higher animals" such as mammals that have complex forms of social organization and highly flexible behavioral interactions. The expression of homosexuality often varies widely between different social contexts, age groups, activities, individuals, and even populations and geographic areas, in ways that transcend any possible genetic "control." We also saw in chapter 2 that there are good reasons to consider homosexual (and other sexual) activity to have a "cultural," social, and/or learned dimension in a number of species, especially primates. Ultimately, then, it is of relatively little importance whether there is an actual homosexual "gene" or whether it is part of a pattern of "superior" bisexual reproduction. Even if homosexuality is shown definitively to have a genetic component (as is likely), it will always remain just that—a *component,* one part of a much larger picture that includes the totality of an animal's biology and social environment.

Homosexuality in the Service of Heterosexuality

If homosexuality does not enhance reproductive performance or act as a population regulator or "helper" system, then how else might it be evolutionarily "useful"? A number of scientists have suggested some other ways that homosexual behavior could contribute, directly or indirectly, to heterosexual activity and/or breeding. In this section, we'll look at several of these proposals, including suggestions that homosexuality is a way of practicing heterosexual mating, a method to attract opposite-sex partners, a form of competition to reduce the heterosexual mating opportunities of rivals, and several other more far-fetched "explanations."

Practicing Homosexuals

Same-sex activity is often claimed to be a way that younger animals practice or "rehearse" heterosexual courtship or mating, or a way for individuals to gain sexual "experience" that will improve their future breeding success.[43] While it is possible that homosexuality could provide this "service," it is unlikely that this is its major function. A serious problem with this proposal—as noted by several scientists—is that in most of these species homosexual behavior is not restricted to younger individuals or those who need to acquire sexual experience.[44] In some cases, it is found in all age groups (Walruses, Northern Elephant Seals, West Indian Manatees), in others it is more prevalent in younger animals but still occurs in older ones (Killer Whales, Giraffes), while in some cases it continues throughout an individual's life (Black-headed Gulls).[45] Moreover, in these and many other species there are frequently other patterns of homosexual or bisexual orientation that stretch the limits of a "practice" interpretation nearly to absurdity. For instance, are animals who engage primarily or exclusively in same-sex activities "practicing" their whole lives for some never-to-be achieved heterosexual opportunity? Or how about individuals who intermingle or alternate between homosexual and heterosexual activity—do they need to constantly return to same-sex behavior for remedial "practice"? And what about animals who "switch" to homosexuality only after they have been involved in heterosexual activity, or who come to homosexuality late in their lives—are their prior heterosexual involvements then "practice" for their homosexual ones?

When faced with these and other examples of homosexuality "persisting" or (re)appearing in adulthood, advocates of the "practice" interpretation have been forced to adopt exactly such unconvincing scenarios. For example, scientists have actually argued that individuals continue to engage in homosexuality as adults so that they can experience differences in sexual behavior across a wide variety of partners (of both sexes) and thereby continue to improve their heterosexual performance—an ironic reversal of the traditional view, in which homosexuality is considered to detract from, or be incompatible with, heterosexuality.[46] According to this "practice" interpretation, heterosexual "competence" is apparently so elusive (or difficult to achieve and maintain) that it requires constant, albeit indirect, reinforcement through the help of homosexuality. When we examine the realities of animal behavior we find that in many cases heterosexual mating *is* far from the automatic or "natural" occurrence that it is commonly assumed to be, and it *does* require some "practice" (see the next section). However, this "explanation" does a disservice to both heterosexuality *and* homosexuality in its view of the former as unnecessarily tenuous and the later as necessarily derivative.

Even in species where homosexuality is limited to younger animals or to adult-juvenile interactions, there are often serious discontinuities in participation that challenge a practice interpretation. For example, in the Guianan Cock-of-the-Rock, homosexual courtship and mating occur between adolescent and adult males and have been classified as "practice" behavior on the part of the younger males. However, there is a curious gap in the age distribution of participating males: primarily one-year-olds are involved, while two-year-olds almost never participate in such activity. Could it be that once they pass the yearling stage, they no longer need any "practice"? Most definitely not: scientists studying this species report that when males first acquire their own territories, between the ages of three and five, they continue to practice their courtship—but largely without homosexual interactions—thereby gaining valuable experience that they will need before embarking on heterosexual mating.[47] Why should birds first "practice" using homosexuality as one-year-olds, then cease such practice as two-year-olds, only to resume practice without homosexuality as three-to-five-year-olds, and then once again participate in homosexual "practice" sessions with younger males when they become older? And what exactly is the role of the adult males who willingly participate in all of these "practice" sessions? It seems unlikely that they need to "improve" their skills, too, or that they are simply serving as "mentors" to (probably unrelated) males, altruistically providing them with the opportunity to rehearse their mating skills. Although younger males may gain sexual and courtship experience as an indirect result of homosexual interactions, this appears to be a relatively minor byproduct of such activity and is highly problematic as an overall "explanation" for the behavior.[48]

There are other questionable aspects to the notion that homosexual behavior is merely rehearsal for heterosexual behavior. In many species where a "practice" interpretation has been suggested, only a small fraction of the population ever engages in same-sex activity, and often an individual participates only a handful of times. It seems highly unlikely that much sexual experience, or any useful "training," could be gained from such activity.[49] Moreover, in many species young animals practice heterosexual behavior by participating directly *in heterosexuality,* either with adults or with each other. In other animals—including ones in which adults engage in homosexuality—heterosexual practice is accomplished without partners or overt same-sex activity. For example, adolescent male Sage Grouse learn the complex courtship display of their species by imitating older males, practicing the "strutting" movements and sounds while they are gathered in their own groups on the periphery of the breeding grounds. Occasionally an older male joins the group so that he can—in the words of one ornithologist—"demonstrate the fine points of strutting to the rapidly maturing novices," but no homosexual courtship or copulation takes place, and in fact homosexual activity in this species is limited to adult females. Among Montagu's harriers and several other species of raptors, young birds are trained in heterosexual courtship displays by parents of both sexes without any homosexual activity.[50] Even if homosexual behavior were training for heterosexuality, this raises more questions than it answers. Why must some animals "resort" to same-sex practice while others can use opposite-sex interactions (or solo practice)? Why does homosexual "training" in some species involve adults "helping" younger animals, while in other species juveniles only "practice" amongst themselves? And why do some individuals apparently not need to "practice" at all? It is clear from these examples that a "practice" interpretation of homosexuality is at best of limited applicability and explanatory value.[51]

Finally, there is a curious gender bias in the application of a "practice" interpretation to homosexuality—in the overwhelming majority of cases, only male animals are said to require such rehearsal.[52] Where complicated courtship displays are performed only by the male of the species, this is perhaps understandable, but why should it be that no female needs to "practice" sexual activity by engaging in same-sex activity? In many animals, especially primates, females are active participants in heterosexual intercourse, often initiating sexual activity and adopting specific postures, positions, or movements as part of a sexual interaction. In most bird species, heterosexual copulation is impossible without the cooperation and active participation of the female: since most male birds do not have a penis, mating can usually only occur if the female actively facilitates the interaction, for example by positioning herself in such a way as to allow genital contact. Yet in none of these species has it been suggested that females "practice" heterosexual mating via homosexual copulation—not because practice isn't required, or because lesbian activity could not serve this "function," but because many scientists still regard the female as an essentially passive participant in sexual activity.[53] This is highly revealing—not only of sexist attitudes in biology, but also of the true "utility" of this explanation. It is not applied systematically and carefully to all potentially relevant cases: more often than not, it is simply introduced when most other "explanations" have failed, a convenient tool with which to discount or dismiss homosexual activity.

Homosexuality as a Breeding Strategy

Some nonsexual interpretations of homosexual activity discussed in the previous chapters hinge on the indirect contribution of homosexuality to heterosexuality. For example, it has been suggested that homosexuality reinforces group cohesion and social bonds between individuals, thereby improving their well-being and allowing them, ultimately, to reproduce more successfully. It has also been claimed that homosexual "alliances" between animals improve their chances of gaining heterosexual copulations.[54] Some scientists have been even bolder in their view of the connection between homosexual and heterosexual activity, regarding the two to be directly related or even essentially continuous: same-sex activity is seen as simply an alternative breeding strategy adopted by some animals, or a way to attract or acquire partners of the opposite sex.[55] It has been proposed, for example, that female Rhesus Macaques sometimes form homosexual consortships to gain access to a male who is himself consorting with their female partner, or that male Bottlenose Dolphins form pairs with each other to seek out female partners.[56]

Another standard "explanation" for homosexual activity among females, especially mammals, is that it attracts males and stimulates them to mate heterosexually. It has also been suggested that female mammals mount each other primarily when one or both partners are in heat, and hence homosexual activity acts as a signal to males of when females are ready to mate. A variation on the notion of homosexuality as a stimulant for heterosexuality is the speculation that males stimulate their *own* libidos by engaging in same-sex activity (rather than attracting female partners). For example, erotic fighting in African Elephants (during which both participants become sexually aroused) is claimed to stimulate the males so that they can then go out and seek female partners, while male homosexuality in Greenshanks and Golden Plovers is claimed to stimulate and strengthen the birds' heterosexual drive.[57]

Most of these rather fanciful speculations are not based on any systematic evidence, and in fact there are many arguments against such interpretations. To begin with, homosexual activity in many species is not restricted to the breeding season (i.e., the time when it could "stimulate" heterosexual mating) or to females who are in heat. In more than a third of the mammals and birds for which information is available concerning the chronology of homosexual activity, it occurs either year-round (i.e., both in the breeding and nonbreeding seasons or in females regardless of whether they are in heat or fertile), or else only during the nonbreeding season.[58] In some cases, the majority of homosexual activity occurs when females cannot conceive, e.g., when they are pregnant (some populations of Japanese Macaques) or during nonfertilizable stages of their cycle (Hanuman Langurs),[59] and therefore it cannot contribute to heterosexual mating.

Furthermore, even when homosexual activity does take place during the breeding season or at times when females can conceive, cases where it attracts members of the opposite sex or stimulates heterosexual mating activity are the exception, not the rule. In most species, other animals are entirely disinterested, nonchalant, or "underwhelmed" by any same-sex activity they may happen to observe (as discussed in chapter 2). Members of the opposite sex are often entirely absent from the vicinity (Hanuman Langurs) or may even stay away or leave when homosexual activity is taking place (Guianan Cock-of-the-Rock) or be chased away or ignored when they attempt to interact with animals engaging in homosexual activity (Japanese Macaques, Hanuman Langurs). Moreover, in many species homosexual alliances do not actually "improve" their participants' chances at gaining opposite-sex partners, and the reproductive advantages of same-sex coalitions are often questionable. Male Calfbird companions who display together, for example, do not attract females, nor are they more successful at acquiring mating territories or overcoming rivals than "single" males. Male Cheetahs living in same-sex bonded coalitions (pairs or trios) are no more likely to encounter females than are single males (even though the standard interpretation of such bonding is that it enhances males' reproductive opportunities and access to opposite-sex partners). They may in fact suffer reduced chances of heterosexual mating (and lowered reproductive output) because of competition or direct interference from their companions. Likewise, although coalitions of male Savanna Baboons sometimes cooperate in obtaining or defending female partners, one researcher points out that this is true for only one-quarter to one-third of all such alliances, concluding that most male partnerships serve many purposes besides obtaining mates and may even lack a recognizable "function."[60]

The case for male pair-bonding being solely a breeding strategy in Bottlenose Dolphins is also far from definitive. Pairing or "coalition" formation between males of this species is often interpreted—and widely cited—as a means whereby the animals obtain heterosexual mates. Although pairs (and trios) of Bottlenose males may cooperatively seek out and herd females for purposes of mating in some populations (e.g., Australia), this is not a uniform aspect of such partnerships, and in many cases it has yet to be documented. Heterosexual matings resulting from such associations have not in fact been observed in the Florida population where the most extensive study of male pairing has been conducted, nor in Ecuador, where it has recently been suggested that paired males may compete for females. In Australia, where herding and mounting of females by paired males have been observed, no "full" copulations involving penetration have actually been documented, so the reproductive status of this behavior is not clear. Moreover, nearly 38 percent of the animals herded by paired males in Australia were not definitvely sexed: researchers simply assumed that they were females. In fact, bonded males in other populations seek out male rather than female sexual partners in at least some contexts. In the Bahamas, pairs or coalitions of adult male Bottlenose Dolphins herd and chase Atlantic Spotted Dolphins; they typically pursue homosexual activities (including full penetrative copulation with other males) during these interspecies encounters. Finally, even if bonded males assist each other in obtaining heterosexual mates, this does not preclude a homosexual dimension to such partnerships—sequential and simultaneous bisexuality are, after all, prominent in this species. Same-sex pairs can form as long as ten to fifteen years before breeding activity commences, and homosexual activity may exist concurrently with heterosexual activity in such pairs once they do reach breeding age.[61]

One species in which same-sex activity among males sometimes does attract females is the Ruff. As already noted, however, homosexual behavior in this bird is not limited to contexts in which it might increase opportunities for heterosexual mating. It also occurs among nonbreeding males, when females are not present, and during the nonbreeding season. Similarly, homosexual activity in female Squirrel Monkeys sometimes does arouse the attentions of males. However, it is clear that the participating females engage in such behavior regardless of whether it draws males and even rebuff the advances of males who approach them during such activity. Sometimes heterosexual behavior serves as a stimulant for homosexual activity and not the other way around. Stumptail Macaques, aroused by the sight of heterosexual activity, often initiate same-sex interactions, while in Wolves, Savanna

(Yellow) Baboons, and Mountain Sheep, animals watching heterosexual mating often become excited and engage in homosexual activities.[62]

Even if males of some species are genuinely aroused by sexual activity between females, the evidence clearly shows that females are unconcerned with the effect of their behavior on males and do not structure their participation in homosexuality to maximize its impact on heterosexuality. Yet in spite of all this counterevidence, biologists still claim that a primary "function" of homosexual activity in females is to arouse males: "The sight of one female mounting another is said to excite males sexually in Squirrel monkeys and it may do so to males of other species too, for example men watching pornographic films of lesbian activity."[63] By drawing an explicit parallel to human sexuality, the author of this statement hopes to argue for the evolutionary "usefulness" of homosexuality—but the analogy actually highlights the fundamental absurdity of this "explanation," as well as its dependence on cultural rather than biological factors. True, many heterosexual men are aroused by the sight of two women having sex together, and lesbian sexuality is often packaged and trivialized as pornography to be consumed by straight men. But it would be ridiculous to conclude, on the basis of this, that lesbians have sex "in order" to arouse heterosexual men—yet this is exactly the type of reductionist thinking that is routinely applied to homosexual behavior in animals. It is also highly revealing that homosexual behavior among male animals is virtually never described as being stimulating for females.[64]

Perhaps the most widespread version of the idea that homosexuality is really just a form of reproductive behavior concerns same-sex pairing in birds. It is frequently asserted that the "function" of such associations is to allow females to successfully raise young when they are unable to obtain a male mate. Not only is the initial premise of this explanation—that homosexual pairs result from the unavailability of members of the opposite sex—incorrect, but first and foremost, birds do not usually form same-sex pairs specifically to undertake parenting.[65] Species in which homosexual pairs never attempt to raise young are nearly as common as

Nests belonging to homosexual pairs of Black-winged Stilts (left) and Red-backed Shrikes (right). Both females in the pair lay eggs, and therefore their nests contain "supernormal clutches" (double the usual number of eggs). Because neither female has mated with males, however, these clutches typically consist entirely of unfertilized eggs. ▼

those in which same-sex parenting does occur. Even in those species where female pairs lay eggs, the proportion of their eggs that are actually fertile is usually low, indicating that the females do not mate with males or "try" to raise a family—fertility rates as low as 0 percent for Kittiwakes, 0–15 percent for Western Gulls, 4–30 percent in Herring Gulls, 33 percent for Silver Gulls, and 8 percent in some populations of Ring-billed Gulls have been documented.[66] In addition, female pairs whose clutches are entirely infertile have been reported for Mute Swans, Black-winged Stilts, Roseate Terns, Blue Tits, Red-backed Shrikes, King and Gentoo Penguins, and Lovebirds (among others). Female Jackdaws who have lost their male partners sometimes pair up with nonbreeding females. However, these associations develop regardless of whether the widow has young, demonstrating that females do not form same-sex associations solely for the purpose of obtaining help in raising offspring. Moreover, only 10 percent of widowed females are involved in homosexual pairs, so even if such partnerships were "reproductively" motivated, it remains to be explained why only some females take advantage of such alternative parenting arrangements.

Furthermore, there are several different forms of same-sex parenting among birds (and other animals). In some cases, individuals develop full pair-bonds with their coparent, including courtship and sexual activity, and the partnership typically exists prior to and extends beyond the duration of parenting (e.g., Western Gulls, Black-winged Stilts). In other species, partners who already have offspring simply enter into a joint-parenting arrangement with no associated courtship or sexual activity between them, often lasting only until the young have been raised (e.g., Lesser Scaup Ducks). In still other cases, animals develop an intermediate arrangement, with "platonic" coparenting between individuals who may nevertheless continue to associate together even when not breeding (e.g., Acorn Woodpeckers, Squirrel Monkeys). And finally, in many species (e.g., Greylag Geese, Oystercatchers), individuals form bisexual trios that parent their offspring together (often contrasting with heterosexual trios and/or homosexual pairs within the same species).[67] All four types of arrangement could be interpreted as "strategies" to raise young, yet the differences between them remain unaddressed if homosexual associations are seen strictly as coparenting arrangements.

The putative benefits of same-sex breeding associations are also generally belied by the fact that not many individuals take advantage of them. The proportion of birds who participate in homosexual pairings or joint parenting arrangements is often relatively small—much smaller, in fact, than would be expected if this were simply an efficient or beneficial reproductive strategy. For example, most male Greater Rheas and Emus raise their young as single parents, but occasionally two males join forces, incubating their eggs in tandem and raising their chicks together. Single parenting can be taxing in these species—partnerless males, for instance, may fast during the entire incubation period, and single Greater Rhea fathers often lose eggs because they can't keep large clutches warm—so it has been suggested that two males may be better equipped to handle the difficulties of parenting by helping each other. However, only a small fraction of nests are tended by two males (less than 3 percent in Greater Rheas): if this were truly a useful parenting strategy, why wouldn't all males—or at least a larger proportion—be using it? Clearly something more—or something else—is involved in associations between males than simply the parenting benefits they may accrue. To further confound the picture, in Greater Rheas both same-sex coparenting and same-sex nest helpers occur. While some males jointly parent the same brood of young, a much higher percentage (about a quarter, still a minority of the population) are assisted by an adolescent male who separately parents one of their nests. Once again, this raises the question of why some males opt for joint parenting, others "choose" to have male helpers, while most do neither. And in many species the supposed advantages of coparenting as opposed to single parenting are in fact illusory. Most female Lesser Scaups raise their young with no help from males, but occasionally two or three females coparent. It is usually assumed that this strategy gives such females an advantage in parenting, but detailed studies of parental investment have shown that same-sex coparents are no more and no less successful than single parents. Moreover, each female in such an arrangement generally spends the same amount of time in parenting duties as do single females, i.e., she is not "relieved" of some of her responsibilities by her companion. In other words, there is essentially no reproductive advantage to joining forces with a parenting partner in this species.[68]

Nor is the occurrence of homosexual pairing in other species correlated with the supposed advantages of having an *opposite-sex* partner to help with parenting. Even in birds where male-female coparenting is typical, there are often significant differences between species in how essential that biparental care is to successful chick-raising. In some birds, females can raise young without the assistance of a male partner, while in other species the male's contribution is indispensable. If homosexual pairing were somehow related to the (in)ability of single birds to raise young on their own, one would expect same-sex associations to occur in species where biparental care is more important, i.e., where single birds cannot raise young on their own—yet the facts do not support this. Consider two parallel examples: Snow Geese and Black-billed Magpies. Homosexual pairing in female Snow Geese is claimed to allow otherwise single birds to raise young. However, biparental care is not essential for successful reproduction in this species: when females in heterosexual pairs have their male partners taken away from them, they are quite capable of raising their young as single parents. On the other hand, biparental care *is* essential in Black-billed Magpies, since females are unable to raise offspring on their own when they lose (or are deprived of) their male mates. Yet homosexual pairs of Magpies do not raise young together, nor do widowed females form same-sex pairs in this species (unlike the closely related Jackdaws). This is exactly the opposite of what would be expected if mateless birds were forming homosexual associations to enable them to parent.[69]

In fact, most pair-bonding birds do not form same-sex couples when heterosexual mates desert them or are experimentally removed, indicating that homosexual pairing is not a widespread mechanism for achieving two-parent care (regardless of whether the latter is "indispensable" or simply preferred). Moreover, in those polygamous (non-pair-bonding) species such as the Superb Lyrebird where females could benefit significantly from male parental assistance (and appear to suffer detrimental effects in its absence, such as slowed growth of their off-spring), female pairing and coparenting is noticeably absent.[70] Conversely, same-sex pairing and/or coparenting *do* occur in many species where single parents routinely raise young successfully. This is true for Hooded Warblers and Mallards (where heterosexual parents almost always separate and become single parents before the young are fledged), and Red Squirrels and Grizzlies (where heterosexual coparenting never occurs as part of these species' polygamous mating systems). In none of these animals is a two-parent family (either heterosexual or homosexual) absolutely *required* for successful parenting.

To take this line of thinking a step further: in a few species homosexual associations may actually be *detrimental* to parenting. Besides providing no apparent parental benefits to each other, Calfbird female companions may in fact increase their risk of predator attacks by nesting so close to each other (thereby drawing attention to their location). Female Japanese Macaques in homosexual consortships also do not typically assist their partner with parenting and are often notably aggressive toward their consort's offspring. Homosexual bonding is reproductively disadvantageous for both Oystercatchers and Jackdaws in bisexual trios, for slightly different reasons. Oystercatchers in such associations typically do not jointly incubate their supernormal clutches (only one bird sits on the nest at a time); because each incubator is unable to cover all the eggs simultaneously, the outsized clutch is often not kept adequately warm. As a result, bisexual trio parents hatch fewer eggs and produce significantly fewer fledglings than heterosexually paired Oystercatchers. Female Jackdaws in bisexual trios, on the other hand, *do* jointly incubate their supernormal clutches. However, because the two females are bonded to each other, both leave the nest together when their male partner arrives to relieve them, and he is unable to cover all their eggs and keep them warm. A parallel effect may occur in Lesser Scaup Ducks: although most female coparents exhibit remarkable cooperative defense of their joint broods, some pairs have been observed flying off together at the approach of a predator, temporarily abandoning their young in the face of danger. Finally, female Canada Geese in homosexual pairs sometimes roll eggs between their adjacent nests, breaking many of them in the process.[71] Clearly, then, successful parenting—and, by extension, reproduction or "perpetuation of the species"—cannot be the whole story behind the formation of same-sex pair-bonds.

Sperm-Swapping and Other Flights of Fancy

Attempts to determine the evolutionary "function" of homosexuality have sometimes led to even more obscure and implausible "explanations," all revolving (predictably) around heterosexual breeding. For example, some scientists have suggested that homosexuality is a form of reproductive "competition": females have sex (or form pair-bonds) with other females to monopolize their partner's time and thereby prevent her from mating heterosexually, while males mount each other to reduce or redirect their rival's sexual drive.[72] However, in many species homosexual interactions are actively initiated by the animals who are mounted rather than by the mounters, and the participants often have a friendly rather than a competitive relationship with each other.[73] Moreover, there is no evidence that participation in homosexual mounting reduces heterosexual activity—indeed, in some species the opposite is true, for the greatest amount of heterosexual mating is accomplished by precisely those individuals who are also the most active homosexually (as discussed previously). And, as already mentioned, in many animals homosexual activity does not even take place during the breeding season or is only exhibited by a small proportion of individuals.

Another version of this competition hypothesis is that homosexuality is a way of directly interfering with the heterosexual activity of a rival. In a number of birds—Pukeko, Guianan Cock-of-the-Rock, Ocher-bellied Flycatchers, and Buff-breasted Sandpipers—homosexual activity is claimed to be a form of "disruption" whereby one male prevents another from mating with a female, while possibly also "usurping" his partner and mating heterosexually himself. The specifics of same-sex courtship and mating in each case, however, do not support this interpretation. In Pukeko, for example, males do sometimes interrupt heterosexual mating attempts by inviting the other male to mount them, but they do not generally take advantage of the situation to mate with the female partner. Moreover, this occurs only infrequently, and males are more likely to ignore heterosexual matings by other males or watch them without interfering than they are to try to prevent them from occurring. In addition, even if a male were trying to use homosexuality as a way to disrupt a heterosexual mating, this strategy would not "work" unless the other male found the prospect of mounting him more appealing than completing his heterosexual copulation. Ironically, then, a "disruption" interpretation of same-sex mounting in this species—typically presented as an example of the primacy of *heterosexual* relations—actually entails the assumption that male Pukeko would prefer *homosexual* activity. In Ocher-bellied Flycatchers, the suggestion that males are trying to disrupt heterosexual matings, or to gain access to females, is entirely speculative. Males have never been seen mating with a female as a result of a homosexual interaction in this species, and in fact females are not even present during the majority of courtship pursuits between males. Homosexual activity is also classified as a form of courtship "disruption" in the Guianan Cock-of-the-Rock, yet there is little evidence in favor of this explanation. As much, if not more, homosexual activity takes place when females are not present on the male's display territories, and males who initiate such "disruptions" almost never gain access to members of the opposite sex as a result and have rarely even been observed mating with females. Furthermore, visits by yearling males involving homosexual activity are distinct from true courtship disruptions, which are performed by rival adult males. Yearling visits are directed toward a wide variety of males, all of whom cooperate in the interaction. In contrast, rival males target only the most successful heterosexual breeders and are violently attacked by the males they try to disrupt. In addition, same-sex interactions are sometimes directed toward adult nonbreeders, who do not participate in heterosexual mating at all, so it is difficult to see how this could be a form of "disruption."[74]

One species in which at least some homosexual activity appears to be genuinely associated with disruption of heterosexual mating is the Buff-breasted Sandpiper, in which rival males often interrupt each other's courtship attempts by mounting and pecking them. However, even in this case, the benefits of such activity are not clear-cut, since "disruption" does not always result in more favorable mating opportunities for the "disrupter" or reduced mating for the "disruptee." Although a "disruptive" male is often able to lure females away from his rival, in other cases he may keep returning to mount the rival without trying to mate with any females. In addition, detailed studies have shown that a male's success at copulating with females is not in fact related to his ability to repel disrupting males. Moreover, not all homosexual mounting is directly involved with disruption of heterosexual courtship, while many disruptions occur without any homosexual activity.[75] This brings up a point that is also relevant for other species. Many animals—including ones that exhibit homosexuality in other contexts—use *direct* tactics to interrupt or harass heterosexual matings. These include threatening or physically attacking couples during copulation, and trying to pull or dislodge the partners from each other. Even if homosexual behavior were in some instances being used as a form of heterosexual disruption, it would still remain to be explained why some species—or only some individuals in a species—resort to this fairly unusual and indirect strategy, when more effective and efficient measures are available.

Other zoologists have proposed, in all seriousness, that homosexual copulation is a way of transferring or "swapping" sperm between same-sex partners for heterosexual (breeding) purposes. For example, one ornithologist has suggested that a male bird might deposit his sperm in another male's genital tract during a homosexual mating so that the latter would then pass it on to a female during heterosexual copulation, indirectly fertilizing the female with the first male's sperm.[76] Not only is this explanation highly implausible, it is factually incorrect for many species. Homosexual matings often take place outside of the breeding season or when females are nonfertilizable. In addition, male birds often solicit homosexual mounts from other males or actively facilitate homosexual interactions, which should not occur if they are a form of "insemination rivalry." Finally, male birds frequently defecate when they are stressed (birds use the same orifice for all excretory and sexual functions)—thus, if same-sex activity were not consensual, males could easily empty their genital tract of any sperm that might have been put there by a rival male.[77]

Even more absurdly, this "explanation" has been proposed for lesbian copulations in Pukeko, where no sperm is directly involved. The claim is that female birds mate with each other to transfer ejaculate between them from *previous heterosexual matings*. And why exactly would they do this? Zoologists have suggested that this is not so they can "fertilize" each other, but so that they can obscure paternity, i.e., confuse several males as to who is the actual father and thereby "trick" more males into caring for their young. This is implausible, however, since a number of other mechanisms already insure that paternity is obscured and parental care is shared in this species. These include multiple copulations by females with all males in the group, variability in timing of ovulation, absence of mate guarding or copulation disruptions, and an independent tendency of males to be "generous" or "indiscriminate" in their parental efforts (i.e., caring for all chicks regardless of whether they fathered them or even mated with any females at all). Another dubious speculation is that female Pukeko copulate with each other so as to "synchronize" their sexual cycles, thereby allowing them to lay their eggs at the same time (which is thought to be a more efficient reproductive strategy). Once again, there is no evidence that lesbian copulations have this effect.[78]

It's hard to imagine more convoluted and conjectural accounts of homosexuality than these. While many of these ideas are highly unlikely and scientifically unsubstantiated, they can actually be traced to misconceptions about human homosexuality that are deeply entrenched in our culture. For example, the belief that lesbian sexual activity serves to transfer semen from heterosexual intercourse can be found in some of the earliest written records concerning same-sex activity in people. A twelfth-century Irish story about Níall Frassach, a king who died in 778 A.D., makes use of this theme:

> A woman came to the king carrying a boy child . . . "find out for me who the carnal father of this boy is, for . . . I have not known guilt with a man for many years now." The king was silent then. "Have you had playful mating with another woman?" said he. "And do not conceal it if you have." "I will not conceal it," said she. "I have." "It is true," said the king. "That woman had mated with a man just before, and the semen which he left with her, she put it into your womb in the tumbling, so that it was begotten in your womb."[79]

In discussing this curious tale, historian John Boswell remarks that it reveals a "preoccupation with women as bearers and conduits of bloodlines rather than as beings with their own erotic lives and needs."[80] It is an alarming comment on the "progress" of history that, nearly 900 years later, almost identical ideas about female animals should reappear under the guise of scientific theories, with scarcely any improvement in the perceptions of same-sex activity (or females, for that matter).

What Is Valuable?

Homosexuality is popularly considered to be nonreproductive or even counter(re)productive. In this section we have considered a wide range of proposals about the possible evolutionary value (and genetics) of homosexuality that challenge this "commonsense" view. These proposals concern ways that homosexuality might somehow contribute to the perpetuation of the species, either directly (for example, by improving an animal's reproductive prowess or increasing its heterosexual mating opportunities) or indirectly (for example, by providing breeding animals with "helpers" or acting as a population regulation mechanism). Implausible as some of these ideas may sound, many aspects of animal homosexuality run counter to preconceived ideas, not the least of which is the widespread participation of breeding animals in homosexual activity. A number of other unexpected phenomena in several species led us to consider further whether some of these proposals might actually have some explanatory value. Examination of deeper patterns within a broader range of animals, however, as well as more rigorous investigation of specific cases, showed that they do not. Thus, in the end we have arrived back at our starting point: homosexuality, whether in breeding or nonbreeding individuals, does not generally contribute to the reproduction of the species. This is an obvious point, perhaps, but one whose very obviousness has usually precluded a serious investigation of its validity. And so once again we are confronted with the evolutionary "paradox" of homosexuality: Why does same-sex activity persist—reappearing in species after species, generation after generation, individual after individual—when it is not "useful"?

Part of the problem is that "usefulness" or "value" in most biological theorizing is narrowly defined to refer only to reproduction. A common thread running through each of the proposals considered in this section is that they view homosexuality only in terms of how it could contribute to breeding or facilitate mating relations between males and females, rather than in terms of any intrinsic value it might have. This brings us to the final, and overarching, problem with all such "evolutionarily valuable" explanations. Scientists have often been led to absurd conclusions about the putative "function" of homosexuality precisely because evolutionary theory cannot readily countenance behaviors that are apparently "useless"—and for a behavior to be "useful" it must contribute in some way to mating and reproduction. Perhaps it is the very notion of "utility" or "value" that needs to be reexamined. In the realm of human culture and biology, the idea that life revolves around heterosexuality and that everything in life can be related, ultimately, to reproduction—a view sometimes known as heterocentrism or heterosexism—is currently being challenged on a number of fronts.[81] Yet this view would appear to be a self-evident truth where animals are concerned, since the passing on of genetic material through reproduction is considered to be the very foundation of biology and evolution. In the next section we will see that, on the contrary, this belief is as incomplete a description of animal biology as it is of human society.

Nonreproductive and Alternative Heterosexualities in Animals

The primary reason that animal homosexuality is considered problematic by many scientists and "abnormal" by many nonscientists—and therefore in need of "explanation"—is that it does not lead to reproduction, and reproduction is considered the be-all and end-all of biological existence. However, animal life and sexuality are *not* organized exclusively around procreation. Just as there is a multiplicity of kinds of homosexuality in the animal world, so, too, there are innumerable ways that males and females interact (sexually and otherwise) with each other, only some of which involve reproduction. In this section we'll explore a wide variety of nonreproductive and alternative heterosexualities: nonbreeding individuals, male-female segregation and hostility, "alternative" parenting and pair-bonding arrangements, and nonprocreative sexual practices. Some form of nonreproductive heterosexuality has been observed in nearly every animal species, far exceeding the incidence of homosexuality. So extensive is this phenomenon that, in the discussion that follows, we will only be able to give the barest indication of its scope and characteristics, as we survey a broad range of behaviors and species. For further details, the reader is referred to the profiles of individual animals in part 2, as well as the references contained in the notes to this section.

Life without Procreation:
Nonbreeders, Celibacy, and Reproductive Suppression

> *It is apparent that in some cases the bulls withdraw entirely from active*
> *participation in sexual competition for the herd.*
>
> —S. K. SIKES, *The Natural History*
> *of the Elephant*[82]

Virtually every animal population includes nonbreeding individuals. There is a tendency to regard the urge to procreate among animals as instinctual, all-pervasive, and unstoppable. While heterosexual interactions often do have this quality, there are just as many examples of animals who do not reproduce: individuals who actively remove themselves from the breeding cycle, whose nonparticipation in reproduction is guaranteed by the overall social organization of the species or by physiological constraints, who produce offspring rarely (if ever), or who lead complete lives after (or without) reproducing. Many nonbreeding animals are still sexually active; on the other hand, celibacy, abstinence, and other kinds of asexuality are also prevalent in the animal kingdom. The proportion of nonbreeding individuals varies widely between different species, and between different populations of the same species. In some cases, only a few lone individuals are not actively reproductive; at the other extreme are species where more than half (American Bison, Right Whales), three-quarters (Blackbuck, Giraffe), or even 80–95 percent (New Zealand Sea Lions, Northern Elephant Seals, naked mole-rats, some dragonfly species) of one or both sexes do not reproduce.[83] Between these extremes, nonbreeders may constitute a quarter (Long-tailed Hermit Hummingbirds) to a third (Common Murres, Kestrels) of the population.[84]

Many types of nonbreeding are found in the animal world, involving individuals of different ages, social circumstances, and varying lifetime reproductive and sexual histories. In hoofed mammals and seals, for example, males often "delay" reproducing for several years after they reach sexual maturity, frequently living in large "bachelor" herds separated from the breeding animals. Although many such animals eventually go on to reproduce, at any given time nonbreeders constitute a large segment of the population, in part because of a preponderance of younger animals in the demographics. In these species as well as others that have polygamous or promiscuous mating systems (where males typically mate with a large number of females without forming pair-bonds with any of them), there are usually also further "mating skews." Only a portion of the male population establishes breeding territories and courts females; of those who do, moreover, only a fraction ever get to actually mate with females and sire offspring. Among Guianan Cock-of-the-Rock, for example, a fifth of the males, on average, do not have courtship territories, while almost two-thirds of the males who do are unable to mate with females. In species that have a ranked form of social organization, it is typically only the higher-ranking males that participate in the most matings. In Squirrel Monkeys and Grizzly Bears (among others) the opposite sometimes happens: the highest-ranking males may fail to obtain any heterosexual copulations at all, due to their greater aggressiveness.[85] In many animals with communal breeding systems, only one or two individuals in each group reproduce while the others are nonbreeders; many of the latter help the breeding animals raise their young, but in a few species such as Red Foxes and Gray-capped Social Weavers, some nonbreeders do not even contribute to other group members' reproductive efforts as helpers.

A period of temporary nonbreeding can sometimes involve an entire population. In one troop of gray-cheeked mangabeys, for instance, all the females stopped cycling for a period of four months, while no reproduction took place in a population of Musk-oxen for several years.[86] In other cases, such as Hanuman Langurs, Northern Fur Seals, Mountain Zebras, Red Deer, Ruffed Grouse, Pied Kingfishers, and red-winged blackbirds,[87] some individuals live their entire lives without ever reproducing, while in some species of mole-rats as well as in Northern Elephant Seals, 90 percent or more of the population never procreates.[88] Entire flocks of Flamingos often abandon or "give up" on breeding in the middle of a season, or forgo reproducing for three to four years at a time, while individual female Silver Gulls may go as long as sixteen years without reproducing. Although most animals have yearly breeding cycles (sometimes even reproducing more than once a year), others have nonyearly or "supra-annual" cycles. King Penguins and Australian Sea Lions, for instance, have 16-to-18-month cycles, while large mammals such as Elephants, Manatees, and Whales typically reproduce only once every several years. Among White-handed Gibbons, males and females are thought to interact sexually with each other only every two years or so, while Siamang females often space their pregnancies by a couple of years, turning over parental duties to males while they assume leadership roles.

One particularly interesting form of nonbreeding involves "postreproductive" animals: individuals who have bred previously during their lifetime but are now "retired" from reproducing. Menopause and/or a period of nonbreeding in old age were long thought to be uniquely human traits. It was assumed that all animals continued to reproduce until they died, or alternatively, died shortly after they were no longer able to reproduce. In chapter 2 we saw the pitfalls of asserting human uniqueness in any area of behavior, and indeed postreproductive animals are now known to occur in several primate, hoofed mammal, seal, and whale species, and even in some birds such as Antbirds.[89] In some cases (e.g., male African Elephants), such animals are loners or peripheral to the social organization of the species. In other instances (e.g., Rhesus Macaques, short-finned pilot whales), they are integrated into the social fabric and may even assume central roles.[90] Among Killer Whales, for example, pods are often led by older, postreproductive matriarchs. Because males remain with their matriarchal group in this species, some pods eventually "die out" (even though they contain breeding-age males) because all of their females are postreproductive. Many postreproductive individuals remain sexually active until their death. Menopausal or old-aged female short-finned pilot whales, Orcas, Japanese

Macaques, and Hanuman Langurs, for example, frequently engage in heterosexual (and in some cases homosexual) activity, sometimes with younger partners.

Sexual activity also occurs in other nonbreeding animals. Among birds that typically form heterosexual pair-bonds, for instance, some individuals remain single yet still court or copulate with members of the opposite sex, often during periods when fertilization is not possible (e.g., Oystercatchers, Humboldt Penguins, Hoary-headed Grebes). In many other cases, birds form heterosexual pairs or trios but do not breed—even though they still continue to be sexually active. Researchers even found that some nonbreeding pairs of Canada Geese had higher copulation rates than breeders.[91] On the other hand, many nonbreeding animals are asexual or "celibate," not courting or interacting with members of the opposite sex at all. An interesting variation of this sort involves Japanese Macaques, who sometimes form "platonic" heterosexual consortships, in which either partner may nevertheless interact sexually with other individuals besides the consort. Similar platonic "friendships" are also found between male and female Savanna Baboons. Paradoxically, the opposite situation to sexually active nonbreeding animals also occurs in a number of animals. In several bird species, pairs that are ostensibly involved in breeding actually *stop* copulating before the female's fertile period has ended, while males of some marine turtles leave the waters where females are located long before the breeding season is over. Although most such animals do reproduce, it appears that in some ways they are not exploiting their reproductive potential to its fullest.[92]

Why do animals not reproduce? Biologists have coined the term *reproductive suppression* to refer to various forms of nonbreeding, implying that all animals would breed if they could, but are somehow "prevented" from doing so. However, the underlying mechanisms involved in nonbreeding are far more complex than this term implies. Numerous social, physiological, environmental, and individual factors are implicated, often interacting in ways that are still poorly understood.[93] In some animals, procreation is indeed actively "suppressed." In Wolves, for example, dominant pack members often physically attack lower-ranking individuals who try to mate; female Savanna Baboons sometimes form coalitions to attack cycling or pregnant females and prevent or terminate their reproductive efforts; while in many hoofed mammals, higher-ranking males prevent other males from gaining access to females. However, in other species the term *suppression* is a misnomer, since no coercion is involved. Young American Bison bulls, for instance, are not "prevented" from mating by older males—they simply do not participate to the same extent (as discussed in chapter 4). In other species—particularly birds with communal breeding systems such as the Pied Kingfisher, as well as primates such as tamarins and marmosets—scientists describe individuals not as "unwillingly" suppressed in their reproductive efforts, but rather as "choosing" to forgo reproduction or exercising "self-restraint" in their reproductive participation.[94] Further evidence that animals are often "voluntary" nonbreeders involves species such as Ocher-bellied Flycatchers and Ruffed Grouse, where prime breeding territories often go unused even though many nonreproducing individuals are in the population. Sometimes physiological mechanisms are involved in nonbreeding, such as lowered hormonal levels, delay of sexual maturation (sometimes indefinitely), inhibition of ovulation, and even blocking of pregnancy following conception (seen in many rodents).[95]

Finally, reproduction is often a physically demanding and exceedingly danger-ous undertaking that some animals may simply "avoid." Nonbreeders are often in better physical condition than breeders, since they do not have to experience the rigors of reproducing and parenting. In fact, breeding could even be considered "suicidal" in some cases, since it may lead to a reduced life expectancy. Male Bighorn Sheep and female Red Deer that breed, for example, have significantly higher mortality rates than nonbreeders. In several species of carnivorous marsupi-als, most males die after mating while nonbreeders generally survive longer. The life expectancy of nonbreeding male Ruffed Grouse often exceeds that of breeding males. And female Western Gulls that breed more often during their lives have a lower survival rate than individuals that are less reproductively "prolific." Sometimes specific biological factors serve to discourage breeding, such as the astonishing phenomenon of clitoral, as opposed to vaginal, births in the Spotted Hyena. Many females of this species die during their first pregnancy or labor because their genital anatomy requires the baby to be born through the clitoris, which ruptures and often results in many other complications for both mother and fetus.[96] Finally, the risk of acquiring sexually transmitted diseases (which are found in a surprising number of animals) may also affect reproductive activity. For example, female Razorbills (a kind of bird) avoid reproductive copulations with males when the risk for infection from STDs is greatest (although they continue to have nonprocreative sex, i.e., mounting without direct genital contact). Heterosexual behavior in a number of other species may also be curtailed by the potential danger of STDs.[97]

In the end, then, there is no single "reason" why animals don't reproduce: non-breeding, like sexuality, is simply a part of the fabric of animals' lives, manifesting itself in many different ways. Heterosexuality constitutes a whole range of behav-iors and life histories, not a single, inalterable template that every animal must fol-low. And nonbreeding is one of the many ways to be "heterosexual." Regardless of the number of nonprocreating animals in a particular population or the "causes" of their nonbreeding, one thing is certain: nonreproducing (heterosexual) animals are a ubiquitous feature of animal life.

Worlds Apart: Sex Segregation, Hostility, and the Dark Side of Heterosexuality

> *Adult males and females [of the Sperm Whale] have lifestyles so distinct that they might be separate species. The males leave tropical waters each summer and voyage into the highest latitudes . . . but females and young seldom venture more than 40° from the equator.*
> —LYALL WATSON, *Sea Guide to Whales of the World*[98]

Heterosexual mating is anything but the "natural," effortless activity that it is often portrayed as. There are many ways that sexual interactions between males and females are avoided, exacerbated, or generally fraught with problems. In numerous animals, for example, it almost seems that the social organization and behaviors of the species have been *designed* to keep males and females apart and prevent reproduction—or at least make it difficult. Consider sex segregation: partial or total separation of males and females is a surprisingly prevalent form of social organization in the animal world. Various forms of sex segregation occur in mammals and birds of all types, though separation of the sexes is especially prevalent in species such as hoofed mammals that have promiscuous or polygamous mating systems (where individuals mate with more than one partner). Often the only time that the two sexes come together is to mate, sometimes for only a few days or months out of the year—the rest of their time is spent living entirely apart. Even "harems"—in which one male associates with a group of females and often prevents other males from gaining access to them—are not the quintessential example of heterosexual mating opportunity that they are usually thought to be. Scientists studying "harems" in a number of species such as sea lions and some hoofed mammals have found that these groups do not always form as a result of heterosexual attraction or males "controlling" females (and thus the term is somewhat inaccurate). Rather, females prefer to associate with each other and therefore they congregate in relatively autonomous groups of their own; males that participate in breeding then end up associating with such groups out of necessity.[99] In addition to social and spatial segregation—living in separate groups or habitats—sex segregation can also be seasonal or migratory. It may occur only during the nonbreeding season, for example, or involve separate migratory journeys or latitudinal destinations for males and females (for example, in Northern Elephant Seals and Kestrels). One of the most extreme forms of "sex segregation" occurs in several species of marsupial mice: all the males die a few days after the mating season, so that when females give birth there are no adult males left in the population.[100]

Sex segregation during the breeding season is often facilitated by a phenomenon known as sperm storage: most female animals have one or more special organs or sites in their reproductive tract that allow them to store a cache of sperm (from a prior mating) for a long time, using it later to "inseminate" themselves while forgoing heterosexual copulations. Birds and reptiles have special glands that allow them to do this. Female Ruffs, for example, often leave the breeding grounds (after having mated with males) and migrate northward, laying their eggs several weeks later by fertilizing them with stored sperm. In some birds such as the fulmar, sperm may be stored by females for up to eight weeks, while in reptiles (as well as insects) sperm stored in females may remain viable for much longer, up to several months or even years. Female garter snakes, for instance, are able to keep sperm for up to three to six months after mating with a male. In fact, females in this species usually do not ovulate until two to five weeks *after* their first mating in the spring. They may even become pregnant without mating at all that season, simply by using sperm from a copulation that took place the previous fall before hibernation. The record for sperm storage is held by the female Javan wart snake, who can store sperm for up to seven years! In mammals, sperm is generally "stored" for shorter periods (although some bats can do so for more than six months) and may be kept in "crypts" on the cervix or inside the uterus. Recent work has also shown that females in most species can control, through behavioral, anatomical, and physiological mechanisms, which portion of the sperm (if any) is stored and/or utilized for fertilization.[101]

A phenomenon known as delayed implantation also enables males and females to spend long periods away from each other. In nearly 50 mammalian species (including seals, bears, other carnivores, marsupials, and some bats) the fertilized egg does not implant right away. It remains in "suspended animation" for several months, after which it implants and begins its regular development. The delay extends the pregnancy by two to five months in seals and up to ten to eleven months in badgers, fishers, stoats, and related small carnivores. In seals, this allows females to spend a longer time out at sea—often completely separate from males—and permits them to optimize the timing of their pregnancies and to take advantage of more favorable times of the year for birthing and pup-raising. Some species of bats also have delayed embryonic development, in which the fertilized egg experiences a temporary cessation of development *after* implantation.[102]

In fact, delayed implantation as well as sperm storage (among a variety of other factors) effectively result in a separation and reordering of key reproductive events in many vertebrates, and consequently an "uncoupling" of male and female reproductive cycles. We are used to thinking of breeding as an ordered progression, one stage leading inevitably to the next: ovulation followed by mating followed by fertilization followed by pregnancy followed by birth (or egg laying). However, there are often significant gaps and rearrangements of these events: sperm storage can temporally separate mating from fertilization, while delayed implantation separates fertilization from fetal development during pregnancy. As noted above, sperm storage can also result in ovulation taking place *after* insemination, and in other animals, further rearrangements occur. In birds, for example, "pregnancy" or the development of the egg inside the mother's body actually precedes fertilization: the egg yolks are already quite large (and may cause a noticeable bulge and weight increase in the female) prior to being fertilized. In fact, the eggs can be laid without ever being fertilized—this is what allows females in homosexual pairs to produce (infertile) eggs. In most fishes, "pregnancy" *ends,* rather than begins, with mating: the eggs develop within the female's body and are then laid or discharged when ready to be fertilized (i.e., fertilization typically takes place outside the female's body).[103] In addition to these delays and reordering of reproductive events, breeding can also, of course, be interrupted or terminated at any of these stages—this will be discussed in the next section, when we look at naturally occurring forms of birth control.

Another common misconception about animal heterosexuality is that only females experience periodic hormonal fluctuations in their reproductive biology. In fact, many male animals also have sexual cycles, entailing considerable periods in which they are sexually inactive and living separate from females. Occasionally, male and female sexual cycles are poorly synchronized or not optimal for breeding, as sometimes happens in Ostriches and Lovebirds. Male cycles are found in a wide range of animals, including primates, deer, seals, and numerous bird species, and usually entail a yearly, rather than a monthly, periodicity. In some instances, dramatic physical and physiological changes are involved. Male Wattled Starlings, for example, undergo regular periods of "balding" (feather loss) and wattle development, and males of many other bird species develop dramatic nuptial plumages associated with breeding. Male Squirrel Monkeys become "fatted" during the peak of their sexual cycle, while male Elephants experience "musth," involving a whole host of changes such as glandular secretions, increased aggression, and rumbling vocalizations.

The significance of male sexual cycles has often been lost or overlooked under sexist biological theories, which tend to emphasize aspects of animal biology that confirm the unflagging "virility" of male animals, to the exclusion of those things that underscore the similarities between male and female sexuality. In fact, reproductive traits that are usually thought to be exclusively male or female can be found in members of the opposite sex in at least some species. Male pregnancy occurs in sea horses, for example, while lactation—milk production from fully functional mammary glands—was recently discovered in male Dayak fruit bats.[104] Females, for their part, can carry sperm within their bodies and "inseminate" themselves (as discussed above) or may possess elongated, phalluslike clitorides that can undergo erections (this is found in numerous mammals, including Spotted Hyenas, moles, and Squirrel Monkeys, as well as several flightless birds).[105] Some animals (e.g., Seals, Bears, Squirrels) even have a clitoral bone, homologous to the male's baculum or penis bone in these species; in female Walruses, this bone may be over an inch long. Female pipefish and Japanese sea ravens (a kind of fish) even have extendable genital organs used to penetrate or retrieve sperm from their male partners.[106]

When males and females do manage to get together, a formidable set of obstacles often stands in the way of achieving sexual contact and, ultimately, reproduction. Refusal or indifference by either the male or female partner is widespread and routine in the animal kingdom, and heterosexual matings are often "incomplete" in the sense that they do not involve erection, genital contact, ejaculation, and/or insemination. In one study of Chaffinch heterosexual copulations, for example, every "complete" and "incomplete" mating attempt was logged: out of 144 attempts, only 75 (52 percent) involved mounting with full genital contact (and therefore could potentially have led to fertilization). Of the remaining "unsuccessful" attempts, 76 percent entailed no mounting at all because one or both partners fled before copulation could take place, 9 percent involved the male mounting without attempting to make genital contact, in 8 percent mounting was terminated when the female refused to continue (in some cases after being pecked by the male), in 5 percent of the mounts the male slipped off the female's back, while in 1 percent of the cases the male mounted in a reversed head-to-tail position and therefore did not make genital contact. In African jacanas, only about one in four sexual solicitations by the female result in the male actually mounting her.[107] In some species, completion of the sexual act is prevented because of interference from other animals, who actively harass males and females while they are copulating. This is typical of many primates, but has also been reported in some birds such as King Penguins, Kittiwakes, and Sage Grouse.[108]

In a number of animals, it appears that male and female anatomy are not ideally suited to heterosexual interactions. The female Elephant's vaginal opening, for example, is much farther forward on her belly than in other mammals. Although the male's penis has a special shape and muscles that allow it to reach the female's genitals, he still often experiences considerable difficulty in achieving penetration and may end up ejaculating on her anus or otherwise outside her body. Moreover, it is not true that male and female genitals always fit together like a "lock and key": in many species the structural "compatibility" of the sex organs is less than perfect. In addition, the female's internal reproductive tract in most animals is—in the words of several zoologists—a tortuous, obstacle-ridden pathway that is "remarkably hostile to sperm." Its structure, chemical composition, and immune response to semen are actually designed to *prevent* most sperm from ever achieving fertilization, in part to protect the female from possible infection (sperm are, after all, "foreign" bodies) and in part to allow her to control paternity.[109] Males and females may be anatomically incompatible in other respects as well. Biologists studying Musk-oxen have observed that the male's build—a deep chest with short legs, and most of his considerable weight concentrated in the front half of his body—is decidedly ill-suited to mounting and clasping the female, and studies have shown that males are able to successfully mount females less than a third of the time.[110] In many other species of hoofed mammals and seals where there is a significant size difference between the sexes, females often fall or are crushed under the weight of the male during copulation and may suffer serious (even lethal) injuries.

Sometimes outright hostility erupts between males and females, including chasing and harassment, as well as actual aggression, violence, and injury (inflicted by males on females and, less commonly, by females on males). Attacks can be as brutal as they are commonplace. Female Savanna (Olive) Baboons, for instance, are liable to be attacked almost daily by males without provocation, and each female is severely wounded about once a year; injuries are sometimes fatal. Sexual coercion (i.e., punishment or intimidation by males of "uncooperative" females) as well as outright rape also occur in a wide variety of animals, occasionally involving "gangs" of males that attack and forcibly mate with females.[111] Heterosexual rape is especially prevalent among birds such as ducks and gulls, but also occurs in mammals such as primates (e.g., Orang-utans, Rhesus Macaques), hoofed mammals (e.g., Bighorn Sheep), and marine mammals (Right Whales, and numerous seal species). In birds that form pair-bonds, rape usually involves mating attempts on females other than the male's partner, but forced copulation within the pair-bond is not unknown, occurring, for example, in Silver Gulls, Lesser Scaup Ducks, and several other duck species.

Throughout the animal kingdom, heterosexual mating can be a dangerous and even lethal undertaking for females. Male Sea Otters often bite females on the nose during aquatic copulations, sometimes resulting in drowning or fatal infections; swarms of a dozen or more woodfrogs often try to mate with the same female, occasionally killing her in the process; female sharks of several species are routinely and severely bitten on the back during heterosexual courtship; while male mink may puncture the base of the female's skull and brain with their teeth during mating.[112] These are just a few examples of how heterosexual mating is often a destructive, rather than a procreative, act.

Animal Family Values: Birth Control, Day Care, Divorce, and Infidelity

> *Although two young are born [in Pronghorn antelope], four to six embryos are implanted in the uterus, where they fight to the death, so to speak, for the limited space. . . . Long projections grow out of the embryonic sheath which puncture other embryos, causing their death. All but two embryos are reabsorbed by the mother's body.*
>
> —VALERIUS GEIST, "Pronghorns"[113]

When most people think about animal families, they imagine a mother deer lovingly tending her fawns, or a father bear diligently protecting mother and babies. The realities of animal heterosexuality are a far cry from this romanticized view. More often than not, a mother deer viciously drives her yearlings away from the family group (when she hasn't aborted her fetuses, that is) while father bear rarely has anything to do with his family—and when he does, it is often to kill and eat them. This section examines some of the stark realities of animal "family life," beginning with a look at the many mechanisms that animals use to avoid having families altogether, by limiting reproduction and eliminating offspring.

In addition to infrequent copulation or mating during times when fertilization cannot occur, several other forms of "birth control"—i.e., ways of preventing pregnancy—occur in animals.[114] In fact, more than 20 different strategies have been identified whereby females are able to limit, control, or prevent insemination. The widespread occurrence of these phenomena throughout the animal world has led one scientist to conclude that "copulation . . . seldom leads directly and inevitably to fertilization."[115] Some female birds such as brown boobies defecate during mating rather than performing the customary genital contractions, thereby preventing insemination, while females of a number of insect, bird, and mammal species actively eject semen after copulating.[116] Among mammals, vaginal or copulatory plugs (sometimes also known as chastity plugs) are found in a number of species. Gelatinous barriers form (or are deposited) in the female's reproductive tract in several different kinds of rodents, bats, insectivores, and wild pigs, as well as in some primates and dolphins. Although their function is not fully understood, it appears that plugs often serve to prevent insemination. In many species a male leaves a copulatory plug in the female following mating (or the semen simply coagulates to form the plug) so that other males will not be able to fertilize her. In Squirrel Monkeys and some bats, hedgehogs, and opossums, however, the female herself produces the plug (often from sloughed vaginal cells), probably to control or prevent inseminations by males. In addition, female squirrels sometimes entirely remove copulatory plugs—which contain all the semen deposited by a male—thereby effectively preventing insemination from their most recent mating.[117] Finally, scientists recently discovered that female Common Chimpanzees employ an extraordinary form of birth control: nipple stimulation. As in a number of other mammals, the regular reproductive cycles of female Chimps are inhibited or interrupted while they are suckling infants (known as lactational amenorrhea). Some females without infants have learned that by stimulating their own nipples they can effectively mimic this physiological effect, thereby preventing themselves from conceiving even though they are not actually lactating. In some cases, Chimps have avoided pregnancy for as long as a decade by employing this ingenious "contraceptive" technique.[118]

Following conception, pregnancy can be blocked (the fertilized egg does not implant, a phenomenon found in many rodents, where it is known as the Bruce effect).[119] Embryos can also *kill each other* by puncturing or strangulation (Pronghorns) or by actively devouring one another inside the uterus (sand sharks, some salamanders).[120] Embryos may also be "eliminated" because the mother has too few nipples to accommodate all of them (some marsupials such as Northern Quolls), while embryos of many hoofed mammals are simply reabsorbed.[121] Actual abortion occurs in many species, including primates (e.g., Hanuman Langurs, Pig-tailed Macaques, Savanna Baboons), marine mammals (Australian Sea Lions, other seals), hoofed mammals (feral Horses, White-tailed Deer), carnivores (Red Foxes), and numerous rodents and insectivores (wood rats, voles, coypus). Abortions may occur either spontaneously, or as a result of stress and harassment from males, or (in primates) possibly also from deliberate ingestion of abortion-inducing plants. Though usually a sporadic or isolated occurrence, abortions may be more commonplace in some species or populations. In California sea lions, for example, large numbers of females routinely abort their fetuses: hundreds of abortions take place each year on some breeding grounds, often as long as four months in advance of the usual birthing period.[122] Many birds and other species that lay eggs practice the equivalent of "abortion"—that is, the termination of an embryo's development—in the form of egg destruction (also known as ovicide), ejection of eggs from the nest, and/or clutch abandonment.[123]

Following birth or hatching, many animals employ strategies of direct or indirect killing to "eliminate" offspring. Infanticide or direct killing of young is a widespread phenomenon in the animal kingdom, reported in all major animal groups.[124] It is also a decidedly heterosexual behavior, often revolving around the creation of new breeding "opportunities." In one common form of infanticide, for instance, a male kills youngsters so that he can mate with their mother and sire his own young.[125] In another type of infanticide, females kills their own or related offspring—in black-tailed prairie dogs, for example, almost 40 percent of all litters suffer partial or complete loss to infanticide by females.[126] Cannibalistic infanticide also occurs, in which young animals are not only killed but also eaten.[127] Offspring are sometimes subjected to neglect and abuse as well, ranging from "absentee" parents who fail to incubate their eggs or properly attend to their youngsters (e.g., storm petrels, Oystercatchers, King Penguins) all the way to abandonment, physical brutality, and/or sexual violence inflicted on youngsters (e.g., Hanuman Langurs, Northern Elephant Seals, Ring-billed Gulls), sometimes resulting in death.[128] In many bird species, the size of families is regulated through a combination of factors. Parents frequently control the sequence of egg laying and hatching so that some offspring routinely perish (usually the last egg to be laid or hatched). In other cases "surplus" or extra young are produced as a "backup" strategy and usually end up fighting and killing each other (a phenomenon sometimes known as siblicide or cainism).[129]

Once animals establish a family, an enormous number of different parenting arrangements can be employed—only a small fraction of which involve a "nuclear family" configuration with a mother and a father both caring for their offspring. In the majority of animals, single parenting (or no parental investment at all) is the rule. In most mammal species, for example, no long-lasting bonds are formed between the sexes, and females raise their young on their own. Even in birds, where "nuclear" two-parent heterosexual families are typical, single parenting also sometimes occurs. In a number of species, male-female pairs routinely separate and one bird takes over parental duties—often the female, but occasionally the male, as in Common Murres and whimbrels. Sometimes this occurs only a few days after hatching (e.g., whimbrels), or even before hatching (e.g., ducks). In other cases the brood is actually split between the two parents, as in some woodpeckers, Hooded Warblers, and many other perching birds.[130] The opposite of single parenting is also found: many birds raise their young in communal breeding groups with multiple parents and caretakers of both sexes, and parenting trios occasionally form in species that otherwise have "nuclear families."[131] A phenomenon that could perhaps be called double-parenting also occurs in Golden Plovers, in which two heterosexual couples join forces and raise their youngsters as a quartet of parents. In general, then, a male-female parenting couple is neither a necessary nor a common arrangement among animals.

In addition to the wide range of family constellations in which animals raise their young, nearly 300 species of mammals and birds have developed adoption, parenting-assistance, and "day-care" systems, in which offspring are raised or cared for by animals other than their biological parents. Sometimes a sort of "baby-sitting" arrangement is formed (also known as alloparenting), in which a male or female assists another individual or couple in the care of their young (including "wet nurses" who suckle another female's young). This helper may be a relative or may not be related at all. In other species, groups of youngsters are pooled together into groups—known variously as crèches (e.g., Botos, Flamingos, Cliff Swallows, and many other bird species), nursery groups or calf pools (e.g., Giraffes, Wapiti), and pods (e.g., Northern Fur Seals). These groups are usually looked after by one or two adult "guardians" while the parents are out foraging or socializing. Such systems can be viewed both as examples of adults being freed from their parenting duties by a natural "day-care" system and instances of animals forgoing a portion of their reproductive "responsibilities" in order to pursue other activities. Outright adoption and various forms of foster-parenting and stepparenting also occur across a wide spectrum of animals. Many Gull chicks actually desert or "run away" from their families as a result of neglect or violence (and are adopted by other families), and White Stork and lesser kestrel chicks also sometimes abandon their nests and switch to neighboring "foster families." In birds, many other types of "adoption" result from eggs being abandoned by other birds, laid in other families' nests, or even transferred to other nests by being carried or swallowed and regurgitated whole.[132] In a few species, "kidnapping" of youngsters or stealing of eggs (with subsequent foster-parenting) may also take place.[133]

Heterosexual mating systems also exhibit a dizzying variety of forms. Pair-bonding between males and females is found in some mammals and most birds, but the majority of animals have polygamous or promiscuous systems, in which animals mate or bond with several different partners. This can take the form of either one male with several females (*polygyny*, the most common form), one female with several males (*polyandry*), a combination of both (each sex mating and bonding with several partners, or *polygynandry*), or mating with multiple partners with no bonding between them (*promiscuity*).[134] Even in species that form male-female pairs, however, there are many different arrangements. Heterosexual pair-bonding was long thought to be a simple and straightforward type of mating system, but biologists now recognize that—as in most other aspects of sexuality and social organization—animals exhibit considerable flexibility and diversity in their pairing arrangements.[135] Many species such as willow warblers, Eleanora's falcons, and sea horses are strictly monogamous. In numerous others, however, "infidelity" or non-monogamous matings occur, among at least a subset of both males and females.[136] Often, such copulations take place at times when females cannot be fertilized, so they are not entirely related to reproduction. In spotted sandpipers and Shags, for instance, almost all "unfaithful" copulations occur outside the females' fertilizable periods, while female Razorbills specifically avoid full genital contact during non-monogamous matings until *after* fertilization is no longer possible.[137] Other, more complex arrangements are also found: the mating system of some species, such as Tasmanian Native Hens, is described as "social polygamy with genetic monogamy." These birds live in polygamous groups, often several males mating with one female, but only one male fathers offspring with the female. This is the opposite of birds that form heterosexual pairs ("social monogamy") but mate and produce offspring with other partners ("genetic polygamy"). In many species where individuals typically form pair-bonds, there is also usually a subset of individuals that form heterosexual trios.[138]

Many pairing systems could probably be characterized as "serial monogamy." Even in birds that tend to form lifelong pair-bonds, divorce occasionally happens, and in many species pairs break up much more frequently, individuals usually then remating with other partners.[139] In Oystercatchers, for instance, divorce and remating is quite common (especially among females), and some individuals have as many as six or seven consecutive mates over their lives. Overall divorce rates vary widely between different individuals and species, from 0 percent in Australian ravens and wandering albatrosses to about a quarter of Kittiwake, a third of natal robin, and two-thirds of Lapland longspur pairs, to nearly 100 percent of all pairs in house martins and Flamingos. Divorce may result from a failure to produce offspring, but in many cases a complex interaction of multiple factors is involved, including general partner incompatibility. Other types of heterosexual family breakup also occur: extended families in Ocellated Antbirds may disintegrate when male-female pairs leave or grandparents isolate themselves; Warthog family units composed of a male and a female with young are generally less stable than female-only families; yearling White-tailed Deer are usually driven away by their own mothers; and Snow Geese family units may break up prematurely when juveniles leave.[140]

Heterosexual mating and parenting arrangements come in a staggering variety of forms—it is simply not the case that one type of "family" configuration is utilized by all species, or even by all individuals within the same species, or by the same individual for all of his or her life. Animal heterosexuality (like homosexuality) is a truly multidimensional, polymorphous phenomenon.

Sex without Purpose: Pleasure and Nonreproduction

> *Suzie stood with her back to Unk and she leaned her upper torso down-ward. He proceeded to manipulate her genitalia. That same day, . . . Suzie allowed subadult Smitty to lick her clitoris. . . . Observations . . . indicated that Suzie may experience orgasm . . . a shudder coursed through her entire body and then she became rigid.*
>
> —GREYSOLYNNE J. FOX, *Social Dynamics in Siamang*[141]

Even when males and females can overcome the considerable hurdles standing in the way of mating, they often engage in sexual activities that do not lead to re-production. Several different forms of such "purposeless" sexual behavior can be identified, the most common being heterosexual sex that involves partners or situations where fertilization is impossible. As previously mentioned, many animals routinely mate (or engage in other sexual activities) outside of the breeding season or when the female is not ovulating—including during menstruation and preg-nancy (or, in birds, during the incubation period). Not only is this found in a wide variety of animals—in mammals, for example, among various primates, hoofed mammals, carnivores, marsupials, rodents, and so on—but such nonprocreative sexual activity frequently constitutes a significant portion of all sexual behavior. In Common Murres, for instance, about half of all copulations in some populations occur during times when fertilization is not possible, while in Proboscis Monkeys and golden lion tamarins, a peak in sexual activity often occurs during preg-nancy.[142] About half of all pregnant or menstruating Rhesus Macaques are sexu-ally active, and some males mate with pregnant females as often as they do with ovulating females. In fact, sexual activity sometimes occurs *during* or *shortly after* birth in this species: males have been observed mounting females who just gave birth, while female attendants occasionally masturbate themselves while watching a female in labor. The birth process itself also stimulates sexual interest (courtship, mounting) in several species of hoofed mammals, including Mountain Goats, ad-dax antelopes, and wildebeest.[143] In none of these cases can the "function" of such sexual activity be procreation. Heterosexual behavior also occurs among sexually immature animals, between adults and juveniles, between genetically related ani-mals, between members of different species, and sometimes even between live and dead animals—all instances in which reproduction is not optimized (if not alto-gether impossible).[144]

Multiple copulations—in which animals mate far in excess of the amount required for fertilization—are also widespread. Several species of wild cats and birds of prey, for instance, have astonishingly high copulation rates. Lions may mate up to 100 times a day during the breeding season (or as much as 1,500 times for each litter produced), while heterosexual pairs of goshawks and American kestrels mate 500–700 times for each clutch of eggs they produce.[145] Oystercatcher pairs also copulate about 700 times each breeding season, while female Kob antelopes may each experience several hundred heterosexual mounts during a 24-hour visit to the mating grounds.[146] In addition, animals of some species (e.g., Spinner Dolphins, Gray and Bowhead Whales, Herons, Swallows) engage in group sexual activity in which only a small subset of the participants (if any) are actually passing on their genes and reproducing.

Specific nonprocreative heterosexual practices in the animal world are many and varied, and they often parallel homosexual behaviors as well as the wide variety of nonreproductive sexual practices found in humans. To begin with, mounting that does not involve full genital contact—sometimes described as "symbolic," "display," or "noncopulatory" mounting—is widespread. For every "full" copulation in Kob antelopes, for instance, an average of three mounts without erection and six mounts with an erection but no penetration are performed by the male.[147] Reverse mounting—in which the female mounts the male, usually without mutual genital contact—also occurs in a wide variety of species, and sometimes involves "reciprocal" mounting or sequential exchange of positions between the male and the female.[148] In some species, males occasionally mount females from the side or in other positions that do not involve penetration or genital contact: for example, Japanese Macaques, Waterbuck, Mountain Sheep, Takhi, Collared Peccaries, Warthogs, Koalas, Ruffs, Hammerheads, and Chaffinches. Many other types of nonprocreative sexual acts occur in mammals: various forms of oral sex (including fellatio, genital licking, and beak-genital propulsion); stimulation of a partner's genitals with the hands or other appendages (such as flippers), including vaginal penetration with the fingers (in primates); anal stimulation, including penetration with fingers or oral-anal contact (e.g., Orang-utans), rump rubbing (e.g., in Bonobos and Common Chimpanzees), and even heterosexual anal intercourse (e.g., in Orang-utans).

Masturbation also occurs widely among animals, both male and female. A variety of creative techniques are used, including genital stimulation using the hand or front paw (primates, Lions), foot (Vampire Bats, primates), flipper (Walruses), or tail (Savanna Baboons), sometimes accompanied by stimulation of the nipples (Rhesus Macaques, Bonobos); auto-fellatio, or licking, sucking, and/or nuzzling by a male of his own penis (Common Chimpanzees, Savanna Baboons, Vervet Monkeys, Squirrel Monkeys, Thinhorn Sheep, Bharal, Aoudad, Dwarf Cavies); stimulation of the penis by flipping or rubbing it against the belly or in its own sheath (White-tailed and Mule Deer, Zebras, and Takhi); spontaneous ejaculations

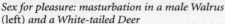

Sex for pleasure: masturbation in a male Walrus (left) *and a White-tailed Deer*

(Mountain Sheep, Warthogs, Spotted Hyenas); and stimulation of the genitals using inanimate objects (found in several primates and cetaceans; see chapter 2 for further discussion). Many birds masturbate by mounting and copulating with tufts of grass, leaves, or mounds of earth, and some mammals such as primates and Dolphins also rub their genitals against the ground or other surfaces to stimulate themselves. One fairly unusual form of (indirect) genital stimulation occurs in some hoofed mammals. Among male Red Deer, Moose, Wapiti, and other species of Deer, the antlers are erotic organs that can result in sexual arousal and even ejaculation when they are rubbed. In addition to occasionally stimulating each other this way, males of these species often stimulate themselves by rubbing their antlers in clumps of vegetation.[149]

Masturbation in female mammals, as well as heterosexual and homosexual intercourse (especially in primates), often involves direct or indirect stimulation of the clitoris (as in the description at the beginning of this section of oral sex among Siamangs, a primate species). This organ is present in the females of all mammalian species and several other animal groups, yet it has generally elicited a uniform reaction throughout much of scientific history: stunned (and embarrassed) silence.[150] This is due not only to the general hush surrounding female sexuality, but because the clitoris poses serious challenges to conventional biological theories. Its only "function" appears to be sexual pleasure, and the notion of pleasure in animals, particularly as it relates to the phenomenon of female orgasm, is a difficult one for biologists to come to terms with. Scientists have been remarkably reticent on the subject, refusing even to believe that female animals can experience orgasm until the phenomenon was "proven" with detailed observations and experimental studies on monkeys.[151]

Even after it was "verified," a debate about the "function" of the female orgasm erupted in the scientific community and continues nearly unabated to this day.[152] When a male animal has an orgasm—i.e., ejaculates—this is typically explained as the "mechanism" that insures sperm is transferred to the female—not as the pursuit of sexual pleasure. But no such mechanistic "explanation" is available for the female orgasm or clitoris. Most current biological discussion of the female orgasmic response attempts to justify its existence in terms of how sexual pleasure might "encourage" or contribute to breeding or social bonding, rather than seeing it as something inherently valuable that requires no further "justification." As always, female sexuality—and sexual pleasure in general—is assumed not to exist until proven otherwise. Once "proven" it requires a "function" or "purpose" rather than having intrinsic worth—a striking echo of the presumption of heterosexuality in biology and the need to find an "explanation" for the occurrence of homosexuality.[153]

Conclusion: Toward a Biology of the Twenty-First Century

The phenomena of nonreproductive and alternative heterosexualities have broad implications for how we look at animal behavior and sexuality in general. Animal social organization and biology do not revolve exclusively around reproduction and, in many cases, appear to be designed specifically to *prevent* procreation. Although heterosexual mating can (and frequently does) lead to reproduction, this is often an incidental consequence rather than an overriding "goal" (or ultimate "purpose"). Sexuality between males and females assumes a wide variety of forms, many of which necessitate recognizing sexual pleasure as a motivating force.[154] Homosexuality is, therefore, not unique in the animal kingdom by virtue of its "failure" to lead to procreation. It is simply one of many animal behaviors that lack the supposed "purpose" of contributing directly to the perpetuation of the species.

Nor is homosexuality unique in being considered a behavioral "anomaly" by scientists. Because they challenge some of biology's most fundamental assumptions about how the natural world is organized—while also reflecting stigmatized human behaviors—nonreproductive and alternative heterosexualities have inspired many of the same negative reactions that homosexuality has. As one ornithologist observes, "Until quite recently, [heterosexual] infidelity among wild birds was written off as aberrant behavior, and males were excused as being 'sick' or having a 'hormone imbalance' or, in one case, being 'dissatisfied at home.'" Likewise, the Brown-headed Cowbird's habit of abdicating parental care (by "parasitizing" other species' nests) was (and in some cases still is) considered particularly "loathsome" by many biologists. As a result, research on this common species has been severely hampered, and information on some basic aspects of its biology and social organization was lacking until fairly recently.[155] Analogously, behaviors such as masturbation, heterosexual trios, interspecies matings, nest desertion, reverse mounting, and forced copulations have all been labeled *abnormal, aberrant, unnatural,* or *anomalous*—in some cases as recently as a decade or so ago.[156]

Unlike homosexuality and transgender, however, most of these phenomena are no longer pathologized by the majority of contemporary biologists, who now recognize that these behaviors are "normal"—that is, a routine aspect of the social and sexual organization of the species in which they occur. Nevertheless, nonreproductive and alternative heterosexualities continue to evoke profound "puzzlement" regarding their function, which closely parallels the ongoing attempt to find "explanations" for homosexuality. In a recent discussion of alloparenting (helping behavior) in Common Murres, for example, the list of possible "causes" or "functions" of this behavior was virtually identical to those currently invoked to "explain" homosexuality: mistaken identity, heterosexual "practice," coercion or "manipulation," hormonal factors, kinship, and nonadaptiveness.[157] Heated debate about their supposed "purpose" continues to envelop the scientific discussion of most other nonreproductive and alternative heterosexualities: adoption, divorce, nonreproductive copulations, infanticide, nest desertion, reverse mountings, sex segregation, nonmonogamous matings, masturbation, multiple copulations, rape, vaginal plugs, reproductive suppression, postreproductive individuals, and harassment of matings all remain "perplexing" and highly contentious phenomena.[158]

Even "kisses" have prompted a barrage of functional "explanations." Unable (or unwilling) to countenance the possibility of pleasure or affection in this "behavior"—much less something more intangible—biologists insist that kisses (even in humans) must be a vestige of ritual food exchange, or olfactory sampling, or have a specific social "function" such as reconciliation or alliance formation.[159] Perhaps this is true—but, as with other social/sexual behaviors, there is so much more to it than this. "The kiss" is a perfect symbol of the limitations of biological reductionism—for even if its origins can ultimately be traced to such functional considerations, something ineffable still remains in the gesture each time it is performed, something that continues to transcend its biological "purpose" and evade "explanation." Poet e. e. cummings once warned of those who, "given the scalpel," would "dissect a kiss."[160] Considering that biologists now have the analytical tools and theoretical frameworks to "dissect" the function of the kiss—and apparently no qualms about doing so—perhaps his admonitions are more than metaphorical.

In 1923 a biologist "explained" polyandry (females mating with more than one male) in phalaropes (a type of sandpiper) as being caused by the "deranged sex organs" of female birds, whose "wanton" behavior was "forcing polygamy on the

The kiss: two male Chimpanzees (left) and female Squirrel Monkeys (right) kissing

race." On discovering this historical "explanation" more than half a century later, one ornithologist conceded that scientific theories often reflect "the passions and prejudices of the time," adding, "Half a century hence, our successors will no doubt find similar amusement in ideas devolving from our present ignorance."[161] Much of the "present ignorance" of biology lies precisely in its single-minded attempt to find reproductive (or other) "explanations" for homosexuality, transgender, and non-procreative and alternative heterosexualities. In the next chapter we'll explore how biology can take a first step into the twenty-first century—by reconciling itself at the most fundamental level to the existence and apparent "purposelessness" of this broad panoply of behaviors, sexualities, and genders in the natural world.

Chapter **6**

A New Paradigm:
Biological Exuberance

What "causes" homosexuality is an issue of importance only to societies which regard gay people as bizarre or anomalous. Most people do not wonder what "causes" statistically ordinary characteristics, like heterosexual desire or right-handedness; "causes" are sought only for personal attributes which are assumed to be outside the ordinary pattern of life.
—historian JOHN BOSWELL

It is our vision, and not what we are viewing, that is limited. . . . We understand nature as source when we understand ourselves as source. We abandon all attempts at an explanation of nature when we see that we cannot be explained, when our own self-origination cannot be stated as fact.
—philosopher and theologian JAMES P. CARSE[1]

Western science has been attempting to explain animal homosexuality for over two hundred years, yet it has run into considerable problems by trying to relate all aspects of animal behavior to reproduction (as the previous chapters have shown). The phenomena of animal homosexuality/transgender—and more generally, nonprocreative heterosexuality—require a rethinking of some of the most fundamental concepts in biology. Where are we to turn for models of animal behavior and evolution that can encompass such seemingly "unproductive" activities as homosexuality, heterosexual oral sex, and reproductive suppression? The key lies in what may at first seem an unlikely source of inspiration: the traditional knowledge of indigenous and tribal cultures. These aboriginal worldviews often regard gender and sexuality (in both animals and people) as inherently multiple and mutable; they are typically part of a larger interpretive framework incorporating sophisticated ideas, observations, learning, and lore about how the natural world works. Indigenous beliefs also show remarkable correspondences with recent scientific discoveries in animal behavior and, more broadly, with a number of ideas emerging from "new" Western scientific and philosophical perspectives such as chaos science, post-Darwinian evolution, Gaia theory, biodiversity studies, and the theory of General Economy.

In this chapter we introduce the concept of Biological Exuberance, which synthesizes these two major strands of thought (indigenous and "modern") into a new way of viewing the natural world. This view is at once consistent with many orthodox ideas about evolution and biology while simultaneously offering a radical shift in perspective. Traditionally, scarcity and functionality have been considered the primary agents of biological change. The essence of Biological Exuberance is that natural systems are driven as much by abundance and excess as they are by limitation and practicality. Seen in this light, homosexuality and nonreproductive heterosexuality are "expected" occurrences—they are one manifestation of an overall "extravagance" of biological systems that has many other expressions.

Left-Handed Bears and Androgynous Cassowaries: Informing Biology with Indigenous Knowledge

To Western science, homosexuality (both animal and human) is an anomaly, an unexpected behavior that above all requires some sort of "explanation" or "cause" or "rationale." In contrast, to many indigenous cultures around the world, homosexuality and transgender are a routine and expected occurrence in both the human and animal worlds. The sporadic attention devoted to animal homosexuality/transgender by Western science spans a little over two centuries, while aboriginal cultures have accumulated a vast storehouse of knowledge about the natural world—including the sexual and gender systems of animals—over thousands of years. It stands to reason, then, that Western science might be able to learn something from indigenous sources. In this section we'll explore some traditional tribal beliefs about animal (and human) homosexuality/transgender from around the world and examine the ways in which these ideas are relevant to contemporary scientific inquiry.

Aboriginal Views of Animal Homosexuality and Transgender

Ideas about animal (and human) homosexuality/transgender figure prominently in three cultural complexes on different continents: native North America, the tribes of New Guinea/Melanesia, and indigenous Siberian/Arctic peoples. Beliefs about sexual and gender variability in animals recur systematically throughout many of the cultures in each of these areas and are paralleled by a corresponding recognition and valuation of human homosexuality and transgender. Although these are by no means the only cultures in the world where such beliefs exist, a relatively extensive body of anthropological research has documented the indigenous views on this subject particularly well for these regions. These cultures offer a useful introduction to aboriginal systems of knowledge concerning gender and sexuality and may be taken as representative of the sorts of worldviews that are likely to be encountered in other indigenous cultures.[2] Moreover, the forms that such beliefs take show striking similarities in each of these regions. Aboriginal ideas about animal homosexuality and transgender are encoded in four principal cultural forms: totemic or symbolic associations of animals with human homosexuality and transgender; beliefs about mutable or nondualistic gender(s) of particular species, often represented in the figure of a powerful cross-gendered animal or in sacred stories ("myths") about sexual and gender variability in animals; ceremonial reenactments or representations of animal homosexuality and transgender, sometimes combined with ritual reversals of ordinary activities; and animal husbandry practices that encourage and value intersexual and/or nonreproductive creatures.[3]

NORTH AMERICA: *Two-Spirit, Shape-Shifter, Trickster-Transformer*

Most Native American tribes formally recognize—and honor—human homosexuality and transgender in the role of the "two-spirit" person (sometimes formerly known as a berdache). The two-spirit is a sacred man or woman who mixes gender categories by wearing clothes of the opposite or both sexes, doing both male and female (or primarily "opposite-gender") activities, and often engaging in same-sex relations. As is true for homosexual and/or transgendered individuals in many other indigenous cultures around the world, two-spirit people are frequently shamans, healers, or intermediaries in their communities, performing religious and/or mediating functions (e.g., between the sexes, or between the human, animal, and spirit realms).[4] In many Native American cultures, certain animals are also symbolically associated with two-spiritedness, often in the form of creation myths and origin legends relating to the first or "supernatural" two-spirit(s). Among the Oto people, for example, Elk (Wapiti) is described as cross-dressing in several origin legends and is considered the original two-spirit; consequently, two-spirits in this culture always belong to the Elk clan.[5] A Zuni creation story relates how the first two-spirits—creatures that were neither male nor female, yet both at the same time—were the twelve offspring of a mythical brother-sister pair. Some of these creatures were human, but one was a bat and another an old buck Deer.[6] In "How the Salmon Were Brought to This World," a Nuxalk (Bella Coola) story that describes the origin of food, the first two-spirit accompanies all the animals (including a Raven, cormorant, crane, osprey, hawk, and mink) on a long canoe journey in their quest for the first salmon. Each of the animals finds a different kind of salmon, while the two-spirit brings back the first berries for people to eat. A mythic journey is also featured in the origin tale of the Kamia (Tipai/Southern Diegueño) people, in which the divine two-spirit and his/her twin sons use the feathers from a number of birds—among them, the crow—to make headdresses.[7] Finally, the Mothway origin story of the Navajo relates the adventures of an extraordinary figure known as *Be'gochídí*. Divine trickster, shape-shifter, world-creator, and two-spirit, *Be'gochídí* is a blond (or red-haired), blue-eyed god who mediates between animals and humans, men and women, and Navajos and non-Navajos. S/he is also intimately associated with Butterflies: born at the ancestral home of Moths and Butterflies, *Be'gochídí* is responsible for raising the Butterfly People and frequently indulges in fondling or masturbation of both male and female Butterflies.[8]

In other Native American cultures, animal associations with transgender and homosexuality take the form of personal vision quests or totemic creatures linked with two-spirits. Among various Siouan peoples such as the Dakota, Lakota, and Ponca, for example, a man or woman becomes two-spirited if his or her sacred dreaming involves Buffalo (especially a hermaphrodite Buffalo or a white Buffalo calf), or if he or she has a vision of Double-Woman, who often appears in the form of a Black-tailed (Mule) Deer. An Omaha (Sioux) man's calling to be a two-spirit might be announced by an owl.[9] Among the Tsistsistas (Cheyenne), a vision quest involving the thunderbird (typically identified with the golden eagle or Harris's hawk) destines an individual to become a member of the Contrary Society, a group of men who are (heterosexually) celibate, do everything the opposite way, and sometimes have relations with two-spirit people. Those manifesting sexual and gender variance (contraries and two-spirits) may also be symbolically associated with birds that have orange-red coloration, such as orioles or tanagers, and possibly also with Dragonflies.[10] The Arapaho people believe that two-spirit is a blessing bestowed as a supernatural gift from birds or mammals, while Hidatsa two-spirits typically wear Magpie feathers in their hair as part of ceremonial dress. This symbolizes their connection to powerful holy women who are associated with Magpies in this culture.[11] In some cases, individual two-spirit shamans may invoke the powers of specific animals, such as a Wolf tutelary for the Tolowa two-spirit shaman *Tsoi'tsoi* and a Grizzly Bear tutelary for the two-spirit shaman *haywič* of the Snoqualmie (Lushootseed/Puget Sound Salish) people.[12]

Bears play a further role in Native American cultures with regard to homosexuality/transgender. A fascinating association between (of all things) left-handed Bears and two-spiritedness reappears in many tribes throughout North America.[13] In a number of First Nations—for example, the Nuu-chah-nulth (Nootka), Kutenai, Keres, and Winnebago—the Bear is seen as a powerful cross-gendered figure. In these tribes, Bears are thought to combine elements of both masculinity and femininity, and they are also seen as mediators between the sexes and between humans and animals (much like the role of the human two-spirit, which is also recognized in all these tribes). Their strength, size, and ferocity are considered quintessentially male attributes, yet Bears are often perceived as female in these cultures and referred to with feminine pronouns and terms of address regardless of their biological sex. In addition, many of the prominent Bear stories and ceremonies concern female Bears, especially the omnipotent, life-giving Bear Mother figure (who often engages in mythic marriage, sexual intercourse, or transformation with humans).[14] There is also a consistent association between Bears and menstruation. A number of Native American peoples have beliefs about the dangers of women going into the forest during their period, since it is thought that they will attract Bears who may try to mate with (or attack) them. Other tribes mythologically connect Bears to menstrual blood or consider Bears to be powerfully drawn to human females in other respects, especially at the onset of puberty.[15]

Most strikingly, Bears of both (biological) sexes are thought to be left-handed—a quality traditionally associated with the feminine in these cultures—and Bear rites often require ceremonial activities to be performed with the left hand. In fact, beliefs about the left-handedness of Bears pervade all aspects of ritual life in some tribes. In the Nuu-chah-nulth culture of Vancouver Island, for example, Bear hunters eat with their left hand (they are the only people allowed to do so) in order to identify with their prey, since Bears are believed to reach for bait with their left paw. In myths and tales such as that told by contemporary Nuu-chah-nulth artist and storyteller George Clutesi, *Chims-meet* the Bear hunts for salmon with his left paw while his mother picks berries with her left paw; Clutesi illustrates one of his tales with a drawing of a Bear using his left paw to swat salmon. Left-handedness is even encoded in the structure of the language: when speaking Nuu-chah-nulth, special affixes can be added to words to indicate that a left-handed person is talking or is being referred to. Of course, this "left-handed speech" is also typical of Bears when speaking in myths, stories, and jokes.[16]

Many First Nations sacred stories and myths, especially those involving a prankish trickster-transformer figure, reveal other associations between animals and homosexuality/transgender. A common theme is that of a male coyote marrying or having sex with a male mountain lion, fox, or other animal—or sometimes even with a man—often by changing sex, mixing gender characteristics, or pretending to be a member of the opposite sex. In the Okanagon story "Coyote, Fox, and Panther," for instance, coyote tricks a panther (mountain lion) into marrying him by pretending to be female; the presence of human two-spirits in this culture is therefore considered to be decreed by coyote. Similar tales are found in many other cultures. In fact, an Arapaho story combines this theme with that of the supernatural two-spirit in the tale of *Nih'a'ca* (the first two-spirit), by having *Nih'a'ca* pretend to be a woman and marry a mountain lion (a symbol of masculinity). The trickster theme takes many other forms as well. The Fox Indians, for example, have a tale in which a male turtle is fooled into having sex with a human trickster figure, who fashions a vulva for himself out of an Elk's spleen and disguises himself as a woman named Doe-Fawn. The Winnebago trickster man also uses the internal organs of an Elk to make female parts for himself, then becomes pregnant by having sex with a number of male animals, including a fox and a blue jay.[17]

Two-spirit is still a living tradition in many First Nations, and there is a continuing association of animals with homosexuality and transgender in the stories, life narratives, and poetry of contemporary Native Americans. Two-spirit Mohawk writer Beth Brant gives the trickster theme a gender spin in her tale "Coyote Learns a New Trick." In this story, a female coyote tries to fool a female fox into sleeping with her by dressing up as a man; the joke is on coyote, however, because fox only pretends to be duped, and the two end up making love without any disguises. In "Coyote and Tehoma," Daniel-Harry Steward of the Wintu nation offers a poetic account of love between a male coyote (accompanied by several animal spirit-guides) and Tehoma, the handsome male "god of the smoking mountain." In this fable, the howling of wild coyotes is attributed to the heartbreak of their mythic coyote ancestor, who calls forlornly to his male lover after Tehoma has been changed into the stars. In "Song of Bear," a contemporary version of a traditional Nuu-chah-nulth tale recorded by Anne Cameron, the human-animal marriage of the Bear Mother myth is given a lesbian retelling. A young woman goes into the forest (disregarding warnings about the attraction of Bears to menstruating women) and draws the attentions of a female Bear; they end up falling in love and living together "forever after" in the Bear's den. Finally, for contemporary two-spirits Terry Tafoya (Taos/Warm Springs), Doyle Robertson (Dakota), and Beth Brant, creatures such as the dragonfly, hawk, eagle, heron, and salmon have powerful personal and symbolic resonance, while the searing poetry of two-spirit writer and activist Chrystos (Menominee) is also replete with bird and other animal imagery.[18]

A tricksterlike figure plays a central role in another manifestation of animal homosexuality/transgender in indigenous cultures, the ritual enactment of same-sex activity during sacred ceremonies. Among the Mandan, a Siouan people of North Dakota, a spectacular religious festival known as the *Okipa* was held annually for at least five centuries (until the late 1800s) to ensure the success of the Buffalo hunt and to ritually dramatize their cosmology.[19] Replete with sacred communal dancing, chanting, and prayers in an ancient liturgical dialect (used only during this festival), the four-day ceremony includes shamanic rites of self-mutilation (such as skewering and suspension of initiates), feats of astounding physical endurance, and graphic sexual imagery. Throughout the festival a special Bull Dance is performed by men representing Bison: cloaked in the entire skins and heads of the animals, they realistically portray the movements of Bison. Surrounding them are dancers dressed as various other animals as well as men impersonating holy women. The dance culminates on the final day with symbolic homosexual activity between the Bison bulls and a clownlike figure called *Okehéede* (known variously as the Foolish One, the Owl, or the Evil Spirit), who is painted entirely black and adorned with a Buffalo tail and Buffalo fur. Wielding an enormous wooden penis, *Okehéede* simulates anal intercourse with the male Bison by mounting them from behind "in the attitude of a buffalo bull in rutting season." He erects and inserts his phallus under

A Bison bull in Wyoming mounting another male. Many Native American peoples have traditional beliefs, ceremonies, and observations regarding sexual and gender variance in this (and other) species. ▼

each dancer's animal hide, even imitating the characteristic thrusting leap that Bison make when ejaculating. The Mandan believe that this ceremonial homosexuality directly ensures the return of the Buffalo in the coming season.[20]

Ceremonial "performances" of sexual and gender variability occur in several other Native American sacred animal rites, such as the *Massaum* ceremony of the Tsistsistas (Cheyenne) people. Also known as the "Crazy" or "Contrary" Animal Dance (from the word *massa'ne* meaning foolish, crazy, or acting contrary to normal), this 2,000-year-old world-renewal festival was performed annually on the Northern Plains until the early 1900s. Timed with key celestial events in the midsummer sky (including the solstitial alignment of three stellar risings), the *Massaum* ceremonial cycle invokes and draws upon the powers of two-spirit and "contrary" shamans in order to reinvigorate the earth and all its inhabitants. The five-day ritual is thought to have been bequeathed to the Tsistsistas people by the prophet *Motseyoef,* an immortal androgynous shaman who presides over each reenactment of the rite in the form of a human representative. A prominent feature of the *Massaum* is a pair of sacred Bison horns, originally taken from a hermaphrodite Buffalo. Among the central participants are a set of sacred male and female canids, all impersonated by men dressed in animal skins and imitating the actions of the creatures: two wolves—a male red (or yellow) wolf and a female white (or Gray) Wolf—as well as a female kit (or blue) fox. As master hunters, game protectors, and messengers from the spirit world, these animals teach humans how to hunt with the proper reverence and skill. The *Massaum* culminates with a ritual hunt of epic proportions, in which nearly a sixth of the Tsistsistas population participates by impersonating all the various creatures of their world. Each species is "led" by someone who has dreamed of that animal acting in a peculiar way. On the final day, the androgynous contrary shamans begin their sacred clowning, doing things backward and generally acting in an eccentric manner. As part of their holy "craziness," they symbolically hunt the animals, "shooting" them with special miniature bows and arrows held in a reversed position. Upon ritually killing each creature, they immediately bring it back to life, thereby assisting in the divine regeneration and fertilization of the earth. By uniting primordial opposites within themselves and in their actions, the two-spirit and contrary shamans are seen by the Tsistsistas as instrumental in restoring wholeness to the world.[21]

Ritual transgender is also enacted in the Buffalo Ceremony of the Oglala Dakota, a girl's puberty rite presided over by a shaman dressed as a Bison. During this ceremony the shaman combines attributes of both male and female Buffalo: he imitates the courting behavior of a Bison bull, but his face is painted with a pattern symbolic of a Bison cow, and he is designated with the word for a female Buffalo. Likewise in a Hopi Buffalo dance, the men portraying Bison wear some articles of women's clothing, while female dancers also don some men's garments. In other sacred *kachina* ceremonies of the Pueblo peoples, some female animal figures are impersonated by male dancers. The Hopi goddess *Talatumsi,* or Dawn Woman, for example, who is the mother of Bighorn Sheep-men, is portrayed by a man dressed as a female Mountain Sheep. The bawdy *kachina* clowns in Hopi ceremonies sometimes simulate sexual intercourse with a burro, one man pretending to be the animal while the other mounts him from behind. The Zuni animal fertility goddess *Chakwena*—mother of rabbits and other game animals—is also impersonated by a man: s/he performs symbolic versions of female reproductive powers, including ritual menstruation in the form of rabbit blood dripped down his/her legs, and a four-day ceremonial enactment of childbirth. Ritual animal birth can also be associated with Wintu two-spirit shamans: one man, for example, was believed to experience menstrual periods and was thought to have given birth to a pair of snakes.[22]

Native American rites and beliefs about sexual and gender diversity sometimes also extend to the sphere of animal husbandry, for example among the Navajo. Consummate shepherds and goatherds, these Southwestern people have developed sophisticated animal-management techniques over the many centuries of tending their domesticated herds. Yet their practical knowledge is also informed by the Navajo recognition and honoring of gender and sexual variability in all creatures. Traditionally, hermaphrodite Sheep and Goats are considered integral and prized members of the flock, since they are thought to increase the other animals' productivity and bring prosperity. For this reason they are never killed, and their presence is further encouraged by several ritual practices. When hunters catch an intersexual Deer, Pronghorn, or Mountain Sheep, for example, they rub its genitals on the tails of their domesticated female herd animals and on the noses of the males, as this is believed to result in more hermaphrodite Sheep and Goats being born into the flocks. In addition, rennet from the stomachs of intersexual animals is rubbed on Sheep to increase their growth and milk production. This convergent valuing of transgender in both wild and domesticated animals is reflected in Navajo mythology and cosmology: *Be'gochidí,* the divine two-spirit described earlier, is regarded as the creator of both game animals and domesticated creatures. S/he is also god of the hunt and a tutelary who instructs humans in stalking techniques and hunting rituals, as well as a prankster who sneaks up on hunters and causes them to lose their aim by grabbing their testicles. Some of the hunting rituals associated with *Be'gochidí* also involve ceremonial reversals. For example, the skin of a slain Deer, after being removed from the animal, is repositioned with the head resting on the carcass's rump, sometimes with the Deer's tail placed in its own mouth.[23]

NEW GUINEA: *Male Mothers and the Living Secret of Androgyny*

In addition to reappearing in the native cultures of North America, beliefs about animal (and human) homosexuality/transgender also feature prominently among the indigenous peoples of New Guinea and Melanesia. Homosexuality is an important aspect of human social and ceremonial interactions in many tribes, while homosexual or transgendered animals are an equally pervasive aspect of their belief systems. In a number of cultures, all males undergo a period of homosexual initiation lasting for several years (from prepuberty to young adulthood). Semen from adult men is considered a vital substance for "masculinizing" boys, and therefore adults "inseminate" younger males through oral or anal intercourse. Other forms of sexual and gender variance also exist: the Sambia and Bimin-Kuskusmin cultures, for instance, recognize a "third sex" among humans (applied to hermaphrodites or intersexed individuals), and the Sambia also have a prominent origin myth involving male parthenogenesis, in which the first people were believed to be created through homosexual fellatio. Ceremonial transvestism occurs in some New Guinean tribes as well, along with beliefs and ceremonies relating to "male menstruation" (often ritualized as bloodletting of the penis).[24] A number of animals are symbolically and ceremonially associated with homosexuality in these cultures as well. Among the Sambia, for example, plumes from several birds, including the Raggiana's Bird of Paradise, the kalanga parrot, and several species of lorikeets (a type of parakeet), are ritually worn by boys and adolescents to mark their various stages of initiation and participation in homosexual activities. Homosexual bonding among the Ai'i people is emblematized by two men sharing a bird of paradise totem, which also connotes the joint land-holding rights of the male couple. And in the Marind-anim tribe, the wallaby, jabiru stork, and cassowary are symbolically associated with homosexuality.[25]

Beliefs about variant gender systems in animals—including all-female offspring and various forms of sex change—also occur in several New Guinean cultures.[26] Opossums, Tree Kangaroos, and other tree-dwelling marsupials are thought by the Sambia to start out life as females, with only some individuals later becoming male once they reach adulthood.[27] Thus, the life cycle of these species, in the indigenous conception, involves a sort of sequential sex change for animals that end up as male. In contrast, the *nungetnyu*—a kind of bird of paradise or bowerbird—is thought to exist only in female form throughout its life. The Sambia liken the communal courtship dances of this species to their own dance ceremonies, except with a gender inversion (all-female bird groups versus all-male human groups).[28] Other birds are thought to go through multiple sex changes: they start out life as female, then some briefly become male birds as adults and develop brightly colored plumage, after which they revert back to a female form (with dull plumage) in their old age. The Bimin-Kuskusmin also believe that several species of birds of paradise go through multiple gender transformations during their lives, but with the opposite sequence: the brightly plumaged individuals are considered to be females and the drably plumaged ones to be males. Likewise, a daily oscillation between genders is attributed to a species of nightjar: these nocturnal birds are thought to be either male or female in the daytime but both male and female at night.[29] Parallel ideas about sex change in sago beetles and their grubs are held by the Bedamini, Onabasulu, and Bimin-Kuskusmin peoples.

Perhaps the most extraordinary example of beliefs about ambiguous or contradictory genders in animals concerns the cassowary. A large, flightless, ostrichlike bird of New Guinea and northern Australia, the cassowary is considered by many New Guinean peoples to be an androgynous or gender-mixing creature, and it often assumes a preeminent mythic status in these cultures. The cassowary possesses many of the physical attributes of strength, audacity, and ferocity that are traditionally considered masculine in these cultures. It has powerful legs, feet, and razor-sharp claws (capable of inflicting serious, even lethal, injuries to people); a dinosaur-like bony helmet or "casque" (used for crashing through the jungle); dangerously sharp spines or quills in place of wing feathers; booming calls (described as "warlike trumpet barks"); bright blue and red neck skin with pendulous, fleshy wattles; and an imposing size (over five feet tall and 100 pounds in some species). Yet numerous New Guinean peoples also regard the cassowary to be an all-female species (or for each bird to be simultaneously male and female) and often associate them with culturally feminine elements.

*The cassowary is considered a powerfully androgynous creature by many indigenous New Guinean peoples. This is the one-wattled cassowary (*Casuarius unappendiculatus*).*

The Sambia, for instance, consider all cassowaries to be "masculinized females," that is, biologically female birds that nevertheless lack a vagina and possess masculine attributes (they're thought to reproduce or "give birth" through the anus). Similarly, the cassowary is perceived as an androgynous figure by the Mianmin people: the bird is thought to have a penis, yet all cassowaries are considered female. One Mianmin tale actually recounts how a woman with a penis was transformed into a cassowary, and this mythological trope is found in the sacred stories of several other New Guinean peoples. Other cultures elevate the cassowary to a prominent position in their traditional cosmologies and origin myths as a generative figure, a powerful female creator of food and human life. The cassowary is believed to combine elements of femininity and masculinity in many other tribes, a number of whom also practice ritualized homosexuality, such as the Kaluli and Keraki. Finally, in a striking parallel to the cross-gendered Bear figure of many Native American cultures, the androgynous cassowary is also considered an intermediary, of sorts, between the animal and human worlds. In addition to mythic transformations and marriages between people and cassowaries, in several tribes this creature is not classified as a bird at all, but is grouped in the same category as human beings because of its size and upright, two-legged gait. Combining images of male-female and bird-mammal, the Waris and Arapesh peoples also believe that cassowaries suckle their young from their neck wattles or wing quills, which are found in both male and female birds.[30]

Ritual performance of the cassowary's gender-mixing also occurs. Among the Umeda people, for example, a central feature of the tribe's *Ida* fertility rite involves two cassowary dancers whose costumes, movements, and symbolism combine both male and female elements. The dancers impersonating the birds are both men and are called by a name that refers to male cassowaries. Yet they are identified with the ancestral mothers of the tribe (who act as female tutelary spirits to the dancers), and the entire ceremony is said to have belonged in mythic times only to women and was performed without men. Each cassowary dancer also has an exaggerated phallus consisting of a large black gourd worn over the head of his penis, but the enormous mask/headdress that he carries (representing the cassowary's plumage as well as a palm tree) is imbued with feminine symbolism (in the form of its inner layer of underbark). The dancing of the cassowary impersonators emphasizes their male sexuality: they rhythmically hop and move their hips in such a way that their penis gourds flip upward and strike their belts in a motion that imitates copulation, and their phallic organs are said to become enormously elongated during the all-night ceremony. At the same time, the two men frequently hold hands and dance as a pair, activities that are otherwise seen only in female dancers among the Umeda.[31]

The figure of the gender-mixing cassowary reaches its greatest elaboration among the Bimin-Kuskusmin people. In the belief system of this remote tribe of the central New Guinea highlands, the cassowary presides over an entire pantheon of androgynous and sex-transforming animals, and it is physically embodied in the form of special human representatives that ritually enact its transgendered characteristics. In addition to the cassowary and sex-changing birds and grubs mentioned previously, numerous other creatures are believed to combine male and female attributes in the worldview and mythology of the Bimin-Kuskusmin. Several species of marsupials, a bowerbird, and a python are all considered androgynous or hermaphroditic. The wild boar is regarded as a feminized male that never breeds but instead fertilizes androgynous plants with its semen and menstrual blood. And a species of centipede is thought to be female on its left side and male on its right, using its venom to bring life to other androgynous centipedes and death to nonandrogynous creatures.

At the pinnacle of this transgendered bestiary stands the creator figures of *Afek*, the masculinized female cassowary, along with her brother/son/consort *Yomnok*, a feminized male fruit bat or echidna (the latter being a spiny anteater, an egg-laying mammal related to the platypus). Both are descended from a powerful double-gendered monitor lizard and are believed to be hermaphrodites possessing breasts and a combined penis-clitoris. *Afek* gives birth through two vaginas (one in each buttock), while *Yomnok* gives birth through his/her penis-clitoris. The gender-mixing of these mythical figures parallels the way they straddle the categories of bird and mammal: the cassowary is a "mammallike" bird—huge, ferocious, flightless, with furlike feathers—while the echidna is a "birdlike mammal"—small, beaked, and egg-laying (the fruit bat is also birdlike, being a flying mammal).

The Bimin-Kuskusmin elect certain people in their tribe to become the sacred representatives and lifelong human embodiments of these primordial creatures: they undergo special initiations and thenceforth ritually reenact and display the intersexuality of their animal ancestors. Two postmenopausal female elders in the clan are chosen to represent *Afek*: they undergo male scarification rituals, experience symbolic veiling or dissolution of their marriages and children, adhere to combined male and female food taboos, receive male names, and are awarded both male and female hunting and gardening tools. During ceremonial functions—in which they are sometimes referred to as "male mothers"—they ornament themselves with cassowary plumes, often cross-dress in male regalia, or wear exaggerated breasts combined with an erect penis-clitoris made of red pandanus fruit. Physically intersexual or hermaphrodite members of the tribe are selected to be the embodiments of *Yomnok*. They are adorned with echidna quills or dried fruit-bat penises, wear both male and female clothing and body decorations, sport an erect penis-clitoris (made from black, salt-filled bamboo tubes) during rituals, and are lifelong celibates.[32] In each case, these living human representatives of the primal animal androgynes become highly revered and powerful figures in the tribe. They apply their sacred double-gendered power in curing, divination, purification, and initiation rites and officiate at ceremonies that require the esoteric manipulation and mediation of both male and female essences. Above all, these transgendered and nonreproductive "animal-people" are symbols of fertility, fecundity, and growth—corporeal manifestations of what one cassowary man-woman calls "the hidden secret of androgyny . . . inside the living center of the life force."[33]

Ritualized "performances" of homosexuality combined with animal imagery are also found in the extraordinary initiation and circumcision rites of several cultures of Vanuatu (formerly the New Hebrides), including the Nduindui and Vao peoples. During these secret ceremonies, symbolic homosexual intercourse is enacted or implied between young male initiates and their elder initiators or ancestral male spirits. Along with other ritual inversions of everyday activities or breaking of taboos during the rites, these ceremonial homosexual activities are thought to imbue the participants with an unusually intense, dangerous, and glorious power. All of these activities coalesce around the image of the shark. The ceremonies are known as shark rites; participants wear elaborate shark headdresses; the initiators/elder partners in actual or symbolic homosexual relationships are referred to as sharks; the rituals are staged in enclosures that symbolize a shark's mouth; and circumcision itself is likened to the bite of a shark. In some cases there is a connection to other gender-mixing creatures. During the enactment of ceremonial homosexual overtures or intercourse, for example, participants sometimes refer to hermaphrodite Pigs, and the story of one Vanuatu culture hero bearing the title of "shark" tells how his son brought intersexual Pigs to several islands. The linked themes of androgyny and Pigs also appear in narratives from outside the Vanuatu region, for example among the Sabarl people. In their tale "The Girl Who Dressed as a Boy," a young woman adopts warrior paraphernalia—and later assumes the full garb of a man—during a heroic encounter with a giant Pig who is the offspring of an androgynous creator god.[34]

Gender-mixing Pigs also feature prominently in another fascinating Vanuatu cultural practice that honors sexual and gender variance in animals (in some cases alongside ritual human homosexuality/transgender). Hermaphrodite Pigs are highly prized in a number of Vanuatu societies, being valued for their uniqueness and relative rarity. Although only a minority of Pigs are intersexual, their husbandry is an esteemed pursuit (especially in the northern and central regions), and animal breeding practices that result in hermaphrodite offspring are encouraged. As a result, nearly every village in some areas has intersexual Pigs, and gender-mixing animals comprise a fairly high proportion of the total domesticated Pig population, perhaps as much as 10–20 percent in some regions. In fact, on these islands there are more hermaphrodite mammals—probably numbering in the thousands—than anywhere else in the world. These intersexual Pigs possess internal male reproductive organs and typically grow tusks like boars (although they are sterile), yet their external genitalia are intermediate between those of males and females, tending toward the female. Behaviorally, they often become sexually aroused in the presence of females and may even mount other females while exhibiting clitoral erections. Among the people of Sakao, seven distinct "genders" of hermaphrodite Pigs are recognized and named, ranging along a continuum from those with the most femalelike genitalia to those that are truly ambiguous to those with the most malelike genitalia. The indigenous classification of these gradations of intersexuality exceeds in completeness any conceptual or nomenclatural system developed by Western science. So precise is this vocabulary, in fact, that the native terminology was actually adopted by the first Western biologist who studied the phenomenon in order to distinguish the various types of gender mixing.

In these Vanuatu cultures, hermaphrodite Pigs are a status symbol of sorts, since their ritual sacrifice is required to achieve progressively higher rank within the society (they are also used in dowries). In some cases, a sophisticated monetary system and trading network has developed in which pigs actually function as a type of currency, complete with forms of "pig credit" and "pig compound interest." In this system, intersexual Pigs (and the sows that produce them) can be worth up to twice as much as nonintersexual Pigs. The prestige of these animals also extends to the domestic sphere: hermaphrodite Pigs are often depicted on finely carved household items such as plates and bowls, and intersexual Pigs are sometimes kept as pets. They may even become highly valued "family members," to the point of being suckled by a woman like one of her own children. Moreover, men who raise tusked Pigs (either boars or hermaphrodites) are in some cases viewed as sexually ambiguous or androgynous themselves, since their intimate tending and nurturing of the Pigs is thought to parallel the mother-child relationship. Simultaneously "father" and "mother" to the creatures, they constitute another example of the indigenous concept of "male motherhood" as it pertains to animals.[35]

SIBERIA/ARCTIC: *Reversal and Renewal, Traversal and Transmutation*

A similar constellation of phenomena concerning animal homosexuality and transgender is found among the numerous indigenous cultures scattered across Siberia and the Arctic (including the Inuit and Yup'ik [Eskimos] of arctic North America).[36] Aboriginal Siberian shamans often harness the power of cross-gender animal spirit guides or assume characteristics of the opposite sex under the direction of spirit animals. The most powerful male shamans among the Sakha (Yakut) people, for example, are believed to undergo a three-year initiation during which they experience aspects of female reproduction, including giving birth to a series of spirit animals (such as a Raven, loon, pike, Bear, or Wolf). Some female shamans also claim to manifest their power by transforming themselves into a male Horse. Gender reversals and recombinations are most prominently expressed in the phenomenon known as the transformed shaman, a sacred man or woman who takes on aspects of an opposite-sex identity. Transformation ranges along a continuum from a simple name-change, to partial or full transvestism during shamanic rituals, to living permanently as a transgendered person (including marrying a husband in the case of a transformed male or marrying a wife for a transformed female).

Among the Chukchi, transformed shamans are sometimes associated with animal powers through spirit-name adoptions and animal transmutations. One such male shaman was named She-Walrus, for instance, while another believed s/he had the ability to change into a Bear when curing patients. Animal gender transformations that parallel those of shamans are also encoded in sacred stories. Among the Koryak, for example, a mythological figure named White-Whale-Woman turns herself into a man and marries another woman. In another story s/he marries a male Raven who has turned himself into a woman (and whose son later gives birth to a boy).[37]

The ornate and beautiful costumes worn by shamans in many Siberian cultures often combine animal impersonation with cross-dressing. The robes, headdresses, and footgear of male shamans among the Yukaghir, Evenk, and Koryak people, for example, are usually women's garments adorned with animal imagery. This may include an "antlered cap" bearing a symbolic representation of Reindeer antlers, or two iron circles representing breasts sewn to the front of the cloak. These sacred vestments—often made from an entire animal skin—are believed to allow the shaman to incarnate an animal or undertake supernatural bird-flight during trance, and s/he often performs dances that closely imitate the movements of a particular species that serves as his/her tutelary spirit. Shamanic ceremonies in a number of Siberian tribes also sometimes involve all-male dances imitating the mating activities of various animals, aimed at promoting sexual activity and a "renewal of life." The word for *shaman* in the Samoyed language actually has the same root as the words *to rut* (of a stag) or *to mate* (of game birds). Chukchi transformed shamans do not generally wear special garments or impersonate animals; however, female-to-male shamans sometimes wear a dildo made from a Reindeer's calf muscle, attached to a leather belt. In addition, Chukchi women and girls who are not shamans often perform all-female dances imitating various species, including white-fronted geese, long-tailed ducks, swans, Walruses, and seals. Some of these dances actually represent the courtship displays of male Ruffs or rutting Reindeer, and dances may also conclude with two girls lying on the ground and simulating sexual intercourse with each other.[38]

Reindeer (known as Caribou in North America) are regarded in the shamanic contexts of some Arctic cultures as powerful transgendered creatures belonging to the supernatural. The Iglulik Inuit (Eskimo), for example, believe in mythical Caribou known as *Silaat* (in their male form) or *Pukit* (in their female form; singular *Pukiq*). These enormous animals are swifter and stronger than ordinary Caribou, can create dangerous weather conditions, and are thought to hatch from giant eggs on the tundra (sometimes identified with actual wild-goose eggs). The males wear female adornments on their robes (such as white pendants) and can transform themselves into females (some *Silaat* also assume the form of bearded seals or Polar Bears). The *Silaat/Pukit* also serve as spirit guides to shamans: one shamanic initiate named *Qingailisaq* tells of encountering a herd of such creatures, one of whom metamorphosed into a woman. The other *Silaat* then instructed him to make a shaman's cloak that resembled her garment. The robe *Qingailisaq* created combines both male and female elements: in pattern and overall style it resembles a man's coat, but in its ornaments and decoration it is similar to women's clothes. The cloak's white pendants evoke the garments of the transgendered Caribou, and an embroidered image of a transformed white Caribou or *Pukiq* adorns each shoulder. These Caribou are thought to be the original male descendants of *Sila*, a powerful deity and life force associated with gender variability. The Iglulik Inuit culture is based on a ternary gender system that recognizes a "third sex" or gender category. This encompasses a number of different cross-gendering phenomena such as "transsexuals" (people believed to have physically changed sex at birth), transvestites (people who adopt or are assigned the clothes, name, and other markers of the opposite sex), and shamans (who may be fully transgendered, or combine various male and female elements, or undergo mythic transformations between sexes and species). *Sila* occupies a central position in the Inuit cosmology as an intermediary between gender poles, and *Sila's* descendants—the transgendered Caribou—are a further manifestation of this bridging and synthesis of "opposites" (male and female, animal and human).[39]

Some Inuit peoples share with Native American tribes the belief that Bears—in this case, Polar Bears—manifest qualities of gender mixing and left-handedness.[40] In Siberian cultures, however, the association of Bears with sexual and gender variability is most notable in the activities generally known as Bear ceremonialism. A pan-Siberian religious complex, Bear ceremonialism involves the ritual killing of a Bear, whose skin and head are then placed on a sacred platform and feted for many days. Among the Ob-Ugrian peoples, these carnivalesque ceremonies involve feasting, dancing, the singing of sacred epics, and the performing of satirical plays. The latter typically include bawdy displays of transvestism: all female roles are played by

A cloak belonging to the Inuit shaman Qingailisaq. *Just below each shoulder is the image of a* Pukiq, *a mythical transgendered Caribou that combines and transforms elements of male and female, animal and human.*▼

men, who often simulate sex acts with one another. In ecstatic ritual dances men may also remove each other's clothes. During Nivkh (Gilyak) Bear festivals, male hunters wearing articles of female clothing (and men's clothing backwards) try to grab a Bear from behind or kiss it. This highlights a fundamental aspect of Siberian Bear ceremonialism: transgressions of gender and sexual boundaries are simply one of many ritual "reversals" that occur during the festivities (others include saying the opposite of what one means, and the breaching of various other social prohibitions). Bear ceremonies thereby serve, in the words of one anthropologist, as a "liminal (mediating) period of ritual excess," believed by these Siberian peoples to be essential for both human and animal fecundity and prosperity.[41]

Dramatic performances of gender reversals and sexual ambiguity are also an integral component of the elaborate animal renewal and fertility ceremonies of the Yup'ik (Alaskan Eskimo) people. Such festivals feature "male mothers," hermaphrodite and androgynous spirits, ritual transvestism, and cross-gender impersonation of animals, among other elements.[42] One of the most important ceremonies is *Nakaciuq* or the Bladder Festival, a ten-day winter-solstice feast in which seals and other sea mammals are honored and invited to return for the next year (so named because the animals' souls are believed to reside in their bladders, which are inflated and displayed during the ceremony). Another important ceremony is *Kelek* or the Masquerade, part of a larger festal cycle in which shamans and others interact with and appease the spirits of game animals. Images of male motherhood, pregnancy, and birth abound during these ceremonies. At the beginning of the Bladder Festival, for example, two men (often shamans) are designated "mothers" and pretend to be married to each other, with a third man playing the part of their "child." In the Masquerade, male participants occasionally enact the part of a nursing woman, wearing a female mask and two wooden breasts carved with nipples. Male shamans dressed in women's clothing also undertake trance journeys to visit animal spirits, symbolically give birth to spirit beings, and observe rituals associated with menstruation and childbirth following their spirit encounters. At the climax of this festival, a young boy dressed in women's clothes acts as a ceremonial staff-carrier. Transvestism occurs in other Yup'ik festivities as well, involving both men and women disguising themselves as the opposite sex or two men dressing as bride and groom and pretending to get married. Men also sometimes wear articles of women's clothing for luck when hunting land mammals.

The *Tuunraat* or spirit helpers visited by shamans—including the powerful guardians of game animals—are often considered to be hermaphrodite beings. During the Masquerade festival they are impersonated by men wearing masks that meld elements of male-female and animal-human. One such mask, for example, combines a downturned mouth—a standard female symbol in Yup'ik art—with labrets at both corners of the mouth (ornaments worn in lower lip piercings by men), symbolic of a male Walrus's tusks. A stylized sea-mammal tail for a nose and other animal imagery also adorn the mask. Masks of androgynous spirits such as *Qaariitaaq*—represented as a sort of "bearded woman"—are used in the Bladder Festival as well. Animal dances, in which people impersonate various creatures using realistic movements, sounds, and costumes, are also a central aspect of the

Yup'ik ceremonial cycle. Most notable of these are a dance in which a man portrays a mother eider duck, wearing a birdlike hunting helmet decorated with female phallic symbols (two young boys play "her" ducklings), and another in which two men impersonating a loon and a murre also wear these gender-mixing helmets as they dance side by side. As in Siberian Bear ceremonialism, all of these activities are part of an overall pattern of reversals and traversals characteristic of Yup'ik fertility ceremonies. Ordinary activities are turned upside down and the boundaries between "opposite" worlds are rendered fluid (e.g., participants walk backwards, invert traditional hospitality rituals, go nude or wear clothing inside out, etc.). In Yup'ik cosmology these sacred inversions are believed to remake, renew, and regenerate the natural world, ultimately insuring a harmonious relationship between humans and animals.

Although Siberian/Arctic peoples do not appear to accord special meaning to intersexuality among domesticated animals (as in some Native American and New Guinean cultures), nonbreeding animals do feature prominently in some Siberian animal husbandry practices. The Chukchi, for example, believe that castrated and nonreproductive animals insure the success of their domesticated Reindeer herds. The largest bucks are always gelded and, along with several "barren" does, allowed to fatten rather than being slaughtered. Castration is often accomplished by the herdsman biting directly through the animal's spermatic ducts or tubules. These "eunuchoid" Reindeer (both male and female) are highly prized, as they are considered essential for the prosperity of the entire herd. Likewise, the Sakha (Yakut) people always donate one mare from their large herds of domesticated Horses to a shaman. This animal is not permitted to breed during its life, and it becomes an embodiment of the cosmic life force and a symbol of fertility for the tribe as a whole.[43]

Despite wide differences in cultural contexts and details, there are a number of remarkable correspondences and continuities between native North America, Melanesia, and Siberia in their perception of alternative systems of gender and sexuality in animals. In numerous indigenous cultures widely separated in space and time, we find recurring variations on five central themes: Animals are totemically or symbolically associated with homosexuality and transgender, often in a shamanic context. Powerful gender-mixing creatures such as the Bear, cassowary, and Caribou/Reindeer occupy a central position in tribal cosmologies and worldviews. Ritual enactments of animal homosexuality and transgender are commonplace and are often directly associated with notions of fertility, growth, or life essence; this is sometimes concretized in the image of a "male mother" figure and may also be part of a larger pattern of sacred reversals or inversions. Among domesticated creatures, hermaphrodite and nonbreeding animals are cultivated and highly valued. And finally, both animals and people that combine aspects of maleness and femaleness or exhibit sexual variation are consistently honored and ceremonialized, and an essential continuity is recognized between homosexuality/transgender in both human and nonhuman creatures. While fascinating in their own right, these cross-cultural parallels are perhaps even more significant in terms of their implications for contemporary scientific thought.

Chimeras, Freemartins, and Gynandromorphs:
The Scientific Reality of Indigenous "Myths"

How accurate are indigenous views about animal homosexuality and transgender? In other words, do the species associated with homosexuality and transgender in these cultures actually exhibit same-sex behavior or intersexuality? If taken literally, the connection is certainly less than systematic: many animals linked in aboriginal cultures with alternate sexualities are not in fact homosexual, bisexual, or transgendered, while many animals in which sexual and gender variance have been scientifically documented do not have symbolic associations with homosexuality/transgender in these cultures. Moreover, many of the more "fanciful" indigenous beliefs about animals are obviously false (at least in their specifics).

Nevertheless, some striking parallels involving particular species suggest a connection that may be more than fortuitous. For example, homosexuality—including full anal penetration between bulls—is common among American Bison, same-sex courtship and pair-bonding occur in Black-billed Magpies, male and female homosexualities are found in Caribou, and same-sex mounting and coparenting also occur among Bears. These species are all directly identified with homosexuality and/or transgender in some Native American tribes. Moreover, in many cases where the exact species that figures in indigenous conceptions of homosexuality is not accurate, a closely related animal (often in another geographic area) does exhibit the behavior. For example, homosexual activity has not been recorded among New Guinean wallabies, yet it does occur in Australian Wallabies. Likewise, although homosexuality is not yet reported for the cassowary, it has been observed in Emus and Ostriches (related species of flightless birds). Other examples are summarized in the table below.

Some Correspondences between Indigenous Beliefs and Western Scientific Observations of Animal Homosexuality/Transgender (TG)

Animal traditionally associated with homosexuality/TG	Homosexuality/TG reported in scientific literature	Homosexuality/TG observed in related species
NORTH AMERICA		
Black-tailed Deer (Lakota, Zuni)	yes (Mule Deer)	yes (White-tailed Deer)
Elk (Oto)	yes (Wapiti)	yes (Red Deer, Moose)
Buffalo (Lakota, Ponca, Mandan, etc.)	yes (American Bison)	yes (other Buffalo species)
Bighorn Sheep (Hopi)	yes (Bighorn Sheep)	yes (other Mountain Sheep)
mountain lion (Okanagon, etc.)	no	yes (African/Asiatic Lion)
fox, coyote (Arapaho, Okanagon, etc.)	yes (Red Fox)	yes (Bush Dog)
Gray Wolf, red wolf (Tsistsistas)	yes (Gray Wolf)	yes (other Canids)

Animal traditionally associated with homosexuality/TG	Homosexuality/TG reported in scientific literature	Homosexuality/TG observed in related species
NORTH AMERICA *(cont.)*		
Bears (Nuu-chah-nulth, Keres, etc.)	yes (Grizzly, Black Bear)	yes (other carnivores)
jackrabbit, cottontail species (Zuni)	no	yes (Eastern Cottontail)
bat species (Zuni)	no	yes (Little Brown Bat, other Bats)
golden eagle, Harris's hawk (Tsistsistas)	no	yes (Kestrel, Steller's Sea Eagle)
owl species (Omaha, Mandan)	yes (Barn Owl)	yes (Powerful Owl)
oriole, tanager species (Tsistsistas)	no	yes (Yellow-rumped Cacique)
Magpie (Hidatsa)	yes (Black-billed Magpie)	yes (other Crows)
blue jay (Winnebago)	no	yes (Mexican Jay)
crow (Kamia)	no	yes (Raven, Jackdaw)
turtle species (Fox)	yes (Wood Turtle)	yes (Desert Tortoise)
salmon species (Nuxalk)	no	yes (European Salmon species)
butterfly/moth species (Navajo)	yes (Monarch, others)	yes (other butterfly species)
dragonfly species (Tsistsistas)	yes (Dragonflies)	yes (Damselflies)
NEW GUINEA		
wild boar, Pig (Bimin-Kuskusmin, Sabarl, etc.)	yes (domestic Pig)	yes (Warthog, Peccaries)
New Guinean wallabies (Marind-anim)	no	yes (Australian Wallabies)
arboreal marsupials (Sambia, Bimin-Kuskusmin)	yes (Tree Kangaroos)	yes (other marsupials)
fruit bat species (Bimin-Kuskusmin)	no	yes (other Fruit Bats)
echidna (Bimin-Kuskusmin)	no	no
jabiru stork (Marind-anim)	no	yes (White Stork)
Raggiana's Bird of Paradise (Sambia)	yes (Raggiana's)	yes (see below)
other birds of paradise (Ai'i, Sambia, etc.)	no	yes (Victoria's Riflebird)
bowerbird species (Sambia, Bimin-Kuskusmin)	no	yes (Regent Bowerbird)
cassowary (Sambia, Bimin-Kuskusmin, etc.)	no	yes (Emu, Ostrich, Rhea)
black-capped & purple-bellied lories (Sambia)	no	yes (several Lorikeet species)
other New Guinean parrots (Sambia)	no	yes (Galah)

Animal traditionally associated with homosexuality/TG	Homosexuality/TG reported in scientific literature	Homosexuality/TG observed in related species
New Guinea *(cont.)*		
nightjar species (Bimin-Kuskusmin)	no	no
python species (Bimin-Kuskusmin)	no	yes (other snake species)
monitor lizard (Bimin-Kuskusmin)	no	yes (other lizard species)
shark species (Nduindui, Vao)	no	no
sago grub (Bedamini, Sambia, Bimin-Kuskusmin)	no	yes (Southern One-Year Canegrub)
centipede species (Bimin-Kuskusmin)	no	yes (Spiders, other arthropods)
Siberia/Arctic		
white whale (Koryak, Inuit)	yes (Beluga)	yes (other Whales & Dolphins)
bearded, ringed, & Spotted Seals (Yup'ik, Inuit)	yes (Spotted Seals)	yes (Harbor Seal, other Seals)
Walrus (Chukchi, Yup'ik, Inuit)	yes (Walrus)	yes (other pinnipeds)
Caribou/Reindeer (Inuit, Yukaghir, Chukchi, etc.)	yes (Caribou)	yes (other Deer)
Horse (Sakha)	yes (Takhi, domestic Horse)	yes (other Equids)
Wolf (Sakha)	yes (Wolf)	yes (other Canids)
Bears (Ob-Ugrian, Nivkh, Chukchi, Inuit, etc.)	yes (Grizzly, Black, Polar Bears)	yes (other carnivores)
eider duck (Yup'ik)	no	yes (Lesser Scaup, other Ducks)
loon (Sakha, Yup'ik)	no	yes (Grebes)
murre species (Yup'ik)	yes (Common Murre)	yes (other diving birds)
Ruff (Chukchi)	yes (Ruff)	yes (other sandpipers)
Raven (Sakha, Koryak)	yes (Raven)	yes (other Crows)
pike species (Sakha)	no	yes (Salmon species)

Some of the most precise correspondences involve transgender (particularly intersexuality) rather than homosexuality per se. Modern science has provided startling confirmation of a number of indigenous "beliefs" about purportedly cross-gendered animals, most notably the left-handed Bear figure of many Native American cultures. Biologists have actually uncovered evidence that some species of Bears probably are left-paw dominant. "Handedness," or laterality, is a widespread phenomenon in the animal kingdom, with species as diverse as primates, cats, parrots, and even whales and dolphins showing preferences for the use of a right or left appendage (or side of the body) in various behaviors and tasks.[44] Although in most species there is considerable variation between individuals as to which side is dominant, it does appear that at least some kinds of Bears are consistently "left-handed." Scientists and naturalists report that Polar Bears, for example, regularly use their left paw for attack and defense as well as for clubbing seals and hauling them out of the water. In many cases they exhibit greater development of the left paw and may also use their left forelimb and shoulder to carry large objects. The consistency of left-paw use is exemplified by an incident in which wildlife biologists set up snare traps to capture and tag Polar Bears (for long-term study of their migrations). The traps were triggered by the bear's reaching for some bait with its paw, and all 21 Bears caught in this way were snared by their left front foot.[45] Incidentally, there also appears to be a correlation between left-handedness and homosexuality/transgender in humans: a higher than average proportion of gays and lesbians are left-handed (or ambidextrous), and one study found the percentage of left-handers among lesbians to be more than four times that among heterosexual women. Left-handedness also appears to be more common among transsexuals, particularly male-to-female transsexuals.[46] As yet, no studies have looked for possible correlations between laterality and homosexuality/transgender in animals.

Another astonishing correspondence concerns the attraction of Bears to human menstrual blood, a widespread belief in many Native American tribes. Implausible as this connection may sound, zoologists decided to conduct experiments to see if there was any truth to these "superstitions." Employing controlled olfactory-preference tests on Polar Bears in both laboratory and field settings, they found that the animals were indeed significantly more attracted to the odors in human menstrual blood than to a number of other smells, including several animal and food odors as well as nonmenstrual blood. In fact, human menstrual odors in many cases elicited a response from the Bears as strong as that prompted by the smell of seals (their primary food in the wild), which they could detect from more than 1,200 feet away.[47]

Even more extraordinary, biologists have found actual cases of physical gender-mixing in Bears. In 1986, Canadian zoologist Marc Cattet made a stunning discovery: the presence of significant numbers of "masculinized females" in wild populations of Grizzly, Black, and Polar Bears. These animals have the internal reproductive anatomy of a female combined with portions of the external genitalia of a male, including "penislike" organs. As many as 10–20 percent of the Bears in some populations may exhibit this phenomenon.[48] Such individuals are able to reproduce, and most adult intersexual Bears are actually mothers that successfully raise cubs. In fact, the reproductive canal in some intersexual Bears extends through the phallus rather than forming a vagina, so that the female actually mates and gives birth through the tip of her "penis"—similar to the way female Spotted Hyenas mate and give birth through their "penile" clitoris. These findings offer striking parallels to the gender-mixing Bear Mother figure of many Native American tribes, as well as the Bimin-Kuskusmin and Inuit beliefs about "male mothers" and androgynous animals that give birth through a penis-clitoris.

Intersexual animals that combine male and female sex organs (and in some cases that are intermediate between males and females in their body proportions and size) also occur spontaneously in other mammals that are not usually hermaphroditic, such as primates (e.g., Common Chimpanzees, Rhesus Macaques, Savanna Baboons), whales and dolphins (e.g., Bowhead and Beluga Whales, Striped Dolphins), marsupials (e.g., Eastern Gray and Red Kangaroos, various Wallabies, Tasmanian Devils), and rodents and insectivores (e.g., moles).[49] In fact, a veritable profusion of different kinds of gender mixing has been uncovered throughout the animal world—so much so that scientists have had to develop a special terminology to refer to the bewildering variety of intersexualities. Fanciful-sounding names such as *chimeras, freemartins, mosaics,* and *gynandromorphs* are actually the technical terms used by biologists to designate animals with various types of chromosomal and anatomical gender mixing.[50] In Greek mythology, a *chimera* is a fantastic creature combining features of a lion, goat, and serpent, while *Hermaphroditus* is the child of the gods Hermes and Aphrodite. It is ironic that Western science uses names with mythological connotations to refer to animals that are actually the living "proof" of indigenous myths about homosexual and transgendered species. The left-handed androgynous Bear may exhibit *chimerism* (scientifically speaking), but it isn't "chimerical" at all—it's alive and well and living in North America!

A *freemartin* is an animal that becomes intersexual as a result of association in the womb (or egg) with a twin of the opposite sex (note the motif of twinning in some Native American two-spirit traditions that involve animals, such as the Kamia and Wintu), while *chimera* refers to an animal with organs that combine genetically male and female elements. Similar to chimerism, a *mosaic* is an individual that has variable chromosomal patterns and a corresponding mixture of male and female traits. Some of the diverse types of chromosome configurations (in addition to the "typical" female and male patterns of XX and XY, respectively) include XXY, XXX, XXYY, XO, and even combinations of these in different cells of the body. Each chromosomal pattern, in turn, manifests itself as a different mixture of male and female sex organs and secondary sexual characteristics, sometimes juxtaposed in separate parts of the body, sometimes combined in the same organ, and sometimes blending together as a gradation of traits or a combination of all of these.[51]

One particularly remarkable type of mosaic is called a *gynandromorph:* an animal that

Myth made real: a transgendered (intersexual) Eastern Gray Kangaroo. This animal has both a penis and a pouch (the latter usually found only in females). Chromosomally, it combines the female pattern (XX) with the male (XY) to yield an XXY pattern. ▶

appears to be literally divided in half, one side (usually the right) male in appearance, the other side female, often with a sharp line of demarcation between them. This occurs in a number of different kinds of animals, such as butterflies, spiders, and small mammals, and bears a noteworthy resemblance to the Bimin-Kuskusmin belief about a centipede that is female on its left side and male on its right. More than 40 cases of gynandromorphism have also been reported in birds such as finches, falcons, and pheasants. In these cases the two halves of the creature differ in plumage (and sometimes even size), usually corresponding to internal reproductive organs of both sexes (an ovary on one side, a testis on the other). Some gynandromorphs have more of a gender mixture in their appearance while still preserving a central dividing line. One warbler, for example, had a left side that was male while its right side was a tapestry of male and female plumage characteristics. Although little information is available on the behavior of gynandromorphs, it appears that some individuals may exhibit a combination of both male and female behavior patterns. One spider gynandromorph, for example, courted and mated with females using its male organs, but also built an egg case as is typical for females. On the other hand, a gynandromorphic chimney swift exhibited primarily (heterosexual) "male" behavior throughout its life, regularly pairing with females and fathering offspring.[52]

Other examples of the correspondences between indigenous views about transgendered animals and scientific observations can be found. Sambia (and other New Guinean) beliefs about all-female offspring, gender mixing, and sex change in marsupials and other animals may seem far-fetched, yet recent discoveries by zoologists studying a variety of species bear remarkable similarities to these ideas. For example, geneticists recently determined that significant numbers of female wood lemmings are actually chromosomally male (having an XY pattern). Moreover, some of these animals only give birth to female offspring, so that the population consists of 80 percent females. Similar phenomena occur in at least 7 other species of rodents, and individual females that only produce female offspring have also been reported as a recurring phenomenon in at least 12 different species of butterflies.[53] True transsexuality is most often found in fishes (and "lower" animals), where the combination of sex change with chronological color change in some coral reef species echoes Sambia beliefs about sequential sex and plumage changes in birds (and other creatures). Cases of chromosomal "sex reversal" (males that have a female chromosomal pattern, or vice versa) occasionally occur in mammals such as moles, mole-voles, and primates (e.g., Orang-utans and Hanuman Langurs).[54] And while Inuit beliefs about gender-mixing Caribou that wear female garments are not literally true, female Caribou often exhibit physical "transvestism" in the sense that they bear antlers (a trait typically associated with males in all other species of Deer).

Striking parallels also exist with regard to the gender transposition, androgyny, and homosexuality that the Sambia and Bimin-Kuskusmin associate with some birds. Female Raggiana's Birds of Paradise, for example, have been observed performing courtship displays to one another. This behavior combines not only same-sex interaction but a gender-role "reversal," since typically only males display in this species. Ornithologists have also determined that males of the king bird of paradise do indeed associate in pairs—recall that among the Ai'i people, some birds of paradise are symbolically related to male couples.[55] As for the cassowary, its polyandrous social system—in which one female mates with several males, who are then left to incubate the eggs and raise the young on their own—shows some correspondence to the notions of "female potency," male motherhood, and gender reversal attributed to this bird by a number of native New Guinean peoples.[56]

Scientists have also discovered some unusual details about the cassowary's genital anatomy that bear an uncanny resemblance to indigenous ideas about the "androgyny" of these creatures, especially the Bimin-Kuskusmin belief about the bird's "penis-clitoris." Unlike most other birds, the cassowary male actually does possess a penis; however, this organ does not transport semen internally as it does in mammals. The cassowary's phallus is described by scientists as being "invaginated," that is, it has a tubular cavity that opens at the tip of the penis but is not connected internally to the male reproductive organs. This vagina-like cavity is in fact used to retract the phallus by turning it "inside out," causing the nonerect penis to resemble the finger of a glove that has been pushed inward. Consequently, although the male cassowary inserts his erect penis into the female during mating, he ejaculates semen through his cloaca, an orifice at the base of the penis that also doubles as the bird's "anus" and urinary organ. Females also mate, lay eggs, defecate, and urinate all through the same orifice, the cloaca (as in all other female birds)—but the cloaca is exceptionally large in this species, being capable of passing eggs weighing up to one and a half pounds. Most amazingly, all female cassowaries *also* possess a phallus, which is essentially identical to the male's in structure but smaller. The "female phallus" is also sometimes referred to as a clitoris, but it would be equally valid to speak of a "male clitoris" in this species (as noted in chapter 5), since the male cassowary's "penis" is not in fact an ejaculatory organ.[57] Thus, the cassowary's genital anatomy exhibits a bewildering juxtaposition of "masculine" and "feminine" traits: both males *and* females possess a penis/clitoris (a phallic organ that nevertheless is "vaginal" in form and nonejaculatory in function), and both sexes also possess another genital orifice that doubles as an anus. Indigenous beliefs about masculinized female cassowaries, the bird's penis-clitoris, anal birth, and women with phalluses being transformed into cassowaries are not nearly as outlandish as they sound.

Another interesting parallel between homosexuality/transgender in animals and indigenous views of these phenomena in people concerns the notion of "hypermasculinity." Contrary to the stereotypical Euro-American view of male homosexuality, in some Native American cultures two-spirit people who are biologically male may manifest (or are considered to manifest) a sort of "excess" or intensification of masculinity (at the same time as they embody a combination of both male and female traits). Among the Coahuiltec, Crow, Keres, and Zuni peoples, for example, male two-spirits are sometimes actually physically larger, taller, and/or stronger than non-two-spirit men, and greater strength has also been attributed to two-spirits among the Luiseño, Hidatsa, and O'odham (Papago). Some male two-spirits are distinguished warriors in their tribes, are notably aggressive, or otherwise fight alongside non-two-spirit males, for example among the Osage, Illinois, Miami, and Hidatsa. The Tsistsistas (Cheyenne) people include male two-spirits in war parties, in part because they are thought to possess a "stored-up virility" that will insure the success of the endeavor, while Lakota and Ojibway male warriors sometimes have sex with male two-spirits in order to partake of the latter's courage, ferocity, and fighting skills. And as already mentioned, in a number of Melanesian cultures (as well as other cultures around the world), homosexuality is thought to have a strengthening or "masculinizing" effect on men and in some cases to express qualities of male potency and even exaggerated virility. Among animals, there are a number of intriguing concordances with this rather unexpected association between male homosexuality/transgender and "extra" masculinity. As discussed in preceding chapters, gander pairs among Greylag Geese and Black Swans are distinguished by their superior strength, courage, aggression, and (in some cases) more intense bonding, while male pairs in a number of other species can be offensively (rather than just defensively) aggressive. In addition, transgendered or homosexual individuals in American Bison, Savanna (Chacma) Baboons, and Hooded Warblers sometimes exceed other males in overall size, weight, or other physical dimensions. Such individuals may also achieve a high-ranking social status (e.g., Greylag Geese, Savanna Baboons), which echoes the honored status of two-spirits in many Native American cultures. Furthermore, transgendered males in several species are often more "virile" or heterosexually active than nontransgendered males (e.g., Northern Elephant Seals, Red Deer, Black-headed Gulls, and Common Garter Snakes). And in Bighorn Sheep, homosexual mounting is more characteristic of "masculine" rams than of "effeminate" rams (i.e., behaviorally transvestite males, who act like females).[58]

Are these various connections between indigenous beliefs and scientific facts merely fortuitous, or do they represent accurate observation of animals on the part of aboriginal cultures? In other words, how likely is it that indigenous peoples could have been aware of the often esoteric details of animal behavior and biology that "corroborate" their beliefs? Although much indigenous thinking about animals is encoded in mythological terms (as we have seen), it is often grounded in a sophisticated framework of direct observation and study of the environment (sometimes known as ethno-science). This is true not only in the area of zoology, but in fields as diverse as botany, geology, geography, oceanography, meteorology, astronomy, and so on. In fact, aboriginal knowledge about the organization of the natural world often mirrors the findings of more "objective" scientific inquiry, sometimes down to the most minute detail. Many tribal cultures, for example, have developed comprehensive classification schemata for plant and animal species that rival the system of scientific nomenclature used by biologists today. The Arfak mountain people of New Guinea identify and name 136 distinct bird species in their environment—almost exactly the number recognized by Western science for the same area.[59] Indigenous knowledge of animal behavior and other aspects of zoology is often remarkably accurate, and in many cases the behavioral, anatomical, or physiological phenomena involved have only been "discovered" or verified by Western science in the last decade or two. As one biologist remarks, "The sum total of the [indigenous] community's empirically based knowledge is awesome in breadth and detail, and often stands in marked contrast to the attenuated data available from scientific studies of these same populations."[60]

The Inuit and Aleutian Islanders, for instance, have an extraordinarily profound understanding of Walrus behavior and social organization, including knowledge of a number of more unusual habits and aspects of the animals' social life that have been verified by zoologists only relatively recently. The use of pharyngeal (throat) pouches in producing metallic sounds, adoption of orphaned pups, all-male summer herds, and mass mortality during huge stampedes were all "unexpected" or "controversial" phenomena when first discovered by Western observers, yet they had been known to indigenous peoples long before their existence was documented by biologists.[61] Western scientists initially considered solitary male Musk-oxen to be older individuals who were "superfluous" to the population; Inuit people, in contrast, believe that these old males are not a surplus component of the population at all. Based on direct observation of Musk-oxen as well as their traditional beliefs about the animals, the Inuit maintain that such animals are a vital element of Musk-ox social structure, serving as a focal point for regathering of the population after the rut as well as functioning as "elders" for the herd. Scientists now know that such males are not in fact superfluous, but serve an important role in the population structure of the species.[62] Moreover, biologists studying other animals have even gone so far as to ascribe "elder"-like roles to postreproductive individuals, suggesting that in short-finned pilot whales, for example, "their principal biological contribution might be to learn, remember, and transmit what pilot whales need to know."[63] Similarly, traditional Cree knowledge about beaver social organization and population regulation rivals the most sophisticated wildlife-management programs developed by Western science—ones that utilize computer modeling, satellite mapping, and complex statistical analysis.[64]

So it is not unreasonable that aboriginal beliefs about animal homosexuality/transgender might represent systematic and careful observations of the natural world, rather than simply the projections of a mythological system. Many indigenous peoples are undoubtedly aware of gender-mixing creatures as part of their natural environment and incorporate them into their belief systems. As anthropologist Jay Miller has observed, "Hunting tribes were also astute enough observers . . . to notice that other Animal People had hermaphroditic members, and often equated these with the berdaches [two-spirits]."[65] Intersexual Bison, for example, are recognized as such when seen by Native Americans in herds of wild animals. The Cree call them *ayekkwe mustus* (*ayekkwe* referring to the quality of being neither male nor female, or both, i.e., hermaphrodite; *mustus* meaning buffalo), while the Lakota and Ponca refer to them as *pte winkte* and *pte mixuga,* respectively—*pte* meaning buffalo and *winkte* or *mixuga* designating two-spirit—thereby drawing an explicit parallel between transgender in animals and people. The sight of a massive hermaphrodite "buffalo ox" towering above its companions undoubtedly reinforces this parallel for many indigenous observers, since it so closely resembles the way two-spirit males in some cultures are taller or stronger than both men and women. (Early white observers, in contrast, erroneously attributed the intersexuality of such Bison to castration, either by people or wolves.)[66]

As mentioned previously, the Navajo also recognize intersexual animals in several game species and are even aware of the "cactus bucks," transgendered Mule

Deer with distinctive antler configurations. They call these creatures *biih nádleeh*— *biih* for deer, *nádleeh* meaning transformed, constantly changing, or hermaphrodite (the same term applied to two-spirit people)—once again establishing the fundamental continuity between animal and human gender/sexual variability.[67] Likewise, although cassowaries are elusive and difficult to observe in the wild, many New Guinean tribes hunt the birds and also keep them in captivity, utilizing semi-domesticated cassowaries for food, trading, harvesting of plumes and other materials, in ceremonial functions, as pets, and even as a form of currency.[68] It is likely, then, that at least some details of this creature's unusual genital anatomy are generally known to indigenous peoples from firsthand observation (rather than simply figuring in mythological contexts). Indeed, at least one tribe, the Mianmin, are aware of the bird's phallus—an organ whose existence and structure in males, let alone females, is not even widely acknowledged among Western ornithologists.[69] Finally, we have already mentioned how indigenous Vanuatu knowledge and terminology relating to intersexuality in domestic Pigs rivals and in some cases even exceeds that of Western science.

In addition to intersexuality, other supposedly mythic traits of animals ritually associated with transgender/homosexuality have also been directly observed by native peoples in the creatures around them. The Inuit report seeing Polar Bears use their left paws to kill seals and throw ice and other objects at Walruses. The Halkomelem (Fraser River Salish) people of British Columbia also describe a curious behavior in Bears that suggests an awareness on their part of the creature's "left-handedness." They tell of Bears staying close to the right perimeter of a cave wall when departing their hibernation dens, thereby leaving the left paw free for defense.[70] White Buffalo—symbolically associated with two-spirit among the Lakota—were also regularly observed by native peoples in wild herds. When reports of these creatures first reached non-Indians, they were usually considered to be a figment of the "native imagination" or else were attributed to "artificial" circumstances, such as an escaped domestic cow, a hybrid offspring of such an animal, or the result of deliberate "whitening" of the hide by Indians. Now, of course, scientists recognize that the indigenous observations were correct: white Bison—both albino and nonalbino—are a recurring, albeit rare, phenomenon in wild populations of this species (as are other colors such as pied and gray).[71]

The discovery of widespread animal homosexuality and transgender by modern science puts a whole new spin on these parallels. Could it be that indigenous cultures actually know *more* about certain aspects of animal sexual and gender variance than zoologists do now? In other words, are species "erroneously" associated with homosexuality/transgender in various indigenous cultures actually genuine examples waiting to be "discovered" by Western science? Certainly when the scientific literature has previously failed to offer corroboration of a particular native belief about the behavior of an animal, it is often the case that the scientific record, and not the aboriginal observation, was in error. Time and again, indigenous beliefs have been dismissed as fanciful "superstitions," only to be confirmed once the technology and observational skills of modern science finally catch up with the age-old teachings of aboriginal peoples. For example, the Hopi have a folktale about hibernation in the poorwill—a bird they call *hölchko,* "the sleeping one"—a belief also shared by the Navajo. This was thought to be purely mythological until scientists discovered a torpid poorwill regularly hibernating during the winter (it was found in a rock crevice in California with a body temperature of about 64 degrees F). Ornithologists now officially recognize the poorwill as the only bird in the world that consistently undergoes long-term hibernation.[72] Likewise, the traditional songs and oral narratives of the O'odham (Pima) people of Arizona refer to moths becoming drunk on the nectar of jimsonweed blossoms. Far from being an inventive anthropomorphization, this "belief" about insects was subsequently verified by Western science. Biologists observed "drunken" behavior in hawkmoths that had consumed jimsonweed nectar (which is now known to contain narcotic alkaloids), including erratic and uncoordinated flight, "crash landings," falling over, and other movements suggesting intoxication.[73] The Kalam people of New Guinea believe that earthworms make croaking noises and can produce various other sounds such as whistles and stridulations. Biologists initially scoffed at these beliefs, yet specialists in worm biology have confirmed that earthworms, particularly some of the larger species found in Southeast Asia and Australia, can make an extraordinary range of sounds, including clicks, rasps, slurps, and even birdlike notes.[74] Numerous references to a giant lizard known as *kawekaweau* occur in the folklore and legends of the Maori (the indigenous people of New Zealand). Initially dismissed by contemporary Western investigators as an imaginary creature, the *kawekaweau* has now been identified by zoologists as corresponding to a recently discovered species of gecko. Though it measures just over a foot long, it is in fact the largest of its kind in the world.[75]

Finally, Navajo legend tells of how Bears taught people about the medicinal properties of a plant known as *na'bi* or bear medicine, instructing them in the proper administration of the drug (including chewing and/or applying a powder or infusion directly to the skin). Scientists recently confirmed the connection of this indigenous pharmaceutical to Bears and also experimentally verified the effectiveness of the plant's active ingredient (ligustilide) as an antibacterial and antiviral. Extraordinary observations have been made of Grizzly Bears actually utilizing the plant as a topical medication on themselves. They chew the root, spit the plant juices and saliva on their paws, then rub the mixture thoroughly into their fur. In fact, this and other examples of "self-medicating" behavior in animals (most notably in Chimpanzees) have recently led to the establishment of a new scientific discipline called *zoopharmacognosy,* the study of animals' use of medicinal plants to treat themselves. Investigators working in this exciting field of inquiry have stumbled upon something that many indigenous peoples have known for an immensely long time, the fact that (in the words of one biologist) "not all pharmacists are human."[76]

Drunken moths, hibernating birds, giant geckos, croaking worms, white Buffalo, self-medicating Bears, left-handedness, menstrual attraction, sex change, gender mixing, homosexuality . . . often the most "preposterous" aboriginal beliefs about animals turn out to have a basis in reality. One could hardly imagine more fantastical creatures than mother Bears with penises or cassowaries of both sexes with vaginal phalluses—yet these "myths" are biological facts. Thus, while many indigenous ideas about animal homosexuality and transgender have yet to be confirmed, scientific "proof" may well be forthcoming—even for the most unlikely sounding of mythological scenarios.

An All-Encompassing Vision

Indigenous "myths," sacred stories, and folk knowledge about animals (including information relating to homosexuality and gender mixing) are part of an oral tradition that is thousands of years old. The Nuu-chah-nulth culture of Vancouver Island, for example, stretches back uninterrupted to at least 3,000 B.C. according to archaeological dating methods, and is by no means a unique example.[77] Contemporary native storytellers are, in a sense, the repositories of a scientific tradition whose continuity can be measured in millennia. It must be remembered that the "accuracy" of indigenous views about animals is being assessed against a Western science that has only recently begun to systematically investigate animal homosexuality/transgender (and that has generally been reluctant even to recognize these phenomena). New cases are being discovered all the time, often in species previously claimed never to exhibit homosexual behavior in the wild. Consequently, animal homosexuality reported in the scientific literature does not represent the sum total of homosexual wildlife in the world—only those cases that scientists happen to have noticed. Undoubtedly many examples have been missed or ignored, especially when the investigator harbors a strong personal distaste for the subject matter or is not prepared to observe same-sex behavior (as discussed in chapter 3).

So rather than simply checking the "correctness" of indigenous beliefs against what Western science has uncovered or currently "knows," perhaps we should also be using the "discoveries" of indigenous science as signposts for where zoology might direct its attentions on this subject. Traditional tribal knowledge about animal homosexuality/transgender can in fact serve as a model for more orthodox scientific investigation of the subject—for example, by leading the way toward study of these phenomena in new species. With thousands of animals remaining to be described in detail by zoologists—and new species being discovered each year—it is a daunting task to know where to begin and which species to focus on when studying homosexuality/transgender in the natural world. If coyotes and mountain lions, for example, or New Guinean birds of paradise, marsupials, and echidnas are consistently singled out by native cultures as being relevant in this area, Western science could do worse than swallow its pride and take these "myths" seriously. In determining once and for all whether these are merely superstitions, it may well discover (once again) that an unexpected kernel of truth in some of these beliefs merits further scientific inquiry.

Both indigenous and Western scientific paradigms have their own particular strengths and weaknesses; by forging a partnership between them, we can achieve a level of knowledge that exceeds the sum of the two. Although the two perspectives would appear to have much to benefit from such an interaction, they have rarely met within the scientific or academic establishment.[78] An intimation of the sort of collaborative effort that is possible is provided by two examples involving indigenous peoples in both North America and New Guinea. The Kalam and other tribes of New Guinea recognize several species of poisonous birds in their environment. One of these is the hooded pitohui, a species they refer to as a "rubbish bird" because it causes burning and numbness of the lips when eaten, and possibly even paralysis and death. Chemical defense through the use of naturally produced toxins in the skin (such as those found in the poisonous frogs of South America) was previously thought not to occur in birds. In 1990, however, scientists confirmed that these birds are indeed poisonous by isolating the chemical compound, homobatrachotoxin, responsible for their toxicity. Their investigations would not have been possible without the help of the aboriginal hunters who shared their traditional knowledge of these species and helped the ornithologists locate specimens of the birds over several field studies. This discovery, in turn, spurred a renewed interest on the part of biologists in the long-neglected topic of avian chemical defense, and subsequent research has revealed a surprisingly large number and variety of poisonous bird species throughout the world.[79]

More than 15 years earlier, Robert Stephenson, a biologist with the Alaska Department of Fish and Game, and Robert Ahgook, a Nunamiut Inuit (Eskimo) hunter, coauthored a scientific report on Wolf ecology and behavior. They pointed out that the indigenous view of Wolves involves a highly developed conception of the creature's behavioral flexibility and individuality, a perspective that zoologists are just beginning to countenance: "As a result of the vast array of behavioral events they have witnessed, the Nunamiut interpret wolf behavior in a broader, yet more intricate theoretical framework than that heretofore used by modern science; their in-depth knowledge gained from patient, on-the-ground observation has taught them that the adaptability and elasticity inherent in wolf behavior rivals that in human behavior."[80] These cross-cultural collaborators suggest that Western scientists should adopt this sort of intellectual framework as a matter of course—a suggestion that rings especially true where animal homosexuality and transgender are concerned, since these phenomena epitomize the behavioral "elasticity" inherent in the natural world.

Important as these findings are, indigenous perspectives on animal (and human) homosexuality/transgender have a significance for Western science that extends far beyond the details of specific behaviors in particular animals. It is striking that in so many cultures that recognize some kind of alternate gender/sexuality system in animals, human homosexuality/transgender are also routinely recognized and even honored. Perhaps, then, what is most valuable about indigenous views of animal homosexuality/transgender is not so much the "accuracy" of beliefs about this species or that, but the overall worldview imparted by these cultures: a view of both animals and people in which sexuality and gender are each realms of multiple possibilities.

In fact, ideas about human and animal homosexuality tend to be mutually reinforcing. When people consider homosexuality/transgender to be an accepted part of human reality, they are not surprised to find gender and sexual variability in animals as well. Similarly, a culture living in intimate association with the natural world will undoubtedly encounter animal homosexuality/transgender on a routine basis; these observations in turn contribute to the culture's view of such things as an integral part of human life. On the other hand, people accustomed to seeing homosexuality/transgender as an aberration will balk at encountering the phenomena in animals. And when a culture no longer lives in close association with wilderness, it will have less opportunity to encounter natural examples of variation in gender and sexual expression.

Consider two contrasting viewpoints on animal homosexuality that epitomize this difference. A man representing the Euro-American cultural tradition of the late twentieth century states that it is impossible for him to even imagine a "queer grizzly bear . . . or a lesbian owl or salmon."[81] In contrast, a contemporary Native American storyteller of the Wintu nation describes coyote as having a homoerotic relationship with another male, guided by the spirits of "grizzly, salmon, and eagle."[82] In a remarkable coincidence, each individual has independently singled out virtually identical animals as somehow emblematic, but with radically different interpretations (neither was aware of the other's words). From the Anglo perspective, homosexuality is an insult to the animals' supposed "purity" or "virility"—sentiments that are echoed, less overtly, throughout the scientific discourse on the subject—while from the native perspective, such homosexuality is an affirmation of nature's plurality, strength, and wholeness.

If taken literally, the Wintu tribesman's account is clearly the more "accurate" of the two: homosexuality and/or transgender occur in Grizzly and Black Bears, Salmon of various species, and several birds of prey (including Barn Owls, Powerful Owls, Kestrels, and Steller's Sea Eagles). But that's almost beside the point: what is most significant is the inclusiveness of his vision, which stands regardless of whether any animals he mentions are "known" by zoology to be homosexual or transgendered. It is not the "accuracy" of individual observations that validates an indigenous perspective, but the expansiveness of that perspective which fosters such "accurate" observations in the first place. What Western science can learn most from aboriginal cultures is precisely this polysexual, polygendered view of the natural world. The next section explores how these ideas can be incorporated in a more concrete fashion into scientific discourse and shows that they are fundamentally compatible with a number of new developments in science and philosophy.

A Revolution Under Way: Contemporary Scientific and Philosophical Perspectives

> We need another and a wiser and perhaps a more mystical concept of animals. . . . They are not brethren, they are not underlings; they are other nations, caught with ourselves in the net of life and time, fellow prisoners of the splendor and travail of the earth.
>
> —naturalist HENRY BESTON

> In effect, chaos is life. All mess, all riot of color, all protoplasmic urgency, all movement—is chaos.
>
> —essayist HAKIM BEY[83]

Biology must reconsider functional explanations based on evolution by natural selection, and it must recognize the inherent multiplicity of all life forms. The existence of a natural phenomenon *is* its function—regardless of how strange, complex, or "unproductive" it may seem. These are just a sample of some of the revisionist ideas that are now being proposed in biology, in conjunction with work in a broad range of other scientific disciplines. Although none of these ideas has yet been applied to the understanding of homosexuality/transgender, they have powerful implications for our ways of seeing these phenomena. The synthesis of these new ideas—and their application to a broad spectrum of natural and cultural phenomena, including systems of gender and sexuality—we will call Biological Exuberance. Biological Exuberance is not a theory or an "explanation" designed to supplant previous ones; rather, it is a fundamental shift in perspective, an alternative vision of something we thought we understood. Through this concept, we seek not so much to add new facts to existing knowledge, but (as Robert Pirsig puts it) to add a new pattern of knowledge to existing facts.[84]

In the following discussion, we will explore the potential inherent in certain contemporary scientific and philosophical perspectives to initiate such a re-visioning. Many of the ideas to be considered here are highly speculative or counter to traditional thinking, and often controversial even within their respective fields. Other, seemingly implausible, concepts will reveal themselves to be compatible with some of the most basic and long-standing concepts of orthodox biological theorizing. Moreover, each of the ideas to be discussed already represents a vast and complex field of knowledge; we can do no more than sketch the merest outlines of a road map for future investigation, suggesting some fruitful paths of inquiry. What the ideas we are conveniently summarizing under the rubric of Biological Exuberance have in common, though, is the capacity to precipitate a breakthrough in understanding (even when, by necessity, they are presented in abbreviated form). Taken together, they offer a new mode of perception, something infinitely more valuable than yet another simplistic "answer." Where basic paradigm shifts are concerned, we should not be puzzled by how firmly we previously held to so many different falsehoods; rather, we should be astounded that there are so many different truths (to paraphrase James Carse).[85]

Post-Darwinian Evolution and Chaotic Order

> Nature . . . is fundamentally erratic, discontinuous, and unpredictable. It is
> full of seemingly random events that elude our models of how things are
> supposed to work.
>
> —DONALD WORSTER, "The Ecology of Chaos and Harmony"[86]

Survival of the fittest, natural selection, random genetic mutations, competition for resources—we all know how evolution works, right? Not quite. Over the past two decades, a quiet revolution has been taking place in biology. Some of the most fundamental concepts and principles in evolutionary theory are being questioned, challenged, reexamined, and (in some cases) abandoned altogether. A new paradigm is emerging: post-Darwinian evolution.[87] "Heretical" ideas are being proposed by post-Darwinian evolutionists, such as the self-organization of life, the notion that the environment can beneficially alter the genetic code, and a suite of evolutionary processes to accompany the once hegemonic principle of natural selection. Moreover, many of the developments in this theorizing reflect surprising convergences with another "new" science, chaos theory.

"Put at its simplest, the new paradigm is an insistence on pluralism in evolutionary studies." That's how scientists Mae-Wan Ho and Peter Saunders characterize the essence of the new thinking on evolution.[88] This paradigm is tackling a number of long-standing puzzles in biology—among them, global patterns of emergence and extinction of species, "mimicry" between animals separated by geography (in which two unrelated butterfly species in different parts of the world, for example, have evolved identical appearances), and convergence between the structure of biological and inorganic forms (in which jellyfish larvae, for instance, closely resemble the patterns made by falling drops of ink in water; or the similarity between animal coat markings and the standing wave patterns that can be generated on thin, vibrating plates). Post-Darwinian evolutionary biologists are synthesizing developments in a number of diverse disciplines such as physics, chemistry, mathematics, and molecular and developmental biology as part of their theorizing on these and other phenomena.

One proposal involves the possibility of the self-organization of life—the notion that the proteins, and in turn the enzymes and the cells, necessary for the first rudimentary life-forms may not have arisen randomly. Rather, experiments have shown that such building blocks can form "spontaneously" through the interaction of chemical and physical processes inherent in the molecules themselves and their watery medium. Similarly, convergences in form between distant species or organic and inorganic matter reveal underlying patterning processes that may actually "direct" evolutionary change. Another revolutionary proposal involves what is known as the "fluid genome": the hypothesis that the environment can beneficially change the genes of an organism. The genetic code was previously thought to be static and inalterable (aside from random mutations), but now biologists are recognizing that a dynamic, complex, two-way interaction between environment and genetics may occur, possibly even leading to the evolution of new species.[89]

Although much of this theorizing is admittedly in its infancy (and even, in a few cases, on the "fringes" of the scientific establishment), some of the most respected names in evolutionary science are participating in the reevaluation of basic tenets of the theory.[90] World-renowned biologist and evolutionist Edward O. Wilson is at the forefront of the discussion, even going so far as to declare that evolution is, in a sense, a form of religion—"The evolutionary epic is probably the best myth we will ever have"[91]—thereby putting an ironic twist on the whole creationism-evolution controversy. Perhaps what is most significant in this entire discussion is not the explanatory power of particular theories (impressive as some of these are), but the spirit of intellectual openness and vision being embraced by many evolutionists, the willingness to reexamine once ironclad principles. Nowhere is this more apparent than in the questioning of the basic principle of natural selection based on random genetic variations. A number of scientists—among them Stephen Jay Gould—have long criticized the attempt to find an adaptive explanation for "every surviving form, structure, or behavior—however bizarre, unnecessarily complex or outright crazy it may appear."[92] Of course, the limitations of such "adaptationist" explanations are precisely the problem that orthodox biology confronts when it looks at the "bizarre" behaviors of homosexuality and nonreproductive heterosexuality. If biology is finally to come to terms with these phenomena, such explanations will need to be seriously reevaluated.

There are a number of parallels between post-Darwinian thought and the emerging science of chaos. Chaos theory is, fundamentally, a recognition of the unpredictability and nonlinearity of natural (and human) phenomena, including apparently destructive or "unproductive" events such as natural catastrophes. Although originally developed in the fields of mathematics, physics, and computer science, chaos science was quickly applied to biological phenomena. In fact, the periodic fluctuations of animal and plant populations were among the first examples of "chaotic behavior" to be uncovered in the natural world. Chaos theory has since been successfully used in the analysis of a wide range of natural and social phenomena, including biological systems (from the ecosystem to the cellular level) and evolutionary processes. Indeed, chaos scientist Joseph Ford has stated that "evolution is chaos with feedback."[93] The fractal or "chaotically ordered" structure of nature has even been revealed in the behavior patterns of individual animals and in the "self-organizing" architecture of honeybee combs.[94]

Arrhythmias, discordant harmonies, and aperiodicities are some of the characterizations of "chaotic" natural phenomena that have been offered. These terms are attempts to convey the idea that fundamental principles of "pattern organization" direct, but do not entirely determine, the development or "shape" of biological (and other) entities. The internal dynamics of such systems generate unpredictable, but not random, patterns.[95] This concept is echoed in recent reappraisals of "adaptationist" explanations for the diversity of plant and animal forms. As one ornithologist studying the proliferation and elaboration of bird plumage has observed, traditional evolutionary theory may be able to account for how a specific pattern, color, or form has developed, but it cannot explain why or how such incredible variety arose in the first place: "Such hypotheses explain a large variety of traits as divergent as a widowbird's tail, a rooster's comb, a peacock's train, or the black bib of a sparrow. While these hypotheses can account for some features of the trait, they cannot account for the enormous diversity in conspicuous traits—why some birds have red heads and others long tails even though the same basic process . . . may be at work."[96] Most current theories of phenomena such as plumage diversity still focus on the putative functional or adaptive role of specific patterns rather than the overall range of variation. However, this is an area where the application of principles from chaos theory might yield fruitful results.[97]

So too for diversity of sexual and gender expression. One of the more important insights to emerge from chaos theory is that the natural world often behaves in seemingly inexplicable or "counterproductive" ways as part of its "normal" functioning. According to Sally Goerner (in her discussion of chaos, evolution, and deep ecology), "Time and again, nonlinear models show that apparently aberrant, illogical behavior is, in fact, a completely lawful part of the system." Similarly, biologist Donald Worster remarks that "scientists are beginning to focus on what they had long managed to avoid seeing. The world is more complex than we ever imagined . . . and indeed, some would add, ever can imagine." More than half a century earlier, evolutionary biologist J. B. S. Haldane presaged these thoughts when he commented that "the universe is not only queerer than we suppose, it is queerer than we *can* suppose"—words we used to open this book.[98] Although none of these scientists is referring specifically to homosexuality, the alternate systems of gender and sexuality found throughout the animal kingdom are exactly the sort of "discontinuities" and "irrational" events that should be generated in a "chaotic" system.

Particularly relevant in this respect is Goerner's statement of one of five basic "principles" of chaos: *"Nonlinear systems may exhibit qualitative transformations of behavior (bifurcations).* The idea is simple: a single system may exhibit many different forms of behavior—all the result of the same basic dynamic. One equation, many faces. A corollary to this idea is that a system may have . . . multiple competing forms of behavior, each perhaps a hairsbreadth away, each representing stable mutual-effect organization."[99] Transposed to the realm of sexuality, this idea offers the potential for intriguing insights: heterosexuality, homosexuality, and all variants in between can be seen as alternative manifestations of a single sexual "dynamic," as it were, which is itself part of a much larger nonlinear system. The "flux" of this system is played out in endless and infinitely varying expressions within individual lives, through various communities, between different species, across sequences of time, and so on and so forth.

Though chaos theory has been applied to various social phenomena, it has yet to be used in the analysis of patterns of sexual behavior. It remains to be seen whether something as relatively elusive as sexual and gender expression could even be quantified to the extent required by the rigorous mathematical models of chaos science. Nevertheless, the broader insights offered by chaos theory are readily apparent: seemingly incoherent or counterintuitive phenomena—whether in the realm of inorganic chemistry or "sexual chemistry"—are components of an overall pattern, regardless of whatever meaning (or lack thereof) they may have individually. In essence, deviation from the norm is part of the norm.

Biodiversity = Sexual Diversity

> Gaia theory . . . has a profound significance for biology. It affects even Darwin's great vision, for it may no longer be sufficient to say that organisms that leave the most progeny will succeed.
>
> —JAMES E. LOVELOCK, "The Earth as a Living Organism"[100]

Nearly two decades ago, British scientist James Lovelock published his book *Gaia: A New Look at Life on Earth,* ushering in a new era in biological thought. What has come to be known as the Gaia hypothesis or Gaia theory has had an immeasurable impact on the way science looks at natural systems in general, and evolution in particular. Gaia theory says that the sum of all living and nonliving matter forms a single self-regulating entity, analogous to a giant living organism. Converging with the results of post-Darwinian evolutionary theory, the Gaia hypothesis has prompted a rethinking of some of the most basic principles of evolution. Cooperation, in addition to competition, is seen as an important force of evolutionary change, while the search for adaptive explanations at the level of the individual has been shifted upward to also include whole species as well as the functioning of the entire biosphere. Although not without controversy, Gaia theory has spawned a number of innovative ideas, many of which are beginning to be empirically and experimentally verified, and has led to important cross-disciplinary collaborations between scientists.[101]

Once again, these new strands of thought have powerful implications for the way animal homosexuality and, more broadly, systems of sexuality and gender are construed. As Lovelock (quoted above) has observed, reproduction is not necessarily a required component of "survival"—in some instances, it may be beneficial for a species or an ecosystem as a whole if some of its members do not procreate. Of course, it is overly simplistic to equate homosexuality with nonreproduction (since, as we saw in previous chapters, many animals that engage in same-sex activity also procreate). There is also little evidence to support the idea that homosexuality operates as a kind of large-scale "population-regulating" mechanism (perhaps the most obvious "function" that would be ascribed to homosexuality in a Gaian interpretation). Nevertheless, one of the fundamental insights of Gaia theory—the value it accords to "paradoxical" phenomena—is directly applicable to homosexuality and transgender. Indeed, the "mosaic" or mixture of male and female characteristics found in intersexual animals such as gynandromorphs is used by some Gaian theorists as a model of multiplicity within oneness, the transformation of disjuncture into wholeness—in other words, the very image of the earth itself.[102]

Like chaos theory, the Gaia hypothesis recognizes that phenomena that appear inexplicable at the level of an individual organism or population may be part of a larger, complex tapestry: a web of seemingly incongruous forces that interact to produce the flow of life, often in ways that are difficult to fathom. Nowhere is this idea better formulated than in the concept of *biodiversity*. Stated simply, this is the principle that *the vitality of a biological system is a direct consequence of the diversity it contains:* "as diversity increases, so does stability and resilience."[103] Traditionally, such diversity is thought of strictly in terms of number and types of species—that is, the *physical* composition of the system, usually expressed in terms of its overall genetic variety. Long-term studies of individual ecosystems have shown, for example, that the health and stability of a natural system is directly linked to the number of different species it contains.[104]

However, variability in number of species is not the only way that biological diversity can be expressed. At all levels of the natural world, social and *sexual* diversity exists—in every type of animal, and between different species, populations, and individuals. As an example, consider just one group of birds, the sandpipers and their relatives.[105] An enormous variety of heterosexual and homosexual mating and social systems are found among the more than 200 species in this group. We find monogamous pairings between birds of the same or opposite sex (Black-winged Stilts, Greenshanks); polygamous associations such as one male mating with more than one female (northern lapwings, curlew sandpipers) or one female mating with more than one male (jacanas), or bisexual trios in which two birds of the same sex bond with each other and with a third individual of the opposite sex (Oystercatchers); and "promiscuous" systems in which birds court and mate with multiple partners of the same or opposite sex without establishing pair-bonds, often involving communal courtship display grounds or leks (Ruffs, Buff-breasted Sandpipers). Even within a particular mating system such as heterosexual "monogamy," there are many different variations: some species form lifelong pair-bonds (e.g., Black Stilts); others are serially monogamous, forming sequential pair-bonds or mating associations with different partners (kentish plovers, sanderlings); others are primarily monogamous but form occasional polygamous trios (Golden Plovers). Some species have largely "faithful" pair-bonds, with birds rarely if ever copulating with individuals other than their mate (Golden Plovers), while in others nonmonogamous matings with birds outside of the pair-bond are routine (Oystercatchers). And even within a given species, there are variations between different geographic areas: lesbian pairs occur in only certain populations of Black-winged and Black Stilts, for example, while snowy plovers exhibit extensive geographic variation in their heterosexual mating patterns, ranging from monogamy to serial polygamy (and numerous versions of each). Within a given population, there is also diversity between individual birds. In Oystercatchers, for example, only some birds participate in homosexual associations, nonmonogamous heterosexual copulations, or serial monogamy, while extensive numbers of nonreproducing birds that do not engage in either heterosexual or homosexual activities are also found in most species. And finally, each individual bird may participate in a variety of sexual and mating behaviors during its lifetime. Among male Ruffs, for example, some birds are exclusively heterosexual for their entire lives, some alternate between periods of heterosexual and homosexual activity or engage in both simultaneously, other individuals participate primarily in same-sex activities for most of their lives, while still others are largely asexual. Similar examples could be furnished from virtually any other animal group, especially now that detailed longitudinal studies are beginning to reveal individual (and idiosyncratic) life-history variations in nearly all organisms.

Scientists are beginning to find evidence that this diversity in social and mating systems contributes directly to the "success" of a species. For instance, among great bustards (a large, storklike bird found in southern Europe and North Africa), flexibility in heterosexual mating systems gives the birds a greater adaptability, enabling them to cope with difficult or variable ecological conditions.[106] And in some species, homosexuality itself appears to be associated with environmental or social changes, in ways that are suggestive but (so far) poorly understood. Male pairing in Golden Plovers, for example, is claimed to be more prevalent in years when severe winter snowstorms have "disrupted" heterosexual pairing, while female coparenting among Grizzlies appears to be characteristic of animals living in conditions of environmental or social flux. In Ostriches, homosexual courtships may be linked to unusually rainy seasons that alter the species' overall sexual and social patterns. Likewise, same-sex pairs in Ring-billed and California Gulls are more common in newly founded colonies that are experiencing rapid expansion, while homosexual activities in Rhesus and Stumptail Macaques (and a number of other primates) are often associated with changes in the composition or dynamics of the social group.[107]

Although the correlations between these factors need to be more systematically investigated—a linear, one-way, cause-and-effect relationship is surely not involved—they do suggest that sexual, social, and environmental variability may be closely allied. Specifically, the capacity for behavioral plasticity—including homosexuality—may strengthen the ability of a species to respond "creatively" to a highly changeable and "unpredictable" world. As primatologist G. Gray Eaton suggests, sexual versatility as both a biological *and* a cultural phenomenon in animals may be directly responsible for a species' success, in ways that challenge conventional views of evolution:

> The macaques' sexual behavior includes both hetero- and homosexual aspects as part of the "normal" pattern. Protocultural variations of some of these patterns have already been discussed but it is well to remember the extreme variation in behavior that characterizes individuals and groups of primates. This plasticity of behavior has apparently played a major role in the evolutionary success of primates by allowing them to adapt to a variety of social and environmental conditions. . . . The variability and plasticity of the behavior . . . suggests an optimistic or *"maximal* view of human potentialities and limitations" . . . rather than a pessimistic or *minimal* view of man as a biological machine functioning on the basis of instinct. This minimal view based on the fang-and-claw school of Darwinism finds little support in the evidence of protocultural evolution in nonhuman primates.[108]

This is not to say that such plasticity always has an identifiable "function" in relation to specific environmental or social factors (even though a few such "functions" can be discerned in specific cases, as we saw in previous chapters). Behavioral versatility is best regarded as a *manifestation* of the larger "chaotic ordering" or nonlinearity of the world, rather than merely a *response* to it. A broader synergy is involved, a pattern of overall adaptability that can be realized in ways that do not necessarily entail any literal "contribution" to reproduction or any straightforward "improvement" in an animal's well-being. In other words, it is the *presence* of behavioral flexibility in a system that is as valuable, if not more so, than its actual concrete "usefulness" or "functionality."

Taken together, these observations—of sexual diversity, and the strength imparted by such sexual variability—lead to an important conclusion. The concept of biodiversity should be extended to include not only the genetic variety, but also the *systems of social organization* found within a species or ecosystem. In other words, sexual and gender systems are an essential measure of biological vitality. The more diverse patterns of social/sexual organization that a species or biological system contains—including homosexuality, transgender, and nonreproductive heterosexuality—the stronger that system will be. Mating and courtship patterns are, after all, as much a part of the "complexity" of an ecosystem as the number of species it contains—and same-sex activity is an integral part of those mating and courtship systems in many animals. It stands to reason, then, that a rich mosaic of different social patterns should increase the vitality of a system, even when such patterns themselves are apparently "unproductive" or are found in only a fraction of the population.

In a rain forest that contains many hundreds of thousands of species of mammals, birds, insects, plants, and so on, the "purpose" of yet one more kind of beetle may be difficult to see—except when understood in terms of its contribution to the overall complexity and vitality of the environment. Similarly, the "function" of a particular social or sexual behavior such as homosexual courtship or heterosexual reverse mounting may seem minimal or even nonexistent at the level of a particular species or individual. But its contribution to the overall strength of the system is independent of such "utility" (or lack thereof) and is also independent of the proportion of the population that participates in it. Every individual, every behavior— whether productive or "counterproductive," comprising 1 percent or 99 percent of the population—has a part to play. Its role is not *in* the tapestry of life, but *as* the tapestry of life: its existence is its "function." Biological diversity is *intrinsically* valuable, and homosexuality/transgender is one reflection of that diversity.

The Extravagance of Biological Systems

> *The history of life on earth is mainly the effect of a wild exuberance . . .*
> —GEORGES BATAILLE, "Laws of General Economy"[109]

There are many points of contact between biodiversity studies, chaos science, and the new evolutionary paradigms, but one of the most significant common threads running through these three disciplines is a recognition of the profound *extravagance* of natural systems. Chaos physicist Joseph Ford speaks of an "exciting variety, richness of choice, a cornucopia of opportunity," in the patterning of physical systems, while fractals and "strange attractors" are described as "bizarre, infinitely tangled abstractions" with "prickly thorns . . . spirals and filaments curling outward and around . . . infinitely variegated."[110] Edward O. Wilson, one of the premier theoreticians on biodiversity, talks about "the engine of tropical exuberance," in which "specialization is . . . pushed to bizarre, beautiful extremes" and where "in the fractal world, an entire ecosystem can exist in the plumage of a bird."[111] Ornithologists studying the complexity of birdsong marvel that "the diversity of modes of singing amongst birds is so great that it defies explanation" and are left "to puzzle over the resulting richness and variety that evolution has created."[112] Entomologists are awed by the "spectacular diversity of complex structures" in the most minute of forms, such as the sperm-reception sites of insect eggs or the "morphological exuberance," "extravagance," and "apparently superfluous complexity" of insect genitalia.[113] Evolutionary theorists grapple with the enigma of "the luxuriant tail feathers of peacocks, the lion's mane, and the flashy dewlaps and throat colorations of many lizards . . . just a few of the extravagant . . . features for which evolutionists have sought explanations ever since Darwin advanced his ideas."[114]

To formally recognize this "extravagance," and also to consolidate some of the converging ideas in these disciplines, we propose the concept of Biological Exuberance, after the work of noted French author and philosopher Georges Bataille.[115] Bataille has presented, in his theory of General Economy, a radical revision in the way we think about the flow of energy in both natural and cultural systems (or "economies"). According to his view, excess and exuberance are primary driving forces of biological systems, as much if not more so than scarcity (competition for resources) or functionality (the "usefulness" of a particular form or behavior). Bataille's fundamental observation is that all organisms are provided with more energy than they need to stay alive—the source of this energy is, ultimately, the sun. This surplus of energy will first be used for the growth of the organism (or larger biological system), but when the system reaches its limits of growth, the excess energy must be spent, expressed in some other form, "used up," or otherwise destroyed. The typical ways that such energy is "squandered," Bataille observes, are through sexual reproduction, consumption by other organisms (eating), and death.

Life on this planet is above all characterized by what Bataille calls "the superabundance of biochemical energy" freely given to it by the sun. The challenge confronting life, then, is not scarcity, but excess—what to do with all this extra energy. Virtually all outpouring of activity, both (pro)creative *and* destructive—the development of baroque ornament and pattern (or its distillation into concentrated minimalism), the wanton consumption of animal and plant foods (or mass starvations in their absence), the extreme elaboration of social systems (encompassing both "complex" and "simple" forms), the florescence of new species and the extinction of others, the cycles of burgeoning and decaying biomass—all of these can be seen, ultimately, as mechanisms that "use up" or express this excess energy. According to this view, life should in fact be full of "wasteful," "extravagant," and "excessive" activities. Bataille also extends his theory to systems of human economy and social organization, including an examination of various attempts to "control" or channel this outpouring of exuberance, often by artificially creating scarcity.[116] Phenomena as diverse as Aztec sacrifice and warfare, potlatch among Northwest Coast Indians, Buddhist monasticism in Tibet, and Soviet industrialization are all revealed to have unexpected properties and interconnections under this analysis.

This theory turns conventional ideas about the world on their head. In spite of its unorthodox perspective, though, it accords startlingly well with a number of observations that scientists have been making for many years (and not just the obvious ones, such as that solar energy is the driving force behind all life and movement on this planet). We have already seen that scientists in such diverse areas as chaos theory, biodiversity studies, and post-Darwinian evolution have been forced to confront the unmitigated extravagance of natural systems, in all their "splendor and squalor."[117] Yet researchers who do not necessarily consider themselves to be part of these "new" streams of thought have independently come to similar conclusions. This is particularly true with regard to the three "expenditures" that Bataille's theory singles out—sexual reproduction, eating, and death.

For instance, biologists have repeatedly remarked that sexual reproduction is costly, draining, dangerous, and yes, even "wasteful." This is true not only for individual animals—who are often reduced to emaciated shadows of their former selves by the end of the breeding season because of the tolls of reproduction—but for entire populations. The insect world, in particular, is famous for its extraordinary orgies of "mating" activity involving hundreds of thousands of individuals at a time, who often perish only a few hours or days after hatching—sometimes without ever mating. So striking is this "costliness" that scientists have questioned why sexual reproduction should exist at all—not all animals reproduce sexually, after all. This is often posed as the long-standing "problem" or "paradox" of sex. Sexual reproduction is generally considered to be more than twice as "expensive" (energetically as well as genetically) as asexual reproduction, because of the "inefficiency" of having each parent contribute only half of the offspring's genetic material, the lack of a male contribution to raising that offspring in many species, as well as the associated risks and energy expenditures of courtship and mating behaviors. Yet exactly this sort of "wastefulness" is expected in a pattern of Biological Exuberance.[118]

Biologists have also observed that eating—the consumption of one organism by another—is not a necessary component of life. Why, for example, don't all species manufacture their own food the way plants do? In fact, compared to the efficiency (and self-sufficiency) of photosynthesis, much more energy is "squandered" when one animal consumes another or consumes plant material. In nature, death itself seems to be elevated to "lavish" proportions, often reaching a "profusion" of its own. Hundreds of baby turtles, after hours of struggling to break through their eggshells, finally reach the sea, only to be picked off by the waiting jaws and beaks of predators—just one of countless examples throughout nature. This "squandering" of life hasn't escaped the attention of biologists, who usually speak of it in terms of the inexorable mechanics of the food chain—otherwise known as the "cruelty" of nature. Yet it, too, is part of an overall pattern of abundance or excess.

In addition to making scientific sense, the concept of Biological Exuberance also makes *common* sense—it is intuitively accessible. We can all think of examples of the "extravagance" of nature in our own lives—maybe it's the overwhelming lushness and beauty of the plants in our garden, the endlessly varied patterns of snowflakes or frost on our window, the infinite and subtle hues of autumn leaves— or perhaps simply our dog or cat, one of many hundreds of different breeds and hybrids. The examples multiply when we turn our attentions to other areas of the natural world, or to human society. Appreciation of the diversity and "exuberance" of life is, of course, nothing new—scientists and artists alike have sung its praises throughout history. The brilliance of Bataille's work lies not so much in his recognition of this concept, but in the *importance* he accords it. Conventional thinking regards the diversity and extravagance of life as the *result* or *by-product* of other, greater forces—evolution, the laws of physics, the progression of history, and so on.

For Bataille, this relation is reversed: exuberance is the *source* and *essence* of life, from which all other patterns flow.

Most importantly, the concept of Biological Exuberance sheds new light on the phenomenon of homosexuality. If, as Bataille suggests, life is characterized by what appear to be "wasteful" activities, then what could be more "wasteful" than homosexuality and nonprocreative heterosexuality (and gender systems)? If sexual reproduction itself is a means of using up excess biochemical energy, then obviously sexual or social activity that does not itself lead to reproduction will be an even greater "squandering" of such energy.[119] Homosexuality/transgender is simply one of the many expressions of the natural intensity or "exuberance" of biological systems. Contrary to what we have all been taught in high school, reproduction is not the ultimate "purpose" or inevitable outcome of biology. It is simply one consequence of a much larger pattern of energy "expenditure," in which the overriding force is the need to use up excess. In the process, many organisms end up passing on their genes, but just as many lead lives in which reproduction figures scarcely at all. Earth's profusion simply will not be "contained" within procreation: it wells up and spills over and beyond this. . . . Lives of intense briefness or sustained incandescence—whether procreative or just creative—each is fueled by the generosity of existence. The equation of life turns on both prodigious fecundity and fruitless prodigality.

Returning to the Source: Indigenous Cosmologies and Fractal Sexualities

> The Ufaina believe in a vital force called fufaka which is . . . present in all living beings. This vital force, whose source is the sun, is constantly recycled among plants, animals, men, and the Earth itself. . . . When a being dies it releases this energy . . . similarly when a living thing consumes another. . . . The sun revolves around the cosmos distributing energy to all equally.
>
> —MARTIN VON HILDEBRAND, "An Amazonian Tribe's View of Cosmology"[120]

> Solar energy is the source of life's exuberant development.
>
> —GEORGES BATAILLE, "Laws of General Economy"[121]

The concept of Biological Exuberance encapsulates a number of converging lines of thought in a wide range of scientific disciplines. In essence, it is a new way of looking at the world—but in a sense, it is not new at all. This "modern" worldview is uncannily similar to the perspectives of indigenous peoples around the globe, whose ancient "cosmologies" often bear striking resemblances to the most sophisticated recent theories of particle physics or deep ecology. Perhaps the most significant aspect of the intersection of chaos science, post-Darwinian evolution, and biodiversity/Gaia theory is its potential to initiate a return to indigenous sources of knowledge.

A number of scientists in each of these "new" scientific disciplines are starting to acknowledge the teachings of aboriginal cultures. Some of the most prominent and respected researchers in biodiversity studies, chaos theory, and the new evolutionary paradigms are waking up to the fact that their innovative ideas are echoed in aboriginal belief systems around the world. For instance, Edward O. Wilson invokes the visionary insights of indigenous Amazonian shamans, as well as the classificatory expertise of native New Guineans, to illustrate the biodiversity and "exuberance of tropical rain-forest life."[122] Pioneering chaos mathematician Ralph Abraham recognizes that ancient and tribal cultures are cut through with "chaotic" patterns, such as the "fractal architecture" of the indigenous peoples of Mali.[123] There is even serious discussion among respected scientists of "respiritualizing" our relationship with nature and looking to indigenous cultures for guidance, faced as we are with the global destruction of ecosystems and massive losses in species diversity.[124] Indigenous knowledge of natural history among the Inupiaq (Eskimo) and Koyukon people of Alaska, the O'odham and Yaqui people of the Southwest, and the Foré and various other New Guinean tribes is offered as a model for Western scientists addressing biodiversity issues.[125] The indigenous concept of an animal's "spirit" is embraced by wildlife biologist Douglas Chadwick, who suggests that a view of animals as "beings with languages and elaborate societies of their own" and perhaps even "some shared quality of consciousness" is useful for an integrated scientific understanding of their behavior and role in the ecosystem. Renowned conservation biologists such as Michael E. Soulé and R. Edward Grumbine also point to Native American spirituality—such as the shamanic Bear ceremonialism of many First Nations (including the Bear Mother myth)—as an important part of the solution to our current biodiversity crisis.[126]

Gaian and post-Darwinian evolutionary theorists such as Peter Bunyard and Edward Goldsmith are also calling for a return to indigenous worldviews as a way of understanding the nonlinear complexities of nature.[127] Many of these aboriginal cosmologies, like that of the Amazonian Ufaina people referred to above, involve sophisticated conceptualizations of the flow of "life energy" that parallel contemporary environmental and economic theories, including Bataille's theory of General Economy. Others are in accordance with some of the basic tenets of chaos and Gaia theory in recognizing the importance of "exceptional," statistically rare, or apparently paradoxical phenomena. Frank LaPena, a traditional poet and artist of the Wintu tribe as well as a native anthropologist, succinctly captures this perspective, which is simultaneously ancient and modern: "The earth is alive and exists as a series of interconnected systems where contradictions as well as confirmations are valid expressions of wholeness."[128]

One of the most powerful symbols of scientists' newfound willingness to listen to indigenous sources took place at the National Forum on BioDiversity in 1986. Organized by the National Academy of Sciences and held at the Smithsonian Institution, this prestigious conference brought together more than 60 distinguished scholars and scientists from around the world. Their task: to discuss the importance of biodiversity as we approach the twenty-first century. One of the most eagerly anticipated speakers was neither a "scholar" nor a "scientist" in the conventional sense. Native American storyteller Larry Littlebird, a member of the Keres nations of New Mexico, was invited to give an indigenous perspective on the natural world. As the audience sat in hushed attention on the final day of the conference, Littlebird treated the biologists to an enigmatic tale of Lizard, who summons forth rain clouds with a song that most ordinary humans can't hear.[129]

Although this unprecedented event is an encouraging sign of a new direction in science, something crucial was missing—the centrality of homosexuality/transgender to indigenous belief systems. How many of the participants at that conference knew that Littlebird's Pueblo tribe, one of the Keresan peoples, recognizes the sacredness of the two-spirit or *kokwimu* (man-woman), and honors homosexuality and transgender in both humans and animals? How many of them realized that the Keresan cosmology includes one of the most noteworthy examples of the left-handed, gender-mixing Bear figure?[130] Or that the "Lizard" of Littlebird's story was most likely a Whiptail Lizard, one of several all-female species of the American Southwest that reproduce by parthenogenesis and engage in lesbian copulation?[131] How many of them knew that some of the animals mentioned in Littlebird's closing words, "The deer, eagle, and butterfly dancers are coming . . . ," exhibit homosexuality and transgender in nature? The answer, unfortunately, is that probably no one in the audience was aware of these connections.

A contemporary Yup'ik two-spirit, Anguksuar (Richard LaFortune), has drawn attention to the recent convergence of Western scientific thought with indigenous perspectives, and the relevance of notions of gender and sexual fluidity: "Modern science emerged [and] linear flight from disorder led directly to quantum theory. This scrambling toward something orderly and manageable has landed right back in the lap of the Great Mystery: chaos, the unknown, and imagination. . . . This is a region of the cosmos familiar to many indigenous taxonomies and to which the Western mind is finally returning. . . . When I read [Fritjof] Capra's description of the 'Crisis of Perception' that appears to be afflicting Western societies, it seemed to make perfect sense that culture, identity, gender, and human sexuality would figure prominently in such a crisis."[132] The fact is that two-spiritedness, homosexuality, bisexuality, and transgender are at the forefront of some of the most significant scientific re-visionings of our time—in which the gap between indigenous and Western perspectives is finally being bridged—yet their contribution is rarely, if ever, acknowledged by Western scientists. When prominent chaos theoreticians, biodiversity experts, and post-Darwinian evolutionists invoke the teachings of tribal peoples, they are usually unaware of the pivotal role played by homosexuality and transgender in these indigenous belief systems, or in the lives of the writers, storytellers, and visionaries who give poetic voice to their scientific concepts.

In the book *Evolution Extended,* for instance—a recent presentation of innovative scientific and philosophical interpretations of evolutionary theory—the words of Native American poet Joy Harjo are featured prominently as a haunting invocation of life's interconnectedness.[133] Of Muscogee Creek heritage, Harjo has received wide acclaim for her writing, which draws heavily on her indigenous roots and often includes powerful images of the natural world, while also juxtaposing references to specific constructs of Western science such as quantum physics or molecular structures. Harjo is also a "lover of women" whose writing has been anthologized in *Gay & Lesbian Poetry in Our Time*. She acknowledges lesbian or bisexual authors such as Audre Lorde, June Jordan, Alice Walker, Beth Brant, and Adrienne Rich, as well as the ideas of lesbian-feminism, as primary influences on her work. She has spoken of the importance of eroticism permeating all aspects of life, and she affirms the power of androgyny and the presence of male and female traits in every individual.[134] Yet these aspects of Harjo's life and work are considered incidental or irrelevant to the perspective she brings to the scientific material—not even worth mentioning as one component of her personal vision, let alone a key feature in the bringing together of seemingly disparate worlds that she achieves through her poetry.

"From Wakan Tanka, the Great Spirit, there came a great unifying life force that flowed in and through all things—the flowers of the plains, blowing winds, rocks, trees, birds, animals—and was the same force that had been breathed into the first man. Thus all things were kindred, and were brought together by the same Great Mystery." So spoke the Oglala (Sioux) chief Luther Standing Bear, whose words grace the pages of *Buffalo Nation,* a recent book by prominent wildlife biologist Valerius Geist. This vision of the life energy connecting the "buffalo nation" (and all of nature) to the "human nation" underscores the parallel that Geist draws between indigenous and contemporary scientific approaches to wildlife conservation. The sophisticated game-management practices developed by many Native Americans—both traditionally and in their current efforts to resurrect Buffalo herds on their lands—are, according to Geist, at the forefront of recent Bison conservation efforts. Beautifully interwoven through his discussion of this species' natural history, behavior, and preservation are evocations of the powerful spiritual role played by the Bison in Native American cultures, including descriptions of Mandan Buffalo Dances and the Lakota legend of the White Buffalo Woman. Yet nowhere in this discussion is there any mention of indigenous views on sexual and gender variability in Bison (or humans), let alone of contemporary scientific findings on these topics. Ironically, though, the book still manages to unintentionally present a vivid picture—literally—of Bison homosexuality. In the section on rutting behavior, a photograph of Buffalo mating activity is identified as a bull mounting a female, when in fact it depicts a bull mounting another bull.[135] In the end, then, perhaps the animals themselves will have the "final say," insuring the representation of homosexuality/transgender and its rightful place in both indigenous and Western scientific thinking.

The importance of this missing link cannot be overemphasized. If Western science is to embrace indigenous perspectives—as it should—then it must do so fully, including views on homosexuality/transgender. It cannot pick and choose among aboriginal "beliefs," salvaging only those that it is most comfortable with while rejecting those that challenge its prejudices. All of us (scientists included) must acknowledge that heeding "aboriginal wisdom" means listening even when—or perhaps, *especially* when—we aren't prepared to hear what it has to say about sexual and gender variance. For too long, native views have been sanitized to make them palatable to nonindigenous people. In a world where Native American spirituality is co-opted to sell bottled water—indeed, is sold directly as a "New Age" commodity—it has become something of a cliché to speak of the environmental "balance" and "harmony" of indigenous cultures.[136] The reality is that homosexuality and transgender—along with many other beliefs and practices that would probably be considered objectionable by large numbers of people—are usually an integral, if not a central, component of such "balance." Consider the cosmology of the Bedamini people of New Guinea, which seems to turn conventional ideas about the natural world upside down:

> It is believed that homosexual activities promote growth throughout nature . . . while excessive heterosexual activities lead to decay in nature. . . . The balance of these forces is dependent on human action. . . . The Bedamini do not . . . experience any inconsistency in the cosmic equation of homosexuality with growth and heterosexuality with decay.
>
> —ARVE SØRUM, "Growth and Decay: Bedamini Notions of Sexuality"[137]

Nor is the association of homosexuality with fecundity unique to this example. As we saw earlier, the renewal and abundance of nature is ensured during Mandan, Yup'ik, and many other cultures' ceremonies by the symbolic reenactment of animal homosexuality and ritual displays of gender mixing. The Bimin-Kuskusmin human-animal androgynes (who are themselves celibate or postreproductive) are seen as embodiments of fertility, life essence, and earth's creative powers, while the presence of transgendered and nonreproductive animals is regarded as vital for the productivity of domesticated herds among the Navajo and Chukchi. Rather than being seen as "barren" or counterproductive, then, homosexuality, transgender, and nonbreeding are considered essential for the continuity of life. This is the fundamental "paradox" at the heart of indigenous thinking on alternate genders and sexualities—something that is not, of course, really considered paradoxical at all in these worldviews. It is important that scientists working in chaos theory, biodiversity/Gaia studies, and post-Darwinian evolution acknowledge their genuine affinities with indigenous perspectives. But this process will be complete only when scientists themselves understand this "paradox" and no longer see any inconsistency in the equation of homosexuality/transgender with the vitality of the natural world.

In his study of the 12,000-year-old shamanic worldview of the Tsistsistas (Cheyenne) people and their ancestors, anthropologist Karl Schlesier makes explicit the concordance between ancient and modern perspectives, and the essence of sexual and gender variability that is at its core. According to Schlesier, "The new scientific paradigm initiated by physics and astronomy during the last decades has not only overthrown the rationalistic description that has dominated science for merely four centuries, but is testing concepts regarded as factual in the Tsistsistas world description. The Tsistsistas world description understands power ('energy') in the universe . . . as cosmic power"—a power that controls quantum phenomena and exhibits paradoxical properties, including being both local and nonlocal, causal and noncausal (among others). Central to this understanding is the figure of the gender-mixing or two-spirit shaman, the "halfman-halfwoman" who is a living exemplar of the reconciliation of contraries, a "traveler in the androgynal quest" uniting within him/herself apparently contradictory categories. This conjunction of opposites is seen as a return to the original and timeless state of all matter—the primordial mystery of totality. Homosexuality/transgender is therefore regarded as a *hierophany,* a manifestation of this sacred oneness and plentitude. "This organic Tsistsistas world description, in which all parts of the universe were interrelated, saw life as wondrous. . . . This is perhaps the greatest achievement of shamanism since its development: . . . to interpret the world with all its manifestations as a place of miracles, transformations, and immortality."[138]

On the eve of the twenty-first century, human beings have begun to reimagine and reconfigure some of the most fundamental aspects of nature and culture. Stepping into a social and biological landscape that could scarcely have been imagined a few decades ago, homosexual, bisexual, and transgendered people are now offering new paradigms of sexuality and gender for all of us to consider. As part of this process, they are looking simultaneously to indigenous and futuristic sources of inspiration:

> In the search for new vocabularies and labels, terms like *shapeshifter* and *morphing* have come to be used to refer to gender identity and sexual style presentations and their fluidity. *Shapeshifter,* originally from Native American culture, was introduced into current popular culture from science fiction, especially a new offshoot of the cyberpunk subgenre made famous by William Gibson and exemplified by the work of Octavia E. Butler, the African-American author of the *Xenogenesis* series. Butler's books are inhabited by genetics-manipulating aliens, a polygendered species whose sexuality is multifarious and who are "impelled to metamorphosis," whose survival in fact depends upon their "morphological change, genetic diversity and adaptations."
>
> —ZACHARY I. NATAF, "The Future: The Postmodern
> Lesbian Body and Transgender Trouble"[139]

Ironically, one need not look into the future or on "alien worlds" to find appropriate models: shape-shifting and morphing creatures are not merely the stuff of fantasy. The animal world—right now, here on earth—is brimming with countless gender variations and shimmering sexual possibilities: entire lizard species that consist only of females who reproduce by virgin birth and also have sex with each other; or the multigendered society of the Ruff, with four distinct categories of male birds, some of whom court and mate with one another; or female Spotted Hyenas and Bears who copulate and give birth through their "penile" clitorides, and male Greater Rheas who possess "vaginal" phalluses (like the females of their species) and raise young in two-father families; or the vibrant transsexualities of coral reef fish, and the dazzling intersexualities of gynandromorphs and chimeras. In their quest for "postmodern" patterns of gender and sexuality, human beings are simply catching up with the species that have preceded us in evolving sexual and gender diversity—and the aboriginal cultures that have long recognized this. The very melding of indigenous cosmologies and fractal sexualities suggested in the passage above is already well under way—but within the realm of science *fact*, not fiction.

The Magnificent Overabundance of Reality

It is early morning in the mountains of Sierra Chincua in central Mexico. Covered with what appear to be the golden and orange leaves of autumn, the forest is aquiver, "her trillion secrets touchably alive"[140]—but these are not leaves, nor is it autumn. The sound of a distant waterfall fills the air—but no cascading rapids are nearby. It is the fluttering of hundreds of thousands of paper-thin wings—for this is the overwintering site of Monarch Butterflies, resting after their epic migration across North America. They cling to the trees in such numbers that the branches are bent toward the ground, and the forest floor is carpeted with their densely packed bodies. Some of the butterflies are in tandem, since mating often takes place at these overwintering sites. And some of this mating is homosexual: one study of an overwintering site revealed that at the peak of mating activity, more than 10 percent of the Monarch pairs were composed of two males, while later in the season, this percentage rose to nearly 50 percent.[141] When the Monarchs take to the air en masse, they form a thick orange cloud that engulfs the trees and requires thirty minutes to pass. Seen from above, their multitude is staggering: the forest seems to be on fire, burning with millions of tiny butterfly-flames. This image is a powerful evocation of the central theme of Biological Exuberance: the glorious multiplicity and bounty of life, what author Hakim Bey has called "the magnificent overabundance of reality."[142]

We conclude this section with a reflection on where this journey through the speculations of post-Darwinian evolution, chaos theory, and biodiversity studies has led us—a journey along circuitous routes, following clues that at times seemed far-flung, straying down paths that never quite lost us (in spite of their tangential meanderings). Our final resting spot—the concept of Biological Exuberance—lies somewhere along the trajectory defined by these three points (chaos, biodiversity, evolution), although its exact location remains strangely imprecise.[143] Seen in the light of Biological Exuberance, animal homosexuality/transgender and other nonreproductive behaviors finally "make sense"—they find an intuitive connection to a larger pattern. Yet they are still, paradoxically, "inexplicable," since they continue to elude conventional definitions of usefulness. Nothing, in the end, has really been "explained"—and rightly so, for it was "sensible explanations" that ran aground in the first place.

Nevertheless, by looking at one particular aspect of animal behavior, we have actually stumbled upon something much larger—a new way of seeing the world, of perceiving broader patterns in nature and human society. Animal homosexuality and transgender may appear far removed from our everyday lives, but through these phenomena, we also arrive at an understanding and appreciation of some of the simplest, most ordinary things around us. Biological Exuberance is *available*, if it is nothing else—at our fingertips, everywhere we turn, in the fibers and textures that surround us, in the spices that fill our nostrils as we walk past the corner store, in the cloud formations above us and the dandelion seeds strewn by the wind about us, in the embrace of a friend and the kiss of a beloved—in all the colors and patterns and sensations that fill our lives. How many of us haven't, at one time or another, been overcome by this variety, this feeling of what poet Louis MacNeice describes as "the drunkenness of things being various," the world as "incorrigibly plural"?[144] Biological Exuberance simply takes our intuitive understanding of the diversity of life and makes it the essence of existence. We needn't be living in material wealth or in an isolated wilderness to experience this lavishness, either. The weeds struggling through a sidewalk crack or choking an abandoned urban plot are every bit as sumptuous as the most refined of rose gardens, the most magnificent of mountain forests—if not more so. Gifted with this heightened understanding, we can now find the intoxication contained in a glass of water, where before even the most sophisticated wine seemed flavorless (to paraphrase Hakim Bey).[145]

Ultimately, the synthesis of scientific views represented by Biological Exuberance brings us full circle—back to a way of looking at the world that is in accordance with some of the most ancient indigenous conceptions of animal (and human) sexual and gender variability. This perspective dissolves binary oppositions, uniting dualities while simultaneously cherishing unlikeness. It suffers difference, honoring the "anomalous" and the "irregular" without reducing them to something familiar or "manageable." And it embraces paradox, recognizing the coexistence of contradictory and seemingly incompatible phenomena. It is about the unspeakable inexplicability of earth's mysteries—which are as immediate as the next heartbeat. Biological Exuberance is, above all, an affirmation of life's vitality and infinite possibilities: a worldview that is at once primordial and futuristic, in which gender is kaleidoscopic, sexualities are multiple, and the categories of male and female are fluid and transmutable. A world, in short, exactly like the one we inhabit.

Part II

A Wondrous Bestiary

Portraits of Homosexual, Bisexual, and Transgendered Wildlife

PIED BEAUTY

Glory be to God for dappled things—
 For skies of couple-color as a brinded cow;
  For rose-moles all in stipple upon trout that swim;
Fresh-firecoal chestnut-falls; finches' wings;
 Landscape plotted and pieced—fold, fallow, and plough;
  And all trades, their gear and tackle and trim.

All things counter, original, spare, strange;
 Whatever is fickle, freckled (who knows how?)
  With swift, slow; sweet, sour; adazzle, dim;
He fathers-forth whose beauty is past change:
      Praise him.

—GERARD MANLEY HOPKINS

Introduction

"*A* Wondrous Bestiary" presents a species-by-species survey of sexual and gender variance in animals. Included are profiles of mammals and birds in which at least some individuals are homosexual, bisexual, and/or transgendered. Only species in which same-sex activities have been scientifically documented are included here; for species excluded from this roster, see p. 673: chapter 1, note 29; for more on the (sometimes controversial) interpretations and categorizations of these behaviors, see chapters 3–5. Because each portrait is self-contained, it can be read either on its own, in sequence as part of a subgrouping of related animals, in conjunction with material in part 1, or browsed at random, according to the particular interests of the reader (the index may be used to investigate particular topics). Each portrait contains the following types of information, arranged sequentially:

Heading: basic identifying information for each profile, including:

- *Name:* common and scientific names of the species, the animal subgrouping, and an icon identifying the major animal type (e.g., primate, marine mammal, etc.).
- *Category:* indicates whether the animal in question exhibits male and/or female homosexuality; the major type of transgender if present (transvestism and/or intersexuality); the types of same-sex behaviors involved; and whether homosexuality/transgender has been observed in the wild, semiwild, and/or captivity (for discussion of these distinctions, see chapters 1 and 4).
- *Ranking:* an informal categorization of each animal in terms of the importance of homosexuality and/or transgender in the species, based on the variety and elaboration of behaviors, the frequency of same-sex activity, and the sexual orientation profiles for the species; categories are "primary," "moderate," and "incidental."
- *Portrait Drawing:* a line drawing identifying one or more of the profiled species.

Ecology: background information about the animal and its environment:

- *Identification:* a brief physical description of the animal.
- *Distribution:* the animal's geographic range, and an indication of the species' endangered status if threatened in the wild (World Conservation Union–designated categories of "critically endangered," "endangered," or "vulnerable"; see pp. 708–9: chapter 5, note 17, for some discussion of these designations).
- *Habitat:* a description of the animal's physical environment.
- *Study Area(s):* specific location(s) and subspecies where homosexuality has been observed and/or studied.

Social Organization: background information about the general social and mating system(s) of the animal, providing a behavioral context for understanding homosexuality/transgender in the species.

Description: detailed information about the particular form(s) of homosexuality and/or transgender found in this animal, including:

- *Behavioral Expression:* the type(s) of behaviors involved, with discussion of courtship, affectionate, sexual, pair-bonding, and/or parenting activities; and the form(s) of transgender, if any (behavioral or physical transvestism, intersexuality, etc.).
- *Frequency:* detailed statistics (where available) or estimates of how often homosexual activity occurs, specified as the proportion of all sexual (or other) activity that is same-sex, and/or frequency rates, time/activity budgets, or other measures.
- *Orientation:* what proportion of the population participates in same-sex activity, where individuals fall along the continuum from homosexual through bisexual to heterosexual, and how this is manifested during individual life histories.
- *Illustrations:* photographs and line drawings of specific activities.

Nonreproductive and Alternative Heterosexualities: summaries of various heterosexual activities that do not lead to reproduction (or that actively suppress it), along with family and pair-bonding configurations that deviate from the species-typical pattern or that are otherwise noteworthy.

Other Species: summary of homosexual activities and/or transgender in related species, where applicable.

Sources: a complete list of references for each animal, with an indication of which ones discuss or mention homosexuality/transgender.

The appendix summarizes the occurrence of homosexuality (and in some cases, transgender) in other major animal groupings: reptiles/amphibians, fish, insects, spiders/other invertebrates, and domesticated animals. Included are tables and a complete list of references.

Mammals

Primates

GREAT APES   *Primate* 🦎

BONOBO or PYGMY CHIMPANZEE
Pan paniscus

HOMOSEXUALITY	TRANSGENDER	BEHAVIORS		RANKING	OBSERVED IN
● Female	○ Intersexuality	● Courtship	● Pair-bonding	● Primary	● Wild
● Male	○ Transvestism	● Affectionate	○ Parenting	○ Moderate	○ Semiwild
		● Sexual		○ Incidental	● Captivity

IDENTIFICATION: Similar to the Common Chimpanzee, but more slender and with longer limbs, a uniformly dark face, and a slight "part" in the hair on top of the head. DISTRIBUTION: Central and western Congo (Zaire); endangered. HABITAT: Tropical lowland rain forest. STUDY AREAS: Wamba and the Lomako Forest, Congo (Zaire); Yerkes Regional Primate Research Center (Georgia); San Diego Zoo; Wild Animal Park (San Diego); Frankfurt and Stuttgart Zoos, Germany.

Social Organization

Bonobos live in communities composed of large mixed-sex and mixed-age groups containing up to 60 or more individuals. These often divide into smaller, temporary subgroups that have a more fluid membership. On reaching adolescence (and becoming sexually mature), female Bonobos typically leave their home group and emigrate to a new one, while males usually remain in their home group for life. Females often form strongly bonded subgroups and are generally dominant to males. The mating system is promiscuous: males and females mate with multiple partners, and males do not generally participate extensively in raising their offspring.

Description

Behavioral Expression: Bonobos have one of the most varied and extensive repertoires of homosexual practices found in any animal. Females engage in an extraordinary form of mutual genital stimulation that, in many aspects, is unique to this species. Sometimes known as GG-RUBBING (for genito-genital rubbing), this behavior is usually performed in a face-to-face embracing position (heterosexual copulation is also sometimes done in this position, but not as often as in lesbian

▲ *Two female Bonobos in Congo (Zaire) "GG (genito-genital) rubbing"*

interactions). One female stands on all fours and literally "carries" or lifts her partner off the ground; the female on the bottom wraps her legs around the other's waist and clings to her as they rapidly rub their genitals against one another, directly stimulating each other's clitoris. Some scientists believe that the particular shape and location of the Bonobo's genitals have evolved specifically for lesbian rather than heterosexual interactions. During GG-rubbing, each female rhythmically swings her pelvis from side to side—precisely timed so that each partner is thrusting in opposite directions—at a rate of about two thrusts per second. This is comparable to the thrusting rate seen in males during heterosexual interactions, but males thrust vertically rather than sideways. In addition, although both homosexual and heterosexual copulations are quite brief, same-sex interactions generally last slightly longer—an average of about 15 seconds (maximum of 1 minute) compared to about 12 seconds (maximum of 45 seconds) for heterosexual matings. Sometimes females GG-rub with the same partner several times in a row.

As shown by their facial expressions, vocalizations, and genital engorgement, females experience intense pleasure—and probably orgasm—during homosexual interactions. Partners gaze intensely into each other's eyes and maintain eye contact throughout the interaction. Sometimes, females grimace or "grin" by baring their teeth wide and also utter screams or squeals that are thought to be associated with

sexual climax. The Bonobo's clitoris is prominent and well-developed; during sexual arousal it undergoes a full erection of both the shaft and glans (in humans, only the glans of the clitoris becomes enlarged), swelling to nearly twice its regular size. Remarkably, clitoral penetration has occasionally been observed between females during homosexual interactions

◀ *Two male Bonobos "rump rubbing"*

(in captivity). When penetration occurs, the females often switch to vertical thrusting (as in heterosexual mating) rather than the usual sideways hip movements.

Genital stimulation between females is sometimes performed in different positions: the two partners may both hang from a branch facing each other; one female may mount the other from behind; one female might lie on her back while the other stands facing away from her, rubbing her genitals on her recumbent partner's vulva; or both females may lie on their

▲ *A male Bonobo mounting another male from behind*

backs or stand rump-to-rump while GG-rubbing. In the face-to-face position, females may alternate between who is on bottom and who is on top; prior to interacting, they often "negotiate" positions by lying down with legs spread to see whether the other partner wants to be on top. GG-rubbing occurs among females of all ages, from adolescent to very old, but if an older and a younger female are interacting, often the younger female will be on top. Sexual activity may also be more common when the females are of different ranks. Homosexual interactions are often initiated with a characteristic series of "courtship" signals: approaching the partner and peering closely, standing on the hind legs and raising the arms over the head while making eye contact, and/or touching the shoulder or knee while staring. Among captive Bonobos, partners may also use a highly developed "lexicon" of manual gestures to help negotiate the position(s) to be used in sexual interactions (see pp. 66–69 for more detailed discussion).

Females may have multiple sexual partners. In one troop containing ten females, each female interacted sexually with five other females on average, and some had as many as nine different partners. Group sexual activity also occasionally takes place, with three to five females simultaneously rubbing their genitals together. Some females are considered especially

Two younger male Bonobos engaging in fellatio ▶

(PHOTO BY FRANS DE WAAL)

(PHOTO BY FRANS DE WAAL)

▲ *An adult male Bonobo* (left) *manually stimulating the penis of a younger male*

"attractive"—usually because of the shape, size, and coloration of their genital swellings—and individuals may have preferred partners that they tend to interact with more often. In fact, females typically form strongly bonded, enduring relationships with one another that are fostered by sexual interactions and include such activities as mutual grooming, play, food-sharing, and alliance-formation (often for challenging males). Females generally prefer each other's company, and their same-sex bonds form the core of social organization. In addition, when new females (usually adolescents) join a troop, they often pair up with an older female with whom they have most of their sexual and affectionate interactions. These bonds need not be exclusive—either party may have sex with other females or males—but such mentorlike pairings can last for a year or more until the newcomer is fully integrated into the troop. In this species, a sort of homosexual "incest taboo" is in effect for these pair-bonds: most females are unrelated to the Bonobos in their new troop, but those who are related are not chosen as special partners. Some homosexual activity does, however, occur between mothers and their daughters.

Male Bonobos also have a wide variety of homosexual interactions. Sometimes, two males mutually stimulate each other's genitals using a face-to-face position similar to GG-rubbing: one male lies on his back and spreads his legs while the other thrusts on him, rubbing their erections together (in this and all other male homosexual activity, anal penetration is not involved). If there is an age difference between partners, often the younger male will be on the bottom. Occasionally, two males hang from a branch facing each other and engage in what is known as PENIS FENCING, swinging their hips from side to side as they rub their erect penises on each other or cross them as if they were fencing with swords. Another activity is RUMP RUBBING, in which two males stand on all fours in opposite directions, pressing their buttocks against each other and mutually rubbing their anal and scrotal regions. Both males often have erections. Males also mount each other from behind and either mountee or mounter may make thrusting movements. Sometimes the males switch positions, and the mounter may scream or grin in sexual arousal as in lesbian or heterosexual interactions. Bonobo males have also been seen standing on their hind legs, one embracing the other from behind. Other sexual activities include oral sex, or fellatio, in which one male sucks another's penis at the initiation of either partner (usually seen only in younger males). Manual stimulation of the genitals by a partner also occurs: typically an adolescent male spreads his legs and presents his erect penis to an adult male, who takes the shaft in his hand and caresses it with up-and-down movements. Younger males (and occasionally females) also sometimes give each other openmouthed kisses, often with extensive mutual tongue stimulation. Although males do not appear to form pairlike bonds with sexual partners (as do some females), occasionally two or three males are intimately associated as companions, constantly accompanying each other and foraging together.

Frequency: Homosexual activity is nearly as common as heterosexual activity in Bonobos, accounting for 40–50 percent of all sexual interactions; two-thirds to three-quarters of this same-sex activity is between females (mostly GG-rubbing). Daily life among Bonobos is characterized by numerous relatively brief episodes of sexual activity scattered throughout the day, and homosexual interactions are frequent. Each female participates in GG-rubbing on average once every two hours or so, and some newcomers to a troop do so even more often, on an hourly basis.

Orientation: Virtually all Bonobos are bisexual, interacting sexually with both males and females. In fact, motherhood and homosexual activity are fully integrated among Bonobos, as a female often GG-rubs with another female while her infant is clinging to her belly. Usually same-sex and opposite-sex activities are interspersed or alternated, although both may occur simultaneously during group sexual interactions. Nevertheless, it appears that—among some females at least—homosexual activity is preferred. Although females vary along a continuum, with one-third to nearly 90 percent of their interactions being with partners of the same sex, overall there is often a predominance of homosexual activity. An average of two-thirds of all sexual interactions among females are with other females, and individuals generally have more female than male sexual partners. In addition, females have sometimes been observed consistently ignoring males who are soliciting them for sex, preferring instead to GG-rub with each other.

Nonreproductive and Alternative Heterosexualities

Variety, flexibility, and frequency of sexual interactions are not limited to contact between Bonobos of the same sex—heterosexual activity is replete with nonreproductive behaviors. Rump rubbing, fellatio, and manual stimulation of the genitals by either sex (including fondling of the scrotum) are all aspects of male-female sexual interactions. In addition, females occasionally mount males from behind (REVERSE mounts), and heterosexual copulation often does not involve penetration and/or ejaculation, but simply mutual rubbing of genitals. Both male and female Bonobos also masturbate (males sometimes using inanimate objects to stimulate themselves). Group sexual activity occurs as well, often with one individual thrusting against a pair who are copulating, and individuals may participate in several bouts of heterosexual activity in rapid succession. Sometimes, because of the frequency and persistence of sexual invitations—often associated with begging for food—individuals (especially males) may even become annoyed and try to avoid further heterosexual interaction. In addition, females occasionally cooperate with one another in harassing and attacking males, in some cases causing severe injuries by holding a male down and biting his ears, fingers, toes, or genitals.

Bonobos mate during all phases of a female's sexual cycle, and about a third of copulations occur during periods when fertilization is unlikely or impossible. Mating also takes place during pregnancy, sometimes as late as one month before delivery. Both adult males and females interact sexually with adolescents and juveniles (three-to-nine-year-olds). In fact, young females go through a five-to-six-year period sometimes referred to as ADOLESCENT STERILITY (although no pathology is involved) during which they actively participate in heterosexual mating (often with adults) but never get pregnant. Sexual behavior between adults and infants of both sexes is also common—about a third of the time it is initiated by the infant and may involve genital rubbing and full copulatory postures (including penetration of an adult female by a male infant). Another form of nonreproductive sexuality involves contact with other species: younger male Bonobos have occasionally been observed engaging in playful sexual interactions with redtail monkeys (*Cercopithecus ascanius*) in the wild.

Sources   *asterisked references discuss homosexuality/transgender*

*Blount, B. G. (1990) "Issues in Bonobo (*Pan paniscus*) Sexual Behavior." *American Anthropologist* 92: 702–14.

*Enomoto, T. (1990) "Social Play and Sexual Behavior of the Bonobo (*Pan paniscus*) With Special Reference to Flexibility." *Primates* 31:469–80.

*Furuichi, T. (1989) "Social Interactions and the Life History of Female *Pan paniscus* in Wamba, Zaire." *International Journal of Primatology* 10:173–97.

*Hashimoto, C. (1997) "Context and Development of Sexual Behavior of Wild Bonobos (*Pan paniscus*) at Wamba, Zaire." *International Journal of Primatology* 18:1–21.

*Hashimoto, C., T. Furuichi, and O. Takenaka (1996) "Matrilineal Kin Relationships and Social Behavior of Wild Bonobos (*Pan paniscus*): Sequencing the D-loop Region of Mitochondrial DNA." *Primates* 37:305–18.

*Hohmann, G. and B. Fruth (1997) "The Function of Genito-Genital Contacts among Female Bonobos (*Pan paniscus*)." In M. Taborsky and B. Taborsky, eds., *Contributions to the XXV International Ethological Conference*, p. 112. Advances in Ethology no. 32. Berlin: Blackwell Wissenschafts-Verlag.

*Idani, G. (1991) "Social Relationships Between Immigrant and Resident Bonobo (*Pan paniscus*) Females at Wamba." *Folia Primatologica* 57:83–95.

*Kano, T. (1992) *The Last Ape: Pygmy Chimpanzee Behavior and Ecology*. Stanford: Stanford University Press. Translated from the Japanese by Evelyn Ono Vineberg.

*———— (1990) "The Bonobos' Peaceable Kingdom." *Natural History* 99(11):62–71.

*———— (1989) "The Sexual Behavior of Pygmy Chimpanzees." In P. G. Heltne and L. A. Marquardt, eds., *Understanding Chimpanzees*, pp.176–83. Cambridge, Mass.: Harvard University Press.

*———— (1980) "Social Behavior of Wild Pygmy Chimpanzees (*Pan paniscus*) of Wamba: A Preliminary Report." *Journal of Human Evolution* 9:243–60.

*Kitamura, K. (1989) "Genito-Genital Contacts in the Pygmy Chimpanzee (*Pan paniscus*)." *African Study Monographs* 10:49–67.

*Kuroda, S. (1984) "Interactions Over Food Among Pygmy Chimpanzees." In R. L. Susman, ed., *The Pygmy Chimpanzee: Evolutionary Biology and Behavior*, pp. 301–24. New York: Plenum Press.

*———— (1980) "Social Behavior of Pygmy Chimpanzees." *Primates* 21:181–97.

*Parish, A. R. (1996) "Female Relationships in Bonobos (*Pan paniscus*): Evidence for Bonding, Cooperation, and Female Dominance in a Male-Philopatric Species." *Human Nature* 7:61–96.

*———— (1994) "Sex and Food Control in the 'Uncommon Chimpanzee': How Bonobo Females Overcome a Phylogenetic Legacy of Male Dominance." *Ethology and Sociobiology* 15:157–79.

*Roth, R. R. (1995) "A Study of Gestural Communication During Sexual Behavior in Bonobos (*Pan paniscus* Schwartz)". Master's thesis, University of Calgary.

Sabater Pi, J., M. Bermejo, G. Illera, and J. J. Vea (1993) "Behavior of Bonobos (*Pan paniscus*) Following Their Capture of Monkeys in Zaire." *International Journal of Primatology* 14:797–804.

*Savage, S. and R. Bakeman (1978) "Sexual Morphology and Behavior in *Pan paniscus*." In D. J. Chivers and J. Herbert, eds., *Recent Advances in Primatology*, vol. 1, pp. 613–16. New York: Academic Press.

*Savage-Rumbaugh, E. S., and R. Lewin (1994) *Kanzi: The Ape at the Brink of the Human Mind*. New York: John Wiley & Sons.

*Savage-Rumbaugh, E. S., and B. J. Wilkerson (1978) "Socio-sexual Behavior in *Pan paniscus* and *Pan troglodytes*. A Comparative Study." *Journal of Human Evolution* 7:327–44.

*Savage-Rumbaugh, E. S., B. J. Wilkerson and R. Bakeman (1977) "Spontaneous Gestural Communication among Conspecifics in the Pygmy Chimpanzee (*Pan paniscus*)." In G. Bourne, ed., *Progress in Ape Research*, pp. 97–116. New York: Academic Press.

*Takahata, Y., H. Ihobe, and G. Idani (1996) "Comparing Copulations of Chimpanzees and Bonobos: Do Females Exhibit Proceptivity or Receptivity?." In W. C. McGrew, L. F. Marchant, and T. Nishida, eds., *Great Ape Societies*, pp. 146–55. Cambridge: Cambridge University Press.

Takeshita, H., and V. Walraven (1996) "A Comparative Study of the Variety and Complexity of Object Manipulation in Captive Chimpanzees (*Pan troglodytes*) and Bonobos (*Pan paniscus*)." *Primates* 37: 423–41.

*Thompson-Handler, N., R. K. Malenky, and N. Badrian (1984) "Sexual Behavior of *Pan paniscus* Under Natural Conditions in the Lomako Forest, Equateur, Zaire." In R.L. Susman, ed., *The Pygmy Chimpanzee: Evolutionary Biology and Behavior*, pp. 347–68. New York: Plenum Press.

*de Waal, F. B. M. (1997) *Bonobo: The Forgotten Ape*. Berkeley: University of California Press.

*——— (1995) "Sex as an Alternative to Aggression in the Bonobo." In P. A. Abramson and S. D. Pinkerton, eds., *Sexual Nature, Sexual Culture*, pp. 37–56. Chicago: University of Chicago Press.

*——— (1989a) *Peacemaking Among Primates*. Cambridge, Mass.: Harvard University Press.

*——— (1989b) "Behavioral Contrasts Between Bonobo and Chimpanzee." In P. G. Heltne and L. A. Marquardt, eds., *Understanding Chimpanzees*, pp. 154–73. Cambridge, Mass.: Harvard University Press.

*——— (1988) "The Communicative Repertoire of Captive Bonobos (*Pan paniscus*), Compared to That of Chimpanzees." *Behavior* 106:184–251.

*——— (1987) "Tension Regulation and Nonreproductive Functions of Sex in Captive Bonobos (*Pan paniscus*)." *National Geographic Research* 3:318–35.

Walraven, V., L. Van Elsacker, and R. F. Verheyen (1993) "Spontaneous Object Manipulation in Captive Bonobos." In L. Van Elsacker, ed., *Bonobo Tidings: Jubilee Volume on the Occasion of the 150th Anniversary of the Royal Zoological Society of Antwerp*, pp. 25–34. Leuven: Ceuterick Leuven.

*White, F., and N. Thompson-Handler (1989) "Social and Ecological Correlates of Homosexual Behavior in Wild Pygmy Chimpanzees, *Pan paniscus*." *American Journal of Primatology* 18:170.

GREAT APES

COMMON CHIMPANZEE

Pan troglodytes

Primate

HOMOSEXUALITY	TRANSGENDER	BEHAVIORS		RANKING	OBSERVED IN
● Female	● Intersexuality	○ Courtship	● Pair-bonding	○ Primary	● Wild
● Male	○ Transvestism	● Affectionate	○ Parenting	● Moderate	○ Semiwild
		● Sexual		○ Incidental	● Captivity

IDENTIFICATION: The familiar small ape, with black, gray, or brownish fur, prominent ears, and variable facial coloring, from black to brown and pink (especially in younger animals). DISTRIBUTION: Western and central Africa, from southeastern Senegal to western Tanzania; endangered. HABITAT: Woodland savanna, grassland, tropical rain forest. STUDY AREAS: Mahale Mountains National Park and the Gombe Stream National Park, Tanzania; Budongo Forest, Uganda; eastern Congo (Zaire); Arnhem Zoo, the Netherlands; Anthropoid Station, Tenerife; Yale University Primate Laboratory and chimpanzee colony (New Haven, Conn., Franklin, N.H., and Orange Park, Fla.); ARL Chimpanzee Colony, N.Mex.; Delta Regional Primate Research Center, La.; subspecies *P.t. schweinfurthii.*

Social Organization

Common Chimpanzees live in groups or communities of 40–60 individuals, usually with twice as many adult females as males. Within each group, smaller subgroups often form, and some individuals form longer-lasting bonds with each other as part of a complex network of social and communicative interactions. The mating system is promiscuous or polygamous: males and females each mate with multiple partners, and males do not generally participate in raising their own offspring.

Description

Behavioral Expression: Female Common Chimpanzees participate in a variety of same-sex activities. One form of mutual genital stimulation is sometimes known as BUMPRUMP: two females, standing on all fours and facing in opposite directions, rub their rumps together (usually in an up-and-down motion), stimulating their genital and anal regions. Sometimes one female lies on top of the other in a face-to-face position—or the two sit facing one another—rubbing their genitals together. Mounting also occurs in the front-to-back position typical of heterosexual mating. Unlike male-female mountings, though, the angle and position of the mounting female's body and arms may be slightly different from that of a male, her pelvic thrusts may be slower or more perfunctory, and she may rub against the other female's genitals with her belly rather than her own genital region. Occasionally female Chimps also engage in cunnilingus: one individual presents her buttocks by crouching in front of the other, who stimulates her external genitalia with her lips and tongue.

Among males, several different kinds of same-sex interactions occur. Manual contact or stimulation of a partner's genitals, for example, can involve fondling, rubbing, or gripping of the penis and/or touching of the scrotum, sometimes while the partner makes pelvic thrusts that "bounce" his genitals on his partner's hand. Chimps occasionally also engage in fellatio, mutual penis-rubbing while sitting face-to-face, mounting in a front-to-back position (sometimes with pelvic thrusts or body shaking), and even insertion of a finger into the partner's anus and oral-anal "grooming" in a 69 position. A number of these activities—notably genital touching, mounting, and anal contact—occur as ritualized sexual gestures in the context of greeting, enlisting of support, reconciliation, and/or reassurance. They are often combined with affectionate gestures between males such as embracing, kissing (including openmouthed contact), grooming, and genital kissing or nuzzling. Males who participate in such activities may be bonded together in a mutually supportive "friendship" or COALITION. Occasionally male Chimpanzees also interact sexually with male Savanna Baboons in the wild. One adolescent Chimp, for example, was observed holding and fondling the penis of an adult male Baboon.

Transgendered or intersexual Common Chimpanzees occasionally occur as well. One individual who was physically and anatomically a male was chromosomally a mosaic, combining both the male (XY) and the female (XX) chromosome types.

Frequency: The prevalence of same-sex activities between male Common Chimpanzees is highly variable. Mounting between males constitutes anywhere from 1–2 percent to one-third or one-half of the behaviors involved in reassurance, enlistment of support, and other activities during or following conflicts. Kissing and embracing between males constitute from 12–30 percent of such interactions (depending on the population). Overall, 29–33 percent of all mounting activity occurs between males. Less detailed information is available for females, but a similar range of frequencies is probably involved. Other homosexual activities such as bumprump and oral or manual stimulation of genitals have so far been observed largely in captivity, where they may be fairly common.

Orientation: Most adult male Chimpanzees that participate in same-sex mounting, genital handling, or other activities also mate with females. Younger (adolescent or juvenile) males, who occasionally engage in such activities as fellatio or mutual genital rubbing, may be less heterosexually involved. In some populations, virtually all adult males participate in same-sex mounting, although such behavior may constitute anywhere from one-fifth to three-quarters of an individual's mounting activity. Females that participate in same-sex activities are also usually functionally bisexual, copulating with males as well. However, a few individuals appear to be more exclusively homosexual: one female, for example, refused to mate with males and was only involved sexually with other females for many years. She even developed a close relationship with another female and her offspring. Socially, she occupied an intermediate position between the male and female subgroups: she often associated with males and "ganged up" with them against other individuals, but she also maintained primary bonds with females and sometimes even defended them against the sexual advances of males. Later, however, she also mated with males.

Nonreproductive and Alternative Heterosexualities

Common Chimpanzees engage in a variety of nonprocreative heterosexual practices that parallel their same-sex behaviors. Heterosexual oral sex involves both cunnilingus (males licking the vagina or mouthing the labia) and fellatio (females sucking or nuzzling the penis). Manual stimulation of the genitals also occurs: males sometimes insert a finger into the vagina (the female may then move his hand in order to stimulate herself), while females occasionally fondle their partner's penis. In addition, bumprump takes place between males and females (sometimes including thrusting and scrotum handling), and both sexes masturbate—stimulating their own genitals manually or with various implements. Some males even perform AUTO-FELLATIO, i.e., they suck their own penis. Male Chimpanzees occasionally mount females without achieving penetration or ejaculate after withdrawing; they may also mate with females who are not in heat. Another form of nonprocreative sex is copulation during pregnancy: some females participate in heterosexual activity for 75–80 percent of the time that they are pregnant. In addition, male Chimps have been observed copulating with female Savanna Baboons in the wild.

When females are in heat, they typically mate numerous times and with multiple partners—as often as six or more times a day (sometimes with two to seven males in quick succession), for a total of several hundred times for each baby conceived. In some cases, though, heterosexual relations are less than amicable: males occasionally try to force females to consort and mate with them by threatening and even violently attacking them, and females often display "blunt refusal" or "abhorrence" reactions toward the advances of older males. Copulations are often interrupted or harassed by other Chimps trying to disrupt the sexual activities. In addition, infanticide and even cannibalism occasionally occur. For example, infants conceived outside the community may be killed by the resident males, and most females mate with males belonging to their own social group. However, in some populations a considerable number of females seek partners outside their group, engaging in "furtive" matings with them. In one community, for example, more than half of all offspring were sired by males living in other groups.

Although incestuous matings between adults are not common, mothers engage in sexual activity with their infant sons fairly often. Young females typically experience a one-to-three-year period of ADOLESCENT STERILITY after their first menstruation, during which time they mate heterosexually without conceiving. Some adult females practice a unique form of birth control: they simulate the contraceptive effect of nursing by stimulating their own nipples, in some cases preventing pregnancy for up to ten years. Females may also experience a postreproductive or "menopausal" period later in their lives, lasting up to two years (about 4–5 percent of the maximum life span). During this time they often continue to mate, accounting for up to 20 percent of all female sexual activity in a group.

Sources   *asterisked references discuss homosexuality/transgender

*Adang, O. M. J., J. A. B. Wensing, and J. A. R. A. M. van Hooff (1987) "The Arnhem Zoo Colony of Chimpanzees *Pan troglodytes:* Development and Management Techniques." *International Zoo Yearbook* 26:236–48.

*Bingham, H. C. (1928) "Sex Development in Apes." *Comparative Psychology Monographs* 5:1–165.

Bygott, J. D. (1979) "Agonistic Behavior, Dominance, and Social Structure in Wild Chimpanzees of the Gombe National Park." In D. A. Hamburg and E. R. McCown, eds., *The Great Apes*, pp. 405–28. Menlo Park, Calif.: Benjamin Cummings.

*———— (1974) "Agonistic Behavior and Dominance in Wild Chimpanzees." Ph.D. thesis, Cambridge University.

Dahl, J. F., K. J. Lauterbach, and C. A. Duffey (1996) "Birth Control in Female Chimpanzees: Self-Directed Behaviors and Infant-Mother Interactions." *American Journal of Physical Anthropology* supp. 22:93.

*Egozcue, J. (1972) "Chromosomal Abnormalities in Primates." In E. I. Goldsmith and J. Moor-Jankowski, eds., *Medical Primatology 1972*, part I, pp. 336–41. Basel: S. Karger.

Gagneux, P. , D. S. Woodruff, and C. Boesch (1997) "Furtive Mating in Female Chimpanzees." *Nature* 387:358–59.

*Goodall, J. (1986) *The Chimpanzees of Gombe: Patterns of Behavior.* Cambridge, Mass.: Belknap Press.

———— (1977) "Infant Killing and Cannibalism in Free Living Chimpanzees." *Folia Primatologica* 28:259–82.

*———— (1965) "Chimpanzees of the Gombe Stream Reserve." In I. deVore, ed., *Primate Behavior: Field Studies of Monkeys and Apes*, pp. 425–73. New York: Holt, Rinehart & Winston.

*Köhler, W. (1925) *The Mentality of Apes.* London: Routledge & Kegan Paul.

*Kollar, E. J., W. C. Beckwith, and R. B. Edgerton (1968) "Sexual Behavior of the ARL Colony Chimpanzees." *Journal of Nervous and Mental Disease* 147:444–59.

*Kollar, E. J., R. B. Edgerton, and W. C. Beckwith (1968) "An Evaluation of the Behavior of the ARL Colony Chimpanzees." *Archives of General Psychiatry* 19:580–94.

*Kortlandt, A. (1962) "Chimpanzees in the Wild." *Scientific American* 206(5):128–38.

*Lawick-Goodall, J. van (1968) "The Behavior of Free-Living Chimpanzees in the Gombe Stream Reserve." *Animal Behavior Monographs* 1:161–311.

*Nishida, T. (1997) "Sexual Behavior of Adult Male Chimpanzees of the Mahale Mountains National Park, Tanzania." *Primates* 38:379–98.

———— (1990) *The Chimpanzees of the Mahale Mountains: Sexual and Life History Strategies.* Tokyo: University of Tokyo Press.

———— (1979) "The Social Structure of Chimpanzees of the Mahale Mountains." In D. A. Hamburg and E. R. McCown, eds., *The Great Apes*, pp. 73–121. Menlo Park, Calif.: Benjamin Cummings.

*———— (1970) "Social Behavior and Relationship Among Wild Chimpanzees of the Mahali Mountains." *Primates* 11:47–87.

*Nishida, T., and K. Hosaka (1996) "Coalition Strategies Among Adult Male Chimpanzees of the Mahale Mountains, Tanzania." In W. C. McGrew, L. F. Marchant, and T. Nishida, eds., *Great Ape Societies*, pp. 114–34. Cambridge: Cambridge University Press.

*Reynolds, V., and F. Reynolds (1965) "Chimpanzees of the Budongo Forest." In I. deVore, ed., *Primate Behavior: Field Studies of Monkeys and Apes*, pp. 368–424. New York: Holt, Rinehart & Winston.

Takahata, Y., N. Koyama, and S. Suzuki (1995) "Do the Old Aged Females Experience a Long Postreproductive Life Span?: The Cases of Japanese Macaques and Chimpanzees." *Primates* 36:169–80.

Tutin, C. E. G., and P. R. McGinnis (1981) "Chimpanzee Reproduction in the Wild." In C.E. Graham, ed., *Reproductive Biology of the Great Apes*, pp. 239–64. New York: Academic Press.

*Tutin, C. E. G., and W. C. McGrew (1973a) "Chimpanzee Copulatory Behavior." *Folia Primatologica* 19:237–56.

———— (1973b) "Sexual Behavior of Group-Living Adolescent Chimpanzees." *American Journal of Physical Anthropology* 38:195–200.

*de Waal, F. B. M. (1982) *Chimpanzee Politics: Power and Sex Among Apes.* New York: Harper & Row.

*de Waal, F. B. M., and J. A. R. A. M. van Hooff (1981) "Side-directed Communication and Agonistic Interactions in Chimpanzees." *Behavior* 77:164–98.

Wrangham, R. W. (1997) "Subtle, Secret Female Chimpanzees." *Science* 277:774–75.

*Yerkes, R. M. (1939) "Social Dominance and Sexual Status in the Chimpanzee." *Quarterly Review of Biology* 14:115–36.

GREAT APES   *Primate*

GORILLA
Gorilla gorilla

HOMOSEXUALITY	TRANSGENDER	BEHAVIORS		RANKING	OBSERVED IN
● Female	○ Intersexuality	● Courtship	● Pair-bonding	● Primary	● Wild
● Male	○ Transvestism	● Affectionate	○ Parenting	○ Moderate	○ Semiwild
		● Sexual		○ Incidental	● Captivity

IDENTIFICATION: A massive ape (adult males generally weigh over 300 pounds) with black fur; old males are called silverbacks because of the silvery-gray fur on their backs. DISTRIBUTION: Central Africa including Congo (Zaire), Uganda, Rwanda; southeastern Nigeria to southern Congo; endangered. HABITAT: Bamboo forests, rain forests. STUDY AREAS: Virunga Mountains, Rwanda and Congo (Zaire), subspecies *G.g. beringei*, the Mountain Gorilla; Basel, Metro Toronto, and St. Louis Zoos, subspecies *G.g. gorilla*, the Lowland Gorilla.

Social Organization

Gorillas live in small groups of eight to fifteen individuals, usually consisting of three to six adult females, one mature male, one to two juvenile males, and five to seven immature offspring. All-male groups also regularly occur. The mating system is polygynous, i.e., the mature male mates with all of the females in the group. The females are usually not related to each other, since they generally leave their family group once reaching adulthood—in many other group-living primates, males leave a core group of (usually related) females.

Description

Behavioral Expression: Within her group, a female Gorilla sometimes forms an intense pairlike friendship with another female, spending as much time with her as with the breeding male of the group. Her interactions with this "favorite" female consist of constant touching while they spend time together, sitting with each other or lying one against the other, and frequent mutual grooming. Female Gorillas also frequently have sex with other females in their group. In a typical lesbian interaction, one female approaches another directly, often making copulatory vocalizations, after which they may sit quietly together for a while. Often they will begin to fondle each other's genitals or bring their face into intimate contact with the other's vulva, smelling or touching with their mouths. This is usually followed by embracing in a face-to-face position (usually lying down) with rubbing of the genitals against each other, sometimes accompanied by growling, grunting, screaming, or pulsing whimpers. The animals may also pause during periods of pelvic thrusting to caress each other, shift their positions, or masturbate themselves.

Lesbian sexual activity is notable for its differences from heterosexual copulation, probably related to an emphasis on achieving mutual pleasure. First of all, the more intimate face-to-face position is rarely used between males and females, who instead mate with the male mounting from behind (and often with the male thrusting and vocalizing significantly less than his partner). Sexual interactions between females are also generally more affectionate, involving much more embracing and grooming, and they usually last longer. One study revealed that sexual interactions between females last on average five times longer than heterosexual ones, and that lesbian activity involves considerably more thrusting and genital stimulation.

Female Gorillas also exhibit clear preferences for particular female sexual partners within their group. Although lesbian activity generally occurs among all members of a group, each female usually has a favorite partner with whom she interacts more often. Homosexuality is also integrated into the general reproductive cycle of the group: breeding females (including mothers) have sex with other females as much as do nonbreeding females, and lesbian sex is common even during pregnancy, sometimes as late as a week or two before birth.

Although male Gorillas (especially younger animals) sometimes mount each other in cosexual groups, homosexuality occurs most commonly in all-male groups, where probably more than 90 percent of all same-sex activity between males takes place. Such groups result when females leave their home group to join another, or when males occasionally leave their own group upon reaching maturity and band together. All-male groups persist for many years and have a complex network of homosexual pairings. Each male has preferred partners whom he courts and has sex with; some interact with only one other male in the group, while others have multiple partners (up to five have been recorded for one individual). Durations of individual pairings can be anywhere from a few months to a year or more. Participants are sometimes related to each other: in one all-male group, about 40 percent of all homosexual activity occurred between half brothers. There is often intense competition among the males for "preferred" partners—often the younger males—and older, higher-ranking males frequently "guard" their favorite males and fight to protect them from the advances of other males. Nevertheless, rates of aggression are significantly lower in all-male groups than in cosexual groups, and some male groups exhibit a high degree of cohesiveness attributable to the sexual bonding and mediating activities of their members.

When one male Gorilla is courting another, he approaches while making intense panting sounds; sometimes contact is initiated with one (or both) males reaching out to touch the other, or one may make a more subtle soliciting approach. Sexual activity involves one male mounting another and thrusting, in either the face-to-face or front-to-back position; both males often emit grumbling, growling, or panting sounds. Orgasm is signaled when the animal emits a deep sigh on dismounting, and often there is direct evidence of ejaculation (e.g., semen spilled on his partner). Most males both mount and are mounted, except for the oldest silverback males, who only mount. Like lesbian interactions, male homosexual encounters tend to last, on average, longer than heterosexual ones and to use the

▲ *Male Gorillas having sex with each other in the mountain rain forests of Rwanda, showing two mounting positions: front-to-back* (left) *and face-to-face*

face-to-face position more often than in male-female interactions (though less often than between females). Male Gorillas also touch and fondle each other's genitals in addition to mounting one another.

Frequency: In cosexual groups, 9 percent of all sexual activity is lesbian and 58 percent of all social/affectionate interactions of females are with other females (mostly with their "favorite" partner); about 2 percent of mounting episodes occur between adult males in such groups. Among younger animals (e.g., adolescents and juveniles) in cosexual groups, 7–36 percent of mounts are between males and 9–14 percent between females. Male homosexual courtship and copulation occur daily in some all-male groups and are thought to exceed the amount of heterosexual activity that takes place in cosexual groups. Some males may engage in homosexual copulation more than 75 times a year in such groups, and homosexual courtship, at its peak, can take place as often as 7 times an hour. Up to 10 percent of groups in some populations are all-male, and Gorillas spend an average of six years in such groups, with some males staying ten or more years (and even occasionally remaining until their death).

Orientation: Many female Gorillas are bisexual, having sexual and affectionate relationships with both males and females, but there are clear differences in the extent to which various individuals participate in homosexual versus heterosexual activity. In general, it appears that a continuum exists from those females who prefer lesbian activity (they account for a large proportion of the same-sex interactions), to those who have a fairly equal amount of interaction with both males and females, to those who interact primarily with males. Many male Gorillas are probably sequentially bisexual, spending portions of their lives having only homosexual encounters (in all-male groups), followed by periods of only heterosexual interactions, and so on. Other males, especially younger ones, may be simultaneously bisexual. Depending on the circumstances (such as the particular group composition), some males may also have primarily or exclusively homosexual interactions throughout their lives, while others may have only heterosexual ones, but it appears that all males at least have the capacity for bisexuality.

Nonreproductive and Alternative Heterosexualities

As noted above, sex (both heterosexual and homosexual) during pregnancy is common among Gorillas. Both males and females also engage in masturbation, and younger animals frequently participate in nonpenetrative sexual activity. Mountings of the latter type are usually incestuous, involving siblings, half siblings, or (more rarely) parents and their offspring (or their siblings' offspring). In captivity, oral sex and manual stimulation of genitals have also been observed in heterosexual interactions. Male Gorillas generally appear to have less interest in sex than females and are sometimes rather reluctant or perfunctory participants in mating. Heterosexual interactions are nearly always initiated by females (whose advances are often initially ignored by the males), males often thrust and vocalize much less than females during copulation (often for no more than 20 seconds), and females generally determine when a particular sexual interaction is finished. However, females may be sexually inactive for up to three or four years while nursing their young. Infanticide is quite common among wild Gorillas: more than 40 percent of all infant deaths in one population were due to infanticide, usually by an adult male trying to gain breeding access to a female (although females have been known to kill infants as well). Probably all adult males commit infanticide at least once during their lifetime, while most females are likely to lose at least one infant during their lifetime to killing by another of their own species.

Sources   *asterisked references discuss homosexuality/transgender*

*Coffin, R. (1978) "Sexual Behavior in a Group of Captive Young Gorillas." *Boletín de Estudios Médicos y Biológicos* 30:65–69.

*Fischer, R. B. and R.D. Nadler (1978) "Affiliative, Playful, and Homosexual Interactions of Adult Female Lowland Gorillas." *Primates* 19:657–64.

*Fossey, D. (1990) "New Observations of Gorillas in the Wild." In *Grzimek's Encyclopedia of Mammals,* vol. 2, pp. 449–62. New York: McGraw-Hill.

——— (1984) "Infanticide in Mountain Gorillas (*Gorilla gorilla beringei*) with Comparative Notes on Chimpanzees." In G. Hausfater and S. B. Hrdy, eds., *Infanticide: Comparative and Evolutionary Perspectives,* pp. 217–35. New York: Aldine.

*——— (1983) *Gorillas in the Mist.* Boston: Houghton Mifflin.

*Harcourt, A. H. (1988) "Bachelor Groups of Gorillas in Captivity: The Situation in the Wild." *Dodo* 25:54–61.

*——— (1979a) "Social Relations Among Adult Female Mountain Gorillas." *Animal Behavior* 27:251–64.

——— (1979b) "Social Relationships Between Adult Male and Female Mountain Gorillas in the Wild." *Animal Behavior* 27:325–42.

*Harcourt, A. H., D. Fossey, K. J. Stewart, and D. P. Watts (1980) "Reproduction in Wild Gorillas and Some Comparisons with Chimpanzees." *Journal of Reproduction and Fertility* suppl. 28:59–70.

Harcourt, A. H., and K. J. Stewart (1978) "Sexual Behavior of Wild Mountain Gorillas." In D. J. Chivers and J. Herbert, eds., *Recent Advances in Primatology,* vol. 1, pp. 611–12. New York: Academic Press.

*Harcourt, A. H., K. J. Stewart, and D. Fossey (1981) "Gorilla Reproduction in the Wild." In C. E. Graham, ed., *Reproductive Biology of the Great Apes,* pp. 265–79. New York: Academic Press.

*Hess, J. P. (1973) "Some Observations on the Sexual Behavior of Captive Lowland Gorillas, *Gorilla g. gorilla* (Savage and Wyman)." In R. P. Michael and J. H. Crook, eds., *Comparative Ecology and Behavior of Primates,* pp. 507–81. London: Academic Press.

*Nadler, R. D. (1986) "Sex-Related Behavior of Immature Wild Mountain Gorillas." *Developmental Psychobiology* 19:125–37.

*Porton, I., and M. White (1996) "Managing an All-Male Group of Gorillas: Eight Years of Experience at the St. Louis Zoological Park." In *AAZPA Regional Conference Proceedings,* pp. 720–28. Wheeling, W.V.: American Association of Zoological Parks and Aquariums.

*Robbins, M. M. (1996) "Male-Male Interactions in Heterosexual and All-Male Wild Mountain Gorilla Groups." *Ethology* 102:942–65.

*——— (1995) "A Demographic Analysis of Male Life History and Social Structure of Mountain Gorillas." *Behavior* 132:21–47.

*Schaller, G. (1963) *The Mountain Gorilla.* Chicago: University of Chicago Press.

*Stewart, K. J. (1977) "Birth of a Wild Mountain Gorilla (*Gorilla gorilla beringei*)." *Primates* 18:965–76.

Watts, D. P. (1990) "Mountain Gorilla Life Histories, Reproductive Competition, and Sociosexual Behavior and Some Implications for Captive Husbandry." *Zoo Biology* 9:185–200.

——— (1989) "Infanticide in Mountain Gorillas: New Cases and a Reconsideration of the Evidence." *Ethology* 81:1–18.

*Yamagiwa, J. (1987a) "Intra- and Inter-Group Interactions of an All-Male Group of Virunga Mountain Gorillas (*Gorilla gorilla beringei*)." *Primates* 28:1–30.

*——— (1987b) "Male Life History and the Social Structure of Wild Mountain Gorillas (*Gorilla gorilla beringei*)." In S. Kawano, J. H. Connell, and T. Hidaka, eds., *Evolution and Coadaptation in Biotic Communities,* pp. 31–51. Tokyo: University of Tokyo Press.

GREAT APES

Primate

ORANG-UTAN
Pongo pygmaeus

HOMOSEXUALITY	TRANSGENDER	BEHAVIORS		RANKING	OBSERVED IN
● Female	● Intersexuality	○ Courtship	● Pair-bonding	○ Primary	○ Wild
● Male	○ Transvestism	● Affectionate	○ Parenting	● Moderate	● Semiwild
		● Sexual		○ Incidental	● Captivity

IDENTIFICATION: A medium-sized ape (adult males generally weigh around 170 pounds) with a long, reddish brown coat; some older males develop prominent cheek pads or "flanges." DISTRIBUTION: Sumatra, Borneo (Indonesia); vulnerable. HABITAT: Swamps, lowland and mountain rain forests. STUDY AREAS: Ketambe region of North Sumatra, Indonesia, subspecies *P.p. abelii,* the Sumatran Orang-utan; Regent's Park Zoo and Singapore Zoological Garden, including subspecies *P.p. pygmaeus,* the Bornean Orang-utan.

Social Organization

Adult Orang-utans are largely solitary—males and females live separate from each other and interact only when the female is ready to mate. Younger Orangs, however, are more sociable and may actively seek each other's company and interact in groups. The mating system is polygynous: males copulate with multiple females and no long-lasting heterosexual pairing occurs, although males and females may "consort" together for shorter periods during mating. Males do not participate in parenting.

Description

Behavioral Expression: Orang-utans engage in a variety of homosexual activities, including a range of different sexual techniques and various affectionate and pairing behaviors. Mounting among male Orangs, especially younger adults (10–15 years old) and adolescents (7–10 years old) often develops into full anal intercourse, with erection of the penis, pelvic thrusting, penetration, and ejaculation. In a more unusual type of homosexual penetration, a male sometimes tries to insert his penis into the small hollow formed when his partner's penis retracts. Another prominent homosexual activity is fellatio (oral-genital contact): one male will lick and suck another's erect penis. In some cases, males take turns fellating each other. Males occasionally also fondle and touch the erect penis of another male, often examining the organ closely by parting the hairs in the genital region. Lesbian activity in Orangs usually involves one female fondling the genitals of another female, often inserting her fingers (thumb or other digits) into the vagina of the other. Sometimes she also masturbates herself with her foot while she is penetrating the other female. Mounting rarely, if ever, occurs among female Orangs, unlike in many other animals in which females engage in homosexual behavior. Female homosexual encounters may last for up to 12 minutes, comparable to the 10–15 minute duration of most heterosexual copulations. Although Orang-utans are usually willing participants in homosexual encounters, sometimes one animal is more reluctant and the partner will then attempt to restrain him or her, for example by using the feet to hold him or her down. However, this contrasts sharply with heterosexual encounters, in which females often scream and struggle violently while males attempt to forcibly mate with them (see the discussion on heterosexualities below).

Sexual behavior often occurs within a "bonding" or special friendship-like pairing between younger animals of the same sex. Two males or two females may become quite attached, following one another over several days, playing together (including play-wrestling), sharing food, and generally spending a great deal of time together and coordinating their activities. One partner may even throw a "temper tantrum" when the other ventures too far away or fails to wait for its companion. Female Orangs have also been known to compete with males for sexual access to a favorite female partner with whom they later develop a bonded relationship. Same-sex companions demonstrate a number of affectionate behaviors toward each other—females, for example, may embrace, cling to one other, walk in tandem, or groom each other, and males may "kiss" each other. While in

▲ *Fellatio in two younger male Orang-utans in Sumatra: the male on the lower right is sucking his partner's penis*

some cases this mouth-to-mouth contact may be for the exchange of food or drink, in other cases it appears to be more of an affectionate or greeting gesture. Such companionships also develop between animals of the opposite sex, and indeed they resemble in many ways the "consortships" that sometimes characterize heterosexual mating relations. Companionships, however, need not involve any sexual contact, whether between animals of the same or opposite sex.

Homosexual interactions sometimes also occur between male Orangs and Crab-eating Macaques. These monkeys often associate with Orangs, feeding in the same areas and interacting nonaggressively. Orangs and Crab-eating Macaques may groom each other, and male Orang-utans will occasionally suck the penis of an adult male Crab-eating Macaque.

Transgendered Orang-utans occasionally occur as well: individuals have been found who are physically male yet have a female (XX) chromosome pattern.

Frequency: Approximately 9 percent of all Orang-utan sexual encounters in some populations involve males mounting each other; the proportion of same-sex activity is probably even higher, since male oral-genital contacts and female homosexual encounters are not included in this figure. Females who engage in lesbian activity may do so frequently and repeatedly over several days, similar to the repeated sexual interactions in a heterosexual consortship.

Orientation: Male homosexual behavior is characteristic of younger Orang-utans. Not all younger males engage in same-sex activity, but those who do are probably bisexual, since most such individuals are also involved in heterosexual pursuits. Mature adult males probably have a bisexual potential: although they rarely engage in homosexual activity in the wild, in captivity they often do (even in the presence of females). Female Orangs are probably also bisexual—for example, one female who was sexually active with another female later mated heterosexually and raised young. However, while engaged in homosexual behavior, she was exclusively lesbian, since she completely ignored males and focused her attentions entirely on other females.

Nonreproductive and Alternative Heterosexualities

A wide variety of nonprocreative heterosexual activities are found in Orang-utans. Both males and females often stimulate their partner's genitals with their mouth or hands, and females may also rub their genitals against the male. The female has a prominent clitoris that is stimulated during intercourse, and she often takes the initiative in heterosexual activity, actually mounting the male, manually guiding his penis into her, and performing pelvic thrusts while he lies on his back. A variety of positions are used for heterosexual copulation, including face-to-face (the most common), front-to-back, and sideways. In almost 30 percent of mounts, vaginal penetration and/or ejaculation do not occur. Anal stimulation can be a component of heterosexual interactions as well: both males and females lick, suck, blow on, insert fingers into, and rub their genitals on their partner's anus; males have also been known to engage in anal intercourse (penetration) with females. Females may consort and copulate with multiple male partners, and copulation can occur throughout pregnancy up to the time of birth. Masturbation is also common among Orangs—females rub their fingers or foot on their clitoris or insert a finger or toe into their vagina, while males rub their penises with their fist or foot. Both males and females also use inanimate objects or "tools" to masturbate. Males sometimes become sexually aroused and spontaneously ejaculate during long-calling (a courtship and territorial vocalization used by mature males). Mothers frequently engage in incestuous contact with their infants, manually or orally stimulating the penis or clitoris (or being stimulated by the infant), and may even mount the infants.

Heterosexual relations are sometimes characterized by aggression and violence rather than pleasure and consensuality. Younger males often chase, harass, and rape females. During such interactions, which may account for the majority of copulations in some populations, the male may grab, slap, bite, and forcibly restrain the female, who struggles violently while screaming or whimpering. Occasionally (about 7–8 percent of the time) she does manage to break free. An unusual form of reproductive suppression also occurs among male Orangs. Although they become sexually mature at seven to ten years old, males generally fail to develop the full range of secondary sexual characteristics (such as the large cheek pads or "flanges," a throat pouch, and a general weight increase) for another seven years, and sometimes this is delayed for as long as two decades. It is thought that this development is suppressed by the presence of a mature male, perhaps through social intimidation or stress, although the exact mechanism is not known. Nonbreeding males have been found to have higher estrogen levels than breeding males, so perhaps a physiological effect is also involved. Interestingly, nonflanged younger males have been observed copulating repeatedly with females without resulting in any pregnancies; perhaps this is related to their arrested sexual development, although it is also possible that they were simply mating during the nonovulatory phase of the female's cycle. In addition, adolescent females experience ADOLESCENT STERILITY, lasting a year or longer, during which they can copulate without becoming pregnant. In fact, adolescent females have higher copulation rates than adult females, accounting for more than 60 percent of heterosexual mating. Adult females breed relatively infrequently, perhaps once every four to eight years. Because females in some populations tend to have synchronized reproductive cycles, there may be periods of up to two years when no adult females are available for mating.

Sources   *asterisked references discuss homosexuality/transgender*

*Dutrillaux, B., M.-O. Rethoré, and J. Lejeune (1975) "Comparaison du caryotype de l'orang-outang (*Pongo pygmaeus*) à celui de l'homme, du chimpanzé, et du gorille [Comparison of the Karyotype of the Orang-utan to Those of Man, Chimpanzee, and Gorilla]." *Annales de Génétique* 18:153–61.

Galdikas, B. M. F. (1995) "Social and Reproductive Behavior of Wild Adolescent Female Orangutans." In R. D. Nadler, B. M. F. Galdikas, L. K. Sheeran, and N. Rosen, eds., *The Neglected Ape*, pp. 183–90. New York: Plenum Press.

——— (1985) "Orangutan Sociality at Tanjung Puting." *American Journal of Primatology* 9:101–19.

——— (1981) "Orangutan Reproduction in the Wild." In C. E. Graham, ed., *Reproductive Biology of the Great Apes*, pp. 281–300. New York: Academic Press.

Harrisson, B. (1961) "A Study of Orang-utan Behavior in the Semi-Wild State." *International Zoo Yearbook* 3:57–68.

Kaplan, G., and L. Rogers (1994) *Orang-Utans in Borneo*. Armidale, Australia: University of New England Press.

Kingsley, S. R. (1988) "Physiological Development of Male Orang-utans and Gorillas." In J.H. Schwartz, ed., *Orang-utan Biology*, pp. 123–31. New York: Oxford University Press.

——— (1982) "Causes of Nonbreeding and the Development of the Secondary Sexual Characteristics in the Male Orang Utan: A Hormonal Study." In L. E. M. de Boer, ed., *The Orang Utan: Its Biology and Conservation*, pp. 215–29. The Hague: Dr W. Junk Publishers.

*MacKinnon, J. (1974) "The Behavior and Ecology of Wild Orang-utans (*Pongo pygmaeus*)." *Animal Behavior* 22:3–74.

*Maple, T. L. (1980) *Orang-utan Behavior*. New York: Van Nostrand Reinhold.

Mitani, J. C. (1985) "Mating Behavior of Male Orangutans in the Kutai Game Reserve, Indonesia." *Animal Behavior* 33:392–402.

*Morris, D. (1964) "The Response of Animals to a Restricted Environment." *Symposia of the Zoological Society of London* 13:99–118.

Nadler, R. D. (1988) "Sexual and Reproductive Behavior." In J. H. Schwartz, ed., *Orang-utan Biology*, pp. 105–16. New York: Oxford University Press.

——— (1982) "Reproductive Behavior and Endocrinology of Orang Utans." In L. E. M. de Boer, ed., *The Orang Utan: Its Biology and Conservation*, pp. 231–48. The Hague: Dr W. Junk Publishers.

*Poole, T. B. (1987) "Social Behavior of a Group of Orangutans (*Pongo pygmaeus*) on an Artificial Island in Singapore Zoological Gardens." *Zoo Biology* 6:315–30.

*Rijksen, H. D. (1978) *A Fieldstudy on Sumatran Orang Utans (*Pongo pygmaeus abelii* Lesson 1827): Ecology, Behavior, and Conservation*. Wageningen, Netherlands: H. Veenman & Zonen b.v.

Rodman, P. S. (1988) "Diversity and Consistency in Ecology and Behavior." In J. H. Schwartz, ed., *Orang-utan Biology*, pp. 31–51. New York: Oxford University Press.

Schürmann, C. (1982) "Mating Behavior of Wild Orang Utans." In L. E. M. de Boer, ed., *The Orang Utan: Its Biology and Conservation*, pp. 269–84. The Hague: Dr W. Junk Publishers.

Schürmann, C., and J. A. R. A. M. van Hooff (1986) "Reproductive Strategies of the Orang-Utan: New Data and a Reconsideration of Existing Sociosexual Models." *International Journal of Primatology* 7:265–87.

*Turleau, C., J. de Grouchy, F. Chavin-Colin, J. Mortelmans, and W. Van den Bergh (1975) "Inversion péricentrique du 3, homozygote et hétérozygote, et translation centromérique du 12 dans une famille d'orangs-outangs. Implications évolutives [Pericentric Inversion of Chromosome 3, Homozygous and Heterozygous, and Transposition of Centromere of Chromosome 12 in a Family of Orang-utans. Implications for Evolution]." *Annales de Génétique* 18:227–33.

Utani, S., and T. M. Setia (1995) "Behavioral Changes in Wild Male and Female Sumatran Orangutans (*Pongo pygmaeus abelii*) During and Following a Resident Male Take-over." In R. D. Nadler, B.M.F. Galdikas, L.K. Sheeran, and N. Rosen, eds., *The Neglected Ape*, pp. 183–90. New York: Plenum Press.

Gibbons or Lesser Apes

WHITE-HANDED GIBBON
Hylobates lar
SIAMANG
Hylobates syndactylus

Primary 🐾

HOMOSEXUALITY	TRANSGENDER	BEHAVIORS		RANKING	OBSERVED IN
○ Female	○ Intersexuality	○ Courtship	○ Pair-bonding	○ Primary	● Wild
● Male	○ Transvestism	● Affectionate	○ Parenting	● Moderate	○ Semiwild
		● Sexual		○ Incidental	● Captivity

WHITE-HANDED GIBBON
IDENTIFICATION: A small ape (weighing up to 13 pounds) with a variable coat color (cream, black, brown, or reddish) and a white face ring, hands, and feet. DISTRIBUTION: China, Thailand, Laos, Burma, Malay Peninsula, Sumatra. HABITAT: Lowland and mountain deciduous and rain forests. STUDY AREAS: Huai Kha Khaeng Wildlife Sanctuary, Thailand.

SIAMANG
IDENTIFICATION: Similar to White-handed Gibbon, but larger (up to 24 pounds), with all-black fur and a prominent throat sac. DISTRIBUTION: Sumatra, Malay Peninsula. HABITAT: Lowland and mountain forests. STUDY AREA: Milwaukee County Zoo, Wisconsin.

Social Organization

Gibbons generally live in family groups consisting of a paired male and female and their offspring. Siamang heterosexual pairs may be more closely bonded than those of White-handed Gibbons. Both males and females perform complex vocal duets as part of bonding and territorial displays, although separate family groups have relatively little interaction with one another.

Description

Behavioral Expression: Within their nuclear family groups, male Gibbons sometimes engage in homosexual activities with each other. This incestuous activity often takes place between an adolescent or younger male and his father (or stepfather, if his parents have divorced and re-paired); in Siamangs it may also occur between brothers. A typical homosexual encounter between father and son in White-handed Gibbons occurs in the trees in the morning or early afternoon, while the family is resting or feeding. The mother is generally nonchalant about such encounters, ignoring the sexual activity even if she is close at hand. The two males often groom each other or engage in playful wrestling or chasing as part of their interaction. During such activities, either male may approach the other to initiate sex, which consists primarily of the two males rubbing their erect penises together, often leading to orgasm. This is done face-to-face (unlike heterosexual copulation, which is typically performed front-to-back). The father lifts up his knees and spreads his legs wide while sitting on a branch or hanging by his arms—this is an invitation to his son to have sex. The adolescent male embraces his father around the waist with his legs, then lowers his body until he is, in effect, sitting in his father's lap, with his legs resting on top of the older male's thighs. This allows their genitals to come into direct contact, and the younger male usually begins rapidly thrusting against his father; sometimes the older male will make pelvic thrusts as well. If his son ejaculates on him, the father may scoop up the semen and eat it. Genital contact can last for up to a minute, although the average is about 20 seconds; in comparison, heterosexual copulations in this species average only about 15 seconds.

A similar form of genital rubbing occurs between father and son in Siamangs. While both males are hanging by their arms, the younger grasps his father around the waist with his legs and both thrust against each other (in this species, heterosexual mating is occasionally also performed face-to-face). Unlike in White-handed Gibbons, this activity is sometimes accompanied by threats between the two males, and it appears that the younger male may on occasion want to terminate the activity before his father does. Sometimes two brothers—juveniles or adolescents, four to nine years old—thrust against each other face-to-face as well. Brothers are also generally affectionate with each other, touching and grooming one another, putting their arms over each other's shoulders, and wrestling together. Fellatio also sometimes occurs in Siamangs: usually an older brother will lick and gently nibble on the penis and groin of his younger brother (who may be only one to three years old) while the latter dangles by his arms above him or sits with his legs spread. The older male may also masturbate the younger by pulling on his erect penis; if ejaculation occurs, the semen may be eaten. Occasionally, a son will lick and groom his father's genital area, or the father might insert one of his fingers into his son's anus.

Frequency: In those Gibbon families where homosexual activity takes place, it occurs quite frequently, and at rates that may equal or exceed heterosexual activity. In one White-handed Gibbon family, the father and son sometimes had sexual encounters as often as 8 times a day—although they averaged about twice a day—and homosexual activity took place on more than a third of the days that the family was observed. In fact, during one 18-day period, 44 homosexual interactions were recorded. In comparison, 23 heterosexual copulations were observed in another family over 18 (different) days, at a rate of 1–3 per day; other studies have found rates of 2 heterosexual matings per day (equivalent to heterosexual activity on about a third of the observation days). In a Siamang family observed in a zoo, about 30 percent of all sexual interactions were between males. It is not yet known, however, in what proportion of families homosexual activity occurs (in either of these species).

Orientation: Male Gibbon sexual life is probably sequentially bisexual, characterized by alternating periods of heterosexual and homosexual activity, with occasional long-term exclusive homosexuality. Younger males may experience entirely homosexual interactions with their fathers, or sexual activity with both parents (see below) while they are growing up, and then go on to mate heterosexually as adults. Once paired with a female, they may engage in incestuous encounters with their offspring of both sexes or have extended periods of exclusive homosexuality. In one White-handed Gibbon family, no heterosexual interactions were observed between the father and his female mate during the entire two years that homosexual activity was taking place between him and his son.

Nonreproductive and Alternative Heterosexualities

Like many other species where homosexual activity occurs between related individuals, heterosexual incest is also prominent among Gibbons. Siamang mothers and fathers both interact sexually with their offspring of the opposite sex, as do siblings. Adult males sometimes perform copulation-like thrusting with their daughters, as well as oral and manual stimulation of their genitals. In one case, a Siamang father was observed fondling his adolescent daughter's vulva with his fingers while her younger brother licked her clitoris. Mothers may invite their juvenile sons—as young as four to five years—to lick and groom their genitals (usually with no hostile reaction from the father). When offspring grow up, mother-son pairs (and occasionally, brother-sister pairs) may sometimes develop in both White-handed Gibbons and Siamangs, often when a father dies and is replaced by his son. Nonreproductive sexual behaviors such as oral sex are also commonly performed in non-incestuous contexts, e.g., between a pair-bonded male and female. Cunnilingus (including direct clitoral licking), manual fondling of the vulva, and vaginal penetration with the fingers have all been observed in mated pairs. Females probably also experience orgasm during heterosexual encounters: in one episode in which a male and female were thrusting against each other, a shudder coursed through the female's body, and she remained still for almost half a minute after a period of intense stimulation. Female White-handed Gibbons sometimes masturbate by rubbing their genitals against a surface, and they may experience orgasm this way; male Siamangs also masturbate, though not necessarily to orgasm.

In White-handed Gibbons, about 6–7 percent of heterosexual copulations occur when the female cannot conceive, e.g., during pregnancy or lactation. Some of these matings may be with males other than her mate. Although most Gibbon pairs are monogamous, it is estimated that 10–12 percent of White-handed Gibbon copulations are promiscuous. Nonmonogamous sexual activity also occurs in Siamangs and may be initiated by the female. Similarly, although many Gibbons (of both species) pair for life, divorce also occurs. In one study that followed 11 Gibbon heterosexual pairs over six years, 5 of them split up—usually when one partner left his or her mate to be with another individual. As a result, many White-handed Gibbon families—perhaps up to a third—involve step-parenting. Interestingly, even though there is a wide variety of possible sexual and pairing activities in these species, heterosexual activity is a relatively rare occurrence in wild Gibbons. For example, sexual behavior between male and female White-handed Gibbons generally occurs only once every two years or so, and for periods of only four or five months at a time when it does (females generally breed only every two to three years). In Siamangs, females go through regular periods of asexuality in which they delay breeding and turn over the care of their young to males. Females of this species look after their young only until they are 12–16 months old; at that time, males assume full responsibility for the offspring, but females do not reproduce again for another year. It is thought that this period of nonreproduction enables them to assume leadership roles in their group.

Sources   *asterisked references discuss homosexuality/transgender

Brockelman, W. Y., U. Reichard, U. Treesucon, and J. J. Raemaekers (1998) "Dispersal, Pair Formation, and Social Structure in Gibbons (*Hylobates lar*)." *Behavioral Ecology and Sociobiology* 42:329–39.

Chivers, D. J. (1974) *The Siamang in Malaya: A Field Study of a Primate in Tropical Rain Forest.* Contributions to Primatology, vol. 4. Basel: S. Karger.

———. (1972) "The Siamang and the Gibbon in the Malay Peninsula." In D. M. Rumbaugh, ed., *Gibbon and Siamang*, vol.1, pp. 103–35. Basel: S. Karger.

Chivers, D. J., and J. J. Raemaekers (1980) "Long-term Changes in Behavior." In D. J. Chivers, ed., *Malayan Forest Primates: Ten Years' Study in Tropical Rain Forest*, pp. 209–60. New York: Plenum.

*Edwards, A.-M. A. R., and J. D. Todd (1991) "Homosexual Behavior in Wild White-handed Gibbons (*Hylobates lar*)." *Primates* 32:231–36.

Ellefson, J. O. (1974) "A Natural History of White-handed Gibbons in the Malayan Peninsula." In D. M. Rumbaugh, ed., *Gibbon and Siamang*, vol. 3, pp. 1–136. Basel: S. Karger.

*Fox, G. J. (1977) "Social Dynamics in Siamang." Ph.D. thesis, University of Wisconsin–Milwaukee.

——— (1972) "Some Comparisons Between Siamang and Gibbon Behavior." *Folia Primatologica* 18:122–39.

Koyama, N. (1971) "Observations on Mating Behavior of Wild Siamang Gibbons at Fraser's Hill, Malaysia." *Primates* 12:183–89.

Leighton, D. R. (1987) "Gibbons: Territoriality and Monogamy." In B. B. Smuts, D. L. Cheney, R. M. Seyfarth, R. W. Wrangham, and T. T. Struhsaker, eds., *Primate Societies*, pp. 135–45. Chicago: University of Chicago Press.

Mootnick, A. R., and E. Baker (1994) "Masturbation in Captive *Hylobates* (Gibbons)." *Zoo Biology* 13:345–53.

Palombit, R. (1996) "Pair Bonds in Monogamous Apes: A Comparison of the Siamang *Hylobates syndactylus* and the White-handed Gibbon *Hylobates lar*." *Behavior* 133:321–56.

——— (1994a) "Dynamic Pair Bonds in Hylobatids: Implications Regarding Monogamous Social Systems." *Behavior* 128:65–101.

——— (1994b) "Extra-pair Copulations in a Monogamous Ape." *Animal Behavior* 47:721–23.

Raemaekers, J. J., and P. M. Raemaekers (1984) "Vocal Interaction Between Two Male Gibbons, *Hylobates lar*." *Natural History Bulletin of the Siam Society* 32:95–106.

Reichard, U. (1995a) "Extra-pair Copulation in Monogamous Wild White-handed Gibbons (*Hylobates lar*)." *Zeitschrift für Säugetierkunde* 60:186–88.

——— (1995b) "Extra-pair Copulation in a Monogamous Gibbon (*Hylobates lar*)." *Ethology* 100:99–112.

LANGURS AND LEAF MONKEYS

<small>LANGURS</small>
HANUMAN LANGUR
Presbytis entellus
NILGIRI LANGUR
Presbytis johnii

Primate

HOMOSEXUALITY	TRANSGENDER	BEHAVIORS		RANKING	OBSERVED IN
● Female	● Intersexuality	○ Courtship	● Pair-bonding	● Primary	● Wild
● Male	○ Transvestism	● Affectionate	○ Parenting	○ Moderate	○ Semiwild
		● Sexual		○ Incidental	● Captivity

HANUMAN LANGUR
IDENTIFICATION: A medium-sized monkey with a silver-gray or brown coat, black face, slender limbs, and a long tail (over 3 feet). DISTRIBUTION: Throughout India, Pakistan, Bangladesh, Sri Lanka, Burma. HABITAT: Scrub and deciduous forests. STUDY AREAS: Numerous locations in India, including Jodhpur, Abu, Kumaon Hills; Polonnaruwa, Sri Lanka; Melemchi, Nepal; University of California—Berkeley.

NILGIRI LANGUR
IDENTIFICATION: Similar to Hanuman Langur but with shiny black fur, a light brown hood, and a prominent brow tuft. DISTRIBUTION: Southwestern India; vulnerable. HABITAT: Montane evergreen rain forests, woodlands. STUDY AREAS: Nilgiri district, Mundanthurai Tiger Reserve, Anaimalai Wildlife Sanctuary, India.

Social Organization

Hanuman Langurs live in cosexual troops (some contain only one male) and also in all-male groups. The latter typically include up to 30 or more individuals; about 20 percent of the population in some areas lives in male-only groups. Nilgiri Langurs live in both cosexual troops (usually eight to nine monkeys each, with one or two adult males) and same-sex groupings (usually two or three males, occasionally more, constituting about a quarter to a third of the adult males). Troop life and stability generally revolve around the females; males are distinctly peripheral to the overall social system, playing only a minor role in defending the troop and minimally helping raise the young.

Description

Behavioral Expression: Homosexual mounting is a prominent feature of female interactions in Hanuman Langurs. One female climbs on the back of another and begins pelvic thrusting. Unlike in other species that engage in homosexual mounting, the female does not rub her genitals on the rump of the other female, but rather thrusts against her buttocks. The mounter may experience indirect stimulation of the area surrounding her clitoris, while the mountee may have direct clitoral stimulation from the female on top of her. In many ways such mounts resemble heterosexual matings, for example in their duration (5 to 10 seconds), number of pelvic thrusts (two to eleven), the grunting and grimacing of the mounter, and the JUMPING DISPLAY, which often precedes or follows. In other respects, however, homosexual mountings are strikingly different. For example, most heterosexual matings are solicited by the female, who lowers her head and shakes it while presenting her hindquarters to the male. While a similar solicitation occurs in about 13 percent of mounts between females, the majority (79 percent) are initiated by the mounting female. In both homosexual and heterosexual mounts, the partners groom following a mount; however, between females the mounter usually grooms the mountee, while in heterosexual mounts, the mountee (female) typically grooms the mounter (male). Finally, only about 30 percent of homosexual mounts are interrupted by other individuals; in contrast, more than 80 percent of heterosexual copulations are harassed by other animals trying to disrupt (and prevent) the mating. Up to seven animals of all ages and sexes may converge on an opposite-sex mating pair and directly attack the animals, slapping them, trying to push the male off or chase the female out from under him, and even kicking the male's testicles.

All females participate freely in homosexual mounting, including lactating, pregnant, menstruating, ovulating, and nonovulating females. This behavior is especially common among mothers, who have developed a special "baby-sitting" system: they transfer their young to other individuals in the troop (usually other females but sometimes males) for a short time (a similar pattern is found in Nilgiri Langurs). This allows them to engage in homosexual (and other) activities. Most mounting between females occurs among adults, though some involves adults with ju-

◄ *A female Hanuman Langur in India mounting another female*

veniles (up to four years old), usually with the younger female mounting the older one. Although most homosexual participants are unrelated, some mountings are incestuous, mostly between half sisters (about 27 percent of all lesbian mounts) and, more rarely, between mother and daughter (about 1 percent of mounts); heterosexual matings are virtually never incestuous.

Male Hanuman Langurs mount each other as well, especially in the all-male bands. Unlike homosexual mounting between females, this activity is typically initiated by the mountee, who invites another male to mount him by performing a head-shaking display (sometimes combined with small jumps and "snoring" vocalizations). Mounting often involves one male rubbing and thrusting his erect penis against the other's body and may be accompanied by a number of affectionate activities that are typically associated with sexual arousal. These include embracing (in which one male buries his head in the chest or shoulder of another), mouth-to-mouth contact or "kissing," reciprocal grooming (often accompanied by erections), and touching or grooming of the genitals with the hand or lips. Males also "cuddle," one sitting closely behind the other, resting his head on the back of the male seated between his legs while touching his loins. Hanuman males sometimes form DUOS—companionships in which two males live and travel together, visiting troops in other areas or sometimes settling into all-male bands. Some duos are short-lived (a month or less), others more long-lasting.

Male Nilgiri Langurs also interact erotically with each other in a number of ways, usually incorporating three types of activities: grooming, embracing, and mounting. These are not entirely separate behaviors, and they often combine with each other. Grooming is an intensely pleasurable, relaxing, and arousing experience: one or both males usually get an erection during grooming, and this may lead to ejaculation (a male may even eat his own semen afterward). In embracing, one monkey runs up to another and gives him a long, clinging hug; this is often combined with grooming and has a noticeable soothing effect. Male Nilgiri Langurs also mount each other, using the same position as for heterosexual intercourse. One male may directly approach another male and present his hindquarters to the other, or else he may go through a more stylized presentation known as REAR-END FLIRTATION, in which he slowly walks by the other male and turns his hindquarters toward him as he passes by. During same-sex mounting, one male climbs on top of the other's rump, grabbing the mounted monkey by the ankles with his feet while making pelvic thrusts (often simultaneously mouthing his back); the mountee sometimes looks or reaches back to grasp the other male. Following mounting, the monkeys often groom or embrace-groom. In cosexual troops with more than one male, two males who engage in homosexual mounting may cooperate in launching attacks against males in neighboring groups.

Intersexual Hanuman Langurs also occur: for example, anatomically male animals with female chromosomes (XX) have been documented. This is thought to be the result of a "sex change" in the chromosomal structure of the animal.

Frequency: Female homosexual mounting is a common and regular occurrence in Hanuman Langurs, comprising 37 percent of all mounting activity in some regions; each female participates in a homosexual mount roughly once every five days, usually in the morning or late afternoon. There is also considerable geographic variation in the frequency of mounting between females. Male homosexual mounting in both Hanuman and Nilgiri Langurs is less common, though it can occur frequently in Hanuman all-male bands (especially during periods of excitement). In one study, more than 40 homosexual interactions between male Hanumans were observed during three months. Among Nilgiri Langur males, nearly half of all grooming sessions are between males and over 10 percent of embracing is between males.

Orientation: All female Hanuman Langurs participate in homosexual activity, to varying degrees; most females are probably bisexual, also mating heterosexually to a greater or lesser extent. Male Hanuman Langurs are probably sequentially bisexual, since each male spends some part of his life in an all-male band, where homosexual activities typically occur—and at any given time, as much as 75–90 percent of the male population may be living in such bands. Some males are exclusively homosexual for long periods, since they may stay in these bands for more than five years—in some cases, decade-long residencies have been documented. In fact, some males never mate heterosexually, since they spend their entire adult lives in all-male bands. The highest-ranking male in a Nilgiri Langur troop engages in both homosexual and heterosexual mounting. The other males in the group, however, participate only in homosexual mounting during their stay in the troop, which can last for four or more years.

Nonreproductive and Alternative Heterosexualities

A large proportion of male Hanuman Langurs are nonbreeding: about a quarter of all males never reproduce during their lifetime, and (as noted above) the majority of the male population lives in all-male bands. Individual females occasionally go through nonbreeding periods in which menstruation may cease for months at a time. Furthermore, both males and females experience a postreproductive or "menopausal" period later in their lives: about 14 percent of the female population consists of nonbreeding older females who nevertheless are still sexually active. This period can last up to nine years, fully one-quarter of the average female's life span. Males frequently rejoin all-male bands after they have bred, where they live out the remainder of their lives (six or more years). Langurs also participate in a variety of nonprocreative sexual behaviors during their breeding prime: in Hanumans, about 8 percent of copulations occur outside of the female's fertilizable period, while sexual activity during pregnancy is common (especially during the second and third months of the six-to-seven-month gestation period). Females also occasionally mount males, while adult-juvenile heterosexual interactions also occur. Male Nilgiri Langurs often masturbate, and in some populations of this species, heterosexual mating is remarkably infrequent. In fact, the sexes often lead largely separate lives: adult males and females hardly ever interact with one another, and most social interactions take place within small subgroups consisting of monkeys of the same sex and age.

The breeding system of Hanuman Langurs is in many ways characterized by hostility and violence between the sexes and toward infants. As mentioned above, group members often harass and disrupt heterosexual matings, with the result that less than half of all copulations are completed (harassment also occurs in more than three-quarters of Nilgiri Langur copulations). Moreover, a systematic pattern of infanticide is prominent in Hanuman Langurs: males attempting to gain sexual access to females often brutally kill their infants. In some populations, infanticide accounts for as many as 30–60 percent of all infant deaths. The stress of male takeover attempts also sometimes results in abortion of fetuses, and in a few cases females even appear to induce the abortions themselves rather than have their babies subsequently killed by a male. For example, pregnant females may press and slide their bellies on the ground or allow other females to climb on or jump forcefully against them. During the raising of infants, abuse and neglect by females is also not uncommon, occurring in 12 percent of mother-infant interactions and 17 percent of "baby-sitter" interactions. Mistreatment includes abandonment; dangling, dropping, or dragging of the baby; shoving it against the ground; biting; and even kicking and throwing infants out of trees. Remarkably, infants are rarely seriously hurt as a result of such behaviors, although a few deaths (including choking) have been documented. In addition, young females from one group sometimes "kidnap" a baby from a neighboring group, keeping the infant for up to 33 hours before its mother is able to retrieve it. Occasionally the stolen baby dies as a result of the mishandling or neglect it experiences during a kidnapping.

Sources   *asterisked references discuss homosexuality/transgender*

Agoramoorthy, G., and S. M. Mohnot (1988) "Infanticide and Juvenilicide in Hanuman Langurs (*Presbytis entellus*) Around Jodhpur, India." *Human Evolution* 3:279–96.

Agoramoorthy, G., S. M. Mohnot, V. Sommer, and A. Srivastava (1988) "Abortions in Free-ranging Hanuman Langurs (*Presbytis entellus*)—a Male Induced Strategy?" *Human Evolution* 3:297–308.

Borries, C. (1997) "Infanticide in Seasonally Breeding Multimale Groups of Hanuman Langurs (*Presbytis entellus*) in Ramnagar (South Nepal)." *Behavioral Ecology and Sociobiology* 41:139–50.

*Dolhinow, P. (1978) "A Behavioral Repertoire for the Indian Langur (*Presbytis entellus*)." *Primates* 19:449–72.

*Egozcue, J. (1972) "XX Male *Presbytis entellus*? A Retrospective Study." *Folia Primatologica* 17:292–96.

*Hohmann, G. (1989) "Group Fission in Nilgiri Langurs (*Presbytis johnii*)." *International Journal of Primatology* 10:441–54.

Hohmann, G., and L. Vogl (1991) "Loud Calls of Male Nilgiri Langurs (*Presbytis johnii*): Age-, Individual-, and Population-Specific Differences." *International Journal of Primatology* 12:503–24.

Hrdy, S. B. (1978) "Allomaternal Care and Abuse of Infants Among Hanuman Langurs." In D. J. Chivers and J. Herbert, eds., *Recent Advances in Primatology*, vol. 1, pp. 169–72. London: Academic Press.

*——— (1977) *The Langurs of Abu: Male and Female Strategies of Reproduction*. Cambridge, Mass.: Harvard University Press.

Johnson, J. M. (1984) "The Function of All-Male Trouping Structure in the Nilgiri Langur, *Presbytis johnii*." In M. L. Roonwal, S. M. Mohnot, and N. S. Rathore, eds., *Current Primate Researches*, p. 397. Jodhpur, India: Jodhpur University.

*Mohnot, S. M. (1984) "Some Observations on All-Male Bands of the Hanuman Langur, Presbytis entellus." In M. L. Roonwal, S. M. Mohnot, and N. S. Rathore, eds., *Current Primate Researches*, pp. 343–59. Jodhpur, India: Jodhpur University.

——— (1980) "Intergroup Infant Kidnapping in Hanuman Langur." *Folia Primatologica* 34:259–77.

*Poirier, F. E. (1970a) "The Nilgiri Langur (*Presbytis johnii*) of South India." In L. A. Rosenblum, ed., *Primate Behavior: Developments in Field and Laboratory Research*, vol .1, pp. 251–383. New York: Academic Press.

*——— (1970b) "The Communication Matrix of the Nilgiri Langur (*Presbytis johnii*) of South India." *Folia Primatologica* 13:92–136.

——— (1969) "Behavioral Flexibility and Intertroop Variation Among Nilgiri Langurs (*Presbytis johnii*) of South India." *Folia Primatologica* 11:119–33.

Rajpurohit, L. S., V. Sommer, and S. M. Mohnot (1995) "Wanderers Between Harems and Bachelor Bands: Male Hanuman Langurs (*Presbytis entellus*) at Jodhpur in Rajasthan." *Behavior* 132:255–99.

Sommer, V. (1989a) "Sexual Harassment in Langur Monkeys (*Presbytis entellus*): Competition for Nurture, Eggs, and Sperm?" *Ethology* 80:205–17.

——— (1989b) "Infant Mistreatment in Langur Monkeys—Sociobiology Tackled from the Wrong End?" In A. E. Rasa, C. Vogel, and E. Voland, eds., *The Sociobiology of Sexual and Reproductive Strategies*, pp. 110–27. London and New York: Chapman and Hall.

*——— (1988) "Female-Female Mounting in Langurs (*Presbytis entellus*)." *International Journal of Primatology* 8:478.

Sommer, V., and L. S. Rajpurohit (1989) "Male Reproductive Success in Harem Troops of Hanuman Langurs (*Presbytis entellus*)." *International Journal of Primatology* 10:293–317.

Sommer, V., A. Srivastava, and C. Borries (1992) "Cycles, Sexuality, and Conception in Free-Ranging Langurs (*Presbytis entellus.*)" *American Journal of Primatology* 28:1–27.

*Srivastava, A., C. Borries, and V. Sommer (1991) "Homosexual Mounting in Free-Ranging Female Hanuman Langurs (*Presbytis entellus.*)" *Archives of Sexual Behavior* 20:487–516.

Tanaka, J. (1965) "Social Structure of Nilgiri Langurs." *Primates* 6:107–22.

Vogel, C. (1984) "Patterns of Infant-Transfer within Two Troops of Common Langurs (*Presbytis entellus*) Near Jodhpur: Testing Hypotheses Concerning the Benefits and Risks." In M. L. Roonwal, S. M. Mohnot, and N. S. Rathore, eds., *Current Primate Researches*, pp. 361–79. Jodhpur, India: Jodhpur University.

*Weber, I. (1973) "Tactile Communication Among Free-ranging Langurs." *American Journal of Physical Anthropology* 38:481–86.

*Weber, I., and C. Vogel (1970) "Sozialverhalten in ein- und zweigeschlechtigen Langurengruppen [Social Behavior in Unisexual and Heterosexual Langur Groups]." *Homo* 21:73–80.

*Weinrich, J. D. (1980) "Homosexual Behavior in Animals: A New Review of Observations From the Wild, and Their Relationship to Human Sexuality." In R. Forleo and W. Pasini, eds., *Medical Sexology: The Third International Congress*, pp. 288–95. Littleton, Mass.: PSG Publishing.

LEAF MONKEYS

Primate 🐒

PROBOSCIS MONKEY
Nasalis larvatus

GOLDEN MONKEY
Pygathrix roxellana

HOMOSEXUALITY	TRANSGENDER	BEHAVIORS		RANKING	OBSERVED IN
● Female	○ Intersexuality	○ Courtship	○ Pair-bonding	○ Primary	● Wild
● Male	○ Transvestism	● Affectionate	○ Parenting	○ Moderate	○ Semiwild
		● Sexual		● Incidental	● Captivity

PROBOSCIS MONKEY

IDENTIFICATION: A long-tailed monkey with a reddish orange to gray coat; males are bigger than females (weighing up to 50 pounds) and develop an enlarged, pendulous nose with age. DISTRIBUTION: Borneo; vulnerable. HABITAT: Coastal swamp forests. STUDY AREAS: Tanjung Puting National Park in Kalimantan Tengah, Indonesia.

GOLDEN MONKEY

IDENTIFICATION: A medium-sized, long-tailed monkey with a dark brown back and tail; golden orange chest, underparts, and elongated shoulder hairs; and a prominent white muzzle and blue face. DISTRIBUTION: South-central China; vulnerable. HABITAT: Mountain coniferous and bamboo forests. STUDY AREAS: Beijing Raising and Training Center for Endangered Animals, China; subspecies *R.r. roxellana*.

Social Organization

Both Proboscis and Golden Monkeys usually live in polygamous groups consisting of one male and several adult females (five on the average), along with several adolescent or juvenile females. Younger male Proboscis Monkeys sometimes join all-male troops before reaching adolescence—such groups contain males of all ages, including adults—while some male Golden Monkeys are solitary or peripheral. Cosexual groups of Proboscis Monkeys are female-centered, since female Proboscis Monkeys direct most of their behaviors—both friendly and aggressive—toward other females rather than males, and such female relationships hold the group together. In many cases females also take the initiative in directing the movement of the group, for example when leaving a sleeping tree or crossing a river. Although little else is known about the social organization of Golden Monkeys, it appears that they may also gather into huge troops—up to 600 animals—which are among the largest groupings found in any tree-living primate.

Description

Behavioral Expression: Both male and female Proboscis Monkeys participate in homosexual mounting. An adult female may mount another from behind, in a position similar to heterosexual mounting, and thrust against her. Such behavior sometimes occurs when two social groups meet each other. Unlike heterosexual copulation, lesbian mounting is usually not preceded by solicitation behavior on the part of the female being mounted (solicitation involves a special "pout" face with pursed lips, shaking of the head from side to side, and presentation of the hindquarters to the mounting animal). Male homosexual mounting—also in the front-to-back position, and with pelvic thrusting—occurs in younger animals (adolescents or juveniles) and is often a part of play-wrestling matches. It may also be interspersed with masturbation, in which the mounting male stimulates his penis with his hand. In some cases, the male being mounted may not be an entirely willing participant and will try to get away as the other male restrains him by biting gently on the neck; similar escape behavior is sometimes shown by females during heterosexual mounting. In Golden Monkeys, homosexual behavior has only been seen among females and again consists of one female mounting another.

Frequency: Same-sex mounting probably occurs only occasionally in Proboscis and Golden Monkeys. However, in Proboscis Monkeys at least, heterosexual mounting is also infrequent: in one study, only 12 mountings were observed over an entire year, and 2 of these (17 percent) were same-sex.

Orientation: Female Proboscis Monkeys that engage in same-sex mounting are probably bisexual, since they may alternate same-sex and opposite-sex activity during the same session. Golden Monkey females probably have a bisexual potential, since same-sex mounting has so far only been observed in captive situations when males are not present, and all such females engaging in same-sex mounting also mated heterosexually and raised young.

Nonreproductive and Alternative Heterosexualities

As noted above, female Proboscis Monkeys often refuse heterosexual copulation by breaking away from the male during mounting. A male may exhibit indifference to a female's solicitations as well, either ignoring her completely or snarling at her to indicate his unwillingness. Male Golden Monkeys also frequently ignore females' sexual invitations: nearly 50 percent of all female invitations fail to result in mounting by the male. Furthermore, many mounts by males do not result in ejaculation: for some Golden males, 18–97 percent of their copulations are nonejaculatory. Females, however, display an intense interest in sexual activity, frequently soliciting the male and copulating repeatedly—one female solicited a male 34 times, and was mounted 23 times, in one day. REVERSE mountings, in which the female mounts the male, are common in Golden Monkeys as well, accounting for anywhere from 3–40 percent of all heterosexual mounts. In such cases, the male usually adopts the prostrate solicitation posture typical of females inviting copulation. Heterosexual copulations in Proboscis Monkeys are sometimes also harassed or interrupted by younger animals, who try to disrupt the mating pair by climbing on the male, pulling at his nose, or making noises and distracting movements. Female Proboscis Monkeys often have sex when they are pregnant, in some cases soliciting copulations from males as late as two weeks before birth. In fact, heterosexual mating may occur more frequently during pregnancy than at any other time of the year. Golden Monkeys engage in sexual mounting outside of the mating season, and females often solicit sexual behavior while they are menstruating. As noted above, masturbation in Proboscis Monkeys may occur in same-sex contexts; similar behavior has been observed in heterosexual contexts among Golden Monkeys. In such cases, both males and females may eat the semen after ejaculation. A number of adult, sexually mature Proboscis males are nonbreeders, since they live in all-male groups: in one population, such individuals comprised 28 percent of all adult males. Solitary male Golden Monkeys are probably also nonbreeders.

Sources   *asterisked references discuss homosexuality/transgender

Clarke, A. S. (1991) "Sociosexual Behavior of Captive Sichuan Golden Monkeys (*Rhinopithecus roxellana*)." *Zoo Biology* 10:369–74.

Gorzitze, A. B. (1996) "Birth-related Behaviors in Wild Proboscis Monkeys (*Nasalis larvatus*)." *Primates* 37:75–78.

Kawabe, M., and T. Mano (1972) "Ecology and Behavior of the Wild Proboscis Monkey, *Nasalis larvatus* (Wurmb) in Sabah, Malaysia." *Primates* 13:213–27.

Poirier, F. E., and H. Hongxhin (1983) "*Macaca mulatta* and *Rhinopithecus* in China: Preliminary Research Results." *Current Anthropology* 24:387–88.

Qi, J.-F. (1988) "Observation Studies on Reproduction of Golden Monkeys in Captivity: I. Copulatory Behavior." *Acta Theriologica Sinica* 8:172–75.

*Ren, R., K. Yan, Y. Su, H. Qi, B. Liang, W. Bao, and F. B. M. de Waal (1995) "The Reproductive Behavior of Golden Monkeys in Captivity." *Primates* 36:135–43.

——— (1991) "The Reconciliation Behavior of Golden Monkeys (*Rhinopithecus roxellanae roxellanae*) in Small Breeding Groups." *Primates* 32:321–27.

Schaller, G. B. (1985) "China's Golden Treasure." *International Wildlife* 15:29–31.

*Yeager, C. P. (1990a) "Notes on the Sexual Behavior of the Proboscis Monkey." *American Journal of Primatology* 21:223–27.

——— (1990b) "Proboscis Monkey (*Nasalis larvatus*) Social Organization: Group Structure." *American Journal of Primatology* 20:95–106.

MACAQUES

Macaques

JAPANESE MACAQUE
Macaca fuscata

Primate 🐾

HOMOSEXUALITY	TRANSGENDER	BEHAVIORS		RANKING	OBSERVED IN
● Female	○ Intersexuality	● Courtship	● Pair-bonding	● Primary	● Wild
● Male	○ Transvestism	● Affectionate	○ Parenting	○ Moderate	● Semiwild
		● Sexual		○ Incidental	● Captivity

IDENTIFICATION: A medium-sized (3-foot-long) monkey with a brownish gray coat, short tail, and red facial skin. DISTRIBUTION: Japan; endangered. HABITAT: Forests, including subalpine and snow-covered terrain. STUDY AREAS: Wild: near Arashiyama, Takasakiyama (Kyushu), Shiga Heights (Jigokundai), Koshima, Miyajima, and other areas of Japan; semiwild: near Laredo, Texas; the Oregon Regional Primate Research Center; captivity: Laboratory of Behavioral Primatology, St-Hyacinthe, Quebec; Calgary Zoo; Cavriglia Park (Italy).

Social Organization

Japanese Macaques live in cosexual troops of 20–100 individuals, subdivided into smaller matrilineal groups composed of numerous related females and several unrelated males. Males usually leave their birth group on reaching maturity and may even periodically transfer to different groups, live in all-male groups, or become solitary or peripheral. Females, in contrast, generally remain in their home group for life. As a result, the group is centered around the kinship and bonding between females (and some groups have only one regular adult male member).

Description

Behavioral Expression: Female Japanese Macaques form intense, exclusive pair-bonds with each other based on mutual sexual attraction. These pairings are known as CONSORTSHIPS and are characterized by a number of distinctive affectionate, sexual, and social activities. Female partners in a consortship typically sit together, huddling or in close physical contact, and often spend long periods grooming one another between their sexual interactions. They also synchronize their movements, including traveling in tandem and persistently following one another, and may make cooing sounds toward one another. Consorts become agitated when another animal approaches or intrudes, responding with threats, shudders, vocalizing, or withdrawal. They sometimes actively separate themselves from their relatives (temporarily forgoing the time they would usually spend grooming their young or other kin), even occasionally withdrawing from the main troop as a whole. Partners in homosexual consortships often forge strong supportive "alliances" with each other, defending their partner and intervening on her behalf when she is threatened by another individual. Most such interventions do not disrupt the traditional hierarchy or rank system of the troop. However, some interventions are termed REVOLUTIONARY because partners challenge higher-ranking individuals, while others are ARBITRATING interventions, involving mediation between individuals with more ambiguous rankings. Females also actively compete with individuals of both sexes for access to other females as sexual partners, sometimes even incurring severe injuries when they challenge intruding males. Consortships typically last from a few days to several weeks during the mating season, although partners often develop a strong friendship as a result of a consortship and remain bonded throughout the year (in contrast to heterosexual consortships, which generally do not extend beyond the mating season). Females are often serially monogamous, forming several exclusive consortships sequentially during the mating season, although they generally have fewer partners than individuals engaging in heterosexual or male homosexual activities. Homosexual pairings occur among females of all ages, from adolescents to the very old, and sometimes an adult female will pair with a pubescent female. Interestingly, an incest taboo is in effect for homosexual but not heterosexual consortships: females never choose close relatives (mothers, sisters, daughters, granddaughters, first cousins) as partners, whereas brother-sister and mother-son consorts do occasionally occur. Aunts and nieces, however, do not generally recognize each other as kin in this species, and so they sometimes consort together.

As signaled by the reddening of their faces and sexual skin (which also swells), females in consortships are usually "in heat," and sexual activity is a regular and prominent feature of these homosexual pairings. Females engage in a variety of behaviors that involve genital stimulation, usually in the form of one female mounting another. Fully seven different positions are used for homosexual copulation (as well as for heterosexual mating, although with different frequencies). Most commonly, one female sits or lies on the back of the other, making pelvic thrusts and rubbing her clitoris against her partner's rump. The mountee's clitoris may be stimulated by her partner's thrusting or with her own tail. Two females also sometimes embrace each other in a face-to-face position and rub their genitals together, either lying down or "sitting in each other's lap"; these postures are more common in lesbian interactions than heterosexual ones. Other positions include the "double-foot-clasp" (one female mounts from behind, grasping her partner's ankles or thighs with her feet); rear mounts where the mounter keeps her feet on the ground; and thrusting against the partner in sideways or variable postures. During mounting, females sometimes make vocalizations such as hoarse, cackling *ko-ko-ko-ko* sounds. Females being mounted commonly reach back and grab their partner, gazing intensely into her eyes and grimacing, while the mounter clutches the fur on her partner's back—indications that both females probably experience orgasm during homosexual interactions. Consorts usually mount each other in a series of three or more consecutive mounts (as in heterosexual mating), and mounting is often reciprocal (partners exchange positions). In addition to mounting and genital rubbing, consorts occasionally suck each other's nipples. Sexual interactions are frequently initiated by the mountee, who performs a number of characteristic "courtship" behaviors to solicit or "demand" a mount from her consort. She may slap the ground while shrieking, run away from her partner and then return while presenting her hindquarters, or else display a variety of other behaviors such as head bobbing, vocalizing, arching of the back, lip quivering, intense staring, and even spasms and more aggressive pushing and grabbing. All of these are distinct from heterosexual courtship, in which the female usually invites the male to mount by slowly inching toward him in a sitting position, with mutual flipping of ears and eyebrows. A wide variety of sounds are used in both heterosexual and homosexual courtships, including cooing, whistling, warbling, squawking, chirping, barking, and squeaking—the latter is more typical of interactions between females, however.

Male Japanese Macaques also mount each other. They do not generally form homosexual consortships—indeed, some males mount up to 24 different partners—but some individuals do have certain preferred partners, usually of the same age. Males also sometimes engage in affectionate and playful activities in combination with their sexual interactions, including touching, huddling, and grooming. Unlike heterosexual and lesbian interactions, most mounts between males are single (rather than series) and are generally briefer, although they can still involve full erections, pelvic thrusting, penetration, and/or ejaculation. The double-foot-clasp is the preferred posture between males, and sitting or lying on the partner's back is rarely if ever used. Young males often make a distinctive purring or churring sound deep in the throat while mounting each other; this vocalization is not heard in heterosexual contexts. Sexual activity between males differs from male-female behavior in a number of other ways: homosexual mounting is more common outside or toward the end of the breeding season, while heterosexual mounts are more frequently interrupted by other individuals than mounts between males.

Frequency: The prevalence of homosexual activity in Japanese Macaques varies greatly between different troops, although it can be found to some degree in virtually every population. In some cases, more than a quarter of all consortships are between females and up to a third or more of all mounting episodes are homosexual; five to ten mounts between males may occur each morning in some troops. In other troops, same-sex behavior is much less frequent.

Orientation: Again, the proportion of the female population participating in homosexual activity is highly variable, ranging from 12–78 percent, averaging 43 percent (in semiwild troops). In some troops, all females that participate in same-sex consortships are bisexual, also consorting with males; however, while paired with a female they remain faithful to their partner, ignoring or rebuffing any advances made by males toward them. In other troops, though, some females are exclusively lesbian, engaging in sexual interactions only with females: in these cases, an average of 9 percent of females are homosexual, 56 percent bisexual, and 35 percent exclusively heterosexual. Among males, there is a similar variation in the proportion of individuals participating in homosexual mounting, from as low as 0–15 percent, to virtually all the males in a troop. Generally, though, males that engage in same-sex activity also participate in opposite-sex mounting. Interestingly, some of the most intense homosexual activity, involving complete mounts with ejaculation, is exhibited by males who are also the most active heterosexually.

Nonreproductive and Alternative Heterosexualities

Nonprocreative heterosexual activities are a prominent feature of Japanese Macaque life. In some populations, as many as three-quarters of all females actively seek sexual interactions while they are pregnant, half do so while menstruating, and individual females may copulate with an average of ten different males during the mating season. Most heterosexual mounts (nearly two-thirds) do not lead to ejaculation. In addition, REVERSE mounting is common, in which a female climbs on top of a male and rubs her genitals on his back. In some troops, about 40 percent of all females engage in this behavior and it occurs in about a third of all heterosexual interactions. Masturbation is also common in both males and females. Females in some troops frequently form consortships with sexually immature (preadolescent) males; as noted above, incestuous pairings also sometimes occur, and up to 15 percent of heterosexual mountings may be between related individuals.

Heterosexual mating occurs year round, but rarely leads to pregnancy when it takes place outside of the breeding season. Nor are females unique in experiencing a sexual cycle with distinct nonreproductive periods: males undergo a yearly seasonal fluctuation in their hormone levels that results in retraction of their testicles, cessation of ejaculation, and loss of color in their sexual skin during the nonmating season. In addition, approximately 10 percent of Japanese Macaques in some troops form nonbreeding or celibate heterosexual pairs: the partners specifically avoid sexual activities with each other, although they may interact sexually with other individuals. Many females also experience a long postreproductive period later in their lives, generally lasting four to five years and constituting about 16 percent of the average life span. Such individuals continue to be sexually active, copulating with males at rates comparable to those of breeding females and also interacting with the same number of female sexual partners as do younger females.

In some troops, a unique form of "baby-sitting" has developed. Although males in this species do not typically participate in parenting, high-ranking males in some populations take care of infants that are not their own for short periods. They groom, carry, embrace, and protect the infants, usually with the consent of their mothers. A few females also act as baby-sitters; however, nonbreeding females have also been known to kidnap infants, sometimes keeping them permanently. In addition, a few male caretakers interact sexually with infants (usually females), masturbating themselves while carrying them or even thrusting against them.

Sources
*asterisked references discuss homosexuality/transgender

*Chapais, B., C. Gauthier, J. Prud'homme, and P. Vasey (1997) "Relatedness Threshold for Nepotism in Japanese Macaques." *Animal Behavior* 53:1089–1101.

*Chapais, B., and C. Mignault (1991) "Homosexual Incest Avoidance Among Females in Captive Japanese Macaques." *American Journal of Primatology* 23:171–83.

*Corradino, C. (1990) "Proximity Structure in a Captive Colony of Japanese Monkeys (*Macaca fuscata fuscata*): An Application of Multidimensional Scaling." *Primates* 31:351–62.

*Eaton, G. G. (1978) "Longitudinal Studies of Sexual Behavior in the Oregon Troop of Japanese Macaques." In T. E. McGill, D. A. Dewsbury, and B. D. Sachs, eds., *Sex and Behavior: Status and Prospectus*, pp. 35–59. New York: Plenum Press.

*Enomoto, T. (1974) "The Sexual Behavior of Japanese Monkeys." *Journal of Human Evolution* 3:351–72.

*Fedigan, L. M. (1982) *Primate Paradigms: Sex Roles and Social Bonds*. Montreal: Eden Press.

*Fedigan, L. M., and H. Gouzoules (1978) "The Consort Relationship in a Troop of Japanese Monkeys." In D. Chivers, ed., *Recent Advances in Primatology*, vol. 1: pp. 493–95. London: Academic Press.

*Gouzoules, H., and R. W. Goy (1983) "Physiological and Social Influences on Mounting Behavior of Troop-Living Female Monkeys (*Macaca fuscata*)." *American Journal of Primatology* 5:39–49.

*Green, S. (1975) "Variation of Vocal Pattern with Social Situation in the Japanese Monkey (*Macaca fuscata*): A Field Study." In L. A. Rosenblum, ed., *Primate Behavior: Developments in Field and Laboratory Research*, vol. 4, pp. 1–102. New York: Academic Press.

*Hanby, J. P. (1974) "Male-Male Mounting in Japanese Monkeys (*Macaca fuscata*)." *Animal Behavior* 22:836–49.

*Hanby, J. P., and C. E. Brown (1974) "The Development of Sociosexual Behaviors in Japanese Macaques *Macaca fuscata*." *Behavior* 49:152–96.

*Hanby, J. P., L. T. Robertson, and C. H. Phoenix (1971) "The Sexual Behavior of a Confined Troop of Japanese Macaques." *Folia Primatologica* 16:123–43.

Itani, J. (1959) "Paternal Care in the Wild Japanese Monkey, *Macaca fuscata fuscata*." *Primates* 2:61–93.

*Lunardini, A. (1989) "Social Organization in a Confined Group of Japanese Macaques (*Macaca fuscata*): An Application of Correspondence Analysis." *Primates* 30:175–85.

*Rendall, D., and L. L. Taylor (1991) "Female Sexual Behavior in the Absence of Male-Male Competition in Captive Japanese Macaques (*Macaca fuscata*)." *Zoo Biology* 10:319–28.

*Sugiyama, Y. (1960) "On the Division of a Natural Troop of Japanese Monkeys at Takasakiyama." *Primates* 2:109–48.

*Takahata, Y. (1982) "The Socio-sexual Behavior of Japanese Monkeys." *Zeitschrift für Tierpsychologie* 59:89–108.

*———— (1980) "The Reproductive Biology of a Free-Ranging Troop of Japanese Monkeys." *Primates* 21:303–29.

Takahata, Y., N. Koyama, and S. Suzuki (1995) "Do the Old Aged Females Experience a Long Postreproductive Life Span?: The Cases of Japanese Macaques and Chimpanzees." *Primates* 36:169–80.

*Tartabini, A. (1978) "An Analysis of Dyadic Interactions of Male Japanese Monkeys (*Macaca fuscata fuscata*) in a Cage-Room Observation." *Primates* 19:423–36.

Tokuda, K. (1961) "A Study on the Sexual Behavior in the Japanese Monkey Troop." *Primates* 3:1–40.

*Vasey, P. L. (1998) "Female Choice and Inter-sexual Competition for Female Sexual Partners in Japanese Macaques." *Behavior* 135:1–19.

*———— (1996–98) Personal communication.

*———— (1996) "Interventions and Alliance Formation Between Female Japanese Macaques, *Macaca fuscata*, During Homosexual Consortships." *Animal Behavior* 52:539–51.

*Vasey, P. L., B. Chapais, and C. Gauthier (1998) "Mounting Interactions Between Female Japanese Macaques: Testing the Influence of Dominance and Aggression." *Ethology* 104:387–98.

*Wolfe, L. D. (1986) "Sexual Strategies of Female Japanese Macaques (*Macaca fuscata*)." *Human Evolution* 1:267–75.

*———— (1984) "Japanese Macaque Female Sexual Behavior: A Comparison of Arashiyama East and West." In M. F. Small, ed., *Female Primates: Studies by Women Primatologists*, pp. 141–58. New York: Alan R. Liss.

*———— (1979) "Behavioral Patterns of Estrous Females of the Arashiyama West Troop of Japanese Macaques (*Macaca fuscata*)." *Primates* 20:525–34.

*Wolfe, L. D., and M. J. S. Noyes (1981) "Reproductive Senescence Among Female Japanese Macaques (*Macaca fuscata fuscata*)." *Journal of Mammology* 62:698–705.

MACAQUES

RHESUS MACAQUE
Macaca mulatta

Primate 🐾

HOMOSEXUALITY	TRANSGENDER	BEHAVIORS		RANKING	OBSERVED IN
● Female	● Intersexuality	● Courtship	● Pair-bonding	● Primary	● Wild
● Male	○ Transvestism	● Affectionate	○ Parenting	○ Moderate	● Semiwild
		● Sexual		○ Incidental	● Captivity

IDENTIFICATION: A brown monkey with a pale reddish face and rump, and a medium-sized tail (up to 1 foot long). DISTRIBUTION: Afghanistan, India, southern China, northern Southeast Asia. HABITAT: Variable, including semidesert, forests, swamps. STUDY AREAS: Wild: near Dehra Dun, India; semiwild: Cayo Santiago Island, Puerto Rico; captivity: Caribbean Primate Research Center, Sabana Seca, Puerto Rico; California, Tulane, Wisconsin, and Yerkes Regional Primate Research Centers; and other locations.

Social Organization

Rhesus Macaques live in troops numbering up to 80–100 individuals. These are composed of several cosexual subgroups (averaging around 18 members) organized along matrilineal kinship. Males typically leave their subgroup during adolescence and establish themselves elsewhere—sometimes in all-male groups—leaving the female line of descent intact.

Description

Behavioral Expression: In female Rhesus Macaques, homosexual behavior usually takes place in a CONSORTSHIP, a type of pair-bond between two females that can last anywhere from a few days to many months (consortships are also characteristic of heterosexual relations). Two consorts follow and spend a great deal of time with each other, and participate in a wide variety of courtship, affectionate, and sexual activities; consort partners also sometimes cooperate in attacking other individuals. Females might consort with several other females, although most have only one partner. Lesbian courtship is highly distinctive, and involves five different playful pursuit games: "hide-and-seek," in which two females peek at each other from around a tree trunk; "kiss and run," in which one female rushes up to another and briefly kisses or nuzzles her before running off, the other in pursuit; "follow the leader," in which the females alternate positions following one another; "lipsmack and circle," where one female circles closer and closer to the other while making lipsmacking noises; and "present and run," where one female invites the other to mount her and then teasingly runs off.

▲ *Courtship between two female Rhesus Macaques: a game of "hide-and-seek"*

Sexual activity usually involves one female mounting the other, either in the position used for heterosexual mounting (with the mounter's feet on the mountee's legs), or in a position unique to homosexual mounting, in which one female climbs directly on the back of the other female and rubs her genitals on the mountee's rump. Mounting from the side also occurs. The mounter sometimes stimulates her clitoris or has her partner do it for her, during a mount. Both females show signs of orgasm: the mounter often "pauses" the way a male does at the moment of ejaculation, while the mountee often reaches back and clutches her partner. In some consortships the two females mount each other reciprocally, although in many cases one female is consistently the mounter and her partner always the mountee. Sometimes two females also participate in sexual hugging, in which they wrap their arms and legs around each other while one or both of them engage in clitoral stimulation (including rubbing their genitals on the ground). Combined with sexual stimulation, females also kiss (touching lips or tongues), caress each other's face, gently bite one another's ears, and groom each other. Pregnant females sometimes participate in homosexual mounting and consortships as well. Occasionally, females behave aggressively toward a female sexual partner, as also seen in heterosexual interactions (see below).

Male Rhesus Macaques also mount each other, sometimes within a consortship as well. Mounting can include full anal penetration and ejaculation, or else simple thrusting against the partner. Sometimes the mounted male masturbates himself or his partner, and reciprocal mounting is also common (in which the two males take turns mounting each other). In some cases, one male performs a series of mounts one after the other on his partner, as in heterosexual mating. Homosexual activity may be accompanied by grooming or play-wrestling, and the two partners in a consortship can be highly affectionate toward each other, with extended periods of touching, holding, and embracing. Mounting between brothers has also been observed, as well as with males of other species such as Crab-eating Macaques (in captivity).

Several types of intersexuality occur spontaneously in Rhesus Macaques, including hermaphrodite monkeys whose internal gonads are a combination of ovaries and testes, as well as individuals that have female external genitalia but are missing a female sex chromosome and have no ovaries.

Frequency: In wild and semiwild populations of Rhesus Macaques, anywhere from 16 percent to 47 percent of mounting is between animals of the same sex; the majority of homosexual mounting (84 percent) is between males. Within a lesbian consortship, two females may mount each other more than 200 times over six months, and one couple participated in homosexual mounting more than 1,000 times during that time.

Orientation: Depending on the population, 20–90 percent of females participate in homosexual mounting and/or consortships. The majority of these females are bisexual, since they also engage in heterosexual activity (either concurrently or during other periods of their lives), and some even alternate between same-sex and opposite-sex activities in the same day. At least some females that are bisexual nevertheless do seem to prefer homosexual activity, since they return to their female consorts even after having mated with males; a female consort may also try to "win" her female partner back whenever she is temporarily with a male. In addition, some individuals participate in lesbian consortships more frequently than others and generally have much less contact with the opposite sex. Females may also be generally more receptive to same-sex advances: in one population, for example, only 6 percent of attempted homosexual mounts were rejected, compared to 29 percent of heterosexual attempts. Many male Rhesus Macaques are bisexual as well, and individuals vary in their participation in homosexual versus heterosexual activity. As with females, though, some individuals do seem to show a "preference" for homosexuality, since males may mount each other while ignoring available females. Moreover, in a detailed study of one male homosexual consortship or sexual "friendship" in captivity, both males preferred each other's company to that of a female and chose each other as sexual partners when given a preference test (even though both were able to perform heterosexually with a female).

Nonreproductive and Alternative Heterosexualities

Rhesus Macaques are noted for their nonprocreative heterosexual behaviors. Half or more of all pregnant females engage in sexual behavior (including mating), and 12 percent of all copulations involve pregnant females. Some individual males even seem to prefer mating with females after they conceive, since nearly half of their copulations are with pregnant females. In fact, parturition itself sometimes stimulates sexual activity in attendants and onlookers, who may masturbate themselves or even mount the mother shortly after she gives birth. More than 40 percent of menstruating females also engage in sexual activity. The typical pattern for heterosexual copulation includes a large number of nonreproductive mounts, since the male may mount the female up to 100 or more times as part of each "copulation." Although penetration may occur during each mount, usually only the final mount in the series involves ejaculation. Females often initiate sexual behavior with males and commonly experience orgasm during heterosexual mating. They may also copulate with several different males—in fact, females typically mate with more consort partners than do males, to such an extent that they experience a phenomenon known as VAGINAL OVERFLOW because of the amount of sperm they receive from such multiple matings. Rhesus Macaque females also sometimes mount males—such REVERSE mountings can account for 2–6 percent of all heterosexual mountings. Males may become sexually stimulated while a female is mounting them, masturbating or ejaculating spontaneously, and she may also achieve orgasm from rubbing during the mount. Males often masturbate to ejaculation on their own, while females have been observed fondling and sucking their own nipples. Sexual activity with nonoptimal partners also occurs: males occasionally mount their mothers or sisters (incestuous activity accounts for 12–15 percent of all sexual interactions in some populations), and adult-juvenile sexuality (primarily mounting but also fellatio, including with infants) may account for more than 15 percent of all sexual activity. A wide range of interspecies sexual interactions in captivity have also been observed (including a female Rhesus soliciting copulations from a Dog).

Male Rhesus Macaques have a yearly hormonal cycle with a distinct non-breeding period. Females also commonly experience a postreproductive or "menopausal" stage later in their lives, in which they no longer breed but may still be sexually active. They may also continue to be valued members of the troop, even contributing to the care and raising of infants and juvenile monkeys. In addition, females of all ages participate in a type of "baby-sitting," in which individuals—including nonbreeding monkeys—look after infants belonging to other females (and also act as "attendants" during their births). These "aunts," as they are sometimes called (though they need not be genetically related to the mother), often protect and take care of the infants. Males (who generally do not participate in parenting) also occasionally engage in similar behavior and may even adopt orphaned infants. In some cases, though, "aunts" engage in aggressive or sexual interactions with the infants as well and may even try to "kidnap" another female's baby. Mothers are also sometimes abusive toward their own infants—shoving, biting, and stepping on the baby's head have all been observed, and one study showed that about 11 percent of infants are abused in their first two years of life. In addition, heterosexual relations are often characterized by aggression: males frequently attack and may severely wound females that they mate or consort with.

Sources   *asterisked references discuss homosexuality/transgender

*Akers, J. S., and C. H. Conaway (1979) "Female Homosexual Behavior in Macaca mulatta." Archives of Sexual Behavior 8:63–80.
*Altmann, S. A. (1962) "A Field Study of the Sociobiology of Rhesus Monkeys, Macaca mulatta." Annals of the New York Academy of Sciences 102:338–435.
*Carpenter, C. R. (1942) "Sexual Behavior of Free Ranging Rhesus Monkeys (Macaca mulatta). I. Specimens, Procedures, and Behavioral Characteristics of Estrus. II. Periodicity of Estrus, Homosexual, Autoerotic, and Non-Conformist Behavior." Journal of Comparative Psychology 33:113–62.
Conaway, C. H., and C. B. Koford (1964) "Estrous Cycles and Mating Behavior in a Free-ranging Band of Rhesus Monkeys." Journal of Mammalogy 45:577–88.

*Erwin, J., and T. Maple (1976) "Ambisexual Behavior with Male-Male Anal Penetration in Male Rhesus Monkeys." *Archives of Sexual Behavior* 5:9–14.

*Fairbanks, L. A., M. T. McGuire, and W. Kerber (1977) "Sex and Aggression During Rhesus Monkey Group Formation." *Aggressive Behavior* 3:241–49.

*Gordon, T. P., and I. S. Bernstein (1973) "Seasonal Variation in Sexual Behavior of All-Male Rhesus Troops." *American Journal of Physical Anthropology* 38:221–26.

*Hamilton, G. V. (1914) "A Study of Sexual Tendencies in Monkeys and Baboons." *Journal of Animal Behavior* 4:295–318.

*Huynen, M. C. (1997) "Homosexual Interactions in Female Rhesus Monkeys, *Macaca mulatta*." In M. Taborsky and B. Taborsky, eds., *Contributions to the XXV International Ethological Conference*, p. 211. Advances in Ethology no. 32. Berlin: Blackwell Wissenschafts-Verlag.

Kaufmann, J. H. (1965) "A Three-Year Study of Mating Behavior in a Free-Ranging Band of Rhesus Monkeys." *Ecology* 46:500–12.

*Kempf, E. J. (1917) "The Social and Sexual Behavior of Infrahuman Primates With Some Comparable Facts in Human Behavior." *Psychoanalytic Review* 4:127–54.

*Lindburg, D. G. (1971) "The Rhesus Monkey in North India: An Ecological and Behavioral Study." In L. A. Rosenblum, ed., *Primate Behavior: Developments in Field and Laboratory Research*, vol. 2, pp. 1–106. New York: Academic Press.

Loy, J. D. (1971) "Estrous Behavior of Free-Ranging Rhesus Monkeys (*Macaca mulatta*)." *Primates* 12:1–31.

———— (1970) "Peri-Menstrual Sexual Behavior Among Rhesus Monkeys." *Folia Primatologica* 13:286–97.

Michael, R. P., M. I. Wilson, and D. Zumpe (1974) "The Bisexual Behavior of Female Rhesus Monkeys." In R. C. Friedman, ed., *Sex Differences in Behavior*, pp. 399–412. New York: John Wiley & Sons.

Missakian, E. A. (1973) "Genealogical Mating Activity in Free-Ranging Groups of Rhesus Monkeys (*Macaca mulatta*) on Cayo Santiago." *Behavior* 45:225–41.

Partch, J. (1978) "The Socializing Role of Postreproductive Rhesus Macaque Females." *American Journal of Physical Anthropology* 48:425.

*Reinhardt, V., A. Reinhardt, F. B. Bercovitch, and R. W. Goy (1986) "Does Intermale Mounting Function as a Dominance Demonstration in Rhesus Monkeys?" *Folia Primatologica* 47:55–60.

Rowell, T. E., R. A. Hinde, and Y. Spencer-Booth (1964) "'Aunt'-Infant Interaction in Captive Rhesus Monkeys." *Animal Behavior* 12:219–26.

*Sade, D. S. (1968) "Inhibition of Son-Mother Mating Among Free-Ranging Rhesus Monkeys." In J. H. Masserman, ed., *Animal and Human*, pp. 18–38. Science and Psychoanalysis, vol. 12. New York: Grune & Stratton.

Schapiro, S. J., and G. Mitchell (1983) "Infant-Directed Abuse in a Seminatural Environment: Precipitating Factors." In M. Reite and N. G. Caine, eds., *Child Abuse: The Nonhuman Primate Data*, pp. 29–48. Monographs in Primatology, vol.1. New York: Alan R. Liss.

*Sullivan, D. J., and H. P. Drobeck (1966) "True Hermaphrodism in a Rhesus Monkey." *Folia Primatologica* 4:309–17.

Tilford, B. (1981) "Nondesertion of a Postreproductive Rhesus Female by Adult Male Kin." *Journal of Mammalogy* 62:638–39.

Vessey, S. H., and D. B. Meikle (1984) "Free-Living Rhesus Monkeys: Adult Male Interactions with Infants and Juveniles." In D. M. Taub, ed., *Primate Paternalism*, pp. 113–26. New York: Van Nostrand Reinhold.

*Weiss, G., R. F. Weick, E. Knobil, S. R. Wolman, and F. Gorstein (1973) "An X-O Anomaly and Ovarian Dysgenesis in a Rhesus Monkey." *Folia Primatologica* 19:24–7.

MACAQUES

Primate 🐒

STUMPTAIL MACAQUE
Macaca arctoides

HOMOSEXUALITY	TRANSGENDER	BEHAVIORS		RANKING	OBSERVED IN
● Female	○ Intersexuality	○ Courtship	● Pair-bonding	● Primary	○ Wild
● Male	○ Transvestism	● Affectionate	○ Parenting	○ Moderate	● Semiwild
		● Sexual		○ Incidental	● Captivity

IDENTIFICATION: A medium-sized (2-foot-long) monkey with dark or reddish brown fur, a short, almost hairless tail, and mottled black-and-red skin on the face. DISTRIBUTION: Southeast Asia and south-central China; vulnerable. HABITAT: Dense forests, including mountainous regions. STUDY AREAS: Semiwild: island of Totogochillo, Lake Catemaco, Mexico; captivity: Stanford University; University of Helsinki; the Netherlands Primate Center; Yerkes and Wisconsin Regional Primate Research Centers; Calcutta and Paris Zoos.

Social Organization

Few field studies of Stumptail Macaques have been conducted, so little is known of their social organization in the wild. They generally appear to live in cosexual groups of 20–50 individuals with a matrilineal organization. The mating system is probably polygamous or promiscuous, involving copulations with multiple partners and little male parental involvement.

Description

Behavioral Expression: Male Stumptails form intense sexual "friendships" with each other, within which an extraordinary variety of homosexual activities are expressed. One male may develop strong affectionate bonds with another male, as shown by their embracing, gently nibbling at each other's mouth, and huddling together. The two partners may even sleep together, one closely hugging the other from behind while holding his partner's penis. The affection between these males is also expressed through sexual activity, ranging from mounting to oral sex to mutual masturbation. Mounting occurs in the typical front-to-back position found in heterosexual mating; pelvic thrusting, anal penetration, and occasionally ejaculation can all be components of this activity. Fellatio or oral-genital activity involves one male licking or sucking on the penis of the other for up to two minutes at a time. This is done in a variety of positions, for example with one male behind the other or mounted on him, sucking his genitals between his legs. The males may even perform mutual fellatio in a 69 position. Males also fondle one another's scrotum and penis, rubbing the hand up and down the shaft; again, a number of positions are used—one male may stand sideways in front of another who is sitting down, spreading his thighs to allow the seated male to masturbate him, or they may engage in mutual masturbation by backing up toward each other and fondling each other's genitals between their legs. Sometimes male Stumptails also sit together and masturbate themselves, stimulated by the sight of one another or of a nearby heterosexual mating. The partners in a sexual friendship may

▲ *Two male Stumptail Macaques in mutual fellatio*

be of the same age, or one may be considerably younger than the other, perhaps even an infant.

Female Stumptails also form sexual friendships with one another. These relationships, which involve a considerable amount of affectionate behavior as well, may be a stable association between just two females, or they may be more fluid, involving a network of three females, for example, or more short-lived pairings. Sexual behavior involves mounting with intense genital stimulation and orgasm. One female will climb onto the back of her favorite partner, using a position slightly different from heterosexual mating that allows her to rub her genitals against the rump of the other female. This may also be accomplished in a front-to-back sitting position, one female pulling the other up against her belly, and the two may even end up lying or leaning back together in this configuration. Lesbian interactions are prolonged, lasting up to two minutes (a duration similar to heterosexual mating), and the females usually make numerous pelvic thrusts comparable to the amount

performed by a male in a heterosexual mounting. Orgasm is striking: the mounting female tenses up, first pausing and then showing a number of body spasms; her fur stands on end, and this is combined with a characteristic frowning and round-mouthed facial expression (also found in ejaculating males) and rapid breathing sounds. She also experiences a number of intense uterine contractions lasting for nearly a minute. The female being mounted does

◀ *A female Stumptail Macaque kissing her female partner at the climax of mounting* not, apparently, have the same sort of orgasmic response, although she is in a state of high sexual arousal, and will often reach back to clutch the mounting female and even kiss her during the climax. Following orgasm, the females usually hug one another and make teeth-chattering or squeaking sounds. In some cases females can apparently reach orgasm without direct genital stimulation, particularly when they are hugging their favorite female partner in great excitement after being reunited.

Frequency: Homosexual activity is common in Stumptail Macaques, accounting for as much as 25–40 percent of all sexual encounters in some captive and semi-wild groups. In one study, nearly two-thirds of this same-sex activity was between females, while in another population, all same-sex mounting was between males.

Orientation: Most Stumptail females are probably simultaneously bisexual, interspersing heterosexual activity with homosexual activity. In fact, females have been known to participate in lesbian activities when they are lactating (and possibly even during pregnancy), indicating an easy compatibility between motherhood and homosexuality in this species. Some males show a decrease in the intensity of their same-sex friendships as they get older and may begin engaging in a larger proportion of heterosexual activity as they mature. Nevertheless, most males probably continue to have some homosexual contact throughout their lives.

Nonreproductive and Alternative Heterosexualities

Male Stumptails engage in both masturbation and nonreproductive heterosexual mating. In the latter case, males have been observed mounting females without full penetration, mating with menstruating females, rubbing their genitals on the female (sometimes to ejaculation), and even stimulating the penis with their own foot while mounted on the female. Both male and female orgasm can be a component of heterosexual mating. Sometimes, however, mating is distinctly less pleasurable—especially for the female, who may collapse under the weight of the male during copulation (he may be up to twice as heavy as she is). Females are also sometimes bitten by males during mounting, resulting in shallow cuts on her shoulders and upper arms (this occurs in about 15–18 percent of copulations). More than half of all matings involve male aggression (including chasing the female, pushing or pulling her, and fighting or biting her) and/or female resistance (including running away from the male, screaming, trying to dislodge him, and fighting with him). In addition, heterosexual copulations are often harassed by other individuals (of both sexes), sometimes in spectacular outbursts of activity that involve an entire social group. This often occurs following ejaculation while the two mating animals remain attached in a "copulatory tie," similar to that of mating Dogs.

Sources   *asterisked references discuss homosexuality/transgender

*Bernstein, I. S. (1980) "Activity Patterns in a Stumptail Macaque Group (*Macaca arctoides*)." *Folia Primatologica* 33:20–45.
*Bertrand, M. (1969) *The Behavioral Repertoire of the Stumptail Macaque: A Descriptive and Comparative Study.* Bibliotheca Primatologica 11. Basel: S. Karger.

*Chevalier-Skolnikoff, S. (1976) "Homosexual Behavior in a Laboratory Group of Stumptail Monkeys (*Macaca arctoides*): Forms, Contexts, and Possible Social Functions." *Archives of Sexual Behavior* 5:511–27.

*——— (1974) "Male-Female, Female-Female, and Male-Male Sexual Behavior in the Stumptail Monkey, with Special Attention to the Female Orgasm." *Archives of Sexual Behavior* 3:95–116.

*Estrada, A., and R. Estrada (1978) "Changes in Social Structure and Interactions After the Introduction of a Second Group in a Free-ranging Troop of Stumptail Macaques (*Macaca arctoides*): Social Relations II." *Primates* 19:665–80.

*Estrada, A., R. Estrada, and F. Ervin (1977) "Establishment of a Free-ranging Colony of Stumptail Macaques (*Macaca arctoides*): Social Relations I." *Primates* 18:647–76.

*Goldfoot, D. A., H. Westerborg-van Loon, W. Groeneveld, and A. K. Slob (1980) "Behavioral and Physiological Evidence of Sexual Climax in the Female Stump-tailed Macaque (*Macaca arctoides*)." *Science* 208:1477–79.

Gouzoules, H. (1974) "Harassment of Sexual Behavior in the Stumptail Macaque, *Macaca arctoides*." *Folia Primatologica* 22:208–17.

*Leinonen, L., I. Linnankoski, M.-L. Laakso, and R. Aulanko (1991) "Vocal Communication Between Species: Man and Macaque." *Language and Communication* 11:241–62.

*Linnankoski, I., and L. M. Leinonen (1985) "Compatibility of Male and Female Sexual Behavior in *Macaca arctoides*." *Zeitschrift für Tierpsychologie* 70:115–22.

Niemeyer, C. L., and A. S. Chamove (1983) "Motivation of Harassment of Matings in Stumptailed Macaques." *Behavior* 87:298–323.

*O'Keefe, R. T., and K. Lifshitz (1985) "A Behavioral Profile for Stumptail Macaques (*Macaca arctoides*)." *Primates* 26:143–60.

*Slob, A. K., and P. E. Schenk (1986) "Heterosexual Experience and Isosexual Behavior in Laboratory-Housed Male Stump-tailed Macaques (*M. arctoides*)." *Archives of Sexual Behavior* 15:261–68.

*de Waal, F. B. M. (1989) *Peacemaking Among Primates*. Cambridge, Mass.: Harvard University Press.

de Waal, F. B. M., and R. Ren (1988) "Comparison of the Reconciliation Behavior of Stumptail and Rhesus Macaques." *Ethology* 78:129–42.

MACAQUES   *Primate*

BONNET MACAQUE
Macaca radiata
CRAB-EATING MACAQUE
Macaca fascicularis

HOMOSEXUALITY	TRANSGENDER	BEHAVIORS		RANKING	OBSERVED IN
● Female	○ Intersexuality	○ Courtship	● Pair-bonding	○ Primary	● Wild
● Male	○ Transvestism	● Affectionate	○ Parenting	● Moderate	○ Semiwild
		● Sexual		○ Incidental	● Captivity

BONNET MACAQUE
IDENTIFICATION: A grayish brown monkey with a circular "cap" of hair on the head, a prominent wrinkled brow and forehead, and a long tail (over 2 feet in males). DISTRIBUTION: Southern India. HABITAT: Forests, scrub, open areas. STUDY AREAS: Near Somanathapur Sandal Reserve and Byrankuppe (Mysore State), Dharwar, Karnataka (Tamil Nadu), and Lal Bagh (Bangalore), India; California Primate Research Center; State University of New York; subspecies *M.r. diluta*.

CRAB-EATING MACAQUE
IDENTIFICATION: A gray-green to reddish brown monkey with a slight pointed crest, pinkish face, and long tail. DISTRIBUTION: Southeast Asia including Indonesia, Philippines, Nicobar Islands, and introduced to Palau. HABITAT: Forests, swamps. STUDY AREAS: Angaur Island, Palau, Micronesia; Yerkes Regional Primate Research Center; University of California—Berkeley.

Social Organization

Both Bonnet and Crab-eating Macaques live in fairly large matriarchal groups containing numerous adult males and females as well as youngsters; males typically emigrate from their home group on becoming adults. Bonnet groups can be as large as 50–60 monkeys, but most average around 18–20 individuals, with four or five each of adult males and females. Male Bonnets demonstrate a strong tendency to interact and cooperate with one another, often forming supportive COALITIONS together. Crab-eating Macaques live in troops containing 40–50 individuals on average; smaller groups each contain 2–9 adult males. Large subgroups of youngsters, as well as some peripheral or solitary animals, also occur.

Description

Behavioral Expression: Male Bonnet Macaques frequently mount one another, using the same front-to-back position found in heterosexual copulation. A male may have anywhere from two to five different partners that he mounts; each male also varies in the proportion of times he mounts or is mounted. One male acted as the mountee in only 9 percent of his homosexual mountings, another did so in all of his mountings, though the average is a roughly equal proportion of mounter-mountee behavior, and reciprocal mounting occurs as well. In addition, male Bonnets engage in a wide variety of other same-sex behaviors, both affectionate and sexual, often within a coalition "bond" between them. Masturbation of another male is common in all age groups, especially younger males—one male holds or fondles the other's penis and may even eat the semen from the resulting ejaculation (mutual masturbation also occurs). Males also sometimes grip and gently tug on each other's scrotum; often, this is accompanied by embracing, nuzzling, grasping of the rump, tongue-clicking, and mouthing of the other's neck or shoulders, all combined into a sort of ritualized "greeting" interaction. Another behavior, unique to homosexual interactions, involves two males rhythmically rubbing their rumps and genital areas together, often reaching back between their legs to fondle each other's genitals. This behavior also occurs between females, as does mounting.

Homosexual mounting also occurs in male Crab-eating Macaques. In addition, male Crab-eaters may mouth and fondle the genitals and anal region of another male, including using their index finger to investigate the area. Males can also develop intense sexual friendships with one another, especially between older and younger males. In one such pair observed in captivity, affectionate embraces frequently led to sexual arousal and homosexual mounting, often accompanied by excited lip-smacking or crooning sounds; the male being mounted sometimes even turned his head to kiss his partner during a mount. Both consensual and nonconsensual mounting occurs in Crab-eating Macaques—in the former (54 percent of mounts between males), the mounted animal fully cooperates by standing still and helps support the weight of the other male (and perhaps even initiates the encounter). In nonconsensual mounts (46 percent of mounts between males), the mounting animal may corner his partner and hold him down (this also occurs in heterosexual mounting). Male Crab-eaters also occasionally engage in homosexual contact with other species. Wild Crab-eating Macaques sometimes allow male Orang-utans to perform fellatio on them, while in captivity they have been known to attempt copulation with males of a number of nonprimate species, including foxes.

An older male Bonnet Macaque in India mounting a younger male ▶

Frequency: Homosexual mounting is very common in male Bonnet Macaques. In some populations, same-sex mounts exceed the number of heterosexual mounts by as much as four to one, and mounting between males can comprise 31 percent to 79 percent of all mounting. Sexual and affectionate behaviors between males occur in about a quarter of their interactions with one another. Female homosexual activity is somewhat less common: one study found that mounting between females occurred at rates that were two to seven times less than male-female or male-male mounting, although the rates of mutual rump rubbing with genital stimulation between females were only slightly less than between males. In Crab-eating Macaques, homosexual mounting accounts for 17–30 percent of mountings, and 10 percent of all interactions between males involve mounting (compared to nearly 50 percent of all interactions between males and females).

Orientation: Nearly all male Bonnet Macaques participate in both homosexual and heterosexual mounting, but it appears that they generally have more different male partners than female ones. Some males participate very little in homosexual activity (accounting for about 10 percent of the same-sex mountings in a population), while others may be involved in more than half of all homosexual mounts. A similar variation occurs with respect to their heterosexual participation. However, males that are the least active heterosexually are not necessarily the most active homosexually: in many cases, those males that participate in a large number of homosexual mountings also have a large number of heterosexual mountings. No quantitative information on sexual orientation is available for Crab-eaters; however, observations in captivity indicate that males engage in both homosexual and heterosexual copulation, sometimes alternating relatively frequently between the two. Furthermore, males that are "bonded" to one another show a preference for their partner that even survives separation and intervening heterosexual activity. If such males are separated from one another, they rush to embrace each other on being reunited, resuming their sexual relationship where they left off.

Nonreproductive and Alternative Heterosexualities

As noted above, heterosexual matings in Crab-eating Macaques are not always consensual: about 19 percent of such mounts are forced by the male on the female. Furthermore, male Crab-eaters sometimes severely attack females with small babies and also occasionally kill those infants to gain sexual and breeding access to the female. Both Bonnets and Crab-eaters participate in a range of nonprocreative sexual behaviors. Among Crab-eaters, females copulate during pregnancy (though usually not during the first two to three weeks), while more than half of male-female copulations do not involve ejaculation. In both species, females have multiple male partners and copulations—during one six-month period, for example, each female Crab-eater mated an average of 45 times, with some copulating more than 110 times. Female Bonnets may mate with up to three different males in succession, and females also sometimes mount males (REVERSE mounting). Male Bonnet Macaques may also manually penetrate females, inserting a finger and then licking or smelling it. Unlike in a number of other primates, in Bonnets this behavior does not appear to be simply a way of testing the sexual receptivity of the female. Both male and female Bonnets also masturbate, females sometimes employing innovative techniques (such as using objects or pulling their tail between their legs and using it to rub their labia while making pelvic thrusting movements). Crab-eating Macaques engage in an interesting form of "infidelity": heterosexual mating typically occurs within a short-term bond or consortship. However, nearly 20 percent of all copulations in some populations are nonmonogamous: half of all females and almost three-quarters of the males "steal" matings with other partners during a consortship, but return to their original partner afterward. The social system of Bonnet Macaques may entail considerable inbreeding, and incestuous mother-son matings that produce viable offspring do occur.

Sources
asterisked references discuss homosexuality/transgender

*Bernstein, S. (1970) "Primate Status Hierarchies." In L. A. Rosenblum, ed., *Primate Behavior: Developments in Field and Laboratory Research*, vol.1, pp. 71–109. New York: Academic Press.

Emory, G. R., and S. J. Harris (1978) "On the Directional Orientation of Female Presents in *Macaca fascicularis*." *Primates* 19:227–29.

*Hamilton, G. V. (1914) "A Study of Sexual Tendencies in Monkeys and Baboons." *Journal of Animal Behavior* 4:295–318.

*Kaufman, I. C., and L. A. Rosenblum (1966) "A Behavioral Taxonomy for *Macaca nemestrina* and *Macaca radiata*: Based on Longitudinal Observation of Family Groups in the Laboratory." *Primates* 7:205–58.

*Makwana, S. C. (1980) "Observations on Population and Behavior of the Bonnet Monkey, *Macaca radiata*." *Comparative Physiology and Ecology* 5:9–12.

Moore, J., and R. Ali (1984) "Are Dispersal and Inbreeding Avoidance Related?" *Animal Behavior* 32:94–112.

*Nolte, A. (1955) "Field Observations on the Daily Routine and Social Behavior of Common Indian Monkeys, with Special Reference to the Bonnet Monkey (*Macaca radiata* Geoffroy)." *Journal of the Bombay Natural History Society* 53:177–84.

Noordwijk, M. A. van (1985) "Sexual Behavior of Sumatran Long-tailed Macaques (*Macaca fascicularis*)." *Zeitschrift für Tierpsychologie* 70:277–96.

*Poirier, F. E., and E. O. Smith (1974) "The Crab-Eating Macaques (*Macaca fascicularis*) of Angaur Island, Palau, Micronesia." *Folia Primatologica* 22:258–306.

Rahaman, H., and M. D. Parthasarathy (1969) "Studies on the Social Behavior of Bonnet Monkeys." *Primates* 10:149–62.

*——— (1968) "The Expressive Movements of the Bonnet Macaque." *Primates* 9:259–72.

*Rasmussen, D. R. (1984) "Functional Alterations in the Social Organization of Bonnet Macaques (*Macaca radiata*) Induced by Ovariectomy: An Experimental Analysis." *Psychoneuroendocrinology* 9:343–74.

*Silk, J. B. (1994) "Social Relationships of Male Bonnet Macaques: Male Bonding in a Matrilineal Society." *Behavior* 130:271–92.

——— (1993) "Does Participation in Coalitions Influence Dominance Relationships Among Male Bonnet Macaques?" *Behavior* 126:171–89.

Sinha, A. (1997) "Complex Tool Manufacture by a Wild Bonnet Macaque, *Macaca radiata*." *Folia Primatologica* 68:23–25.

*Simonds, P. E. (1996) Personal communication.

*——— (1965) "The Bonnet Macaque in South India." In I. DeVore, ed., *Primate Behavior: Field Studies of Monkeys and Apes*, pp. 175–96. New York: Holt, Rinehart, & Winston.

*Sugiyama, Y. (1971) "Characteristics of the Social Life of Bonnet Macaques (*Macaca radiata*)." *Primates* 12:247–66.

*Thompson, N. S. (1969) "The Motivations Underlying Social Structure in *Macaca irus*." *Animal Behavior* 17:459–67.

*——— (1967) "Some Variables Affecting the Behavior of Irus Macaques in Dyadic Encounters." *Animal Behavior* 15:307–11.

MACAQUES   *Primate* 🐾

PIG-TAILED MACAQUE
Macaca nemestrina

CRESTED BLACK MACAQUE
Macaca nigra

HOMOSEXUALITY	TRANSGENDER	BEHAVIORS		RANKING	OBSERVED IN
● Female	○ Intersexuality	○ Courtship	○ Pair-bonding	○ Primary	● Wild
● Male	○ Transvestism	● Affectionate	○ Parenting	● Moderate	○ Semiwild
		● Sexual		○ Incidental	● Captivity

PIG-TAILED MACAQUE
IDENTIFICATION: A medium-sized monkey (up to 30 pounds in males) with olive-brown fur and a short, curly, nearly naked tail. DISTRIBUTION: Southeast Asia from Burma to Sumatra; vulnerable. HABITAT: Forests. STUDY AREAS: Near Mt. Kerinci, West Sumatra, Indonesia, subspecies *M.n. nemestrina;* Bernam River, western Malaysia; Washington and Yerkes Regional Primate Research Centers; State University of New York; Turin Zoo.

CRESTED BLACK MACAQUE
IDENTIFICATION: An entirely black monkey with a striking crest, long muzzle, high cheek and brow ridges, and a short tail. DISTRIBUTION: Sulawesi, Indonesia; endangered. HABITAT: Tropical forest. STUDY AREAS: Tangkoko-DuaSudara Nature Reserve, North Sulawesi, Indonesia; Oregon and Yerkes Regional Primate Research Centers; Woodland Park Zoological Gardens, Seattle, Washington.

Social Organization

Both of these species live in cosexual groups, containing 15–40 individuals in Pigtails and up to 40–90 individuals in Crested Black Macaques. Pig-tailed groups are matrilineal clans, in which females remain in their home (natal) group while males emigrate on reaching maturity. The heterosexual mating system is promiscuous: both sexes mate with multiple partners. Smaller Crested Black Macaque groups (6–15 individuals) may contain only one adult male.

Description

Behavioral Expression: Pig-tailed Macaques engage in homosexual mounting as well as kissing. Male Pigtails mount each other using the posture found in heterosexual mating (one male behind the other, hands grasping the loins, and feet clasping the calves of the other), sometimes with erection of the penis and pelvic thrusting (though anal penetration does not occur). Females use the same position and also occasionally thrust against their female partner; the mounting female is usually in heat. Some Pigtails engage in same-sex mounting with only one partner, while other Pigtails have several partners (as many as seven for females, though the average is three). Individuals differ as to whether they prefer mounting or being mounted in homosexual encounters: a few animals engage in only one or the other, while most exhibit a range of mounter/mountee behavior. Males may even mount each other reciprocally, exchanging positions in different mounting sessions. Same-sex kissing (mouth-to-mouth contact) occurs more often than heterosexual kissing and is most common between females. However, many female homosexual mounts are "forced" in the sense that the animal being mounted does not solicit the mounting, and aggression may also be involved: 48 percent of mounts between females are nonconsensual, compared to 18 percent of heterosexual ones. No mounts between males are forced. Some same-sex mounting is incestuous, e.g., between mother and daughter or siblings of either sex.

Crested Black Macaques also participate in male and female homosexual mounting, similar in many ways to that found in Pigtails. Females often reach back to clasp the leg of a female mounting them, which is believed to be a sign of orgasm (it also occurs in heterosexual mounting), while males being mounted by other males frequently fondle their own penis. Mounting is sometimes preceded by a gesture of invitation known as REAR-END FLIRTATION, in which one male walks by another and presents his hindquarters. Younger males often mount older ones in this species. A number of other homosexual activities occur in Crested Black Macaques. Females engage in a form of mutual masturbation unique to homosexual interactions, in which they stand side by side facing in opposite directions (draping an arm over the other's waist) and fondle and sniff each other's vulva, sometimes with direct clitoral stimulation. Males often participate in a form of erotic grooming: one male uses his hands, lips, and tongue to groom the other, who usually has an erection and may masturbate himself by rolling his penis between his palms or licking it (he usually eats his own semen if he ejaculates). The grooming animal is often sexually aroused, too, as evidenced by his erection. Males (especially younger ones) also use a number of ritualized erotic "greeting" gestures with one another, including embracing, face-licking or kissing, fondling or grabbing of the erect penis, mounting, and rump fingering.

Frequency: Same-sex mounting occurs frequently in Pig-tailed Macaques, accounting for 7–23 percent of all mounting activity. More than three-quarters of all kissing occurs between females. In Crested Black Macaques, about 5–8 percent of all mounting activity occurs between males. Ritualized penis-grabbing or fondling between males takes place regularly

Two female Crested Black Macaques masturbating each other ▶

in this species and can be observed weekly, and perhaps even daily, in some wild populations.

Orientation: Most Pig-tailed Macaques are probably bisexual, engaging in both same- and opposite-sex mounting. However, individuals vary along a continuum of what percentage of their activity is homosexual: for some males, it is as little as 8 percent, while for others, nearly two-thirds of their mounting activity is same-sex. Although less specific information is available for Crested Black Macaques, they appear to have a similar orientation profile: in one wild troop, for example, all males participated in both heterosexual and homosexual mounting (as well as penis-grabbing) to varying degrees.

Nonreproductive and Alternative Heterosexualities

As noted above, some heterosexual copulations in Pig-tailed Macaques are forced, in that the female is an unwilling participant. Furthermore, more than a third of aggressive interactions in Pigtails are between males and females (73 percent are directed by the male against the female). Infanticide has also been seen among captive Pigtails—in some cases, infants as young as one day old have died from head and neck injuries inflicted by adult males. In addition, a 33–year study of this species (spanning seven generations and nearly 400 individuals) found that one in eight infants is physically abused or neglected by its mother. This includes being dragged across the ground, having its fingers or tail chewed, suffering severe eye damage or blindness as a result of compulsive grooming around the eyes, having its head or body crushed on the ground, and/or being rejected, abandoned, or starved by its mother. Physical abuse accounts for about a third of all Pigtail infant injuries or deaths; it appears to run in families and is usually repeated with successive offspring. Infants can also be injured when they are kidnapped, which occurs only occasionally and is typically carried out by a nonbreeding female.

A number of nonreproductive sexual behaviors occur in these two species as well. In Pigtails, males mount nonovulating females 8–15 percent of the time, and 1–2 percent of heterosexual behavior involves females mounting males (REVERSE mounts). In addition, female Pigtails may mate with up to five different males during a single period of heat. Mother-son mountings occur as well. In both of these species, heterosexual mounting sometimes does not involve penetration: nearly a fifth of Crested Black male-female mounts, for example, are "ritualized" or noncopulatory. Male Pigtails and Crested Black Macaques also masturbate, occasionally eating their own semen, while female Crested Black Macaques sometimes masturbate by inserting a finger into the vagina and simultaneously slapping their rump with one hand. Infant and very young male Crested Blacks often mount adult females, performing pelvic thrusts and even achieving penetration. Finally, spontaneous abortions occur among Pigtails, often associated with a number of physiological changes in the female's blood chemistry: one study in captivity found that 14 percent of pregnancies terminated in abortions (among females not otherwise at high risk for miscarriages).

Other Species

Homosexual mounting occurs in three other species of Macaques: Lion-tailed (*Macaca silenus*), Tonkean (*Macaca tonkeana*), and Moor (*Macaca maurus*) Macaques. In the latter two species, 11–13 percent of mounting activity is same-sex.

Sources   *asterisked references discuss homosexuality/transgender*

*Bernstein, I. S. (1972) "Daily Activity Cycles and Weather Influences on a Pigtail Monkey Group." *Folia Primatologica* 18:390–415.

*———— (1970) "Primate Status Hierarchies." In L. A. Rosenblum, ed., *Primate Behavior: Developments in Field and Laboratory Research,* vol.1, pp. 71–109. New York: Academic Press.

*———— (1967) "A Field Study of the Pigtail Macaque (*Macaca nemestrina*)." *Primates* 8:217–28.

Bernstein, I. S., and S. C. Baker (1988) "Activity Patterns in a Captive Group of Celebes Black Apes (*Macaca nigra*)." *Folia Primatologica* 51:61–75.

*Bound, V., H. Shewman, and J. Sievert (1988) "The Successful Introduction of Five Male Lion-tailed Macaques (*Macaca silenus*) at Woodland Park Zoo." In *AAZPA Regional Conference Proceedings,* pp. 122–31. Wheeling, W.Va.: American Association of Zoological Parks and Aquariums.

*Caldecott, J. O. (1986) *An Ecological and Behavioral Study of the Pig-Tailed Macaque.* Basel: Karger.

*Dixson, A. F. (1977) "Observations of the Displays, Menstrual Cycles, and Sexual Behavior of the 'Black Ape' of Celebes (*Macaca nigra*)." *Journal of Zoology, London* 182:63–84.

*Giacoma, C., and P. Messeri (1992) "Attributes and Validity of Dominance Hierarchy in the Female Pigtail Macaque." *Primates* 33:181–89.

*Kaufman, I. C., and L. A. Rosenblum (1966) "A Behavioral Taxonomy for *Macaca nemestrina* and *Macaca radiata*: Based on Longitudinal Observation of Family Groups in the Laboratory." *Primates* 7:205–58.

Kyes, R. C., R. E. Rumawas, E. Sulistiawati, and N. Budiarsa (1995) "Infanticide in a Captive Group of Pig-tailed Macaques (*Macaca nemestrina*)." *American Journal of Primatology* 36:135–36.

Maestripieri, D., K. Wallen, and K. A. Carroll (1997) "Infant Abuse Runs in Families of Group-Living Pigtail Macaques." *Child Abuse & Neglect* 21:465–71.

*Matsumura, S., and K. Okamoto (1998) "Frequent Harassment of Mounting After a Takeover of a Group of Moor Macaques (*Macaca maurus*)." *Primates* 39:225–30.

*Nickelson, S. A., and J. S. Lockard (1978) "Ethogram of Celebes Monkeys (*Macaca nigra*) in Two Captive Habitats." *Primates* 19:437–47.

Oi, T. (1996) "Sexual Behavior and Mating System of the Wild Pig-tailed Macaque in West Sumatra." In J. E. Fa and D. G. Lindburg, eds., *Evolution and Ecology of Macaque Societies,* pp. 342–68. Cambridge: Cambridge University Press.

*———— (1991) "Non-copulatory Mounting of Wild Pig-tailed Macaques (*Macaca nemestrina nemestrina*) in West Sumatra, Indonesia." In A. Ehara, T. Kimura, O. Takenaka, and M. Iwamoto, eds., *Primatology Today,* pp. 147–50. Amsterdam: Elsevier Science Publishers.

*———— (1990a) "Patterns of Dominance and Affiliation in Wild Pig-tailed Macaques (*Macaca nemestrina nemestrina*) in West Sumatra." *International Journal of Primatology* 11:339–55.

———— (1990b) "Population Organization of Wild Pig-tailed Macaques (*Macaca nemestrina nemestrina*) in West Sumatra." *Primates* 31:15–31.

*Poirier, F. E. (1964) "The Communicative Matrix of the Celebes Ape (*Cynopithecus niger*): A Study of Sixteen Male Laboratory Animals." Master's thesis, University of Oregon.

*Reed, C. (1997) Personal communication.

*Reed, C., T. G. O'Brien, and M. F. Kinnaird (1997) "Male Social Behavior and Dominance Hierarchy in the Sulawesi Crested Black Macaque (*Macaca nigra*)." *International Journal of Primatology* 18:247–60.

Schiller, H. S., G. P. Sackett, W. T. Frederickson, and L. J. Risler (1983) "Maintenance of High-density Lipoprotein Blood Levels Prior to Spontaneous Abortion in Pig-tailed Macaques (*Macaca nemestrina*)." *American Journal of Primatology* 4:127–33.

*Skinner, S. W., and J. S. Lochard (1979) "An Ethogram of the Liontail Macaque (*Macaca silenus*) in Captivity." *Applied Animal Ethology* 5:241–53.

*Thierry, B. (1986) "Affiliative Interference in Mounts in a Group of Tonkean Macaques (*Macaca tonkeana*)." *American Journal of Primatology* 11:89–97.

*Tokuda, K., R. C. Simons, and G. D. Jensen (1968) "Sexual Behavior in a Captive Group of Pigtailed Monkeys (*Macaca nemestrina*)." *Primates* 9:283–94.

OTHER PRIMATES

BABOONS

SAVANNA BABOON
Papio cynocephalus

HAMADRYAS BABOON
Papio hamadryas

GELADA BABOON
Theropithecus gelada

Primate

HOMOSEXUALITY	TRANSGENDER	BEHAVIORS		RANKING	OBSERVED IN
● Female	● Intersexuality	○ Courtship	● Pair-bonding	○ Primary	● Wild
● Male	○ Transvestism	● Affectionate	○ Parenting	● Moderate	○ Semiwild
		● Sexual		○ Incidental	● Captivity

SAVANNA BABOON

IDENTIFICATION: The familiar baboon, with variable coat color (greenish to yellowish brown to grayish black), doglike head with a black face, and long tail (over 2 feet in males). DISTRIBUTION: Equatorial, eastern, and southern Africa. HABITAT: Scrub, savanna, woodland. STUDY AREAS: Gombe Stream National Park, Tanzania; Ishasha Forest and Queen Elizabeth National Park, Uganda; Amboseli National Park, and near Gilgil and the Athi River, Kenya; Cape of Good Hope Nature Reserve, South Africa; Namibia; subspecies *P.c. anubis*, the Olive Baboon; *P.c. ursinus*, the Chacma Baboon; and *P.c. cynocephalus*, the Yellow Baboon.

HAMADRYAS BABOON

IDENTIFICATION: A gray baboon with a striking silver-gray "cape" or shoulder mane in adult males. DISTRIBUTION: Somalia, Ethiopia, southern Saudi Arabia, Yemen. HABITAT: Semidesert, steppe, savanna woodlands, rocky terrain. STUDY AREAS: Erer-Gota region, eastern Ethiopia; Brookfield (Illinois) and London Zoos.

GELADA BABOON

IDENTIFICATION: A brown baboon with a thick "cape" of fur in adult males; both sexes have an hourglass-shaped patch of bare skin on the chest, encircled by fleshy "beads" in estrous females. DISTRIBUTION: Northern and central Ethiopia. HABITAT: Mountain grasslands, rocky gorges. STUDY AREAS: Simien Mountain National Park, Ethiopia; Yerkes Regional Primate Research Center, Georgia; San Antonio Zoo, Texas.

Social Organization

Savanna Baboons live in groups of 30–100 containing both adult males and females. Females form the matriarchal core of each group since they remain for life, whereas males often emigrate to a new group on reaching adulthood. However, some troops are strongly inbred because individuals rarely leave. In contrast, both Gelada and Hamadryas Baboons live in large troops that include so-called HAREM groups—bands that have a single male and several females. In Geladas, the primary social bonds are between the females in such groups (most of whom are related to each other, as in Savanna Baboons), whereas in Hamadryas Baboons the primary bonds are between the male and the females. Unlike many other primates, Hamadryas females emigrate from the group while males remain (hence, most of the females in a group are not related to each other). Geladas also have "bachelor" troops of nonbreeding males, and "bachelor" Hamadryas or Gelada males sometimes associate with a harem group and may develop a close relationship with its male leader.

Description

Behavioral Expression: Homosexual mounting among both males and females occurs in all three of these Baboon species. The position used is similar to that for heterosexual copulation: the mounting animal places its hands on the mountee's lower back and clasps the mountee's ankles or thighs with its feet. Both sexes often make pelvic thrusts during homosexual mounts; males usually have an erection, and ejaculation does occur in at least some Savanna Baboon mounts between males. Male Savanna Baboons also sometimes fondle their own or their partner's genitals during a same-sex mount, and male mounters may gently bite or nuzzle

their partner's neck after a homosexual mount. Male Geladas have also been observed masturbating other males. Females of these species mount each other both during "heat" and outside of their sexual cycle; about 9 percent of lesbian mounts in Savanna Baboons involve a pregnant female mounting another female. Because most Savanna and Gelada females are related to the other females in a troop, at least some homosexual activity is incestuous.

In Savanna Baboons, homosexual mounting occurs in a variety of contexts, including during good-natured play-fighting. However, in this species (and to some extent also in the other species) same-sex mounting is most prominent as

Mounting between two male Savanna Baboons in Tanzania ▶

▲ *A male Savanna Baboon in southern Africa nuzzling the genitals of another male*

part of a unique form of male "greeting" interaction. Whenever two males meet each other, they exchange a series of ritualized sexual behaviors that may include homosexual mounting and invitations to mount, as well as a wide variety of other sexual and affectionate contacts. One such behavior is known as DIDDLING, in which the males fondle each other's genitals, including touching or pulling on the penis and fondling the scrotum. Males also embrace and kiss each other on the head or mouth, and may even bend down to kiss, lick, or nibble on another male's penis or nuzzle his groin and thighs. Sometimes a male will also nuzzle another male's back with his nose, especially during a mount. Another "greeting" behavior involves one male patting, grabbing, and sometimes even nuzzling or fingering another male's rear end. Although most "greeting" interactions are relatively brief and one-sided, occasionally two males develop a closer, bonded relationship sometimes known as a COALITION, in which the "greeting" interactions are more extensive, frequent, and reciprocated. Partners in such an alliance take turns exchanging sexual behaviors with each other (especially diddling), and they often protect and help one another. Such coalitions between males may become stable, long-lasting associations that persist for many years.

Several different kinds of intersexual or hermaphrodite individuals occur spontaneously in Savanna Baboons. In South African populations, gender-mixing individuals sometimes become high-ranking and powerful members of their troops. Such animals have female genitalia and internal organs, yet their mammary glands are not developed. Genetically, they are male (having XY chromosomes), physically they are large—exceeding the proportions of nonintersexed females and sometimes even males—and sexually, they often interact with males. Another type of intersexuality found in Baboons involves animals that have male external genitals (including testes) combined with some female internal reproductive organs (such as a uterus and fallopian tubes).

Frequency: Among Savanna Baboons, homosexual activity is common: 13–24 percent of all mounting behavior is between males and up to 9 percent is between females. "Greetings" interactions—which include ritualized homosexual activities—occur more than twice as often as any other form of interaction between males and take place roughly once every 50 minutes in some troops. Approximately 10 percent of males in some areas develop closely bonded coalitions, which comprise about 2 percent of all dyads between males. In Hamadryas Baboons, sexual behavior (homosexual or heterosexual) is overall less frequent, but approximately one-third to two-thirds of sexual activity may occur between animals of the same sex (mostly between males). In a study of wild Geladas, homosexual interactions between males were observed about once every two hours; in captivity, 14–25 percent of mounts are between males and 2–3 percent are between females.

Orientation: In Savanna Baboons, the entire male population participates in homosexual "greetings" interactions; all adult males (including those in bonded coalitions with other males) are sexually active with females as well, indicating a high degree of bisexuality. This is probably true to a lesser extent for females, not all of whom engage in homosexual mounting. In Hamadryas Baboons, harem-holding males may be bisexual, mounting both males and females, while Gelada harem males are usually exclusively heterosexual (although they may engage in same-sex genital handling if there is another male in the troop). Some "bachelor" males in both species may engage solely in homosexual activity, especially adolescents and younger adults. In Geladas, for example, an average of 12 percent of the population lives in all-male groups (where same-sex activity usually occurs). In some areas this proportion is as high as 40 percent or more, although most such males do eventually breed.

Nonreproductive and Alternative Heterosexualities

A large proportion of Baboon heterosexual behavior involves nonprocreative activities. Savanna and Hamadryas Baboons commonly mate when the female cannot get pregnant. In some Savanna populations, up to 18 percent of heterosexual copulations occur during the nonfertilizable stages of her ovulatory cycle (including during menstruation) and 2–12 percent of matings happen when the female is already pregnant. Most females, however, remain abstinent while pregnant or lactating and in fact are (hetero)sexually active for less than 10 percent of their adult lives. Heterosexual mountings do not always involve penetration and/or ejaculation, either—in fact, one study found that less than 40 percent of opposite-sex matings in Savanna Baboons involved "full" copulation. Moreover, a variety of sexual behaviors besides vaginal intercourse occur, including fondling of genitals and females mounting males (REVERSE mounts). Masturbation is also a component of Baboon sexual expression, with several innovative techniques employed: Savanna males stimulate the penis with their hand, lick their own penis, rub their genitals against the ground, or stroke the penis with the tip of the tail, while females stimulate the clitoris and perineal area with the tail or fingers. Both males and females show evidence of sexual arousal during grooming, in the form of rhythmic erection of the genitals—"penis flicking" in males, and labial and clitoral swelling and "pulsation" in females. Sexual activity in Baboons also sometimes includes copulation with partners that are not optimal for breeding: adult male Savanna Baboons and adult female Hamadryas Baboons may mate with juvenile animals, incestuous matings are common in inbred troops, while "friendships" and sexual behavior between Baboons and Common Chimpanzees have been observed in the wild (and between female Savanna Baboons and male Macaques in captivity). In addition, infertile females are occasionally found in Savanna Baboons—in one troop they constituted 10 percent of all the adult females—and these individuals continue to engage in sexual behavior.

Several other forms of nonbreeding are found among Baboons. Geladas and Hamadryas Baboons, for example, have significant "bachelor" populations that do not generally participate in reproduction (constituting 20 percent of the male Hamadryas population). Older female Savanna Baboons sometimes experience a postreproductive period, while breeding-age males and females in this species often form platonic "friendships" with each other. Sexual relations in Baboons, however, are sometimes characterized by significant antagonism and even violence between the sexes. Hamadryas harem males often threaten and attack females, biting them on the neck to prevent them from leaving the group. Adult male Savanna Baboons sometimes rape younger females, often seriously injuring them, and adult females avoid or refuse a third of all mating attempts by males. In some populations, abortions and infanticide result when an outside male takes over a troop. He viciously attacks both mothers and infants to maximize his breeding opportunities; as a result of this brutality, females may suffer severe injuries as well as miscarriages, and infants may be killed. There is also evidence that females attack other females in order to suppress their reproduction. Infanticide was recently discovered among Geladas as well. Many male Baboons act as "baby-sitters" for infants, although occasionally the youngsters become injured during fights between their "baby-sitter" (or "kidnapper") and other males.

Other Species

Homosexual activity occurs in several other species of African monkeys. Same-sex mounting (in both males and females) is found in Vervets (*Cercopithecus aethiops*) (comprising about 11 percent of all mounting), Sooty Mangabeys (*Cercocebus torquatus*) (about 18 percent of mounting), and Talapoins (*Miopithecus talapoin*). Mounting between male Vervets is often accompanied by grooming, embracing from behind, fondling and displaying of the genitals, and nuzzling of the perineum and scrotum. Same-sex mounting in Talapoins may also include embracing or play-wrestling. Among Patas Monkeys (*Erythrocebus patas*), adolescent and younger males often fondle and nuzzle the scrotum and genitals of adult males.

Sources   *asterisked references discuss homosexuality/transgender*

*Abegglen, J.-J. (1984) *On Socialization in Hamadryas Baboons: A Field Study*. Lewisburg, Pa.: Bucknell University Press.

*Bernstein, I. S. (1975) "Activity Patterns in a Gelada Monkey Group." *Folia Primatologica* 23:50–71.

*——— (1970) "Primate Status Hierarchies." In L. A. Rosenblum, ed., *Primate Behavior: Developments in Field and Laboratory Research*, vol.1, pp. 71–109. New York: Academic Press.

*Bielert, C. (1985) "Testosterone Propionate Treatment of an XY Gonadal Dysgenetic Chacma Baboon." *Hormones and Behavior* 19:372–85.

*——— (1984a) "The Social Interactions of Adult Conspecifics with an Adult XY Gonadal Dysgenetic Chacma Baboon (*Papio ursinus*)." *Hormones and Behavior* 18:42–55.

*———— (1984b) "Estradiol Benzoate Treatment of an XY Gonadal Dysgenetic Chacma Baboon." *Hormones and Behavior* 18:191–205.

*Bielert, C., R. Bernstein, G. B. Simon, and L. A. van der Walt (1980) "XY Gonadal Dysgenesis in a Chacma Baboon (*Papio ursinus*)." *International Journal of Primatology* 1:3–14.

*Dixson, A. F., D. M. Scruton, and J. Herbert (1975) "Behavior of the Talapoin Monkey (*Miopithecus talapoin*) Studied in Groups, in the Laboratory." *Journal of Zoology, London* 176:177–210.

Dunbar, R. (1984) *Reproductive Decisions: An Economic Analysis of Gelada Baboon Social Strategies.* Princeton: Princeton University Press.

*Dunbar, R., and P. Dunbar (1975) *Social Dynamics of Gelada Baboons.* Basel: S. Karger.

*Fedigan, L. M. (1972) "Roles and Activities of Male Geladas (*Theropithecus gelada*)." *Behavior* 41:82–90.

*Gartlan, J. S. (1974) "Adaptive Aspects of Social Structure in *Eryhthrocebus patas*." In S. Kondo, M. Kawai, A. Ehara, and S. Kawamura, eds., *Proceedings from the Symposia of the Fifth Congress of the International Primatological Society*, pp. 161–71. Tokyo: Japan Science Press.

*———— (1969) "Sexual and Maternal Behavior of the Vervet Monkey, *Cercopithecus aethiops*." *Journal of Reproduction and Fertility*, supplement 6:137–50.

*Hall, K. R. L. (1962) "The Sexual, Agonistic, and Derived Social Behavior Patterns of the Wild Chacma Baboon, *Papio ursinus*." *Proceedings of the Zoological Society of London* 139:283–327.

*Hausfater, G., and D. Takacs (1987) "Structure and Function of Hindquarter Presentations in Yellow Baboons (*Papio cynocephalus*)." *Ethology* 74:297–319.

*Kummer, H. (1968) *Social Organization of Hamadryas Baboons.* Chicago: University of Chicago Press.

*Kummer, H., and F. Kurt (1965) "A Comparison of Social Behavior in Captive and Wild Hamadryas Baboons." In H. Vagtborg, ed., *The Baboon in Medical Research*, pp. 65–80. Austin: University of Texas Press.

*Leresche, L. A. (1976) "Dyadic Play in Hamadryas Baboons." *Behavior* 57:190–205.

*Marais, E. N. (1926) "Baboons, Hypnosis, and Insanity." *Psyche* 7:104–10.

*———— (1922/1969) *The Soul of the Ape.* New York: Atheneum.

*Maxim, P. E., and J. Buettner-Janusch (1963) "A Field Study of the Kenya Baboon." *American Journal of Physical Anthropology* 21:165–80.

Mori, A., T. Iwamoto, and A. Bekele (1997) "A Case of Infanticide in a Recently Found Gelada Population in Arsi, Ethiopia." *Primates* 38:79–88.

*Mori, U. (1979) "Individual Relationships within a Unit; Development of Sociability and Social Status." In M. Kawai (ed.) *Ecological and Sociological Studies of Gelada Baboons*, pp. 93–154. Basel: S. Karger.

*Noë, R. (1992) "Alliance Formation Among Male Baboons: Shopping for Profitable Partners." In A. H. Harcourt and F. B. M. de Waal, eds., *Coalitions and Alliances in Humans and Other Animals*, pp. 284–321. Oxford: Oxford University Press.

*Owens, N. W. (1976) "The Development of Sociosexual Behaviour in Free-Living Baboons, *Papio anubis*." *Behavior* 57:241–59.

Packer, C. (1980) "Male Care and Exploitation of Infants in *Papio anubis*." *Animal Behavior* 28:512–20.

Pereira, M. E. (1983) "Abortion Following the Immigration of an Adult Male Baboon (*Papio cynocephalus*)." *American Journal of Primatology* 4:93–98.

*Ransom, T. W. (1981) *Beach Troop of the Gombe.* Lewisburg, Pa.: Bucknell University Press; London and Toronto: Associated University Presses.

*Rowell, T. E. (1973) "Social Organization of Wild Talapoin Monkeys." *American Journal of Physical Anthropology* 38:593–98.

*———— (1967a) "Female Reproductive Cycles and the Behavior of Baboons and Rhesus Macaques." In S. A. Altmann, ed., *Social Communication Among Primates*, pp. 15–32. Chicago: University of Chicago Press.

*———— (1967b) "A Quantitative Comparison of the Behavior of a Wild and a Caged Baboon Group." *Animal Behavior* 15:499–509.

Saayman, G. S. (1970) "The Menstrual Cycle and Sexual Behavior in a Troop of Free Ranging Chacma Baboons (*Papio ursinus*)." *Folia Primatologica* 12:81–110.

Smuts, B. B. (1987) "What Are Friends For?" *Natural History* 96(2):36–45.

———— (1985) *Sex and Friendship in Baboons.* New York: Aldine.

*Smuts, B. B., and J. M. Watanabe (1990) "Social Relationships and Ritualized Greetings in Adult Male Baboons (*Papio cynocephalus anubis*)." *International Journal of Primatology* 11:147–72.

*Struhsaker, T. T. (1967) *Behavior of Vervet Monkeys (Cercopithecus aethiops).* University of California Publications in Zoology, vol. 82. Berkeley: University of California Press.

*Wadsworth, P. F., D. G. Allen, and D. E. Prentice (1978) "Pseudohermaphroditism in a Baboon (*Papio anubis*)." *Toxicology Letters* 1:261–66.

Wasser, S. K., and S. K. Starling (1988) "Proximate and Ultimate Causes of Reproductive Suppression Among Female Yellow Baboons at Mikumi National Park, Tanzania." *American Journal of Primatology* 16:97–121.

*Weinrich, J. D. (1980) "Homosexual Behavior in Animals: A New Review of Observations From the Wild, and Their Relationship to Human Sexuality." In R. Forleo and W. Pasini, eds., *Medical Sexology: The Third International Congress*, pp. 288–95. Littleton, Mass.: PSG Publishing.

*Wolfheim, J. H., and T. E. Rowell (1972) "Communication Among Captive Talapoin Monkeys (*Miopithecus talapoin*)." *Folia Primatologica* 18:224–55.

*Zuckerman, S. (1932) *The Social Life of Monkeys and Apes*. New York: Harcourt, Brace and Company.

NEW WORLD MONKEYS   Primate

(COMMON) SQUIRREL MONKEY
Saimiri sciureus

RUFOUS-NAPED TAMARIN
Saguinus geoffroyi

HOMOSEXUALITY	TRANSGENDER	BEHAVIORS		RANKING	OBSERVED IN
● Female	○ Intersexuality	● Courtship	● Pair-bonding	○ Primary	○ Wild
● Male	○ Transvestism	● Affectionate	● Parenting	● Moderate	● Semiwild
		● Sexual		○ Incidental	● Captivity

SQUIRREL MONKEY
IDENTIFICATION: A small (9–14 inch) monkey with a long tail; a pinkish white, heart- or skull-shaped facial pattern; and dense, yellowish or gray-green fur. DISTRIBUTION: Throughout most of northeastern South America, including Brazil, Colombia. HABITAT: Forests, swamps. STUDY AREAS: Monkey Jungle, Miami, Florida; Washington and California Regional Primate Research Centers; University of California—Santa Barbara; Max-Planck Institute of Psychiatry, Munich, Germany.

RUFOUS-NAPED TAMARIN
IDENTIFICATION: A squirrel-sized monkey with a mottled black and golden coat, a reddish tail and head, and a white crown. DISTRIBUTION: Northwestern Colombia through central Panama and Costa Rica. HABITAT: Tropical forests. STUDY AREA: Barro Colorado Island, Panama.

Social Organization

Squirrel Monkeys live in troops of 20–70 animals containing a majority of females. Younger males leave their home troops and live for several years in all-male "bachelor" bands of 2–10 monkeys, after which they join the cosexual troops as peripheral members. Females generally remain in their home troops for life, where they develop strong bonds with each other. Rufous-naped Tamarins live in cosexual groups of 3–9 individuals in which usually only one male-female pair breeds; the remainder of the group consists of their offspring and unrelated adult non-breeders.

Description

Behavioral Expression: Female Squirrel Monkeys court and mount each other. Homosexual courtship is initiated by one female facing the other, tilting her head, and making a "purring" noise (a series of soft, guttural clicklike sounds). This may be accompanied by a GENITAL DISPLAY, in which the soliciting female positions herself in front of the other, spreading her thighs to expose her vulva and engorged or erect clitoris. As an invitation to the other female to mount, she turns around and presents her hindquarters, looking over her shoulder with her feet spread apart. This may be repeated several times, the presenting female moving away each time as the other female approaches in a sort of courtship "chase." Mounting is done in the same position used for heterosexual mating: one female grasps the other's waist with her hands and her calves with her feet, making thrusting movements with her pelvis. The mounted female frequently purrs during the sexual interaction. Sometimes two females take turns mounting each other, but often one female is more typically a mounter and the other a mountee.

Female Squirrel Monkeys may develop a short, consortlike bond (also seen in heterosexual interactions) during which they interact sexually with one another. In addition, a number of other types of female bonding occur in this species. Females frequently have one close female "friend" with whom they travel and rest; often this relationship develops into a highly affectionate one and may even include coparenting. The two females frequently touch hands, kiss each other on the mouth, and huddle together. If one of them is a mother, the other helps her raise her infant; if both are mothers, they help each other with parenting, including nuzzling and carrying each other's infant and protecting them from predators. Often the infant develops a strong bond with the comother, although some females act as coparents for infants belonging to several different mothers. The coparenting female is sometimes known as an "aunt," although she need not be genetically related to the mother; her relationship with the mother often outlasts the duration of parenting.

A female Squirrel Monkey mounting another female ▶

▲ *A male Squirrel Monkey* (right) *performing a "genital display" toward another male*

Homosexual mounting sometimes occurs between male Squirrel Monkeys, especially younger individuals or between an older and a younger (adult) partner; adult males also perform the genital display to each other. During intense displays, one male will thrust his erect penis into the face of the other male while holding him down with his hands and may even climb onto the back of the other male (who also sometimes develops an erection). Several males can be drawn into the activity, forming a ball or "pile-up" of three or four individuals all twisting and climbing on one another as they try to perform genital displays.

In Rufous-naped Tamarins, homosexual behavior takes the form of same-sex mounting, including pelvic thrusting (as in heterosexual copulation); both males and females participate in homosexual mounts.

Frequency: In captivity, homosexual mounting can occur quite frequently in both mixed-sex and single-sex groups of Squirrel Monkeys: one study recorded mounts between females roughly once every 40 minutes, with homosexual activity taking place over three to seven days each month. An average of about 40 percent of genital displays occur between animals of the same sex; more than one-quarter are between females. In Rufous-naped Tamarins, homosexual mounting occurs sporadically.

Orientation: Some female Squirrel Monkeys that bond with other females as coparents are themselves nonbreeders; to this extent, then, they are involved exclusively in same-sex activities. Most other Squirrel Monkeys, as well as Rufous-naped Tamarins, that participate in homosexual behavior are probably simultaneously bisexual. In a group of three Squirrel Monkeys (two females and one male) whose sexual activities were briefly sampled, for example, heterosexual and homosexual encounters alternated continuously for a half hour, and more than 25 percent of the courtship and sexual activities were between the females. However, no detailed long-term studies have been conducted to verify the extent of individuals' same-sex versus opposite-sex activity throughout their entire lives.

Nonreproductive and Alternative Heterosexualities

In addition to the female coparenting arrangement described above, other alternative family configurations and nonbreeding individuals occur in Squirrel Monkeys. One female sometimes adopts another female's infant, raising it along with her own baby, while some male Squirrel Monkeys never copulate at all during the mating season. Interestingly, these may be the highest-ranking males in the troop, who are often more aggressive and less patient than other males and therefore more likely to disturb females or fail to attract willing mates. Several other aspects of Squirrel Monkey heterosexual life reveal considerable antagonism and separation between the sexes. Females often form groups or COALITIONS during the mating season to chase off males who are pursuing unwilling females; females (and occasionally males) may also directly disrupt heterosexual copulations in progress. At other times, females persistently harass males so that they remain spatially segregated from the females, either on the periphery of the troop or closer to the ground. When a willing female is found and the sexual interaction is not disturbed, often several males will participate, all joining in by kissing the female on the mouth, displaying their genitals, and sniffing or nuzzling her genitalia. The mating system is promiscuous, as both males and females copulate with multiple partners.

A number of nonprocreative sexual activities are also found in these New World monkeys. Male Squirrel Monkeys masturbate by either sucking their own penis or rubbing it with one or both hands, while females may copulate when not in heat or during pregnancy (up to the fourth month). In addition, females sometimes produce a vulvar plug from sloughed vaginal cells when they are in heat, which may serve to limit inseminations. Rufous-naped Tamarin males on occasion mount females without thrusting or penetration. Male Squirrel Monkeys also have a pronounced sexual cycle: for three to four months out of the year, they are sexually active, more aggressive, and develop a characteristic appearance—heavier, with more fluffed fur—during which time they are known as FATTED MALES. For the remainder of the year, however, their testes are essentially dormant; they lose their "fatted" appearance and live largely separate from the females. In Rufous-naped Tamarins, as in other tamarins and marmosets, all but the highest-ranking female in a group forgo reproduction, perhaps through a complex mechanism of "self-restraint" that is mediated by pheromones from the lead female. As a result, only about half of all mature females actually reproduce at one time; however, nonbreeding individuals often continue to copulate. In addition, most matings outside the breeding season do not result in offspring, and it is thought that many embryos may be reabsorbed, aborted, or the young die soon after birth.

Other Species

Same-sex activity occurs in several other species of Central and South American monkeys. Homosexual mounting (in both males and females), including pelvic thrusting and genital rubbing on the partner, has been observed in a variety of Tamarin species, including the Saddle-back Tamarin (*Saguinus fuscicollis*), the Mustached Tamarin (*S. mystax*), and the Cotton-top Tamarin (*S. oedipus*). Both male and female Lion Tamarins (*Leontopithecus rosalia*) sometimes mount their own offspring of both sexes, including adolescents and younger individuals. Approximately 1 percent of mounting activity in Common Marmosets (*Callithrix jacchus*) occurs among adolescent and younger males (brothers living in the same family group). In White-fronted Capuchin Monkeys (*Cebus albifrons*), young males occasionally suck and fondle the scrotum of older males, while homosexual activity among females (including mounting) also occurs in Brown Capuchins (*C. apella*) and Weeper Capuchins (*C. olivaceous*). More than half of all mounting in White-faced Capuchins (*C. capucinus*) is between same-sex partners, often preceded by a type of courtship activity known as WHEEZE DANCING, involving contorted postures, wheezing vocalizations, and "slow-motion" chases.

Sources
asterisked references discuss homosexuality/transgender

*Akers, J. S., and C. H. Conaway (1979) "Female Homosexual Behavior in *Macaca mulatta*." *Archives of Sexual Behavior* 8:63–80.

*Anschel, S., and G. Talmage-Riggs (1978) "Social Structure Dynamics in Small Groups of Captive Squirrel Monkeys." In D. J. Chivers and J. Herbert, eds., *Recent Advances in Primatology*, vol. 1, pp. 601–4. New York: Academic Press.

*Baldwin, J. D. (1969) "The Ontogeny of Social Behavior of Squirrel Monkeys (*Saimiri sciureus*) in a Seminatural Environment." *Folia Primatologica* 11:35–79.

——— (1968) "The Social Behavior of Adult Male Squirrel Monkeys (*Saimiri sciureus*) in a Seminatural Environment." *Folia Primatologica* 9:281–314.

Baldwin, J. D., and J. I. Baldwin (1981) "The Squirrel Monkeys, Genus *Saimiri*." In A. F. Coimbra-Filho and R. A. Mittermeier, eds., *Ecology and Behavior of Neotropical Primates*, vol. 1, pp. 277–330. Rio de Janeiro: Academia Brasileira de Ciências.

*Castell, R., and B. Heinrich (1971) "Rank Order in a Captive Female Squirrel Monkey Colony." *Folia Primatologica* 14:182–89.

Dawson, G. A. (1976) "Behavioral Ecology of the Panamanian Tamarin, *Saguinus oedipus* (Callitrichidae, Primates)." Ph.D. thesis, Michigan State University.

*Defler, T. R. (1979) "On the Ecology and Behavior of *Cebus albifrons* in Eastern Colombia: II. Behavior." *Primates* 20:491–502.

*Denniston, R. H. (1980) "Ambisexuality in Animals." In J. Marmor, ed., *Homosexual Behavior: A Modern Reappraisal*, pp. 25–40. New York: Basic Books.

*DuMond, F. V. (1968) "The Squirrel Monkey in a Seminatural Environment." In L. A. Rosenblum and R. W. Cooper, eds., *The Squirrel Monkey*, pp. 87–145. New York: Academic Press.

DuMond, F. V., and T. C. Hutchinson (1967) "Squirrel Monkey Reproduction: The 'Fatted' Male Phenomenon and Seasonal Spermatogenesis." *Science* 158:1067–70.

*Hoage, R. J. (1982) *Social and Physical Maturation in Captive Lion Tamarins, Leontopithecus rosalia rosalia (Primates: Callitrichidae)*. Smithsonian Contributions to Zoology no. 354. Washington, D.C.: Smithsonian Institution Press.

Hopf, S., E. Hartmann-Wiesner, B. Kühlmorgen, and S. Mayer (1974) "The Behavioral Repertoire of the Squirrel Monkey (*Saimiri*)." *Folia Primatologica* 21:225–49.

Latta, J., S. Hopf, and D. Ploog (1967) "Observation of Mating Behavior and Sexual Play in the Squirrel Monkey (*Saimiri sciureus*)." *Primates* 8:229–46.

*Linn, G. S., D. Mase, D. LaFrançois, R. T. O'Keeffe, and K. Lifshitz (1995) "Social and Menstrual Cycle Phase Influences on the Behavior of Group-Housed *Cebus apella*." *American Journal of Primatology* 35:41–57.

*Manson, J. H., S. Perry, and A. R. Parish (1997) "Nonconceptive Sexual Behavior in Bonobos and Capuchins." *International Journal of Primatology* 18:767–86.

*Mendoza, S. P., and W. A. Mason (1991) "Breeding Readiness in Squirrel Monkeys: Female-Primed Females are Triggered by Males." *Physiology & Behavior* 49:471–79.

Mitchell, C. L. (1994) "Migration Alliances and Coalitions Among Adult Male South American Squirrel Monkeys (*Saimiri sciureus*)." *Behavior* 130:169–90.

*Moynihan, M. (1970) *Some Behavior Patterns of Platyrrhine Monkeys. II.* Saguinus geoffroyi *and Some Other Tamarins.* Smithsonian Contributions to Zoology no. 28. Washington, D.C.: Smithsonian Institution Press.

*Perry, S. (1998) "Male-Male Social Relationships in Wild White-faced Capuchins, *Cebus capucinus.*" *Behavior* 135:139–72.

Peters, M. (1970) "Mouth to Mouth Contact in Squirrel Monkeys (*Saimiri sciureus*)." *Zeitschrift für Tierpsychologie* 27:1009–10.

Ploog, D. W. (1967) "The Behavior of Squirrel Monkeys (*Saimiri sciureus*) as Revealed by Sociometry, Bioacoustics, and Brain Stimulation." In S. Altmann, ed., *Social Communication Among Primates*, pp. 149–84. Chicago: University of Chicago Press.

*Ploog, D. W., J. Blitz, and F. Ploog (1963) "Studies on the Social and Sexual Behavior of the Squirrel Monkey (*Saimiri sciureus*)." *Folia Primatologica* 1:29–66.

Ploog, D. W., and P. D. Maclean (1963) "Display of Penile Erection in Squirrel Monkey (*Saimiri sciureus*)." *Animal Behavior* 11:32–39.

Rosenblum, L.A. (1968) "Mother-Infant Relations and Early Behavioral Development in the Squirrel Monkey." In L. A. Rosenblum and R. W. Cooper, eds., *The Squirrel Monkey*, pp. 207–33. New York: Academic Press.

*Rothe, H. (1975) "Some Aspects of Sexuality and Reproduction in Groups of Captive Marmosets (*Callithrix jacchus*)." *Zeitschrift für Tierpsychologie* 37:255–73.

*Shadle, A. R., E. A. Mirand, and J. T. Grace Jr. (1965) "Breeding Responses in Tamarins." *Laboratory Animal Care* 15:1–10.

Skinner, C. (1985) "A Field Study of Geoffroy's Tamarin (*Saguinus geoffroyi*) in Panama." *American Journal of Primatology* 9:15–25.

Snowdon, C. T. (1996) "Infant Care in Cooperatively Breeding Species." *Advances in the Study of Behavior* 25:643–89.

Srivastava, P. K., F. Cavazos, and F. V. Lucas (1970) "Biology of Reproduction in the Squirrel Monkey (*Saimiri sciureus*): I. The Estrus Cycle." *Primates* 11:125–34.

*Talmage-Riggs, G., and S. Anschel (1973) "Homosexual Behavior and Dominance in a Group of Captive Squirrel Monkeys (*Saimiri sciureus*)." *Folia Primatologica* 19:61–72.

*Travis, J. C., and W. N. Holmes (1974) "Some Physiological and Behavioral Changes Associated with Oestrus and Pregnancy in the Squirrel Monkey (*Saimiri sciureus*)." *Journal of Zoology, London* 174:41–66.

*Vasey, P. L.(1995) "Homosexual Behavior in Primates: A Review of Evidence and Theory." *International Journal of Primatology* 16:173–204.

LEMURS AND BUSHBABIES   *Primate*

VERREAUX'S SIFAKA
Propithecus verreauxi

LESSER BUSHBABY or MOHOL GALAGO
Galago moholi

HOMOSEXUALITY	TRANSGENDER	BEHAVIORS		RANKING	OBSERVED IN
● Female	○ Intersexuality	○ Courtship	○ Pair-bonding	○ Primary	● Wild
● Male	● Transvestism	○ Affectionate	○ Parenting	○ Moderate	○ Semiwild
		● Sexual		● Incidental	● Captivity

VERREAUX'S SIFAKA
IDENTIFICATION: A long-legged lemur with a plush white coat, black face, black or brown crown and under-parts, and long tail (nearly 2 feet). DISTRIBUTION: Western and southern Madagascar; vulnerable. HABITAT: Forests. STUDY AREA: Near Hazafotsy, Madagascar; subspecies *P.v. verreauxi*.

LESSER BUSHBABY
IDENTIFICATION: A small, squirrel-like primate (7 inches, plus a foot-long tail) with silky, grayish-yellow fur, a broad face, and enormous eyes and ears. DISTRIBUTION: Sub-Saharan Africa. HABITAT: Woodland, savanna, scrub. STUDY AREA: Witwatersrand University, South Africa.

Social Organization

Verreaux's Sifakas live in cosexual groups of up to 12 individuals and some-times associate as male-female pairs. As in most Lemurs, females are generally dominant to males in this species. Females typically remain in their birth group for life, while males leave their group on maturing and transfer between groups several times throughout their lives. The mating system has elements of POLYGYNANDRY, that is, females generally mate with more than one male and vice versa. Lesser Bushbabies generally live in family groups consisting of females and their offspring along with peripheral males. They are often found singly or in pairs and may form sleeping groups of up to seven individuals.

Description

Behavioral Expression: Male Verreaux's Sifakas sometimes mount other males during the mating season. The mounted animal—usually a younger adult or ado-lescent male—often snaps at the mounter and tries to wriggle free (as do females trying to escape from unwelcome heterosexual advances). In Lesser Bushbabies, fe-males occasionally mount and thrust against each other when in heat. Like other species of Bushbabies, the genitals of female Lesser Bushbabies are unusual in sev-eral respects. The clitoris is long and pendulous, greatly resembling the male's pe-nis, and the urethra extends through to the tip of the organ, so that females urinate through the clitoris rather than through a urethral opening near the vagina. Females do not menstruate, and in fact the vagina remains closed at all times except during the mating season (which lasts no more than two to three weeks and occurs twice a year).

Frequency: Homosexual mounting probably occurs only occasionally in Verreaux's Sifakas and Lesser Bushbabies. However, one study of wild Verreaux's Sifakas found that 3 out of 21 mountings (14 percent) were between males.

Orientation: Lemurs and Bushbabies that participate in same-sex mounting probably also engage in heterosexual activity, although too little is known about individual life histories in these species to draw any firm conclusions.

Nonreproductive and Alternative Heterosexualities

Many heterosexual copulations in Verreaux's Sifakas—more than two-thirds in some populations—do not involve penetration or ejaculation, often because the female resists the mating attempt and wriggles free. Females sometimes also mate when they are not in heat: for some individuals, 30–80 percent of their sexual activity is nonprocreative, taking place during times when they cannot conceive. In some populations females also delay reproducing for several years, and only slightly more than half of all adult females reproduce each year. Infanticide occurs occasionally in this species, and possibly also abortions. In Lesser Bushbabies, heterosexual copulations can be lengthy—more than nine minutes in some cases—and a female often bites the male, "boxes" him with her hands, and tries to push him off or get away from him during mating.

Sources   *asterisked references discuss homosexuality/transgender*

*Andersson, A. B. (1969) "Communication in the Lesser Bushbaby (*Galago senegalensis moholi*)." Master's thesis, Witwatersrand University.

Bearder, S. K., and G. A. Doyle (1974) "Field and Laboratory Studies of Social Organization in Bushbabies (*Galago senegalensis*)." *Journal of Human Evolution* 3:37–50.

Brockman, D. K., and P. L. Whitten (1996) "Reproduction in Free-Ranging *Propithecus verreauxi*: Estrus and the Relationship Between Multiple Partner Matings and Fertilization." *American Journal of Physical Anthropology* 100:57–69.

Butler, H. (1967) "The Oestrus Cycle of the Senegal Bush Baby (*Galago senegalensis senegalensis*) in the Sudan." *Journal of Zoology, London* 151:143–62.

Dixson, A. F. (1995) "Sexual Selection and the Evolution of Copulatory Behavior in Nocturnal Prosimians." In L. Alterman, G. A. Doyle, and M. K. Izard, eds., *Creatures of the Dark: The Nocturnal Prosimians*, pp. 93–118. New York: Plenum Press.

*Doyle, G. A. (1974a) "Behavior of Prosimians." *Behavior of Nonhuman Primates* 5:154–353.

——— (1974b) "The Behavior of the Lesser Bushbaby." In R. D. Martin, G. A. Doyle, and A. C. Walker, eds., *Prosimian Biology*, pp. 213–31. Pittsburgh: University of Pittsburgh Press.

Doyle, G. A., A. Pelletier, and T. Bekker (1967) "Courtship, Mating, and Parturition in the Lesser Bushbaby (*Galago senegalensis moholi*) Under Semi-Natural Conditions." *Folia Primatologica* 7:169–97.

Kubzdela, K. S., A. F. Richard, and M. E. Pereira (1992) "Social Relations in Semi-Free-Ranging Sifakas (*Propithecus verreauxi verreauxi*) and the Question of Female Dominance." *American Journal of Primatology* 28:139–45.

Lipschitz, D. L. (1996) "Male Copulatory Patterns in the Lesser Bushbaby (*Galago moholi*) in Captivity." *International Journal of Primatology* 17:987–1000.

Lowther, F. D. L. (1940) "A Study of the Activities of a Pair of *Galago senegalensis moholi* in Captivity, Including the Birth and Postnatal Development of Twins." *Zoologica* 25:433–65.

Richard, A., (1992) "Aggressive Competition Between Males, Female-Controlled Polygyny, and Sexual Monomorphism in a Malagasy Primate, *Propithecus verreauxi.*" *Journal of Human Evolution* 22:395–406.

———— (1978) *Behavioral Variation: Case Study of a Malagasy Lemur.* Lewisburg, Pa.: Bucknell University Press.

*———— (1974a) "Intra-specific Variation in the Social Organization and Ecology of *Propithecus verreauxi.*" *Folia Primatologica* 22:178–207.

*———— (1974b) "Patterns of Mating in *Propithecus verreauxi verreauxi.*" In R. D. Martin, G. A. Doyle, and A. C. Walker, eds., *Prosimian Biology,* pp. 49–74. London: Duckworth; Pittsburgh: University of Pittsburgh Press.

Richard, A., P. Rakotomanga, and M. Schwartz (1991) "Demography of *Propithecus verreauxi* at Beza Mahafaly, Madagascar: Sex Ratio, Survival, and Fertility, 1984–1988." *American Journal of Physical Anthropology* 84:307–22.

Marine Mammals

DOLPHINS AND WHALES

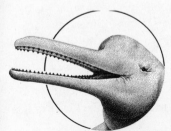

River Dolphins

Marine Mammal

BOTO or AMAZON RIVER DOLPHIN
Inia geoffrensis

HOMOSEXUALITY	TRANSGENDER	BEHAVIORS		RANKING	OBSERVED IN
○ Female	○ Intersexuality	○ Courtship	○ Pair-bonding	○ Primary	○ Wild
● Male	○ Transvestism	● Affectionate	○ Parenting	● Moderate	○ Semiwild
		● Sexual		○ Incidental	● Captivity

IDENTIFICATION: An 8-foot-long dolphin with a long, toothed beak and light blue or even bright pink skin. DISTRIBUTION: The Amazon and Orinoco River systems; vulnerable. HABITAT: Slow-moving streams and tributaries, flooded forests, lakes. STUDY AREAS: Duisburg Zoo, Germany; Aquarium of Niagara Falls, New York; subspecies *I.g. humboldtiana*, the Orinoco Dolphin.

Social Organization

Although little is known about their social organization, it appears that Botos are largely solitary animals that occasionally associate in groupings of up to a dozen or more individuals. Larger aggregations generally occur at feeding areas, and Botos may even coordinate their fishing attempts with other species, such as the giant river otter (*Pteronura brasiliensis*). The Boto mating system is probably polygamous.

Description

Behavioral Expression: Male Botos participate in a wide variety of homosexual interactions, including mating with each other using fully three different types of penetration: one male may insert his erect penis into the genital slit of the other, into his anus, or into his blowhole. When engaging in anal or genital-slit intercourse, one male swims upside down beneath the other one as in heterosexual copulation; blowhole mating occurs with the inserting male above the other one. If there is an age difference between the males, typically the older one penetrates the younger one. Males also rub their genital openings or erect penises against each other; alternatively, one male might rub his head against the other's genitals, stimulating an erection. Pairs of males who interact sexually also display a great deal of affection toward one another, caressing each other with their beak or flippers, brushing against one another, swimming side by side while touching each other's body, flippers, or flukes, surfacing to breathe simultaneously, or playing and resting together. Male homosexual encounters can be quite lengthy—continuing for a whole afternoon, for example—although if mating occurs, the actual penetration lasts for only about one minute (in anal intercourse).

Male Botos also engage in homosexual activity with another species of Amazon River dolphin, the Tucuxi (*Sotalia fluviatilis*). In these interspecies encounters, genital slit intercourse between males involves the same belly-to-belly position described above, but sometimes the penetrating animal twists around so that his head faces in the opposite direction (while still remaining inserted in the other male). In addition to caresses and genital rubbing, homosexual activity sometimes includes more unusual behavior: a male Boto was once seen gently taking a Tucuxi's entire head into his mouth, in an apparently affectionate gesture.

Frequency Homosexual activity is common in captive Botos; its prevalence among wild animals is not known. Similarly, sexual behavior between Botos and Tucuxis has only been seen in captivity, but these two species do occasionally interact with one another in the wild.

Orientation: Because homosexual behavior has been studied in detail only in captive male Botos without access to females, it is not known whether this behavior occurs in other contexts, or if it is simply an expression of a latent or "situational" bisexual potential. However, given the varied and generally plastic nature of dolphin sexuality, it is likely that homosexual or bisexual expression is a basic component of Boto social life for at least some individuals.

Nonreproductive and Alternative Heterosexualities

Male and female Botos sometimes engage in nonreproductive matings: heterosexual blowhole copulations have been observed, and a male will also sometimes rub his penis against the female's fins or flukes, especially if she does not permit him to copulate vaginally. In addition, heterosexual matings can be remarkably frequent and prolonged affairs:

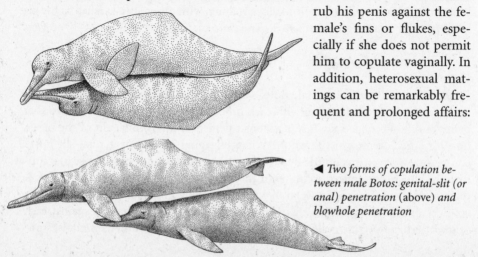

◄ *Two forms of copulation between male Botos: genital-slit (or anal) penetration* (above) *and blowhole penetration*

▲ *Two male Botos, a younger and an older individual who are sexually involved with one another, swimming side by side while touching*

one male and female were seen to mate once every four minutes for a virtually continuous period of over three hours. However, females are not always willing participants in such repeated copulations, often fleeing into shallow waters to avoid males that are harassing them. Females that cannot escape may be attacked and bitten around the genital area by males. Masturbation is also common in Botos: males rub the penis with one of their fins, females sometimes try to insert objects into the genital slit, while both sexes rub their genitals against underwater objects or surfaces. Botos have also developed an alternate parenting or "baby-sitting" arrangement of communal nursery groups. Young Botos gather together in shallow water, forming what are sometimes known as CRÈCHES that contain both calves and older juveniles; these groups offer them safety in numbers while their parents feed on their own.

Sources
asterisked references discuss homosexuality/transgender

*Best, R. C., and V. M. F. da Silva (1989) "Amazon River Dolphin, Boto, *Inia geoffrensis* (de Blainville, 1817)." In S. H. Ridgway and R. Harrison, eds., *Handbook of Marine Mammals, vol. 4: River Dolphins and the Larger Toothed Whales*, pp. 1–23. London: Academic Press.

*Caldwell, M. C., D. K. Caldwell, and R. L. Brill (1989) "*Inia geoffrensis* in Captivity in the United States." In W. F. Perrin, R. L. Brownell, Jr., Z. Kaiya, and L. Jiankang, eds., *Biology and Conservation of the River Dolphins*, pp. 35–41. Occasional Papers of the IUCN Species Survival Commission no. 3. Gland, Switzerland: International Union for Conservation of Nature and Natural Resources.

*Caldwell, M. C., D. K. Caldwell, and W. E. Evans (1966) "Sounds and Behavior of Captive Amazon Freshwater Dolphins, *Inia geoffrensis*." *Los Angeles County Museum Contributions in Science* 108:1–24.

Layne, J. N. (1958) "Observations on Freshwater Dolphins in the Upper Amazon." *Journal of Mammology* 39:1–22.

*Layne, J. N, and D. K. Caldwell (1964) "Behavior of the Amazon Dolphin, *Inia geoffrensis* (Blainville), in Captivity." *Zoologica* 49:81–108.

*Pilleri, G., M. Gihr, and C. Kraus (1980) "Play Behavior in the Indus and Orinoco Dolphin (*Platanista indi* and *Inia geoffrensis*)." *Investigations on Cetacea* 11:57–107.

*Renjun, L., W. Gewalt, B. Neurohr, and A. Winkler (1994) "Comparative Studies on the Behavior of *Inia geoffrensis* and *Lipotes vexillifer* in Artificial Environments." *Aquatic Mammals* 20:39–45.

*Spotte, S. H. (1967) "Intergeneric Behavior Between Captive Amazon River Dolphins *Inia* and *Sotalia*." *Underwater Naturalist* 4:9–13.

*Sylvestre, J.-P. (1985) "Some Observations on Behavior of Two Orinoco Dolphins (*Inia geoffrensis humboldtiana* [Pilleri and Gihr 1977]), in Captivity, at Duisburg Zoo." *Aquatic Mammals* 11:58–65.

Trujillo, F. (1996) "Seeing Fins." *BBC Wildlife* 14:22–28.

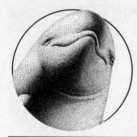

DOLPHINS   *Marine Mammal*

BOTTLENOSE DOLPHIN
Tursiops truncatus
SPINNER DOLPHIN
Stenella longirostris

HOMOSEXUALITY	TRANSGENDER	BEHAVIORS		RANKING	OBSERVED IN
● Female	○ Intersexuality	● Courtship	● Pair-bonding	● Primary	● Wild
● Male	○ Transvestism	● Affectionate	○ Parenting	○ Moderate	○ Semiwild
		● Sexual		○ Incidental	● Captivity

BOTTLENOSE DOLPHIN

IDENTIFICATION: The familiar gray, 10–13-foot-long dolphin. DISTRIBUTION: Worldwide oceans and seas. HABITAT: Coastal, temperate-to-tropical waters. STUDY AREAS: Near Sarasota, Florida; Grand Bahama Island, the Bahamas; Marineland, Florida; Marine World Africa, California; Marineland of the Pacific, California; Port Elizabeth Oceanarium, South Africa; Harderwijk Dolphinarium, the Netherlands; subspecies, *T.t. truncatus*, the Atlantic Bottlenose; *T.t. gilli*, the Pacific Bottlenose; and *T.t. aduncus*, the Indian Ocean Bottlenose.

SPINNER DOLPHIN

IDENTIFICATION: A 6-foot-long dolphin with a long, slender beak; steep, triangular dorsal fin; dark upperparts and light underparts. DISTRIBUTION: Tropical oceans worldwide. HABITAT: Often in deep, offshore waters. STUDY AREAS: Kealake'akua Bay, Hawaii; Sea Life Park Oceanarium, Hawaii; subspecies *S.l. longirostris*, the Hawaiian Spinner Dolphin.

Social Organization

Bottlenose Dolphins have a highly developed social system characterized by four basic social units: mother-calf pairs, groups of adolescents (often male-only, or with a preponderance of males), bands of up to a dozen adult females and their young, and adult males in pair-bonds (and less commonly, on their own). Spinner Dolphins may have a more fluid social organization, although coalitions of males can sometimes be recognized, as well as schools of a thousand or more individuals. The heterosexual mating system is poorly understood; however, there are no strong male-female bonds, and animals probably mate with multiple partners.

Description

Behavioral Expression: In both Bottlenose and Spinner Dolphins, animals of the same sex frequently engage in affectionate and sexual activities with each other that have many of the elements of heterosexual courtship and sexuality. For example, two males or two females often rub their bodies together, mouthing and nuzzling one another, and may caress and stroke each other—simultaneously or alternately—with their fins, flukes, snouts (or "beaks"), and heads. Sometimes this is accompanied by playful rolling, chasing, pushing, and leaping. During this activity—which can last anywhere from several minutes to several hours—males may display erect penises. More overt homosexual activity takes a variety of forms. One animal might stroke or gently probe the other's genital area with the soft tips of its flukes or flippers. Female Spinner Dolphins sometimes even "ride" on each other's dorsal fin—one inserts her fin into the other's vulva or genital slit, then the two swim together in this position. Among Bottlenose females, direct stimulation of the clitoris is a prominent feature of homosexual interactions. Two females often take turns rubbing each other's clitoris, using either the snout, flippers, or flukes, or else actively masturbate against their partner's appendages. Females may also clasp one another in a belly-to-belly position (as in heterosexual mating) and thrust against each other.

Homosexual interactions also involve a form of "oral" sex in which one animal rubs and nuzzles the other's genitals with its snout or beak; because both males and females have a genital slit or opening, penetration is also possible in this fashion for both sexes. One animal might insert the tip of its beak into the other's genitals or perhaps just use its lower jaw to penetrate and stimulate his or her partner. Sometimes this develops into a sexual activity known as BEAK-GENITAL PROPULSION, in which one partner inserts its beak into the other's genitals and gently propels the two of them forward, maintaining penetration while they swim together. The lower animal may also turn on its side or rotate belly up during this activity. Male Dolphins sometimes rub their erect penises on one another's body or genital area. This may lead to copulation, in which one male swims upside down underneath the other, pressing his genitals against the other and even inserting his penis into the genital slit (or less commonly, anus) of the male above him (this same position is used in heterosexual intercourse). The two partners may switch positions, alternating during the same session, or perhaps exchanging "roles" over a longer period. If there is an age difference between male partners, either may penetrate the other, and Bottlenose adolescents have even been observed penetrating much older males. Groups of three or four males may engage in homosexual activity together, or one male may masturbate (by rubbing his penis on rocks or sand) while other males are coupling nearby. Homosexual activity is sometimes accompanied by aggressive behaviors, but these can also occur during heterosexual interactions (males and females have been observed diving forcefully at each other, for example, and violently ramming their foreheads together as a prelude to mating). In Spinner Dolphins, groups of a dozen or more dolphins of both sexes sometimes gather together in near "orgies" of caressing and sexual behavior (both same-sex and opposite-sex); these groups are known as WUZZLES.

◀ *"Beak-genital propulsion" between two female Spinner Dolphins*

Male Bottlenose Dolphins often form life-long pair-bonds with each other. Adolescent and younger males typically live in all-male groups in which homosexual activity is common; within these groups, a male begins to develop a strong bond with a particular partner (usually of the same age)

▲ *Homosexual copulation in Bottlenose Dolphins: the male in an upside-down position is penetrating the male above him*

with whom he will spend the rest of his life. The two Dolphins become constant companions, often traveling widely; although sexual activity probably declines as they get older, it may continue to be a regular feature of such partnerships. Paired males sometimes take turns guarding or remaining vigilant while their partner rests. They also defend their mates against predators such as sharks and protect them while they are healing from wounds inflicted during predators' attacks. Sometimes three males form a tightly bonded trio. On the death of his partner, a male may spend a long time searching for a new male companion—usually unsuccessfully, since most other males in the community are already paired and will not break their bonds. If, however, he can find another "widower" whose male partner has died, the two may become a couple.

Male Bottlenose Dolphins also sometimes aggressively pursue and copulate with male Atlantic Spotted Dolphins (*Stenella frontalis*), both adults and juveniles. After an initial chase, the Bottlenose male typically arches and rubs his body and erect penis against the Spotted male, then mounts (and often penetrates) him from an upright, sideways position. This mounting position is distinct from the upside-down, belly-to-belly approach generally used for within-species sexual encounters. Sometimes a pair of Bottlenose males pursue a Spotted male and both partners mount him at the same time. Though often playful, this high-energy interspecies homosexual activity may also be accompanied by aggressive behaviors such as tail slaps, threatening postures, and squawking vocalizations (also part of heterosexual interactions between these two species and among Bottlenose Dolphins, as noted above). In fact, groups of Spotted Dolphins—sometimes accompanied by Bottlenose males—may band together to chase off Bottlenose males that are engaging in these more aggressive sexual interactions. However, even when this activity is accompanied by overt aggression, Bottlenose and Spotted males that interact sexually with one another may later also band together and cooperate with one another. Male Atlantic Spotted Dolphins also engage in homosexual activity with each other, and adults sometimes even copulate with male calves of their own species. In one such instance, adult-juvenile homosexual activity was preceded by a vocalization known as a GENITAL BUZZ, in which the adult male directed a stream of low-pitched, rapid buzzing clicks toward the genital area of the male calf. This sound, which is also a component of heterosexual courtship in this species, may serve as a form of acoustic "foreplay," actually stimulating the genitals of the recipient via the strongly pulsed sound waves. Bottlenose and Spinner Dolphins of both sexes have also been observed participating in homosexual activity with other species of dolphins in captivity, such as Pacific

▲ *Two male Bottlenose Dolphins with erections both attempting to mount a male Atlantic Spotted Dolphin (in the Bahamas) using an upright, sideways mounting position*

Striped Dolphins (*Lagenorhynchus obliquidens*), Common Dolphins (*Delphinus delphis*), Bridled Dolphins (*Stenella attenuata*), and False Killer Whales (*Pseudorca crassidens*).

Frequency: Homosexual interactions are a frequent and regular occurrence in wild Dolphins, particularly among groups of younger Bottlenose males. In mixed-sex groups in captivity, homosexual behavior occurs with equal frequency—and in some cases, more often—than heterosexual activity. Male couples are a ubiquitous feature of many Bottlenose communities; in some cases, more than three-quarters of all males live in same-sex pair-bonds. About 30 percent of interactions between wild Bottlenose and Atlantic Spotted Dolphins include homosexual activities (often accompanied by aggressive behaviors).

Orientation: The lives of male Bottlenose Dolphins are characterized by extensive bisexuality, combined with periods of exclusive homosexuality. As adolescents and young males, they have regular homosexual interactions in all-male groups, sometimes alternating with heterosexual activity. From age 10 onward, most male Dolphins form pair-bonds with another male, and because they do not usually father calves until they are 20–25 years old, this can be an extended period—10–15 years—of principally same-sex interaction. Later, when they begin mating heterosexually, they still retain their primary male pair-bonds, and in some populations male pairs and trios cooperate in herding females or in interacting homosexually with Spotted Dolphins. Because only five or six calves are born to a community each year, however, probably no more than half of the adult males are heterosexually active each mating season (and perhaps far fewer if, as some biologists have suggested, only two or three males monopolize all copulations). Males that do not form same-sex pairs may have a more exclusively heterosexual orientation. Female Bottlenose Dolphins probably have a similar pattern of bisexual interactions overlaid on a largely female-centered social framework. Spinner Dolphins seem to be more uniformly bisexual without extensive periods of exclusive homosexuality, often alternating between same-sex and opposite-sex interactions in quick succession (this sort of concurrent bisexuality has also been observed in Bottlenose and Atlantic Spotted Dolphins). In captivity, though, Spinners exhibit a continuum, with homosexual activity making up only 10 percent of some individuals' behavior, half to two-thirds for other animals, while some Dolphins interact almost exclusively with animals of the same sex.

Nonreproductive and Alternative Heterosexualities

Nonprocreative activities are a hallmark of Dolphin heterosexual interactions. Virtually all of the nonreproductive behaviors described above for same-sex interactions also occur between males and females, including beak-genital propulsion and stimulation of the genitals with the flippers, flukes, and snout. Group sexual activity—much of it heterosexual but nonreproductive—occurs in Spinner wuzzles, and courtship and sexual activity in Bottlenose Dolphins sometimes involves up to ten animals at a time. Female sexuality in Dolphins is often pleasure-oriented, focusing on stimulation of the clitoris as much if not more so than vaginal penetration and insemination. Bottlenose Dolphins mate and interact sexually at all times of the year, not just during the mating season; in Spinner Dolphins (and other species as well), males have a yearly sexual cycle, with significant periods when they are probably unable to fertilize females. In addition, masturbation is a prominent feature of Bottlenose sexual life: both males and females rub their genitals against inanimate objects or other animals, sometimes even developing the activity into a playful "game." Females have even been observed using the muscles of their vaginal region to carry small rubber balls, which they then rub their genitals against. Young Dolphins are sexually precocious, and incestuous copulations have been observed between males a few months old and their mothers. Both male and female Bottlenose Dolphins also interact heterosexually with Atlantic Spotted Dolphins, often using the same sideways mounting position and aggressive behaviors described above for interspecies homosexual encounters. Adults often direct sexual behaviors toward juveniles during these interactions, and female Bottlenose Dolphins have even been seen sideways mounting younger male Spotteds (REVERSE mounting). Many heterosexual interactions in captivity also take place between Dolphins of different species.

Interestingly, this broad variety of heterosexual expression takes place in a larger social framework of primarily separate spheres of activity for males and females, at least in Bottlenose Dolphins. As described above, the two sexes are largely segregated for most of their lives, often socializing in same-sex groups. Furthermore, many animals spend a large portion of their lives uninvolved in breeding: most males do not begin mating until they are at least 20 years old (well beyond the time they become sexually mature), and many Dolphins of that age still do not participate in heterosexual mating. Females breed only once every three to six years, and nearly a quarter of the adult female population may not be involved in reproductive activities at any time. When females do bear calves, they are often assisted by another adult—usually a female—who acts as a "baby-sitter," taking care of the calf while she feeds. Males do not generally parent, and indeed, most Bottlenose calves are sired by males from outside the community. In Spinner Dolphins, "helpers" may be of both sexes, and parental helping behavior has also been observed between Dolphins of different species, for example, by adult Common, Spotted, and Spinner Dolphins toward Bottlenose calves. At times, however, this behavior (within the same species) may be less than "helpful," especially when it involves males. In captivity, "baby-sitting" males have been observed harassing mothers, trying to "kidnap" their calves, and even behaving sexually toward the infants (including trying to mate with them). Pairs and trios of males in some Bottlenose populations also occasionally harass adult females, chasing, herding, and even "kidnapping" and attacking them (e.g., with charges, bites, tail slaps, and body slams) in an attempt to mate with them. Recently, infanticide has also been discovered in some wild Bottlenose communities.

Other Species

Homosexual activity has also been reported in (captive) male Harbor Porpoises (*Phocoena phoecena*) and Commerson's Dolphins (*Cephalorhynchus commersoni*), among others. Intersexual or hermaphrodite individuals (possessing external female genitals along with testes and other internal male reproductive organs) occasionally occur in Striped Dolphins (*Stenella coeruleoalba*).

Sources   *asterisked references discuss homosexuality/transgender*

*Amudin, M. (1974) "Some Evidence for a Displacement Behavior in the Harbor Porpoise, *Phocoena phocoena* (L.). A Causal Analysis of a Sudden Underwater Expiration Through the Blow Hole." *Revue du comportement animal* 8:39–45.

*Bateson, G. (1974) "Observations of a Cetacean Community." In J. McIntyre, ed., *Mind in the Waters,* pp. 146–65. New York: Charles Scribner's Sons.

*Brown, D. H., D. K. Caldwell, and M. C. Caldwell (1966) "Observations on the Behavior of Wild and Captive False Killer Whales, With Notes on Associated Behavior of Other Genera of Captive Delphinids." *Contributions in Science (Los Angeles County Museum of Natural History)* 95:1–32.

*Brown, D. H., and K. S. Norris (1956) "Observations of Captive and Wild Cetaceans." *Journal of Mammalogy* 37:311–26.

*Caldwell, M. C., and D. K. Caldwell (1977) "Cetaceans." In T.A. Sebeok, ed., *How Animals Communicate,* pp. 794–808. Bloomington: Indiana University Press.

*——— (1972) "Behavior of Marine Mammals." In S. H. Ridgway, ed., *Mammals of the Sea: Biology and Medicine,* pp. 419–65. Springfield: Charles C. Thomas.

*——— (1967) "Dolphin Community Life." *Quarterly of the Los Angeles County Museum of Natural History* 5(4):12–15.

*Connor, R. C., and R. A. Smolker (1995) "Seasonal Changes in the Stability of Male-Male Bonds in Indian Ocean Bottlenose Dolphins (*Tursiops* sp.)." *Aquatic Mammals* 21:213–16.

*Connor, R. C., R. A. Smolker, and A. F. Richards (1992) "Dolphin Alliances and Coalitions." In A. H. Harcourt and F. B. M. de Waal, eds., *Coalitions and Alliances in Humans and Other Animals,* pp. 415–43. Oxford: Oxford University Press.

*Dudok van Heel, W. H., and M. Mettivier (1974) "Birth in Dolphins (*Tursiops truncatus*) in the Dolfinarium, Harderwijk, Netherlands." *Aquatic Mammals* 2:11–22.

*Félix, F. (1997) "Organization and Social Structure of the Coastal Bottlenose Dolphin *Tursiops truncatus* in the Gulf of Guayaquil, Ecuador." *Aquatic Mammals* 23:1–16.

*Herzing, D. L. (1996) "Vocalizations and Associated Underwater Behavior of Free-ranging Atlantic Spotted Dolphins, *Stenella frontalis* and Bottlenose Dolphins, *Tursiops truncatus.*" *Aquatic Mammals* 22:61–79.

*Herzing, D. L., and C. M. Johnson (1997) "Interspecific Interactions Between Atlantic Spotted Dolphins (*Stenella frontalis*) and Bottlenose Dolphins (*Tursiops truncatus*) in the Bahamas, 1985–1995." *Aquatic Mammals* 23:85–99.

*Irvine, A. B., M. D. Scott, R. S. Wells, and J. H. Kaufmann (1981) "Movements and Activities of the Atlantic Bottlenose Dolphin, *Tursiops truncatus*, Near Sarasota, Florida." *Fishery Bulletin, U.S.* 79:671–88.

*McBride, A. F., and D. O. Hebb (1948) "Behavior of the Captive Bottle-Nose Dolphin, *Tursiops truncatus*." *Journal of Comparative and Physiological Psychology* 41:111–23.

*Nakahara, F., and A. Takemura (1997) "A Survey on the Behavior of Captive Odontocetes in Japan." *Aquatic Mammals* 23:135–43.

*Nishiwaki, M. (1953) "Hermaphroditism in a Dolphin (*Prodelphinus caeruleo-albus*)." *Scientific Reports of the Whales Research Institute* 8:215–18.

*Norris, K. S., and T. P. Dohl (1980a) "Behavior of the Hawaiian Spinner Dolphin, *Stenella longirostris*." *Fishery Bulletin, U.S.* 77:821–49.

*——— (1980b) "The Structure and Functions of Cetacean Schools." In L. M. Herman, ed., *Cetacean Behavior: Mechanisms and Functions*, pp. 211–61. New York: Wiley-InterScience.

*Norris, K. S., B. Würsig, R. S. Wells, and M. Würsig (1994) *The Hawaiian Spinner Dolphin*. Berkeley: University of California Press.

*Östman, J. (1991) "Changes in Aggressive and Sexual Behavior Between Two Male Bottlenose Dolphins (*Tursiops truncatus*) in a Captive Colony." In K. Pryor and K. S. Norris, eds., *Dolphin Societies: Discoveries and Puzzles*, pp. 304–17. Berkeley: University of California Press.

Patterson, I. A. P., R. J. Reid, B. Wilson, K. Grellier, H. M. Ross, and P. M. Thompson (1998) "Evidence for Infanticide in Bottlenose Dolphins: an Explanation for Violent Interactions with Harbor Porpoises?" *Proceedings of the Royal Society of London*, Series B 265:1167–70.

*Saayman, G. S., and C. K. Tayler (1973) "Some Behavior Patterns of the Southern Right Whale, *Eubalaena australis*." *Zeitschrift für Säugetierkunde* 38:172–83.

Samuels, A., and T. Gifford (1997) "A Quantitative Assessment of Dominance Among Bottlenose Dolphins." *Marine Mammal Science* 13:70–99.

Shane, S. H. (1990) "Behavior and Ecology of the Bottlenose Dolphin at Sanibel Island, Florida." In S. Leatherwood and R. R. Reeves, eds., *The Bottlenose Dolphin*, pp. 245–65. San Diego: Academic Press.

Shane, S. H., R. S. Wells, and B. Würsig (1986) "Ecology, Behavior, and Social Organization of the Bottlenose Dolphin: A Review." *Marine Mammal Science* 2:34–63.

*Tavolga, M. C. (1966) "Behavior of the Bottlenose Dolphin (*Tursiops truncatus*): Social Interactions in a Captive Colony." In K. S. Norris, ed., *Whales, Dolphins, and Porpoises*, pp. 718–30. Berkeley: University of California Press.

*Tayler, C. K., and G. S. Saayman (1973) "Imitative Behavior by Indian Ocean Bottlenose Dolphins (*Tursiops aduncus*) in Captivity." *Behavior* 44:286–98.

*Wells, R. S. (1995) "Community Structure of Bottlenose Dolphins Near Sarasota, Florida." Paper presented at the 24th International Ethological Conference, Honolulu, Hawaii.

*——— (1991) "The Role of Long-Term Study in Understanding the Social Structure of a Bottlenose Dolphin Community." In K. Pryor and K. S. Norris, eds., *Dolphin Societies: Discoveries and Puzzles*, pp. 199–225. Berkeley: University of California Press.

*——— (1984) "Reproductive Behavior and Hormonal Correlates in Hawaiian Spinner Dolphins, *Stenella longirostris*." In W. F. Perrin, R. L. Brownell, Jr., and D. P. DeMaster, eds., *Reproduction in Whales, Dolphins, and Porpoises*, pp. 465–72. Report of the International Whaling Commission, Special Issue 6. Cambridge, UK: International Whaling Commission.

*Wells, R. S., K. Bassos-Hull, and K. S. Norris (1998) "Experimental Return to the Wild of Two Bottlenose Dolphins." *Marine Mammal Science* 14:51–71.

*Wells, R. S., M. D. Scott, and A. B. Irvine (1987) "The Social Structure of Free-ranging Bottlenose Dolphins." In H. Genoways, ed., *Current Mammalogy*, vol. 1, pp. 247–305. New York: Plenum Press.

Dolphins

Marine Mammal

ORCA or KILLER WHALE
Orcinus orca

HOMOSEXUALITY	TRANSGENDER	BEHAVIORS		RANKING	OBSERVED IN
○ Female	○ Intersexuality	● Courtship	● Pair-bonding	● Primary	● Wild
● Male	○ Transvestism	● Affectionate	○ Parenting	○ Moderate	○ Semiwild
		● Sexual		○ Incidental	○ Captivity

IDENTIFICATION: The largest member of the dolphin family (16–26 feet in length); a tall dorsal fin and distinctive black-and-white markings. DISTRIBUTION: Seas and oceans worldwide. HABITAT: Often found in coastal waters. STUDY AREAS: Johnstone Strait, Vancouver Island, British Columbia, Canada; Puget Sound, Washington.

Social Organization

Killer Whales live in a complex society based on a female-centered social unit called the MATRILINEAL GROUP. This is made up of an adult female, the matriarch—usually reproductively active, but sometimes older and postreproductive—her young, and any adult sons of hers. Sometimes her mother or grandmother is also present, and possibly her brothers or uncles. Matrilineal groups usually contain three or four Orcas (although some have up to nine); these groups are organized into larger social units known as PODS, which tend to socialize together and share a common dialect in their vocalizations. Some populations of Killer Whales are TRANSIENTS, who travel widely in smaller groups (occasionally singly) and are less vocal. Unlike nontransient or RESIDENT Orcas, they feed primarily on marine mammals rather than fish.

Description

Behavioral Expression: Homosexual interactions are an integral and important part of male Orca social life. During the summer and fall—when resident pods join together to feast on the salmon runs—males of all ages often spend the afternoons in sessions of courtship, affectionate, and sexual behaviors with each other. A typical homosexual interaction begins when a male Killer Whale leaves his matrilineal group to join a temporary male-only group; a session can last anywhere from a few minutes to more than two hours, with the average length being just over an hour. Usually only two Orcas participate at a time, although groups of three or four males are not uncommon, and even five participants at one time have been observed. The males roll around with each other at the surface, splashing and making frequent body contact as they rub, chase, and gently nudge one another. This is usually accompanied by acrobatic displays such as vigorous slapping of the water with the tail or flippers, lifting the head out of the water (SPYHOPPING), arching the body while floating at the surface or just before a dive, and vocalizing in the air. Particular attention is paid by the males to each other's belly and genital region, and often they initiate a behavior known as BEAK-GENITAL ORIENTATION, which is also seen in heterosexual courtship and mating sequences. Just below the surface of the water, one male swims underneath the other in an upside-down position, touching and nuzzling the other's genital area with his snout or "beak." The two males swim together in this position, maintaining beak-genital contact as the upper one surfaces to breathe; then they dive together, spiraling down into the depths in an elegant double-helix formation. As a variation on this sequence, sometimes one male will arch his tail flukes out of the water just before a dive, allowing the other male to rub his beak against his belly and genital area. When the pair resurfaces after three to five minutes, they repeat the sequence, but with the positions of the two males reversed. In fact, almost 90 percent of all homosexual behaviors are reciprocal, in that the males take turns touching or interacting with one another. During all of these interactions, the Orcas frequently display their erect penises, rolling at the surface or underwater to reveal the distinctive yard-long, pink organs. One male may even attempt to insert his penis—which has a prehensile tip that can be independently moved—into the genital slit of another male (although this has yet to be fully verified).

Although males of all ages participate in homosexual activities, this behavior is most prevalent among "adolescent" Orcas (sexually mature individuals 12–25 years old). More than three-quarters of all sessions involve males who are more than five years apart in age, although age-mates also interact together (especially among adolescents). Occasionally, adult-only homosexual activity (i.e., between males 25 years and older) takes place. Some males have favorite partners with whom they interact year after year, and they may even develop a long-lasting "friendship" or pairing with one particular male. Other males seem to interact with a wide variety of different partners. Most participants in homosexual activity come from different matrilineal groups and are therefore not related; however, more than a third of the sessions include brothers or half brothers (along with other participants), while 9 percent are entirely incestuous.

Frequency: Homosexual interactions are common in Orcas during the summer and fall, especially during August and September—anywhere from 6 to 30 or more sessions of same-sex activity may occur each season in some populations. On average, each male participates in one or two sessions each season, spending about 10 percent of his time in this activity; however, some males may be involved in as many as seven or eight sessions and devote more than 18 percent of their time to this behavior. Overall, in some populations more than three-quarters of observed sexual activity occurs between males.

Orientation: Anywhere from one-third to more than one-half of all males engage in homosexual interactions. This behavior is especially prevalent among younger Orcas: adolescent males participate four times more often than adults.

Many males that engage in this behavior are probably bisexual, since they also court and mate with females. However, there are clear differences between individual males in their affinity or "preference" for homosexual interactions: some Killer Whales participate often and actively seek out male partners, while others are much less involved.

Nonreproductive and Alternative Heterosexualities

Orca communities contain a sizable population of older, nonbreeding females. With an average life span of 50 years—and a possible maximum longevity of 80 years—female Killer Whales can experience a postreproductive period of up to 30 years. In some populations, one-third to one-half of adult females are postreproductive, and it is estimated that a stable population can support as many as two-thirds postreproductive females. Many such females are the matriarchs of their group, and their leadership continues even if it means the ultimate demise of the pod: if there are only male offspring, a pod will eventually disappear upon the death of its matriarch, since there are no breeding females to continue the matriline. Many postreproductive females, while not breeding themselves, act as "babysitters" or helpers in an elaborate communal parenting system. They, along with breeding females, nonbreeding adult and adolescent females, and adult males, frequently take care of calves when their mothers are away or attending to a sibling. Since most breeding females reproduce only once every five years, there is a large pool of potential helpers who are not themselves parents in the population. It is estimated that each calf may be baby-sat as often as once a day during particularly busy times. Although postreproductive females no longer procreate, they may still participate in sexual activity, often with younger males. Several other types of nonprocreative heterosexuality also occur among Orcas: pregnant females have been observed engaging in courtship and sexual behavior with males, while heterosexual interactions also occur between adults or adolescents (of both sexes) and youngsters (juveniles as well as calves). Some incestuous sexual activity has also been documented, for example between an adolescent male and his juvenile sister. Finally, heterosexual interactions do not always involve just two individuals—sometimes a trio of two males and a female will engage in courtship activity together, and one male may even touch and hold the female while the other copulates with her.

Other Species

Same-sex activity occurs in several other species of toothed whales. Pairs of male Sperm Whales (*Physeter macrocephalus*) that may be homosexually bonded occur in some populations. In the waters surrounding New Zealand, for example, 3–5 percent of males are found in such pairs, probably belonging to a semiresident population. These male couples travel together closely and are usually composed of two adults or one older and one younger male. Sexual interactions leading to orgasm also take place in groups of (primarily younger) male Sperm Whales off the coast of Dominica. Homosexual activity has been seen in captive male Beluga Whales (*Delphinapterus leucas*) as well. In addition, hermaphrodite individuals occasionally occur in Belugas and possibly also in Sperm Whales. One Beluga, for example, had male external genitalia combined with a complete set of both male and female internal reproductive organs (i.e., two ovaries and two testes).

Sources

asterisked references discuss homosexuality/transgender

*Ash, C. E. (1960) "Hermafrodite spermhval/Hermaphrodite Sperm Whale." *Norsk Hvalfangst-Tidende* 49:433.

*Balcomb, K. C. III, J. R. Boran, R. W. Osborne, and N. J. Haenel (1979) "Observations of Killer Whales (*Orcinus orca*) in Greater Puget Sound, State of Washington." Unpublished report, Moclips Cetological Society, Friday Harbor, Wash.; 46 pp. (available at National Marine Mammal Laboratory Library, Seattle, Wash.).

*De Guise, S., A. Lagacé, and P. Béland (1994) "True Hermaphroditism in a St. Lawrence Beluga Whale (*Delphinapterus leucas*)." *Journal of Wildlife Diseases* 30:287–90.

Ford, J. K. B., G. M. Ellis, and K. C. Balcomb (1994) *Killer Whales: The Natural History and Genealogy of Orcinus orca in British Columbia and Washington State*. Vancouver: UBC Press; Seattle: University of Washington Press.

*Gaskin, D. E. (1982) *The Ecology of Whales and Dolphins*. London: Heinemann.

*——— (1971) "Distribution and Movements of Sperm Whales (*Physeter catodon* L.) in the Cook Strait Region of New Zealand." *Norwegian Journal of Zoology* 19:241–59.

*——— (1970) "Composition of Schools of Sperm Whales *Physeter catodon* Linn. East of New Zealand." *New Zealand Journal of Marine and Freshwater Research* 4:456–71.

*Gewalt, W. (1976) *Der Weisswal*, Delphinapterus leucas [The Beluga]. Wittenberg: A. Ziemsen-Verlag

*Gordon, J., and R. Rosenthal (1996) "Sperm Whales: The Real Moby Dick." BBC-TV productions, UK.

*Haenel, N. J. (1986) "General Notes on the Behavioral Ontogeny of Puget Sound Killer Whales and the Occurrence of Allomaternal Behavior." In B. C. Kirkevold and J. S. Lockard, eds., *Behavioral Biology of Killer Whales*, pp. 285–300. New York: Alan R. Liss.

*Jacobsen, J. K. (1990) "Associations and Social Behaviors Among Killer Whales (*Orcinus orca*) in the Johnstone Strait, British Columbia, 1979–1986." Master's thesis, Humboldt State University.

*——— (1986) "The Behavior of *Orcinus orca* in the Johnstone Strait, British Columbia." In B.C. Kirkevold and J. S. Lockard, eds., *Behavioral Biology of Killer Whales*, pp. 135–85. New York: Alan R. Liss.

Martinez, D. R., and E. Klinghammer (1978) "A Partial Ethogram of the Killer Whale (*Orcinus orca* L.)." *Carnivore* 1:13–27.

Olesiuk, P. F., M. A. Bigg, and G. M. Ellis (1990) "Life History and Population Dynamics of Resident Killer Whales (*Orcinus orca*) in the Coastal Waters of British Columbia and Washington State." In P. S. Hammond, S. A. Mizroch, and G. P. Donovan, eds., *Individual Recognition of Cetaceans: Use of Photo-Identification and Other Techniques to Estimate Population Parameters*, pp. 209–43. Report of the International Whaling Commission, Special Issue 12. Cambridge, UK: International Whaling Commission.

*Osborne, R. W. (1986) "A Behavioral Budget of Puget Sound Killer Whales." In B. C. Kirkevold and J. S. Lockard, eds., *Behavioral Biology of Killer Whales*, pp. 211–49. New York: Alan R. Liss.

Reeves, R. R., and E. Mitchell (1988) "Distribution and Seasonality of Killer Whales in the Eastern Canadian Arctic." In J. Sigurjónsson and S. Leatherwood, eds., *North Atlantic Killer Whales*, pp. 136–60. *Rit Fiskideildar (Journal of the Marine Research Institute, Reykjavík)*, vol. 11. Reykjavík: Hafrannsóknastofnunin.

*Rose, N.A. (1992) "The Social Dynamics of Male Killer Whales, *Orcinus orca*, in Johnstone Strait, British Columbia." Ph.D. thesis, University of California–Santa Cruz.

*Saulitis, E. L. (1993) "The Behavior and Vocalizations of the 'AT' Group of Killer Whales (*Orcinus orca*) in Prince William Sound, Alaska." Master's thesis, University of Alaska.

*Utrecht, W. L. van (1960) "Notat om den hermafrodite spermhval/Note on the 'Hermaphrodite Sperm Whale.'" *Norsk Hvalfangst-Tidende* 49:520.

BALEEN WHALES

Marine Mammal

GRAY WHALE
Eschrichtius robustus

HOMOSEXUALITY	TRANSGENDER	BEHAVIORS		RANKING	OBSERVED IN
● Female	○ Intersexuality	○ Courtship	● Pair-bonding	● Primary	● Wild
● Male	○ Transvestism	● Affectionate	○ Parenting	○ Moderate	○ Semiwild
		● Sexual		○ Incidental	○ Captivity

IDENTIFICATION: A baleen whale (fringed plates in the mouth are used to filter food) reaching 38–48 feet in length and 27–37 tons (males are slightly smaller than females); characterized by its grayish color, tufts of bristly hairs on its head, and distinctive white splotches and bumps on skin surface that differ like "fingerprints" for each individual. DISTRIBUTION: West coast of North America from Baja California to Arctic Ocean; from southern Korea and Japan to Sea of Okhotsk. HABITAT: Shallow coastal waters, fjordlike inlets, open oceans. STUDY AREA: Wickaninnish Bay, Vancouver Island, British Columbia, Canada.

Social Organization

For eight months of the year—during the migration and summering periods—Gray Whales generally travel and socialize in sex-segregated groups (sometimes known as PODS), while for the remaining time the two sexes are together. Gray Whales have one of the longest migration routes of any mammal: they spend their summers feeding in northern waters, while in the fall they head south on a four-month journey to the mangrove lagoons off Baja California where they mate and their calves are born, only to return to their northern waters in the early spring. A few populations of Gray Whales are nonmigratory, remaining year-round in northern waters.

Description

Behavioral Expression: Male homosexual interactions among Gray Whales occur frequently in the northern summering waters and during the northward migration. Sexual and affectionate activities occur close to the surface of the water in long sessions lasting anywhere from 30 minutes to more than an hour and a half. Often more than two males are involved, sometimes as many as four or five. The whales begin by rolling around each other and onto their sides, with much splashing of water, flailing of fins and flukes at the surface, and occasional slapping of the surface and blowing; sometimes two males rise out of the water several feet in a throat-to-throat position. The whales rub their bellies together and position themselves so that their genital areas are in contact, and usually one or more has an arching, erect or semi-erect penis (which is a distinctive light pink in color and may be

▲ *Penis intertwining between two male Gray Whales off the coast of Vancouver Island (the erect organs of the two males are visible above the surface of the water)*

three to five feet in length and a foot in circumference at its base). Often two or more males intertwine their penises above the water surface, or one male may lay his erect penis on another male's belly or perhaps nudge the other's penis with his head. Female homosexual interactions may also occur.

Gray Whales also frequently form same-sex companionships (pairs and trios) that travel and feed together throughout the summer (without necessarily engaging in sexual activity with one another). They swim in an intimate side-by-side position, often with their side fins touching, and travel back and forth along the length of coastal inlets for hours at a time, apparently with no particular purpose other than to be together. Such companions also perform synchronized blowing and diving maneuvers, including feeding and BREACHING (an acrobatic leap two-thirds out of the water, landing with a dramatic splash on their sides or backs). Two whales also often roll over and under each other, rubbing bellies. Both short-term and long-term (recurring) pair- and trio-bonds occur: some last only for a few hours or days, with the whales changing partners several times over the summer; other companionships endure from year to year.

Frequency: Homosexual activity is fairly common in Gray Whales outside the breeding season, and can be seen perhaps as often as five or six times a month in the early spring in some populations. Actual frequencies may be higher, since much sexual activity probably occurs underwater or in locations that are otherwise difficult to observe. At least a quarter of all companion pairs and trios are same-sex.

Orientation: The majority of male Gray Whales are probably bisexual, interacting primarily with each other on the migrations and during the summer in the north, and interacting heterosexually in the calving waters in the south.

Nonreproductive and Alternative Heterosexualities

Migration in Gray Whales involves a lot more than travel: feeding, nursing, and sexual activity all take place during the journey. Migrating pods are separated according to the sex, age, and reproductive status of the whales: on the northward migration, for example, newly pregnant females usually leave first, then adult males, followed by nonovulating and immature females, immature males, and lastly females with new calves. Male Gray Whales also have a distinct seasonal sexual cycle related to sperm production: during the northward migration and the summer months, their testes are essentially inactive, producing little or no sperm; on the southward migration sperm production resumes and peaks in late fall and early winter in preparation for heterosexual breeding. Consequently, for two-thirds of the year, male Gray Whales are nonfertile, even though they sometimes engage in heterosexual copulation during these times. Combined with the female sexual cycle (with its infertile period in the spring and summer) and the fact that sexual interactions outside the mating season may involve groups of whales (not all of whom copulate), this means that a significant portion of heterosexual activity is nonreproductive. Furthermore, heterosexual courtship and copulation during the mating season also sometimes involves groups of up to 18 whales interacting at the same time, and both males and females mate with multiple partners. During the actual mating act, trios consisting of two males and one female are sometimes involved: one male is a "helper" who does not interact sexually with the female, but seems instead to assist the other two to align their bodies and maintain their position during copulation. Females usually breed only every other year, and some wait two years between calves.

Sources   *asterisked references discuss homosexuality/transgender*

Baldridge, A. (1974) "Migrant Gray Whales with Calves and Sexual Behavior of Gray Whales in the Monterey Area of Central California, 1967–1973." *Fishery Bulletin, U.S.* 72:615–18.

Darling, J. D. (1984) "Gray Whales Off Vancouver Island, British Columbia." In M. L. Jones, S. L. Swartz, and S. Leatherwood, eds., *The Gray Whale*, Eschrichtius robustus, pp. 267–87. Orlando: Academic Press.

*———— (1978) "Aspects of the Behavior and Ecology of Vancouver Island Gray Whales, *Eschrichtius glaucus* Cope." Master's thesis, University of Victoria.

*———— (1977) "The Vancouver Island Gray Whales." *Waters: Journal of the Vancouver Public Aquarium* 2:4–19.

Fay, F. H. (1963) "Unusual Behavor of Gray Whales in Summer." *Psychologische Forschung* 27:175–76.

Hatler, D. F., and J. D. Darling (1974) "Recent Observations of the Gray Whale in British Columbia." *Canadian Field-Naturalist* 88:449–59.

Houck, W. J. (1962) "Possible Mating of Gray Whales on the Northern California Coast." *Murrelet* 43:54.

Rice, D. W., and A. A. Wolman (1971) *The Life History and Ecology of the Gray Whale* (Eschrichtius robustus). American Society of Mammalogists Special Publication no. 3. Stillwater, Okla.: American Society of Mammalogists.

Samaras, W. F. (1974) "Reproductive Behavior of the Gray Whale, *Eschrichtius robustus,* in Baja, California." *Bulletin of the Southern California Academy of Sciences* 73(2):57–64.

Sauer, E. G. F. (1963) "Courtship and Copulation of the Gray Whale in the Bering Sea at St. Lawrence Island, Alaska." *Psychologische Forschung* 27:157–74.

Swartz, S. L. (1986) "Demography, Migration, and Behavior of Gray Whales *Eschrichtius robustus* (Lilljeborg, 1861) in San Ignacio Lagoon, Baja California Sur, Mexico and in Their Winter Range." Ph.D. thesis, University of California–Santa Cruz.

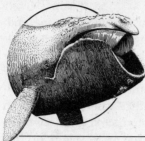

RIGHT WHALES

Marine Mammal 🐬

BOWHEAD WHALE

Balaena mysticetus

RIGHT WHALE

Balaena glacialis

HOMOSEXUALITY	TRANSGENDER	BEHAVIORS		RANKING	OBSERVED IN
● Female	● Intersexuality	○ Courtship	○ Pair-bonding	○ Primary	● Wild
● Male	○ Transvestism	● Affectionate	○ Parenting	● Moderate	○ Semiwild
		● Sexual		○ Incidental	○ Captivity

BOWHEAD WHALE
IDENTIFICATION: A black, 50–65-foot whale with a huge head and arched jaw comprising 40 percent of its total length. DISTRIBUTION: Arctic waters of Canada and Greenland; Barents Sea. HABITAT: Ice-edge waters, bays, straits, estuaries. STUDY AREA: Isabella Bay, Baffin Island, Canada.

RIGHT WHALE
IDENTIFICATION: A 50–60-foot whale weighing up to 104 tons, whose enormous jaws are often encrusted with barnacles and callosites. DISTRIBUTION: Temperate and subarctic waters worldwide; vulnerable. HABITAT: Primarily oceangoing, but closer to land during breeding season. STUDY AREA: Near Valdés Peninsula, Argentina; subspecies *B.g. australis,* the Southern Right Whale.

Social Organization

Bowhead Whales socialize and travel in small groups of 2–7 animals as well as larger herds of 50–60 individuals; many animals are solitary as well. During much of the year—e.g., the spring migrations, and the socializing and feeding periods of summer and early fall—males and females (as well as different age groups) generally associate separately from each other. Right Whales may form aggregations of 100 or more individuals, although most social interactions occur during the mating period.

Description

Behavioral Expression: Intensive sexual encounters between male Bowhead Whales take place in shallow waters, involving three to six males at a time. Amid much splashing and churning of water, the males roll over each other with erect (unsheathed) penises, caress one another, slap the surface of the water with their tails or flippers, chase each other, and perform TAIL LOFTS, in which the tail is lifted high above the water while the whale sinks vertically down. Generally there is one central whale that the others are trying to copulate with, although this whale often rolls belly up in the water, perhaps attempting to avoid their advances (similar to the behavior of females during heterosexual mating activity). Nevertheless, a male

Bowhead sometimes inserts his penis into the genital slit of another male. Sessions of homosexual activity can last for 40 minutes or more, during which males often produce loud and complex vocalizations that resemble roars, screams, or trumpetings. Both male and female Right Whales also engage in homosexual activity, involving such behaviors as caressing, rolling and pushing, and flipper and fluke slaps.

Bowhead Whales also have a relatively high incidence of intersexual or hermaphrodite individuals with female external genitalia and mammary glands combined with male chromosomes and internal sexual organs such as testes (which are contained within the body cavity in this species, as in other cetaceans).

Frequency: Homosexual activity is characteristic of certain times of the year: in Bowheads, it generally occurs during the late summer and fall, while in Right Whales, it occurs early in the season for females, and late in the season for males. Beyond this, it is difficult to quantify the frequency of same-sex interactions. Among Bowheads, social activity is common during the fall, and about 40 percent of all socializing groups include three or more whales (the configuration typical of sexual interactions). Although the exact percentage of these interactions that are homosexual is not known, in two out of three such groups in which the sexes of all the animals could be determined, the sexual activity involved only males. It is possible, therefore, that a significant proportion of fall sexual activity—perhaps even a majority—is homosexual. Intersexuality in Bowheads is relatively common, occurring in about 1 in 4,000 individuals (compared with a rate of 1 in 62,400 humans for the same type of intersexuality).

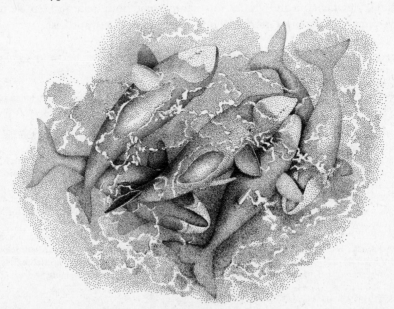

▲ *Aerial view of six male Bowhead Whales participating in intensive homosexual activity at the surface of the water; some of the males are displaying erections*

Orientation: In Bowhead Whales, homosexual behavior appears to be typical of adolescent or younger adult males, so it may be that individuals engage in sequential or chronological bisexuality over their lives, with an initial period of homosexuality followed by heterosexuality. This is speculative, however, because the life histories of individual whales have not been tracked. In Right Whales, homosexual behavior is not restricted to younger animals, but in fact occurs among whales of all ages; the extent of heterosexual activities (if any) of such individuals are not fully known.

Nonreproductive and Alternative Heterosexualities

Because Bowhead and Right Whales generally mate throughout the year—and in particular outside of the female's fertilizable period—a large proportion of heterosexual activity is nonprocreative. In both species, heterosexual copulation usually involves a group of several males trying to mate with one female, who often tries to escape their attentions. At times, the interaction can become violent: groups of male Right Whales searching for females have been described as "rape gangs," and sometimes two or more males cooperate in forcing a female underwater so that they can take turns mating with her. In some cases, calves get caught in the middle of a heterosexual mating attempt and are hit, crushed, and perhaps even killed. Females of these two species generally do not breed every year. In Right Whales, for example, five or more years may elapse between calves, with the result that sometimes less than half of the adult females in an area are breeding.

Other Species

Intersexuality or transgender has also been reported among Fin Whales (*Balaenoptera physalus*): one individual, for example, had both male and female reproductive organs, including a uterus, vagina, elongated clitoris, and testes.

Sources   *asterisked references discuss homosexuality/transgender*

*Bannister, J. L. (1963) "An Intersexual Fin Whale *Balaenoptera physalus* (L.) from South Georgia." *Proceedings of the Zoological Society of London* 141:811–22.

Clark, C. W. (1983) "Acoustic Communication and Behavior of the Southern Right Whale (*Eubalaena australis*)." In R. S. Payne, ed., *Communication and Behavior of Whales*, pp. 163–98. American Association for the Advancement of Science Selected Symposium 76. Boulder: Westview Press.

Everitt, R. D., and B. D. Krogman (1979) "Sexual Behavior of Bowhead Whales Observed Off the North Coast of Alaska." *Arctic* 32:277–80.

Finley, K. J. (1990) "Isabella Bay, Baffin Island, an Important Historical and Present-Day Concentration Area for the Endangered Bowhead Whale (*Balaena mysticetus*) of the Eastern Canadian Arctic." *Arctic* 43:137–52.

*Koski, W. R., R. A. Davis, G.W. Miller, and D. E. Withrow (1993) "Reproduction." In J. J. Burns, J. J. Montague, and C. J. Cowles, eds., *The Bowhead Whale*, pp. 239–74. Lawrence, Kans.: Society for Marine Mammalogy.

*Moore, S. E., and R. R. Reeves (1993) "Distribution and Movement." In J. J. Burns, J. J. Montague, and C. J. Cowles, eds., *The Bowhead Whale*, pp. 313–86. Lawrence, Kans.: Society for Marine Mammalogy.

*Östman, J. (1991) "Changes in Aggressive and Sexual Behavior Between Two Male Bottlenose Dolphins (*Tursiops truncatus*) in a Captive Colony." In K. Pryor and K. S. Norris, eds., *Dolphin Societies: Discoveries and Puzzles*, pp. 304–17. Berkeley: University of California Press.

Payne, R. (1995) *Among Whales*. New York: Scribner.

*Richardson, W. J., and K. J. Finley (1989) *Comparison of Behavior of Bowhead Whales of the Davis Strait and Bering/Beaufort Stocks*. Report from LGL Ltd., King City, Ontario, for U.S. Minerals Management

RIGHT WHALES | 359

Service, Herndon, Va.; OCS Study MMS 88–0056, NTIS no. PB89–195556/AS. Springfield, Va.: National Technical Information Service.

*Richardson, W. J., K. J. Finley, G.W. Miller, R. A. Davis, and W. R. Koski (1995) "Feeding, Social, and Migration Behavior of Bowhead Whales, *Balaena mysticetus,* in Baffin Bay *vs.* the Beaufort Sea—Regions with Different Amounts of Human Activity." *Marine Mammal Science* 11:1–45.

Saayman, G. S., and C. K. Tayler (1973) "Some Behavior Patterns of the Southern Right Whale, *Eubalaena australis." Zeitschrift für Säugetierkunde* 38:172–83.

*Tarpley, R. J., G. H. Jarrell, J. C. George, J. Cubbage, and G. G. Stott (1995) "Male Pseudohermaphroditism in the Bowhead Whale, *Balaena mysticetus." Journal of Mammalogy* 76:1267–75.

*Würsig, B., and C. Clark (1993) "Behavior." In J. J. Burns, J. J. Montague, and C. J. Cowles, eds., *The Bowhead Whale,* pp. 157–99. Lawrence, Kans.: Society for Marine Mammalogy.

*Würsig, B., J. Guerrero, and G. Silber (1993) "Social and Sexual Behavior of Bowhead Whales in Fall in the Western Arctic: A Re-examination of Seasonal Trends." *Marine Mammal Science* 9:103–11.

SEALS AND MANATEES

Common or "Earless" Seals   Marine Mammal

GRAY SEAL
Halichoerus grypus
NORTHERN ELEPHANT SEAL
Mirounga angustirostris
HARBOR SEAL
Phoca vitulina

HOMOSEXUALITY	TRANSGENDER	BEHAVIORS		RANKING	OBSERVED IN
● Female	○ Intersexuality	● Courtship	○ Pair-bonding	● Primary	● Wild
● Male	● Transvestism	● Affectionate	● Parenting	○ Moderate	○ Semiwild
		● Sexual		○ Incidental	● Captivity

GRAY SEAL
IDENTIFICATION: A large seal (up to 7 feet in males) with an elongated muzzle and a spotted coat. DISTRIBU-
TION: North Atlantic waters, including northeastern North America (especially Newfoundland), Iceland,
British Isles, Norway, Kola Peninsula, Baltic Sea. HABITAT: Temperate and subarctic waters; breeds and molts
on rocky coasts and islands. STUDY AREA: Ramsay Island, England.

NORTHERN ELEPHANT SEAL
IDENTIFICATION: One of the largest seals, reaching a length of up to 16 feet and a weight of 5,500 pounds
(in males); adult males have a prominent proboscis. DISTRIBUTION: North Pacific waters from Alaska to Baja
California. HABITAT: Oceangoing; breeds and molts on islands and coasts. STUDY AREA: Año Nuevo State Reserve, California.

HARBOR SEAL
IDENTIFICATION: A smaller, round-headed seal with grayish brown, usually spotted fur. DISTRIBUTION: Waters of
the North Atlantic and North Pacific. HABITAT: Coastal reefs, sandbars, rocks. STUDY AREAS: Otter Island, Pribilof Islands, Alaska; Nanvak Bay, Cape Newenham National Wildlife Refuge, Alaska; Seaside Aquarium,
Oregon; subspecies *P.v. richardsi,* the Pacific Harbor Seal.

Social Organization

Gray Seals are highly gregarious, congregating in large colonies for mating and
molting, and in large groups to feed. In some populations the mating system is primarily polygynous, meaning that males mate with multiple partners, do not form

heterosexual pair-bonds, and do not participate in parenting. However, some individuals in these areas are "monogamous" in that they mate with the same partner year after year, while in other populations the majority of individuals mate with only one partner but not necessarily the same one each year. Northern Elephant Seals are more solitary when at sea, although they form large breeding and molting aggregations in traditional areas known as ROOKERIES and also have a polygynous mating system. Harbor Seals generally congregate in mixed-sex groups on land, anywhere from a dozen to several thousand animals; they often mate in the water, however, and appear to have a polygynous mating system as well.

Description

Behavioral Expression: During the nonbreeding season, both Gray Seal and Northern Elephant Seal bulls engage in homosexual activity. Gray Seals come ashore to molt their fur, gathering in groups of up to 150 animals, no more than half a dozen of which are females. Those males who have completed their molt often roll around in pairs near the water and mount each other; bulls of all ages participate in this activity. Both adolescent and younger adult Northern Elephant Seal bulls also engage in homosexual mounting during the molt period. This occurs in shallow waters near the shore, often as a part of extended bouts of harmless play-fighting among clusters of males. Prior to and during the mating season, this activity continues among adolescent males, though it is usually no longer aquatic. Adolescent males often spend time in male-only areas that are separate from the breeding grounds. Males are attracted to the play-fighting and mounting activity in these areas and may travel up to 100 yards through the rookery to join in. Adult bulls do not participate in this activity. However, they do sometimes mount younger adolescent and juvenile males (two to four years old). Typically the older male approaches a younger male at rest, moving up alongside him and sometimes placing his front flipper over his back in the position characteristic of heterosexual copulations. Usually the younger male struggles to escape, and the mounter may try to subdue him by pressing or bouncing his neck down on top of him or biting his neck. The older male may have an erection and attempt to penetrate the younger male, but he rarely if ever succeeds. Although they prefer juveniles or adolescents, a few bulls also try to copulate with much younger animals, such as weaned pups of both sexes (who strongly resist their advances).

Homosexual activity is also prevalent in male Harbor Seals in the form of PAIR-ROLLING: two males embrace and mount each other in the water, continuously twisting and writhing about one another while maintaining full body contact. Rolling can become quite vigorous as the two animals spiral synchronously underwater and at the surface (often in a vertical position), sometimes gently mouthing or biting each other's neck, chasing each other's flippers, yelping and snarling, blowing streams of bubbles underwater, or slapping the surface of the water. One male usually has an erection, and the bout of courtship rolling typically ends when he mounts the other male, grasping him from behind and maintaining this position for up to 3 minutes (sometimes sinking to the bottom in shallow waters). The two males may also take turns mounting each other. Heterosexual copulations,

▲ *Two male Harbor Seals "pair-rolling"*
(a courtship and sexual behavior)

in contrast, can occur both in the water and on land; they are not usually accompanied by pair-rolling and can last for up to 15 minutes. Although males of all ages engage in pair-rolling, most participants are adults (sexually mature individuals over six years old) or adolescents.

Same-sex courtship or sexual behavior is not found among females in these species, but two cow Seals occasionally co-parent a pup. In Northern Elephant Seals, for example, two females who have each lost their own pup sometimes adopt an orphan and raise it together, or (more commonly) a cow who has lost her pup associates with a mother and shares parenting duties with her, including nursing the pup.

Finally, some adolescent Northern Elephant Seal males are transvestite, acting and looking like females. They have the body proportions of cow Seals, and they also deliberately pull in their noses so that they resemble females (who do not have the enlarged snouts that bulls do) and keep their heads low so as not to attract attention. Moving stealthily through the breeding grounds, these younger males try to copulate with females, who, nevertheless, are usually not fooled by their attempts to disguise themselves and usually do not allow them to mate. However, because most adult males are not able to mate with females, some transvestite males are actually more successful at breeding than non-transvestite males.

Frequency: Homosexual activity occurs frequently in Harbor Seals during the late spring, summer, and fall (except during the pupping season). In one two-month study period, for example, pair-rolling between males occurred daily and in total nearly 285 same-sex rolling pairs were observed (during the same period, no heterosexual matings were seen). Homosexual behavior is also common among male Gray Seals during the molting period, less frequent among Elephant Seal bulls (though it occurs at more times of the year in the latter species). Among females, approximately 2 percent of Elephant Seal adoptive families involve two pupless females coparenting orphaned pups, and another 14 percent involve one female sharing the care of a pup with its mother. Overall, these two-mother families probably represent about 2–3 percent of all families (the remainder are single-mother).

Orientation: Male Gray Seals exhibit seasonal bisexuality: during the molting period, many bulls participate in preferential homosexual activity—generally ignoring any females present in the herd—while during the mating season heterosexual behavior is the norm. However, only about a third of older males actually copulate with females, while less than 2 percent of younger males (up to eight years old) regularly have access to females. Thus, many bulls probably engage exclusively in homosexual activity for at least part of their lives. Younger adolescent male Elephant Seals—who make up 25–55 percent of the male population—may participate primarily in homosexual mounting, since few actually mate with females. At the other extreme, the highest-ranking bulls are probably exclusively heterosexual, since their attentions are usually directed toward mating (often with hundreds of females each season). Some older adolescent males (40–55 percent of the population) or younger adults may be bisexual, mounting both males and females. However, since less than 9 percent of all males ever mate with females during their lifetime, and less than half of those males surviving to breeding age ever mate, a large number may participate only in same-sex activity (as in Gray Seals). Bulls who mount pups—only a fraction of the male population—do so with equal frequency on both male and female pups. In Harbor Seals, males participate in pair-rolling activity with one another even in the presence of receptive females and generally do so for several months each year (heterosexual mating is usually restricted to a shorter period, perhaps a month or so). Similar patterns of sexual orientation among different age classes probably occur in this species as in the other two.

Nonreproductive and Alternative Heterosexualities

Gray, Harbor, and Northern Elephant Seals engage in a wide variety of nonprocreative heterosexual behaviors. Sexual activity during pregnancy is not uncommon. When female Gray Seals come ashore just before their pups are born, for example, they often participate in heterosexual copulation and other sexual interactions with males, including REVERSE mountings (in which they mount the male rather than vice versa). Male Northern Elephant Seals also mate with pregnant females, including cows who are leaving the breeding grounds after having already been inseminated. Gray and Harbor Seals sometimes copulate outside of the mating season when fertilization is impossible—not only because the females are pregnant, but because (in Grays) males have their own sexual cycle that renders their testes inactive at that time. Heterosexual matings also occasionally occur between these two species. In addition, females in all three species may copulate with multiple male partners.

As noted above, some male Northern Elephant Seals try to copulate with weaned pups—about half of all pups are subjected to such forced mating or rape attempts, which they usually violently resist. In some cases the pups are severely injured by the bulls, with deep gashes and punctures from neck bites. Aggressive sexual behavior by bulls is the leading cause of mortality among pups on the breeding grounds, accounting for the deaths of about 1 in 200 pups each year. Male Northern Elephant Seals also sometimes aggressively mount pups of other species such as Harbor Seals. Similar aggression, violence, and attempted rape—sometimes lethal—is also directed by bulls toward adult females and adolescents. During mating, male Northern Elephant Seals routinely bite, pin down, and slam the full weight of their bodies against females (bulls are 5–11 times heavier than females). A female may be pursued by groups of males as she leaves the rookery, sometimes being raped three to seven times as she tries to escape. Some bulls even try to mate with dead females that have been killed during such attacks (and even with dead seals of other species). Mating in Harbor Seals may also involve aggressive attacks by males, female refusal, and even "gangs" of two or three males forcibly copulating with a female. In addition, Gray and Harbor Seal pups are sometimes killed by adults (accounting for about 7 percent of Gray Seal pup deaths), while roughly 6 percent of Harbor Seal pups are abandoned by their mothers shortly after birth.

For much of the year, the two sexes lead largely segregated lives: Northern Elephant Seal males and females, for example, each embark on their own epic migratory journeys twice a year. Males travel farther north to Alaska while females journey out into the central Pacific, remaining separate for up to 300 days as they traverse more than 13,000 miles in their double migrations. Male Gray Seals are at sea (or molting on land) essentially separate from females for nine to ten months of the year. This segregation is facilitated in part by the phenomenon of DELAYED IM- PLANTATION, in which a female's fertilized embryo remains in "suspended animation" for three to four months, extending the duration of the pregnancy to eleven or more months. Even during the breeding season, many males do not copulate or reproduce: usually only 14–35 percent of the males present on the breeding grounds mate each season. Likewise, more than 90 percent of male Elephant Seals never copulate during their entire lives (most delay breeding until fairly late and simply perish before reaching the age when reproduction usually begins). Because a small number of individuals often monopolize mating opportunities, some populations may experience high levels of inbreeding. In addition, about 20 percent of females skip breeding each year in some populations.

Separation of the sexes continues through pup-rearing: like most polygamous animals, male Seals do not participate in any parenting duties. Females, however, engage in an assortment of fostering activities, often after they have lost their own pup (although some take care of other pups in addition to their own). More than half of all Northern Elephant Seal pups become separated from their mothers each season, and about 18 percent of all females adopt pups. Besides the female coparenting arrangements mentioned above, many females adopt orphan pups on their own, some female Elephant Seals nurse several orphans at once, while others nurse already weaned pups (who become bloated from the extra milk, turning into gigantic "superweaners," as they are called). Some females even try to "kidnap" or steal pups away from their own mothers, and females who have not lost their own pup often threaten, attack, and even kill stray pups. As many as a quarter to three-quarters of female Gray Seals and 10 percent of female Harbor Seals participate in foster-parenting in some populations.

Other Species

Pairs of female Spotted or Larga Seals (*Phoca largha*) occasionally coparent their pups together, even sharing in nursing their offspring. Male Sea Otters (*Enhydra lutris*) have been observed clasping and mounting other males (in the water) in the position usually seen during heterosexual matings. Male Sea Otters also sometimes mount and attempt to mate with Seals, including Harbor Seals and Northern Elephant Seals, and some of these interactions may be same-sex.

Sources   *asterisked references discuss homosexuality/transgender*

Allen, S. G. (1985) "Mating Behavior in the Harbor Seal." *Marine Mammal Science* 1:84–87.

Amos, B., S. Twiss, P. Pomeroy, and S. Anderson (1995) "Evidence for Mate Fidelity in the Gray Seal." *Science* 268:1897–99.

Anderson, S. S. (1991) "Gray Seal, *Halichoerus grypus*." In G. B. Corbet and S. Harris, eds., *The Handbook of British Mammals*, pp. 471–80. Oxford: Blackwell Scientific Publications.

Anderson, S. S., and M. A. Fedak (1985) "Gray Seal Males: Energetic and Behavioral Links Between Size and Sexual Success." *Animal Behavior* 33:829–38.

*Backhouse, K. M. (1960) "The Gray Seal (*Halichoerus grypus*) Outside the Breeding Season: A Preliminary Report." *Mammalia* 24:307–12.

——— (1954) "The Gray Seal." *University of Durham Medical Gazette* 48:9–16.

*Backhouse, K. M., and H. R. Hewer (1957) "Behavior of the Gray Seal (*Halichoerus grypus* Fab.) in the Spring." *Proceedings of the Zoological Society of London* 129:450.

Baker, J. R. (1984) "Mortality and Morbidity in Gray Seal Pups (*Halichoerus grypus*)." *Journal of Zoology, London* 203:23–48.

Bishop, R. H. (1967) "Reproduction, Age Determination, and Behavior of the Harbor Seal, *Phoca vitulina* L. in the Gulf of Alaska." Master's thesis, University of Alaska.

Boness, D. J., D. Bowen, S. J. Iverson, and O. T. Oftedal (1992) "Influence of Storms and Maternal Size on Mother-Pup Separations and Fostering in the Harbor Seal, *Phoca vitulina*." *Canadian Journal of Zoology* 70:1640–44.

Boness, D. J., and H. James (1979) "Reproductive Behavior of the Gray Seal (*Halichoerus grypus*) on Sable Island, Nova Scotia." *Journal of Zoology, London* 188:477–500.

Burton, R. W., S. S. Anderson, and C. F. Summers (1975) "Perinatal Activities in the Gray Seal (*Halichoerus grypus*)." *Journal of Zoology, London* 177:197–201.

Coulson, J. C., and G. Hickling (1964) "The Breeding Biology of the Gray Seal, *Halichoerus grypus* (Fab.), on the Farne Islands, Northumberland." *Journal of Animal Ecology* 33:485–512.

*Deutsch, C. J. (1990) "Behavioral and Energetic Aspects of Reproductive Effort of Male Northern Elephant Seals (*Mirounga angustirostris*)." Ph.D. thesis, University of California–Santa Cruz.

*Hatfield, B. B., R. J. Jameson, T. G. Murphey, and D. D. Woodard (1994) "Atypical Interactions Between Male Southern Sea Otters and Pinnipeds." *Marine Mammal Science* 10:111–14.

Hewer, H. R. (1960) "Behavior of the Gray Seal (*Halichoerus grypus* Fab.) in the Breeding Season." *Mammalia* 24:400–21.

Hewer, H. R., and K. M. Backhouse (1960) "A Preliminary Account of a Colony of Gray Seals, *Halichoerus grypus* (Fab.) in the Southern Inner Hebrides." *Proceedings of the Zoological Society of London* 134:157–95.

Hoover, A. A. (1983) "Behavior and Ecology of Harbor Seals (*Phoca vitulina richardsi*) Inhabiting Glacial Ice in Aialik Bay, Alaska." Master's thesis, University of Alaska.

*Johnson, B.W. (1976) "Studies on the Northernmost Colonies of Pacific Harbor Seals, *Phoca vitulina richardsi*, in the Eastern Bering Sea." Unpublished report, University of Alaska Institute of Marine Science and Alaska Department of Fish and Game; 67 pp. (available at National Marine Mammal Laboratory Library, Seattle, Wash.).

*——— (1974) "Otter Island Harbor Seals: A Preliminary Report." Unpublished report, University of Alaska Institute of Marine Science and Alaska Department of Fish and Game; 20 pp. (available at National Marine Mammal Laboratory Library, Seattle, Wash.).

*Johnson, B. W., and P. Johnson (1977) "Mating Behavior in Harbor Seals?" *Proceedings (Abstracts) of the Conference on the Biology of Marine Mammals* (San Diego) 2:30.

Kroll, A. M. (1993) "Haul Out Patterns and Behavior of Harbor Seals, *Phoca vitulina*, During the Breeding Season at Protection Island, Washington." Master's thesis, University of Washington.

*Le Boeuf, B. J. (1974) "Male-Male Competition and Reproductive Success in Elephant Seals." *American Zoologist* 14:163–76.

——— (1972) "Sexual Behavior in the Northern Elephant Seal, *Mirounga angustirostris*." *Behavior* 41:1–26.

Le Boeuf, B. J., and R. M. Laws (eds.) (1994) *Elephant Seals: Population Ecology, Behavior, and Physiology*. Berkeley: University of California Press.

Le Boeuf, B. J., and S. Mesnick (1991) "Sexual Behavior of Male Northern Elephant Seals: I. Lethal Injuries to Adult Females." *Behavior* 116:143–62.

Le Boeuf, B. J., and J. Reiter (1988) "Lifetime Reproductive Success in Northern Elephant Seals." In T. H. Clutton-Brock, ed., *Reproductive Success: Studies of Individual Variation in Contrasting Breeding Systems,* pp. 344–62. Chicago: University of Chicago Press.

Mortenson, J., and M. Follis (1997) "Northern Elephant Seal (*Mirounga angustirostris*) Aggression on Harbor Seal (*Phoca vitulina*) Pups." *Marine Mammal Science* 13:526–30.

Perry, E. A., and W. Amos (1998) "Genetic and Behavioral Evidence That Harbor Seal (*Phoca vitulina*) Females May Mate with Multipile Males." *Marine Mammal Science* 14:178–82.

Riedman, M. L. (1990) *The Pinnipeds: Seals, Sea Lions, and Walruses.* Berkeley: University of California Press.

*Riedman, M. L., and B. J. Le Boeuf (1982) "Mother-Pup Separation and Adoption in Northern Elephant Seals." *Behavioral Ecology and Sociobiology* 11:203–15.

*Rose, N. A., C. J. Deutsch, and B. J. Le Boeuf (1991) "Sexual Behavior of Male Northern Elephant Seals: III. The Mounting of Weaned Pups." *Behavior* 119:171–92.

Smith, E. A. (1968) "Adoptive Suckling in the Gray Seal." *Nature* 217:762–63.

Stewart, B. S., and R. L. DeLong (1995) "Double Migrations of the Northern Elephant Seal, *Mirounga angustirostris.*" *Journal of Mammalogy* 76:196–205.

Sullivan, R. M. (1981) "Aquatic Displays and Interactions in Harbor Seals, *Phoca vitulina*, with Comments on Mating Systems." *Journal of Mammalogy* 62:825–31.

Thompson, P. (1988) "Timing of Mating in the Common Seal (*Phoca vitulina*)." *Mammal Review* 18:105–12.

Tinker, M. T., K. M. Kovacs, and M. O. Hammill (1995) "The Reproductive Behavior and Energetics of Male Gray Seals (*Halichoerus grypus*) Breeding on a Land-Fast Ice Substrate." *Behavioral Ecology and Sociobiology* 36:159–70.

Venables, U. M., and L. S. V. Venables (1959) "Vernal Coition of the Seal *Phoca vitulina* in Shetland," *Proceedings of the Zoological Society of London* 132:665–69.

——— (1957) "Mating Behavior of the Seal *Phoca vitulina* in Shetland." *Proceedings of the Zoological Society of London* 128:387–96.

Wilson, S. C. (1975) "Attempted Mating Between a Male Gray Seal and Female Harbor Seals." *Journal of Mammalogy* 56:531–34.

EARED SEALS

Marine Mammal

AUSTRALIAN SEA LION
Neophoca cinerea
NEW ZEALAND SEA LION
Phocarctos hookeri
NORTHERN FUR SEAL
Callorhinus ursinus

HOMOSEXUALITY	TRANSGENDER	BEHAVIORS		RANKING	OBSERVED IN
● Female	○ Intersexuality	○ Courtship	○ Pair-bonding	○ Primary	● Wild
● Male	○ Transvestism	● Affectionate	○ Parenting	● Moderate	○ Semiwild
		● Sexual		○ Incidental	● Captivity

AUSTRALIAN, NEW ZEALAND SEA LIONS
IDENTIFICATION: Males reach 8–10 feet in length and have a massive neck, shoulders, and mane; females are smaller. DISTRIBUTION: Southern coast of Australia; South Island of New Zealand and other subantarctic islands to the south; New Zealand species is vulnerable. HABITAT: Coastal waters and adjacent shores. STUDY AREAS: Dangerous Reef Island, South Australia; Enderby Island, Auckland Islands, New Zealand.

NORTHERN FUR SEAL
IDENTIFICATION: A dark brown or gray seal with extremely dense fur, large flippers and a relatively small head; females reach 4½ feet in length, males nearly 7 feet. DISTRIBUTION: North Pacific waters; vulnerable. HABITAT: Oceangoing; breeds onshore. STUDY AREAS: St. Paul and St. George Islands, Pribilof Islands, Alaska.

Social Organization

During the breeding season, females of each of these species aggregate into groups among which a smaller number of males are distributed. This organization has been mistakenly interpreted as a "harem" structure, when in fact males have little control over the movements of the females, whose grouping is often the result of seeking each other out while simultaneously avoiding males. In the Northern Fur Seal, nonbreeding animals and juveniles gather on islands separate from the breeding areas; nonbreeding groups also occur in the other two species. Outside of the mating season, Australian and New Zealand Sea Lions form smaller mixed-sex groups, while Northern Fur Seals are oceangoing, relatively solitary, and sex-segregated during the fall, winter, and spring.

Description

Behavioral Expression: In Australian and New Zealand Sea Lions, homosexual mounting is common: one male mounts the other from behind (as in heterosexual copulation) and makes pelvic thrusts against the other male. Homosexual copulation can take place either on the beach or in the surf (the latter especially among older males). All age groups participate, although there is often an age difference between the two males, with the younger male typically mounting the older one (especially in New Zealand Sea Lions). Among younger males, homosexual behavior is often a component of play-fighting, in which the two males stand chest to chest and push against one another, with each trying to grab the other's neck in his mouth. Female Australian Sea Lions also occasionally mount one another, but lesbian mounting is more common in Northern Fur Seals. During the mating season, one female sometimes copulates with another by mounting her and performing pelvic thrusts; the mounted female often facilitates the homosexual mount by arching her back and extending her flippers, thereby making her genital region more accessible to the other female.

Frequency: Homosexual behavior occurs fairly frequently in Australian and New Zealand Sea Lions and occasionally in Northern Fur Seals.

Orientation: In Australian and New Zealand Sea Lions, younger males that do not associate with female groups may engage exclusively in homosexual activity, since many such individuals (which make up 81 percent of the New Zealand, and 33 percent of the Australian, male population) do not mate heterosexually. In New Zealand Sea Lions, adult breeding males also sometimes participate in homosexual mounting, making them bisexual, whereas in Australian Sea Lions, adult breeding males are exclusively heterosexual. In Northern Fur Seals, all females that participate in lesbian mounting are also active heterosexually, since they also mate with males. In fact, nearly all females involved in homosexual activity are mothers, although not all mothers engage in homosexual mounting. However, the amount of homosexual activity that an individual female participates in may be equal to or greater than her heterosexual activity, since females usually mate with a male only once during the entire breeding season.

Nonreproductive and Alternative Heterosexualities

In all three of these eared seals, significant proportions of the population do not breed. As noted above, more than 80 percent of New Zealand Sea Lion adult males, and a third of Australian Sea Lion males, do not participate in heterosexual mating. In Northern Fur Seals most males younger than nine years old do not mate because they cannot compete with older males for access to females, while most breeding males actually participate in reproduction for only a single season out of their entire lives. The average male copulates with females only 3–4 times during his life, and many males never do so. In addition, 8–17 percent of females on the breeding grounds do not get pregnant each year, and females generally reproduce only once every five years or so. In fact, nearly 60 percent of the total population consists of nonbreeders that do not even attempt to reproduce. Australian Sea Lions are unusual among mammals in having an exceptionally long or "supra-annual" breeding cycle, 17–18 months from mating to birth (most mammals complete the cycle in less than a year, allowing them to breed annually). Part of the reason for this extended cycle is because of the phenomenon known as DELAYED IMPLANTATION (also found in other seals), in which the fertilized egg fails to develop and instead remains in "suspended animation"—in this species, for as long as eight or nine months. In addition, late-term abortions are relatively common in Australian Sea Lions. Implantation in Northern Fur Seals is delayed for four to five months, but about 11 percent of embryos fail to implant or are aborted or reabsorbed.

In addition to this separation between insemination and fetal development, there is notable spatial and temporal segregation of the sexes in these species. In Northern Fur Seals, males and females go separate ways once the mating season is over: females range widely over the north Pacific Ocean while males remain in the Bering Sea. Since males and females interact for only two months out of the year, this means that the majority of their lives are spent apart from each other. Furthermore, the two sexes are often antagonistic when they are together. Male Northern Fur Seals sometimes try to prevent females from leaving their territory by throwing them bodily and flipping them into the air. Two males may also try to claim the same female by grabbing her with their jaws—sometimes while she is actually giving birth—causing severe lacerations or even death in the resulting tug-of-war. In Australian Sea Lions, "gangs" of younger males roam through the colony, sexually harassing females and attacking those that try to get away. A male New Zealand Sea Lion was once observed trying to copulate with a dead female New Zealand fur seal (*Arctocephalus forsteri*), which he may have killed during a previous mating attempt. Sexual interactions between Northern Fur Seals and California sea lions (*Zalophus californianus*) also occur, and male Northern Fur Seals have also been known to attempt forcible copulation with pups of their own species. In addition, male (and occasionally female) Australian Sea Lions often savagely attack pups, shaking, tossing, and biting them. Death from the resulting injuries is the primary cause of mortality for pups on land, accounting for nearly a fifth of all pup deaths in this species. About 17 percent of all Northern Fur Seal pup fatalities are due to attacks from adults (usually females).

In spite of these severe obstacles facing adult females and young seals, a number of innovative shared parenting or "day-care" arrangements have developed in these species. In Australian Sea Lions, females take turns watching over and defending a group of pups. In Northern Fur Seals, pups gather in nursery groups or PODS for protection while their mothers are away at sea—which is most of the time, since mothers usually spend only one day ashore each week and may be gone for up to 16 days at a time. Female Australian Sea Lions that have lost their own pup also sometimes try to abduct another female's youngster.

Sources   *asterisked references discuss homosexuality/transgender*

*Bartholomew, G. A. (1959) "Mother-Young Relations and the Maturation of Pup Behavior in the Alaska Fur Seal." *Animal Behavior* 7:163–71.

Bartholomew, G. A., and P. G. Hoel (1953) "Reproductive Behavior of the Alaska Fur Seal, *Callorhinus ursinus*." *Journal of Mammalogy* 34:417–36.

Gales, N. J., P. D. Shaughnessy, and T. E. Dennis (1994) "Distribution, Abundance, and Breeding Cycle of the Australian Sea Lion *Neophoca cinerea* (Mammalia: Pinnipedia)." *Journal of Zoology, London* 234:353–70.

*Gentry, R. L. (1998) *Behavior and Ecology of the Northern Fur Seal.* Princeton: Princeton University Press.

——— (1981) "Northern Fur Seal, *Callorhinus ursinus* (Linnaeus, 1758)." In S. H. Ridgway and R. J. Harrison, eds., *Handbook of Marine Mammals, vol. 1*, pp. 143–60. London: Academic Press.

Higgins, L. V. (1993) "The Nonannual, Nonseasonal Breeding Cycle of the Australian Sea Lion, *Neophoca cinerea*." *Journal of Mammalogy* 74:270–74.

Higgins, L. V., and R. A. Tedman (1990) "Attacks on Pups by Male Australian Sea Lions, *Neophoca cinerea*, and the Effect on Pup Mortality." *Journal of Mammalogy* 71:617–19.

Kenyon, K. W., and F. Wilke (1953) "Migration of the Northern Fur Seal, *Callorhinus ursinus*." *Journal of Mammalogy* 34:86–89.

*Marlow, B. J. (1975) "The Comparative Behavior of the Australasian Sea Lions *Neophoca cinerea* and *Phocarctos hookeri* (Pinnipedia: Otariidae)." *Mammalia* 39:159–230.

——— (1972) "Pup Abduction in the Australian Sea-lion, *Neophoca cinerea*." *Mammalia* 36:161–65.

Miller, E. H., A. Ponce de Léon, and R. L. DeLong (1996) "Violent Interspecific Sexual Behavior by Male Sea Lions (Otariidae): Evolutionary and Phylogenetic Implications." *Marine Mammal Science* 12:468–76.

Peterson, R. S. (1968) "Social Behavior in Pinnipeds with Particular Reference to the Northern Fur Seal." In R. J. Harrison, R. C. Hubbard, R. S. Peterson, C. E. Rice, and R. J. Schusterman, eds., *The Behavior and Physiology of Pinnipeds*, pp. 3–53. New York: Appleton-Century-Crofts.

Walker, G. E., and J. K. Ling (1981) "New Zealand Sea Lion, *Phocarctos hookeri* (Gray, 1844)" and "Australian Sea Lion, *Neophoca cinerea* (Péron, 1816)." In S. H. Ridgway and R. J. Harrison, eds., *Handbook of Marine Mammals, vol. 1*, pp. 25–38, 99–118. London: Academic Press.

Wilson, G. J. (1979) "Hooker's Sea Lions in Southern New Zealand." *New Zealand Journal of Marine and Freshwater Research* 13:373–75.

York, A. E., and V. B. Scheffer (1997) "Timing of Implantation in the Northern Fur Seal, *Callorhinus ursinus*." *Journal of Mammalogy* 78:675–83.

OTHER PINNIPEDS   Marine Mammal
WALRUS
Odobenus rosmarus

HOMOSEXUALITY	TRANSGENDER	BEHAVIORS		RANKING	OBSERVED IN
○ Female	○ Intersexuality	● Courtship	● Pair-bonding	● Primary	● Wild
● Male	○ Transvestism	● Affectionate	○ Parenting	○ Moderate	○ Semiwild
		● Sexual		○ Incidental	● Captivity

IDENTIFICATION: A huge pinniped (up to 12 feet and 3,500 pounds in males) with sparsely furred, brownish orange skin, a bristled "mustache," and prominent tusks in adult males. DISTRIBUTION: Throughout the Arctic. HABITAT: Shallow waters, coastal areas, ice floes. STUDY AREAS: Round Island, Bristol Bay, Alaska; Coats Island, Hudson Bay; Bathurst and Dundas Islands, Northwest Territories, Canada; New York Aquarium; subspecies *O.r. divergens*, the Pacific Walrus, and *O.r. rosmarus*, the Atlantic Walrus.

Social Organization

From January to March (the breeding season), Walruses congregate far from shore on and around pack ice, where heterosexual courtship and mating take place. The mating system is polygynous—males generally copulate with several different females without forming long-lasting heterosexual bonds. During the summer and early fall, males gather together in large aggregations numbering in the thousands, typically on islands that are used for this purpose year after year. These "haul-outs" are sex-segregated. When both males and females are present, they tend to occupy distinct areas; more typically, however, the haul-outs are entirely male, since females and their young migrate to the far north to spend the summer.

Description

Behavioral Expression: In the shallow waters off the coast of the summer haul-out grounds, male Walruses engage in homosexual courtship, sexual, and affectionate activities. Pairs of males—sometimes as many as 50 animals at a time—float at the surface of the water by inflating special pouches in their throat, which act like the buoyant sacs in a life vest. While floating and swimming, the males—especially younger ones—rub their bodies against one another, clasp and embrace each other with their front flippers, touch noses, and loll together in groups. Males sometimes even sleep together in the water—pairs or groups of males float vertically at the surface (a posture known as BOTTLING), one behind the other, each male clasping the one in front of him in a "sleeping line." Males also perform courtship displays for other males, employing a number of extraordinary behaviors and sounds that are also used in heterosexual courtship. Typically a younger male displays to an older one, and courting males often situate themselves in the water near a favorite cliff face, boulder, or rock formation at the water's edge. The spectacular display consists of inflation of the throat pouches—often preceded by head bowing—interspersed with dives by one or both males, and an incredible series of vocalizations that form a courtship "song." In homosexual encounters, at least three types of calls are used: KNOCKS, which are rapid, clicklike sounds resembling castanets, produced by "chattering" the cheek teeth underwater; a metallic BELL call, which is an eerie gonglike sound thought to be produced underwater by striking the throat pouches with the flippers to generate air pulses; and a short, piercing WHISTLE made through pursed lips when the Walrus resurfaces.

During same-sex courtship, sometimes one male rubs his erect, arm-sized penis with his front flipper. Overt sexual behavior between males takes the form of mounting (in the shallow water): one male clasps

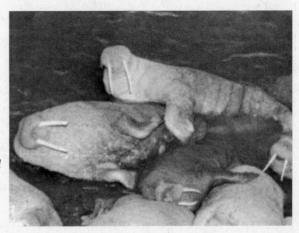

A male Walrus mounting another male off the coast of Round Island (Alaska), thrusting his erect penis (visible below the surface of the water) in the other male's anal region. ▶

another with his flippers from behind, thrusting his pelvis and erect penis against the other male's anal region. Younger males mount older ones and vice versa. Although most homosexual behavior is confined to the summer haul-outs, younger males sometimes also mount adults or other younger males during the breeding season. Groups of younger Walruses may crowd around an older male and roll on top of him; in addition, an adult male occasionally sings his courtship songs to a group of younger males or is accompanied by a younger attendant male while he sings. Often, his companion surfaces and dives in synchrony with him. This behavior, known as SHADOWING, occurs regardless of whether the adult is courting a female. In addition, male Walruses in captivity have been observed participating in cross-species homosexual encounters with male Gray Seals.

Frequency: Male Walruses engage in homosexual activities frequently during the summer haul-outs: approximately a quarter of all social interactions of males in the shallow waters involve same-sex affectionate, courtship, or sexual behaviors. On average, each male participates in such activity roughly five times per hour when in the water, and contact between males (including sexual and courtship behaviors) occupies about 3 percent of the overall time spent by males in the water. Up to 2 percent of the total male population (which may number more than 3,000) may be in the water at any one time engaging in homosexual activities. During the breeding season, up to a third of mounting activity may take place between younger males or between adult and younger males, and 2–19 percent of singing males have a younger male companion "shadowing" them.

Orientation: Because male Walruses achieve sexual maturity approximately four years before they actually participate in breeding, a sizable segment of the population is probably involved exclusively in homosexuality for a portion of their lives. Roughly 40–60 percent of all sexually mature males—those between the ages of 10 and 14—do not participate in heterosexual mating, and a large proportion of these males engage in homosexual activity. Most older males are seasonally bisexual, courting and copulating with females during the mating season and participating in same-sex activities during the summer and early fall; some, however, also engage in same-sex mounting and companionships during the breeding season.

Nonreproductive and Alternative Heterosexualities

As noted above, male Walruses experience a delay in their breeding careers, since most do not begin mating heterosexually until about 40 percent of their maximum life span is over. Once they do begin procreating, often only a small percentage of males actually copulates each year, perhaps as few as a quarter of all adult males. Among females, about half of the individuals over the age of 23 are nonbreeding, experiencing a postreproductive or "menopausal" period that may last for 7 or more years. Because of their long pregnancy (15–16 months) and nursing period (two years), most breeding-age females do not reproduce every year. One reason the pregnancy is so long is because of DELAYED IMPLANTATION: after mating, the fertilized egg fails to implant in the uterus and temporarily stops developing, remaining in "suspended animation" for four to five months. This results in an even longer period that the two sexes are separated. Although their social life is characterized by extensive sex segregation (see above), Walruses do sometimes copulate outside of the mating season. Heterosexual matings have been recorded in nearly all months of the spring and fall. Because males have a distinct yearly sexual cycle—their testes are essentially dormant except during January–March—most of this sexual activity is nonreproductive. Even during the mating season, Walruses participate in a variety of nonprocreative sexual behaviors. REVERSE mountings, in which the female clasps the male from behind and mounts him, occur in the water, and younger, sexually immature individuals also mount each other. Females have also been observed sucking their partner's penis as well as kneading it to erection with their flippers prior to copulation (in captivity). Outside of the breeding season, males masturbate by stroking their penis with a front flipper, sometimes accompanied by knock sounds or STRUM calls (the latter sounding something like the strumming of fingers on a guitar or zither). In addition, interspecies heterosexual copulations with Gray Seals have been seen in captivity.

As in many other polygamous mammals, female Walruses generally raise their young on their own, occasionally supplemented with a number of alternative parenting arrangements. Unrelated females and males may assist in the care and protection of calves, nursery groups of youngsters sometimes play together while their mothers are occupied during the mating season, and orphaned calves are commonly adopted by other mothers or nonbreeding females. Occasionally, females even try to steal or "kidnap" calves from other females. Unfortunately, the lives of a calf and its mother are often endangered by male violence. Calves are sometimes gored by a male's tusks, while mass tramplings may occur on haul-out sites—often triggered by belligerent Walrus bulls roaming through groups of females and their young. In some locations, such stampedes occur regularly, littering the beach with hundreds and even thousands of carcasses each year. Nearly a quarter of all fatalities are calves less than six months old, while 15 percent are aborted fetuses.

Sources   *asterisked references discuss homosexuality/transgender*

Born, E. W., and L. Ø. Knutson (1997) "Haul-out and Diving Activity of Male Atlantic Walruses (*Odobenus rosmarus rosmarus*) in NE Greenland." *Journal of Zoology, London* 243:381–96.

Dittrich, L. (1987) "Observations on Keeping the Pacific Walrus *Odobenus rosmarus divergens* at Hanover Zoo." *International Zoo Yearbook* 26:163–70.

Eley, T. J., Jr. (1978) "A Possible Case of Adoption in the Pacific Walrus." *Murrelet* 59:77–78.

Fay, F. H. (1982) *Ecology and Biology of the Pacific Walrus,* Odobenus rosmarus divergens *Illiger*. North American Fauna, no. 74. Washington, D.C.: U.S. Department of the Interior, Fish and Wildlife Service.

———— (1960) "Structure and Function of the Pharyngeal Pouches of the Walrus (*Odobenus rosmarus* L.)." *Mammalia* 24:361–71.

*Fay, F. H., G. C. Ray, and A. A. Kibal'chich (1984) "Time and Location of Mating and Associated Behavior of the Pacific Walrus, *Odobenus rosmarus divergens* Illiger." In F. H. Fay and G. A. Fedoseev, eds., *Soviet-American Cooperative Research on Marine Mammals, vol. 1: Pinnipeds,* pp. 89–99. NOAA Technical Report NMFS 12. Washington, D.C.: U.S. Department of Commerce.

Fay, F. H., and B. P. Kelly (1980) "Mass Natural Mortality of Walruses (*Odobenus rosmarus*) at St. Lawrence Island, Bering Sea, Autumn 1978." *Arctic* 33:226–45.

*Mathews, R. (1983) "The Summer-Long Bachelor Party on Round Island." *Smithsonian* 14:68–75.

Miller, E. H., (1985) "Airborne Acoustic Communication in the Walrus *Odobenus rosmarus*." *National Geographic Research* 1:124–45.

*———— (1976) "Walrus Ethology. II. Herd Structure and Activity Budgets of Summering Males." *Canadian Journal of Zoology* 54:704–15.

*———— (1975) "Walrus Ethology. I. The Social Role of Tusks and Applications of Multidimensional Scaling." *Canadian Journal of Zoology* 53:590–613.

*Miller, E. H., and D. J. Boness (1983) "Summer Behavior of Atlantic Walruses *Odobenus rosmarus rosmarus* (L.) at Coats Island, N.W.T. (Canada)." *Zeitschrift für Säugetierkunde* 48:298–313.

Nowicki, S. N., I. Stirling, and B. Sjare (1997) "Duration of Stereotyped Underwater Vocal Displays by Male Atlantic Walruses in Relation to Aerobic Dive Limit." *Marine Mammal Science* 13:566–75.

Ray, G. C., and W. A. Watkins (1975) "Social Function of Underwater Sounds in the Walrus *Odobenus rosmarus*." *Rapports et Procès-Verbaux des Réunions, Conseil International pour l'Exploration de la Mer* 169:524–26.

*Salter, R. E. (1979) "Observations on Social Behavior of Atlantic Walruses (*Odobenus rosmarus* [L.]) During Terrestrial Haul-Out." *Canadian Journal of Zoology* 58:461–63.

Schevill, W. E., W. A. Watkins, and C. Ray (1966) "Analysis of Underwater *Odobenus* Calls with Remarks on the Development and Function of the Pharyngeal Pouches." *Zoologica* 51:103–6.

*Sjare, B., and I. Stirling (1996) "The Breeding Behavior of Atlantic Walruses, *Odobenus rosmarus rosmarus*, in the Canadian High Arctic." *Canadian Journal of Zoology* 74:897–911.

Stirling, I., W. Calvert, and C. Spencer (1987) "Evidence of Stereotyped Underwater Vocalizations of Male Atlantic Walruses (*Odobenus rosmarus rosmarus*)." *Canadian Journal of Zoology* 65:2311–21.

MANATEES   *Marine Mammal*

WEST INDIAN MANATEE
Trichechus manatus

HOMOSEXUALITY	TRANSGENDER	BEHAVIORS		RANKING	OBSERVED IN
○ Female	○ Intersexuality	○ Courtship	○ Pair-bonding	● Primary	● Wild
● Male	○ Transvestism	● Affectionate	○ Parenting	○ Moderate	○ Semiwild
		● Sexual		○ Incidental	○ Captivity

IDENTIFICATION: A large (8–14 foot), streamlined, seal-like animal with a rounded tail, foreflippers but no hind legs, and a thick, hairless skin. DISTRIBUTION: Coastal waters and rivers of southeastern United States, the Caribbean, and northeastern Brazil; vulnerable. HABITAT: Shallow tropical and subtropical waters with abundant aquatic plants. STUDY AREAS: Crystal and Homosassa Rivers, Florida; subspecies *T.m. latirostris*, the Florida Manatee.

Social Organization

West Indian Manatees are generally solitary and only moderately social; however, they may congregate in loose herds of two to six animals. Some herds are cosexual, while others are "bachelor" groups of younger males.

Description

Behavioral Expression: Male West Indian Manatees of all ages regularly engage in intense homosexual activities. In a typical encounter, two males embrace, rub their genital openings against each other, and then unsheathe or erect their penises and rub them together, often to ejaculation. During a homosexual mating, the two males often tumble to the bottom, thrusting against each other and wallowing in the mud as they clasp each other tightly. A wide variety of positions are used, including embracing in head-to-tail and sideways positions, often with interlocking penises or flipper-penis contact. All of these are distinct from the position used for heterosexual copulation, in which the male typically swims underneath the female on his back and mates with her upside down. Lasting for up to two minutes, homosexual copulations are generally four to eight times longer than heterosexual ones. Before they engage in sexual activity, males often "kiss" each other by touching their muzzles at the surface of the water. In addition, several other types of affectionate and tactile activities are a part of homosexual interactions, including mouthing and caressing of each other's body, nibbling or nuzzling of the genital region, and riding by one male on the back of the other (a behavior also seen in heterosexual interactions). Sometimes a male emits vocalizations indicating his pleasure during homosexual activity, variously described as high-pitched squeaks, chirp-squeaks, or snort-chirps. If, however, he is not interested in participating, he may emit a squealing sound, slapping his tail as he flees from the other male (just the way females do when trying to escape from unwanted advances of males).

Often several males participate at the same time in homosexual interactions: groups of up to four animals have been seen kissing, embracing each other in an interlocked "hug," thrusting, and rubbing their penises against one another. These homosexual "orgies" can last for hours as new males arrive to join the group, subgroups form and re-form, and participants leave and return. Homosexual behavior is often part of a social activity known as CAVORTING, in which animals travel and splash about in groups, nuzzling, grabbing, chasing, rubbing, and rolling against one another. Cavorting groups can be mixed-sex or all-male.

Frequency: Homosexuality is common among West Indian Manatees. In addition, males spend on average about 11 percent of their time in cavorting groups.

Orientation: Most male Manatees are probably bisexual, since homosexual behavior is sometimes interspersed with, or develops out of, heterosexual interactions when more than one male is involved. However, much homosexual activity occurs independent of heterosexual activity, and some males may engage primarily in same-sex interactions.

Nonreproductive and Alternative Heterosexualities

Heterosexual interactions in West Indian Manatees often involve considerable harassment and coercion of females by males. Large, jostling herds containing as many as 17–22 males relentlessly pursue females in heat as well as nonfertilizable females, attempting to copulate with them and often following them for weeks at a time. In her attempts to escape from the males, the female may violently slap her tail, twisting and turning as she dives away, or else tear through the underwater vegetation, even plunging into the mud or stranding herself onshore. Calves whose mothers are being pursued sometimes get lost or are fatally fatigued or injured. Female Manatees generally reproduce only once every three years, and at any given time, only about 30–40 percent of all females are reproducing. Most male Manatees have a distinct seasonal sexual cycle as well, with their testes generally dormant and not producing sperm during the winter months. Females raise their young on their own with no help from the males. However, a mother will occasionally allow another female to nurse her calf or may leave her calf in the company of other mothers and/or their calves while she goes off to feed on her own.

Other Species

Homosexual activity has also been observed in Dugongs (*Dugong dugon*), a species of Manatee that inhabits Australasian waters. A pair of captive males, for example, engaged in courtship and sexual behaviors with each other, including rolling, nudging, gentle biting, and splashing, often with erect penises. Although same-sex activity has yet to be documented in wild Dugongs, most observations of mating activity for this species in the wild involve individuals whose sex has not been unequivocally determined.

Sources   *asterisked references discuss homosexuality/transgender*

*Anderson, P. K. (1997) "Shark Bay Dugongs in Summer. I: Lek Mating." *Behavior* 134:433–62.

*Bengtson, J. L. (1981) "Ecology of Manatees (*Trichechus manatus*) in the St. Johns River, Florida." Ph.D. thesis, University of Minnesota.

*Hartman, D. S. (1979) *Ecology and Behavior of the Manatee* (Trichechus manatus) *in Florida*. American Society of Mammalogists Special Publication no. 5. Pittsburgh: American Society of Mammalogists.

*——— (1971) "Behavior and Ecology of the Florida Manatee, *Trichechus manatus latirostris* (Harlan), at Crystal River, Citrus County." Ph.D. thesis, Cornell University.

Hernandez, P. , J. E. Reynolds, III, H. Marsh, and M. Marmontel (1995) "Age and Seasonality in Spermatogenesis of Florida Manatees." In T. J. O'Shea, B. B. Ackerman, and H. F. Percival, eds., *Population Biology of the Florida Manatee*, pp. 84–97. Information and Technology Report 1. Washington, D.C.: U.S. Department of the Interior.

Husar, S. L. (1978) "*Trichechus manatus*." *Mammalian Species* 93:1–5.

*Jones, S. (1967) "The Dugong *Dugong dugon* (Müller): Its Present Status in the Seas Round India with Observations on Its Behavior in Captivity." *International Zoo Yearbook* 7:215–20.

Marmontel, M. (1995) "Age and Reproduction in Female Florida Manatees." In T. J. O'Shea, B. B. Ackerman, and H. F. Percival, eds., *Population Biology of the Florida Manatee*, pp. 98–119. Information and Technology Report 1. Washington, D.C.: U.S. Department of the Interior.

Moore, J. C. (1956) "Observations of Manatees in Aggregations." *American Museum Novitates* 1811:1–24.

*Nair, R. V., R. S. Lal Mohan, and K. Satyanarayana Rao (1975) *The Dugong* Dugong dugon. ICAR Bulletin of the Central Marine Fisheries Research Institute no. 26. Cochin, India: Central Marine Fisheries Research Institute.

Preen, A. (1989) "Observations of Mating Bevavior in Dugongs (*Dugong dugon*)." *Marine Mammal Science* 5:382–87.

*Rathbun, G. B., J. P. Reid, R. K. Bonde, and J. A. Powell (1995) "Reproduction in Free-ranging Florida Manatees." In T. J. O'Shea, B. B. Ackerman, and H. F. Percival, eds., *Population Biology of the Florida Manatee,* pp. 135–56. Information and Technology Report 1. Washington, D.C.: U.S. Department of the Interior.

Reynolds, J. E., III. (1981) "Aspects of the Social Behavior and Herd Structure of a Semi-Isolated Colony of West Indian Manatees, *Trichechus manatus.*" *Mammalia* 45:431–51.

——— (1979) "The Semisocial Manatee." *Natural History* 88(2):44–53.

Reynolds, J. E., III., and D. K. Odell (1991) *Manatees and Dugongs.* New York: Facts on File.

*Ronald, K., L. J. Selley, and E. C. Amoroso (1978) *Biological Synopsis of the Manatee.* Ottawa: International Development Research Center.

Hoofed Mammals

DER

WHITE-TAILED DEER
Odocoileus virginianus

MULE or BLACK-TAILED DEER
Odocoileus hemionus

Hoofed Mammal

HOMOSEXUALITY	TRANSGENDER	BEHAVIORS		RANKING	OBSERVED IN
● Female	● Intersexuality	● Courtship	● Pair-bonding	○ Primary	● Wild
● Male	● Transvestism	○ Affectionate	○ Parenting	● Moderate	○ Semiwild
		● Sexual		○ Incidental	● Captivity

WHITE-TAILED DEER

IDENTIFICATION: A medium-sized deer (approximately 3 feet tall at shoulder) with a white undertail and multipronged antlers that sweep forward. DISTRIBUTION: Southern Canada, United States except Southwest, Mexico south to Bolivia and northeastern Brazil. HABITAT: Varied, from thickets to open country. STUDY AREAS: Welder Wildlife Refuge, Sinton, Texas; Edwards Plateau, Llano County, Texas; subspecies *O.v. texanus*, the Texas White-tailed Deer.

MULE DEER

IDENTIFICATION: A stocky, grayish deer with a black-tipped tail and antlers that branch into two equal portions. DISTRIBUTION: Western North America, northern Mexico. HABITAT: Semiarid forest, brushlands. STUDY AREAS: Waterton and Banff National Parks, Alberta, Canada; University of British Columbia, Canada; near Fort Collins, Colorado; subspecies *O.h. hemionus*, the Rocky Mountain Mule Deer, and *O.h. columbianus*, the Black-tailed Deer.

Social Organization

During most of the year, White-tailed and Mule Deer live in sex-segregated groups: females form groups with other does and their offspring, while males (bucks) live in "bachelor" groups or on their own. During the rutting season, males form short-lived, consecutive "tending bonds" with multiple females—a form of polygamy or "serial monogamy." Larger cosexual groups may also form during the winter.

Description

Behavioral Expression: Adult male White-tailed Deer sometimes mount each other, as do yearling males (especially during the nonbreeding season); occasionally a younger male mounts an older one during this activity. Homosexual mounts (like heterosexual ones) are usually preceded by one male nuzzling the other's rear end, and sometimes one male mounts another twice in a row; occasionally the mounting buck has an erection. The mount may be briefer than a male-female copulation, but the same duration as heterosexual nonreproductive mounts (5–15 seconds, as opposed to 15–20 seconds). Yearling Mule Deer occasionally mount each other during SPARRING MATCHES—ritualized, nonviolent contests in which the bucks lock horns. During this activity one male might assume a stiffened posture, similar to a female's before copulation. The other male—sometimes younger or smaller than the first—then mounts him, after first licking and smelling the special scent-producing glands on his hind legs. Female Mule Deer also sometimes mount each other when in heat; in addition, some does court other females using a chasing sequence known as RUSH COURTSHIP. In this behavior (which also occurs in heterosexual contexts), they race toward another female, stopping abruptly and sometimes pawing the ground, pacing, leaping into the air while twisting their body, or running in circles or figure eights; this causes the other doe to become excited and aroused. Adult male White-tailed Deer frequently develop "companionships" or bonds with one (or occasionally two) other adult males in their buck groups; male companions are generally not related to each other. These strong attachments constitute the stable "core" of each buck group, and although male companions typically separate during the breeding season, they usually resume their bonds once mating is over.

An extraordinary form of transgendered deer occurs in some populations of White-tails. These animals, which are genetically male but actually combine characteristics of both males and females, are sometimes called VELVET-HORNS because their antlers are permanently covered with the special "velvet" skin that in most males is shed after the antlers have grown. Their antlers are usually only spikes (without the extensive branching of other males' antlers) and they slope backward and sometimes have enlarged bases. Physically, velvet-horns often have body proportions and facial features more typical of does, while their testes are small and undeveloped (and in fact the

A male White-tailed Deer mounting another buck ▶

▲ *A transgendered "velvet-horn"*
White-tailed Deer

animals are infertile). A similar type of trans-gender is found among Mule Deer, where the animals are known as CACTUS BUCKS owing to the distinctive shape of their antlers (which sometimes also have elaborate spikes, prongs, and asymmetrical growths). Velvet-horns usually form their own social groups of three to seven animals and live separately from both does and nontransgendered males. In fact, they are often harassed and attacked by other deer. Nontransgendered White-tails (both does and bucks of all ages, even fawns) threaten velvet-horns who try to approach them—forcing them to remain no less than ten feet away—while bucks may actively charge velvet-horns to drive them away. When threatened, velvet-horns flee without giving the standard alarm signals of other deer (stamping their feet, snorting or whistling, and raising their tails). Some-times, groups of up to six bucks "gang up" on a velvet-horn, chasing and even vio-lently attacking it by gashing its rump with their antlers. As a result, velvet-horns are extremely wary around other deer, venturing near feeding areas cautiously and always remaining in groups on the periphery, or else refusing to approach at all when other deer are present. Interestingly, velvet-horns are almost always in supe-rior physical condition compared to nontransgendered males, precisely because they do not breed. The rutting season is extremely taxing on breeding bucks, who rarely eat and may lose up to a quarter of their body weight. In contrast, velvet-horns consistently have excellent body fat deposits and are in prime shape.

Two types of gender-mixing females also occur among White-tailed and Mule Deer, both bearing antlers (females in these species do not usually have antlers). In one type, the antlers are similar to those of velvet-horns: they are permanently cov-ered in velvet, are never shed, and are either spikes or asymmetrically branching. Unlike velvet-horns, such females are usually fertile, mating heterosexually and be-coming mothers. The other type is a more complete form of intersexuality: the antlers are hard and polished, more closely resemble those of males in their branch-ing structure, and may even be shed seasonally. These individuals usually combine both male and female sexual traits, such as having the genitalia and/or reproductive organs of both sexes, or partial organs of each sex, or chromosomes of one sex com-bined with the genitalia of the other.

Frequency: Homosexual mounting probably occurs only occasionally among White-tailed and Mule Deer; however, in one study of White-tails, two out of ten observed mountings were same-sex. Up to 10 percent of males are velvet-horns in some areas, although their incidence fluctuates. In some years they may constitute as many as 40–80 percent of all males in a given population. One study of a White-tailed Deer population over 14 years found that 1–2 percent of the females had antlers; overall, approximately 1 in every 1,000–1,100 does is antlered.

Orientation: Most Deer that participate in same-sex mounting probably also engage in heterosexual courtship and copulation. Gender-mixing Deer that are fertile (almost always genetically female) are usually heterosexual (i.e., they mate with genetic males), while nonfertile transgendered Deer (e.g., velvet-horns) are probably asexual or associate only with other transgendered Deer.

Nonreproductive and Alternative Heterosexualities

Deer participate in a variety of nonprocreative sexual behaviors besides homosexuality. White-tails sometimes engage in heterosexual mountings outside of the mating season, which are nonreproductive for two reasons. They often do not involve penetration, and bucks have a seasonal sexual cycle, so that during the spring and summer their testes are small and produce little, if any, sperm. Mating episodes among Mule Deer during the breeding season often involve the male performing extensive non-insertive sexual activity prior to actual copulation: in this activity he mounts the female with his penis erect (unsheathed) but without penetration. These mounts may be fairly lengthy—up to 15 seconds—and frequent (anywhere from 5 to more than 40 in one session). Bucks of both species sometimes masturbate in a unique fashion: the penis is first unsheathed and licked, then stimulated by moving it back and forth (via pelvic rotations and thrusts) in its sheath or against the belly until orgasm is reached. Because their antlers are actually sensitive—even erotic—organs (as in several other species of Deer), buck Mule Deer also sometimes sexually stimulate themselves by rubbing their antlers on vegetation. Incestuous activity—including fawns mounting their mothers—also occurs in these species.

As mentioned above, sex segregation is a notable feature of White-tailed Deer society. This pattern usually begins during the fawning period, when does become aggressive toward adult males and may even kick and chase them away. When their male fawns become yearlings, females also drive them away in the same violent fashion. In addition to nonbreeding transgendered animals, other nonreproducing individuals occur. White-tail bucks often do not mate until they are three to five years old; because of the physical stresses of reproduction, bucks that delay breeding may actually grow larger than those that reproduce earlier. When breeding does occur, females of both species sometimes terminate their pregnancies by aborting the fetus or reabsorbing the embryo. This probably occurs in 1–10 percent of Mule Deer pregnancies, but is more likely to happen when unfavorable climate and forage would make it difficult for mothers to feed and care for their young.

Sources *asterisked references discuss homosexuality/transgender*

*Anderson, A. E. (1981) "Morphological and Physiological Characteristics." In O. C. Wallmo, ed, *Mule and Black-tailed Deer of North America*, pp. 27–97. Lincoln and London: University of Nebraska Press.
*Baber, D. W. (1987) "Gross Antler Anomaly in a California Mule Deer: The 'Cactus' Buck." *Southwestern Naturalist* 32:404–6.

Brown, B. A. (1974) "Social Organization in Male Groups of White-tailed Deer." In V. Geist and F. Walther, eds., *The Behavior of Ungulates and Its Relation to Management*, vol. 1, pp. 436–46. IUCN Publication no. 24. Morges, Switzerland: International Union for Conservation of Nature and Natural Resources.

*Cowan, I. McT. (1946) "Antlered Doe Mule Deer." *Canadian Field-Naturalist* 60: 11–12.

*Crispens, C. G., Jr., and J. K. Doutt (1973) "Sex Chromatin in Antlered Female Deer." *Journal of Wildlife Management* 37:422–23.

*Donaldson, J. C., and J. K. Doutt (1965) "Antlers in Female White-tailed Deer: A 4-Year Study." *Journal of Wildlife Management* 29:699–705.

*Doutt, J. K., and J. C. Donaldson (1959) "An Antlered Doe With Possible Masculinizing Tumor." *Journal of Mammalogy* 40:230–36.

*Geist, V. (1981) "Behavior: Adaptive Strategies in Mule Deer." In O. C. Wallmo, ed., *Mule and Black-tailed Deer of North America*, pp. 157–223. Lincoln and London: University of Nebraska Press.

*Halford, D. K., W. J. Arthur III, and A. W. Alldredge (1987) "Observations of Captive Rocky Mountain Mule Deer Behavior." *Great Basin Naturalist* 47:105–9.

*Hesselton, W. T., and R. M. Hesselton (1982) "White-tailed Deer." In J. A. Chapman and G. A. Feldhamer, eds., *Wild Mammals of North America: Biology, Management, and Economics*, pp. 878–901. Baltimore and London: Johns Hopkins University Press.

*Hirth, D. H. (1977) "Social Behavior of White-Tailed Deer in Relation to Habitat." *Wildlife Monographs* 53:1–55.

Jacobson, H. A. (1994) "Reproduction." In D. Gerlach, S. Atwater, and J. Schnell, eds., *Deer*, pp. 98–108. Mechanicsburg, Pa.: Stackpole Books.

Marchinton, R. L., and D. H. Hirth (1984) "Behavior." in L. K. Halls, ed., *White-tailed Deer: Ecology and Management*, pp. 129–68. Harrisburg, Pa.: Stackpole Books; Washington, DC: Wildlife Management Institute.

Marchinton, R. L. and W. G. Moore (1971) "Auto-erotic Behavior in Male White-tailed Deer." *Journal of Mammalogy* 52:616–17.

*Rue, L. L., III (1989) *The Deer of North America*. 2nd ed. Danbury, Conn.: Outdoor Life Books.

Sadleir, R. M. F. S. (1987) "Reproduction of Female Cervids." In C. M. Wemer, ed., *Biology and Management of the Cervidae*, pp. 123–44. Washington, D.C.: Smithsonian Institution Press.

Salwasser, H., S. A. Holl, and G. A. Ashcraft (1978) "Fawn Production and Survivial in the North Kings River Deer Herd." *California Fish and Game* 64:38–52.

*Taylor, D. O. N., J. W. Thomas, and R. G. Marburger (1964) "Abnormal Antler Growth Associated with Hypogonadism in White-tailed Deer of Texas." *American Journal of Veterinary Research* 25:179–85.

*Thomas, J. W., R. M. Robinson, and R. G. Marburger (1970) *Studies in Hypogonadism in White-tailed Deer of the Central Mineral Region of Texas*. Texas Parks and Wildlife Department Technical Series no. 5. Austin: Texas Parks and Wildlife Department.

*——— (1965) "Social Behavior in a White-tailed Deer Herd Containing Hypogonadal Males." *Journal of Mammalogy* 46:314–27.

*——— (1964) "Hypogonadism in White-tailed Deer in the Central Mineral Region of Texas." In J. B. Trefethen, ed., *Transactions of the North American Wildlife and Natural Resources Conference* 29:225–36. Washington, D.C.: Wildlife Management Institute.

*Wishart, W. D. (1985) "Frequency of Antlered White-tailed Does in Camp Wainright, Alberta." *Journal of Mammalogy* 35:486–88.

*Wislocki, G. B. (1956) "Further Notes on Antlers in Female Deer of the Genus *Odocoileus*." *Journal of Mammalogy* 37:231–35.

*——— (1954) "Antlers in Female Deer, With a Report on Three Cases in *Odocoileus*." *Journal of Mammalogy* 35:486–95.

*Wong, B, and K. L. Parker (1988) "Estrus in Black-tailed Deer." *Journal of Mammalogy* 69:168–71.

Deer

Hoofed Mammal

WAPITI, ELK, or RED DEER
Cervus elaphus
BARASINGHA or SWAMP DEER
Cervus duvauceli

HOMOSEXUALITY	TRANSGENDER	BEHAVIORS		RANKING	OBSERVED IN
● Female	○ Intersexuality	○ Courtship	● Pair-bonding	○ Primary	● Wild
● Male	● Transvestism	○ Affectionate	○ Parenting	○ Moderate	● Semiwild
		● Sexual		● Incidental	○ Captivity

WAPITI/RED DEER

IDENTIFICATION: A large deer (standing 4–5 feet at the shoulder) with brownish red fur and a pale rump patch; males generally have enormous antlers and a long mane. DISTRIBUTION: Southern Canada, United States, northern Mexico; Eurasia, northwest Africa. HABITAT: Varied, including forests, meadows, chaparral, highlands. STUDY AREAS: Prairie Creek Redwoods State Park, California, subspecies *C.e. roosevelti*, the Roosevelt Elk; Isle of Rhum, Scotland, subspecies *C.e. scoticus*, the British Red Deer.

BARASINGHA

IDENTIFICATION: A 3–4-foot-tall deer with a brownish coat and large antlers (3 feet long) in males. DISTRIBUTION: India, Nepal; vulnerable. HABITAT: Meadows, woodland, marshy grassland. STUDY AREA: Kanha National Park, Madhya Pradesh, India; subspecies *C.d. brannderi*, the South Indian Barasingha.

Social Organization

Male Wapiti/Red Deer live for nine to ten months of the year in bachelor groups, while females (cows or hinds) associate with each other and their offspring in matriarchal groups. During the rut, which lasts for one to two months, males herd females and mate polygamously with them. Barasingha generally live in groups of 3–13 animals, although toward the end of the rutting season aggregations of up to 70 Deer may form. During most of the year Barasingha herds are sex-segregated.

Description

Behavioral Expression: In both of these Deer species, homosexual mounting occurs outside of the breeding season—in females among Barasingha, and in both sexes among Wapiti and Red Deer (Wapiti or Elk is the name for this species in North America, Red Deer is the European name). In addition, Red Deer hinds sometimes mount one another when they are in heat during the breeding season. Female homosexual mounting in Wapiti generally takes place in the cow groups. Usually the two animals engaging in same-sex activity are fully grown adults, but in male Wapiti homosexual mounting may occur between adult bulls and spikehorns (yearlings

▲ *A female Red Deer mounting another female*

whose antlers are spikes, having yet to develop prongs). Red Deer yearlings also participate in same-sex activity, including occasional incestuous homosexual mountings by females of their mothers. Homosexual mounting is done in the same position as heterosexual mating, with one animal behind the other; Red Deer stags have been observed with a full erection when mounting another male. In Wapiti and female Red Deer, same-sex (and opposite-sex) mounting may also be preceded by CHIN-RESTING, in which one animal rests its chin on the rump of the other, signaling his or her intention to mount. About a third of all Red Deer females participate in homosexual mounting as both mounter and mountee, while another third only participate as mounters, and another third only as mountees. Reciprocal mounting—in which two animals take turns mounting each other— sometimes occurs in male Wapiti. A type of same-sex, "platonic" pair-bonding is also found in this species. Both males and females may form "companionships" with an animal of the same sex; female companions are usually of the same age, while male companions may be two adult bulls, or an adult male with a younger male. Occasionally bulls will try to separate female companions during the breeding season. Their bond is strong, however, and the females travel great distances to rejoin each other, calling toward their companion until they are reunited.

Among Red Deer, gender-mixing individuals with various antler configurations are occasionally found. In this species, the vast majority of males have antlers; however, some stags, known as HUMMELS, physically resemble females in that they do not have this secondary sexual characteristic. Interestingly, hummels are in many ways more successful than antlered stags. Many become "master stags," that is, the highest ranking males, because they are generally in better physical condition, more resourceful, better fighters (in spite of not having antlers), and more successful at mating with females than antlered males. In addition, a few males are PERUKES, that is, their antlers are spikes and permanently covered in velvet. Such males are generally nonreproductive, having undeveloped testes. Antlered females also sometimes occur.

Frequency: Same-sex mounting occurs occasionally among Wapiti; in Barasingha, approximately 2–3 percent of sexual activity is between females. Homosexual mounting makes up about a third of all mounting behavior outside the breeding season in Red Deer, with the majority of this activity (64 percent) taking place between females.

Orientation: In Red Deer, about 70 percent of all females engage in some homosexual activity outside the rutting season; of these, about 30 percent participate exclusively in homosexuality, while the remainder are bisexual. The proportion of homosexual activity in bisexual females ranges from 6 percent to 80 percent, with the average being about 48 percent same-sex mounting per individual. Life histories of individual Wapiti and Barasingha that participate in homosexual mounts have not been compiled, so it is not known whether they also engage in heterosexual behavior.

Nonreproductive and Alternative Heterosexualities

Significant portions of the Wapiti and Red Deer population do not participate in reproduction. Only about a third of adult male Wapiti and half of adult male Red Deer mate with females each year. In fact, some Red Deer males (and a few females) are lifetime nonbreeders, never fathering offspring; others may have a postreproductive period in their old age. Moreover, about 30 percent of females, on average, are nonreproductive each season; individuals that do not breed generally have a lower mortality rate than breeders. As described above, Deer society is largely sex-segregated: males and females live mostly separate from each other except for one month out of the year (during the breeding season). Nevertheless, some heterosexual activity does take place outside of the rutting season: younger male Wapiti—often those that did not breed the previous season—may try to court and mount females, and heterosexual mounting also occurs outside the rut in Red Deer. Interestingly, some female Wapiti come into heat outside of the breeding season, but they are usually ignored by most adult males.

Even during the breeding season, heterosexual relations are sometimes strained: female Wapiti often refuse to be mounted by adult males, and they bite or kick yearling males that try to mate with them. On the other hand, a variety of nonprocreative sexual behaviors also make up the heterosexual repertoire: male Wapiti and Red Deer may lick and nuzzle the female's genitals, while REVERSE mounts (in which the female mounts the male) make up more than a quarter of all heterosexual activity outside the breeding season in Red Deer (they also occur in Wapiti). Both Red Deer and Wapiti males also masturbate, using a fairly unusual method: antlers in these species are actually erotic zones, and males derive sexual stimulation by rubbing them against vegetation. Red Deer stags have regularly been observed developing an erection and ejaculating from this activity. Sexual behavior by calves—including adult-calf interactions—also occurs in these species. Wapiti/Red Deer calves sometimes mount adults (including their mothers, in Red Deer), while female Red Deer occasionally mount calves. More than half of all mounting by yearling Red Deer is incestuous, with the younger animal mounting its mother. Finally, Wapiti females have developed a communal parenting or "day-care" system of CALF POOLS or CRÈCHES. These nursery groups, containing up to 50 or more calves, form in late summer to early fall, with one or two females watching over the youngsters while the other mothers go off on their own.

Other Species

In Père David's Deer (*Elaphurus davidianus*) stags sometimes mount each other, with younger males typically mounting older ones. Male Reeve's Muntjacs (*Muntiacus reevesi*), a small Chinese Deer, sometimes court other males. Transgendered peruke stags occasionally occur in other species such as Sika Deer (*Cervus nippon*)—where they may have a female coat color—Roe Deer (*Capreolus capreolus*), and Fallow Deer (*Dama dama*). In addition, intersexual individuals combining the genitals or reproductive organs of both sexes also occur in these species, as do Indian Muntjacs (*Muntiacus muntjak*) with a combined male-female chromosomal pattern (XXY).

Sources   *asterisked references discuss homosexuality/transgender*

Altmann, M. (1952) "Social Behavior of Elk, *Cervus canadensis nelsoni,* in Jackson Hole Area of Wyoming." *Behavior* 4:116–43.

*Barrette, C. (1977) "The Social Behavior of Captive Muntjacs *Muntiacus reevesi* (Ogilby 1839)." *Zeitschrift für Tierpsychologie* 43:188–213.

*Chapman, D. I., N. G. Chapman, M. T. Horwood, and E. H. Masters (1984) "Observations on Hypogonadism in a Perruque Sika Deer (*Cervus nippon*)." *Journal of Zoology, London* 204:579–84.

Clutton-Brock, T. H., F. E. Guiness, and S. D. Albon (1983) "The Costs of Reproduction to Red Deer Hinds." *Journal of Animal Ecology* 52:367–83.

——— (1982) *Red Deer: Behavior and Ecology of Two Sexes.* Chicago: University of Chicago Press.

*Darling, F. F. (1937) *A Herd of Red Deer.* London: Oxford University Press.

*Donaldson, J. C., and J. K. Doutt (1965) "Antlers in Female White-tailed Deer: A 4-Year Study." *Journal of Wildlife Management* 29:699–705.

Franklin, W. L., and J. W. Lieb (1979) "The Social Organization of a Sedentary Population of North American Elk: A Model for Understanding Other Populations." In M. S. Boyce and L. D. Hayden-Wing, eds., *North American Elk: Ecology, Behavior, and Management,* pp. 185–98. Laramie: University of Wyoming.

Graf, W. (1955) *The Roosevelt Elk.* Port Angeles, Wash.: Port Angeles Evening News.

*Guiness, F., G. A. Lincoln, and R. V. Short (1971) "The Reproductive Cycle of the Female Red Deer, *Cervus elaphus* L." *Journal of Reproduction and Fertility* 27:427–38.

*Hall, M. J. (1983) "Social Organization in an Enclosed Group of Red Deer (*Cervus elaphus* L.) on Rhum. II. Social Grooming, Mounting Behavior, Spatial Organization, and Their Relationships to Dominance Rank." *Zeitschrift für Tierpsychologie* 61:273–92.

*Harper, J. A., J. H. Harn, W. W. Bentley, and C. F. Yocom (1967) "The Status and Ecology of the Roosevelt Elk in California." *Wildlife Monographs* 16: 1–49.

*Lieb, J. W. (1973) "Social Behavior in Roosevelt Elk Cow Groups." Master's thesis, Humboldt State University.

*Lincoln, G. A., R. W. Youngson, and R. V. Short (1970) "The Social and Sexual Behavior of the Red Deer Stag." *Journal of Reproduction and Fertility* suppl. 11:71–103.

Martin, C. (1977) "Status and Ecology of the Barasingha (*Cervus duvauceli branderi*) in Kanha National Park (India)." *Journal of the Bombay Natural History Society* 74:60–132.

Morrison, J. A. (1960) "Characteristics of Estrus in Captive Elk." *Behavior* 16:84–92.

Prothero, W. L., J. J. Spillett, and D. F. Balph (1979) "Rutting Behavior of Yearling and Mature Bull Elk: Some Implications for Open Bull Hunting." In M. S. Boyce and L. D. Hayden-Wing eds., *North American Elk: Ecology, Behavior, and Management,* pp. 160–65. Laramie: University of Wyoming.

*Schaller, G. B. (1967) *The Deer and the Tiger.* Chicago: University of Chicago Press.

*Schaller, G. B., and A. Hamer (1978) "Rutting Behavior of Père David's Deer, *Elaphurus davidianus.*" *Zoologische Garten* 48:1–15.

*Wurster-Hill, D. H., K. Benirschke, and D. I. Chapman (1983) "Abnormalities of the X Chromosome in Mammals." In A. A. Sandberg, ed., *Cytogenetics of the Mammalian X Chromosome,* Part B, pp. 283–300. New York: Alan R. Liss.

DEER

Hoofed Mammal 🐏

CARIBOU or REINDEER
Rangifer tarandus

MOOSE
Alces alces

HOMOSEXUALITY	TRANSGENDER	BEHAVIORS		RANKING	OBSERVED IN
● Female	● Intersexuality	● Courtship	● Pair-bonding	○ Primary	● Wild
● Male	● Transvestism	○ Affectionate	○ Parenting	○ Moderate	○ Semiwild
		● Sexual		● Incidental	● Captivity

CARIBOU
IDENTIFICATION: A medium-sized deer typically with a grayish brown coat and white underparts, and antlers in both sexes. DISTRIBUTION: Circumboreal, including northern North America and Eurasia. HABITAT: Tundra, taiga, coniferous forest. STUDY AREA: Badger, Newfoundland, Canada; subspecies *R.t. caribou*, the Woodland Caribou, and *R.t. groenlandicus*, the Barren-Ground Caribou.

MOOSE
IDENTIFICATION: The largest species of deer (weighing up to 1,300 pounds); has slender legs, a pendulous nose, and (in males) prominent palmate antlers and a dewlap or "bell" beneath the throat. DISTRIBUTION: Northern Eurasia and North America. HABITAT: Moist woodland. STUDY AREAS: Jackson Hole, Wyoming; Kenai Peninsula, Alaska; Badger, Newfoundland; Wells Gray Provincial Park, British Columbia, Canada; subspecies *A.a. shirasi*, the Wyoming Moose; *A.a. gigas*, *A.a. americana*, and *A.a. andersoni*.

Social Organization

Caribou are highly gregarious, sometimes forming herds of tens or even hundreds of thousands of animals (although most groups contain 40–400 animals). They typically associate in all-male, mother-calf, and juvenile/adolescent bands. Moose, on the other hand, are more solitary, although they form aggregations of up to several dozen animals during the fall rutting period. Groups of bulls and cosexual herds may also coalesce after the mating season. In both species, animals mate with multiple partners rather than forming long-term heterosexual bonds, and males do not participate in raising their young (i.e., they have a polygamous mating system).

Description

Behavioral Expression: Caribou and Moose occasionally participate in a variety of same-sex courtship and sexual activities. Male Moose, for example, sometimes direct courtship behaviors toward other males, including sniffing the anal and genital region, and approaching another bull while making the characteristic rutting sound, the CROAK (a grunting call that combines a deep, resonant syllable with a popping or suction noise). Younger male Caribou may also court other

▲ *A male Caribou* (left) *courting another male by "vacuum licking"*

males by making a similar sound, sometimes known as SLURPING or VACUUM LICKING, in which the animal flicks or smacks his tongue against the upper palate while approaching the other male with his head outstretched. Female Caribou sometimes mount each other, as do younger males, while yearling male Moose have been observed trying to mount adult bulls. In addition, bull Moose sometimes associate with younger male companions —known as SATELLITES—that travel together in pairs or small groups, usually outside the breeding season. Another homosexual activity among males in both Moose and Caribou is antler rubbing. In these two species (as in several other types of Deer), antlers are highly sensitive organs and genuine erotic zones, and males may become sexually aroused when they rub their antlers together. Among Moose, this is done rather gently as a sort of "play-fighting" (the antlers may still have their velvet covering), while Caribou males rattle their antlers against each other when they are free of velvet.

Several types of gender-mixing occasionally occur in Moose, often involving unusual antler configurations. Intersexed males lacking a scrotum or testes sometimes develop what are known as VELERICORN antlers, which are covered in velvet and festooned with various ridges and knobs; such antlers are permanent, unlike regular antlers, which are shed and regrown each season. Other males—sometimes known as PERUKES—have elaborate, misshapen antlers covered with baroque nodule-like growths. Occasionally, females develop antlers, which may be single; spiked (without branches); covered in velvet; or lacking the flat, palmated structure typical of male Moose antlers. Caribou are the only deer in which females regularly sport antlers: depending on the population, anywhere from 8–95 percent of females may be antlered.

Frequency: Homosexual activity occurs only sporadically in Moose and Caribou. About a quarter of all male Moose associate together in pairs during at least part of the year.

Orientation: Adult animals that participate in homosexual activities in these two species are, in all likelihood, predominantly heterosexual, albeit with some bisexual capability. Some younger animals—especially among male Moose—may tend toward a less heterosexually oriented bisexuality, since many do not fully participate in heterosexual mating.

Nonreproductive and Alternative Heterosexualities

In both Moose and Caribou, many animals do not procreate. Caribou males are physiologically capable of breeding when they are a year old, yet most do not mate until they are at least four years old since they cannot successfully compete with the older bucks; a similar pattern is found in Moose. Among Caribou, females without calves may associate with a breeding female as an "assistant mother," and some even try to "kidnap" or lure a calf away from its biological mother. During severe food shortages, pregnant Caribou females may terminate the breeding process by reabsorbing their embryos, since they would not be able to successfully raise their calves under such conditions. Approximately 8 percent of the male Caribou population consists of older (10+ years), postreproductive males that do not participate in breeding. However, many stags never reach this age, since the life expectancy of males is considerably shorter than that of females, at least in part because of the stresses associated with breeding. In Moose, breeding is also a taxing activity for bulls, who fast completely during the rutting period. Mating can also be a traumatic activity for females: because males in these two species are considerably larger, females often suffer injuries from copulation, sometimes literally collapsing under the weight of a male mounting them. As a result, female Caribou often strongly resist mating attempts and struggle to escape (less than two-thirds of matings are completed), while males may strike them with their antlers to make them submit to mounting. Females, in turn, may use their own antlers to fight back. Female Moose often strike males with their front hooves during the rutting season as well, and are capable of inflicting serious injury. In both species, there is significant segregation of the sexes outside of the breeding season: in Moose, for example, only 10–20 percent of winter groups are cosexual.

Moose and Caribou also participate in a variety of nonreproductive sexual behaviors. Males of both species sometimes try to mount calves, and female Caribou sometimes REVERSE mount males. Heterosexual interactions often involve oral-genital contact—male Moose and Caribou lick the female's vulva, while female Caribou sometimes lick the male's penis. About 45 percent of heterosexual mounts in Moose do not involve penetration or ejaculation, and males sometimes mount females up to 14 times in a sequence. In addition, both male Moose and Caribou "masturbate" by rubbing their antlers against vegetation, which often results in sexual stimulation (including erection of the penis and possibly ejaculation).

Sources   *asterisked references discuss homosexuality/transgender*

Altmann, M. (1959) "Group Dynamics in Wyoming Moose During the Rutting Season." *Journal of Mammalogy* 40:420–24.

*Bergerud, A. T. (1974) "Rutting Behavior of Newfoundland Caribou." In V. Geist and F. Walther, eds., *The Behavior of Ungulates and Its Relation to Management*, vol. 1, pp. 395–435. IUCN Publication no. 24. Morges, Switzerland: International Union for Conservation of Nature and Natural Resources.

*Bubenik, A. B., G. A. Bubenik, and D. G. Larsen (1990) "Velericorn Antlers on a Mature Male Moose *(Alces a. gigas)." Alces* 26:115–28.

Bubenik, A. B., and H. R. Timmerman (1982) "Spermatogenesis of the Taiga-Moose—a Pilot Study." *Alces* 18:54–93.

*Denniston, R. H., II (1956) "Ecology, Behavior, and Population Dynamics of the Wyoming or Rocky Mountain Moose, *Alces alces shirasi." Zoologica* 41: 105–18.

de Vos, A. (1958) "Summer Observations on Moose Behavior in Ontario." *Journal of Mammalogy* 39:128–39.

*Dodds, D. G. (1958) "Observations of Pre-Rutting Behavior in Newfoundland Moose." *Journal of Mammalogy* 39:412–16.

*Geist, V. (1963) "On the Behavior of the North American Moose (*Alces alces andersoni* Peterson 1950), in British Columbia." *Behavior* 20:377–416.

Houston, D. B. (1974) "Aspects of the Social Organization of Moose." In V. Geist and F. Walther, eds., *The Behavior of Ungulates and Its Relation to Management,* vol. 2, pp. 690–96. IUCN Publication no. 24. Morges, Switzerland: International Union for Conservation of Nature and Natural Resources.

Kojola, I. (1991) "Influence of Age on the Reproductive Effort of Male Reindeer." *Journal of Mammalogy* 72:208–10.

*Lent, P. C. (1974) "A Review of Rutting Behavior in Moose." *Naturaliste canadien* 101:307–23.

——— (1966) "Calving and Related Social Behavior in the Barren-Ground Caribou." *Zeitschrift für Tierpsychologie* 23:701–56.

Miquelle, D. G., J. M. Peek, and V. Van Ballenberghe (1992) "Sexual Segregation in Alaskan Moose." *Wildlife Monographs* 122:1–57.

*Murie, O. J. (1928) "Abnormal Growth of Moose Antlers." *Journal of Mammalogy* 9:65.

*Pruitt, W. O., Jr. (1966) "The Function of the Brow-Tine in Caribou Antlers." *Arctic* 19:111–13.

——— (1960) "Behavior of the Barren-Ground Caribou." *Biological Papers of the University of Alaska* 3:1–44.

*Reimers, E. (1993) "Antlerless Females Among Reindeer and Caribou." *Canadian Journal of Zoology* 71:1319–25.

Skogland, T. (1989) *Comparative Social Organization of Wild Reindeer in Relation to Food, Mates, and Predator Avoidance.* Advances in Ethology no. 29. Berlin and Hamburg: Paul Parey Scientific Publishers.

Van Ballenberghe, V., and D. G. Miquelle (1993) "Mating in Moose: Timing, Behavior, and Male Access Patterns." *Canadian Journal of Zoology* 71:1687–90.

*Wishart, W. D. (1990) "Velvet-Antlered Female Moose (*Alces alces)." Alces* 26:64–65.

GIRAFFES, ANTELOPES, AND GAZELLES

GIRAFFES AND ANTELOPES   *Hoofed Mammal*

GIRAFFE
Giraffa camelopardalis

HOMOSEXUALITY	TRANSGENDER	BEHAVIORS		RANKING	OBSERVED IN
● Female	○ Intersexuality	● Courtship	○ Pair-bonding	● Primary	● Wild
● Male	○ Transvestism	● Affectionate	○ Parenting	○ Moderate	○ Semiwild
		● Sexual		○ Incidental	○ Captivity

IDENTIFICATION: The tallest mammal (up to 19 feet), with a sloping back, enormously long neck, bony, knobbed "horns" in both sexes, and the familiar reddish brown spotted patterning. DISTRIBUTION: Sub-Saharan Africa. HABITAT: Savanna. STUDY AREAS: Tsavo East and Nairobi National Parks, Kenya; Serengeti, Arusha, and Tarangire National Parks, Tanzania; eastern Transvaal, South Africa; subspecies *G.c. tippelskirchi*, the Masai Giraffe, and *G.c. giraffa*.

Social Organization

Female Giraffes tend to congregate in groups of up to 15 individuals, including their calves and perhaps a few younger males. Males generally associate in all-male "bachelor" groups, but tend to become solitary as they get older. The mating system is polygamous: mostly a few older males mate with more than one female, but take no part in raising their offspring.

Description

Behavioral Expression: Male Giraffes have a unique "courtship" or affectionate activity called NECKING, which is often associated with homosexual mounting. When necking, two males stand side by side, usually facing in opposite directions, while they gently rub their necks on each other's body, head, neck, loins, and thighs, sometimes for as long as an hour. Necking sessions are usually initiated with one male assuming a formal posture with his neck held rigid and upright. One male

▲ *Two male Giraffes engaging in "necking" behavior*

may also affectionately lick the other's back or sniff his genitals during necking. Necking Giraffes also sometimes swing their necks at each other in what has been described as a "stately dance" or a form of play-fighting (although they rarely hit, and virtually never injure, each other). Necking usually leads to sexual arousal: one or both males develop erections, and occasionally one might exhibit a curling of the lip similar to the FLEHMEN response seen in heterosexual courtships (associated with sexual arousal and testing sexual "readiness"). Sometimes after necking for 15 minutes or so, one male suddenly stops and "freezes" with his neck stretched forward, which is thought to indicate intense sexual excitement approaching orgasm. Males also commonly mount each other with erect penises during or following bouts of necking and probably reach orgasm (sometimes liquid—presumably semen—can be seen streaming from their penises). At times, groups of four or five males will gather to neck and mount each other, and males may mount several individuals in quick succession or the same male as many as three times in a row. Females also occasionally mount each other, but they do not participate in necking.

Frequency: Homosexual activity is common in Giraffes and in many cases is actually more frequent than heterosexual behavior (which may be quite rare): in one study area, mountings between males accounted for 94 percent of all observed sexual activity. Anywhere from a third to three-quarters of all courtship sessions are homosexual (i.e. they involve necking between males), and at any given time, about 5 percent of all males are participating in necking. Among females, less than 1 percent of interactions involving body contact consist of homosexual mounting.

Orientation: Homosexual activity is characteristic of younger adult males, who may constitute more than 80 percent of the male population. As they get older, males participate less in homosexual courtship and mounting and more in heterosexual activity. Among younger males, it is likely that all of their mounting behavior is homosexual, although a small percentage also court (but do not mount) females. Males participating in homosexual mounting and necking frequently disregard any females present in the herd, perhaps indicating a "preference" for same-sex activity.

Nonreproductive and Alternative Heterosexualities

Only a relatively small percentage of adult Giraffes breed: in some populations, less than a quarter of the females reproduce in any year, while usually only one or two males actually mate with females. A number of factors contribute to this infrequency of breeding: pregnancies last 15 months, and there is a minimum of 20 months between calves. Males are unable to compete successfully for matings until they are at least eight years old, even though they mature sexually at under four years. And as mentioned above, actual copulations can be remarkably rare—in one population, only a single heterosexual mating was observed during more than 3,200 hours of detailed observation over an entire year. In some areas, it also appears that a small class of old, postreproductive males are generally solitary and do not court or mate with females. Giraffes engage in a few forms of nonprocreative heterosexual activity as well: younger females in heat occasionally mount male calves, while calves sometimes mount their mothers. As in most polygamous animals, males do not participate in calf-raising. Females, however, often leave their young in nursery groups or CALVING POOLS containing as many as nine other calves, attended by one or more of the other mothers. This "day-care" arrangement allows a female to feed on her own without having to constantly look after her calf.

Sources   *asterisked references discuss homosexuality/transgender*

*Coe, M. J. (1967) "'Necking' Behavior in the Giraffe." *Journal of Zoology, London* 151:313–21.

*Dagg, A. I., and J. B. Foster (1976) *The Giraffe: Its Biology, Behavior, and Ecology.* New York: Van Nostrand Reinhold.

*Innis, A. C. (1958) "The Behavior of the Giraffe, *Giraffa camelopardalis,* in the Eastern Transvaal." *Proceedings of the Zoological Society of London* 131:245–78.

Langman, V. A. (1977) "Cow-Calf Relationships in Giraffe (*Giraffa camelopardalis giraffa*)." *Zeitschrift für Tierpsychologie* 43:264–86.

*Leuthold, B. M. (1979) "Social Organization and Behavior of Giraffe in Tsavo East National Park." *African Journal of Ecology* 17:19–34.

*Pratt, D. M., and V. H. Anderson (1985) "Giraffe Social Behavior." *Journal of Natural History* 19:771–81.

———— (1982) "Population, Distribution, and Behavior of Giraffe in the Arusha National Park, Tanzania." *Journal of Natural History* 16:481–89.

———— (1979) "Giraffe Cow-Calf Relationships and Social Development of the Calf in the Serengeti." *Zeitschrift für Tierpsychologie* 51:233–51.

*Spinage, C. A. (1968) *The Book of the Giraffe.* London: Collins.

GIRAFFES AND ANTELOPES
PRONGHORN
Antilocapra americana

Hoofed Mammal

HOMOSEXUALITY	TRANSGENDER	BEHAVIORS		RANKING	OBSERVED IN
● Female	○ Intersexuality	● Courtship	○ Pair-bonding	○ Primary	● Wild
● Male	● Transvestism	○ Affectionate	○ Parenting	● Moderate	○ Semiwild
		● Sexual		○ Incidental	○ Captivity

IDENTIFICATION: A deer-sized mammal with distinctive, sharply forked horns in males and reddish brown fur with white patches. DISTRIBUTION: West-central United States, adjacent areas of Canada and Mexico. HABITAT: Prairies, deserts. STUDY AREAS: Yellowstone National Park and National Bison Range, Montana; subspecies *A.a. americana*.

Social Organization

Pronghorn society is characterized by a distinction between territorial males, who establish territories and mate with females, and nonterritorial males, who live primarily in bachelor herds of seven to ten individuals throughout the spring and into early fall. Females associate in groups of up to two dozen individuals, often accompanied by a territorial male. After the breeding season—during which males copulate with multiple partners and do not assist in parenting—most Pronghorns join large mixed-sex herds for the winter.

Description

Behavioral Expression: Male Pronghorns court and mount each other in their bachelor herds from April to October, using many of the same behavior patterns found in heterosexual courtship and mating. As a prelude to sexual behavior, one male follows another, sometimes sniffing his anal region. The courting male might then touch his chest to the other male's rump, a signal that he wants to mount. Usually this leads to a full mount, in which the courting male rises on his hind legs and, with erect penis, slides up onto the other male from behind. Sometimes a whole string or "chain" of courting males forms as each follows and tries to mount the male in front of him. Males of all age groups participate in homosexual courtship and mounting, although adult males usually direct their attentions to adolescent males. Mounting between males sometimes occurs during sparring or play-fighting as well. Female Pronghorns also rump-sniff and mount each other when they are in heat, though less frequently than males.

Male Pronghorns shed their horns after the breeding season and some researchers have suggested that this allows them to "pass" as females in mixed-sex herds during the winter. Since males are usually physically exhausted after the rut, they make easier targets for predators than females: by engaging in a form of female mimicry or transvestism, they may avoid being singled out.

Frequency: Overall, about 7 percent of all courtship/sexual behavior is between animals of the same sex, and about 10 percent of all mounts are homosexual (roughly two-thirds of these are between males). Among animals of the same sex, approximately 3–4 percent of their interactions involve some sort of sexual behavior.

▲ *A male Pronghorn mounting another male*

Orientation: Anywhere from two-thirds to three-quarters of the male population does not participate in breeding; many of these animals are exclusively homosexual. Two-year-old males, for example, never mount females, yet bachelors participate in nearly a third of all homosexual mounts. At the other end of the scale, territorial males are exclusively heterosexual. In between, various forms of bisexuality occur. About 7 percent of adult bachelor males are able to mate with females, yet they also account for 18 percent of homosexual interactions. Some males transfer from the bachelor herds to territorial status, thereby participating in sequential bisexuality over the course of their lives. Many males, however, never become territorial, and though they may try to court females, most of their sexual behavior will continue to be homosexual for the majority—if not the totality—of their lives.

Nonreproductive and Alternative Heterosexualities

As described above, the majority of the male population is not involved in procreation, living as they do in bachelor herds or as loners, and Pronghorn social life is characterized by sex segregation for six to seven months of the year. Some bachelors, however, do try to court females; although their advances are consistently rebuffed, the males often persist and may harass the females relentlessly by chasing them, horning and roaring at them, and sometimes even knocking them down during a chase. Reproduction in the Pronghorn is also characterized by aggression *within the womb:* procreation routinely involves embryos killing each other. As many as seven embryos may initially be present in the female's uterus, but only two of these will survive. The remainder are killed by the other developing fetuses, which grow long projections out of their fetal membranes that fatally puncture the others and drag them out of the uterus back up into the female's oviduct. Some embryos also die earlier because they get strangled in the ropelike bodies of the other embryos. The female reabsorbs any embryos that die.

Sources   *asterisked references discuss homosexuality/transgender*

Bromley, P. T. (1991) "Manifestations of Social Dominance in Pronghorn Bucks." *Applied Animal Behavior Science* 29:147–64.

Bromley, P. T., and D. W. Kitchen (1974) "Courtship in the Pronghorn *Antilocapra americana.*" In V. Geist and F. Walther, eds., *The Behavior of Ungulates and Its Relation to Management,* vol. 1, pp. 356–74. IUCN Publication no. 24. Morges, Switzerland: International Union for Conservation of Nature and Natural Resources.

Geist, V. (1990) "Pronghorns." In *Grzimek's Encyclopedia of Mammals,* vol. 5, pp. 282–85. New York: Mc-Graw-Hill.

Geist, V., and P. T. Bromley (1978) "Why Deer Shed Antlers." *Zeitschrift für Säugetierkunde* 43:223–31.

*Gilbert, B. K. (1973) "Scent Marking and Territoriality in Pronghorn (*Antilocapra americana*) in Yellowstone National Park." *Mammalia* 37:25–33.

*Kitchen, D. W. (1974) "Social Behavior and Ecology of the Pronghorn." *Wildlife Monographs* 38:1–96.

O'Gara, B. W. (1978) *"Antilocapra americana."* *Mammalian Species* 90:1–7.

——— (1969) "Unique Aspects of Reproduction in the Female Pronghorn (*Antilocapra americana* Ord.)" *American Journal of Anatomy* 125:217–32.

KOB ANTELOPES

KOB
Kobus kob

WATERBUCK
Kobus ellipsiprymnus

LECHWE
Kobus leche

PUKU
Kobus vardoni

Hoofed Mammal

HOMOSEXUALITY	TRANSGENDER	BEHAVIORS		RANKING	OBSERVED IN
● Female	● Intersexuality	● Courtship	○ Pair-bonding	● Primary	● Wild
○ Male	○ Transvestism	○ Affectionate	○ Parenting	○ Moderate	○ Semiwild
		● Sexual		○ Incidental	○ Captivity

KOB
IDENTIFICATION: A large grazing antelope with a reddish coat, white underparts, and black markings on the legs; males have lyre-shaped horns, while females are more slender. DISTRIBUTION: Western Kenya to Senegal. HABITAT: Open savanna near water. STUDY AREA: Toro Game Reserve, Uganda; subspecies *K.k. thomasi*, the Uganda Kob.

WATERBUCK
IDENTIFICATION: A 4-foot-tall (shoulder height) antelope with long, straggly brown or grayish hair and a white rump; males have sickle-shaped, ridged horns. DISTRIBUTION: Sub-Saharan Africa. HABITAT: Grassland, savanna, forest near water. STUDY AREA: Queen Elizabeth Park, Uganda; subspecies *K.e. defassa*, the Defassa Waterbuck.

LECHWE
IDENTIFICATION: Similar to Kob, but horns longer and thinner, and coat yellowish brown to black. DISTRIBUTION: Southeastern Zaire, Zambia, Botswana. HABITAT: Wetlands. STUDY AREAS: Chobe Game Reserve and Lochinvar National Park, Zambia; subspecies *K.l. kafuensis*, the Kafue Lechwe.

PUKU
IDENTIFICATION: Similar to Kob but with shorter horns. DISTRIBUTION: Scattered locations throughout south-central Africa. HABITAT: Moist savanna, floodplains, woodland. STUDY AREAS: Kafue National Park and Luangwa Game Reserve, Zambia.

Social Organization

Kob society is complex and is organized around two types of social systems: sex-segregated herds and LEKS. Outside of the breeding grounds, the antelopes congregate in same-sex herds: bachelor herds contain 400–600 males, while female herds usually have 30–50 adults (as well as young of both sexes), though they can contain as many as 1,000 antelopes. On the breeding grounds, the population is structured into a dozen or more small territories known as leks. These are small arenas that the males—and occasionally females—use for performing intricate courtship displays, and which they defend against the intrusion of other males. Females leave their herds to visit these leks, where they choose males to mate with and also interact sexually with other females. The other Kob antelopes also live in sex-segregated female and bachelor herds, although some Lechwe herds are cosexual. In addition, a few males—who do the most mating—are territorial, while some Waterbuck males are SATELLITES, associating with territorial males and occasionally mating with females.

Description

Behavioral Expression: Virtually all Kob females engage in some form of homosexual activity, ranging from simple sexual mounting of other females all the way up to elaborate courtship displays. These interactions usually take place when the females are in heat and may occur either in the female herds or on the leks. Homosexual courtship and sexual interactions consist of a rich array of stylized movements in a fixed sequence, which are all used in heterosexual courtship as well. Individual females vary in how many of these display behaviors they employ when courting another female—some use only one or two, while others employ the full repertoire. A

female usually begins her homosexual courtship by PRANCING: she approaches another female with short, stiff-legged steps, her head held high and tail raised. This is followed by an action known as FLEHMEN or LIP-CURLING: she sniffs the vulva of the other female, who crouches and urinates while her partner places her nose in the stream of urine. While doing this, she retracts her upper lip in a curling gesture, exposing a special sexual scenting organ that allows her to sample the odor of the urine. Her courtship dance continues with a stylized gesture known as FORELEG KICKING: she raises her foreleg and gently touches the other female between her legs from behind. The other female responds with ritual MATING-CIRCLING, in which she circles tightly around the courting female, sometimes nipping or butting her hindquarters. This leads to mounting, in which the first female stands on her hind legs and climbs on top of the other from behind, as in heterosexual

◀ *Courtship and sexual activity between female Kob: "prancing" (above), "foreleg kicking" (middle), and mounting*

▲ *Following homosexual mounting, female Kob may engage in "inguinal nuzzling"* (left) *and "pincers movement"* (right)

mating. Sometimes the mounting female gives a single vigorous pelvic thrust, similar to the thrusting that a male makes when he reaches orgasm.

Homosexual copulation may be followed by a further display of stylized behaviors. The courting female, for example, might make a distinctive whistling sound by forcing air loudly through her nostrils with her mouth closed (also made by males in heterosexual courtship). The two females may also engage in what is known as INGUINAL NUZZLING: the female who was mounted adopts a special posture with her hind legs spread wide, tail raised, back arched, and her neck extended in a graceful swanlike position. The other Kob licks her partner's vulva and udder from behind and then concentrates on nuzzling and licking two special "inguinal glands" located in the same region, which secrete a pungent, waxy substance. Finally, the interaction concludes with what is known as the PINCERS MOVEMENT, in which one female gently holds the other in a "pincers" position with her head on the other Kob's back and her leg raised underneath her belly. Occasionally, a female Kob will herd other females and even defend her display territory against courting males by attacking the males head-on—no small feat, considering that she does not have the horns that most males use for such purposes. The majority of Kob that participate in homosexual mounting also become pregnant and raise young—and in all cases, this is done in the female-only herds, with little or no participation from males beyond insemination.

Female homosexual mounting also occurs in three other closely related species of antelopes, the Waterbuck, Lechwe, and Puku. Interestingly, Waterbuck females that mount each other are not usually in heat, unlike Kob. Occasionally, a Waterbuck female will perform courtship flehmen with another female as well. Hermaphrodite or intersexual individuals also sometimes occur in Kob: one animal, for example, was chromosomally male and had testes and large horns, combined with a vagina, uterus, and enlarged clitoris.

Frequency: Homosexual mounting is common among Kob. Each female participates in same-sex mounting about twice an hour (on average) during the mating season, and over an entire mating season a female might mount other females 60 or more times (although most females probably engage in this activity a dozen or so times). However, because heterosexual mounting rates are extraordinarily high—more than seven times higher than homosexual rates—same-sex mounting accounts for only about 9 percent of all sexual activity. Homosexual courtship displays are less common than same-sex mounting in this species. In Puku and Lechwe, mounting between females is also common, but it occurs only occasionally among Waterbuck.

Orientation: Most, if not all, female Kob are bisexual, participating in both heterosexual and homosexual mounting, but individuals vary along a continuum in their orientation. For some, same-sex mounting makes up nearly 60 percent of their sexual activity, while for others it constitutes only 1–3 percent, but the average is about 11 percent. Fewer Kob females use courtship displays with other females, but there is a parallel range in variation. About 7 percent of females employ a significant portion of the full courtship repertoire when interacting with other females. In the other species of Kob antelopes, females that engage in homosexual mounting probably also participate in heterosexual activities as well.

Nonreproductive and Alternative Heterosexualities

As described above, Kob society is sex-segregated, and there are large numbers of nonbreeding animals, particularly among males. Only a relatively small proportion of males (about 5 percent) have access to lek territories at one time, and only some of these will be selected by females to mate with. In some populations of Waterbuck, large numbers of males are also nonbreeders: at any given time, only 7 percent of males are territory holders, 9 percent are satellites, and the remainder live in bachelor herds. In fact, only 20 percent of males in this species become territorial during their lives. Although a few satellite and bachelor males mate with females, the majority do not. Female Kob usually mate repeatedly with their chosen males—generally many more times than is required to become pregnant—and may copulate with up to nine different males when they visit the lek. Waterbuck females also mate repeatedly when in heat, usually with the same male each time. Kob heterosexual copulations are often preceded by numerous nonreproductive mounts in which the male does not have an erection. Furthermore, full penetration may not occur during copulation, and often the male does not ejaculate even when he does achieve penetration. Waterbuck males sometimes mount females from the side or other positions where penetration cannot occur. When all types of mounts are considered, the rate of heterosexual activity in Kob is staggering: during a 24-hour visit to the lek, each female may engage in several hundred mountings, 40 of which will be full copulations. Female Lechwe are often chased and harassed by males (especially nonterritorial ones) trying to mate with them. Sometimes several males will disrupt a heterosexual copulation, and only 8 percent of matings in cosexual herds and 42 percent on leks result in ejaculation.

Sources   *asterisked references discuss homosexuality/transgender*
Balmford, A., S. Albon, and S. Blakeman (1992) "Correlates of Male Mating Success and Female Choice in a Lek-Breeding Antelope." *Behavioral Ecology* 3:112–23.

*Benirschke, K. (1981) "Hermaphrodites, Freemartins, Mosaics, and Chimaeras in Animals." In C. R. Austin and R. G. Edwards, eds., *Mechanisms of Sex Differentiation in Animals and Man,* pp. 421–63. London: Academic Press.

Buechner, H. K., J. A. Morrison, and W. Leuthold (1966) "Reproduction in Uganda Kob, with Special Reference to Behavior." In I. W. Rowlands, ed., *Comparative Biology of Reproduction in Mammals,* pp. 71–87. Symposia of the Zoological Society of London no. 15. London: Academic Press.

Buechner, H. K., and H. D. Roth (1974) "The Lek System in Uganda Kob." *American Zoologist* 14:145–62.

*Beuchner, H. K., and R. Schloeth (1965) "Ceremonial Mating Behavior in Uganda Kob (*Adenota kob thomasi* Neumann)." *Zeitschrift für Tierpsychologie* 22:209–25.

*DeVos, A., and R. J. Dowsett (1966) "The Behavior and Population Structure of Three Species of the Genus *Kobus.*" *Mammalia* 30:30–55.

Leuthold, W. (1966) "Variations in Territorial Behavior of Uganda Kob *Adenota kob thomasi* (Neumann 1896)." *Behavior* 27:214–51.

Morrison, J. A., and H. K. Buechner (1971) "Reproductive Phenomena During the *Post Partum*–Preconception Interval in the Uganda Kob." *Journal of Reproduction and Fertility* 26:307–17.

Nefdt, R. J. C. (1995) "Disruptions of Matings, Harassment, and Lek-Breeding in Kafue Lechwe Antelope." *Animal Behavior* 49:419–29.

Rosser, A. M. (1992) "Resource Distribution, Density, and Determinants of Mate Access in Puku." *Behavioral Ecology* 3:13–24.

*Spinage, C. A. (1982) *A Territorial Antelope: The Uganda Waterbuck.* London: Academic Press.

——— (1969) "Naturalistic Observations on the Reproductive and Maternal Behavior of the Uganda Defassa Waterbuck *Kobus defassa ugandae* Neumann." *Zeitschrift für Tierpsychologie* 26:39–47.

Wirtz, P. (1983) "Multiple Copulations in the Waterbuck (*Kobus ellipsiprymnus*)." *Zeitschrift für Tierpsychologie* 61:78–82.

——— (1982) "Territory Holders, Satellite Males, and Bachelor Males in a High-Density Population of Waterbuck (*Kobus ellipsiprymnus*) and Their Associations with Conspecifics." *Zeitschrift für Tierpsychologie* 58:277–300.

GAZELLES   *Hoofed Mammal*

BLACKBUCK
Antilope cervicapra
THOMSON'S GAZELLE
Gazella thomsoni
GRANT'S GAZELLE
Gazella granti

HOMOSEXUALITY	TRANSGENDER	BEHAVIORS		RANKING	OBSERVED IN
● Female	○ Intersexuality	● Courtship	● Pair-bonding	● Primary	● Wild
● Male	○ Transvestism	● Affectionate	○ Parenting	○ Moderate	● Semiwild
		● Sexual		○ Incidental	○ Captivity

BLACKBUCK
IDENTIFICATION: A medium-sized gazelle; males have distinctive spiral horns and a black-and-white coat; females and juvenile males are tan colored. DISTRIBUTION: India; vulnerable. HABITAT: Semidesert to open woodland. STUDY AREAS: Kanha National Park, Madhya Pradesh, India; Clères Park, Rouen, France.

THOMSON'S, GRANT'S GAZELLES
IDENTIFICATION: Smaller gazelles (2–3 feet at shoulder height) with ringed, slightly S-shaped horns in both sexes; Thomson's have a conspicuous black flank band, and Grant's horns may bend sharply outward. DISTRIBUTION: East Africa, especially Kenya, Tanzania, Sudan. HABITAT: Grassy steppes. STUDY AREAS: Serengeti National Park and Ngorogoro Crater, Tanzania; subspecies *G.g. robertsi*, the Wide-horned Grant's Gazelle.

Social Organization

Blackbucks live in small, same-sex herds containing 10–50 individuals. Female herds circulate within the territory of one or several adult males who mate with them; the remaining males live in "bachelor" herds on the periphery of the breeding territories. Thomson's and Grant's Gazelles have a similar social organization, except that mixed herds containing both males and females also form, especially during migration.

Description

Behavioral Expression: The majority of male Blackbucks have homosexual interactions: among all age groups, mounting of one male by another occurs in the position used for heterosexual intercourse. Usually mounting happens during play-fighting—friendly sparring matches with erotic overtones, sometimes involving three males at a time. In addition, adult males often perform courtship displays toward adolescent males (one-to-two-year-olds) prior to mounting them. These displays, which also occur in heterosexual interactions, begin with the older male DISPLAY WALKING: he stands some distance away from the object of his attentions, lowering his ears and curling his tail up to touch his back. He walks in this posture parallel to the younger male so that the younger one has to walk in a circle. This is followed by PRESENTING THE THROAT: the older male raises his nose high in the air so that his spiral horns touch the back of his neck. This exposes the striking black-and-white pattern of his neck. While doing this, he briskly kicks first one foreleg, then the other, in front of him several times in a row, sometimes reaching under the other male's belly or between his thighs. Occasionally, the older male makes a distinctive barking sound as he does this. This is then followed by mounting of the younger male by the older. Occasionally, female Blackbucks mount other females.

▲ *A male Blackbuck mounting another male during a bout of play-fighting*

In Thomson's Gazelles, male homosexual mounting may occur in a variety of contexts, including during migration and in encounters between two nonterritorial males. Males also occasionally direct courtship displays toward one another, including the NECK-STRETCH, FORELEG KICK, and NOSE-UP POSTURE, as well as the PURSUIT MARCH (the latter similar to heterosexual courtships). Homosexual courtship displays are preceded by one or both males displaying their horns to the other (often interpreted as a threatening gesture). Homosexual mounting in Grant's Gazelles typically occurs as part of a formalized display in which two males march toward one another, lifting their heads high and showing their white throat patches when they are next to each other. The mounted male, if an adult, often attacks the male trying to mount him (females also sometimes respond aggressively to a male's advances, see below).

Frequency: Male homosexual activity is common among Blackbucks: at any given time, fully three-quarters of the male population lives in the bachelor herds, where most homosexual interactions take place. Among Thomson's and Grant's Gazelles, homosexual behavior is much less frequent: 12 percent of encounters between male Grant's involve mounting, while 1–8 percent of encounters between male Thomson's involve sexual behavior.

Orientation: All Blackbuck males over three years old leave the bachelor herd temporarily to attempt mating with females. However, this usually occurs only once or twice in each male's lifetime; for the remainder of his life, he interacts homosexually. Technically, then, all male Blackbuck are bisexual, though in practice they are predominantly homosexual. In Thomson's Gazelles, homosexual mounting typically occurs among males in bachelor or migratory groups, not territorial males (who are involved principally in heterosexual activities). Although these males occasionally court and attempt to mount females, the majority of their sexual interactions may be with other males. In Grant's males, homosexual behavior does occur in some territorial males; since these males direct sexual behaviors toward both males and females, they are functionally bisexual (although males generally do not consent to being mounted by other males).

Nonreproductive and Alternative Heterosexualities

Because of the organization of Blackbuck society into sex-segregated herds and the small number of active breeding males, only a fraction of the male population is ever involved in heterosexual activity. Furthermore, although all males attempt to leave the bachelor herds and mate with females, most are unable to do so because of the males already defending the breeding territories; consequently life in the bachelor herd is preferable for many males. Among Grant's and Thomson's Gazelles, there are similar patterns of sex segregation and nonparticipation in heterosexuality—in fact, more than 90 percent of the male Grant's population may be composed of nonbreeders at any given time. In addition, female Grant's Gazelles often behave aggressively toward males during heterosexual courtship, performing threat displays and sometimes even fighting bucks to fend off unwanted advances. Female Blackbucks sometimes engage in nonreproductive mounts of fawns or young animals.

Sources   *asterisked references discuss homosexuality/transgender*

*Dubost, G., and F. Feer (1981) "The Behavior of the Male *Antilope cervicapra* L., Its Development According to Age and Social Rank." *Behavior* 76:62–127.

*Schaller, G. B. (1967) *The Deer and the Tiger.* Chicago: University of Chicago Press.

Walther, F. R. (1995) *In the Country of Gazelles.* Bloomington: Indiana University Press.

*———— (1978a) "Quantitative and Functional Variations of Certain Behavior Patterns in Male Thomson's Gazelle of Different Social Status." *Behavior* 65:212–40.

*———— (1978b) "Forms of Aggression in Thomson's Gazelle; Their Situational Motivation and Their Relative Frequency in Different Sex, Age, and Social Classes." *Zeitschrift für Tierpsychologie* 47:113–72.

*———— (1974) "Some Reflections on Expressive Behavior in Combats and Courtship of Certain Horned Ungulates." In V. Geist and F. Walther, eds., *Behavior in Ungulates and Its Relation to Management,* vol. 1, pp. 56–106. IUCN Publication no. 24. Morges, Switzerland: International Union for Conservation of Nature and Natural Resources.

———— (1972) "Social Grouping in Grant's Gazelle (*Gazella granti* Brooke, 1827[sic])" in the Serengeti National Park." *Zeitschrift für Tierpsychologie* 31:348–403.

*———— (1965) "Verhaltensstudien an der Grantgazelle (Gazella granti Brooke, 1872) im Ngorogoro-Krater [Behavioral Studies on Grant's Gazelle in the Ngorogoro Crater]." *Zeitshcrift für Tierpsychologie* 22:167–208.

WILD SHEEP, GOATS, AND BUFFALO

MOUNTAIN SHEEP

Hoofed Mammal

BIGHORN SHEEP
Ovis canadensis
THINHORN or DALL'S SHEEP
Ovis dalli
ASIATIC MOUFLON or URIAL
Ovis orientalis

HOMOSEXUALITY	TRANSGENDER	BEHAVIORS		RANKING	OBSERVED IN
● Female	○ Intersexuality	● Courtship	○ Pair-bonding	● Primary	● Wild
● Male	● Transvestism	● Affectionate	○ Parenting	○ Moderate	● Semiwild
		● Sexual		○ Incidental	● Captivity

BIGHORN SHEEP
IDENTIFICATION: A large wild sheep (weighing up to 300 pounds) with massive spiral horns in males; coat is brown with a white muzzle, underparts, and rump patch. DISTRIBUTION: Southwestern Canada, Rocky Mountains to northern Mexico. HABITAT: Mountain and desert rocky terrain. STUDY AREAS: Banff National Park, Alberta; Kootenay National Park and the Chilcotin-Cariboo Region, British Columbia, Canada; National Bison Range, Montana; subspecies *O.c. canadensis,* the Rocky Mountain Bighorn, and *O.c. californiana,* the California Bighorn Sheep.

THINHORN SHEEP
IDENTIFICATION: Similar to Bighorn, except smaller and with thinner horns; coat is all white or brownish black to gray. DISTRIBUTION: Alaska, northwestern Canada. HABITAT: Rocky alpine and arctic terrain. STUDY AREAS: Kluane Lake, the Yukon; Cassiar Mountains, British Columbia, Canada; subspecies *O.d. dalli,* Dall's Sheep, and *O.d. stonei,* Stone's Sheep.

ASIATIC MOUFLON
IDENTIFICATION: Similar to N. American wild sheep, except coat varies from reddish brown or black-brown to light tan, and males may have a light saddle patch and a "bib" or chest mane; horns can be up to 4 feet long, spiral or arching back. DISTRIBUTION: Southwest Asia (including Iran, Afghanistan, Pakistan); Corsica, Sardinia, Cyprus; vulnerable. HABITAT: Hilly or steep terrain, from deserts to mountains. STUDY AREAS: Bavella, Island of Corsica, France; Salt Range near Kalabagh, Pakistan; Johnson City, Texas; subspecies *O.o. musimon,* the European Mouflon, and *O.o. punjabiensis,* the Punjab Urial.

Social Organization

Mountain Sheep live in sex-segregated bands, usually numbering 5–15 individuals. During the rutting season, the sexes intermingle and mate promiscuously (males copulate with multiple partners and do not form long-term pair-bonds or participate in parenting).

Description

Behavioral Expression: In Bighorn and Thinhorn Sheep, males live in what one zoologist has described as "homosexual societies" where same-sex courtship and sexual activity occur routinely among all rams. Typically an older, higher-ranking male will court a male younger than him, using a sequence of stylized movements. Same-sex courtship is often initiated when one male approaches the other in the LOW-STRETCH posture, in which the head and neck are lowered and extended far forward. This might be combined with the TWIST, where the male sharply rotates his head and points his muzzle toward the other male, often while flicking his tongue and making growling or grumbling sounds. The courting ram often performs a FORELEG KICK, stiffly snapping his front leg up against the other male's belly or between his hind legs. He also occasionally sniffs and nuzzles the other male's genital area and may perform LIP-CURLING or FLEHMEN, in which he samples the scent of the other male's urine by retracting his upper lip to expose a special olfactory organ. Thinhorn rams may even lick the penis of the male they are courting. The male being courted sometimes rubs his forehead and cheeks on the other ram's face—even licking and nibbling him—and may also rub his horns on the other male's neck, chest, or shoulders, occasionally developing an erection. Similar courtship behaviors occur among male Asiatic Mouflons.

In addition to genital licking (in Thinhorns), sexual activity between rams usually involves mounting and anal intercourse: typically the larger male rears up on his hind legs and mounts the smaller male, placing his front legs on the other's

flanks. The mountee assumes a characteristic posture known as LORDOSIS, in which he arches his back to facilitate the copulation (this posture is also seen in many female mammals during heterosexual mating). Usually the mounting male has an erect penis and achieves full anal penetration, performing pelvic thrusts that probably lead to ejaculation in many

◄ *A male Bighorn Sheep in the Rocky Mountains mounting another male*

cases. Mounting and courtship interactions between males sometimes also take place in groups known as HUDDLES: three to ten rams cluster together in a circle, rubbing, nuzzling, licking, horning, and mounting each other. Usually huddles are non-aggressive interactions in which all males are willing participants; occasionally, though, several rams in a huddle focus all their attentions on the same (usually smaller) male, taking turns mounting him and even chasing him if he tries to get away. Female Mountain Sheep also occasionally participate in sexual activity with one another, including licking each other's genitals, mounting, and occasional courtship activities.

So pervasive and fundamental is same-sex courtship and sexuality in Bighorns and Thinhorns that females are said to "mimic" males in order to mate with them. They adopt the behavior patterns typical of younger males being courted by older males, thereby sparking sexual interest on the part of rams because, ironically, they now resemble males. In another twist on gender roles and sexuality, there are also occasionally "female-mimicking" males in some populations—but notably, such males do *not* typically participate in homosexual mounting and courtship. Trans-gendered males are physically indistinguishable from other rams, but behaviorally they resemble females. They remain in the sex-segregated ewe herds year-round, they often adopt the crouching urination posture typical of females, and they are lower-ranking and less aggressive than most males and even many females (even though they are often larger in body and horn size, the typical criteria used to es-tablish rank). Most significantly, transgendered rams do not usually allow other males to court or mount them. Again, this is a typically female pattern, since ewes in these species generally do not permit rams to court or mount them except for the few days out of each year when they are in heat.

Frequency: In Bighorns and Thinhorns, homosexual mounting occurs com-monly throughout the year, but is especially frequent during the rut when hetero-sexual activity is also taking place, accounting for about a quarter of all sexual activity at that time (and occurring in up to 69 percent of males' interactions with each other). Outside of the rut, all mounting activity is homosexual, but mounting only accounts for 2–3 percent of males' interactions with each other. Among fe-males, 1–2 percent of interactions include mounting. At least 70 percent of males' interactions with one another involve courtship behaviors. Homosexual activity appears to be less frequent in Asiatic Mouflons: it is seen sporadically in wild ani-mals, while in captivity about 10 percent of mounting and some courtship be-haviors occur between animals of the same sex, mostly females. Behavioral transvestism occurs in approximately 5 percent of rams in some populations of Bighorn Sheep.

Orientation: Virtually all male Bighorn and Thinhorn Sheep participate in ho-mosexual courtship and mounting; the extent to which they also engage in hetero-sexual pursuits during the rut varies with their age and rank. Younger, lower-ranking rams—close to half of the male population—rarely get to mate with females at all, and some of these males have only homosexual relations. Among older, higher-ranking rams, heterosexual behavior is much more common—but even when they are courting and mounting females, it is often because of the malelike behavior patterns that the females are using (as described above). In other words, even in their heterosexuality, Mountain Sheep may be decidedly "homosexual."

Nonreproductive and Alternative Heterosexualities

Large portions of the male population in Bighorn and Thinhorn Sheep do not breed (as mentioned above). Although many younger and lower-ranking males try to mount females, they are able to mate less than 20 percent of the time because both females and higher-ranking rams will not usually allow them to complete their copulations. However, nonbreeding rams actually have a much lower mortality rate than breeding males—nearly six times lower—owing to the stresses of reproduction (including fasting during the breeding season, fights and chases, and other major energy expenditures). Ewes often reject the advances of older, higher-ranking rams as well (nearly 65 percent of the time in Bighorns), and this may lead to harassment and even forced copulations or rapes. In fact, rams employ three distinct strategies to try to mate with females, only one of which entails courtship and consensual copulations. TENDING involves a ram following a particular female for short periods of time, during which he courts her and is usually permitted to mate. COURSING consists of a ram chasing and sometimes butting a female, who is usually forced to copulate under threat of further punishment from the ram. BLOCK-ING involves forcefully cornering and trapping females with threats and more violent actions such as horn butts; ewes may be knocked down or bounced against trees if they try to escape and have been sequestered for up to nine days at a time by blocking rams. Almost half of all ewes in heat, on average, experience the trauma of blocking. Rams also sometimes mount lambs as well as females who are not in heat—in all, about 15 percent of heterosexual mounts are on such nonfertilizable partners. Male Mountain Sheep "masturbate" by crouching, protruding the penis sideways past the front legs, and ejaculating (sometimes after nuzzling the penis or rubbing it against the front leg). As described above, Mountain Sheep society is strongly sex-segregated for the majority of the year. Since rams and ewes only associate for two months or so during the rut, females usually raise their young on their own with no help from males. Occasionally, however, a ewe who has lost her own lamb will help another mother suckle her young. Such "helpers" are more common among higher-ranking females, where up to 30 percent of mothers who have lost their lambs may foster-nurse other youngsters.

Other Species

Same-sex courtship and mounting occur in several other species of wild sheep and goats, involving similar behavior patterns to those found in North American and European wild sheep. Among Bharal or Blue Sheep (*Pseudois nayaur*) of the Himalayas, 36–57 percent of mounting occurs between males (sometimes in huddles), while approximately 11 percent of courtship displays such as the low-stretch, twist, and foreleg kick are performed between males. Males also perform a "penis display" toward other males, in which the animal sometimes licks or sucks his own organ. Male Markhor (*Capra falconeri*) and Wild Goats or Bezoar (*Capra aegagrus*), two Central Asian species, also occasionally court and mount other males, as do male and female Aoudad or Barbary Sheep (*Ammotragus lervia*) of North Africa.

Sources

asterisked references discuss homosexuality/transgender

*Berger, J. (1985) "Instances of Female-Like Behavior in a Male Ungulate." *Animal Behavior* 33:333–35.

Demarchi, D. A., and H. B. Mitchell (1973) "The Chilcotin River Bighorn Population." *Canadian Field-Naturalist* 87:433–54.

Festa-Bianchet, M. (1991) "The Social System of Bighorn Sheep: Grouping Patterns, Kinship, and Female Dominance Rank." *Animal Behavior* 42:71–82.

*Geist, V. (1975) *Mountain Sheep and Man in the Northern Wilds*. Ithaca, N.Y.: Cornell University Press.

*———— (1971) *Mountain Sheep: A Study in Behavior and Evolution*. Chicago: University of Chicago Press.

*———— (1968) "On the Interrelation of External Appearance, Social Behavior, and Social Structure of Mountain Sheep." *Zeitschrift für Tierpsychologie* 25:199–215.

*Habibi, K. (1987a) "Behavior of Aoudad (*Ammotragus lervia*) During the Rutting Season." *Mammalia* 51:497–513.

*———— (1987b) "Overt Sexual Behavior Among Female Aoudads." *Southwestern Naturalist* 32:148.

Hass, C. C. (1991) "Social Status in Female Bighorn Sheep (*Ovis canadensis*): Expression, Development, and Reproductive Correlates." *Journal of Zoology, London* 225:509–23.

*Hass, C. C., and D. A. Jenni (1991) "Structure and Ontogeny of Dominance Relationships Among Bighorn Rams." *Canadian Journal of Zoology* 69:471–76.

*Hogg, J. T. (1987) "Intrasexual Competition and Mate Choice in Rocky Mountain Bighorn Sheep." *Ethology* 75:119–44.

———— (1984) "Mating in Bighorn Sheep: Multiple Creative Male Strategies." *Science* 225:526–29.

*Katz, I. (1949) "Behavioral Interactions in a Herd of Barbary Sheep (*Ammotragus lervia*)." *Zoologica* 34:9–18.

*McClelland, B. E. (1991) "Courtship and Agonistic Behavior in Mouflon Sheep." *Applied Animal Behavior Science* 29:67–85.

*Pfeffer, P. (1967) "Le mouflon de Corse (*Ovis ammon musimon*, Schreber 1782). Position systématique, écologie, et éthologie comparées [The Mouflon of Corsica: Comparative Systematics, Ecology, and Ethology]." *Mammalia* (suppl.) 31:1–262.

*Schaller, G. B. (1977) *Mountain Monarchs: Wild Sheep and Goats of the Himalaya*. Chicago: University of Chicago Press.

*Schaller, G. B., and Z. B. Mirza (1974) "On the Behavior of Punjab Urial (*Ovis orientalis punjabiensis*)." In V. Geist and F. Walther, eds. *The Behavior of Ungulates and Its Relation to Management*, vol. 1, pp. 306–23. IUCN Publication no. 24. Morges, Switzerland: International Union for Conservation of Nature and Natural Resources.

*Shackleton, D. M. (1991) "Social Maturation and Productivity in Bighorn Sheep: Are Young Males Incompetent?" *Applied Animal Behavior Science* 29: 173–84.

Valdez, R. (1990) "Oriental Wild Sheep." In *Grzimek's Encyclopedia of Mammals*, vol. 5, pp. 544–48. New York: McGraw-Hill.

*Wilson, P. (1984) "Aspects of Reproductive Behavior of Bharal (*Pseudois nayaur*) in Nepal." *Zeitschrift für Säugetierkunde* 49:36–42.

OTHER GOAT-ANTELOPES   *Hoofed Mammal*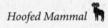
MUSK-OX
Ovibos moschatus
MOUNTAIN GOAT
Oreamnos americanus

HOMOSEXUALITY	TRANSGENDER	BEHAVIORS		RANKING	OBSERVED IN
● Female	○ Intersexuality	● Courtship	● Pair-bonding	○ Primary	● Wild
● Male	○ Transvestism	○ Affectionate	○ Parenting	○ Moderate	○ Semiwild
		● Sexual		● Incidental	● Captivity

MUSK-OX
IDENTIFICATION: A large (6—8-foot-long) mammal with long, shaggy fur, humped shoulders, and massive, down-sweeping horns. DISTRIBUTION: Arctic North America and Greenland. HABITAT: Tundra and meadows. STUDY AREAS: Nunivak Island, Alaska; Thelon Game Sanctuary, Northwest Territories, Canada; University of Saskatchewan; subspecies *O.m. wardi* and *O.m. moschatus*.

MOUNTAIN GOAT
IDENTIFICATION: A stocky, 3-foot-tall, goatlike mammal with shaggy white fur and sharp horns in both sexes. DISTRIBUTION: Western North America from southeastern Alaska to western Montana. HABITAT: Steep mountain slopes, cliffs. STUDY AREAS: Cassiar Mountain Range, British Columbia, Canada; Swan Mountains and Glacier National Park, Montana; Olympic National Park, Washington; subspecies *O.a. americanus, O.a. columbiae,* and *O.a. missoulae.*

Social Organization
Musk-oxen are generally social animals, living in mixed-sex herds (usually 10–20 animals) or smaller all-male groups; some males are also solitary. Male and female Mountain Goats remain largely segregated from each other for most of the summer, females associating in groups of usually less than 15 (including their offspring). Female Mountain Goats are generally dominant to males, who are solitary or peripheral to the female groups except during the rut. The mating system for both species is polygamous or promiscuous: animals copulate with several partners, and males do not participate in parenting.

Description
Behavioral Expression: Male Musk-oxen sometimes court and mount each other. Homosexual courtship involves several of the same patterns used in heterosexual interactions: POSITIONING, in which one male stands next to the other in a standard position such as perpendicular, parallel, or head-to-tail; SNIFFING OF THE REAR, in which one male smells and inspects the anal and genital region of the other; FORELEG KICKING, where one male gently swings his front leg against another male; and CHIN-RESTING, in which the courting male places his lower jaw on

top of the other male's body. Males also mount other males from behind (as in heterosexual mating). The mounted male sometimes resists (as do females when mounted by males) but may also permit himself to be mounted. Homosexual courtship and mounting both occur among younger males, while during the rutting season adult males sometimes court younger males as well (occasionally even

▲ *A male Musk-ox courting another male with "foreleg kicking" and "chin-resting"*

juveniles). Adult male Musk-oxen may also form pairlike companionships that travel, graze, and spend time together (sometimes also fighting with one another), although overt courtship and sexual behavior probably does not occur between them.

Adult male Mountain Goats court younger (yearling) males, again using the species-typical heterosexual behavior patterns: the courting male approaches the other in a crouching position, creeping on his stomach with his head stretched forward (a posture referred to as the LOW-STRETCH). He may also flick his tongue in and out of his mouth while making a soft buzzing sound, jerk his head sideways, and attempt to lick the other male's flanks. Typically the yearling male responds aggressively to the courting adult. Adult males occasionally perform this display toward other adult males as well. In addition, yearling females sometimes mount their own mothers.

Frequency: In captive Musk-oxen, about 40 percent of courtship behavior and 10 percent of mounting activity is homosexual. A little over a quarter of wild, nonbreeding males associate in pairs. In Mountain Goats, nearly 18 percent of courtships during the breeding season are between adult and yearling males; about 8 percent of courtship displays outside of the breeding season occur between two adult males. In one study, 1 out of 14 mounts (7 percent) performed by yearling Mountain Goats on their mothers were same-sex, involving a female offspring.

Orientation: Some younger male Musk-oxen may participate exclusively in homosexual activity, since most males do not breed until they are older than six. In contrast, yearling male Mountain Goats that are courted by older males are primarily heterosexual, since they reject most same-sex advances. The majority of adult male Musk-oxen and Mountain Goats that court other males are probably bisexual, since they also court females and usually do so more often than they court males.

Nonreproductive and Alternative Heterosexualities

Although male Musk-oxen become sexually mature by the time they are two years old, most bulls do not mate heterosexually for another five years because older males generally monopolize breeding opportunities in the female herds. Even among older males, less than half—and often as few as one-quarter—actually participate in procreation. The remainder are nonbreeding bulls that are often solitary or associate with other males in pairs or small groups, sometimes wandering far from the herds. The rate of calf production is low in this species (females usually reproduce every other year), and entire populations may forgo breeding in some years. Even during breeding years, some heterosexual courtship activity in both Musk-oxen and Mountain Goats may be directed by males toward nonprocreating individuals such as yearlings and calves. In addition, male Mountain Goats sometimes court and mount females outside the breeding season or even court females in the act of giving birth. Females in this species have been observed mounting adult males and courting, mounting, or being mounted by their own yearlings or kids. Relations between the sexes are often marked by strife. Females of both species sometimes reject the courtship and mounting attempts of males. Male Musk-oxen may become violent during their courtship kicks of females (the impact of the blow against a female's spine or pelvis can be considerable). As many as two-thirds of Musk-ox mounts may not culminate in ejaculation, because the male is anatomically unsuited to remaining mounted on the female (he is considerably heavier than her and unable to clasp with his forelegs during a mount). Among Mountain Goats, females are often notably aggressive toward males, attacking and sometimes viciously wounding them with stabs from their sharp horns. In addition, violence toward calves has been observed among Musk-oxen: females sometimes flip calves other than their own into the air with their horns, while males have been known to gore calves.

Other Species

In the distantly related Himalayan Tahr (*Hemitragus jemlahicus*), intersexuality sometimes occurs: one individual, for example, had testes and the general appearance of a male combined with a vulva, enlarged clitoris, and a female chromosome pattern.

Sources   *asterisked references discuss homosexuality/transgender*

*Benirschke, K. (1981) "Hermaphrodites, Freemartins, Mosaics, and Chimaeras in Animals." In C. R. Austin and R. G. Edwards, eds., *Mechanisms of Sex Differentiation in Animals and Man*, pp. 421–63. London: Academic Press.

Chadwick, D. H. (1983) *A Beast the Color of Winter: The Mountain Goat Observed*. San Francisco: Sierra Club Books.

*——— (1977) "The Influence of Mountain Goat Social Relationships on Population Size and Distribution." In W. Samuel and W. G. Macgregor, eds., *Proceedings of the First International Mountain Goat Symposium*, pp. 74–91. Victoria, B. C.: Fish and Wildlife Branch.

*Geist, V. (1964) "On the Rutting Behavior of the Mountain Goat." *Journal of Mammalogy* 45:551–68.

Gray, D. R. (1979) "Movements and Behavior of Tagged Muskoxen (*Ovibos moschatus*) on Bathurst Island, N.W.T." *Musk-ox* 25:29–46.

——— (1973) "Social Organization and Behavior of Muxkoxen (*Ovibos moschatus*) on Bathurst Island, N.W.T." Ph.D. thesis, University of Alberta.

*Hutchins, M. (1984) "The Mother-Offspring Relationship in Mountain Goats (*Oreamnos americanus*)." Ph.D. thesis, University of Washington.

Jingfors, K. (1984) "Observations of Cow-Calf Behavior in Free-Ranging Muskoxen." In D. R. Klein, R. G.

White, and S. Keller, eds., *Proceedings of the First International Muskox Symposium,* pp. 105–9. Biological Papers of the University of Alaska Special Report no. 4. Fairbanks: University of Alaska.

Lent, P. C. (1988) *"Ovibos moschatus." Mammalian Species* 302:1–9.

*Reinhardt, V. (1985) "Courtship Behavior Among Musk-ox Males Kept in Confinement." *Zoo Biology* 4:295–300.

*Smith, T. E. (1976) "Reproductive Behavior and Related Social Organization of the Muskox on Nunivak Island." Master's thesis, University of Alaska.

*Tener, J. S. (1965) *Muskoxen in Canada: A Biological and Taxonomic Review.* Ottawa: Canadian Wildlife Service.

BUFFALO   *Hoofed Mammal*

AMERICAN BISON
Bison bison
WISENT or EUROPEAN BISON
Bison bonasus
AFRICAN BUFFALO
Synceros caffer

HOMOSEXUALITY	TRANSGENDER	BEHAVIORS		RANKING	OBSERVED IN
● Female	● Intersexuality	○ Courtship	● Pair-bonding	● Primary	● Wild
● Male	○ Transvestism	● Affectionate	○ Parenting	○ Moderate	● Semiwild
		● Sexual		○ Incidental	● Captivity

AMERICAN BISON

IDENTIFICATION: An enormous buffalo (up to 6½ feet high) with massive forequarters, humped shoulders, and (in males) a beard. DISTRIBUTION: Formerly throughout north-central North America, now only in protected areas. HABITAT: Grassland, forests. STUDY AREAS: National Bison Range, Montana; Catalina Island, California; Wind Cave National Park, South Dakota; Wichita Mountain Wildlife Refuge, Oklahoma; Yellowstone National Park, Wyoming; Mackenzie Bison Sanctuary, Northwest Territories, Canada; Waterhen Wood Bison Ranch, Manitoba, Canada; Steel Rose Ranch, Saskatchewan, Canada; subspecies *B.b. bison,* the Plains Bison, and *B.b. athabascae,* the Wood Bison.

WISENT

IDENTIFICATION: Similar to American Bison but more slender, less hunched, and with longer legs. DISTRIBUTION: Formerly throughout Europe and central Asia, now only in protected areas; endangered. HABITAT: Forests. STUDY AREAS: Białowieża Primeval Forest and Reserve, and Niepołomice Reserve, Poland; Polish Academy of Sciences.

AFRICAN BUFFALO

IDENTIFICATION: A huge (11-foot-long), usually black buffalo with massive, upward-curving horns in both sexes. DISTRIBUTION: Sub-Saharan Africa. HABITAT: Savannas, forests. STUDY AREA: Serengeti National Park, Tanzania; subspecies *S.c. caffer,* the Cape, or Steppe, Buffalo.

Social Organization

Adult males (bulls) in American and European Bison generally live separately from females in groups that may contain up to 12 animals, or else solitarily. Females, their calves, and younger males (generally less than three or four years old) all live together in their own groups. For two months out of the year, female groups aggregate and adult males join these larger herds (which may contain hundreds of animals) for the rutting season. The mating system is "serial monogamy" within an overall framework of polygamy, i.e., males mate with several females, but remain exclusively with each female for a short period. African Buffalo have a similar social organization, living in herds ranging in size from 40–1,500 animals, mostly composed of females and their young in family groups, along with some adult males for part of the year. In addition, about 15 percent of adult males live in smaller bachelor herds, and older males may form peripheral groups.

Description

Behavioral Expression: Male American Bison participate in a variety of homosexual activities. Among younger bulls (less than five years old, particularly one-to-three-year-olds), anal intercourse is common. One male mounts the other with an erect penis and achieves anal penetration; the animal being mounted often facilitates the sexual interaction by positioning his hips or backing toward the other male with his tail lifted to the side. Homosexual copulation lasts on average nearly twice as long as heterosexual mating. The same bull may be mounted several times in succession by one or several other males, but reciprocal mounting is less common, since bulls that mount other males often do not allow themselves to be mounted (although males that are mounted do try to mount their partners). Males that are frequently mounted by other males often exhibit tears in the skin on their back where the mounting bull's hooves rub on either side of their spine. An identical type of skin abrasion is seen in female Bison that are frequently mounted by males. Homosexual mounting also occurs in a number of other contexts: in Bison (both American and European) and African Buffalo, males sometimes mount each other during play-fighting. An adult male American Bison may also mount another bull at the conclusion of an aggressive interaction. In these two contexts, mounting usually does not involve penetration, although erection of the penis and pelvic thrusting may occur. Sometimes one male will rest his chin on the other's rump as a prelude to mounting, often while making a soft panting sound. Female homosexual mounting and CHIN-RESTING also occur in Wisent and African Buffalo.

American Bison bulls—especially younger males—also sometimes form a TENDING BOND or consortship with another male. This paired association resembles the temporary (from a few hours to several days) monogamous bond formed between males and females during the rutting season. In a homosexual tending, one male closely follows and defends another male and may mount him as well. In some pairs mounting is reciprocal; in others, only one partner mounts or is mounted. In addition, younger males sometimes form "tending groups" of four to five individuals who take turns mounting one another or the same individual. Homosexual tending groups are unique in the joint participation of all the males in sexual activity: although several males often accompany heterosexual tending pairs, they never participate in sexual activity with either member of the pair.

Among American Bison, various types of intersexuality or hermaphroditism occasionally occur spontaneously in nature. Some transgendered individuals are known as BUFFALO OX and grow to be extraordinarily tall—they may be one and a half times bigger than a nontransgendered bull and generally have shorter fur as well. Other intersexual individuals are intermediate in size between males and females, possess malelike horns, and have female external genitalia and a uterus combined with testes. During tending bonds, these animals interact with both males and females: one individual tended females the way a (heterosexual) male would, but was also tended by other bulls as in heterosexual and homosexual interactions.

Frequency: Homosexual mounting is very prevalent among American Bison bulls, especially during the rutting season, when it may be seen several times a day. In fact, homosexual mounting is more common than heterosexual mounting in this species, since each female rarely mates with a male more than once a year, while each male may engage in same-sex mounting many times a day. The behavior is especially frequent among younger males, peaking in three-year-olds. Studies of semiwild populations have found that more than 55 percent of mounting in younger males is same-sex, and for some age categories all mounting behavior may be homosexual. It is less common among older adult males and three-to-four-year-olds. Female homosexual mounting in Bison and African Buffalo, as well as male same-sex mounting in African Buffalo, occurs occasionally.

Orientation: In American (and probably also European) Bison, younger bulls—nearly two-thirds of the male population—are functionally bisexual, although many actually participate exclusively in homosexual activity. It was once thought that such males only engage in homosexual mounting because older bulls prevent their access to females; however, studies on captive herds have shown that bulls still participate extensively in homosexual activity even when older bulls are not present. Older bulls, as well as females in Wisent and African Buffalo, are probably functionally bisexual but primarily heterosexual, with many individuals never engaging in homosexual activity.

Nonreproductive and Alternative Heterosexualities

As noted above, large portions of the Bison bull population do not breed: males of both the American and European species are sexually mature by the time they are three years old, yet they do not get a chance to breed until they are six and large enough to compete with older males. Even among older bulls, more than a quarter do not copulate heterosexually during the rutting period, and as many as 15 percent of females may not breed in a given year. In Wisent and African Buffalo, there are some postreproductive males and females as well—older individuals who have ceased breeding in the later years of their life. Nonprocreative sexual activities also figure in the social lives of heterosexual Bison: female American Bison often mount the male during tending, for example, and male Wisent occasionally ejaculate by rubbing the penis against the female's flanks. More than 20 percent of American Bison females engage in repeated copulations (only a single mating is necessary for procreation), and Wisent females have been observed mating with the same male eight times within half an hour. Wisent females also occasionally copulate during pregnancy (as late as three to four days before birth), and heterosexual activity sometimes occurs outside the breeding season. In American Bison, a notable separation and even hostility often exists between the sexes. As mentioned above, males and females live apart from one another for most of the year; during the rutting season, females frequently refuse the advances of males, and females often bear the scars of repeated heterosexual matings (described above). Wisent family life is occasionally marked by violence or abuse: calves have been killed by rutting bulls, and females sometimes desert their calves (especially those born late in the calving season).

Other Species

Among feral Water Buffalo (*Bubalus bubalis*) in Australia, female homosexual mounting is common: all cows mount other cows in heat, with 15–20 percent of the female population participating at any given time.

Sources   *asterisked references discuss homosexuality/transgender*

Caboń-Raczyńska, K., M. Krasińska, and Z. Krasiński (1983) "Behavior and Daily Activity Rhythm of European Bison in Winter." *Acta Theriologica* 28:273–99.

*Caboń-Raczyńska, K., M. Krasińska, Z. Krasiński, and J. M. Wójcik (1987) "Rhythm of Daily Activity and Behavior of European Bison in the Białowieża Forest in the Period without Snow Cover." *Acta Theriologica* 32:335–72.

*Jaczewski, Z. (1958) "Reproduction of the European Bison, *Bison bonasus* (L.), in Reserves." *Acta Theriologica* 1:333–76.

*Komers, P. E., F. Messier, and C. C. Gates (1994) "Plasticity of Reproductive Behavior in Wood Bison Bulls: When Subadults Are Given a Chance." *Ethology Ecology & Evolution* 6:313–30.

*——— (1992) "Search or Relax: The Case of Bachelor Wood Bison." *Behavioral Ecology and Sociobiology* 31:195–203.

Krasińska, M., and Z. A. Krasiński (1995) "Composition, Group Size, and Spatial Distribution of European Bison Bulls in Białowieża Forest." *Acta Theriologica* 40:1–21.

*Krasiński, Z., and J. Raczyński (1967) "The Reproduction Biology of European Bison Living in Reserves and in Freedom." *Acta Theriologica* 12:407–44.

*Lott, D. F. (1996–7) Personal communication.

*——— (1983) "The Buller Syndrome in American Bison Bulls." *Applied Animal Ethology* 11:183–86.

——— (1981) "Sexual Behavior and Intersexual Strategies in American Bison (*Bison bison*)." *Zeitschrift für Tierpsychologie* 56:115–27.

*——— (1974) "Sexual and Aggressive Behavior of Adult Male American Bison (*Bison bison*)." In V. Geist and F. Walther, eds., *Behavior in Ungulates and Its Relation to Management*, vol. 1, pp. 382–94. Morges, Switzerland: International Union for Conservation of Nature and Natural Resources.

*Lott, D. F., K. Benirschke, J. N. McDonald, C. Stormont, and T. Nett (1993) "Physical and Behavioral Findings in a Pseudohermaphrodite American Bison." *Journal of Wildlife Diseases* 29: 360–63.

*McHugh, T. (1972) *The Time of the Buffalo*. New York: Knopf.

*——— (1958) "Social Behavior of the American Buffalo (*Bison bison bison*)." *Zoologica* 43:1–40.

*Mloszewski, M. J. (1983) *The Behavior and Ecology of the African Buffalo*. Cambridge: Cambridge University Press.

*Reinhardt, V. (1987) "The Social Behavior of North American Bison." *International Zoo News* 203:3–8.

*——— (1985) "Social Behavior in a Confined Bison Herd." *Behavior* 92:209–26.

*Roe, F. G. (1970) *The North American Buffalo: A Critical Study of the Species in Its Wild State*. Toronto: University of Toronto Press.

*Rothstein, A., and J. G. Griswold (1991) "Age and Sex Preferences for Social Partners by Juvenile Bison Bulls." *Animal Behavior* 41:227–37.

*Sinclair, A. R. E. (1977) *The African Buffalo: A Study of Resource Limitation of Populations.* Chicago: University of Chicago Press.

*Tulloch, D. G. (1979) "The Water Buffalo, *Bubalus bubalis,* in Australia: Reproductive and Parent-Offspring Behavior." *Australian Wildlife Research* 6:265–87.

OTHER HOOFED MAMMALS

WILD HORSES (EQUIDS)　　　　　　　*Hoofed Mammal*

MOUNTAIN ZEBRA
Equus zebra

PLAINS ZEBRA
Equus quagga

TAKHI or PRZEWALSKI'S HORSE
Equus przewalskii

HOMOSEXUALITY	TRANSGENDER	BEHAVIORS		RANKING	OBSERVED IN
● Female	○ Intersexuality	● Courtship	○ Pair-bonding	○ Primary	● Wild
● Male	○ Transvestism	● Affectionate	○ Parenting	○ Moderate	○ Semiwild
		● Sexual		● Incidental	● Captivity

MOUNTAIN, PLAINS ZEBRAS
IDENTIFICATION: The familiar wild horse with a black-and-white-striped pattern; Mountain Zebras usually have a distinctive dewlap. DISTRIBUTION: Southern and eastern Africa; Mountain species is endangered. HABITAT: Mountainous slopes and plateaus; grassland, desert, semidesert. STUDY AREAS: Mountain Zebra National Park, South Africa, subspecies *E.z. zebra*, the Cape Mountain Zebra; Burgers Zoo, the Netherlands, subspecies *E.q. chapman*, Chapman's Zebra, and *E.q. boehmi*, Grant's Zebra.

TAKHI
IDENTIFICATION: The wild ancestor of domestic horses; coat usually tan or chestnut colored, with an erect mane, black tail and lower legs, white muzzle, and thin black stripe along back and several on upper forelegs. DISTRIBUTION: Formerly in central Asia (Mongolia, Kazakhstan, Sinkiang, Transbaikal); now extinct in the wild. HABITAT: Steppes. STUDY AREA: Bronx Zoo, New York.

Social Organization

Mountain and Plains Zebras have two main social units: breeding groups containing a herd stallion and three to five females with their offspring, and nonbreeding or "bachelor" groups. Groups combine to form herds numbering in the tens of thousands in Plains Zebras. Little is known of Takhi social organization in the wild, where the species is extinct (although it is beginning to be reintroduced from captive populations). It is likely that they have a system similar to that of Mountain and Plains Zebras, including both bachelor and "harem" herds.

Description

Behavioral Expression: Mounting between male Mountain Zebras is prefaced by a special ritualized display or "greeting" ceremony performed by two herd stallions, combining elements of courtship and sexual behaviors similar to those in heterosexual interactions. When two herd stallions meet, they approach each other with a stiff, high-stepping walk, holding their heads erect and ears forward as a friendly gesture. The males then rub first their noses and then their bodies together. Body rubbing is done either with the stallions facing in the same direction, or with one male's head at the other's rump. In the latter position, one male may nuzzle and sniff the other's genitals with his muzzle. Finally, one stallion sometimes mounts the other, or they may take turns mounting each other; occasionally a male will walk a few steps while another stallion is mounted on him. Plains Zebra males have also been observed placing their head on the rump of another male, a ritualized movement (also found in heterosexual courtship) that is thought to indicate an intention to mount the other male. When a herd stallion meets a bachelor male, many of the same behaviors occur, except for mounting. In addition, the bachelor male displays a distinctive facial expression resembling that used by Zebra mares in heat—lowering the head and pulling the lips and mouth corners back to expose the teeth—combined with a high-pitched call. Bachelor males also "greet" each other this way, often leading to play-fighting in which the males gently bite at each other and rear up on their hind legs. Bachelor males also sometimes mount each other as part of play-fighting.

Takhi mares occasionally mount each other; in some cases, pregnant females perform this sexual behavior with other females. Mares may mount each other from a sideways position in addition to from behind (the usual position for heterosexual mounting, although younger males also sometimes use a lateral mounting position). Such females may be among the highest-ranking mares in the herd and can also be noticeably aggressive toward males, kicking or biting them when the latter try to court other females.

Frequency: In captivity, mounting and attempted mounting occur in about 60 percent of interactions between male Plains Zebras. Among wild Mountain Zebras, homosexual interactions are less frequent. About 20 percent of play interactions between bachelor males involve mounting, while herd stallions associate with bachelor males (including "greetings" interactions) about 5 percent of the time. Female homosexual mounting in Takhi occurs occasionally (in captivity).

Orientation: In Zebras, herd stallions that engage in homosexual mounting and courtshiplike "greeting" behavior with other males also court and mate with females. Bachelor males, on the other hand, are exclusively homosexual to the extent that they engage in such behaviors, since they do not generally participate in heterosexual activity while in the bachelor herds. A little more than half of the male population of Mountain Zebras consists of bachelor males; most males join bachelor herds when they are just under two years old and stay for an average of two and a half years. About half of all bachelor males go on to become herd stallions and therefore are sequentially bisexual. However, some males remain in the bachelor herds for their entire lives, never mating heterosexually. Among Takhi, at least some females that mount other females are functionally bisexual, since they may be pregnant when they engage in homosexual activity.

Nonreproductive and Alternative Heterosexualities

As noted above, a large proportion of the male population in Zebras, as well as in Takhi, are nonbreeding bachelors. Some female Zebras also join the bachelor herds and do not participate in heterosexual activity while there (they remain for just under a year, on average). Wild equids also engage in an assortment of nonprocreative heterosexual activities. Males of all three species sometimes perform heterosexual mounts without an erection or penetration, while Takhi mares sometimes REVERSE mount stallions. Male Mountain Zebras and Takhi also frequently masturbate by erecting the penis and flipping it against the belly. Female equids sometimes also engage in an activity known as CLITORAL WINKING as part of courtship, in which the clitoris is rhythmically erected and wetted against the labia (often in conjunction with urination). Mountain Zebras occasionally participate in incestuous copulations: both father-daughter and brother-sister matings have been documented, although generally such pairings are avoided because females leave their family's herd before they reach sexual maturity. In addition, male Plains Zebras often try to mate with unrelated juvenile females that are not yet sexually mature. In fact, stallions—alone or in groups of up to 18 at a time—may "abduct" adolescent females by separating and chasing them from their family groups, after which they will try to copulate with the young mares. Interestingly, the female shows the behavioral signs of being "in heat" before she reaches the age when she can actually conceive; usually an "abducted" mare returns to her family group after her period of "heat" is finished. In contrast, in Takhi it is often the females who behave aggressively toward males (as mentioned above).

In these equids, a number of violent behaviors are also directed toward young foals. Mountain Zebra and Takhi stallions occasionally kill foals; in the latter species, infanticide occurs when the male grabs the youngster by its neck, shaking it and tossing it into the air. Female Mountain Zebras also sometimes accidentally kill their foals by kicking them; ironically, this may occur when they are trying to defend them from other mares, who are often aggressive toward unrelated foals. However, in a few cases females have adopted an unrelated foal, and in one instance a female even rejected her own offspring and adopted another. A Plains Zebra mare may also allow another mare's foal to suckle from her. Although Takhi males are not as involved in parenting as mares, a stallion may act as a "surrogate mother" to his own foal if it has lost its mother, even allowing the foal to "suckle" on his penis sheath.

Sources   *asterisked references discuss homosexuality/transgender*

*Boyd, L. E. (1991) "The Behavior of Przewalski's Horses and Its Importance to Their Management." *Applied Animal Behavior Science* 29:301–18.

*——— (1986) "Behavior Problems of Equids in Zoos." In S. L. Crowell-Davis and K. A. Houpt, eds., *Behavior,* pp. 653–64. The Veterinary Clinics of North America: Equine Practice 2(3). Philadelphia: W. B. Saunders.

*Boyd, L. E. and K. A. Houpt (1994) "Activity Patterns." In L. Boyd and K. A. Houpt, eds., *Przewalski's Horse: The History and Biology of an Endangered Species,* pp. 195–227. Albany: State University of New York Press.

Houpt, K. A., and L. Boyd (1994) "Social Behavior." In L. Boyd and K. A. Houpt, eds. *Przewalski's Horse: The History and Biology of an Endangered Species,* pp. 229–54. Albany: State University of New York Press.

Klingel, H. (1990) "Horses." In *Grzimek's Encyclopedia of Mammals,* vol. 4, pp. 557–94. New York: McGraw-Hill.

——— (1969) "Reproduction in the Plains Zebra, *Equus burchelli boehmi:* Behavior and Ecological Factors." *Journal of Reproduction and Fertility,* suppl. 6:339–45.

Lloyd, P. H., and D. A. Harper (1980) "A Case of Adoption and Rejection of Foals in Cape Mountain Zebra, *Equus zebra zebra." South African Journal of Wildlife Research* 10:61–62.

Lloyd, P. H., and O. A. E. Rasa (1989) "Status, Reproductive Success, and Fitness in Cape Mountain Zebra (*Equus zebra zebra*)." *Behavioral Ecology and Sociobiology* 25:411–20.

*McDonnell, S. M., and J. C. S. Haviland (1995) "Agonistic Ethogram of the Equid Bachelor Band." *Applied Animal Behavior Science* 43:147–88.

Monfort, S. L., N. P. Arthur, and D. E. Wildt (1994) "Reproduction in the Przewalski's Horse." In L. Boyd and K. A. Houpt, eds., *Przewalski's Horse: The History and Biology of an Endangered Species,* pp. 173–93. Albany: State University of New York Press.

*Penzhorn, B. L. (1984) "A Long-term Study of Social Organization and Behavior of Cape Mountain Zebras *Equus zebra zebra." Zeitschrift für Tierpsychologie* 64:97–146.

Rasa, O. A. E., and P. H. Lloyd (1994) "Incest Avoidance and Attainment of Dominance by Females in a Cape Mountain Zebra (*Equus zebra zebra*) Population." *Behavior* 128:169–88.

Ryder, O. A., and R. Massena (1988) "A Case of Male Infanticide in *Equus przewalskii." Applied Animal Behavior Science* 21:187–90.

*Schilder, M. B. H. (1988) "Dominance Relationships Between Adult Plains Zebra Stallions in Semi-Captivity." *Behavior* 104:300–319.

Schilder, M. B. H., and P. J. Boer (1987) "Ethological Investigations on a Herd of Plains Zebra in a Safari Park: Time-Budgets, Reproduction, and Food Competition." *Applied Animal Behavior Science* 18:45–56.

van Dierendonck, M. C., N. Bandi, D. Batdorj, S. Dügerlham, and B. Munkhtsag (1996) "Behavioral Observations of Reintroduced Takhi or Przewalski Horses (*Equus ferus przewalskii*) in Mongolia." *Applied Animal Behavior Science* 50:95–114.

WILD PIGS

Hoofed Mammal 🐏

WARTHOG
Phacochoerus aethiopicus

COLLARED PECCARY or JAVELINA
Tayassu tajacu

HOMOSEXUALITY	TRANSGENDER	BEHAVIORS		RANKING	OBSERVED IN
● Female	○ Intersexuality	○ Courtship	● Pair-bonding	○ Primary	● Wild
● Male	○ Transvestism	○ Affectionate	● Parenting	● Moderate	● Semiwild
		● Sexual		○ Incidental	● Captivity

WARTHOG
IDENTIFICATION: A 3–5-foot-long wild pig with a large head, prominent tusks, and distinctive warts in front of the eyes and on the jaw. DISTRIBUTION: Sub-Saharan Africa. HABITAT: Steppe, savanna. STUDY AREA: Andries Vosloo Kudu Reserve, South Africa.

COLLARED PECCARY
IDENTIFICATION: A piglike mammal with grayish, speckled, or salt-and-pepper fur and a light-colored collar. DISTRIBUTION: Arizona, New Mexico, Texas, southward to northern Argentina. HABITAT: Varied, including desert, woodland, rain forest. STUDY AREAS: In the Tucson Mountains and near Tucson, Arizona; University of Arizona; National Institute of Agronomic Research, French Guiana; subspecies *T.t. sonoriensis*.

Social Organization

Warthogs tend to associate in matriarchal groups (also known as SOUNDERS) of several females and their offspring, and in all-male "bachelor" groups. Only 3 percent of groups contain both males and females, and many Warthog males are solitary. Males join female groups only briefly for mating, which is usually promiscuous—both males and females copulate with multiple partners—and the only long-lasting bonds that form are between animals of the same sex, primarily females. Collared Peccaries live in herds of 5–15 individuals, containing animals of both sexes.

Description

Behavioral Expression: Homosexual mounting occurs in both Collared Peccaries and Warthogs. In Collared Peccaries, females in heat often RIDE or mount other females, and males occasionally mount one another as well. In Warthogs, homosexual mounting also takes place among females in heat, though it is less common. Sometimes a female Warthog will mount another female from the side, a position that is also occasionally used in heterosexual mounting. Warthog females often develop long-lasting bonds with each other, and same-sex mounting can be a part of these pairings (stable male-female pairs do not occur in this species). The two females associate together for many years and may even jointly raise their young, combining their litters and suckling each other's offspring. In addition, when one female is injured or temporarily unable to look after her young, the other female will take over parental duties. One such pair was seen consistently chasing away males who tried to get close to them. Biolo-

▲ *A female Collared Peccary "riding" another female*

gists studying Warthogs call these pairs or groups of adult females without any males or offspring SPINSTER groups—they typically contain an older female with a younger one. Some of these pairings involve related females, such as sisters or mother and daughter—in which case some same-sex mounting may be incestuous—although nonrelated pairings also occur. Occasionally two male Warthogs pair off as well, though no sexual behavior has been observed between them.

Frequency: Homosexual mounting occurs commonly in Collared Peccaries during heat; it is less frequent in Warthogs, probably comprising 1–3 percent of all mountings. About 5 percent of all Warthog groups are "spinster" (female-only) groups.

Orientation: Females that participate in same-sex mounting are probably bisexual, since most also engage in heterosexual relations. Warthog female companions, for example, may mate with males and reproduce, even if they do not consistently socialize with males. More than a quarter of Warthog females do not get pregnant each season, however, so it is possible that some are involved exclusively in same-sex (bonding and/or sexual) activities.

Nonreproductive and Alternative Heterosexualities

Significant portions of Warthog populations do not procreate. In addition to the nonbreeding females and sex-segregated groups mentioned above, one- and two-year-olds that are sexually mature may remain with their mother's group to help raise additional litters (rather than breeding themselves). These two species also participate in a variety of nonreproductive sexual behaviors. About 6 percent of heterosexual activity in Collared Peccaries, for example, involves females mounting males (REVERSE mounts), while another 22 percent of copulations are incomplete mounts by males. Males also frequently mount females during nonfertile periods, including pregnancy. Male Warthogs have been observed spontaneously ejaculating, including in their sleep. In addition, opposite-sex mountings in both species sometimes consist of a male mounting the female from the side without actual penetration (about 2 percent of sexual behavior in Collared Peccaries). Moreover, insemination does not necessarily occur even if penetration does, due to VAGINAL PLUGS. In both Collared Peccaries and Warthogs, a gelatinous barrier in the female's reproductive tract is deposited by a male when he copulates, very likely insuring that sperm from any subsequent matings cannot impregnate the female. Since female Warthogs usually copulate with more than one male, and female Collared Peccaries often mate repeatedly with the same male (as many as 18 times in three hours), a large proportion of copulations are therefore probably nonprocreative. Females can also refuse copulations by covering the vulva with their tail and tightening their leg muscles upward. In Collared Peccaries, offspring are cared for not only by their biological mothers, but also by "nursemaids," usually older sisters of the youngsters, that defend and nurse them. Often the nursemaids are not in fact sexually mature—they may be as young as six months old—with the amazing consequence that many nursemaids are themselves still nursing from their own "nursemaids." It is thought that they are able to produce milk because they consume the mother's placenta when she gives birth, perhaps thereby receiving some sort of hormonal influence from the mother. In Warthogs a number of violent counterreproductive activities also occur: adult males occasionally kill their younger brothers or cannibalize other males that they kill.

Other Species

Same-sex mounting also occurs in White-lipped Peccaries (*Tayassu pecari*) among both males and females.

Sources   asterisked references discuss homosexuality/transgender

Bissonette, J. A. (1982) *Ecology and Social Behavior of the Collared Peccary in Big Bend National Park, Texas.* Scientific Monograph Series no. 16. Washington, D.C.: U.S. National Park Service.

Byers, J. A., and M. Bekoff (1981) "Social, Spacing, and Cooperative Behavior of the Collared Peccary, *Tayassu tajacu.*" *Journal of Mammalogy* 62:767–85.

Child, G., H. H. Roth, and M. Kerr (1968) "Reproduction and Recruitment Patterns in Warthog (*Phacochoerus aethiopicus*) Populations." *Mammalia* 32:6–29.

Cumming, D. H. M. (1975) *A Field Study of the Ecology and Behavior of Warthog.* Salisbury, Rhodesia: Trustees of the National Museums and Monuments of Rhodesia.

*Dubost, G. (1997) "Comportements comparés du Pécari à levres blanches, *Tayassu pecari,* et du Pécari à collier, *T. tajacu* (Artiodactylea, Tayassuidés) [Comparative Behaviors of the White-lipped Peccary and of the Collared Peccary (Artiodactyla, Tayassuidae)]." *Mammalia* 61:313–43.

Frädrich, H. (1965) "Zur Biologie und Ethologie des Warzenschweines (*Phacochoerus aethiopicus* Pallas), unter Berücksichtigung des Verhaltens anderer Suiden [On the Biology and Ethology of Warthogs, in View of the Behavior of Other Suidae]." *Zeitschrift für Tierpsychologie* 22:328–93.

Packard, J. M., K. J. Babbitt, K. M. Franchek, and P. M. Pierce (1991) "Sexual Competition in Captive Collared Peccaries (*Tayassu tajacu*)." *Applied Animal Behavior Science* 29:319–26.

Schmidt, C. R. (1990) "Peccaries." In *Grzimek's Encyclopedia of Mammals,* vol. 5, pp. 48–55. New York: McGraw-Hill.

*Somers, M. J., O. A. E. Rasa, and B. L. Penzhorn (1995) "Group Structure and Social Behavior of Warthogs *Phacochoerus aethiopicus.*" *Acta Theriologica* 40:257–81.

*Sowls, L. K. (1997) *Javelinas and Other Peccaries: Their Biology, Management, and Use.* College Station: Texas A&M University Press.

*———— (1984) *The Peccaries.* Tucson: University of Arizona Press.

*———— (1974) "Social Behavior of the Collared Peccary *Dicotyles tajacu* (L.)." In V. Geist and F. Walther, eds., *The Behavior of Ungulates and Its Relation to Management,* vol. 1, pp. 144–65. IUCN Publication no. 24. Morges, Switzerland: International Union for Conservation of Nature and Natural Resources.

*———— (1966) "Reproduction in the Collared Peccary (*Tayassu tajacu*)." In I. W. Rowlands, ed., *Comparative Biology of Reproduction in Mammals,* Symposia of the Zoological Society of London no. 15, pp. 155–72. London and New York: Academic Press.

Torres, B. (1993) "Sexual Behavior of Free-Ranging Amazonian Collared Peccaries (*Tayassu tajacu.*)" *Mammalia* 57:610–13.

LLAMAS

Hoofed Mammal

VICUÑA
Vicugna vicugna

HOMOSEXUALITY	TRANSGENDER	BEHAVIORS		RANKING	OBSERVED IN
● Female	○ Intersexuality	○ Courtship	○ Pair-bonding	○ Primary	● Wild
● Male	○ Transvestism	● Affectionate	○ Parenting	○ Moderate	○ Semiwild
		● Sexual		● Incidental	○ Captivity

IDENTIFICATION: A small (3 feet at shoulder), camel-like animal with a slender body and a long, thin neck; coat is tawny brown or sandy-colored with white underparts and a chest mane. DISTRIBUTION: Andes Mountains of Peru, Bolivia, Argentina, Chile. HABITAT: High-elevation grasslands, plains. STUDY AREAS: Aricoma and Huaylarco, Peru.

Social Organization

Vicuñas live in cosexual groups usually containing 1 male, 3–10 females, and their offspring. In addition, all-male groups are a regular feature of Vicuña populations; they usually contain 5–10 animals, but may swell to include more than 150 individuals.

Description

Behavioral Expression: Female Vicuñas sometimes mount each other, with one animal straddling the other's back with her forelegs. This is similar to a heterosexual mating, except the mounted animal does not typically lie down as she would if a male were mounting her (even in heterosexual interactions, though, the female does not always cooperate by lying down). In one case, a pregnant female chased another female and mounted her. Adolescent males also sometimes mount one another during play-fighting, remaining astraddle for up to a quarter of a minute. Play-fights are gentle frolics in which the two males push and wrestle each other with their heads and long necks, interspersed with chasing or rearing on the hind legs.

Frequency: Same-sex mounting probably occurs only sporadically in Vicuñas. However, heterosexual mating is also infrequent: during a seven-month study period, for example, 5–11 heterosexual matings were observed compared to one mount between females.

Orientation: At least some females that mount other females are bisexual, since this behavior occurs in pregnant females. During the time that adolescent males are living in bachelor groups, the majority of their mounting activities are same-sex. Many of these males will go on to mate heterosexually, although about 10 percent of the nonbreeding animals in male bands are adults.

Nonreproductive and Alternative Heterosexualities

About 40 percent of Vicuñas do not breed: many are younger males living in the sex-segregated male groups (although these also include some adults), and there are solitary older animals as well. Among breeding animals, there is often considerable antagonism between the sexes: males have been known to fight with pregnant females, and territorial males are openly hostile toward females from neighboring bands, often chasing and attacking them. During copulation, females sometimes refuse to lie down; a male may try to force his partner to mate by bringing his full weight onto her back, causing the female to stagger underneath him. Adult males occasionally copulate with yearling females, who are probably not old enough to breed. Sexual activity between males and females also takes place outside of the breeding season.

Sources   asterisked references discuss homosexuality/transgender

Bosch, P. C., and G. E. Svendsen (1987) "Behavior of Male and Female Vicuña (*Vicugna vicugna* Molina 1782) as It Relates to Reproductive Effort." *Journal of Mammalogy* 68:425–29.

Carwardine, M. (1981) "Vicuña." *Wildlife* 23:8–11.

Franklin, W. L., (1983) "Contrasting Socioecologies of South America's Wild Camelids: The Vicuña and Guanaco." In J. F. Eisenberg and D. G. Kleiman, eds., *Advances in the Study of Mammalian Behavior*, pp. 573–629. American Society of Mammalogists Special Publication no. 7. Stillwater, Okla.: American Society of Mammalogists.

——— (1974) "The Social Behavior of the Vicuña." In V. Geist and F. Walther, eds., *The Behavior of Ungulates and Its Relation to Management*, vol. 1, pp. 477–87. IUCN Publication no. 24. Morges, Switzerland: International Union for Conservation of Nature and Natural Resources.

Franklin, W. L., and W. Herre (1990) "South American Tylopods." In *Grzimek's Encyclopedia of Mammals*, vol. 5, pp. 96–111. New York: McGraw-Hill.

*Koford, C. B. (1957) "The Vicuña and the Puna." *Ecological Monographs* 27:153–219.

ELEPHANTS
AFRICAN ELEPHANT
Loxodonta africana
ASIATIC ELEPHANT
Elephas maximus

Hoofed Mammal

HOMOSEXUALITY	TRANSGENDER	BEHAVIORS		RANKING	OBSERVED IN
● Female	○ Intersexuality	○ Courtship	● Pair-bonding	○ Primary	● Wild
● Male	○ Transvestism	● Affectionate	○ Parenting	● Moderate	○ Semiwild
		● Sexual		○ Incidental	○ Captivity

AFRICAN ELEPHANT
IDENTIFICATION: The familiar large (up to 7½ tons), trunked mammal with enormous ears and tusks in both sexes. DISTRIBUTION: Sub-Saharan Africa; endangered. HABITAT: Varied, including forest, savanna, marsh, semidesert, mountains. STUDY AREAS: Several locations in Africa, including Uganda and the Zambezi Valley of Zimbabwe; Kronberg Zoo, Germany; subspecies *L.a. africana*, the Bush Elephant, and *L.a. cyclotis*, the Forest Elephant.

ASIATIC ELEPHANT
IDENTIFICATION: Similar to African Elephant, but smaller, with tusks only in males, face and ears often mottled, forehead more convex and back more sloping, ears much smaller, and trunk with two fingerlike tips. DISTRIBUTION: India, Sri Lanka, Southeast Asia, China; endangered. HABITAT: Savanna, forest. STUDY AREAS: Periyar Tiger Reserve, Manakkavala, India; Lahugala, Sri Lanka; Pinnawala Elephant Orphanage, Sri Lanka; subspecies *E.m. indicus*, the Indian Elephant, and *E.m. maximus*, the Sri Lankan Elephant.

Social Organization

Elephants have a complex and highly organized community life. Females usually live in matriarchal herds of up to 50 individuals (loosely organized into family groups) led by an older female and generally containing no permanent adult males. Bulls often form male-only herds of 7–15 individuals (particularly in the African species), but may also be loners. Breeding males associate only temporarily with the female herds and mate with several different females.

Description

Behavioral Expression: Both African and Asiatic Elephant males participate in homosexual mounting. Among African Elephants, same-sex activity—which often takes place at or in watering holes—may be preceded by a great deal of caressing and affectionate behaviors. Two males intertwine their trunks, gently nudge each other, touch mouths in a "kiss," place their trunk tips in each other's mouth, roll over one another, and generally frolic together (sometimes with erections). One male often signals his intention to mount by extending his trunk along the other male's back, sometimes pushing him forward with his tusks (a gesture also used in sexual interactions between males and females). Homosexual mounting may also be preceded by one male sniffing or touching the other's penis with the tip of his trunk. Mounting occurs in the typical heterosexual position, with one male behind the other, and often the mounting male has an erect penis. Homosexual mating lasts about the same length as heterosexual mating, generally less than a minute, although one male may mount another several times in succession. Both older bulls and younger males participate in this activity. Among Asiatic Elephants, same-sex mounting sometimes occurs as part of play-fighting, in which two males swing their trunks at each other and gently rush at and butt one another. In African Elephants living in bull herds, a form of "erotic combat" also occurs. Two males push against each other while locking tusks and intertwining trunks; this activity stimulates the males sexually, and they develop full erections over up to half a dozen such bouts. Mounting sometimes occurs following a bout.

Although female homosexual activity has not yet been observed among wild Elephants, in captivity females sometimes masturbate one another with their trunks (the female's clitoris is nearly 17 inches long when erect or engorged). In addition, both female and male Asiatic Elephants in captivity engage in a variety of same-sex interactions with one another, including mounting activities and touching of the genitals with the trunk. Pregnant females also sometimes participate in these interactions.

Male Elephants also form "companionships," usually composed of an older bull and an attendant younger male (in contrast, there are no long-lasting heterosexual bonds in these species). In African Elephants, the younger male often helps the older one by guarding him or pulling down branches for him; in other cases, the older bull may help a younger male (or vice versa) who is injured or suffering from blindness or paralysis. The two males are constant companions and generally isolated from other Elephants; occasionally, an older bull will have two younger attendants. Among Asiatic Elephants, such male companionships appear not to be as

◀ *A male African Elephant mounting another male*

▲ *"Erotic combat" between two male African Elephants*

long-lasting as in African Elephants. Younger attendant males in African Elephants are sometimes reported to have enlarged genitalia.

Frequency: In wild African and Asiatic Elephants, homosexual mounting is a fairly common and regular occurrence, especially among younger bulls. In addition, Asiatic Elephant males spend an average of 10 percent of their time in play-fighting (which can also include mounting between males), while individual African Elephant bulls may participate in erotic combat up to four or five times a day during some times of the year. Approximately 18 percent of male Asiatic Elephants (not living in mixed-sex herds) have a male companion. In captivity, about 11 percent of social interactions between male Asiatic Elephants involve sexual activities, compared to approximately a quarter of social interactions between females; overall, roughly 45 percent of sexual interactions involve same-sex participants.

Orientation: Some younger Asiatic Elephant males that participate in homosexual activity are bisexual, since they may direct their sexual attentions to both males and females. However, some males are probably exclusively homosexual for at least part of their lives, since many Asiatic and African males do not participate in heterosexual activity until they are much older. African Elephant males in companionships are also exclusively same-sex oriented, since they are nonbreeders.

Nonreproductive and Alternative Heterosexualities

Elephant heterosexual life is frequently characterized by segregation and even antagonism between the sexes. In Asiatic Elephants, males and females often live separate from each other: males only associate with female herds about 25–30 percent of the time, and approximately 60 percent of the herds are not accompanied by males. As described above, Elephant herd structure is matriarchal, and females have even developed alternative parenting and "baby-sitting" arrangements without the contribution of males. Asiatic Elephant mothers often leave their calves in "nursery groups" that the adult females take turns looking after, allowing the other mothers to forage on their own. Female African Elephants often look after and occasionally even suckle other calves in their matriarchal groups. Furthermore, in both African and Asiatic species males experience a periodic sexual cycle known as MUSTH. A male in musth exhibits a number of characteristic physiological and behavioral changes, including increased aggression, ear waving and head swinging, infrasonic rumbling calls, continuous urinal discharge, and secretion by the temporal glands (located on either side of the head). Musth can last a few days to several months; during this time, males tend to associate more with females, but once musth is over, they usually return to male-only groups. Among African Elephants, bulls frequent special areas during their nonmusth time where their interactions are exclusively with other males, and they sometimes form a stronger association with one or two particular bulls.

During heterosexual interactions, females may be overtly aggressive toward males. Female Asiatic Elephants, for example, often charge and chase males (especially younger ones) that are trying to mate with them. In addition, males sometimes try to mount juvenile animals, who squeal in protest until a female intervenes. In fact, it is thought that female aggressiveness toward males contributes to a significant delay in breeding in this species: although males are sexually mature when about 10 years old, most do not begin breeding until about 17 years old. Reproductive suppression also occurs in African Elephants: the onset of puberty in females may be delayed by up to 10 years in some populations as a result of social, nutritional, or physiological stresses. In addition, most males do not father their first offspring until they are 30–35 years old—this is 15–20 years after they become sexually mature. Other nonbreeding individuals include loner males (often older, postreproductive individuals), males in "companionships" (described above), as well as postreproductive or menopausal females (generally older than 50 years), and females who have nonbreeding intervals of up to 13 years between calves. Heterosexual relations are further complicated by an apparent mismatch in the structure of the male and female genitalia: unlike in most mammals, the female's vaginal opening is placed far forward on her belly, hampering access by the male. Although his penis is able to assume a special S-shape to reach her vulva, penetration is often difficult to achieve. On occasion, the male's organ will contact the female's anus rather than her vulva, and he may ejaculate before achieving penetration. Heterosexual relations also often include touching and stimulation of the genitals with the trunk; in captivity, one female was also observed rubbing her clitoris against a male's side while he was lying down.

Other Species

Homosexual mounting in the Indian Rhinoceros (*Rhinoceros unicornis*) has been observed among adolescent females in captivity.

Sources

asterisked references discuss homosexuality/transgender

*Buss, I. O. (1990) *Elephant Life: Fifteen Years of High Population Density.* Ames, Iowa: Iowa State University Press.

Buss, I. O,. and N. S. Smith (1966) "Observations on Reproduction and Breeding Behavior of the African Elephant." *Journal of Wildlife Management* 30:375–88.

*Dixon, A., and M. MacNamara (1981) "Observations on the Social Interactions and Development of Sexual Behavior in Three Sub-adult, One-horned Indian Rhinoceros (*Rhinoceros unicornis*) Maintained in Captivity." *Zoologische Garten* 51:65–70.

Douglas-Hamilton, I., and O. Douglas-Hamilton (1975) *Among the Elephants.* London: Collins & Harvill.

Eisenberg, J. F., G. M. McKay, and M. R. Jainudeen (1971) "Reproductive Behavior of the Asiatic Elephant (*Elephas maximus maximus* L.)." *Behavior* 38:193–225.

*Grzimek, B. (1990) "African Elephant." In *Grzimek's Encyclopedia of Mammals,* vol. 4, pp. 502–20. New York: McGraw-Hill.

*Jayewardene, J. (1994) *The Elephant in Sri Lanka.* Colombo, Sri Lanka: Wildlife Heritage Trust of Sri Lanka.

*Kühme, W. (1962) "Ethology of the African Elephant (*Loxodonta africana* Blumenbach 1797) in Captivity." *International Zoo Yearbook* 4:113–21.

Laws, R. M. (1969) "Aspects of Reproduction in the African Elephant, *Loxodonta africana.*" *Journal of Reproduction and Fertility,* suppl. 6:193–217.

Lee, P. C. (1987) "Allomothering Among African Elephants." *Animal Behavior* 35:278–91.

*McKay, G. M. (1973) *Behavior and Ecology of the Asiatic Elephant in Southeastern Ceylon.* Smithsonian Contributions to Zoology no. 125. Washington, D.C.: Smithsonian Institution Press.

*Morris, D. (1964) "The Response of Animals to a Restricted Environment." *Symposia of the Zoological Society of London* 13:99–118.

Moss, C. (1988) *Elephant Memories: Thirteen Years in the Life of an Elephant Family.* New York: William Morrow and Co.

Moss, C., and J. H. Poole (1983) "Relationships and Social Structure of African Elephants." In R. A. Hinde, ed., *Primate Social Relationships: An Integrated Approach,* pp. 315–25. Oxford: Blackwell Scientific Publications.

Poole, J. H. (1994) "Sex Differences in the Behavior of African Elephants." In R. V. Short and E. Balaban, eds., *The Differences Between the Sexes,* pp. 331–46. Cambridge: Cambridge University Press.

——— (1987) "Rutting Behavior in African Elephants: The Phenomenon of Musth." *Behavior* 102:283–316.

*Poole, T. B., V. J. Taylor, S. B. U. Fernando, W. D. Ratnasooriya, A. Ratnayeke, G. Lincoln, A. McNeilly, and A. M. V. R. Manatunga (1997) "Social Behavior and Breeding Physiology of a Group of Asian Elephants *Elephas maximus* at the Pinnawala Elephant Orphanage, Sri Lanka." *International Zoo Yearbook* 35:297–310.

*Ramachandran, K. K. (1984) "Observations on Unusual Sexual Behavior in Elephants." *Journal of the Bombay Natural History Society* 81:687–88.

*Rosse, I. C. (1892) "Sexual Hypochondriasis and Perversion of the Genetic Instinct." *Journal of Nervous and Mental Disease* 19(11): 795–811.

*Shelton, D. J. (1965) "Some Observations on Elephants." *African Wild Life* 19: 161–64.

*Sikes, S. K. (1971) *The Natural History of the Elephant.* New York: Elsevier Publishing Co.

Other Mammals

CARNIVORES

WILD CATS (FELIDS)
LION
Panthera leo
CHEETAH
Acinonyx jubatus

Carnivore

HOMOSEXUALITY	TRANSGENDER	BEHAVIORS		RANKING	OBSERVED IN
● Female	○ Intersexuality	● Courtship	● Pair-bonding	○ Primary	● Wild
● Male	○ Transvestism	● Affectionate	● Parenting	● Moderate	● Semiwild
		● Sexual		○ Incidental	● Captivity

LION
IDENTIFICATION: A large wild cat (up to 550 pounds) with a prominent mane in males. DISTRIBUTION: Throughout Africa and in Gujarat, northwest India; vulnerable. HABITAT: Plains, savannas, scrub, open forest. STUDY AREAS: Serengeti National Park, Tanzania; Gir Wildlife Sanctuary, India; Gay Lion Farm, California; subspecies *P.l. massaieus*, the Masai Lion, and *P.l. persica*, the Asiatic Lion.

CHEETAH
IDENTIFICATION: A medium-sized wild cat with a sleek, greyhoundlike physique and a spotted coat. DISTRIBUTION: Throughout Africa and sporadically in central Asia and the Middle East; vulnerable. HABITAT: Semi-desert, grassland, steppes. STUDY AREAS: Serengeti National Park, Tanzania; Lion Country Safari, California; National Zoological Park, Washington, D.C.; Hogle Zoo, Utah; subspecies *A.j. jubatus*, the African Cheetah.

Social Organization

Lions have two distinct forms of social organization. Some individuals are RESIDENTS, living in prides of up to a dozen or more adult females (usually all related to one another) with their cubs, along with an associated COALITION of one to nine adult males. Other Lions are NOMADS, ranging widely as solitary individuals or pairs. Female Cheetahs are largely solitary, while males are either RESIDENTS (with their own territories) or FLOATERS (without resident territories). Some males associate in groups of two to three (occasionally four) animals, often littermates (see below). The mating system is promiscuous or polygamous: males and females generally mate with multiple partners, no long-term heterosexual bonds are formed, and males do not typically participate in parenting.

Description

Behavioral Expression: In female Lions, homosexual interactions are often initiated by one female pursuing another and crawling under her to encourage the other female to mount her. When mounting another female, a Lioness displays a number of behaviors also associated with heterosexual mating, including gently biting the mountee on the neck, growling, making pelvic thrusts, and rolling on her own back afterward. Sometimes Lionesses take turns mounting each other. Because most females in a pride are related to each other—on average about as closely related as cousins—homosexual behavior among Lionesses may be largely incestuous. Among male pride-mates (residents), homosexual activity often begins with a great deal of affectionate activity (which is also a component of "greetings" interactions between males). This includes mutual head-rubbing (often accompanied by a low moaning or humming noise), presentation of the hindquarters to the other male, sliding and rubbing against each other, circling one another, and rolling on the back with an erect penis. This may lead to intense caressing and eventually mounting of one male by the other, including pelvic thrusting. Sometimes three males rub and roll together, mounting each other in turn. A male Lion sometimes courts a particular individual, keeping company with him for several days while engaging in sexual behavior. He typically defends his partner against intruding males, and often other males in the group will join him in attacking any interfering males. Because male pride-mates, like females, are usually related to each other (as close, on average, as half brothers), this activity may also be incestuous. Nomadic male Lions also form long-lasting platonic "companionships" with other males, spending nearly all of their time together; these male pair- or trio-bonds are generally stronger than heterosexual bonds between residents. Companions are usually close in age; some are pride-mates or brothers, although about half of all companionships include unrelated individuals. Females occasionally form companionships with each other as well.

Female Cheetahs sometimes court other females, including participation in courtship chases, play-fighting, and mating circles. Courtship chases take place in the early morning, late afternoon, or on moonlit nights; groups of Cheetahs—including females—chase a female in heat for up to 150 yards. This may be interspersed with mock-fighting, in which the courting animals (females or males) rise up on their hind legs and drop their forelegs on the female being courted. Females also sometimes join MATING CIRCLES, where the animals lie in a circle around the courted female, often while the males take turns fighting each other. A female Cheetah may also mount another female who is in heat, clasping her with the forelegs, gently biting the scruff of her neck (as in heterosexual copulation), and thrusting against her. Male homosexual mounting occurs as well, and one male may also mount another male that is in the act of mating with a female. During courtship interactions, males also sometimes lick and nuzzle another male's genitals while closely following him, occasionally exhibiting an erection while doing so.

Male Cheetahs often live in permanent partnerships or COALITIONS, consisting of a pair or trio of animals; about 30 percent of these associations include animals that are not related to each other, while the remainder consist of brothers. Partners are strongly bonded to one another and probably remain together for life. Spending almost all (93 percent) of their time in each other's company, male pair-mates frequently groom one another (licking each other's face and neck), defend each other in fights, and prefer resting together in close contact (even if this means that one of them will not be fully shaded against the harsh midday sun). Bonded males also become strongly distressed when separated, continuously searching and calling loudly for one another with birdlike *yip* or *chirp* calls. On being reunited, they may engage in a variety of affectionate and/or sexual activities, including reciprocal mounting with erections, face-rubbing, and STUTTERING (a purrlike vocalization often associated with sexual excitement). These activities appear to be more common between siblings than nonsiblings. Very rarely, paired males may temporarily adopt or look after lost cubs (most other parents in this species—foster or otherwise—are single mothers).

Frequency: In Lions, homosexual behavior in females is fairly common in captivity, while in the wild, two Lionesses were observed to mount each other three times over two days. In males, homosexual mounting may account for up to 8 percent of all mounting episodes. About 47 percent of all companionships involving adult nomadic Lions are between males, and about 37 percent are between females. Homosexual behavior (courtship and sexual) in Cheetahs is also quite frequent (at least in captive or semiwild conditions). In the wild, 27–40 percent of males live in same-sex pairs while 16–19 percent live in same-sex trios. In one study, 1 out of 11 instances of foster-parenting involved a pair of males looking after a cub (representing perhaps less than 1 percent of all families, adoptive and nonadoptive).

Orientation: Female Lions and male Cheetahs that participate in homosexual mounting may be bisexual, since same-sex activity sometimes alternates (or co-occurs) with heterosexual mounting in the same session. Some female Lions react aggressively to homosexual overtures and therefore these individuals are probably predominantly heterosexual. However, other females engage in same-sex mounting even in the presence of males, indicating more of a "preference" for homosexual activity. Many male Cheetahs living in partnerships do court and mate with females. However, pairs or trios of males are only with females 9 percent of the time, and they may experience reduced heterosexual mating opportunities compared to single males. Moreover, same-sex coalitions usually constitute life-long pair-bonds (which are not found between males and females in this species). Only about half of all companionships of two to three male Lions ever become residents that mate with a pride of females. Those that don't may associate exclusively with other males for most of their lives, while some that do join a pride may participate in both same-sex and opposite-sex activities.

Nonreproductive and Alternative Heterosexualities

More than 60 percent of male Lions (solitary or in companionships) do not become residents during their lives and therefore do not participate in breeding. When they do, however, Lions have extraordinarily high heterosexual copulation rates. The female may mate as often as four times an hour when she is in heat over a continuous period of three days and nights (without sleeping), and sometimes with up to five different males—far in excess of the amount required if mating were simply for fertilization. A number of other nonprocreative sexual behaviors occur in these felids as well. Lions sometimes mate during pregnancy (up to 13 percent of all sexual activity), and as much as 80 percent of all heterosexual mating in some populations may not result in reproduction. In fact, following arrival of new males in a pride, females often increase their sexual activity while reducing their fertility (by failing to ovulate). In addition, "oral sex" is a feature of heterosexual foreplay— female Lions may lick and rub the male's genitals, while Cheetahs of both sexes lick their partner's genitals as part of heterosexual courtship. Male Lions have also been observed masturbating in captivity: an unusual technique is used, in which the Lion lies on his back and rolls his hindquarters up above his head, so that he can rub his penis with one of his front legs.

In wild Cheetahs, incestuous activity occasionally takes place when adult males try to mount their mothers, who typically react aggressively to such advances. In fact, heterosexual relations are in general characterized by a great deal of aggression between the sexes. In Lions, heterosexual copulation is often accompanied by snarling, biting, growling, and threats, and sometimes the female actually wheels around and slaps the male. During Cheetah courtship chases, males often knock females down and slap them, and these interactions may develop into full-scale fights. When the female is not in heat, the two sexes live largely segregated lives. Family life in these species may also be fraught with violence: infanticide occurs in Lions (where it accounts for more than a quarter of all cub deaths) as well as Cheetahs. Abandonment by their mother is the second highest cause of cub mortality in Cheetahs, and mother Cheetahs occasionally eat their own cubs if they have been killed by a predator (adult male Cheetahs also sometimes cannibalize each other). However, female Lions often participate in productive alternative family arrangements amongst themselves, such as communal care and suckling of young, as well as the formation of CRÈCHES or nursery groups. Female Cheetahs also sometimes (reluctantly) adopt orphaned or lost cubs.

Sources   *asterisked references discuss homosexuality/transgender*

*Benzon, T. A., and R. F. Smith (1975) "A Case of Programmed Cheetah *Acinonyx jubatus* Breeding." *International Zoo Yearbook* 15:154–57.

Bertram, B. C. R. (1975) "Social Factors Influencing Reproduction in Wild Lions." *Journal of Zoology, London* 177:463–82.

*Caro, T. M. (1994) *Cheetahs of the Serengeti Plains: Group Living in an Asocial Species.* Chicago: University of Chicago Press.

*——— (1993) "Behavioral Solutions to Breeding Cheetahs in Captivity: Insights from the Wild." *Zoo Biology* 12:19–30.

*Caro, T. M., and D. A. Collins (1987) "Male Cheetah Social Organization and Territoriality." *Ethology* 74:52–64.

*———— (1986) "Male Cheetahs of the Serengeti." *National Geographic Research* 2:75–86.

*Chavan, S. A. (1981) "Observation of Homosexual Behavior in Asiatic Lion *Panthera leo persica.*" *Journal of the Bombay Natural History Society* 78:363–64.

*Cooper, J. B. (1942) "An Exploratory Study on African Lions." *Comparative Psychology Monographs* 17:1–48.

Eaton, R. L. (1978) "Why Some Felids Copulate So Much: A Model for the Evolution of Copulation Frequency." *Carnivore* 1:42–51.

*———— (1974a) *The Cheetah: The Biology, Ecology, and Behavior of an Endangered Species.* New York: Van Nostrand Reinhold.

*———— (1974b) "The Biology and Social Behavior of Reproduction in the Lion." In R. L. Eaton, ed., *The World's Cats, vol. 2: Biology, Behavior, and Management of Reproduction,* pp. 3–58. Seattle: Woodland Park Zoo.

*Eaton, R. L., and S. J. Craig (1973) "Captive Management and Mating Behavior of the Cheetah." In R. L. Eaton, ed., *The World's Cats, vol. 1: Ecology and Conservation,* pp. 217–254. Winston, Oreg.: World Wildlife Safari.

Herdman, R. (1972) "A Brief Discussion on Reproductive and Maternal Behavior in the Cheetah *Acinonyx jubatus.*" In *Proceedings of the 48th Annual AAZPA Conference (Portland, OR),* pp. 110–23. Wheeling, W. Va.: American Association of Zoological Parks and Aquariums.

Laurenson, M. K. (1994) "High Juvenile Mortality in Cheetahs (*Acinonyx jubatus*) and Its Consequences for Maternal Care." *Journal of Zoology, London* 234:387–408.

Morris, D. (1964) "The Response of Animals to a Restricted Environment." *Symposia of the Zoological Society of London* 13:99–118.

Packer, C., L. Herbst, A. E. Pusey, J. D. Bygott, J. P. Hanby, S. J. Cairns, and M. B. Mulder (1988) "Reproductive Success of Lions." In T. H. Clutton-Brock, ed., *Reproductive Success: Studies of Individual Variation in Contrasting Breeding Systems,* pp. 363–83. Chicago and London: University of Chicago Press.

Packer, C., and A. E. Pusey (1983) "Adaptations of Female Lions to Infanticide by Incoming Males." *American Naturalist* 121:716–28.

———— (1982) "Cooperation and Competition Within Coalitions of Male Lions: Kin Selection or Game Theory?" *Nature* 296:740–42.

Pusey, A. E., and C. Packer (1994) "Non-Offspring Nursing in Social Carnivores: Minimizing the Costs." *Behavioral Ecology* 5:362–74.

*Ruiz-Miranda, C. R., S. A. Wells, R. Golden, and J. Seidensticker (1998) "Vocalizations and Other Behavioral Responses of Male Cheetahs (*Acinonyx jubatus*) During Experimental Separation and Reunion Trials." *Zoo Biology* 17:1–16.

*Schaller, G. B. (1972) *The Serengeti Lion.* Chicago: University of Chicago Press.

Subba Rao, M. V., and A. Eswar (1980) "Observations on the Mating Behavior and Gestation Period of the Asiatic Lion, *Panthera leo,* at the Zoological Park, Trivandrum, Kerala." *Comparative Physiology and Ecology* 5:78–80.

Wrogemann, N. (1975) *Cheetah Under the Sun.* Johannesburg: McGraw-Hill.

WILD DOGS (CANIDS)
RED FOX
Vulpes vulpes
(GRAY) WOLF
Canis lupus
BUSH DOG
Speothos venaticus

Carnivore

HOMOSEXUALITY	TRANSGENDER	BEHAVIORS		RANKING	OBSERVED IN
● Female	○ Intersexuality	○ Courtship	○ Pair-bonding	○ Primary	○ Wild
● Male	○ Transvestism	○ Affectionate	● Parenting	○ Moderate	○ Semiwild
		● Sexual		● Incidental	● Captivity

RED FOX
IDENTIFICATION: A small canid (body length up to 3 feet) with a bushy tail and a reddish brown coat (although some variants are silvery or black). DISTRIBUTION: Throughout most of Eurasia, northern Africa, and North America. HABITAT: Variable, including forest, tundra, prairie, farmland. STUDY AREA: Oxford University, England.

WOLF
IDENTIFICATION: The largest wild canid (reaching up to 7 feet in length) with a gray, brown, black, or white coat. DISTRIBUTION: Throughout most of the Northern Hemisphere. HABITAT: Widely varied, excepting tropical forests and deserts. STUDY AREAS: Bavarian Forest National Park, Germany; Basel Zoo, Switzerland; subspecies *C.l. lupus,* the Common Wolf.

BUSH DOG
IDENTIFICATION: A small (3 foot long), reddish brown, bearlike canid with short legs and tail. DISTRIBUTION: Northern and eastern South America; vulnerable. HABITAT: Forest, savanna, swamp, riverbanks. STUDY AREA: London Zoo, England.

Social Organization

Red Fox society is characterized by highly complex and flexible living arrangements and social interactions, varying both between and within populations. Many Foxes live in groups with several, often related, adult females and one male (or rarely several); mated pairs are characteristic of other populations. Mating systems range from monogamy to polygamy. Wolves have a highly developed social system revolving around the pack, a group of usually a dozen or so individuals consisting of a mated pair and up to two generations of their offspring; occasionally, a few unrelated adults also live in the pack. Much less is known of the social life of wild Bush Dogs, although it appears that they, too, live in groups (possibly also pairs) and hunt in packs of usually a dozen individuals (though much larger groups containing hundreds of dogs have also been reported).

Description

Behavioral Expression: When the breeding female in a Red Fox group comes into heat, both the male and female group members become sexually interested in her. Homosexual interactions involve the younger females—usually her daughters—running up to the vixen in heat, sniffing her genitals and mounting her. Although the mounting female clasps the other tightly, the older vixen usually responds to these sexual advances aggressively (as she does to most sexual approaches by males) by rearing up and "boxing" with the other female while gaping her mouth. The younger females may also mount each other, all the while making staccato, rasping click sounds known as GECKERING and SNIRKING. Pairs of Red Fox females also sometimes coparent their young, sharing a den, rearing their cubs together, bringing food for each other, and even suckling each other's young. Although coparents are often related to each other, some may not be relatives.

Male Wolves often mount each other when the highest-ranking female in their pack comes into heat (a time when heterosexual activity also reaches its peak). As in Red Foxes, homosexual activity may be incestuous, since the males in the pack are often related to each other. A male Wolf sometimes also mounts another male when the latter is mounting a female. Male Bush Dogs have also been observed mounting each other, often accompanied by playful nipping of the legs or hindquarters.

Frequency: In captivity, homosexual mounting in Red Foxes and Wolves occurs frequently when the breeding female is in heat. In Bush Dogs, mounting between males is less common. The prevalence of these behaviors in the wild is not known.

Orientation: Many female Red Foxes that mount other females may be exclusively same-sex oriented, since such younger or lower-ranking individuals usually do not mate with males. For some females, this homosexual orientation may be long-lasting—perhaps even continuing for a female's entire life—since as many as 50–70 percent of vixens never leave their home groups to begin breeding on their own. Male Wolves that mount each other are bisexual, also showing sexual interest in females. However, their heterosexual activity is limited to the highest-ranking breeding female: males routinely ignore lower-ranking females in favor of homosexual activity.

A female Red Fox mounting another female ▶

Nonreproductive and Alternative Heterosexualities

In all three of these wild dog species, reproductive suppression is a prominent feature of the social system. For example, only a fraction of female Red Foxes reproduce—a third or more of all vixens (depending on the population) are nonbreeders, and in some areas as many as 95 percent of adult females do not reproduce. There are multiple mechanisms for this "birth control." In some cases, nonbreeding females simply do not mate, or else they fail to come into heat (a similar phenomenon occurs in Bush Dogs). In other cases, females become pregnant, but routinely abort their fetuses or abandon their young once they are born. Neglect or abuse of pups (leading to their death) has been documented in both Red Foxes and Bush Dogs, as well as cannibalism (in Red Foxes). In Wolves, the highest-ranking individuals (especially males) often prevent other animals from mating by intruding or attacking them directly; as many as a quarter of all mounts may be interrupted this way. In other cases, Wolves simply show no sexual interest in the opposite sex; these and other factors function to curtail reproduction in 40–80 percent of all packs. Breeding may also be inhibited by an incest taboo when a pack comes to consist entirely of closely related individuals (usually siblings). However, mother-son and brother-sister matings have occasionally been observed, and some packs may be highly inbred. In both Red Foxes and Wolves (and to a lesser extent, Bush Dogs), nonbreeding animals sometimes help the breeding female raise her young, including feeding, guarding, and "baby-sitting" them. There are even cases of a female Red Fox adopting an entire litter after their biological mother has died or been killed. However, some nonbreeding Red Foxes do not contribute any such care, and there is actually some evidence that more offspring may be successfully reared when there are fewer such "helpers" present in the group. Nonbreeding lone Wolves that do not act as helpers may constitute as much as 28 percent of some populations.

In addition to patterns of reproductive suppression, a number of nonreproductive heterosexual activities also occur in these canids. About 8 percent of female Red Foxes mate outside of the breeding period (this practice occurs in Wolves as well). Because males also have a sexual cycle ensuring that they cannot produce sperm during this time, such matings are definitively nonprocreative. About half of all heterosexual mounts in Wolves do not involve thrusting, penetration, or ejaculation; female Wolves also sometimes mount and thrust against males (REVERSE mounting). When mating does occur in Wolves and Red Foxes, individuals often engage in multiple copulations (i.e., more than the number of times simply required for fertilization). These often involve "copulatory ties" that keep the partners joined at the genitals for long periods. Heterosexual relations are sometimes fraught with difficulty, for example when female Red Foxes aggressively gape at males trying to mount them, or when both male and female Wolves display indifference or aggression toward animals trying to mate with them. In fact, one study showed that less than 3 percent of all heterosexual courtships in Wolves actually result in copulation.

Other Species

Several forms of intersexuality or transgender occasionally occur in Raccoon Dogs (*Nyctereutes procyonoides*). Some individuals, for example, combine female genitals with testes, while others have a mosaic chromosome pattern that combines a male pattern (XY) with a joint male-female pattern (XXY).

Sources   *asterisked references discuss homosexuality/transgender*

Creel, S., and D. Macdonald (1995) "Sociality, Group Size, and Reproductive Suppression Among Carnivores." *Advances in the Study of Behavior* 24:203–57.

Derix, R., J. van Hooff, H. de Vries, and J. Wensing (1993) "Male and Female Mating Competition in Wolves: Female Suppression vs. Male Intervention." *Behavior* 127:141–74.

Drüwa, P. (1983) "The Social Behavior of the Bush Dog (*Speothos*)." *Carnivore* 6:46–71.

*Fentener van Vlissengen, J. M., M. A. Blankenstein, J. H. H. Thijssen, B. Colenbrander, A. J. E. P. Verbruggen, and C. J. G. Wensing (1988) "Familial Male Pseudohermaphroditism and Testicular Descent in the Raccoon Dog (*Nyctereutes*)." *Anatomical Record* 222:350–56.

van Hooff, J. A. R. A. M., and J. A. B. Wensing (1987) "Dominance and Its Behavioral Measures in a Captive Wolf Pack." In H. Frank, ed., *Man and Wolf: Advances, Issues, and Problems in Captive Wolf Research*, pp. 219–52. Dordrecht: Dr W. Junk.

*Kleiman, D. G. (1972) "Social Behavior of the Maned Wolf (*Chrysocyon brachyurus*) and Bush Dog (*Speothos venaticus*): A Study in Contrast." *Journal of Mammalogy* 53:791–806.

Lloyd, H. G. (1975) "The Red Fox in Britain." In M. W. Fox, ed., *The Wild Canids: Their Systematics, Behavioral Ecology, and Evolution*, pp. 207–15. New York: Van Nostrand Reinhold.

Macdonald, D. W. (1996) "Social Behavior of Captive Bush Dogs (*Speothos venaticus*)." *Journal of Zoology, London* 239:525–43.

*——— (1987) *Running with the Fox*. New York: Facts on File.

*——— (1980) "Social Factors Affecting Reproduction Amongst Red Foxes (*Vulpes vulpes* L., 1758)." In E. Zimen, ed., *The Red Fox: Symposium on Behavior and Ecology*, pp. 123–75. Biogeographica no. 18 . The Hague: Dr W. Junk.

——— (1979) "'Helpers' in Fox Society." *Nature* 282:69–71.

——— (1977) "On Food Preference in the Red Fox." *Mammal Review* 7:7–23.

Macdonald, D. W., and P. D. Moehlman (1982) "Cooperation, Altruism, and Restraint in the Reproduction of Carnivores." In P. P. G. Bateson and P. H. Klopfer, eds., *Perspectives in Ethology, vol. 5: Ontogeny*, pp. 433–67.

Mech, L. D. (1970) *The Wolf: The Ecology and Behavior of an Endangered Species*. New York: Natural History Press.

Packard, J. M., U. S. Seal, L. D. Mech, and E. D. Plotka (1985) "Causes of Reproductive Failure in Two Family Groups of Wolves." *Zeitchrift für Tierpsychologie* 68:24–40.

Packard, J. M., L. D. Mech, and U. S. Seal (1983) "Social Influences on Reproduction in Wolves." In L. N. Carbyn, ed., *Wolves in Canada and Alaska: Their Status, Biology, and Management*, pp. 78–85. Canadian Wildlife Service Series no. 45. Ottawa: Canadian Wildlife Service.

Porton, I. J., D. G. Kleiman, and M. Rodden (1987) "Aseasonality of Bush Dog Reproduction and the Influence of Social Factors on the Estrous Cycle." *Journal of Mammalogy* 68:867–71.

Schantz, T. von (1984) "'Non-Breeders' in the Red Fox *Vulpes vulpes*: A Case of Resource Surplus." *Oikos* 42:59–65.

——— (1981) "Female Cooperation, Male Competition, and Dispersal in the Red Fox *Vulpes vulpes*." *Oikos* 37:63–68.

*Schenkel, R. (1947) "Ausdrucks-Studien an Wölfen: Gefangenschafts-Beobachtungen [Expression Studies of Wolves: Captive Observations]." *Behavior* 1:81–129.

Schotté, C. S., and B. E. Ginsburg (1987) "Development of Social Organization and Mating in a Captive Wolf Pack." In H. Frank, ed., *Man and Wolf: Advances, Issues, and Problems in Captive Wolf Research*, pp. 349–74. Dordrecht: Dr W. Junk.

Sheldon, J. W. (1992) *Wild Dogs: The Natural History of the Nondomestic Canidae*. San Diego: Academic Press.

Smith, D., T. Meier, E. Geffen, L. D. Mech, J. W. Burch, L. G. Adams, and R. K. Wayne (1997) "Is Incest Common in Gray Wolf Packs?" *Behavioral Ecology* 8:384–91.

Storm, G. L., and G. G. Montgomery (1975) "Dispersal and Social Contact Among Red Foxes: Results From Telemetry and Computer Simulation." In M. W. Fox, ed., *The Wild Canids: Their Systematics, Behavioral Ecology, and Evolution*, pp. 237–46. New York: Van Nostrand Reinhold.

*Wurster-Hill, D. H., K. Benirschke, and D. I. Chapman (1983) "Abnormalities of the X Chromosome in Mammals." In A. A. Sandberg, ed., *Cytogenetics of the Mammalian X Chromosome*, Part B, pp. 283–300. New York: Alan R. Liss.

*Zimen, E. (1981) *The Wolf: His Place in the Natural World*. London: Souvenir Press.

*———— (1976) "On the Regulation of Pack Size in Wolves." *Zeitschrift für Tierpsychologie* 40:300–341.

BEARS

Carnivore

GRIZZLY or BROWN BEAR
Ursus arctos

(AMERICAN) BLACK BEAR
Ursus americanus

HOMOSEXUALITY	TRANSGENDER	BEHAVIORS		RANKING	OBSERVED IN
● Female	● Intersexuality	○ Courtship	● Pair-bonding	○ Primary	● Wild
● Male	○ Transvestism	○ Affectionate	● Parenting	● Moderate	○ Semiwild
		● Sexual		○ Incidental	○ Captivity

GRIZZLY BEAR
IDENTIFICATION: A huge bear (7–10 feet tall) with dark brown, golden, cream, or black fur. DISTRIBUTION: Northern North America, Europe, central Asia, Middle East, north Africa. HABITAT: Tundra, forests. STUDY AREA: Yellowstone National Park, Wyoming; subspecies *U.a. horribilis*.

BLACK BEAR
IDENTIFICATION: A smaller bear (4–6 feet) with coat color ranging from black to gray, brown, and even white. DISTRIBUTION: Canada and northern, eastern, and southwestern United States. HABITAT: Forest. STUDY AREA: Jasper National Park, Alberta; Prince Albert National Park, Saskatchewan, Canada; subspecies *U.a. altifrontalis*.

Social Organization

Grizzly Bears and Black Bears are largely solitary animals. Some Grizzly populations, however, tend to aggregate around abundant food sources such as salmon, marine mammals (stranded onshore), garbage dumps, and even insect swarms; fairly complex social interactions may develop in these contexts. The heterosexual mating system is polygamous, as both males and females generally mate with multiple partners; males do not contribute to parenting.

Description

Behavioral Expression: Female Grizzly Bears sometimes bond with each other and raise their young together as a single family unit. The two mothers become inseparable companions, traveling and feeding together throughout the summer and fall seasons as they share in the parenting of their cubs. Female companions have not been observed interacting sexually with one another, however. A bonded pair jointly defends their food (such as Elk or Bison carcasses), and the two females also protect one another and their offspring (including defending them against attacks by Grizzly males). The cubs regard both females as their parents, following and responding to either mother equally; bonded females occasionally also nurse each other's cubs. If one female dies, her companion usually adopts her cubs and looks after them along with her own.

As winter approaches and Grizzlies prepare for hibernation, female coparents often continue to associate with one another. Foraging together late into the fall, they are apparently reluctant to end their relationship and may even delay the onset of their own sleep. Although paired females do not hibernate together, they frequently visit each other (with their cubs) prior to hibernation, staying nearby while their partner prepares her den. They also sleep together outside their denning sites during this final preparatory period and only retreat to their separate dens once the snow gets too deep. Most Grizzlies seek solitude prior to hibernating and locate their winter dens miles away from each other (and with rugged terrain separating them), but bonded females often hibernate relatively close to one another. Such females have even been known to abandon their traditional denning locations to be nearer to their coparent. One female, for example, moved her usual den site more than 14 miles to be closer to her companion. Pair-bonds are not usually resumed after hibernation, although one female may adopt her companion's yearling offspring the following season. The average age of a bonded female is about 11 years, although Grizzlies as young as 5 and as old as 19 have formed bonds with other females. Companions may be of the same age, or one female might be several years older than the other. Sometimes more than two females are involved: three Grizzlies may form a strongly bonded "triumvirate," and groups of four or five females have even been known to associate (sometimes also forming pair- or trio-bonds within such a group).

Younger male Black Bears (adolescents and cubs) sometimes mount their siblings, both male and female. One male approaches another with his ears in a

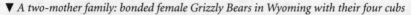

▼ *A two-mother family: bonded female Grizzly Bears in Wyoming with their four cubs*

CRESCENT configuration (facing forward and perpendicular from the head), then rears up on his hind legs in a STANDING OVER position, in which he places his front paws on the other male's back. This develops into sexual mounting as he clasps his partner and gently bites the loose skin on his shoulder, sometimes making pelvic thrusts. The other male often rolls over and begins play-fighting with the mounting male, pawing and biting at him.

Intersexual or hermaphrodite Black and Grizzly Bears occur in some populations. These individuals are genetically female and have female internal reproductive organs, combined with various degrees of external male genitalia. In some cases, they have a penislike organ (complete with a penis bone or BACULUM) that is not connected to the internal reproductive organs, while in others the penis is more fully developed, serving as both a genital orifice connected to the womb as well as a urinary organ. Most transgendered Bears are mothers, mating with males and bearing offspring. Some individuals actually copulate and give birth through their "penis": their male partner inserts his organ into the tip of the intersexual Bear's phallus, and the resulting offspring emerge through the penis as well.

Frequency: Female bonding and coparenting among Grizzly Bears occur sporadically. In a 12-year study of one population, for example, bonds between females were observed during 4 of those years (a third of the study period), with roughly 20 percent of all females participating in same-sex bonding and coparenting at some point in their life (usually 1–2 years out of an adult life span of 7–12 years). About 9 percent of all Grizzly cubs are raised in families headed by two (or more) pair-bonded mothers (constituting about 4 percent of all families). Sexual activity between younger male Black Bears occurs only occasionally, comprising perhaps less than 2 percent of their play. The incidence of intersexual Bears is probably sporadic as well, although some populations appear to have fairly high proportions: in Alberta, for example, researchers found that 4 out of 38 Black Bears (11 percent) and 1 of 4 Grizzlies were transgendered.

Orientation: Extended heterosexual pair-bonding and parenting by male-female couples do not occur among Grizzly Bears; however, only a subset of females bond with each other and coparent their young. Thus, some individuals are probably more inclined to form same-sex attachments than others, and these females may even develop same-sex bonds on more than one occasion. Although such females mate with males (and may not bond with females in other years), their primary social relationship during the time they are bonded is with their female companion (and their young). Male Black Bears participate in homosexual mounting only as youngsters and adolescents; most probably go on to mate heterosexually as adults.

Nonreproductive and Alternative Heterosexualities

Some Grizzly and Black Bear populations have significant numbers of non-breeding animals. Each year, perhaps as many as one-third to one-half of all female Grizzlies do not mate or are otherwise nonreproductive (including copulating with males without becoming pregnant), and some individuals do not breed during their whole lives. In some Black Bear populations, only 16–50 percent of the adult females reproduce each year, and many skip breeding for several years. Female Bears who do become pregnant exhibit DELAYED IMPLANTATION—the fertilized embryo ceases development for about five months before implanting in the uterus. In some cases embryos are reabsorbed, aborted, or prevented from implanting rather than carried to term (e.g., when food supplies are inadequate). In addition, many female Grizzlies and Black Bears delay reproducing anywhere from one to four years after they become sexually mature. Juvenile (sexually immature) Black and Grizzly Bears also engage in sexual activity with each other, including mounting and licking of the vulva. Among adult male Grizzly Bears, higher-ranking individuals often have lower copulation rates because of their preoccupation with aggressive interactions, and in some populations top-ranked males may actually go entire breeding seasons without mating at all. When mating does take place, one partner may display indifference or refusal, and as many as 47 percent of all copulations are incomplete in that they do not involve full penetration or ejaculation. Occasionally, a particularly aggressive male will force a female to mate with him, although females usually have control of the interaction. In fact, females often mate with multiple partners—as many as eight males in a single breeding season for Grizzlies, four to six for Black Bears—and cubs belonging to the same litter may be fathered by different males. Nevertheless, male Black and Grizzly Bears can become violent toward females and cubs, occasionally even killing and cannibalizing adults and/or young. Female Black Bears also sometimes kill cubs that are not their own (especially during the nursing period), although it is not uncommon for mothers of both species to adopt cubs that have been orphaned or abandoned.

Other Species

Intersexual or transgendered individuals also occur among Polar Bears (*Ursus maritimus*), comprising about 2 percent of some populations.

Sources   *asterisked references discuss homosexuality/transgender*

Alt, G. L. (1984) "Cub Adoption in the Black Bear." *Journal of Mammalogy* 65:511–12.

Brown, G. (1993) *The Great Bear Almanac.* New York: Lyons and Beuford.

*Cattet, M. (1988) "Abnormal Sexual Differentiation in Black Bears (*Ursus americanus*) and Brown Bears (*Ursus arctos*)." *Journal of Mammalogy* 69:849–52.

*Craighead, F. C., Jr. (1979) *Track of the Grizzly.* San Francisco: Sierra Club Books.

*Craighead, F. C., Jr., and J. J. Craighead (1972) "Grizzly Bear Prehibernation and Denning Activities as Determined by Radiotracking." *Wildlife Monographs* 32:1–35

*Craighead, J. J., J. S. Sumner, and J. A. Mitchell (1995) *The Grizzly Bears of Yellowstone: Their Ecology in the Yellowstone Ecosystem, 1959–1992.* Washington, D.C. and Covelo, Calif.: Island Press.

*Craighead, J. J., M. G. Hornocker, and F. C. Craighead Jr. (1969) "Reproductive Biology of Young Female Grizzly Bears." *Journal of Reproduction and Fertility,* suppl. 6:447–75.

Egbert, A. L. (1978) "The Social Behavior of Brown Bears at McNeil River, Alaska." Ph.D. thesis, Utah State University.

Egbert, A. L., and A. W. Stokes (1976) "The Social Behavior of Brown Bears on an Alaskan Salmon Stream." In M. R. Pelton, J. W. Lentfer, and G. E. Folk, eds., *Bears—Their Biology and Management: Papers from the Third International Conference on Bear Research and Management,* pp. 41–56. Morges, Switzerland: International Union for Conservation of Nature and Natural Resources.

Erickson, A. W., and L. H. Miller (1963) "Cub Adoption in the Brown Bear." *Journal of Mammalogy* 44:584–85.

Goodrich, J. M., and S. J. Stiver (1989) "Co-occupancy of a Den by a Pair of Great Basin Black Bears." *Great Basin Naturalist* 4:390–91.

*Henry, J. D., and S. M. Herrero (1974) "Social Play in the American Black Bear: Its Similarity to Canid Social Play and an Examination of Its Identifying Characteristics." *American Zoologist* 14:371–89.

Jonkel, C. J., and I. McT. Cowan (1971) "The Black Bear in the Spruce-Fir Forest." *Wildlife Monographs* 27:1–57.

Rogers, L. (1976) "Effects of Mast and Berry Crop Failures on Survival, Growth, and Reproductive Success of Black Bears." *Transactions of the North American Wildlife and Natural Resources Conference* 41:431–38.

Schenk, A., and K. M. Kovacs (1995) "Multiple Mating Between Black Bears Revealed by DNA Fingerprinting." *Animal Behavior* 50:1483–90.

Stonorov, D., and A. W. Stokes (1972) "Social Behavior of the Alaska Brown Bear." In S. Herrero, ed., *Bears—Their Biology and Management: Papers from the Second International Conference on Bear Research and Management*, pp. 232–42. Morges, Switzerland: International Union for Conservation of Nature and Natural Resources.

Tait, D. E. N. (1980) "Abandonment as a Reproductive Tactic—the Example of Grizzly Bears." *American Naturalist* 115:800–808.

Wimsatt, W. A. (1969) "Delayed Implantation in the Ursidae, with Particular Reference to the Black Bear (*Ursus americanus* Pallus)." In A. C. Enders, ed., *Delayed Implantation*, pp. 49–76. Chicago: University of Chicago Press.

HYENAS

Carnivore

SPOTTED HYENA
Crocuta crocuta

HOMOSEXUALITY	TRANSGENDER	BEHAVIORS		RANKING	OBSERVED IN
● Female	○ Intersexuality	○ Courtship	○ Pair-bonding	○ Primary	● Wild
○ Male	● Transvestism	○ Affectionate	○ Parenting	● Moderate	○ Semiwild
		● Sexual		○ Incidental	● Captivity

IDENTIFICATION: A yellowish brown hyena with spotted flanks and back, a strongly sloping body profile, and rounded ears; females typically heavier than males. DISTRIBUTION: Sub-Saharan Africa. HABITAT: Open country, including plains, semidesert, savanna. STUDY AREAS: Kalahari and Gemsbok National Parks, South Africa and Botswana; University of California—Berkeley.

Social Organization

Spotted Hyenas live in matrilineal clans of 30–80 individuals. Females are dominant to males and remain in their home group for life, while males emigrate to single-sex groups during adolescence and then join other clans (usually for only a few years at a time) on reaching adulthood. This species has a highly organized social system, engaging in cooperative hunting and communal denning. The breeding system is polygynous: generally only one male in each clan mates with several females. Spotted Hyenas are largely nocturnal.

Description

Behavioral Expression: Female Spotted Hyenas have an extraordinary genital configuration that makes them superficially resemble males: the clitoris is 90 percent of the length of the male's penis (nearly seven inches long) and equal to it in diameter; it can be fully erected. In addition, the labia are fused to resemble a "scrotum" containing fat and connective tissue that give the appearance of testes. There is no vaginal opening—instead, the female mates and gives birth (as well as urinates) through the tip of her clitoris. Heterosexual mating is accomplished by retracting the clitoris inside the abdomen, essentially turning it inside out to form a passageway within which the male can insert his penis. Homosexual mounting between females also occurs in this species; in some cases, an adolescent or younger adult mounts an older one. During a homosexual encounter, one female approaches another with her clitoris erect, often "flipping" it up against her abdomen as a sign of sexual arousal (also seen in males preparing to mate). She may lick her partner on the back, then mount by rising up and clasping her front paws

around the other female, resting her head on the other's neck, and thrusting against her. Clitoral penetration may also occur, though it is not common. Sometimes the female being mounted appears to be disinterested or nonchalant and may even wander off during mounting. These are also characteristics of heterosexual courtship and copulation—females often walk away or do not permit males to achieve penetration and, indeed, may be overtly aggressive toward them. Spotted Hyenas also perform a MEETING CEREMONY involving clitoral

◄ (Above) *A younger female Spotted Hyena (with erect clitoris) mounting an older female in southern Africa.* (Below)*Two female Spotted Hyenas nuzzling and sniffing each other's genitals during the "meeting ceremony"; note the erect clitoris and "scrotum" of the female raising her leg.*

erection and genital licking: two females stand parallel to each other but with their heads in opposite directions so that they can access each other's genitals. One or both of them lifts up her hind leg and allows the other to sniff, nuzzle, and lick her erect clitoris and "scrotum"—sometimes for as long as half a minute at a time— occasionally accompanied by a soft groaning or whining sound. Although meeting ceremonies are sometimes performed between males and females (or between two males), they are most common between females.

Frequency: All adult female Spotted Hyenas have enlarged clitorides and labial "scrota." Homosexual mounting probably occurs only occasionally in this species, although the majority of meeting ceremonies—anywhere from 55–95 percent— occur between adult females.

Orientation: Most females who participate in same-sex mounting and meeting ceremonies are probably functionally bisexual (if not predominantly heterosexual), also mating with males.

Nonreproductive and Alternative Heterosexualities

The female Spotted Hyena's reproductive anatomy and genitals are not optimal for breeding. Heterosexual mating is often difficult, as males have trouble locating and penetrating the clitoral opening. In addition, many females (and infants) suffer severe trauma and even death during the birth process. Because there is no vaginal canal, Hyenas must give birth through the clitoris itself—an extraordinarily painful process, considering that the newborn's head is significantly larger than the diameter of the clitoris. This causes the clitoral head to rupture in all females during their first birth, and it is estimated that about 9 percent of females in the wild die during their first labor. In addition, the newborn must travel an extraordinarily long way through the female's birth canal, which makes a 180-degree turn to exit through the clitoris. Because the umbilical cord is less than a third of the length of the birth canal, many babies suffocate during birth—perhaps as many as 60 percent of infants are stillborn to first-time mothers, and a female's lifetime production of offspring may be reduced by as much as 25 percent because of such complications. Once born, up to a quarter of Spotted Hyena youngsters may be killed by their siblings, who are fiercely competitive and aggressive toward one another. Infanticide (usually by females) also occasionally occurs among Spotted Hyenas, and cannibalism has been reported as well. Most males do not breed at all: the social system of Spotted Hyenas is such that only one male in a clan gets to mate with the females (although other males may participate in heterosexual courtship). Some males, unable to copulate, engage in a form of "masturbation" in which they thrust their penis in the air and spontaneously ejaculate; others have been seen mounting cubs.

Other Species

Homosexual mounting occurs in Dwarf Mongooses (*Helogale undulata*), a weasel-like carnivore of Africa. In one group studied in captivity, 16 percent of mounting took place between animals of the same sex (mostly males, including brothers), and some individuals had preferred partners with whom they interacted most frequently. Homosexual behavior has also been observed in two other species of small carnivores: male Common Raccoons (*Procyon lotor*) and female Martens (*Martes* sp.).

Sources   *asterisked references discuss homosexuality/transgender

*Burr, C. (1996) *A Separate Creation: The Search for the Biological Origins of Sexual Orientation*. New York: Hyperion.

*East, M. L., H. Hofer, and W. Wickler (1993) "The Erect 'Penis' Is a Flag of Submission in a Female-Dominated Society: Greetings in Serengeti Spotted Hyenas." *Behavioral Ecology and Sociobiology* 33:355–70.

*Frank, L. G. (1996) "Female Masculinization in the Spotted Hyena: Endocrinology, Behavioral Ecology, and Evolution." In J. L. Gittleman, ed., *Carnivore Behavior, Ecology, and Evolution*, vol. 2, pp. 78–131. Ithaca: Cornell University Press.

———— (1986) "Social Organization of the Spotted Hyena (*Crocuta crocuta*). I. Demography. II. Dominance and Reproduction." *Animal Behavior* 34:1500–1527.

Frank, L. G., J. M. Davidson, and E. R. Smith (1985) "Androgen Levels in the Spotted Hyena *Crocuta crocuta*: The Influence of Social Factors." *Journal of Zoology, London* 206:525–31.

Frank, L. G., and S. E. Glickman (1994) "Giving Birth Through a Penile Clitoris: Parturition and Dystocia in the Spotted Hyena (*Crocuta crocuta*)." *Journal of Zoology, London* 234:659–90.

Frank, L. G., S. E. Glickman, and P. Licht (1991) "Fatal Sibling Aggression, Precocial Development, and Androgens in Neonatal Spotted Hyenas." *Science* 252:702–04.

*Frank, L. G., S. E. Glickman, and I. Powch (1990) "Sexual Dimorphism in the Spotted Hyena (*Crocuta crocuta*)." *Journal of Zoology, London* 221:308–13.

*Frank, L. G., M. L. Weldele, and S. E. Glickman (1995) "Masculinization Costs in Hyenas." *Nature* 377:584–85.

*Glickman, S. E., C. M. Drea, M. Weldele, L. G. Frank, G. Cunha, and P. Licht (1995) "Sexual Differentiation of the Female Spotted Hyena (*Crocuta crocuta*)." Paper presented at the 24th International Ethological Conference, Honolulu, Hawaii.

*Glickman, S. E., L. G. Frank, K. E. Holekamp, L. Smale, and P. Licht (1993) "Costs and Benefits of 'Androgenization' in the Female Spotted Hyena: The Natural Selection of Physiological Mechanisms." In P. P. G. Bateson, N. Thompson, and P. Klopfer, eds., *Perspectives in Ethology, vol. 10: Behavior and Evolution*, pp. 87–117. New York: Plenum Press.

*Hamilton, W. H., III, R. L. Tilson, and L.G. Frank (1986) "Sexual Monomorphism in Spotted Hyenas (*Crocuta crocuta*)." *Ethology* 71:63–73.

*Harrison Mathews, L. (1939) "Reproduction in the Spotted Hyena, *Crocuta crocuta* (Erxleben)." *Philosophical Transactions of the Royal Society of London*, Series B 230:1–78.

*Hofer, H., and M. L. East (1995) "Virilized Sexual Genitalia as Adaptations of Female Spotted Hyenas." *Revue Suisse de Zoologie* 102:895–906.

*Kinsey, A. C., W. B. Pomeroy, C. E. Martin, and P. H. Gebhard (1953) *Sexual Behavior in the Human Female*. Philadelphia: W. B. Saunders.

Kruuk, H. (1975) *Hyena*. Oxford: Oxford University Press.

———— (1972) *The Spotted Hyena, a Study of Predation and Social Behavior*. Chicago: University of Chicago Press.

*Mills, M. G. L. (1990) *Kalahari Hyenas: Comparative Behavioral Ecology of Two Species*. London: Unwin Hyman.

*Neaves, W. B., J. E. Griffin, and J. D. Wilson (1980) "Sexual Dimorphism of the Phallus in Spotted Hyena (*Crocuta crocuta*)." *Journal of Reproduction and Fertility* 59:509–13.

*Rasa, O. A. E. (1979a) "The Ethology and Sociology of the Dwarf Mongoose (*Helogale undulata rufula*)." *Zeitschrift für Tierpsychologie* 43:337–406.

*———— (1979b) "The Effects of Crowding on the Social Relationships and Behavior of the Dwarf Mongoose (*Helogale undulata rufula*)." *Zeitschrift für Tierpsychologie* 49:317–29.

KANGAROOS AND WALLABIES   *Marsupial*

EASTERN GRAY KANGAROO
Macropus giganteus

RED-NECKED WALLABY
Macropus rufogriseus

WHIPTAIL or PRETTY-FACED WALLABY
Macropus parryi

HOMOSEXUALITY	TRANSGENDER	BEHAVIORS		RANKING	OBSERVED IN
● Female	● Intersexuality	● Courtship	● Pair-bonding	○ Primary	● Wild
● Male	○ Transvestism	● Affectionate	○ Parenting	● Moderate	○ Semiwild
		● Sexual		○ Incidental	● Captivity

EASTERN GRAY KANGAROO
IDENTIFICATION: A large (over 3 foot tall) kangaroo with a gray coat and a hair-covered muzzle. DISTRIBUTION: Eastern Australia. HABITAT: Open grasslands, forest, woodland. STUDY AREAS: Nadgee Nature Reserve, New South Wales, Australia; Cowan Field Station (Muogammarra Nature Reserve) of the University of New South Wales.

RED-NECKED WALLABY
IDENTIFICATION: A smaller kangaroo (2½ feet tall) with a reddish brown wash on its neck. DISTRIBUTION: Coastal southeastern Australia. HABITAT: Forest, brush areas. STUDY AREAS: Michigan State University; Cowan Field Station (Muogammarra Nature Reserve) of the University of New South Wales; subspecies *M.r. rufogriseus*, Bennett's Wallaby, and *M.r. banksianus*.

WHIPTAIL WALLABY
IDENTIFICATION: A light gray kangaroo standing up to 3 feet tall, with a white facial stripe and a long, slender tail. DISTRIBUTION: Northeastern Australia. HABITAT: Open forest, savanna. STUDY AREA: Near Bonalbo, New South Wales, Australia.

Social Organization

Eastern Gray Kangaroos often associate in large groups of 40–50 animals—sometimes known as MOBS. These comprise smaller cosexual groups of up to 15 individuals, largely females and their young along with a few males. Some individuals are solitary. No pair-bonding occurs between males and females, and the mating system is polygamous or promiscuous. Whiptail Wallabies have a similar social organization, while Red-necked Wallabies are largely solitary (although groups of 8–30 animals may form at times).

Description

Behavioral Expression: Pair-bonds occasionally develop between female Eastern Gray Kangaroos, involving frequent mutual grooming in which the partners affectionately lick, nibble, and rake the fur on each other's head and neck with their paws. Females in such associations also sometimes court and mount each other, and sexual activity may occur as well between females who are not necessarily bonded to one another. Significantly, heterosexual pair-bonds are not found in this species. In Red-necked Wallabies, females frequently mount each other: one female grabs the other from behind, wrapping her forearms around her partner's abdomen and tucking her forepaws inside her partner's thighs. This position resembles heterosexual copulation except that the mounting female is higher up on her partner's body. Sexual activity is often accompanied by grooming, fur-nibbling and licking, pawing, and nosing of the partner. Males also sometimes mount one another, usually during play-fights in which the partners gently push, wrestle, or "box" one another with their forearms. Occasional affectionate activities such as grooming, licking, embracing, and touching also take place during these sessions, and sometimes one male will sniff or nuzzle the other's scrotum. Courtship and sexual interactions between male Whiptail Wallabies involve TAIL-LASHING, a sinuous, sideways movement of the tail indicating sexual arousal, often accompanied by an erection. Mounting also occurs; one male sometimes presents his hindquarters to the other by crouching with his chest on the ground and raising his rump. Prior to mounting, a male frequently sniffs the other male's scrotum (as in Red-necked Wallabies). Swept up in a courtship frenzy, males also sometimes embark on homosexual chases as a part of heterosexual interactions. A group of males will be pursuing a female in heat, furiously circling and dashing after her at breakneck speed; occasionally, other males on the sidelines are then drawn into the excitement of these wild chases—and they are as likely to pursue other males as they are the female.

In Eastern Gray Kangaroos, intersexual or hermaphrodite individuals also occur: some animals, for example, have both a penis and a pouch (the latter usually found only in females), mammary glands, and testes, all combined with body proportions that are generally intermediate between male and female. Chromosomally, these individuals have a mosaic of the female type (XX), the male type (XY), and a combined XXY pattern.

Frequency: Female homosexual mounting in Eastern Gray Kangaroos occurs sporadically: in one study of a wild population, for example, homosexual behavior was recorded eight times during four months of observation. It should be noted, however, that heterosexual mating is rarely observed as well: only one male-female mating was seen during that same time, while only three heterosexual copulations were recorded during a three-year study of Red-necked Wallabies. In captivity, mounting between female Red-necked Wallabies is quite common, but male homosexual activity is much less frequent. During play-fights between males, courtship and sexual behaviors occur roughly once every five hours of activity. Homosexual mounts between males account for about 1 percent of all mounting activity in Whiptail Wallabies.

Orientation: In captive groups of Eastern Gray Kangaroos (which included both sexes), four of six females formed same-sex pairs. Male Red-necked Wallabies may be sequentially bisexual during their lifetimes: sexual interactions during play-fighting are most common among adolescent males, while adults are probably more heterosexually oriented. Male Whiptail Wallabies that court or mount other males also interact sexually with females, i.e., they are simultaneously bisexual.

Nonreproductive and Alternative Heterosexualities

Female Red-necked Wallabies are often harassed by males trying to mate with them (similar behavior also occurs in Whiptail Wallabies). As many as seven males at a time may pursue a single female, and they may injure her while mating: females have been seen limping or with cuts on their backs after heterosexual copulations. Mating attempts may also be interrupted by other males charging the pair and trying to dislodge the mounting male. In this species, only about 18 percent of the males ever mate with females. Nonbreeding females also occur in Eastern Gray Kangaroos. In addition, females who are pregnant (including late-term), not in heat, or sexually immature also occasionally participate in heterosexual activity, and males also regularly masturbate by thrusting the erect penis into the paws. DELAYED IMPLANTATION is another notable feature of the reproductive cycle in these species.

Other Species

Homosexual mounting occurs in a number of other Kangaroo and Wallaby species: between females in Yellow-footed Rock Wallabies (*Petrogale xanthopus*), and between males in Western Gray Kangaroos (*Macropus fuliginosus*), Red Kangaroos (*Macropus rufus*), Agile Wallabies (*Macropus agilis*), Black-footed Rock Wallabies (*Petrogale lateralis*), and Swamp Wallabies (*Wallabia bicolor*). Transgendered or intersexual individuals of various types are also found in several species, including Red Kangaroos, Euros (*Macropus robustus*), Tammar Wallabies (*Macropus eugenii*), and Quokkas (*Setonix brachyurus*). Some of these individuals have female body proportions and external genitalia, female or combined male-female internal reproductive organs, a scrotum, and absence of a pouch and mammary glands. Others have male reproductive organs, intermediate or female body proportions, and a pouch and mammary glands.

Sources   *asterisked references discuss homosexuality/transgender*

Coulson, G. (1997) "Repertoires of Social Behavior in Captive and Free-Ranging Gray Kangaroos, *Macropus giganteus* and *Macropus fuliginosus* (Marsupialia: Macropodidae)." *Journal of Zoology, London* 242:119–30.

*—— (1989) "Repertoires of Social Behavior in the Macropodoidea." In G. C. Grigg, P. J. Jarman, and I. D. Hume, eds., *Kangaroos, Wallabies, and Rat-Kangaroos*, pp. 457–73. Chipping Norton, NSW: Surrey Beatty and Sons.

*Grant, T. R., (1974) "Observations of Enclosed and Free-Ranging Gray Kangaroos *Macropus giganteus*." *Zeitschrift für Säugetierkunde* 39:65–78.

*—— (1973) "Dominance and Association Among Members of a Captive and a Free-Ranging Group of Gray Kangaroos (*Macropus giganteus*)." *Animal Behavior* 21:449–56.

Jarman, P. J., and C. J. Southwell (1986) "Grouping, Association, and Reproductive Strategies in Eastern Gray Kangaroos." In D. I. Rubenstein and R. W. Wrangham, eds., *Ecological Aspects of Social Evolution*, pp. 399–428. Princeton: Princeton University Press.

Johnson, C. N. (1989) "Social Interactions and Reproductive Tactics in Red-necked Wallabies (*Macropus rufogriseus banksianus*)." *Journal of Zoology, London* 217:267–80.

Kaufmann, J. H. (1975) "Field Observations of the Social Behavior of the Eastern Gray Kangaroo, *Macropus giganteus*." *Animal Behavior* 23:214–21.

*—— (1974) "Social Ethology of the Whiptail Wallaby, *Macropus parryi*, in Northeastern New South Wales." *Animal Behavior* 22:281–369.

*LaFollette, R. M. (1971) "Agonistic Behavior and Dominance in Confined Wallabies, *Wallabia rufogrisea frutica*." *Animal Behavior* 19:93–101.

Poole, W. E. (1982) "*Macropus giganteus* Shaw 1790, Eastern Gray Kangaroo." *Mammalian Species* 187:1–8.

—— (1973) "A Study of Breeding in Gray Kangaroos, *Macropus giganteus* Shaw and *M. fuliginosus* (Desmarest), in Central New South Wales." *Australian Journal of Zoology* 21:183–212.

Poole, W. E., and P. C. Catling (1974) "Reproduction in the Two Species of Gray Kangaroos, *Macropus giganteus* Shaw and *M. fuliginosus* (Desmarest). I. Sexual Maturity and Oestrus." *Australian Journal of Zoology* 22:277–302.

*Sharman, G. B., R. L. Hughes, and D. W. Cooper (1990) "The Chromosomal Basis of Sex Differentiation in Marsupials." *Australian Journal of Zoology* 37:451–66.

*Stirrat, S. C., and M. Fuller (1997) "The Repertoire of Social Behaviors of Agile Wallabies, *Macropus agilis*." *Australian Mammalogy* 20:71–78.

*Watson, D. M., and D. B. Croft (1993) "Playfighting in Captive Red-Necked Wallabies, *Macropus rufogriseus banksianus*." *Behavior* 126:219–245.

TREE AND RAT KANGAROOS   *Marsupial*

RUFOUS BETTONG or RAT KANGAROO
Aepyprymnus rufescens

DORIA'S TREE KANGAROO
Dendrolagus dorianus

MATSCHIE'S TREE KANGAROO
Dendrolagus matschiei

HOMOSEXUALITY	TRANSGENDER	BEHAVIORS		RANKING	OBSERVED IN
● Female	○ Intersexuality	● Courtship	○ Pair-bonding	○ Primary	○ Wild
○ Male	○ Transvestism	○ Affectionate	○ Parenting	○ Moderate	○ Semiwild
		● Sexual		● Incidental	● Captivity

RUFOUS BETTONG
IDENTIFICATION: A small (6–7 pound), rodentlike kangaroo with reddish brown fur. DISTRIBUTION: Eastern and southern Australia. HABITAT: Grassy woodlands. STUDY AREAS: National Parks and Wildlife Service Center, Townsville, Australia; Zoological Garden of West Berlin, Germany.

DORIA'S, MATSCHIE'S TREE KANGAROOS
IDENTIFICATION: Stocky, tree-dwelling kangaroos; chestnut or chocolate brown fur with lighter patches. DISTRIBUTION: Interior New Guinea; Doria's is vulnerable, Matschie's is endangered. HABITAT: Mountainous rain forests. STUDY AREAS: Karlsruhe Zoo, Germany; Woodland Park Zoological Gardens, Seattle, Washington.

Social Organization

Tree Kangaroos and Rufous Bettongs are largely solitary, although they sometimes associate in pairs, trios, or small groups of adults and young. About 15 percent of Bettong groups are same-sex. The mating systems of these species, though poorly understood, may involve polygamy or promiscuity, perhaps combined with monogamous pair-bonding in some populations of Rufous Bettongs. Males do not generally participate in raising their own offspring.

Description

Behavioral Expression: Female Rufous Bettongs sometimes court and mount other females, using behavior patterns also found in opposite-sex contexts. A homosexual interaction begins with one female approaching another and then sniffing and nuzzling her genital and pouch openings as well as her anus. The courting female becomes sexually aroused and exhibits a sinuous TAIL-LASHING, in which she moves her tail rapidly from side to side. The other female may initially react with hostility (as do females in heterosexual courtships), lying on her side and kicking with her hind feet while softly growling. As a result, the courting female might perform FOOT-DRUMMING, in which she stands upright on her hind legs near the other female and stamps one foot on the ground in front of her. If the other female calms down, the courting female may mount her by clasping her waist from behind and thrusting against her. Some homosexual interactions in Rufous Bettongs involve adult females courting and mounting juvenile females and vice versa. Sometimes males in the vicinity try to intervene or disrupt females engaging in homosexual behavior, but in other cases they simply ignore the activity. Mountings between females, including pelvic thrusting, also occur in Doria's and Matschie's Tree Kangaroos.

Frequency: Homosexual interactions occur fairly frequently in captive Rufous Bettongs: in one study, 3 out of 8 mounts were between females, and homosexual activity was observed a total of 19 times over one month. In Tree Kangaroos samesex mounting occurs only occasionally.

Orientation: Female Rufous Bettongs that participate in homosexual behavior are probably bisexual, since most mate with males and become successful mothers. Most female Tree Kangaroos that mount other females probably also participate in heterosexual activity, although at least one Matschie's Tree Kangaroo that participated in same-sex mounts was a nonbreeder.

Nonreproductive and Alternative Heterosexualities

Heterosexual interactions in Rufous Bettongs have a number of nonprocreative aspects. For example, adult males generally make more sexual approaches to juvenile females than to sexually mature adult females. Nonreproductive REVERSE mounts—in which females mount males—occasionally occur in this species, as do attempted mounts by males on females who are not in heat. In the latter case, the female typically responds aggressively, vigorously kicking and growling at him. Although most heterosexual interactions occur between pairs of animals, sometimes two female Rufous Bettongs consort simultaneously with the same male as part of a trio. Among Matschie's Tree Kangaroos, a form of infanticide known as POUCH-ROBBING has occasionally been observed in captivity, in which females are severely aggressive toward other infants and may actually pull joeys from their mother's pouch and kill them.

Other Species

Female Tasmanian Rat Kangaroos (*Bettongia gaimardi*) also engage in homosexual mounting.

Sources   *asterisked references discuss homosexuality/transgender*

*Coulson, G. (1989) "Repertoires of Social Behavior in the Macropodoidea." In G. C. Grigg, P. J. Jarman, and I. D. Hume, eds., *Kangaroos, Wallabies, and Rat-Kangaroos*, pp. 457–73. Chipping Norton, NSW: Surrey Beatty and Sons.

Dabek, L. (1994) "Reproductive Biology and Behavior of Captive Female Matschie's Tree Kangaroos, *Dendrolagus matschiei*." Ph.D. thesis, University of Washington.

Frederick, H., and C. N. Johnson (1996) "Social Organization in the Rufous Bettong, *Aepyprymnus rufescens*." *Australian Journal of Zoology* 44:9–17.

Ganslosser, U. (1993) "Stages in Formation of Social Relationships—an Experimental Investigation in Kangaroos (Macropodoidea: Mammalia)." *Ethology* 94:221–47.

*———— (1979) "Soziale Kommunikation, Gruppenleben, Spiel- und Jugendverhalten des Doria-Baumkänguruhs (*Dendrolagus dorianus* Ramsay, 1833) [Social Communication, Group Life, and Play Behavior of Doria's Tree Kangaroo]." *Zeitschrift für Säugetierkunde* 44:137–53.

*Ganslosser, U., and C. Fuchs (1988) "Some Quantitative Data on Social Behavior of Rufous Rat-Kangaroos (*Aepyprymnus rufescens* Gray, 1837 (Mammalia: Potoroidae)) in Captivity." *Zoologischer Anzeiger* 220:300–312.

George, G. G. (1977) "Up a Tree with Kangaroos." *Animal Kingdom* 80(2):20–24.

*Hutchins, M., G. M. Smith, D. C. Mead, S. Elbin, and J. Steenberg (1991) "Social Behavior of Matschie's Tree Kangaroo (*Dendrolagus matschiei*) and Its Implications for Captive Management." *Zoo Biology* 10:147–64.

Jarman, P. J. (1991) "Social Behavior and Organization in the Macropodoidea." *Advances in the Study of Behavior* 20:1–50.

*Johnson, P. M. (1980) "Observations of the Behavior of the Rufous Rat-Kangaroo, *Aepyprymnus rufescens* (Gray), in Captivity." *Australian Wildlife Research* 7:347–57.

OTHER MARSUPIALS   *Marsupial*

KOALA
Phascolarctos cinereus

HOMOSEXUALITY	TRANSGENDER	BEHAVIORS		RANKING	OBSERVED IN
● Female	○ Intersexuality	○ Courtship	○ Pair-bonding	○ Primary	○ Wild
● Male	○ Transvestism	○ Affectionate	○ Parenting	● Moderate	○ Semiwild
		● Sexual		○ Incidental	● Captivity

IDENTIFICATION: A bearlike marsupial with woolly brown or gray fur, large black nose, white chest, and long claws. DISTRIBUTION: Eastern and southeastern Australia. HABITAT: Eucalyptus forests. STUDY AREAS: Lone Pine Sanctuary, Brisbane, Australia; San Diego Zoo; subspecies *P.c. adustus*.

Social Organization

Koalas are largely nocturnal and solitary, although in some populations they tend to live in scattered clusters of two to six females with several males. The mating system is probably promiscuous or polygamous (animals mate with multiple partners), and males take no part in raising their young.

Description

Behavioral Expression: Female Koalas in heat sometimes mount each other in the trees: while one female clings vertically to the trunk, another climbs behind her and reaches around to simultaneously hold on to the tree. She begins to make

▲ *A female Koala mounting another female*

pelvic thrusts against the other female, while also typically gripping the other female's neck in her teeth (as does the male during heterosexual mounting). Occasionally one female mounts another from the side (a position sometimes also used by younger males). Usually the mounted female does not display the receptive posture (which involves arching her back while throwing her head back), and homosexual mounts are generally briefer than heterosexual ones. Like male-female copulations, homosexual interactions sometimes involve aggression between the participants: one female may attack the other or pin her to the ground following a mounting. Sometimes two females take turns mounting each other, and homosexual mounting is often interspersed with other signs of intense sexual arousal, including chasing, bellowing, and jerking. BELLOWING is an extraordinary call (also made by males) that has been described as a combination of rasping, growling, wheezing, grunting, rumbling, and braying. It consists of a long series of in-drawn, snorelike breaths alternating with exhaled, belchlike sounds. JERKING is a display resembling the hiccups, in which the female simultaneously jerks her body upward and flicks her head backward repeatedly. Male Koalas also sometimes mount each other, and a few even perform the jerking display like females in heat.

Frequency: In captivity, same-sex mounting accounts for 11 percent of all copulatory activity, with the majority of this being mounts between females.

Orientation: Koalas that participate in homosexual mounting are probably bisexual, since females that mount other females have also been observed mating with males.

Nonreproductive and Alternative Heterosexualities

Heterosexual relations in Koalas are marked by a striking amount of aggression and violence: more than two-thirds of fights are between males and females rather than between males. Females are sometimes "pestered" by males that persistently follow, touch, bite, or snap at them; if the female returns the bites, the encounter can escalate into a severe fight. Males have been known to brutally attack females—including pregnant and nursing mothers—knocking them from the trees and savagely mauling them. In fact, it is typical for males to nip females on the neck during mating, and for heterosexual copulations to end with the male attacking the female.

Females also fight with males (though less violently), and aggressiveness toward males is considered to be a defining feature of estrus for female Koalas. Occasionally adults are also abusive toward babies: mothers sometimes bite their young, while males have been observed attacking infants that interrupt them during a mating with their mother. Many heterosexual interactions are nonprocreative, since males often try to mount females who are not in heat. Although the females typically rebuff their advances, in some cases the males are able to mount them, often thrusting against the female and ejaculating on her without any penetration. Females in heat also sometimes mount males (REVERSE mounts).

Many wild populations of Koalas have particularly high rates of female infertility (and significantly reduced reproductive rates) due to venereal disease. More than half of all females in some areas are infected with genital chlamydia, a bacteria that causes a number of reproductive tract diseases and, ultimately, sterility. This pathogen has apparently been present in Koala populations for a relatively long time, as records of the associated diseases date back to at least the 1890s. Although the exact mode of its transmission is not yet fully understood, two routes have been implicated: sexual and mother-to-young. The latter may be due to the infant Koala's habit of eating its mother's feces directly from her anus during weaning, since she produces a special form of excrement known as PAP especially for feeding her young (this practice is also found in a number of other marsupials).

Other Species

In another marsupial, the Common Brushtail Possum (*Trichosurus vulpecula*), intersexuality or hermaphroditism occasionally occurs: one individual, for example, had male body proportions, coloring, and genitals combined with mammary glands and a pouch.

Sources   *asterisked references discuss homosexuality/transgender*

Brown, A. S., A. A. Girjes, M. F. Lavin, P. Timms, and J. B. Woolcock (1987) "Chlamydial Disease in Koalas." *Australian Veterinary Journal* 64:346–50.

*Gilmore, D. P. (1965) "Gynandromorphism in *Trichosurus vulpecula*." *Australian Journal of Science* 28:165.

Lee, A., and R. Martin (1988) *The Koala: A Natural History*. Kensington, Australia: New South Wales University Press.

Phillips, K. (1994) *Koalas: Australia's Ancient Ones*. New York: Macmillan.

*Sharman, G. B., R. L. Hughes, and D. W. Cooper (1990) "The Chromosomal Basis of Sex Differentiation in Marsupials." *Australian Journal of Zoology* 37:451–66.

Smith, M. (1980a) "Behavior of the Koala, *Phascolarctos cinereus* (Goldfuss) in Captivity. III. Vocalizations." *Australian Wildlife Research* 7:13–34.

*——— (1980b) "Behavior of the Koala, *Phascolarctos cinereus* (Goldfuss) in Captivity. V. Sexual Behavior." *Australian Wildlife Research* 7:41–51.

——— (1980c) "Behavior of the Koala, *Phascolarctos cinereus* (Goldfuss) in Captivity. VI. Aggression." *Australian Wildlife Research* 7:177–90.

——— (1979) "Behavior of the Koala, *Phascolarctos cinereus* (Goldfuss) in Captivity. I. Non-Social Behavior." *Australian Wildlife Research* 6:117–29.

*Thompson, V. D. (1987) "Parturition and Development in the Queensland Koala *Phascolarctos cinereus adustus* at San Diego Zoo." *International Zoo Yearbook* 26:217–22.

Weigler, B. J., A. A. Girjes, N. A. White, N. D. Kunst, F. N. Carrick, and M. F. Lavin (1988) "Aspects of the Epidemiology of *Chlamydia psittaci* Infection in a Population of Koalas (*Phascolarctos cinereus*) in Southeastern Queensland, Australia." *Journal of Wildlife Diseases* 24:282–91.

CARNIVOROUS MARSUPIALS

Marsupial

FAT-TAILED DUNNART
Sminthopsis crassicaudata

NORTHERN QUOLL
Dasyurus hallucatus

HOMOSEXUALITY	TRANSGENDER	BEHAVIORS		RANKING	OBSERVED IN
● Female	○ Intersexuality	○ Courtship	○ Pair-bonding	○ Primary	○ Wild
● Male	○ Transvestism	○ Affectionate	○ Parenting	○ Moderate	○ Semiwild
		● Sexual		● Incidental	● Captivity

FAT-TAILED DUNNART
IDENTIFICATION: A small, mouselike marsupial with a thick, conical, fat-storing tail. DISTRIBUTION: Inland southern Australia. HABITAT: Varied, including rocky areas. STUDY AREA: University of Adelaide, Australia.

NORTHERN QUOLL
IDENTIFICATION: A catlike marsupial, up to 2 feet long, with grayish brown fur and white splotches. DISTRIBUTION: Northern and eastern Australia. HABITAT: Woodland, rocky areas. STUDY AREA: Monash University, Australia.

Social Organization
Fat-tailed Dunnarts often live together in small groups or pairs that share nests; these groupings are temporary and may consist of individuals of the same sex (especially outside of the breeding season). Although little is known about the social system of Northern Quolls, it appears that most individuals are largely solitary. Both species are nocturnal.

Description
Behavioral Expression: Homosexual mounting occurs among female, and to a lesser extent male, Northern Quolls. One animal climbs on top of another as in a heterosexual mating, grasping its chest with its front paws, and sometimes even riding on the back of the mounted animal as it walks around. In Fat-tailed Dunnarts, females in heat sometimes mount other females.

Frequency: In captivity, homosexual mounting among Northern Quolls occurs in almost two-thirds of encounters between females and 10 percent of encounters between males, though this may not reflect the frequency of its occurrence in the wild. Same-sex mounting only happens occasionally in female Fat-tailed Dunnarts.

Orientation: It is possible that some individuals in these species engage exclusively in same-sex behavior, while others may be bisexual, but little is known about the life histories of specific individuals. In one study, none of the Northern Quolls that participated in homosexual behavior engaged in heterosexual mounting (although observations were not made during the breeding season), while one female Fat-tailed Dunnart that mounted another female did not breed during a nearly yearlong study.

Nonreproductive and Alternative Heterosexualities

Reproduction in Northern Quolls is characterized by an extraordinary phenomenon sometimes known as MALE DIE-OFF. In many areas, virtually the entire male population perishes following the breeding season, while females typically survive to breed for another couple of seasons (some variation occurs between geographic locations and years in the proportion of males and females surviving). This complete annihilation of males is also a feature of a number of other carnivorous marsupial social systems and is found to a much lesser extent in Fat-tailed Dunnarts. Although the exact mechanism responsible for male mortality is not fully understood, it is thought to result from a number of stress-induced factors, perhaps directly related to participation in procreation. There is some evidence that nonbreeding males with lower testosterone levels—essentially "lower-ranking" males—have a higher survival rate than males that reproduce. Female Northern Quolls also routinely practice "abortion" or elimination of unborn young. As many as 17 embryos may begin developing in the female's uterus, but because females typically have no more than 8 nipples in their pouch, most of the embryos and/or newborn young will not survive. In Fat-tailed Dunnarts, breeding females can be noticeably aggressive toward males, attacking them when they attempt to mount; in captivity, females have even been known to kill their mates. In this species, heterosexual copulation can be a remarkably long affair, with the male remaining mounted on the female for hours at a time (sometimes as long as 11 hours); the female may struggle and attempt to escape during such arduous matings. In Northern Quolls, females often have neck and chest wounds inflicted by the male during mating. Adult male Fat-tailed Dunnarts sometimes display sexual interest in juvenile females, and incestuous matings have also been recorded. In addition, females in heat occasionally mount males (REVERSE mountings).

Other Species

Male Stuart's Marsupial Mice (*Antechinus stuartii*) mount individuals of both sexes during the mating period. Transgendered (intersexual) Tasmanian Devils (*Sarcophilus harrisii*) have also been reported: one individual had female genitalia and internal reproductive organs, combined with a scrotum and a pouch with mammary glands on only one side.

Sources   *asterisked references discuss homosexuality/transgender*

Begg, R. J. (1981) "The Small Mammals of Little Nourlangie Rock, N.T. III. Ecology of *Dasyurus hallucatus*, the Northern Quoll (Marsupialia: Dasyuridae)." *Australian Wildlife Research* 8:73–85.

Croft, D. B. (1982) "Communication in the Dasyuridae (Marsupialia): A Review." In M. Archer, ed., *Carnivorous Marsupials*, vol. 1, pp. 291–309. Chipping Norton, Australia: Royal Zoological Society of New South Wales.

*Dempster, E. R. (1995) "The Social Behavior of Captive Northern Quolls, *Dasyurus hallucatus.*" *Australian Mammalogy* 18:27–34.

Dickman, C. R., and R. W. Braithwaite (1992) "Postmating Mortality of Males in the Dasyurid Marsupials, *Dasyurus* and *Parantechinus.*" *Journal of Mammalogy* 73:143–47.

*Ewer, R. F. (1968) "A Preliminary Survey of the Behavior in Captivity of the Dasyurid Marsupial, *Sminthopsis crassicaudata* (Gold)." *Zeitschrift für Tierpsychologie* 25:319–65.

*Lee, A. K., and A. Cockburn (1985) *Evolutionary Ecology of Marsupials.* Cambridge: Cambridge University Press.

Morton, S. R. (1978) "An Ecological Study of *Sminthopsis crassicaudata* (Marsupialia: Dasyuridae). II. Behavior and Social Organization. III. Reproduction and Life History." *Australian Wildlife Research* 5:163–211.

Schmitt, L. H., A. J. Bradley, C. M. Kemper, D. J. Kitchener, W. F. Humphreys, and R. A. How (1989) "Ecology and Physiology of the Northern Quoll, *Dasyurus hallucatus* (Marsupialia, Dasyuridae), at Mitchell Plateau, Kimberley, Western Australia." *Journal of Zoology, London* 217:539–58.

*Sharman, G. B., R. L. Hughes, and D. W. Cooper (1990) "The Chromosomal Basis of Sex Differentiation in Marsupials." *Australian Journal of Zoology* 37:451–66.

RODENTS, INSECTIVORES, AND BATS

SQUIRRELS   *Rodent/Insectivore*

(AMERICAN) RED SQUIRREL
Tamiasciurus hudsonicus

GRAY SQUIRREL
Sciurus carolinensis

LEAST CHIPMUNK
Tamias minimus

HOMOSEXUALITY	TRANSGENDER	BEHAVIORS		RANKING	OBSERVED IN
● Female	○ Intersexuality	○ Courtship	● Pair-bonding	○ Primary	● Wild
● Male	○ Transvestism	● Affectionate	● Parenting	● Moderate	● Semiwild
		● Sexual		○ Incidental	● Captivity

RED SQUIRREL
IDENTIFICATION: A medium-sized (10–15 inch), primarily tree-dwelling squirrel with a reddish brown or tawny coat and white underparts, often with a dark stripe on the side. DISTRIBUTION: Canada, Alaska, Rocky and Appalachian Mountains, northeastern United States. HABITAT: Coniferous or mixed forests. STUDY AREAS: Near Ithaca, New York; Manning Provincial Park, British Columbia, Canada; Saint-Hippolyte, Quebec (University of Montreal); subspecies *T.h. gymnicus, T.h. streatori, T.h. laurentianus.*

GRAY SQUIRREL
IDENTIFICATION: A large (20 inch), tree-dwelling squirrel with a long, bushy tail and gray, grizzled, or buff fur. DISTRIBUTION: Eastern United States, southeastern Canada. HABITAT: Hardwood forests and parks. STUDY AREA: University of Maryland; subspecies *S.c. pennsylvanicus.*

LEAST CHIPMUNK
IDENTIFICATION: A small, ground-dwelling squirrel with alternating dark and light stripes on back and face. DISTRIBUTION: Yukon to Ontario; upper Midwest; mountainous western United States. HABITAT: Conifer forests, sagebrush. STUDY AREA: Lake Superior State College, Sault Ste. Marie, Michigan; subspecies *T.m. neglectus.*

Social Organization

Squirrels are generally not gregarious, especially outside of the breeding season, and Red Squirrels defend individual territories throughout the year. The mating system is promiscuous: males and females mate with multiple partners, and only females raise the offspring. Chipmunks are territorial and live in elaborate underground burrows.

Description

Behavioral Expression: Male and female Red Squirrels participate in homosexual mountings, particularly outside of the breeding season. One animal approaches the other from behind (either on the ground or vertically, as on a tree trunk) and grasps it around the waist. The posture used is the same as for heterosexual mating, with the mounter tucking its paws in front of the haunches of the other animal. The mounter often licks, nibbles, or grooms the fur on its partner's nape or the sides of its neck, and mounting may also be accompanied by play-fighting (harmless "boxing" or kicking sessions) and soft buzzing or *mok*-calls. Sometimes three animals of the same sex participate in mounting activities at the same time, either alternating among themselves or else all three mounting each other simultaneously (one behind the other in a row). Unlike heterosexual copulation, mounting may be reciprocal, and the mounted animal is usually a more willing participant (although occasionally it will turn around to bite the mounter). In addition, same-sex mounting does not usually involve penetration or pelvic thrusting and is not usually preceded by a courtship chase. Among Gray Squirrels, a similar type of same-sex mounting (with grooming) occurs primarily among younger animals (sometimes among siblings). The activity is often initiated differently from in Red Squirrels, in that the mounting animal makes a bouncing leap onto its partner before grasping it around the waist. Adult male Least Chipmunks also sometimes mount one another.

Occasionally two female Red Squirrels form a bond with each other that includes sexual and affectionate activities and joint parenting. The two share a den (tree hole), follow one another about, and often touch noses or gently nudge each other's flanks. They also take turns mounting each other, each female nibbling or stroking her partner's fur while mounting. The pair may raise a single litter of young together; although the youngsters are probably the biological offspring of only one female, both partners may nurse them. In addition, they cooperate in taking care of the young: one such pair was observed leading their lost baby out of a tree and across a street back to their home den. Notably, heterosexual pairs do not usually form in this species: females typically raise young on their own and are generally very aggressive toward other adults that try to approach them.

Frequency: In Red Squirrels, 18 percent of mounts are homosexual; the majority of these are between males, who may participate in same-sex mounts as often as once every half hour outside the breeding season. This is slightly higher than the rate of heterosexual mounting at the same time of the year. Young Gray Squirrels participate in same-sex mounting at a much higher rate, up to 10 times an hour for some age groups (which is more than three times the rate of heterosexual mounting). Among Least Chipmunks, one male was seen to mount another 20 times over four days. Pair-bonding with joint parenting between female Red Squirrels probably occurs only occasionally, although no systematic study has been conducted to determine its prevalence.

Orientation: Adult Red Squirrels that participate in same-sex mounting are seasonally bisexual: outside of the breeding season they engage in both same-sex and opposite-sex mounting, while during the breeding season they mostly participate in heterosexual matings. Females that pair with each other, however, are involved primarily in same-sex activity for the duration of their association, although one and possibly both are functionally bisexual (since they reproduce). Younger animals tend to mount their siblings and therefore are either simultaneously bisexual (interacting with both males and females) or, among males, primarily homosexual (interacting mostly with other males even if females are present). Gray Squirrels are sequentially or chronologically bisexual: juveniles, adolescents, and young adults (up to one and a half years old) show a preference for homosexual activity, while older adults generally exhibit much less same-sex mounting. Since males do not usually start breeding until they are 18 months old, some individuals may be involved exclusively in homosexuality up to that time.

Nonreproductive and Alternative Heterosexualities

Both Red and Gray Squirrels participate in a wide range of nonprocreative sexual activities. Heterosexual mounting outside of the breeding periods is common, especially in Red Squirrels. Since males have a seasonal sexual cycle that renders them infertile at these times (like females), this activity is definitely nonreproductive. In addition, penetration is usually not involved, and REVERSE mounting (females mounting males) also occurs, accounting for about 5 percent of such activity in Red Squirrels. Sexual behavior also takes place in other situations that are not optimal for breeding. Juveniles of both species often participate in mounting or sexual chases, for example, including incestuous activity between siblings or mothers being mounted by their offspring. Interspecies mating chases have also been observed between Gray Squirrels and fox squirrels (*Sciurus niger*).

During Gray Squirrel mating chases, as many as 34 males may pursue and harass a single female; when cornered in a tree cavity or at the end of a limb, she often defends herself by screaming and lunging at the males and then bolting away. Mounts by males result in full copulation only 40 percent of the time, as females usually escape from a male that is trying to mate with them. In addition, mating pairs are often attacked by other males, who knock the couple from the tree or savagely bite them, in some cases inflicting fatal wounds on the female. When mating does occur, the female copulates with several different males, sometimes as many as eight during a single mating bout. Mating chases in Red Squirrels can involve up to seven males at a time; a female may drag a male mounted on her for some distance or vigorously try to shake him off her back. In Gray Squirrels, sperm coagulates in the female's vagina following copulation, forming a plug that may prevent inseminations by other males. However, females often remove these plugs, thereby preventing insemination by the male who just mated with them (while at the same time possibly allowing subsequent males to impregnate them). Young Red Squirrels sometimes suffer fatal injuries from adults during territory takeovers, and Gray Squirrel youngsters are occasionally attacked and/or cannibalized by adults. Not all adults participate in breeding: about a third of female Red Squirrels in some populations are nonbreeders, while about 30 percent of male Gray Squirrels do not mate with females each breeding season (and some individuals skip entire seasons).

Sources   *asterisked references discuss homosexuality/transgender

Barkalow, F. S., Jr., and M. Shorten (1973) *The World of the Gray Squirrel*. Philadelphia and New York: J.B. Lippincott.

Ferron, J. (1981) "Comparative Ontogeny of Behavior in Four Species of Squirrels (Sciuridae)." *Zeitschrift für Tierpsychologie* 55:193–216.

*——— (1980) "Le comportement cohésif de l'Écureuil roux (*Tamiasciurus hudsonicus*) [Cohesive Behavior of the Red Squirrel]." *Biology of Behavior* 5:118–38.

*Horwich, R. H. (1972) *The Ontogeny of Social Behavior in the Gray Squirrel* (Sciurus carolinensis). Berlin and Hamburg: Paul Parey.

Koprowski, J. L. (1994) "*Sciurus carolinensis.*" *Mammalian Species* 480:1–9.

——— (1993) "Alternative Reproductive Tactics in Male Eastern Gray Squirrels: 'Making the Best of a Bad Job.'" *Behavioral Ecology* 4:165–71.

——— (1992a) "Do Estrous Female Gray Squirrels, *Sciurus carolinensis*, Advertise Their Receptivity?" *Canadian Field-Naturalist* 106:392–94.

——— (1992b) "Removal of Copulatory Plugs by Female Tree Squirrels." *Journal of Mammalogy* 73:572–76.

——— (1991) "Mixed-species Mating Chases of Fox Squirrels, *Sciurus niger,* and Eastern Gray Squirrels, *S. carolinensis.*" *Canadian Field-Naturalist* 105:117–18.

*Layne, J. C. (1954) "The Biology of the Red Squirrel, *Tamiasciurus hudsonicus loquax* (Bangs), in Central New York." *Ecological Monographs* 24:227–67.

Moore, C. M. (1968) "Sympatric Species of Tree Squirrels Mix in Mating Chase." *Journal of Mammalogy* 49:531–33.

Price, K., and S. Boutin (1993) "Territorial Bequeathal by Red Squirrel Mothers." *Behavioral Ecology* 4:144–49.

*Reilly, R. E. (1972) "Pseudo-Copulatory Behavior in *Eutamias minimus* in an Enclosure." *American Midland Naturalist* 88:232.

Smith, C. C. (1978) "Structure and Function of the Vocalizations of Tree Squirrels (*Tamiasciurus*)." *Journal of Mammalogy* 59:793–808.

*——— (1968) "The Adaptive Nature of Social Organization in the Genus of Tree Squirrels *Tamiasciurus.*" *Ecological Monographs* 38:31–63.

Thompson, D. C. (1978) "The Social System of the Gray Squirrel." *Behavior* 64:305–28.

——— (1977) "Reproductive Behavior of the Gray Squirrel." *Canadian Journal of Zoology* 55:1176–84.

——— (1976) "Accidental Mortality and Cannibalization of a Nestling Gray Squirrel." *Canadian Field-Naturalist* 90:52–53.

MARMOTS
OLYMPIC MARMOT
Marmota olympus
HOARY MARMOT
Marmota caligata

Rodent/Insectivore

HOMOSEXUALITY	TRANSGENDER	BEHAVIORS		RANKING	OBSERVED IN
● Female	○ Intersexuality	○ Courtship	○ Pair-bonding	○ Primary	● Wild
○ Male	○ Transvestism	● Affectionate	○ Parenting	● Moderate	○ Semiwild
		● Sexual		○ Incidental	○ Captivity

IDENTIFICATION: Woodchucklike rodents with gray, brown, reddish, or black fur. DISTRIBUTION: Olympic Peninsula, Washington; Alaska south to northwestern United States. HABITAT: Alpine slopes. STUDY AREAS: Olympic National Park, Washington; Glacier National Park, Montana; subspecies *M.c. nivaria*.

Social Organization

These two species of Marmots are highly social creatures that live in clusters of colonies; each colony is a series of underground burrows that is home to one male, one to three females, and their offspring. Males are generally not involved directly in parental care of their young. Occasionally an additional SATELLITE male is peripherally associated with an Olympic Marmot colony.

Description

Behavioral Expression: Olympic and Hoary Marmot females often mount other females and participate in other same-sex affectionate and sexual behaviors, especially when they are in heat. A homosexual encounter often begins with a "greeting" interaction in which the two females touch noses or mouths, or one female nuzzles her nose on the other's cheek or mouth. She may also gently chew on the ear or neck of the other female, who then responds by raising her tail. The first female sometimes also sniffs or nuzzles the other's genitals with her mouth. At this point she may mount the other female, gently biting her neck fur while she thrusts against her partner. The

A female Olympic Marmot mounting another female ▶

female being mounted arches her back and holds her tail to the side to facilitate the sexual interaction.

Frequency: Homosexual behavior is quite common in Marmots: in one study of Hoary Marmots, for example, three of five observed mounts by adults were between females.

Orientation: Many female Marmots that participate in same-sex mountings also mate with males. However, some nonbreeding females in Hoary Marmots (see below) probably also participate, which means that they may be involved only in homosexual interactions for those seasons that they do not breed.

Nonreproductive and Alternative Heterosexualities

Although many Marmots form monogamous heterosexual pair-bonds, in some populations the majority (two-thirds) of Hoary Marmots actually live in trios consisting of one male and two females. Occasionally a "quartet" of one male and three females live together as well. Some male Hoary Marmots also seek promiscuous matings with females outside their colony, a behavior that has been termed GALLIVANTING. A form of reproductive suppression occurs in this species as well: females usually procreate every other year, but 11 percent of the time, a female "skips" breeding for two consecutive years. This is especially common in trios, where the two females alternate their skipping patterns. Males sometimes still try to mount females that are not breeding, however. Sexual activity also occurs among juveniles, including mounting of adults.

Sources   *asterisked references discuss homosexuality/transgender*
*Barash, D. P. (1989) *Marmots: Social Behavior and Ecology.* Stanford: Stanford University Press.
———— (1981) "Mate Guarding and Gallivanting by Male Hoary Marmots (*Marmota caligata*)." *Behavioral Ecology and Sociobiology* 9:187–93.
*———— (1974) "The Social Behavior of the Hoary Marmot (*Marmota caligata*)." *Animal Behavior* 22:256–61.
*———— (1973) "The Social Biology of the Olympic Marmot." *Animal Behavior Monographs* 6:171–245.
Holmes, W. G. (1984) "The Ecological Basis of Monogamy in Alaskan Hoary Marmots." In J. O. Murie and G. R. Michener, eds., *The Biology of Ground-Dwelling Squirrels: Annual Cycles, Behavioral Ecology, and Sociality,* pp. 250–74. Lincoln: University of Nebraska Press.
Wasser, S. K., and D. P. Barash (1983) "Reproductive Suppression Among Female Mammals: Implications for Biomedicine and Sexual Selection Theory." *Quarterly Review of Biology* 58:513–38.

CAVIES   Rodent/Insectivore

DWARF CAVY
Microcavia australis
CUI or YELLOW-TOOTHED CAVY
Galea musteloides
APEREA or WILD CAVY
Cavia aperea

HOMOSEXUALITY	TRANSGENDER	BEHAVIORS		RANKING	OBSERVED IN
● Female	○ Intersexuality	● Courtship	● Pair-bonding	○ Primary	● Wild
● Male	○ Transvestism	● Affectionate	● Parenting	● Moderate	● Semiwild
		● Sexual		○ Incidental	● Captivity

IDENTIFICATION: Small, guinea-pig-like rodents with coarse fur; Dwarf Cavies have a distinctive white eye-ring, Cuis have yellow-colored incisors. DISTRIBUTION: South America, primarily Peru, Bolivia, Argentina, Chile. HABITAT: Savanna, brushlands. STUDY AREAS: Several locations in Buenos Aires Province, Argentina, including near Magdalena, Tornquist, and Carmen de Patagones.

Social Organization

Cavies live in colonies of 20–50 animals, although no permanent social groupings occur. Each female usually has her own home bush under which she lives and raises her young. The mating system is promiscuous, in that males and females copulate with multiple partners without forming any long-lasting heterosexual bonds.

Description

Behavioral Expression: Cavies participate in a wide variety of homosexual courtship patterns, most of which are also found in heterosexual interactions. In Dwarf Cavies, adult males are sexually attracted to juveniles of both sexes. A typical homosexual courtship begins with the adult and the younger male sitting quietly together—often right in front of the youngster's mother, who is not visibly disturbed. The two males may kiss and nuzzle their noses together, after which the adult begins to gently nibble on the juvenile's rump. This is often followed by a courtship behavior known as CHIN-RUMP FOLLOWING, in which the adult male places his chin on the youngster's rump, as they move together in circles or figure-eight patterns. If the male is especially aroused, he will stop to pat his paws against the juvenile's rump, sometimes even licking, sniffing, and nuzzling the anus and perineal region while lifting up the youngster's hindquarters with his nose. Adult males frequently have favorite males that they prefer to court and interact sexually with, actively seeking them out while ignoring others.

Female Cuis (and less commonly, males) use a stylized form of same-sex courtship chasing known as REARING: one female approaches another and rears up

▲ *A female Aperea* (right) *performing a "rumba" while courting another female*

on her hind legs as if to mount, then drops back down on all fours to follow the other female, repeating the pattern. Sometimes this leads to actual mounting, in which one female thrusts against the other in an upright posture behind her. Homosexual mounting is common when females are in heat, but it also occurs when a female is not in heat and even when she is pregnant. Among Apereas, females court one another by performing the RUMBA: in this courtship dance, one female rhythmically sways her hindquarters from side to side while slowly approaching, circling, or following another female. She may also make a burbling sound known as RUMBLING. Adult male Apereas sometimes court younger males by RUMPING, in which the adult throws one or both hind legs over the rump of the youngster.

Dwarf Cavies also sometimes bond with animals of the same sex, forming "companionships" and even coparenting arrangements. Two (occasionally three) females share a home (living under the same bush), frequently sitting together and kissing each other with prolonged nuzzling of mouths and noses. They may even cooperate in nursing one another's young. Adult males occasionally also have an adolescent companion; the two males sit and feed together, as well as kiss one another. Sometimes the adult male will permit his companion to mount a female that he is courting.

Frequency: Homosexual courtships make up 58 percent of adult-juvenile chin-rump follows (and 44 percent of all such courtships) in Dwarf Cavies, while more than a third of Cui courtship rearing interactions are between females. Homosexual courtships are less common in Apereas.

Orientation: Most Cavies that participate in same-sex activities are probably bisexual. For example, adult male Apereas and Dwarf Cavies court both juvenile males and females, while same-sex bonded Dwarf Cavies usually copulate with members of the opposite sex. However, there are differences in the apparent preference of adult males for homosexual interactions: in Dwarf Cavies, the majority of adult-juvenile courtships are homosexual, while in Apereas fewer such courtships are same-sex.

Nonreproductive and Alternative Heterosexualities

Cavies regularly engage in heterosexual courtship and sexual behaviors that are nonprocreative. Female Cuis that are pregnant, lactating, or not in heat, for example, often participate in mounting and courtship (including REVERSE mounting of males). Aperea males commonly court pregnant females. Male Dwarf Cavies occasionally masturbate by sitting back on their haunches, making pelvic thrusts, and then licking and nuzzling the erect penis. As mentioned above, sexual behavior between adults and juveniles is widespread in these species. About a quarter of all courtship and sexual interactions in Dwarf Cavies involve adult males and juveniles, while adult male Cuis also chin-rump follow juvenile females as young as two weeks old. In addition, juvenile Cuis of both sexes sometimes mount both adult males and females, including their own mothers. Cui mothers have also developed an extensive system of communal nursing in which most females suckle young other than their own. Although each female nurses her own offspring for longer periods than she does others', she may actually spend a greater total amount of time nursing nonoffspring.

Other Species

Homosexual behavior occurs in several other rodents. Adult males in two other species of Cavies, the Mocó or Rock Cavy (*Kerodon rupestris*) and the Préa (*Galea spixii*), also sometimes court juvenile males, and youngsters mount adults of both sexes as well. Male Spinifex Hopping Mice (*Notomys alexis*) from Australia sometimes engage in sexual and bonding activities with one another, including mounting, nesting together, feeding and resting beside each other, and burrowing in tandem. Both male and female Brown Rats (*Rattus norvegicus,* the wild ancestor of the common household and laboratory rat) participate in homosexual mounting. Male Eastern Cottontail Rabbits (*Sylvilagus floridanus*) sometimes court other males in addition to females, including displays such as the JUMP SEQUENCE, in which the two males alternately rush toward and jump over one another. Males in this species also occasionally attempt to mount other males. In some populations of Kangaroo Rats (*Dipodomys ordii*), as many as 16 percent of the animals are intersexed, having the reproductive organs of both sexes (including a vagina, penis, uterus, and testes).

Sources   *asterisked references discuss homosexuality/transgender*

*Barnett, S. A. (1958) "An Analysis of Social Behavior in Wild Rats." *Proceedings of the Zoological Society of London* 130:107–52.

*Happold, M. (1976) "Social Behavior of the Conilurine Rodents (Muridae) of Australia." *Zeitschrift für Tierpsychologie* 40:113–82.

Künkele, J., and H. N. Hoeck (1995) "Communal Suckling in the Cavy *Galea musteloides.*" *Behavioral Ecology and Sociobiology* 37:385–91.

*Lacher, T. E., Jr. (1981) "The Comparative Social Behavior of *Kerodon rupestris* and *Galea spixii* and the Evolution of Behavior in the Caviidae." *Bulletin of the Carnegie Museum of Natural History* 17:1–71.

*Marsden, H. M., and N. R. Holler (1964) "Social Behavior in Confined Populations of the Cottontail and the Swamp Rabbit." *Wildlife Monographs* 13:1–39.

*Pfaffenberger, G. S., F. W. Weckerly, and T. L. Best (1986) "Male Pseudohermaphroditism in a Population of Kangaroo Rats, *Dipodomys ordii.*" *Southwestern Naturalist* 31:124–26.

*Rood, J. P. (1972) "Ecological and Behavioral Comparisons of Three Genera of Argentine Cavies." *Animal Behavior Monographs* 5:1–83.

*——— (1970) "Ecology and Social Behavior of the Desert Cavy (*Microcavia australis*)." *American Midland Naturalist* 83:415–54.

Stahnke, A., and H. Hendrichs (1990) "Cavy Rodents." In *Grzimek's Encyclopedia of Mammals*, vol. 3, pp. 325–37. New York: McGraw-Hill.

HEDGEHOGS   *Rodent/Insectivore*

LONG-EARED HEDGEHOG
Hemiechinus auritus

HOMOSEXUALITY	TRANSGENDER	BEHAVIORS		RANKING	OBSERVED IN
● Female	○ Intersexuality	● Courtship	○ Pair-bonding	○ Primary	○ Wild
○ Male	○ Transvestism	● Affectionate	○ Parenting	● Moderate	○ Semiwild
		● Sexual		○ Incidental	● Captivity

IDENTIFICATION: A small (less than 1 foot long) insectivore with sandy-colored spines, white underparts, and prominent ears. DISTRIBUTION: Central Asia, Middle East. HABITAT: Steppe, desert. STUDY AREA: Research Laboratory for Comparative Insectivorology, Vienna, Austria; subspecies *H.a. syriacus.*

Social Organization

Long-eared Hedgehogs live in burrows and are largely nocturnal and solitary, although small groups of animals may gather at feeding or resting sites. Males take no part in parental care.

Description

Behavioral Expression: Homosexual interactions in female Long-eared Hedgehogs involve a great deal of courtship and affectionate behaviors as well as direct sexual encounters, frequently consisting of oral sex. A typical lesbian interaction begins, often at dusk, with two females rubbing each other, sliding along each other's body, and cuddling. One female might also crawl directly under the other, sliding back from her throat to her belly. Another courtship display involves one female stretching out full length and pressing her belly against the ground with a concave "arch" in her back. During sexual contact, females intensively lick, sniff, and nibble on each other's genitals. Sometimes, to have better access, one female will raise the hindquarters of the other high into the air with her paws and lower jaw, lifting her partner's hind legs clear off the ground while she continues licking. At other times, one or both females will present their raised hindquarters as an invitation for the other to mount as in heterosexual copulation. Often the presenting female is in such a state of arousal that her hindquarters are actually lifted too high for the other female to fully mount her, although she may try. In captivity, homosexual encounters have been observed between adult sisters, that is, members of the same litter.

Frequency: Homosexual interactions occur frequently between females paired together in
captivity, but the incidence of this
activity in wild Long-eared
Hedgehogs is not known.

Orientation: Female Long-
eared Hedgehogs may have a
latent capacity for bisexual or homosexual behavior, since same-
sex activity surfaces when females
are kept together without males.
However, it appears that this may
initiate a preference for homosexuality that can be long-lasting: one pair of females who
courted and had sex with each
other refused to participate in
heterosexual activity for more
than two years after they were
separated, although eventually
both did mate with males and reproduce.

▲ *Courtship and sexual activity between female
Long-eared Hedgehogs:* sliding *(above),* arching posture *(middle), and* cunnilingus

Nonreproductive and Alternative Heterosexualities

Oral-genital stimulation is a frequent component of heterosexual encounters in
Long-eared Hedgehogs, with males licking and sniffing the female's genitals. Cannibalism also occurs in this species: animals may eat already dead hedgehogs or else
kill them directly and then devour them.

Other Species

Homosexual activity occurs in several species of Tree Shrews, a group of animals found in Southeast Asia and thought to have affiliations with insectivores
(and possibly also primates). In Common Tree Shrews (*Tupaia glis*), for example,
about a third of all sexual activity occurs between females, including sexual approaches and following, genital licking and sniffing, and mounting. Same-sex
mounting has also been observed in Slender Tree Shrews (*T. gracilis*), Mountain
Tree Shrews (*T. montana*), and Long-footed Tree Shrews (*T. longipes*). In the latter
species, mounting between females accounts for about 9 percent of all mounting
activity. Female Long-footed Tree Shrews sometimes form consortships with one
another as well; these typically last longer than heterosexual consortships (several
months as opposed to several hours) and involve mutual grooming, lying on or

next to each other, and sleeping together. Male and female homosexuality also occur in North American Porcupines (*Erethizon dorsatum*), including periods of exclusive homosexual activity among males.

Sources   *asterisked references discuss homosexuality/transgender

*Kaufmann, J. H. (1965) "Studies on the Behavior of Captive Tree Shrews (*Tupaia glis*)." *Folia Primatologica* 3:50–74.

*Kinsey, A. C., W. B. Pomeroy, C. E. Martin, and P. H. Gebhard (1953) *Sexual Behavior in the Human Female*. Philadelphia: W. B. Saunders.

Maheshwari, U. K. (1984) "Food of the Long Eared Hedgehog in Ravine Near Agra." *Acta Theriologica* 29:133–37.

*Poduschka, W. (1981) "Abnormes Sexualverhalten Zusammengehaltener, Weiblicher *Hemiechinus auritus syriacus* (Insectivora: Erinaceinae) [Abnormal Sexual Behavior of Confined Female *Hemiechinus auritus syriacus*]." *Bijdragen tot de Dierkunde* 51:81–88.

Prakash, I. (1953) "Cannibalism in Hedgehogs." *Journal of the Bombay Natural History Society* 51:730–31.

Reeve, N. (1994) *Hedgehogs*. London: T. and A. D. Poyser.

Schoenfeld, M., and Y. Yom-Tov (1985) "The Biology of Two Species of Hedgehogs, *Erinaceus europaeus concolor* and *Hemiechinus auritus aegyptus*, in Israel." *Mammalia* 49:339–55.

*Sorenson, M.W., and C. H. Conaway (1968) "The Social and Reproductive Behavior of *Tupaia montana* in Captivity." *Journal of Mammalogy* 49:502–12.

*———— (1966) "Observations on the Social Behavior of Tree Shrews in Captivity." *Folia Primatologica* 4:124–45.

BATS

GRAY-HEADED FLYING FOX

Pteropus poliocephalus

LIVINGSTONE'S FRUIT BAT

Pteropus livingstonii

(COMMON) **VAMPIRE BAT**

Desmodus rotundus

Rodent/Insectivore

HOMOSEXUALITY	TRANSGENDER	BEHAVIORS		RANKING	OBSERVED IN
● Female	○ Intersexuality	○ Courtship	● Pair-bonding	○ Primary	● Wild
● Male	○ Transvestism	● Affectionate	○ Parenting	● Moderate	○ Semiwild
		● Sexual		○ Incidental	● Captivity

GRAY-HEADED FLYING FOX

IDENTIFICATION: A large bat with an enormous wingspan (up to 4 feet), a doglike face, dark brown fur, a light gray head, and a reddish yellow mantle. DISTRIBUTION: Eastern Australia. HABITAT: Tropical and subtropical forests; roosts in trees. STUDY AREA: Near Brisbane, southeastern Queensland, Australia.

LIVINGSTONE'S FRUIT BAT

IDENTIFICATION: Similar to Gray-headed, except coat is black with tawny shoulders and groin; wingspan over 3 feet. DISTRIBUTION: Anjouan and Mohéli Islands, Comoros Archipelago (Indian Ocean); critically endangered. HABITAT: Upland forests. STUDY AREA: Jersey Wildlife Preservation Trust, England.

VAMPIRE BAT

IDENTIFICATION: A small bat with grayish brown fur and pointed ears. DISTRIBUTION: Northern Mexico through central Chile, Argentina, Uruguay; Trinidad. HABITAT: Forests, open areas; roosts in caves, tree hollows. STUDY AREAS: Hacienda La Pacifica and Parque Nacional de Santa Rosa, Costa Rica; University of the West Indies, Trinidad.

Social Organization

Gray-headed Flying Foxes live in groups known as CAMPS, which may contain many thousands of individuals. These camps are segregated by sex for most of the year: males and females roost in separate trees—or in separate locations within the same tree—except during the breeding season (generally March–April). Some individuals become nomadic, solitary, or much less gregarious following the breeding season. Livingstone's Fruit Bats appear to have a polygamous mating system, in which males mate with multiple female partners but do not participate in raising their offspring. Vampire Bat colonies may contain up to 2,000 individuals, although most have 20–100. The female group is the primary social unit, consisting of 8–12 females (many of whom are related to each other) and their young. Males sometimes form "bachelor" groups of up to 8 individuals, or they may roost in the same tree with female groups.

Description

Behavioral Expression: Gray-headed Flying Foxes of both sexes engage in a form of mutual homosexual grooming and caressing when they are in their separate camps. One animal wraps its wings around another of the same sex in an embrace, licking and gently biting the chest and wings of its partner, rubbing its head on the other's chest, and grooming it with its claws. Males may have an erection while they do this, and individuals generally utter a continuous pulsed, grating call while engaged in this activity. Livingstone's Fruit Bats participate in similar forms of grooming and other homosexual activities. Combined with bouts of intense body licking—either mutual or one-sided—both males and females in this species sometimes lick, nuzzle, and sniff the genitals of a same-sex partner (one male was even seen to drink another's urine as part of this activity). Clasping, play-wrestling, and gentle mouthing or biting of the partner occur as well. This may lead to homosexual mounting, in which one Bat grips the other from behind, holding the scruff of its neck in its mouth (as in heterosexual mating, although males do not usually experience erections or penetration during same-sex activity). Females sometimes mount their adult daughters and vice versa. In one instance, a daughter repeatedly approached, pursued, and mounted her mother for extended periods, and even successfully fought off males who were interested in mating with her mother.

Male Vampire Bats also participate in sexual grooming and licking of one another. Two males hang belly to belly, each with an erect penis. One male then works his tongue over the entire body of the other male, paying particular attention to licking the other male's genitals. Sometimes one male will masturbate himself while licking his partner, using his free foot to rub his own penis. Although overt sexual behavior has not been observed among female Vampire Bats, females do form long-lasting bonds with one another. Companions share the same roost, groom one another, huddle together, and go foraging with each other. Another important aspect of these female companionships is blood-sharing: one female feeds the other by "donating" or regurgitating blood for her to consume (males also occasionally engage in reciprocal blood-sharing). Associations like these can last for five to ten years or more, and some females bond with several different female companions simultaneously.

Frequency: Overt sexual behavior among Gray-headed Flying Foxes and Vampire Bats probably occurs only occasionally (and is more common in male Flying Foxes than in females), but various same-sex activities occur regularly in Livingstone's Fruit Bats (in captivity). In Vampire Bats, between one-half and three-quarters of all companionships or close associations are between females.

Orientation: Little is known of the individual life histories of these Bats, and so it is difficult to draw any definitive conclusions regarding the orientation of their sexual behavior. Nevertheless, it is likely that many Gray-headed Flying Foxes are seasonally bisexual, since they participate in homosexual activities when they are in the sex-segregated camps during the nonbreeding season. Among Vampire Bats in captivity, some males seem to show what amounts to a preference for homosexual activity, since they bypass females in order to interact sexually with another male (although it is not known whether this "preference" is temporary or long-lasting). Livingstone's Fruit Bats may be simultaneously bisexual, able to alternate between same-sex and opposite-sex activities in a relatively short span of time.

Nonreproductive and Alternative Heterosexualities

Heterosexuality in all three of these species of Bats is characterized by a variety of nonreproductive sexual behaviors. Gray-headed Flying Foxes copulate throughout the year, including outside the breeding season when females cannot get pregnant, and mating also takes place during pregnancy. In addition, males have a distinct annual hormonal cycle that affects sperm production, with the result that many of their matings are nonprocreative. Male Livingstone's Fruit Bats sometimes participate in heterosexual mounting without an erection or penetration, and females may REVERSE mount males as well. A prominent feature of Gray-headed Flying Fox sexual behavior is oral sex, in which the male deeply tongues the female's genitalia for long periods. Both male and female Livingstone's Fruit Bats also lick the genitals of their partners during heterosexual interactions. In Vampire Bats, masturbation occurs among younger males, while male Livingstone's Fruit Bats have been observed licking their own penises to erection. Female Vampire Bats sometimes mate with several different males in succession. In this species, a vaginal plug forms in the female's reproductive tract following copulation, which may prevent insemination from subsequent matings. When not in heat females frequently refuse to mate with males altogether, especially aggressive ones. Heterosexual relations in Livingstone's Fruit Bats are also less than amicable: females sometimes cuff males or otherwise refuse their advances, and partners may threaten, wrestle, cuff, and bite each other during actual courtship and mating. Vampire Bats have developed an alternative form of parenting behavior in their female groups known as FOOD SHARING: females sometimes help each other feed infants by regurgitating blood for young that are not their own.

Other Species

Male Serotine Bats (*Eptesicus serotinus*), a Eurasian species, have been observed making sexual advances toward other males in captivity. While suspended upside down, one male approaches another with his penis erect and mounts him from behind, grasping him above the neck and thrusting his penis between the other male's legs (under the membrane that stretches between his limbs). Homosexual activity in several species of British Bats is also common among wild males during the spring and summer (i.e., outside of the breeding season). These include Noctules (*Nyctalus noctula*), Common Pipistrelles (*Pipistrellus pipistrellus*), Brown Long-eared Bats (*Plecotus auritus*), Daubenton's Bats (*Myotis daubentonii*), and Natterer's Bats (*Myotis nattereri*) (including interspecies encounters between the latter two). Among wild Little Brown Bats (*Myotis lucifugus*) in North America, males often mount other males (as well as females) during the late fall, when many of the mounted individuals are semitorpid. These same-sex copulations usually include ejaculation, and the mounted animal often makes a squawking vocalization.

Homosexual behavior also occurs in several other species of Fruit Bats: male Rodrigues Fruit Bats (*Pteropus rodricensis*) participate in same-sex mounting, while younger male Indian Fruit Bats (*Pteropus giganteus*) often mount one another (with erections and thrusting) while play-wrestling.

Sources   *asterisked references discuss homosexuality/transgender*

*Barclay, R. M. R., and D. W. Thomas (1979) "Copulation Call of *Myotis lucifugus*: A Discrete Situation-Specific Communication Signal." *Journal of Mammalogy* 60:632–34.

*Courts, S. E. (1996) "An Ethogram of Captive Livingstone's Fruit Bats *Pteropus livingstonii* in a New Enclosure at Jersey Wildlife Preservation Trust." *Dodo* 32:15–37.

DeNault, L. K., and D. A. McFarlane (1995) "Reciprocal Altruism Between Male Vampire Bats, *Desmodus rotundus*." *Animal Behavior* 49:855–56.

*Greenhall, A. M. (1965) "Notes on the Behavior of Captive Vampire Bats." *Mammalia* 29:441–51.

Martin, L., J. H. Kennedy, L. Little, H. C. Luckhoff, G. M. O'Brien, C. S. T. Pow, P. A. Towers, A. K. Waldon, and D. Y. Wang (1995) "The Reproductive Biology of Australian Flying-Foxes (Genus *Pteropus*)." In P. A. Racey and S. M. Swift, eds., *Ecology, Evolution, and Behavior of Bats*, pp. 167–84. Oxford: Clarendon Press.

*Nelson, J. E. (1965) "Behavior of Australian Pteropodidae (Megachiroptera)." *Animal Behavior* 13:544–57.

*——— (1964) "Vocal Communication in Australian Flying Foxes (Pteropodidae; Megachiroptera)." *Zeitschrift für Tierpsychologie* 21:857–70.

*Neuweiler, V. G. (1969) "Verhaltensbeobachtungen an einer indischen Flughundkolonie (*Pteropus g. giganteus* Brünn) [Behavioral Observations on a Colony of Indian Fruit-Bats]." *Zeitschrift für Tierpsychologie* 26:166–99.

*Rollinat, R., and E. Trouessart (1896) "Sur la reproduction des chauves-souris [On the Reproduction of Bats]." *Mémoires de la Société Zoologique de France* 9:214–40.

*——— (1895) "Deuxième note sur la reproduction des Chiroptères [Second Note on the Reproduction of the Chiroptera]." *Comptes Rendus Hebdomadaires des Séances et Mémoires de la Société de Biologie* 47:534–36.

Schmidt, C. (1988) "Reproduction." In A. M. Greenhall and U. Schmidt, eds., *Natural History of Vampire Bats*, pp. 99–109. Boca Raton, Fla.: CRC Press.

*Thomas, D. W., M. B. Fenton, and R. M. R. Barclay (1979) "Social Behavior of the Little Brown Bat, *Myotis lucifugus*. I. Mating Behavior. II. Vocal Communication." *Behavioral Ecology and Sociobiology* 6:129–46.

Trewhella, W. J., P. F. Reason, J. G. Davies, and S. Wray (1995) "Observations on the Timing of Reproduction in the Congeneric Cómoro Island Fruit Bats, *Pteropus livingstonii* and *P. seychellensis comorensis*." *Journal of Zoology, London* 236:327–31.

Turner, D. C. (1975) *The Vampire Bat: A Field Study in Behavior and Ecology*. Baltimore: Johns Hopkins University Press.

*Vesey-Fitzgerald, B. (1949) *British Bats*. London: Methuen.

Wilkinson, G. S. (1988) "Social Organization and Behavior." In A. M. Greenhall and U. Schmidt, eds., *Natural History of Vampire Bats*, pp. 85–97. Boca Raton, Fla.: CRC Press.

*——— (1985) "The Social Organization of the Common Vampire Bat. I. Pattern and Cause of Association. II. Mating System, Genetic Structure, and Relatedness." *Behavioral Ecology and Sociobiology* 17:111–34.

——— (1984) "Reciprocal Food Sharing in the Vampire Bat." *Nature* 308:181–84.

Birds

Waterfowl and Other Aquatic Birds

GEESE, SWANS, AND DUCKS

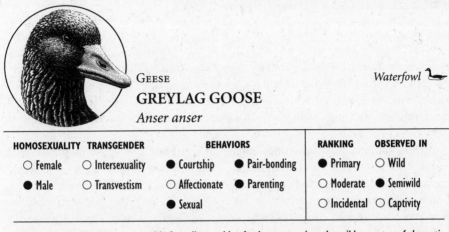

GEESE

GREYLAG GOOSE
Anser anser

Waterfowl

HOMOSEXUALITY	TRANSGENDER	BEHAVIORS		RANKING	OBSERVED IN
○ Female	○ Intersexuality	● Courtship	● Pair-bonding	● Primary	○ Wild
● Male	○ Transvestism	○ Affectionate	● Parenting	○ Moderate	● Semiwild
		● Sexual		○ Incidental	○ Captivity

IDENTIFICATION: A dark gray goose with fine silvery-white feather patterning; the wild ancestor of domestic geese. DISTRIBUTION: Northern and central Eurasia, from Iceland to northeast China. HABITAT: Variable, including marshes, swamps, lakes, lagoons. STUDY AREAS: Konrad-Lorenz Institute, Grünau, Austria; Max-Planck Institute, Seewiesen, Germany; Wörlitzer Park, Dessau, Germany; subspecies *A.a. anser*, the Western Greylag Goose.

Social Organization

Greylag Geese usually associate in flocks containing a complex mixture of pairs, families with offspring, single birds, and subgroups of juveniles. Following the breeding season, migratory flocks sometimes contain thousands of birds. The mating system generally involves long-term, monogamous pair-bonding.

Description

Behavioral Expression: Homosexual pairs made up of two ganders are a prominent form of pair-bonding in Greylag Geese. Male couples are stable and long-lasting: some have been documented as persisting for more than fifteen years, and most homosexual pairs (like heterosexual ones) are probably lifelong partnerships (Greylag Geese can live to be more than 20 years old). "Widower" ganders may even exhibit signs of "grief," becoming despondent and defenseless upon the loss of their male partner. Most heterosexual pairs are also lifelong (and partners

▲ *Two pair-bonded male Greylag Geese performing the "triumph ceremony"*

grieve the loss of their mates), but in many cases gander pairs are actually more closely bonded than male-female pairs, due in part to the intensity of their displays. One of these is the TRIUMPH CEREMONY, a pair-bonding behavior in which the two partners approach each other with extended necks and spread wings while making loud gabbling calls. Gander pairs spend significantly more time in this activity than do heterosexual pairs. They are also generally more vocal than male-female pairs—they often utter PRESSED CACKLING calls (rapid syllables produced with a high-pressure airstream) together in a cheek-to-cheek position and may even perform extended duets with ROLLING calls (deeper and louder notes).

Paired ganders also sometimes engage in courtship and sexual behavior with one another. Pair-bonding is often initiated with the BENT-NECK DISPLAY, in which one male approaches and follows another with a distinct "kink" in his neck, his bill pointing downward. As a prelude to mating, both males perform aquatic displays such as NECK-DIPPING or NECK-ARCHING, in which the head is dipped below the surface while the neck is held in an elegant curve, its feathers ruffled to reveal their distinctive patterning. Following these displays, one male may mount the other as in heterosexual copulation. If there is a size difference between the two males, often the larger male mounts the smaller one. If the two ganders are equal in size, either bird may mount the other, and they often exchange positions when they copulate on different days. Following mating, the male who mounted his partner performs a display in which he lifts his head up and arches his folded wings almost vertically above his back. Sometimes, during homosexual activity one male may "masturbate" by mounting a log or some other object (a common form of masturbation in birds). In addition, a third bird—either male or female—occasionally joins a homosexual pair in their courtship activities, and may even be mounted by one of the ganders. In all cases, though, the

A Greylag gander mounting his male partner ▶

concluding display takes place between the members of the male pair rather than with the third bird. Some gander pairs do not regularly engage in full mounting behavior, in part because both males prefer to mount each other without either one permitting himself to be mounted.

Gander pairs often assume a powerful, high-ranking position within their flock, owing to their superior strength and courage. They are notably more aggressive than heterosexual pairs, frequently threatening, charging, chasing, and jointly attacking predators as well as other geese (especially unpaired males) and often appear to "terrorize" other birds. Paradoxically, each individual gander in a homosexual pair is significantly less aggressive than a male in a heterosexual pair—it is their combined strength that gives the couple its advantage. Homosexual pairs also differ from heterosexual ones in spending significantly more time on the periphery of the flock or away from it, especially during the spring breeding season. This, combined with the gander pair's greater vigilance behavior (as well as the pair's aggressiveness), has led some researchers to suggest that homosexual pairs may act as "guardians" for the flock as a whole. Sometimes a female is attracted to a gander pair—perhaps because of their strength and high standing—and tries to establish a bond with one or both of them. Often the males simply ignore such a female, but in some cases she is allowed to join them to form a trio. When this happens, one or both ganders may copulate with the female, although their homosexual bond usually remains primary. The trio may raise a family together, with the two ganders often searching for a nest site together and jointly defending their eggs and goslings. Occasionally, three ganders bond with each other as a same-sex trio, which may also later be joined by a female to form a "quartet"; again, goslings can be raised by all four birds together.

Although most gander pairs are stable partnerships, occasionally one or both birds may behave antagonistically toward his partner. Fights sometimes erupt when one male tries to mount the other, while occasionally the aggression aimed at an intruder is turned back on one of the partners. Bonded ganders (especially in trios) may also become "jealous" of the attentions their partner shows toward another bird. Some gander pairs are incestuous, as brothers may form long-term homosexual bonds. In addition, interspecies same-sex pairs also occur, for example between Greylag Geese and Mute Swans. Like Greylag-only gander pairs, these partnerships are long-lasting and distinguish themselves by their aggressiveness, with the two males frequently defending their territory against intruders.

Frequency: Homosexual couples constitute a significant proportion of pairs in Greylag Geese: an average of 14 percent of pairs in some populations are same-sex, and in some years this proportion can be even higher, with more than 20 percent of all pair-bonds consisting of ganders.

Orientation: Some Greylag males in gander pairs are exclusively homosexual, since they remain in a monogamous same-sex pair-bond for their entire lives (or re-pair with another gander on the death of their partner). Other males, however, are bisexual: some copulate with a female while remaining primarily bonded to a male (as described above), while others are involved in bisexual trios. Still other males alternate or switch between female and male partners over their lives—for example, ganders in heterosexual pairs sometimes find a male partner after the death of their mate. More than half of all widowers re-pair with a bird of the opposite sex, less than a third remain single, while the remainder form homosexual bonds.

Nonreproductive and Alternative Heterosexualities

Several variations on the monogamous, lifelong pair-bond occur in this species. Divorce occasionally happens: in some populations as many as a quarter of all females, for example, may abandon their mates and find a new gander, and overall, 5–8 percent of pairs divorce. Greylag Geese also sometimes form polygamous heterosexual trios, in which bonding occurs primarily between birds of the opposite sex—two males with a female or, more rarely, two females with a male. In addition, some families foster-parent chicks by combining broods with another family, while widowed ganders occasionally adopt goslings. Birds in heterosexual pairs may engage in promiscuous courtship and mating. Ganders sometimes try to mount females other than their mate, while females may pursue other males—much to the consternation of their mates, who often try to physically prevent them from engaging in "extramarital" activities. Although Greylag Geese become sexually mature in their third year, some one-year-olds form pair-bonds and even engage in courtship and sexual activity long before they begin breeding. Like homosexual pairs, heterosexual associations may also occur between related birds (especially parent-offspring), or birds of different species (e.g., with Canada Geese). However, sibling pairings are much less common among birds of the opposite sex.

Sources   *asterisked references discuss homosexuality/transgender*

Ens, B. J., S. Choudhury, and J. M. Black (1996) "Mate Fidelity and Divorce in Monogamous Birds." In J. M. Black, ed., *Partnerships in Birds: The Study of Monogamy*, pp. 344–401. Oxford: Oxford University Press.

Huber, R. (1988) "Sex-Specific Behavior in Greylag Geese, *Anser anser* L." *Texas Journal of Science* 40:107–9.

*Huber, R., and M. Martys (1993) "Male-Male Pairs in Greylag Geese (*Anser anser*)." *Journal für Ornithologie* 134:155–64.

*Lorenz, K. (1991) *Here Am I—Where Are You? The Behavior of the Greylag Goose*. New York: Harcourt Brace Jovanovich.

*——— (1979) *The Year of the Greylag Goose*. New York: Harcourt Brace Jovanovich.

Olsson, H. (1978) "Probable Polygamy in the Greylag Goose, *Anser anser*, and an Instance of Combined Broods." *Vår Fågelvärld* 37:257–58.

*Schönfeld, M. (1985) "Beitrag zur Biologie der Schwäne: 'Männchenpaar' zwischen Graugans und Höckerschwan [Contribution to the Biology of Swans: 'Male Pairing' Between a Greylag Goose and a Mute Swan]." *Der Falke* 32:208.

GEESE

Waterfowl

CANADA GOOSE
Branta canadensis

SNOW GOOSE
Anser caerulescens

HOMOSEXUALITY	TRANSGENDER	BEHAVIORS		RANKING	OBSERVED IN
● Female	○ Intersexuality	● Courtship	● Pair-bonding	○ Primary	● Wild
● Male	○ Transvestism	○ Affectionate	● Parenting	● Moderate	● Semiwild
		● Sexual		○ Incidental	● Captivity

CANADA GOOSE
IDENTIFICATION: A brown-plumaged goose with a distinctive black neck and white cheek patch; varies widely in size, from 2–24 pounds. DISTRIBUTION: Mostly throughout North Amerca. HABITAT: Lakes, rivers, marshes, meadows, and tundra. STUDY AREAS: Horicon Marsh Wildlife Refuge, Wisconsin; Holkham Park, England; in captivity in Ithaca, New York; subspecies *B.c. interior*, the Hudson Bay Canada Goose, and *B.c. canadensis*, the Atlantic Canada Goose.

SNOW GOOSE
IDENTIFICATION: A pinkish-red-billed goose with two major color phases: all-white and "blue" (grayish plumage with a white head and neck). DISTRIBUTION: Alaska and north-central Canada, northwestern Greenland; winters in southern United States and northern Mexico. HABITAT: Tundra, marshes, floodlands. STUDY AREAS: La Pérouse Bay, Churchill, Manitoba, Canada; Carver Park, Minnesota; subspecies *A.c. caerulescens*, the Lesser Snow Goose.

Social Organization

Snow Geese are extremely gregarious and nest in dense colonies that can number in the thousands of birds; Canada Goose breeding grounds are generally less dense. In both species, birds usually pair in long-term, monogamous bonds (albeit with a number of variations—see below), and outside of the mating season they gather in large flocks.

Description:

Behavioral Expression: In Canada Geese, two birds of the same sex sometimes form a pair-bond. Both male and female homosexual pairs occur, and the partners may be either adults or juveniles; homosexual pair-bonds often persist for many years (as do heterosexual ones). Courtship behavior in the form of HEAD-DIPPING is frequently a part of such bonds: in this display, one bird ritually splashes water over the back of its head and neck by dipping its head deep into the water and then lifting it back up. In heterosexual contexts, this is often a prelude to copulation, but homosexual copulation is not a prominent feature of same-sex pairs. One exception involves trios: occasionally a bond forms between three birds—two females

▲ *Courtship between a pair of female Canada Geese: the "head-dipping" display*

and one male—and sometimes one of the females will mount and copulate with the other female. Some lesbian pairs try to raise a family: one female in a homosexual pair, for example, built a nest and laid eggs while her partner stood guard, then the other female built her own nest next to the first and also laid eggs. None of the eggs hatched, however, because the females constantly rolled the eggs (which were probably not fertile in any case) back and forth between their nests and broke all of them.

More successful lesbian parenting occurs in Snow Geese. Pair-bonds between females are strong: when one member of the pair is absent from her mate, the other begins loudly calling to her until she returns. The pair builds a single nest in which each female lays eggs; as a result, such nests may have SUPERNORMAL CLUTCHES containing double the number of eggs found in heterosexual nests (8 eggs versus 4–5). Both birds take turns incubating the eggs (in heterosexual pairs, the male does not incubate). Since one or both females sometimes copulate with males, some of their eggs may be fertile. When they hatch, both females raise the goslings, including defending them against intruders and predators (such as Herring Gulls) by standing over them with cupped wings. Male homosexual pairs are not found in Snow Geese, although occasionally a cross-species pairing does develop between a male Snow Goose and a male Canada Goose. The two birds become constant companions, following one another and roosting close together, although nest-building and copulation do not usually take place. However, same-sex mounting does sometimes occur between male Snow Geese who take part in heterosexual "gang rapes." In this species, males often sexually harass females, chasing them and forcing them to copulate. In some cases, other males gather together in large "spectator" groups—sometimes containing as many as 20–80 males—to watch and perhaps even join in. Occasionally, one male mounts another male as part of the group sexual activity that ensues.

Frequency: In Canada Geese, up to 12 percent of pairs in some (semiwild) populations are homosexual. The proportion is much smaller in Snow Geese: about 1 in 200 nests belongs to a pair of females. Approximately 4 percent of all mountings during Snow Goose rape attempts are between males.

Orientation: In one study of Canada Geese, 18 percent of the males formed homosexual pair-bonds while 6–12 percent of females did. Some birds in same-sex pairs appear to "prefer" their homosexual association, even if they have the opportunity for heterosexual interactions. In one case, a male harassed a female who was part of a long-lasting lesbian pair and separated her from her companion, mating with her. However, the next year she returned to her female partner and their pair-bond resumed. On the other hand, some birds have a preference for heterosexual pairings: many males remain unpaired if there are no available females rather than forming homosexual pair-bonds with each other. In Snow Geese, females in homosexual pairs may be functionally bisexual—they sometimes copulate with up to three different males to fertilize their eggs—although their same-sex pair-bond remains primary. Males who mount other males are otherwise primarily heterosexual, since most are paired with females and the majority of their sexual interactions are probably not with males.

Nonreproductive and Alternative Heterosexualities

As mentioned above, heterosexual rape is common in Snow Geese: during some mating seasons, each female is subjected to a rape attempt every five days (on average). Females are occasionally successful in thwarting such attacks, but males who rape can be very aggressive and may attack in groups. Sometimes the female's mate can successfully chase an intruder off, but often he is not around to defend her because he is also raping another female. Significantly, most rapes are nonreproductive: more than 80 percent of all rape attempts are directed toward females that are nonfertilizable, such as incubating birds, and only about 2 percent of goslings are actually fathered this way. Rape is much less common among Canada Geese. However, ganders frequently harass and attack neighboring females when their mates are gone, often leading to abandonment of their eggs—as many as one-quarter of all nests may end up being deserted this way.

Several other variations on the heterosexual nuclear family occur in these species. Although most male-female pairs remain together for life, divorce and re-mating do occasionally take place in both Canada and Snow Geese. In addition, although most Snow Goose families remain together until the next breeding season, in some populations as many as 20 percent of the families split or break up before that time, usually because of separation of juveniles. Polygamous, heterosexual trios consisting of one male and two females sometimes form in Canada Geese (these differ from the bisexual trios described above in that the females are not bonded to each other). Some birds pair outside of their species, and Snow and Canada Geese may in fact mate with each other. CRÈCHES or combined broods—containing as many as 60 young—are sometimes found in Canada Geese, attended by one or several heterosexual pairs. In addition, families often "trade" goslings, caring for young other than their own on either a temporary or permanent basis. Up to 46 percent of Canada Goose broods and at least 13 percent of Snow Goose broods may contain adopted young, and over 60 percent of Canada broods in some populations experience a loss and/or gain of goslings due to adoption. Egg "adoption" is also common in Snow Geese because females often lay eggs in nests other than their own: 15–22 percent of all nests contain such eggs (although in some colonies more than 80 percent of nests may be affected), and more than 5 percent of all goslings are raised by a female other than their biological mother. Females that lay eggs in others' nests are often aided by their mates, who distract the nest-owning gander by acting as a decoy for him to attack, allowing the female to approach the nest and lay her eggs. Sometimes the intruding female actually helps with building or repairing the nest; for her part, the nest-owning female often actively adopts foreign eggs that have not been laid directly in the nest by rolling them into her own clutch.

Snow Goose females also occasionally "abandon" their eggs by laying in what are known as DUMP NESTS, which contain large numbers of unincubated eggs from many different females. Abandonment of nests may also be triggered by the stresses of reproduction: females can lose up to a third of their body mass while incubating, and some individuals desert their clutches or even starve on the nest as a result of such hardships. Most Snow Goose nesting colonies also have a nonbreeding flock on their peripheries. In some years, the proportion of nonbreeding adults is sizable—as much as 40 percent of the population—and occasionally an entire colony will forgo breeding (for example if the weather is particularly adverse). Many Canada Goose heterosexual pairs are nonbreeding as well: in some populations, for example, more than a quarter of all male-female pairs do not procreate, although they may copulate frequently. In fact, some nonbreeders have sexual activity rates that are almost twice as high as pairs that do reproduce.

Sources   *asterisked references discuss homosexuality/transgender

*Allen, A. A. (1934) "Sex Rhythm in the Ruffed Grouse (*Bonasa umbellus* Linn.) and Other Birds." *Auk* 51:180–99.

Ankney, C. D., and C. D. MacInnes (1978) "Nutrient Reserves and Reproductive Performance of Female Lesser Snow Geese." *Auk* 95:459–71.

*Collias, N. E., and L. R. Jahn (1959) "Social Behavior and Breeding Success in Canada Geese (*Branta canadensis*) Confined Under Semi-Natural Conditions." *Auk* 76:478–509.

*Conover, M. R. (1989) "What Are Males Good For?" *Nature* 342:624–25.

Cooke, F., and D. S. Sulzbach (1978) "Mortality, Emigration, and Separation of Mated Snow Geese." *Journal of Wildlife Management* 42:271–80.

Cooke, F., M. A. Bousfield, and A. Sadura (1981) "Mate Change and Reproductive Success in the Lesser Snow Goose." *Condor* 83:322–27.

*Diamond, J. M. (1989) "Goslings of Gay Geese." *Nature* 340:101.

Ewaschuk, E., and D. A. Boag (1972) "Factors Affecting Hatching Success of Densely Nesting Canada Geese." *Journal of Wildlife Management* 36:1097–106.

*Grether, G. F., and A. M. Weaver (1990) "What Are Sisters Good For?" *Nature* 345:392.

*Klopman, R. B. (1962) "Sexual Behavior in the Canada Goose." *Living Bird* 1:123–29.

Lank, D. B., P. Mineau, R. F. Rockwell, and F. Cooke (1989) "Intraspecific Nest Parasitism and Extra-Pair Copulation in Lesser Snow Geese." *Animal Behavior* 37:74–89.

Luekpe, K. (1984) "A Strange Goose: Canada-Snow Hybrid?" *Passenger Pigeon* 46:92.

MacInnes, C. D., R. A. Davis, R. N. Jones, B. C. Lieff, and A. J. Pakulak (1974) "Reproductive Efficiency of McConnell River Small Canada Geese." *Journal of Wildlife Management* 38:686–707.

Martin, K., F. G. Cooch, R. F. Rockwell, and F. Cooke (1985) "Reproductive Performance in Lesser Snow Geese: Are Two Parents Essential?" *Behavioral Ecology and Sociobiology* 17:257–63.

*Mineau, P., and F. Cooke (1979) "Rape in the Lesser Snow Goose." *Behavior* 70:280–91.

Nastase, A. J., and D. A. Sherry (1997) "Effect of Brood Mixing on Location and Survivorship of Juvenile Canada Geese." *Animal Behavior* 54:503–7.

Prevett, J. P. and C. D. MacInnes (1980) "Family and Other Social Groups in Snow Geese." *Wildlife Monographs* 71:1–46.

*Quinn, T. W., J. C. Davies, F. Cooke, and B. N. White (1989) "Genetic Analysis of Offspring of a Female-Female Pair in the Lesser Snow Goose (*Chen c. caerulescens*)." *Auk* 106:177–84.

*Starkey, E. E. (1972) "A Case of Interspecific Homosexuality in Geese." *Auk* 89:456–57.

Syroechkovsky, E. V. (1979) "Podkladyvaniye byelymi gusyami yaits v chuzhiye gnyezda [The Laying of Eggs by White Geese into Strange Nests]." *Zoologichesky Zhurnal* 58:1033–41.

Williams, T. D. (1994) "Adoption in a Precocial Species, the Lesser Snow Goose: Intergenerational Conflict, Altruism, or a Mutually Beneficial Strategy?" *Animal Behavior* 47:101–7.

Zicus, M. C. (1984) "Pair Separation in Canada Geese." *Wilson Bulletin* 96:129–30.

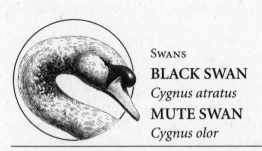

SWANS

Waterfowl

BLACK SWAN
Cygnus atratus
MUTE SWAN
Cygnus olor

HOMOSEXUALITY	TRANSGENDER	BEHAVIORS		RANKING	OBSERVED IN
● Female	○ Intersexuality	● Courtship	● Pair-bonding	● Primary	● Wild
● Male	○ Transvestism	○ Affectionate	● Parenting	○ Moderate	● Semiwild
		● Sexual		○ Incidental	● Captivity

BLACK SWAN
IDENTIFICATION: The only swan with fully black plumage; wing feathers are white, bill is bright red, and the neck is especially long. DISTRIBUTION: Australia, Tasmania, New Zealand. HABITAT: Lakes, lagoons, swamps, bays, floodlands. STUDY AREAS: Lake George and Lake Bathurst, New South Wales, Australia; in captivity at the Division of Wildlife Research, Canberra, Australia.

MUTE SWAN
IDENTIFICATION: A large swan (up to 33 pounds) with a black knob at the base of its reddish orange bill (less prominent in females). DISTRIBUTION: Europe and temperate Asia. HABITAT: Marshes, ponds, lakes, slow-moving rivers, lagoons, coastal areas. STUDY AREAS: Abbotsbury (Dorset) and Rainworth Lodge (Notts), England; Renfrewshire, Scotland.

Social Organization
Black Swans sometimes flock by the thousands and usually form mated pairs (although with numerous variations—see below) that nest either colonially or in separate territories. Mute Swans also generally develop long-term, monogamous

bonds and nest in widely spaced territories, although some pairs form nesting colonies. Outside of the breeding season, they often associate in flocks.

Description

Behavioral Expression: Some male Black Swans form stable, long-lasting homosexual pairs. Like heterosexual mates, same-sex partners often remain together for many years. The two males frequently perform the GREETING CEREMONY, a pair-bonding display that helps solidify and reinforce their partnership: the birds face one another, raise their wings (sometimes flapping them to expose the white feathers), and call repeatedly while extending their necks and lifting their bills up. Males in homosexual pairs also perform a courtship behavior known as HEAD-DIPPING. In this display—a prelude to copulation—the two birds repeatedly immerse first the head, then the neck, and finally the body in a wavelike fashion, sometimes for extended periods of 20–25 minutes. This can lead to homosexual mounting, although if one male does not want to participate in sexual activity he may respond aggressively to his partner's overtures.

Male pairs of Black Swans fiercely defend territories that, during the mating season, are often significantly larger than those of heterosexual pairs. Because two males are able to pool their strength, they are more successful at chasing away other swans and can often annex a major portion of a pond (1,500–3,300 square feet) into their territory. In contrast, heterosexual pairs are often relegated to less favorable nesting areas and smaller territories (15–60 square feet). Homosexual pairs are also successful parents, acquiring nests and eggs in two different ways. Some male pairs associate temporarily with a female, building a nest together, mating with her, and then chasing her away once the eggs are laid, after which they begin parenting as a male couple. Other homosexual pairs chase heterosexual pairs from their nests and adopt the eggs that have already been laid. The two males incubate the clutch, hatch the eggs, and raise the chicks together. In fact, homosexual pairs are often more successful than heterosexual ones at raising chicks, in part because they have access to the best nesting sites and the largest territories, and probably because they also share incubation duties more equally. On average, 80 per-

◀ *A homosexual pair of male Black Swans performing the "greeting ceremony"*

cent of homosexual parenting efforts are successful, compared to only about 30 percent of heterosexual ones.

Both male and female homosexual pairs occur in Mute Swans. In female pairs, both birds build a nest, lay eggs (which are usually infertile), and incubate the eggs. Sometimes one female stands guard over both the nest and her mate (as does the male in heterosexual pairs) and defends their territory. If their nest is disturbed by intruders, the females may begin a second nest and lay a new clutch of eggs, while still attending to the first as best they can. Male pairs also annually build nests together on which they take turns sitting, although unlike Black Swans they do not acquire any eggs. Male Mute Swans also sometimes form homosexual pair-bonds with other species, including Trumpeter Swans (*Cygnus buccinator*) and Greylag Geese.

Frequency: Overall, male couples constitute 5–6 percent of all pairings in Black Swans; in a given year, an average of 13 percent of the birds are in homosexual pair-bonds, and sometimes this proportion is as high as 20 percent. Homosexual parents account for 20–25 percent of all successful families. Same-sex bonds probably occur only sporadically in Mute Swans.

Orientation: Many Black and Mute Swans in same-sex pairs are probably exclusively homosexual, since they do not engage in heterosexual copulation or pairing and usually ignore unpaired birds of the opposite sex. However, some male pairs of Black Swans—while primarily homosexual—form short-lived bisexual trios in order to mate with females and thereby father their own offspring.

Nonreproductive and Alternative Heterosexualities

Populations of both Mute and Black Swans contain large proportions of nonbreeding birds. More than half of all Mute Swans are nonbreeders (as much as 89 percent of some populations), often gathering into their own flocks separate from breeding pairs. Many birds nest only once or twice during their entire lifetimes (which can last for 15–20 years), and a few never breed. Overall, only about a fifth of all Black Swans nest in any year, and in some populations more than 90 percent of the adults do not breed. Young, sexually mature Black Swans may remain with their parents and delay their breeding for many years (as long as three to eight years in some cases). On occasion, such a youngster will form an incestuous pair-bond with its parent: male Swans have been known to mate with their mother on the death of their father. Brother-sister and parent-offspring matings also occur in Mute Swans, as do heterosexual pairings with other species of swans (such as bewick's, whistling, whooper, and Trumpeter) and geese (e.g., Snow, Canada, and Greylag). In fact, Black and Mute Swans may pair with each other, and trios of a male Black with two female Mute Swans have also been observed. Heterosexual trios within the same species are also common: about 14 percent of all bonds in Black Swans involve two males with a female, while Mute Swan trios are usually made up of two females with a male.

In addition to such polygamous associations, several other alternative family arrangements occur. "Foster parenting" or adoption takes place frequently among Black Swans (and occasionally in Mute Swans). In some colonies, more than two-thirds of all cygnets are raised in broods that combine offspring from 2–4 families (and occasionally from as many as 30 different families). Such BROOD AMALGAMA-TIONS—which may have up to 40 youngsters—are attended by a single pair of adults, who are not necessarily the biological parents of any of the cygnets. Adoption also occasionally occurs when adults "steal" eggs laid near their nest by other birds, rolling them into their own nest. Single parenting is a prominent feature of Black Swan social life: often a male or female deserts its mate during incubation, and in some colonies the majority of nests are attended by single parents. Occasionally, a pair "separates" rather than divorces, with one bird taking the newly hatched young while the other remains to incubate the rest of the eggs. Among Mute Swans, the divorce rate is 3–10 percent of all pairs, and about a fifth of all birds have two to four mates during their lifetime. Some Mute Swans are non-monogamous, courting or mating with another bird while remaining paired with their partner; some of this activity may involve REVERSE copulations (in which the female mounts the male). Many within-pair copulations are nonprocreative, since most pairs mate far more often than is required for fertilization of the eggs. Swans also sometimes engage in behaviors that are counterreproductive. A third of all Black Swan eggs, for example, are lost through abandonment of the nest by the parent(s), while 3 percent of Mute Swan parents desert their nests, and birds often attack and even kill youngsters that stray into their territory. Eggs are sometimes also destroyed during territorial disputes, and adult birds may be killed as a direct result of such attacks as well (accounting for 3 percent of all deaths).

Sources   *asterisked references discuss homosexuality/transgender*

Bacon, P. J., and P. Andersen-Harild (1989) "Mute Swan." In I. Newton, ed., *Lifetime Reproduction in Birds*, pp. 363–86. London: Academic Press.

Braithwaite, L. W. (1982) "Ecological Studies of the Black Swan. IV. The Timing and Success of Breeding on Two Nearby Lakes on the Southern Tablelands of New South Wales." *Australian Wildlife Research* 9:261–75.

*——— (1981) "Ecological Studies of the Black Swan. III. Behavior and Social Organization." *Australian Wildlife Research* 8:135–46.

*——— (1970) "The Black Swan." *Australian Natural History* 16:375–79.

Brugger, C., and M. Taborsky (1994) "Male Incubation and Its Effect on Reproductive Success in the Black Swan, *Cygnus atratus*." *Ethology* 96:138–46.

Ciaranca, M. A., C. C. Allin, and G. S. Jones (1997) "Mute Swan (*Cygnus olor*)." In A. Poole and F. Gill, eds., *The Birds of North America: Life Histories for the 21st Century*, no. 273. Philadelphia: Academy of Natural Sciences; Washington, D.C.: American Ornithologists' Union.

Dewer, J. M. (1942) "Ménage à Trois in the Mute Swan." *British Birds* 30:178.

Huxley, J. S. (1947) "Display of the Mute Swan." *British Birds* 40:130–34.

*Kear, J. (1972) "Reproduction and Family Life." In P. Scott, ed., *The Swans*, pp. 79–124. Boston: Houghton Mifflin.

*Low, G. C. and Marquess of Tavistock (1935) "The Extent to Which Captivity Modifies the Habits of Birds." *Bulletin of the British Ornithologists' Club* 55:144–54.

Mathiasson, S. (1987) "Parents, Children, and Grandchildren—Maturity Process, Reproduction Strategy, and Migratory Behavior of Three Generations and Two Year-Classes of Mute Swans *Cygnus olor*." In M. O. G. Eriksson, ed., *Proceedings of the Fifth Nordic Ornithological Congress, 1985*, pp. 60–70. Acta Regiae Societatis Scientiarum et Litterarum Gothoburgensis Zoologica no. 14. Göteborg: Kungl. Vetenskaps- och Vitterhets-Samhället.

Minton, C. D. T. (1968) "Pairing and Breeding of Mute Swans." *Wildfowl* 19:41–60.

*O'Brien, R. M. (1990) "Black Swan, *Cygnus atratus*." In S. Marchant and P. J. Higgins, eds., *Handbook of Australian, New Zealand, and Antarctic Birds,* vol. 1, pp. 1178–89. Melbourne: Oxford University Press.

Ogilvie, M. A. (1972) "Distribution, Numbers, and Migration." In P. Scott, ed., *The Swans,* pp. 29–55. Boston: Houghton Mifflin.

Rees, E. C., P. Lievesley, R. A. Pettifor, and C. Perrins (1996) "Mate Fidelity in Swans: An Interspecific Comparison." In J. M. Black, ed., *Partnerships in Birds: The Study of Monogamy,* pp. 118–37. Oxford: Oxford University Press.

*Ritchie, J. P. (1926) "Nesting of Two Male Swans." *Scottish Naturalist* 159:95.

*Schönfeld, M. (1985) "Beitrag zur Biologie der Schwäne: 'Männchenpaar' zwischen Graugans und Höckerschwan [Contribution to the Biology of Swans: 'Male Pairing' Between a Greylag Goose and a Mute Swan]." *Der Falke* 32:208.

Sears, J. (1992) "Extra-Pair Copulation by Breeding Male Mute Swan." *British Birds* 85:558–59.

*Whitaker, J. (1885) "Swans' Nests." *The Zoologist* 9:263–64.

Williams, M. (1981) "The Demography of New Zealand's *Cygnus atratus* Population." In G. V. T. Matthews and M. Smart, eds., *Proceedings of the 2nd International Swan Symposium (Sapporo, Japan),* pp. 147–61. Slimbridge: International Waterfowl Research Bureau.

DUCKS   *Waterfowl* 🦆

MALLARD DUCK
Anas platyrhynchos

BLUE-WINGED TEAL
Anas discors

HOMOSEXUALITY	TRANSGENDER	BEHAVIORS		RANKING	OBSERVED IN
● Female	○ Intersexuality	● Courtship	● Pair-bonding	○ Primary	● Wild
● Male	○ Transvestism	○ Affectionate	● Parenting	● Moderate	● Semiwild
		● Sexual		○ Incidental	● Captivity

MALLARD DUCK
IDENTIFICATION: A familiar duck with a blue wing patch, an iridescent green head and white collar in males, and brown, mottled plumage in females. DISTRIBUTION: Throughout the Northern Hemisphere; Australia and New Zealand. HABITAT: Wetlands. STUDY AREAS: J. Rulon Miller Wildlife Refuge, McDonogh, New Jersey; Haren and Middleburg, the Netherlands; Augsburg, Germany, and the Max-Planck Institute, Seewiesen, Germany; Delta Waterfowl Research Station, Lake Manitoba, Canada; subspecies *A.p. platyrhynchos,* the Common Mallard.

BLUE-WINGED TEAL
IDENTIFICATION: A grayish brown duck with a light blue upper-wing patch, tawny spotted underparts, and white, crescent-shaped facial stripes in males. DISTRIBUTION: Northern and central North America; winters in Central America and northern South America. HABITAT: Marshes, lakes, streams. STUDY AREA: Delta Waterfowl Research Station, Lake Manitoba, Canada.

Social Organization

Mallard Ducks and Blue-winged Teals are highly sociable birds, usually congregating in their own flocks of hundreds or (in Mallards) even thousands for most of the year. During the breeding season, they typically form monogamous pairs, although many variations exist. As in many other duck species, heterosexual pairs usually separate soon after incubation begins. Females then incubate the eggs and raise their families on their own.

Description

Behavioral Expression: Female Mallards sometimes mount and copulate with other females in the early fall, when ducks congregate in groups and begin to establish pair-bonds. Two females may engage in the PUMPING display, a prelude or invitation to mating in which the head is bobbed up and down so that the bill touches the water in a horizontal position. Following this, one female flattens her body on the water and extends her neck, allowing the other female to mount. While copulating, the mounting female may grab her partner's neck feathers in her bill or gently peck at her head. After dismounting, she performs a concluding display (also shown by females in heterosexual interactions) in which she dips her head in the water and then shakes the drops down her back while beating her wings. Homosexual mountings occasionally occur later in the season between heterosexually paired females and single females.

Homosexual pair-bonds also occur in both male and female Mallards. As in heterosexual pairs, the two partners keep close company, swimming together as well as resting, preening, and feeding in perfect synchrony. Same-sex partners also "defend" their mate from the approach of other Mallards. Females use a special INCITING display for this, in which they trail their partner while looking back over their shoulder and making a trembling call. Overt sexual activity is not generally a feature of same-sex partnerships, however: drake pairs, for example, engage in mutual head pumping and feather ruffling (which are preludes to copulation) but neither partner mounts the other or invites his mate to mount him. Interestingly, though, some males in homosexual pairs have been observed attempting to rape or forcibly copulate with males outside their pair-bond (just the way drakes in heterosexual pairs often participate in nonmonogamous raping of females—see below). Among females, homosexual pair-bonds are more ephemeral, generally occurring only in the pre- and postbreeding seasons.

◀ A homosexual pair of male Mallard Ducks engaging in synchronized preening

▲ *A flock, or "club," of homosexual Mallard drakes*

Some drake pairs are also temporary, while others are long-lasting, persisting for years and possibly even for life.

Male Mallards that have been raised together also frequently develop homosexual bonds of great strength and longevity. When large numbers of such birds are present, they often form their own groups, known as CLUBS. They flock together for hours or even days at a time, excitedly running about and swimming together while quacking continuously. Sometimes a female associates herself with a drake pair to form a bisexual trio; although one or both males may mate with her, their homosexual bond remains primary. Less commonly, females that have been raised together may also form a pair-bond, jointly incubating a nest and coparenting any ducklings that may result from promiscuous matings with males.

Blue-winged Teal drakes will court each other in the absence of females, even competing and fighting with one another for the attentions of another male.

Frequency: Homosexual copulations and pairings between female Mallard Ducks occur sporadically and are most common during the fall. In one study, roughly a quarter of the days on which sexual activity was observed included same-sex mountings. The proportion of male homosexual pairs varies between populations, anywhere from 2–19 percent of all pairs.

Orientation: Several forms of bisexuality occur among Mallards: females may participate in homosexual copulations while paired with a male, and both sexes may form seasonal homosexual attachments prior to or following a period of heterosexual mating. Some males are probably more exclusively homosexual, forming ongoing same-sex bonds that last for many years. In addition, most males probably have a bisexual potential: when raised in all-male groups, Mallards usually form lifelong homosexual pairs, and re-pair with other males on being "widowed." Nevertheless, even among males that have not been raised together, approximately 13–17 percent still participate in homosexual pairing for at least a portion of their lives. In Blue-winged Teals, homosexual behavior appears to be primarily a manifestation of a bisexual potential, since same-sex pairing or courtship have so far only been observed in males isolated from females.

Nonreproductive and Alternative Heterosexualities

Mallard pairs regularly engage in nonreproductive matings. For example, copulation is common during the five-to-seven months that heterosexual couples are together prior to the breeding season (when males are not producing sperm). Later in the breeding season, however, male-female relations are often marked by hostility, since forced copulation or rape is a common feature of opposite-sex interactions in both Mallards and Blue-winged Teals. Following egg-laying, male Mallards regularly abandon their female mates (who thus become "single parents"), congregate in all-male groups, and begin pursuing other females to try to forcibly mate with them. Rapes also occur between paired ducks in both species. As many as a 12–40 males may chase a single female in aerial or aquatic pursuit; drakes have even been known to grab and mount females underwater when they dive (attempting to escape), or to knock females to the ground in midflight. In some populations, as many as 7–10 percent of all females die each year as a result of drownings or other injuries incurred during rapes. Occasionally, males even try to mate with dead females. Even while they are still paired earlier in the breeding season, males frequently court and attempt to mate with (or rape) females other than their mate. About 3–7 percent of offspring are a result of such nonmonogamous matings, and in some populations multiple parentage occurs in at least 17–25 percent of all broods.

Mallards also sometimes form trio-bonds, either one male with two females (2–4 percent of all heterosexual bonds) or, more commonly, two males with one female (3–6 percent of all bonds). Paired males sometimes switch mates during the breeding season as well, and at least 9 percent of all heterosexual couples divorce between breeding seasons. Overall, long-term male-female pair-bonds (lasting two or more seasons) are rare in this species. Mallard mothers can be extremely aggressive in defense of their young, even killing other youngsters that stray from their own broods. In some populations the greatest cause of mortality among ducklings is attacks from other mothers. Occasionally, however, two broods join together and are defended by a single mother for short periods.

Other Species

Homosexual pairs also form among male Wood Ducks (*Aix sponsa*) that are raised together; such pairs are lifelong, and the two males may even search for nesting sites together each year. Female Chiloe Wigeons (*Anas sibilatrix*) have also been known to pair together in captivity; the partners remain bonded for many years and each lays eggs in their nest.

Sources   *asterisked references discuss homosexuality/transgender

Bailey, R. O., N. R. Seymour, and G. R. Stewart (1978) "Rape Behavior in Blue-winged Teal." *Auk* 95:188–90.

Barash, D. P. (1977) "Sociobiology of Rape in Mallards (*Anas platyrhynchos*): Responses of the Mated Male." *Science* 197:788–89.

Boos, J. D., T. D. Nudds, and K. Sjöberg (1989) "Posthatch Brood Amalgamation by Mallards." *Wilson Bulletin* 101:503–5.

*Bossema, I., and E. Roemers (1985) "Mating Strategy, Including Mate Choice, in Mallards." *Ardea* 73:147–57.

Cheng, K. M., J. T. Burns, and F. McKinney (1983) "Forced Copulation in Captive Mallards. III. Sperm Competition." *Auk* 100:302–10.

Evarts, S., and C. J. Williams (1987) "Multiple Paternity in a Wild Population of Mallards." *Auk* 104:597–602.

*Geh, G. (1987) "Schein-Kopula bei Weibchen der Stockente *Anas platyrhynchos* [Pseudo-Copulation of Female Mallard Ducks]." *Anzeiger der Ornithologischen Gesellschaft in Bayern* 26:131–32.

*Hochbaum, H. A. (1944) *The Canvasback on a Prairie Marsh*. Washington, D.C.: American Wildlife Institute.

Huxley, J. S. (1912) "A 'Disharmony' in the Reproductive Habits of the Wild Duck (*Anas boschas* L.)." *Biologisches Centralblatt* 32:621–23.

*Lebret, T. (1961) "The Pair Formation in the Annual Cycle of the Mallard, *Anas platyrhynchos* L." *Ardea* 49:97–157.

*Lorenz, K. (1991) *Here Am I—Where Are You? The Behavior of the Greylag Goose*. New York: Harcourt Brace Jovanovich.

*——— (1935) "Der Kumpan in der Umwelt des Vögels." *Journal für Ornithologie* 83:10–213, 289–413. Reprinted as "Companions as Factors in the Bird's Environment." In K. Lorenz (1970), *Studies in Animal and Human Behavior*, vol. 1, pp. 101–258. Cambridge, Mass.: Harvard University Press.

Losito, M. P., and G. A. Baldassarre (1996) "Pair-bond Dissolution in Mallards." *Auk* 113:692–95.

McKinney, F., S. R. Derrickson, and P. Minneau (1983) "Forced Copulation in Waterfowl." *Behavior* 86:250–94.

Mjelstad, H., and M. Sætersdal (1990) "Reforming of Resident Mallard Pairs *Anas platyrhynchos,* Rule Rather Than Exception?" *Wildfowl* 41:150–51.

Raitasuo, K. (1964) "Social Behavior of the Mallard, *Anas platyrhynchos,* in the Course of the Annual Cycle." *Papers on Game Research (Helsinki)* 24:1–72.

*Ramsay, A. O. (1956) "Seasonal Patterns in the Epigamic Displays of Some Surface-Feeding Ducks." *Wilson Bulletin* 68:275–81.

*Schutz, F. (1965) "Homosexualität und Prägung: Eine experimentelle Untersuchung an Enten [Homosexuality and Developmental Imprinting: An Experimental Investigation of Ducks]." *Psychologische Forschung* 28:439–63.

*Titman, R. D., and J. K. Lowther (1975) "The Breeding Behavior of a Crowded Population of Mallards." *Canadian Journal of Zoology* 53:1270–83.

Weston, M. (1988) "Unusual Behavior in Mallards." *Vogeljaar* 36:259.

Williams, D. M. (1983) "Mate Choice in the Mallard." In P. Bateson, ed., *Mate Choice*, pp. 33–50. Cambridge: Cambridge University Press.

OTHER DUCKS

Waterfowl 🦆

LESSER SCAUP DUCK
Aythya affinis

AUSTRALIAN SHELDUCK
Tadorna tadornoides

MUSK DUCK
Biziura lobata

HOMOSEXUALITY	TRANSGENDER	BEHAVIORS		RANKING	OBSERVED IN
● Female	○ Intersexuality	● Courtship	● Pair-bonding	○ Primary	● Wild
● Male	○ Transvestism	○ Affectionate	● Parenting	● Moderate	○ Semiwild
		● Sexual		○ Incidental	○ Captivity

LESSER SCAUP DUCK
IDENTIFICATION: A broad-billed duck with a purplish black head and breast and white underparts in males, and a dark head and brownish plumage in females. DISTRIBUTION: Northern and central North America; winters in southern United States and Mexico. HABITAT: Lakes, marshes, lagoons. STUDY AREAS: Lake Manitoba (Delta Marsh) and near Erickson, Manitoba; Cariboo region of British Columbia, Canada, including Watson and 150 Mile Lakes.

AUSTRALIAN SHELDUCK
IDENTIFICATION: Cinnamon breast, dark green head and back, and white collar; adult females have white eye and bill rings. DISTRIBUTION: Southern Australia, Tasmania. HABITAT: Marshes, lakes, lagoons. STUDY AREA: Rottnest Island, Western Australia.

MUSK DUCK
IDENTIFICATION: A large, grayish duck with a prominent lobe hanging from the lower bill, and a spike-fan tail. DISTRIBUTION: Southern Australia, Tasmania. HABITAT: Swamps, lakes, other wetlands. STUDY AREA: Kangaroo Lake, Victoria, Australia.

Social Organization

Lesser Scaup Ducks are highly social, gathering into large waterborne flocks or "rafts" that may number in the tens of thousands. They form pair-bonds during the mating season, but males typically leave their mates following egg-laying (see below) and join large all-male groups. Australian Shelducks also form mated pairs during the breeding season (both parents care for the young) but otherwise associate in flocks. Musk Ducks are largely solitary except during the mating season; adult males are territorial, and they are probably polygamous or promiscuous (copulating with more than one female).

Description

Behavioral Expression: Male Lesser Scaup Ducks occasionally try to copulate with one another; drakes who participate in such homosexual activity are usually unpaired birds. While same-sex mounting does not occur among females, coparenting does. In this species, males usually abandon their female mates shortly after incubation of the eggs begins. Most females take care of their young entirely on their own as single parents, but sometimes two females join together and help each other raise their families. Accompanying their combined broods of 20 or more ducklings, the two females cooperate in all parental duties, including coordinated defense of the youngsters. If a predator or intruder approaches, one female distracts it by boldly approaching and feigning an injury, while her partner stealthily leads all of the ducklings away to safety. As their offspring get older, however, female coparents show less of this "distraction" behavior; at the approach of a predator, they may depart with one another and temporarily leave their ducklings behind to fend for themselves. Occasionally three females join forces and raise their combined broods—as many as 50 or more ducklings—as a parenting trio. Interestingly, duckling survival rates are not significantly different in families with one as opposed to two (or three) female parents.

Female Australian Shelducks often court one another and form homosexual pair-bonds. In December, when birds begin pairing, females display to each other with ritual preening movements and chases. These often develop into a full-fledged WATER-THRASHING DISPLAY, in which one female swims toward the other while making sideways pointing movements with her outstretched head and neck. She may also dive and resurface, then chase after the other female. Her partner is frequently captivated by this performance and responds by enthusiastically diving and chasing in return; the two females may, as a result, form a bond that lasts until the next pairing season. Females that engage in homosexual courtship and pairing are usually younger adult or juvenile birds.

Male Musk Ducks perform an extraordinary courtship display that attracts both males and females. The male arches his back and lifts his head up, engorging his large throat pouch; at the same time he fans and bends his tail forward over his back at an extreme angle—a feat made possible because of two extra vertebrae. This gives the bird an astounding, reptilian appearance. While in this posture, he kicks both feet backward or to the side, producing enormously loud splashes or jets of water. Multiple kicks of various types are given in series, and often the courting male rapidly back-paddles in between kicks. He also produces a wide variety of sounds during these PADDLE-KICK, PLONK-KICK, and WHISTLE-KICK displays, including *ker-plonks* (made by the kicking and splashing) combined with *whirr* or *cuc-cuc* vocalizations, and even a distinctive whistling sound. In addition, many displaying males emit a musky odor (hence the name of the bird), and their plumage

A male Musk Duck (left) *attracted to another male performing a "whistle-kick" as part of an elaborate courtship display sequence* ▼

is so oily that puddles of oil may form on the surface of the water around them. This display—perhaps one of the most dramatic of all birds—draws both males and females, who crowd around the courting male. Often, males appear to be more attracted than females, swimming much closer to the displaying male and sometimes even making physical contact by gently and repeatedly nudging their breast against his shoulder. The displaying male is in a trancelike state and rarely responds directly to any of the onlookers. Indeed, although homosexual copulation has not been observed as a part of these displays, heterosexual mating has hardly ever been seen during such courtship sessions either.

Frequency: Homosexual mounting probably occurs only occasionally in Lesser Scaup Ducks, but female coparenting is a regular feature of some populations. As many as a quarter to a third or more of all families have two mothers, although in other populations they are less frequent, comprising about 12 percent of families in some years. Same-sex courtship and pairing occur frequently among younger Australian Shelduck females, while male Musk Ducks are routinely attracted to displaying males.

Orientation: A variety of bisexual arrangements characterize these duck species. Some female Lesser Scaups (perhaps 30–40 percent of the population) have a seasonal alternation between same-sex and opposite-sex pairings: they begin the breeding season in heterosexual pairs, but end it in same-sex associations. Most Australian Shelduck females in homosexual pairs probably go on to form heterosexual bonds as adults. Musk Duck males display to both females and other males; some of the males they attract are also interested in females, but others are apparently only attracted to the courting male.

Nonreproductive and Alternative Heterosexualities

As described above, separation of heterosexual pairs with subsequent female single-parenting (or coparenting) is the usual pattern for Lesser Scaup families. Several other alternative parenting and pairing arrangements occur in these species. Occasionally a female Lesser Scaup associates with a mated pair and even lays eggs in their nest. Musk Ducks often lay their eggs in other birds' nests, where they are foster-parented by both their own and foreign species—including many other kinds of ducks (e.g., the blue-billed duck, *Oxyura australis*) as well as Dusky Moorhens. For their part, Lesser Scaups occasionally raise ducklings of other species of ducks, hatched from eggs that have been laid in their nests by, for example, redheads (*Aythya americana*). Australian Shelducks sometimes foster-parent chicks as well: about 5 percent of all broods contain "extra" ducklings from other families, and about 1 percent of all ducklings are adopted or "exchanged" between families. Lesser Scaup ducklings are occasionally abandoned by their mothers and may be adopted into other families. Abandonment of eggs is also prevalent: female Lesser Scaups may desert entire clutches, while egg DUMPING is common in Australian Shelducks. Many Shelduck pairs copulate but then lay or abandon the resulting eggs in caves, along the shore, or on islands, never incubating or hatching them. Most of these pairs—who may constitute close to half the population—have been unable to secure a breeding territory of their own. Many other birds are non-breeders as well: a large proportion of Lesser Scaups of both sexes, and Australian Shelduck females, are younger birds that are sexually mature but unpaired. In addition, it is thought that reproduction in younger male Musk Ducks may be suppressed by the presence of older males.

Although many Australian Shelducks form long-lasting heterosexual bonds, about 10 percent of breeding pairs divorce, and many more juvenile pairs separate. In Lesser Scaups, nonmonogamous copulations are common, accounting for more than half of all heterosexual activity. Many of these are rapes or forced copulations performed by paired males on females other than their mate; occasionally groups of up to eight males will pursue a female and try to mate with her. Only about 20 percent of such rapes involve penetration—the male Lesser Scaup, like most waterfowl (but unlike most other birds), does have a penis. More than a quarter of all such attempts are nonreproductive, occurring too early in the breeding season, during incubation, after breeding, or on nonbreeding females. In fact, the highest rates of attempted rape occur on females just before their molting period, when they are nonfertilizable. In Australian Shelducks, it is the females who vigorously pursue males, often courting already paired drakes in dramatic aerial chases. One or several females may try to maneuver in between a mated pair to separate the male, even grabbing at the tail feathers of his mate to force her to change direction. Females frequently suffer broken wings and may even be killed when they hit obstacles during such high-speed chases.

Other Species

Interspecies homosexual pairs involving several other kinds of ducks and geese have been observed in captive birds. A pair consisting of a female Common Shelduck (*Tadorna tadorna*) with a female Egyptian Goose (*Alopochen aegyptiacus*), for example, both laid eggs in a shared nest and jointly incubated them.

Sources   **asterisked references discuss homosexuality/transgender*

*Afton, A. D. (1993) "Post-Hatch Brood Amalgamation in Lesser Scaup: Female Behavior and Return Rates, and Duckling Survival." *Prairie Naturalist* 25:227–35.

——— (1985) "Forced Copulation as a Reproductive Strategy of Male Lesser Scaup: A Field Test of Some Predictions." *Behavior* 92:146–67.

——— (1984) "Influence of Age and Time on Reproductive Performance of Female Lesser Scaup." *Auk* 101:255–65.

Attiwell, A. R., J. M. Bourne, and S. A. Parker (1981) "Possible Nest-Parasitism in the Australian Stiff-Tailed Ducks (Anatidae: Oxyurini)." *Emu* 81:41–42.

Bellrose, F. C. (1976) *Ducks, Geese, and Swans of North America*. Harrisburg, PA: Stackpole.

Fullagar, P. J., and M. Carbonell (1986) "The Display Postures of the Male Musk Duck." *Wildfowl* 37:142–50.

Gehrman, K. H. (1951) "An Ecological Study of the Lesser Scaup Duck (*Aythya affinis* Eyton) at West Medical Lake, Spokane County, Washington." Master's thesis, State College of Washington (Washington State University).

*Hochbaum, H. A. (1944) *The Canvasback on a Prairie Marsh*. Washington, D.C.: American Wildlife Institute.

*Johnsgard, P. A. (1966) "Behavior of the Australian Musk Duck and Blue-billed Duck." *Auk* 83:98–110.

*Low, G. C., and Marquess of Tavistock (1935) "The Extent to Which Captivity Modifies the Habits of Birds." *Bulletin of the British Ornithologists' Club* 55:144–54.

*Lowe, V. T. (1966) "Notes on the Musk Duck *Biziura lobata*." *Emu* 65:279–89.

*Munro, J. A. (1941) "Studies of Waterfowl in British Columbia: Greater Scaup Duck, Lesser Scaup Duck." *Canadian Journal of Research*, section D 19:113–38.

O'Brien, R. M. (1990) "Musk Duck, *Biziura lobata*." In S. Marchant and P. J. Higgins, eds., *Handbook of Australian, New Zealand, and Antarctic Birds*, vol. 1, pp. 1152–60. Melbourne: Oxford University Press.

Oring, L. W. (1964) "Behavior and Ecology of Certain Ducks During the Postbreeding Period." *Journal of Wildlife Management* 28:223–33.

*Riggert, T. L. (1977) "The Biology of the Mountain Duck on Rottnest Island, Western Australia." *Wildlife Monographs* 52:1–67.

Rogers, D. I. (1990) "Australian Shelduck, *Tadorna tadornoides*." In S. Marchant and P. J. Higgins, eds., *Handbook of Australian, New Zealand, and Antarctic Birds*, vol. 1, pp. 1210–18. Melbourne: Oxford University Press.

OTHER AQUATIC BIRDS

Other Aquatic

COMMON MURRE or GUILLEMOT
Uria aalge

LAYSAN ALBATROSS
Diomedea immutabilis

HOMOSEXUALITY	TRANSGENDER	BEHAVIORS		RANKING	OBSERVED IN
● Female	○ Intersexuality	● Courtship	○ Pair-bonding	○ Primary	● Wild
● Male	○ Transvestism	○ Affectionate	○ Parenting	● Moderate	○ Semiwild
		● Sexual		○ Incidental	○ Captivity

COMMON MURRE

IDENTIFICATION: A gull-sized, web-footed bird with contrasting black upperparts and white underparts; some individuals have a white eye ring. DISTRIBUTION: Northern oceans and adjacent coasts. HABITAT: Marine coasts, bays, islands. STUDY AREAS: Gannet Islands, Labrador, Canada; Skomer Island, Wales; subspecies *U.a. aalge* and *U.a. albionis*.

LAYSAN ALBATROSS

IDENTIFICATION: A large, white-plumaged, gull-like bird with an enorous wingspan (over 6½ feet), a dark back, and a grayish black wash on the face. DISTRIBUTION: Northern Pacific Ocean. HABITAT: Oceangoing; breeds on oceanic islands. STUDY AREA: Eastern Island in the Midway Atoll.

Social Organization

Common Murres and Laysan Albatrosses spend eight to nine months of the year at sea (often in large flocks for Murres). The remainder of the time, they gather at traditional nesting sites in extraordinary densities—Murre colonies, for example, can contain hundreds of thousands of pairs. The mating system is a combination of long-term pair-bonds and promiscuous copulations.

Description

Behavioral Expression: Male Common Murres—usually heterosexually paired—often try to copulate with birds other than their mate, including other

▲ *A male Common Murre attempting to forcibly copulate with another male*

males. Homosexual mountings—like heterosexual promiscuous mountings—are usually performed on birds returning to the colony after having been away (for example, while feeding). Immediately upon spotting an arriving male (or female), another male runs toward him, making a harsh, yodel-like crowing sound. He then hooks his neck around the other male and attempts to copulate with him. The other male usually prevents or resists the mounting attempt by standing upright, running away, or directly attacking him. Homosexual mountings also take place during "gang rape" attempts, which occur in 20–30 percent of all promiscuous matings. Groups of males—sometimes as many as ten at a time—gather to try to forcibly copulate with the same female, and occasionally males also mount each other during the ensuing sexual activity.

A similar form of rape occurs among Laysan Albatrosses. Early in the breeding season, males often leave their partner's side to try to copulate with males or females that are passing through the breeding colony. This is especially true if they momentarily and inadvertently spread and droop their wings (signals usually given by a female before copulation). Groups of five or six males often pursue the same individual, all jostling to mount him or her; typically a male will hook his bill across the neck of the bird being pursued, to throw it off balance. Homosexual mountings are common in these group rape attempts, and "pile-ups" or stacks of up to four males mounted on top of each other have been observed. Rape attempts—whether on males or females—never result in ejaculation, since the bird being mounted always resists the advances of the pursuing bird. A completely different homosexual activity also occurs in this species: occasionally two birds of the same sex perform an elaborate courtship dance with one another. This complex synchronized display involves more than 25 different postures. The two birds stand facing each other, stretching their heads upward during SKY CALLS and SKY MOOS, clap their bills, and bow, strut, and circle around their partner, all the while making a cacophony of clicking, whinnying, wailing, and grunting sounds.

Frequency: At least 5–6 percent of all promiscuous mating attempts on arriving Common Murres are homosexual, and one out of ten arriving males is mounted by another male (compared to three out of four arriving females). Homosexual copulation attempts probably represent 1 percent or less of all mountings (both promiscuous and between pair-bonded birds). In Laysan Albatrosses, rape attempts are frequent before egg laying and probably occur with equal regularity on males and females. Approximately 9 percent of courtship dances take place between two females and 4 percent between two males.

Orientation: About two-thirds of all male Common Murres participate in promiscuous copulations; only a fraction of these engage in homosexual mountings. Male Laysan Albatrosses are as likely to pursue and mount other males as females in their rape attempts. Although it is difficult to draw firm conclusions without detailed study of individual birds, most males that engage in homosexual behavior in these species are probably functionally bisexual, since they are usually already paired with a female (although a few Common Murres who participate in such activity may be unpaired). However, their primary orientation is probably heterosexual since relatively few of their sexual interactions are with other males. The same probably holds for males being mounted by other males: because they usually resist forced mounts by other males, it is likely that most such males are heterosexually oriented. However, most females also resist forced mountings by males, so it is possible that males are reacting negatively to the forced nature of the copulation attempt, as much if not more so than to the sex of the bird mounting them.

Nonreproductive and Alternative Heterosexualities

As mentioned above, promiscuous copulations occur frequently in these species. About 10 percent of all Common Murre matings are forced copulations between a male and a female other than his partner, and on some days each female is subjected to such a rape attempt nearly every hour. Females usually respond aggressively to such attacks, and their mates also try to defend them, although sometimes an intruding male will actually disrupt a copulation between a mated pair by knocking the male off his partner's back. In about 15 percent of all promiscuous matings, the female does not react aggressively and appears to cooperate in allowing the male to make genital contact. Female Laysan Albatrosses always resist rape attempts and may be severely injured in the process: one female was attacked by four different gangs of males in ten minutes, losing an eye and sustaining severe wing injuries. However, forced copulations in this species are always nonprocreative since sperm is never transferred. Many promiscuous matings in Common Murres are nonreproductive as well: cloacal contact often does not occur (less than 1 in 200 such matings result in insemination), and during group promiscuous matings, males often mount on any part of the female's body, including her head. In addition, about 15–30 percent of promiscuous copulations occur outside of the female's fertile period. The same is also true for sexual activity between mated partners: copulation begins as long as four to five months before the start of egg laying, and half of all heterosexual matings in some populations occur during nonfertilizable periods. In addition, almost a quarter of pair copulations do not involve genital contact. In Common Murres—as in most other birds—females have the remarkable ability to store sperm in special ducts in their reproductive tract, allowing them to inseminate their eggs even when not directly engaging in reproductive copulations.

Other forms of nonprocreative sexuality also occur. Nonbreeding female Common Murres often solicit promiscuous matings from males, while nonbreeding pairs or those who have lost their young (which can make up as much as a third of all pairs) frequently continue to copulate throughout the season. Nonbreeding Laysan Albatross pairs also sometimes engage in copulation. Birds in this species do not reproduce until they are 6–16 years old, even though they mature at one year old and may form pairs fully two years before they actually breed. Similarly, younger Common Murres usually delay breeding until they are five years old, congregating in CLUBS on the tidal rocks beneath the breeding colonies. Such nonbreeders make up approximately 13 percent of the population; among adults, 5–10 percent of birds do not breed each year, and more than a third skip breeding for at least one season during their life. In addition, masturbatory activity—birds mount and "copulate" with clumps of grass—was recently discovered in a closely related species, the thick-billed murre (*Uria lomvia*); it is likely that similar behavior also occurs in Common Murres.

A variety of alternative parenting arrangements are also found in these species. About 8 percent of all Common Murre chicks have "baby-sitters"—a pair of birds other than their parents who help brood (keep warm), protect, and sometimes feed the chick (even when the youngster's parents are not away). Most such helpers are nonbreeders; others have tried but failed to breed, while some have finished raising their family or are also taking care of their own chick. In addition, pair separation and single parenting is routine in Common Murres: when a chick is old enough to leave the colony, only its father accompanies it to sea, feeding and chaperoning it for up to 12 weeks without his female partner. In Laysan Albatrosses, heterosexual parents are together at the nest for a remarkably short time—only 5–10 days out of the 230-day breeding season. Eggs are sometimes temporarily "adopted" by other birds who incubate them when the parents are away from the nest. Nonbreeding females have even been known to "join" existing pairs and regularly take turns with the parents incubating their egg. Sometimes females also lay a second egg in a stranger's nest. Reproduction in this species is often fraught with difficulties, however. More than 20 percent of parents (both males and females) desert their nests—often when their partner fails to return for an incubation shift on time—and couples also occasionally divorce (2 percent of all pairs). Once the chicks have hatched, they are often subjected to abuse from neighboring birds, who may savagely peck, stab, bite, and occasionally even kill the youngsters if they stray too close.

Other Species

Homosexual copulations are common in another species of auk, the Razorbill (*Alca torda*), where 41 percent of nonmonogamous mountings (about 18 percent of all mountings) are between males. Up to 200 or more such mountings have been observed each season in some populations. Nearly two-thirds of all males mount other males (an average of 5 partners, sometimes as many as 16) and more than 90 percent of males receive mounts from other males. Older males participate more often than younger ones, and mountings are occasionally reciprocal. Like females, males usually resist such promiscuous mating attempts: although the mounter usually tries to achieve cloacal (genital) contact, only about 1 percent of same-sex mountings include genital contact or ejaculation (compared to 12 percent of promiscuous heterosexual mounts).

Sources   *asterisked references discuss homosexuality/transgender

Birkhead, T. R. (1993) *Great Auk Islands*. London: T. and A.D. Poyser.

*———— (1978a) "Behavioral Adaptations to High Density Nesting in the Common Guillemot *Uria aalge*." *Animal Behavior* 26:321–31.

———— (1978b) "Attendance Patterns of Guillemots *Uria aalge* at Breeding Colonies on Skomer Island." *Ibis* 120:219–29.

Birkhead, T. R., and P. J. Hudson (1977) "Population Parameters for the Common Guillemot *Uria aalge*." *Ornis Scandinavica* 8:145–54.

Birkhead, T. R., S. D. Johnson, and D. N. Nettleship (1985) "Extra-pair Matings and Mate Guarding in the Common Murre *Uria aalge*." *Animal Behavior* 33:608–19.

Birkhead, T. R., and D. N. Nettleship (1984) "Alloparental Care in the Common Murre (*Uria aalge*)." *Canadian Journal of Zoology* 62:2121–24.

Fisher, H. I. (1975) "The Relationship Between Deferred Breeding and Mortality in the Laysan Albatross." *Auk* 92:433–41.

*———— (1971) "The Laysan Albatross: Its Incubation, Hatching, and Associated Behaviors." *Living Bird* 10:19–78.

———— (1968) "The 'Two-Egg Clutch' in the Laysan Albatross." *Auk* 85:134–36.

Fisher, H. I., and M. L. Fisher (1969) "The Visits of Laysan Albatrosses to the Breeding Colony." *Micronesica* 5:173–221.

Fisher, M. L. (1970) *The Albatross of Midway Island: A Natural History of the Laysan Albatross*. Carbondale, Ill.: Southern Illinois University Press.

*Frings, H., and M. Frings (1961) "Some Biometric Studies on the Albatrosses of Midway Atoll." *Condor* 63:304–12.

Gaston, T., and K. Kampp (1994) "Thick-billed Murre Masturbating on Grass Clump." *Pacific Seabirds* 21:30.

Harris, M. P., and S. Wanless (1995) "Survival and Non-Breeding of Adult Common Guillemots *Uria aalge*." *Ibis* 137:192–97.

*Hatchwell, B. J. (1988) "Intraspecific Variation in Extra-pair Copulation and Mate Defence in Common Guillemots *Uria aalge*." *Behavior* 107:157–85.

Hudson, P. J. (1985) "Population Parameters for the Atlantic Alcidae." In D. N. Nettleship and T. R. Birkhead, eds., *The Atlantic Alcidae*, pp. 233–61. London: Academic Press.

Johnson, R. A. (1941) "Nesting Behavior of the Atlantic Murre." *Auk* 58:153–63.

Meseth, E. H. (1975) "The Dance of the Laysan Albatross, *Diomedea immutabilis*." *Behavior* 54:217–57.

Rice, D. W., and K. W. Kenyon (1962) "Breeding Cycles and Behavior of Laysan and Black-footed Albatrosses." *Auk* 79:517–67.

Tuck, L. M. (1960) *The Murres: Their Distribution, Populations, and Biology*. Ottawa: Canadian Wildlife Service.

*Wagner, R. H. (1996) "Male-Male Mountings by a Sexually Monomorphic Bird: Mistaken Identity or Fighting Tactic?" *Journal of Avian Biology* 27:209–14.

———— (1991) "Evidence That Female Razorbills Control Extra-Pair Copulations." *Behavior* 118:157–69.

CORMORANTS

Other Aquatic 🦆

GREAT CORMORANT
Phalacrocorax carbo

EUROPEAN SHAG
Phalacrocorax aristotelis

HOMOSEXUALITY	TRANSGENDER	BEHAVIORS		RANKING	OBSERVED IN
○ Female	○ Intersexuality	● Courtship	● Pair-bonding	○ Primary	● Wild
● Male	○ Transvestism	○ Affectionate	○ Parenting	○ Moderate	○ Semiwild
		○ Sexual		● Incidental	● Captivity

GREAT CORMORANT
IDENTIFICATION: A large (3 foot), black, web-footed bird with a white throat and white filamentary plumes on the nape. DISTRIBUTION: Throughout Europe, Australasia, Africa, and Atlantic North America. HABITAT: Seacoasts, lakes, rivers. STUDY AREAS: Shinobazu Pond, Tokyo, Japan; Amsterdam Zoo, the Netherlands; subspecies *P.c. sinensis* and *P.c. hanedae*.

EUROPEAN SHAG
IDENTIFICATION: Similar to Great Cormorant, but smaller and uniformly black, with a prominent forehead crest. DISTRIBUTION: Northwestern Europe, Mediterranean basin. HABITAT: Coastal waters; nests on cliffs. STUDY AREA: Lundy Island in the Bristol Channel, England; subspecies *P.a. aristotelis*.

Social Organization

Great Cormorants and Shags form mated pairs and generally nest in colonies, which may contain as many as 20,000 pairs in some populations of Great Cormorants. Outside of the mating season, these species are moderately gregarious, wandering solitarily but sometimes forming flocks.

Description

Behavioral Expression: Homosexual pairs consisting of two males sometimes form in Great Cormorants and last for up to five years (heterosexual pairs in this species are usually seasonal but may also last for several years). Male pairs often build oversize nests because both birds contribute to the construction of the nest. They often sit on the nest as if incubating eggs; similar behavior is also seen in heterosexual pairs prior to egg laying. In some homosexual pairs, one partner may use vocalizations that are typical of females (such as panting or purring sounds), or else calls that are intermediate between male and female vocal patterns. Male pairs are sometimes incestuous, composed of two brothers.

In European Shags, males occasionally court other males. As one male approaches—hopping along the rocks, pausing every now and then in an erect pose known as the UPRIGHT-AWARE POSTURE—the other male performs two displays. In the DART-GAPE, he pulls his head back and then darts it forward, at the same time opening his bill to expose the yellow interior and fanning his tail. In the THROW-BACK, he arches his neck along his back and points his beak upward while quivering his throat pouch. Sometimes the courting male will become aggressive and attack another male that approaches too closely, which also happens frequently when females approach courting males.

Frequency: Homosexual pairs and courtship occur only occasionally in these two species: no more than perhaps 1 in 500 pairs of Great Cormorants, for example, is composed of two males.

Orientation: Great Cormorants in homosexual pairs are sometimes sequentially bisexual, divorcing their male partners and going on to breed in heterosexual pairs. However, others re-pair with another male partner, and some homosexual pairs appear to last much longer, perhaps even for the birds' lifetimes—in which case such individuals are exclusively (or extensively) homosexual for a significant part of their lives. Male European Shags that court other males are simultaneously bisexual, alternating heterosexual courtship with homosexual interactions (and probably a greater proportion of the former).

Nonreproductive and Alternative Heterosexualities

Several forms of nonprocreative sexual behavior are exhibited by these Cormorants. REVERSE mountings constitute 8 percent of European Shag heterosexual copulations (and also occur in Great Cormorants), while at least a quarter of all sexual activity takes place prior to the female's fertilizable period. Great Cormorants sometimes copulate during the incubation period, while heterosexual copulations may continue even after Shag chicks have hatched. In a few cases, adult male Shags have been observed mating with their own chicks, and incestuous pairings sometimes also develop between brothers and sisters when they are still young.

Courtship between two male European Shags: the "upright-aware" posture (left) *and the "throw-back" display* ▼

Nearly half of all heterosexual copulations in Shags involve mounting without genital contact, often because the female will not permit it. In addition, males are frequently hostile to females during the early phases of courtship (as noted above).

Nonmonogamous matings and courting of birds other than one's partner occur in both of these species. In European Shags, for example, 14 percent of all copulations are promiscuous. Almost 18 percent of all chicks are fathered by a male other than their mother's mate, but nearly 80 percent of all nonmonogamous matings are nonreproductive, taking place before females can be fertilized. At least 4 percent of all chicks are related to neither of the parents caring for them; this results from adoption and from females' laying eggs in nests other than their own. About 3–5 percent of male Shags bond polygamously with two females; in addition, 30–40 percent of heterosexual Shag pairs divorce and re-pair with new mates the next season. Individuals may also change mates during the season. Some Shag parents are severely neglectful, refusing to feed their offspring, who may, as a result, die of starvation. In addition, about a third of all eggs lost through breakage result from interference by the females in polygamous associations. Finally, nonbreeding is a regular feature of some Shag populations: on average, 12–25 percent of all adults skip breeding at least once during their lifetime, and in some years as many as 60 percent of all birds forgo reproduction.

Sources   *asterisked references discuss homosexuality/transgender

Aebischer, N. J., G. R. Potts, and J. C. Coulson (1995) "Site and Mate Fidelity of Shags *Phalacrocorax aristotelis* at Two British Colonies." *Ibis* 137:19–28.

Aebischer, N. J., and S. Wanless (1992) "Relationships Between Colony Size, Adult Non-Breeding, and Environmental Conditions for Shags *Phalacrocorax aristotelis* on the Isle of May, Scotland." *Bird Study* 39:43–52.

*Fukuda, M. (1992) "Male-Male Pairing of the Great Cormorant (*Phalacrocorax carbo hanedae*)." *Colonial Waterbird Society Bulletin* 16:62–63.

Graves, J., R. T. Hay, M. Scallan, and S. Rowe (1992) "Extra-Pair Paternity in the Shag, *Phalacrocorax aristotelis* as Determined by DNA Fingerprinting." *Journal of Zoology, London* 226:399–408.

Harris, M. P. (1982) "Promiscuity in the Shag as Shown by Time-Lapse Photography." *Bird Study* 29:149–54.

Johnsgard, P. A. (1993) *Cormorants, Darters, and Pelicans of the World.* Washington and London: Smithsonian Institution Press.

*Kortlandt, A. (1995) "Patterns of Pair-Formation and Nest-Building in the European Cormorant, *Phalacrocorax carbo sinensis*." *Ardea* 83:11–25.

*——— (1949) "Textuur en structuur van het broedvoorbereidingsgedrag bij de aalscholver [Texture and Structure of Brooding-Preparatory Behavior in the Cormorant]." Ph.D. thesis, University of Amsterdam.

*Snow, B. K. (1963) "The Behavior of the Shag." *British Birds* 56:77–103, 164–86.

GREBES
SILVERY GREBE
Podiceps occipitalis
HOARY-HEADED GREBE
Poliocephalus poliocephalus

HOMOSEXUALITY	TRANSGENDER	BEHAVIORS		RANKING	OBSERVED IN
● Female	○ Intersexuality	● Courtship	○ Pair-bonding	○ Primary	● Wild
● Male	○ Transvestism	○ Affectionate	○ Parenting	○ Moderate	○ Semiwild
		● Sexual		● Incidental	○ Captivity

SILVERY GREBE
IDENTIFICATION: A ducklike bird with grayish white plumage, bright red eyes, and yellow facial tufts. DISTRIBUTION: Western and southern South America. HABITAT: Lakes, marshy ponds. STUDY AREA: Laguna Nevada, southern Patagonia, Argentina.

HOARY-HEADED GREBE
IDENTIFICATION: Similar to Silvery Grebe, but with a buff or chestnut wash on the breast, white streaks on the head, and black-and-white eyes. DISTRIBUTION: Australia, Tasmania. HABITAT: Wetlands, estuaries, bays. STUDY AREA: Lake Bathurst, New South Wales, Australia.

Social Organization

Both of these species of Grebes commonly socialize in pairs or small groups, and sometimes in dense flocks (which may contain several thousand birds in Hoary-headed Grebes). They nest in large colonies—up to 400 nests for Hoary-headed Grebes—and most form monogamous pair-bonds.

Description

Behavioral Expression: Male Silvery Grebes occasionally mount other males, using the same position as for heterosexual copulation—although unlike in heterosexual mating, the mounted bird does not typically invite the other male to mount him. In Hoary-headed Grebes, two birds of the same sex sometimes perform courtship displays to each other on floating platforms, constructed of water vegetation, and perhaps also mount one another. One such display is called REARING, in which

The "rearing" display of a Hoary-headed Grebe, used during courtships with same- and opposite-sex partners ▶

the bird lifts its body up, ruffles its mantle feathers, and extends its head and neck downward. This may lead to one bird INVITING the other to mount, by settling back down, kinking its neck, and resting its throat on the nest.

Frequency: Same-sex activity occurs only occasionally in these two species of Grebes. In Silvery Grebes, for example, about 1 in 300 mounts is between two males.

Orientation: Grebes that participate in same-sex activities are probably bisexual, also courting and mating with members of the opposite sex.

Nonreproductive and Alternative Heterosexualities

In these two species of Grebes—as in many other Grebes—REVERSE heterosexual mountings are common. An average of 27 percent of all copulations among Silvery Grebes involve females mounting males, and early in the breeding season as many as 40 percent of all mounts are reverse. Their prevalence during this period—long before fertilization is possible—indicates that this form of behavior is decidedly nonprocreative. In addition, ejaculation does not usually occur during reverse mounts, although females typically perform tail thrusting, perhaps to facilitate genital contact. Grebes also copulate repeatedly, with as many as five or six matings taking place over 15–20 minutes; during this activity, males and females may also alternate positions in a form of reciprocal mounting. Although most heterosexual mating in Hoary-headed Grebes takes place between pair members, some copulations also occur between unpaired birds.

Sources   *asterisked references discuss homosexuality/transgender*
*Fjeldså, J. (1983) "Social Behavior and Displays of the Hoary-headed Grebe *Poliocephalus poliocephalus.*" *Emu* 83:129–40.
Fjeldså, J. and N. Krabbe (1990) *Birds of the High Andes.* Copenhagen: Zoological Museum, University of Copenhagen; Svendborg, Denmark: Apollo Books.
Johnsgard, P. A. (1987) *Diving Birds of North America.* Lincoln and London: University of Nebraska Press.
*Nuechterlein, G. L., and R. W. Storer (1989) "Reverse Mounting in Grebes." *Condor* 91:341–46.
*O'Brien, R. M. (1990) "Hoary-headed Grebe, *Poliocephalus poliocephalus.*" In S. Marchant and P. J. Higgins, eds., *Handbook of Australian, New Zealand, and Antarctic Birds,* vol. 1, pp. 100–107. Melbourne: Oxford University Press.
Storer, R. W. (1969) "The Behavior of the Horned Grebe in Spring." *Condor* 71:180–205.

HERONS

Wading Bird

BLACK-CROWNED NIGHT HERON
Nycticorax nycticorax

HOMOSEXUALITY	TRANSGENDER	BEHAVIORS		RANKING	OBSERVED IN
● Female	○ Intersexuality	● Courtship	● Pair-bonding	○ Primary	○ Wild
● Male	○ Transvestism	○ Affectionate	○ Parenting	● Moderate	● Semiwild
		● Sexual		○ Incidental	● Captivity

IDENTIFICATION: A stocky, medium-sized (2 foot long) heron with a black crown and back, white underparts, gray wings, and white ribbon plumes at the nape of the neck. DISTRIBUTION: Over much of Europe, Asia, Africa, and North and South America. HABITAT: Wetlands. STUDY AREAS: American Museum of Natural History, New York; Altenberg, Austria; subspecies *N.n. hoactli* and *N.n. nycticorax*.

Social Organization

Black-crowned Night Herons are fairly gregarious birds, gathering in colonies that may contain hundreds or thousands of individuals nesting close together. Monogamous pairs predominate during the mating season.

Description

Behavioral Expression: Male Black-crowned Night Herons sometimes court other males and form homosexual pairs. To attract other birds, a male performs the SNAP-HISS CEREMONY, in which he repeatedly extends and lowers his head with erected plumes while "treading" with his feet and making a combined snapping (or clicking) and hissing sound. This courtship display is directed at both males and females. While most males are not interested in the ceremony, those that are attracted go on to participate in an OVERTURE display (also used in heterosexual courtship). In this activity, one or both males stretch their heads forward (again with feathers raised) while making their eyeballs protrude from their sockets, at the same time clicking and touching their bills. Males in homosexual pairs also mount each other (in the position used for heterosexual copulation); both males may mount or be mounted, although individuals sometimes show a preference for one or the other activity. Typically one male in the pair also builds a nest out of twigs; sometimes this occurs before the pair-bond forms, or else the two males may search for a nest site together. Once the pair-bond is established, both males vigorously defend their territory against other birds. Homosexual pair-bonds are strong, lasting the entire breeding season (as do heterosexual bonds). Young Black-crowned Night Herons also form pairs or "companionships," some of which are between birds of the same sex (although no overt sexual behavior occurs between companions of either the same or opposite sex). Occasionally, a juvenile female may even bond with more than one other female. Although no homosexual pairs between adult females have yet been observed, females sometimes perform typically male courtship displays such as the snap-hiss ceremony.

Frequency: In some captive populations, homosexual pairs in adult males make up more than 20 percent of all pairings, while 38 percent of juvenile pairs are same-sex (three-quarters of which are between females). The incidence of same-sex pairs in the wild is not known. Adult male partners may mount each other more than 30 times over the mating season.

Orientation: During the courtship phase, males exhibit a form of simultaneous bisexuality by displaying to both sexes, and some males are clearly more attracted to same-sex courtships than others. Most young females that form same-sex bonds only pair with other females, while a few form both homosexual and heterosexual pair-bonds. Once a homosexual bond has formed, though, the birds will maintain the bond even if they are separated for several weeks, returning to their same-sex partner when reunited rather than establishing a heterosexual bond. Some birds may exhibit sequential bisexuality, forming heterosexual bonds after participating in homosexual pairings when young (or vice versa).

Nonreproductive and Alternative Heterosexualities

Male-female relations in Black-crowned Night Herons are sometimes fraught with complications. Heterosexual copulation is often incomplete because females refuse to cooperate in mating. In addition, in the early stages of courtship, males are often aggressive toward any bird that approaches them, including females. Chicks that hatch late are usually deserted by their parents after their siblings have fledged; they often move to other nests and are adopted by those families. Adults also occasionally accept eggs laid in their nest by other herons such as great egrets (*Casmerodius albus*) or lay their own eggs in nests of other species such as snowy egrets (*Egretta thula*).

Sources   *asterisked references discuss homosexuality/transgender*

Allen, R. P., and F. P. Mangels (1940) "Studies of the Nesting Behavior of the Black-crowned Night Heron." *Proceedings of the Linnaean Society of New York* 50–51:1–28.

Cannell, P. F., and B. A. Harrington (1984) "Interspecific Egg Dumping by a Great Egret and Black-crowned Night Herons." *Auk* 101:889–91.

Davis, W. E., Jr. (1993) "Black-crowned Night Heron (*Nycticorax nycticorax*)." In A. Poole and F. Gill, eds., *The Birds of North America: Life Histories for the 21st Century*, no. 74. Philadelphia: Academy of Natural Sciences; Washington, D.C.: American Ornithologists' Union.

Gross, A. O. (1923) "The Black-crowned Night Heron (*Nycticorax nycticorax naevius*) of Sandy Neck." *Auk* 40:1–30, 191–214.

Kazantzidis, S., V. Goutner, M. Pyrovetsi, and A. Sinis (1997) "Comparative Nest Site Selection and Breeding Success in 2 Sympatric Ardeids, Black-Crowned Night-Heron (*Nycticorax nycticorax*) and Little Egret (*Egretta garzetta*) in the Axios Delta, Macedonia, Greece." *Colonial Waterbirds* 20:505–17.

*Lorenz, K. (1938) "A Contribution to the Comparative Sociology of Colonial-Nesting Birds." In F. C. R. Jourdain, ed., *Proceedings of the Eighth International Ornithological Congress, Oxford (July 1934)*, pp. 207–21. Oxford: Oxford University Press.

McClure, H. E., M. Yoshii, Y. Okada, and W. F. Scherer (1959) "A Method for Determining Age of Nestling Herons in Japan." *Condor* 61:30–37.

*Noble, G. K., and M. Wurm (1942) "Further Analysis of the Social Behavior of the Black-crowned Night Heron." *Auk* 59:205–24.

*Noble, G. K., M. Wurm, and A. Schmidt (1938) "Social Behavior of the Black-crowned Night Heron." *Auk* 55:7–40.

Schorger, A. W. S. (1962) "Black-crowned Night Heron." In R. S. Palmer, ed., *Handbook of North American Birds, vol. 1: Loons through Flamingos*, pp. 472–84. New Haven and London: Yale University Press.

HERONS AND EGRETS

Wading Bird

CATTLE EGRET
Bubulcus ibis
LITTLE EGRET
Egretta garzetta
LITTLE BLUE HERON
Egretta caerulea
GRAY HERON
Ardea cinerea

HOMOSEXUALITY	TRANSGENDER	BEHAVIORS		RANKING	OBSERVED IN
○ Female	○ Intersexuality	○ Courtship	○ Pair-bonding	○ Primary	● Wild
● Male	○ Transvestism	○ Affectionate	○ Parenting	● Moderate	○ Semiwild
		● Sexual		○ Incidental	○ Captivity

CATTLE EGRET, LITTLE EGRET
IDENTIFICATION: Long-legged, typically white herons with ornamental, filamentous plumes on the back, breast, and nape; these are golden-buff-colored in the Cattle Egret. DISTRIBUTION: Throughout Africa, southern Europe, Australasia, and (in Cattle Egret) North and South America. HABITAT: Variable, including swamps, marshes, rivers, lakes, meadows. STUDY AREA: Near Tsu City, Japan; subspecies *B.i. coromanda* and *E.g. garzetta*.

LITTLE BLUE HERON
IDENTIFICATION: Similar to Little Egret but with slaty-gray plumage and a reddish brown head and neck. DISTRIBUTION: Southeastern United States to northern South America. HABITAT: Lakes, marshes, streams. STUDY AREA: Swan Lake, Arkansas; Cliftonville, Massachusetts.

GRAY HERON
IDENTIFICATION: A large (3 foot long) heron with a gray back, white head and neck, and black "eyebrow" stripe and nape plumes. DISTRIBUTION: Throughout Eurasia and Africa. HABITAT: Wetlands. STUDY AREA: Doñana National Park, Spain; subspecies *A.c. cinerea*.

Social Organization

Herons and Egrets are highly social birds, nesting in dense colonies that may include birds of several different species. During the mating season the primary social unit is the monogamous pair, although several alternative mating systems occur (see below). Outside of the breeding season, they may be found either singly or in flocks.

Description

Behavioral Expression: In all four of these Heron and Egret species, males that are paired to females often copulate with birds other than their mates; in some cases, these involve homosexual copulations with other males who are themselves also paired to females. Homosexual mountings always take place during the mating season. In Little Egrets, mountings between males are most common during the early stages of heterosexual pair formation (before nest-building begins), while in Little Blue Herons at least some homosexual activity takes place during the incubation period, since males have been seen mounting other males that are sitting on eggs. Typically, males mount birds in neighboring nests, although in Little Egrets and Little Blue Herons males may travel to other areas of the breeding colony to engage in "extramarital" or promiscuous copulations (both homosexual and heterosexual).

In Cattle Egrets (and probably the other species as well), homosexual mountings always take place on the mountee's nest. In a typical encounter, the male seeking an "extramarital" liaison approaches another male, uttering RICK RACK calls (a harsh double croaking sound, also used in heterosexual encounters). The first male then mounts the other bird and crouches on his back; some males only act as mounters in homosexual copulations, others only as mountees, while some males perform both roles. In Little Blue Herons and Cattle Egrets, homosexual mountings may also occur when one male mounts another male who is himself attempting to copulate with a female; sometimes, "pile-ups" of three or four males on top of each other may develop in this way. Usually the mountee is aggressive toward the male mounting him and does not permit cloacal contact. Similarly, male-female "extramarital" copulations are rarely completed, owing to resistance by the female or defense by her mate. In Cattle Egrets, nearly a quarter of all such heterosexual mounting attempts do not involve cloacal contact, while in Little Egrets more than 85 percent of such opposite-sex copulations are "incomplete."

Frequency: Homosexual mountings can be quite common: in Little Egrets, for example, more than 100 mounts between males were recorded over four months in one colony, with such copulations comprising 5–6 percent of all "extramarital" sexual activity. In Little Blue Herons, homosexual mountings make up 3–6 percent of all copulations outside the pair-bond. Mounts between males represent 5 percent of "extramarital" copulations and 3 percent of all copulations in Cattle Egrets, while in Gray Herons they constitute 8 percent of all promiscuous mountings and 1 percent of the total number of copulations. In 18 percent of "extramarital" copulation attempts on female Cattle Egrets, additional males mount other males in a pile-up.

Orientation: Since males that participate in homosexual activity almost always have female mates, they are technically bisexual (and some birds may even participate in "group" sexual activity involving both males and females simultaneously, as in Little Blue Herons). In Little Egrets, about one-quarter of the male population engages in homosexual mounting, in Gray Herons 5–7 percent of males are involved in such activity, while in Cattle Egrets six out of ten males in one colony participated in same-sex mounting. Some individuals seem to show more of a "predilection" for homosexual behavior than others. In Cattle Egrets, for example, certain males engage in "extramarital" mountings only with males rather than females, while in both this species and in Little Egrets, some individuals participate in same-sex activity noticeably more often than others. In addition, homosexual activity comprises a greater proportion of "extramarital" sexual activity for some males than for others.

Nonreproductive and Alternative Heterosexualities

As described above, "extramarital" or promiscuous heterosexual copulations occur commonly in all four of these species. In Cattle Egrets, as many as 60 percent of all mountings are by males on females other than their mates, while in Gray Herons such matings account for more than 12 percent of all sexual activity. Nearly a third of all Little Egret copulations are promiscuous. In fact, many copulations in this context are actually rapes, since the female is not a willing participant (although in both Cattle and Little Egrets, females may also consent to such matings). Up to 7 percent of Cattle Egret eggs may be fertilized by a male other than the bird's (social) father; however, many "extramarital" copulations are nonprocreative, since almost a quarter of all such matings take place when the female is already incubating her eggs. In addition to stepparenting of birds fathered by other males, several other alternative family arrangements occur: Cattle Egret trios of two females and one male may raise a family, while foster-parenting sometimes occurs when females lay their eggs in nests of other birds, including other species of Egrets and Herons.

Several mating behaviors in these species indicate that not all aspects of heterosexuality revolve around breeding. Cattle Egrets sometimes mate when fertilization is not possible, for example during incubation or chick-raising. And up to 14 percent of copulations between pair members may be "incomplete" in the sense that no genital contact or sperm transfer occurs—sometimes because the male is apparently not "interested" in mating even though his female partner is. In Little Blue Herons, some males copulate with females and yet remain "single" (i.e., do not pairbond with them), while other males never pair with a female during the entire mating season. REVERSE mounts (females mounting males) also occur in Cattle Egrets, and in polygamous trios this sometimes results in a "pile-up" of three birds (one female mounting the second female who is mounting the male).

A number of violent and counterreproductive behaviors can make life harsh for young Egrets and Herons. In Little Blue Herons, infidelity often leads to abandonment of the nest by one or both partners (in part because eggs may be broken during the promiscuous sexual activity). Following a partner's injury, male Cattle Egrets have been known to destroy their own eggs and desert their mates for a new female. Male Gray Herons also occasionally destroy their eggs by stabbing at them. Nest and mate desertion (especially by females) are common in Little Egrets as well. Often the remaining bird will successfully raise the chicks as a single parent; sometimes, though, the chicks die as a result of desertion. If a single father pairs with a new female, she may kill his nestlings by pecking them to death, so that she can mate with him and raise her own offspring. Cannibalism by siblings or parents sometimes occurs in Gray Herons. In addition, Heron and Egret families are often systematically "pared down" because the youngest nestlings starve to death when they are unable to compete for food; more than three-quarters of all nestling deaths in Little Blue Herons are the result of such "brood reduction."

Sources   *asterisked references discuss homosexuality/transgender

Blaker, D. (1969) "Behavior of the Cattle Egret *Ardeola ibis*." *Ostrich* 40:75–129.

*Fujioka, M. (1996) Personal communication.

——— (1989) "Mate and Nestling Desertion in Colonial Little Egrets." *Auk* 106:292–302.

*——— (1988) "Extrapair Copulations in Little Egrets (*Egretta garzetta*)." Paper presented at the annual meeting of the Animal Behavior Society, University of Montana.

——— (1986a) "Infanticide by a Male Parent and by a New Female Mate in Colonial Egrets." *Auk* 103:619–21.

——— (1986b) "Two Cases of Bigyny in the Cattle Egret *Bubulcus ibis*." *Ibis* 128:419–22.

——— (1985) "Sibling Competition and Siblicide in Asynchronously-Hatching Broods of the Cattle Egret *Bubulcus ibis*." *Animal Behavior* 33:1228–42.

*Fujioka, M., and S. Yamagishi (1981) "Extramarital and Pair Copulations in the Cattle Egret." *Auk* 98:134–44.

Lancaster, D. A. (1970) "Breeding Behavior of the Cattle Egret in Colombia." *Living Bird* 9:167–94.

McKilligan, N. G. (1990) "Promiscuity in the Cattle Egret (*Bubulcus ibis*)." *Auk* 107:334–41.

*Meanley, B. (1955) "A Nesting Study of the Little Blue Heron in Eastern Arkansas." *Wilson Bulletin* 67:84–99.

Milstein, P. le S., I. Prestt, and A. A. Bell (1970) "The Breeding Cycle of the Gray Heron." *Ardea* 58:171–257.

*Ramo, C. (1993) "Extra-Pair Copulations of Gray Herons Nesting at High Densities." *Ardea* 81:115–20.

Rodgers, J. A., Jr. (1980a) "Little Blue Heron Breeding Behavior." *Auk* 97:371–84.

——— (1980b) "Breeding Ecology of the Little Blue Heron on the West Coast of Florida." *Condor* 82:164–69.

Rodgers, J. A., Jr., and H. T. Smith (1995) "Little Blue Heron (*Egretta caerulea*)." In A. Poole and F. Gill, eds., *The Birds of North America: Life Histories for the 21st Century,* no. 145. Philadelphia: Academy of Natural Sciences; Washington, D.C.: American Ornithologists' Union.

Telfair, R. C., II (1994) "Cattle Egret (*Bubulcus ibis*)." In A. Poole and F. Gill, eds., *The Birds of North America: Life Histories for the 21st Century,* no. 113. Philadelphia: Academy of Natural Sciences; Washington, D.C.: American Ornithologists' Union.

*Werschkul, D. F. (1982) "Nesting Ecology of the Little Blue Heron: Promiscuous Behavior." *Condor* 84:381–84.

——— (1979) "Nestling Mortality and the Adaptive Significance of Early Locomotion in the Little Blue Heron." *Auk* 96:116–30.

RAILS AND COOTS   *Wading Bird*

PUKEKO or PURPLE SWAMPHEN
Porphyrio porphyrio
TASMANIAN NATIVE HEN
Tribonyx mortierii
DUSKY MOORHEN
Gallinula tenebrosa

HOMOSEXUALITY	TRANSGENDER	BEHAVIORS		RANKING	OBSERVED IN
● Female	○ Intersexuality	● Courtship	○ Pair-bonding	● Primary	● Wild
● Male	○ Transvestism	○ Affectionate	○ Parenting	○ Moderate	○ Semiwild
		● Sexual		○ Incidental	○ Captivity

PUKEKO
IDENTIFICATION: A large (nearly 20 inch) wading bird with bluish purple plumage, a red shield on its fore-head, and red feet with long toes. DISTRIBUTION: From the western Mediterranean through Middle East, east-ern and southern Africa, and throughout Australasia. HABITAT: Wetlands, especially swamps and marshy areas. STUDY AREA: Shakespear Regional Park, North Island, New Zealand; subspecies *P.p. melanotus*.

TASMANIAN NATIVE HEN
IDENTIFICATION: Similar to Pukeko, but flightless, and with grayish brown plumage, no red frontal shield, and shorter legs. DISTRIBUTION: Tasmania. HABITAT: Pasture, marshes, lakes, rivers. STUDY AREA: Hunting Ground, near Hobart, Tasmania.

DUSKY MOORHEN
IDENTIFICATION: Similar to Pukeko, but with black plumage and shorter legs. DISTRIBUTION: Australia, New Guinea, Indonesia. HABITAT: Wetlands. STUDY AREAS: Sullivan's Creek and Gungahlin, near Canberra, Australia; subspecies *G.t. tenebrosa*.

Social Organization

Pukeko are notable for their "communal" breeding system: they form stable groups of 4–14 birds during the mating season, usually with an equal number of males and females. All adult members (except nonbreeding "helpers," see below) generally mate with each other, and both sexes take turns incubating the eggs, which are often laid by several females in the same nest. Some birds pair off into (heterosexual) couples or remain single rather than forming communal groups, and outside the breeding season Pukeko usually live in flocks. Many Tasmanian Na-tive Hens and Dusky Moorhens also live in (generally smaller) communal breeding groups that have various types of polygamous or promiscuous mating arrange-ments. Some Tasmanian Native Hens also form monogamous pair-bonds.

Description

Behavioral Expression: Both male and female Pukeko engage in homosexual courtship and copulation with members of their communal group; these activities are more common and more developed between female birds. Lesbian courtship consists of a sequence of three activities (which also occur during heterosexual interactions). The courtship begins with ALLOPREENING: one female approaches another and bows to her, initiating mutual preening (stylized stroking of each other's feathers with their bill). This is followed by COURTSHIP-FEEDING, in which the two females exchange ritual gifts of food (usually small leaves or shoots). Finally, copulation takes place: the courting bird approaches the other in a distinctive upright posture while making a nasal HUMMING CALL; her partner adopts a hunched posture and mounting takes place, often with full cloacal (genital) contact. Sometimes, the two females reciprocate, the bird who was mounted first then mounting the other. Male homosexuality usually involves copulation and sometimes allopreening, but not courtship-feeding. Unlike female same-sex interactions, sexual activity between males is often initiated by the bird that is mounted, who places himself in the hunched posture in front of the other male as an invitation to mount.

In Tasmanian Native Hens, homosexual copulations occur in both males (often younger birds) and females living in the same group, while only male Dusky Moorhens participate in same-sex mounting. In all three species, birds that engage in homosexual activities also undertake parental duties such as nest-building, egg laying, incubation, and care of chicks (whether their own or those of other members of their communal group). In fact, homosexual activity (like heterosexual activity) in female Pukeko is most frequent just before or during the period of egg laying. Because birds in Pukeko and Tasmanian Native Hen groups are often related to each other, at least some homosexual—as well as heterosexual—activity is incestuous (mostly between brothers in Tasmanian Native Hens).

Frequency: Homosexuality is common among Pukeko, occurring in nearly

Courtship and sexual activity between female Pukeko: "allopreening" (above), approach with "humming call" (middle), and mounting ▶

45 percent of all communal groups. In this species, 7 percent of all copulations are homosexual, while 24 percent of courtship-feedings and 59 percent of courtship allopreening interactions are same-sex. In Tasmanian Native Hens and Dusky Moorhens, homosexual copulations make up 1–2 percent of all matings.

Orientation: Many adult breeding Pukeko are probably bisexual, capable of engaging in both homosexual and heterosexual courtship and copulation. In some cases, birds alternate between same-sex and opposite-sex copulations in quick succession. Nonbreeding helpers, however, generally participate in neither heterosexual nor homosexual activity. Far fewer individuals participate in homosexual behavior in Tasmanian Native Hens and Dusky Moorhens, but those that do are probably also bisexual.

Nonreproductive and Alternative Heterosexualities

As described above, the most common social unit among these species of Rails is not the heterosexual pair or nuclear family, but the communal group. In some populations of Pukeko, each such group may include up to seven nonbreeding individuals who assist in parental duties. These "helpers" (offspring from previous years) may delay their own reproductive careers for up to three years (one or two years in Tasmanian Native Hens, where an average of 18 percent of adults are nonbreeders). Some physiological mechanisms may be involved in this breeding suppression, since Pukeko helpers often have underdeveloped reproductive organs. In addition to several forms of polygamy, a number of other mating and parenting arrangements are found in these groups. For example, although all birds in Tasmanian Native Hen groups usually mate with each other, sometimes only one male-female pair actually has reproductive copulations. This social system has been called GENETIC MONOGAMY (because only one couple breeds) within SOCIAL POLYGAMY (since multiple partners mate with each other). In addition, although most group members remain together for life, females occasionally "divorce" their mates and join a new group; in some cases, this may lead to a female parenting young that are not her own. Occasionally, she may behave aggressively toward these foster chicks, even expelling them from the group. More violent confrontations sometimes occur when chicks stray into neighboring territories, where they may be killed by the resident group members. However, chicks are sometimes also adopted by neighboring groups, in both Tasmanian Native Hens and Dusky Moorhens.

Among breeding Pukeko, up to five males may court and mount the same female in quick succession—in these cases, one or more of the males may not actually inseminate the female. In fact, the "success" rate for most heterosexual copulations is not high: between one-half and two-thirds do not involve full genital contact and so do not result in insemination. Females often resist heterosexual advances by refusing to allow males to mount them, pecking at males who do mount them, preventing genital contact by not raising their tails, and prematurely terminating mating attempts. Copulations also occur when females are not fertile—for example, long before egg laying begins—and some males mate repeatedly without ever fathering any offspring that year. Similarly, mounts without genital contact account for more than a third of Dusky Moorhen matings and 60 percent of Tasmanian Native Hen matings. A few of these are REVERSE copulations, in which the female mounts the male. Incestuous matings are also common in Pukeko and Tasmanian Native Hens. Nearly two-thirds of all Pukeko heterosexual copulations in some populations are between related individuals, including mother-son, father-daughter, and brother-sister matings. More than 40 percent of Tasmanian Native Hen breeding groups contain related adults that mate with each other (mostly siblings); in addition, about 10 percent of copulations involve parents mounting their own offspring, including young chicks.

Other Species

Stable homosexual pairs often form among Cranes (e.g., *Grus* spp.) in captivity, a group of birds that is closely related to Rails.

Sources   *asterisked references discuss homosexuality/transgender*
*Archibald, G. W. (1974) "Methods for Breeding and Rearing Cranes in Captivity." *International Zoo Yearbook* 14:147–55.
Craig, J. L. (1990) "Pukeko: Different Approaches and Some Different Answers." In P. B. Stacey and W. D. Koenig, eds., *Cooperative Breeding in Birds: Long-term Studies of Ecology and Behavior*, pp. 385–412. Cambridge: Cambridge University Press.
*——— (1980) "Pair and Group Breeding Behavior of a Communal Gallinule, the Pukeko, *Porphyrio p. melanotus.*" *Animal Behavior* 28:593–603.
——— (1977) "The Behavior of the Pukeko." *New Zealand Journal of Zoology* 4:413–33.
Craig, J. L., and I. G. Jamieson (1988) "Incestuous Mating in a Communal Bird: A Family Affair." *American Naturalist* 131:58–70.
*Derrickson, S. R., and J. W. Carpenter (1987) "Behavioral Management of Captive Cranes—Factors Influencing Propagation and Reintroduction." In G. W. Archibald and R. F. Pasquier, eds., *Proceedings of the 1983 International Crane Workshop*, pp. 493–511. Baraboo, Wis.: International Crane Foundation.
Garnett, S. T. (1980) "The Social Organization of the Dusky Moorhen, *Gallinula tenebrosa* Gould (Aves: Rallidae)." *Australian Wildlife Research* 7:103–12.
*——— (1978) "The Behavior Patterns of the Dusky Moorhen, *Gallinula tenebrosa* Gould (Aves: Rallidae)." *Australian Wildlife Research* 5:363–84.
Gibbs, H. L., A. W. Goldizen, C. Bullough, and A. R. Goldizen (1994) "Parentage Analysis of Multi-Male Social Groups of Tasmanian Native Hens (*Tribonyx mortierii*): Genetic Evidence for Monogamy and Polyandry." *Behavioral Ecology and Sociobiology* 35:363–71.
Goldizen, A. W., A. R. Goldizen, and T. Devlin (1993) "Unstable Social Structure Associated with a Population Crash in the Tasmanian Native Hen, *Tribonyx mortierii.*" *Animal Behavior* 46:1013–16.
Goldizen, A. W., A. R. Goldizen, D. A. Putland, D. M. Lambert, C. D. Millar, and J. C. Buchan (1998) "'Wifesharing' in the Tasmanian Native Hen (*Gallinula mortierii*): Is It Caused By a Male-biased Sex Ratio?" *Auk* 115:528–32.
*Jamieson, I. G., and J. L. Craig (1987a) "Male-Male and Female-Female Courtship and Copulation Behavior in a Communally Breeding Bird." *Animal Behavior* 35:1251–53.
——— (1987b) "Dominance and Mating in a Communal Polygynandrous Bird: Cooperation or Indifference Towards Mating Competitors?" *Ethology* 75:317–27.
Jamieson, I. G., J. S. Quinn, P. A. Rose, and B. N. White (1994) "Shared Paternity Among Non-Relatives Is a Result of an Egalitarian Mating System in a Communally Breeding Bird, the Pukeko." *Proceedings of the Royal Society of London*, Series B 257:271–77.
*Lambert, D. M., C. D. Millar, K. Jack, S. Anderson, and J. L. Craig (1994) "Single- and Multilocus DNA Fingerprinting of Communally Breeding Pukeko: Do Copulations or Dominance Ensure Reproductive Success?" *Proceedings of the National Academy of Sciences* 91:9641–45.
*Ridpath, M. G. (1993) "Tasmanian Native-hen, *Tribonyx mortierii.*" In S. Marchant and P. J. Higgins, eds., *Handbook of Australian, New Zealand, and Antarctic Birds*, vol. 2, pp. 615–24. Melbourne: Oxford University Press.

*———— (1972) "The Tasmanian Native Hen, *Tribonyx mortierii*. I. Patterns of Behavior. II. The Individual, the Group, and the Population." *CSIRO Wildlife Research* 17:1–90.

*Swengel, S. R., G. W. Archibald, D. H. Ellis, and D. G. Smith (1996) "Behavior Management." In D. H. Ellis, G. F. Gee, and C. M. Mirande, eds., *Cranes: Their Biology, Husbandry, and Conservation*, pp. 105–22. Washington, D.C.: National Biological Service; Baraboo, Wis.: International Crane Foundation.

OTHER WADERS

Wading Bird

HAMMERHEAD
Scopus umbretta

HOMOSEXUALITY	TRANSGENDER	BEHAVIORS		RANKING	OBSERVED IN
● Female	○ Intersexuality	● Courtship	○ Pair-bonding	○ Primary	● Wild
● Male	○ Transvestism	○ Affectionate	○ Parenting	● Moderate	○ Semiwild
		● Sexual		○ Incidental	○ Captivity

IDENTIFICATION: A medium-sized brown, storklike bird with a prominent crest almost as long as its bill. DISTRIBUTION: Throughout tropical Africa from Senegal to east and south Africa; Madagascar. HABITAT: Wetlands, including savanna and woodland near water. STUDY AREAS: Near Niono, Mali; Karen and Nairobi, Kenya.

Social Organization

Hammerheads are usually found in pairs or groups of 8–10 individuals, though larger groups of up to 50 birds may also congregate. The mating system is believed to involve monogamous pair-bonds. Couples build extraordinarily large domed nests, often several on the same territory.

Description

Behavioral Expression: Hammerheads engage in a striking group courtship ceremony that includes same-sex mounting. Gathering at dawn on open lawns, riverbanks, rocks, or in the trees, groups of 3–20 birds (including some mated pairs) begin calling in unison. Their calls consist of a series of loud, high-toned *yips* that develop into a rapid sequence of "purring" or trilling notes, often accompanied by wing-flapping. Two or more birds may then perform a NODDING DISPLAY to each other, in which the bill is rapidly bobbed up and down (sometimes still accompanied by the YIP-PURR chorus), or pairs of birds may run in circles side by side. The culmination of this spectacular display is a series of mountings, initiated when one bird runs up to another, drooping and flicking its wings while it raises and lowers its crest, or when one bird solicits another by crouching, cocking its tail, and partially opening its wings. One Hammerhead then hops onto the back of the other, similar to a heterosexual copulation—except that, in addition to males mounting females, males also mount other males, and females mount females as well as males. The mounting bird beats its wings and gives the YIP-PURR call, while the mounted bird presses its tail up against the lowered tail of the mounter (though no cloacal [genital] contact takes place). Sometimes the two birds face in opposite directions; reciprocal mounting (mounter and mountee exchanging positions, often several times in succession) also occurs, as do "pile-ups" of three or four birds all mounted on each other.

▲ A female Hammerhead mounting another female

Frequency: Social courtship displays including ritualized same-sex mounting occur commonly in Hammerheads throughout the year. Each mounting session lasts for 10–40 minutes and may include dozens of mountings.

Orientation: All Hammerheads participate in group courtship displays, and most birds are probably involved in both same-sex and opposite-sex mountings.

Nonreproductive and Alternative Heterosexualities

As noted above, nonprocreative heterosexual activity (REVERSE mounting, as well as male-female mounts without cloacal contact) is common in Hammerheads. In addition, a significant percentage of heterosexual pairs may be nonbreeders. In some populations, as many as three-quarters of nests go unused (although some of these are "extra" nests built by breeding pairs), and couples may forgo breeding for four or more years at a time.

Other Species

Same-sex pairing occurs in semiwild White Storks (*Ciconia ciconia*) in some European populations, and individuals form homosexual pairs even when numerous opposite-sex partners are available. Same-sex pairs are able to successfully incubate, hatch, and raise foster young.

Sources   *asterisked references discuss homosexuality/transgender*

*Brown, L. H. (1982) "Scopidae, Hamerkop." In L. H. Brown, E. K. Urban, and K. Newman, eds., *The Birds of Africa*, vol. 1, pp. 168–72. London and New York: Academic Press.
*Campbell, K. (1983) "Hammerkops." *E.A.N.H.S. (East Africa Natural History Society) Bulletin* (January-April):11.

Cheke, A. S. (1968) "Copulation in the Hammerkop *Scopus umbretta.*" *Ibis* 110:201–3.

*Elliott, A. (1992) "Scopidae (Hamerkop)." In J. del Hoyo, A. Elliott, and J. Sargatal, eds., *Handbook of the Birds of the World, vol. 1, Ostrich to Ducks*, pp. 430–35. Barcelona: Lynx Ediciós.

Goodfellow, C. F. (1958) "Display in the Hamerkop, *Scopus umbretta.*" *Ostrich* 29:1–4.

Kahl, M. P. (1967) "Observations on the Behavior of the Hamerkop *Scopus umbretta* in Uganda." *Ibis* 109:25–32.

*King, C. E. (1990) "Reproductive Management of the Oriental White Stork *Ciconia boyciana* in Captivity." *International Zoo Yearbook* 29:85–90.

Stowell, R. F. (1954) "A Note on the Behavior of *Scopus umbretta.*" *Ibis* 96:150–51.

Wilson, R. T., and M. P. Wilson (1984) "Breeding Biology of the Hamerkop in Central Mali." In J. Ledger, ed., *Proceedings of the 5th Pan-African Ornithological Congress*, pp. 855–65. Johannesburg: Southern African Ornithological Society.

OTHER WADERS

Wading Bird

FLAMINGO
Phoenicopterus ruber

HOMOSEXUALITY	TRANSGENDER	BEHAVIORS		RANKING	OBSERVED IN
● Female	○ Intersexuality	○ Courtship	● Pair-bonding	○ Primary	○ Wild
● Male	○ Transvestism	○ Affectionate	● Parenting	● Moderate	○ Semiwild
		● Sexual		○ Incidental	● Captivity

IDENTIFICATION: The largest flamingo species (4–5 feet tall) with plumage ranging from pale whitish pink to bright orange-pink. DISTRIBUTION: The Mediterranean, sub-Saharan Africa, western Asia, Caribbean and Galapagos Islands, temperate South America. HABITAT: Shallow lakes, lagoons, mudflats, salt pans. STUDY AREAS: Zoo Atlanta, Georgia; Audubon Park Zoo, New Orleans, Louisiana; Chester Zoo, United Kingdom; Rotterdam Zoo, the Netherlands; Basel Zoo, Switzerland; subspecies *P.r. roseus*, the Greater Flamingo; *P.r. ruber*, the Caribbean Flamingo; and *P.r. chilensis*, the Chilean Flamingo.

Social Organization

Flamingos are extremely gregarious birds, often congregating in enormous groups numbering in the tens of thousands. During the mating season pair-bonds are formed and the birds nest in large colonies.

Description

Behavioral Expression: Flamingos form both male and female homosexual pairs. The bond is similar to that of a heterosexual mated pair and is expressed through shared activities, such as feeding and traveling together, calling in unison, helping each other in aggressive encounters against other birds, and sleeping side by side. The pair-bond is also reinforced through a number of stylized displays, such as ritual preening and feeding, as well as standing in an alert posture (with a graceful S-curve in the neck) in front of each other. Partners may even engage in mounting and copulations with one another; full genital contact occurs in matings between females, but not usually between males. Early in the breeding season, single male Flamingos also sometimes pursue other males in an attempt to mount them; this is known as DRIVING. Single males seeking male partners have even been known to harass heterosexual pairs, following them around and disrupting their copulations and incubation shifts in an attempt to gain access to the male.

Once formed, homosexual pair-bonds are strong and may persist from one breeding season to the next. Most pairings are monogamous; however, in some male pairs the partners also attempt to mount other birds (usually incubating birds of either sex). Sometimes two males and a female will even form a TRIAD in which the two males are bonded or sexually interested in each other as much, if not more so, than they are in the female. Homosexual partners sometimes also build nests together; in the case of male pairs, the nest (a pedestal-shaped mud platform) may become exceptionally large because of the contributions of both partners. Some male pairs, rather than building their own nest, "steal" or take over the nest of a heterosexual pair, occasionally breaking eggs in the process (this behavior also occurs between heterosexual pairs).

Homosexual couples often engage in parenting behavior. Male pairs incubate, hatch, and successfully raise foster chicks (for example, from a nest they have taken over, or from eggs supplied in captivity). Described as "model" parents, male partners may even "nurse" their chick. Flamingo parents (of either sex) typically feed their chicks a blood-red "milk" produced in their crops, and both males in homosexual couples feed their chicks with this crop milk. Some male pairs, however, do not attempt to acquire eggs, even if they have their own nest, while others do not appear to be interested in parenting at all, since they may roll the egg out of a nest they have acquired. Female pairs take turns incubating eggs on their nest; such eggs may be infertile, having been laid by the females themselves rather than acquired from another nest. As in heterosexual pairs, variation exists in the amount of incubation time contributed by each partner. Some share incuba-

Two pair-bonded male Flamingos feeding crop milk to their foster chick ▶

tion duties equally, while in other pairs, one female puts in more incubation shifts than the other. Overall, though, females in lesbian pairs contribute about five to six incubation shifts each, which is comparable to the average for heterosexual partners.

Frequency: In most captive populations with same-sex couples, about 5–6 percent of pairs are homosexual, although some populations have more than a quarter same-sex couples. Although homosexual pairs have not yet been observed in the wild, oversize nests similar to those built by male pairs are found in most colonies and may in fact belong to homosexual pairs (especially considering that most field studies have not systematically and unambiguously determined the sexes of all paired birds).

Orientation: Overall, only a fraction of the population is involved in same-sex pairing. Among these, some individuals have no prior heterosexual experience. However, other Flamingos in homosexual pairs were previously members of a heterosexual pair (and vice versa). Some males in same-sex pairs also occasionally attempt to mount other birds, including females. These are both examples of different types of bisexuality (sequential and simultaneous). In populations with more males than females (or vice versa), many birds remain single rather than forming same-sex pair-bonds, perhaps indicating more of a heterosexual orientation on their part.

Nonreproductive and Alternative Heterosexualities

Although the standard social unit in Flamingos is the breeding monogamous pair, a number of alternative heterosexual pairing and family arrangements occur. Trios or triads—one male with two females or one female with two males—are fairly common (at least in zoos). Typically all three birds share incubation and chick-raising duties (although no same-sex bonding occurs); in trios with two females, there may be two separate nests, or both females may share a nest. Flamingo pairs also sometimes engage in nonreproductive copulations, mating far in advance of the female's fertile (or the male's sperm-producing) period. Many mated pairs are nonmonogamous, with both males and females seeking copulations with outside partners. In one zoo population, 47 percent of the females and 79 percent of the males participated in such "infidelity," and about 8 percent of all copulations were with outside partners; at another zoo, 25–60 percent of all pairs were nonmonogamous. Furthermore, divorce is extremely common in wild Flamingos: nearly all birds change partners between breeding seasons, and about 30 percent of males even switch mates during the season (in contrast, most pairings in zoos are long-lived).

Once chicks are hatched, a number of social systems are available to relieve the biological mother and father of some of their parenting duties. For example, nonbreeding birds sometimes produce crop milk and "nurse" other birds' chicks or else act as foster feeders for orphaned chicks. In addition, as they get older, Flamingo chicks typically gather into large nursery groups or CRÈCHES, which may contain several thousand youngsters. These groups provide them with safety in the absence of direct parental supervision, and adults also sometimes feed youngsters other than their own in these crèches. Chicks are often forced into crèches as a result of attacks from adult birds, including their own parents, and crèches may therefore also provide refuge from aggression by other Flamingos. Breeding in wild Flamingos can be irregular, with entire colonies sometimes forgoing reproduction for three or four years at a time—one colony in France failed to produce chicks for 13 out of 34 years (38 percent of the time). In addition, even if breeding is undertaken, it may be abruptly halted, with all or a large portion of the colony—often as many as half of all pairs—abandoning their eggs. Usually it is the female who initiates desertion of a nest.

Other Species

Homosexual pairs occur among both male and female Lesser Flamingos (*Phoeniconaias minor*) in captivity, including nest-building and (in females) egg-laying. Both sexes participate in homosexual mounting, though only males generally achieve full genital contact. Males in same-sex pairs also sometimes chase and mount females in lesbian pairs, while the latter may mate with males to fertilize their eggs. Most females in homosexual pairs, however, show no interest in males. Pairs of female Scarlet Ibises (*Eudocimus ruber*) have also been observed in captive flocks, nesting together and sometimes laying fertile eggs.

Sources   *asterisked references discuss homosexuality/transgender*

Allen, R. P. (1956) *The Flamingos: Their Life History and Survival.* National Audubon Society Research Report no. 5. New York: National Audubon Society.

*Alraun, R., and N. Hewston (1997) "Breeding the Lesser Flamingo *Phoeniconaias minor*," *Avicultural Magazine* 103:175–81.

Bildstein, K. L., C. B. Golden, and B. J. McCraith (1993) "Feeding Behavior, Aggression, and the Conservation Biology of Flamingos: Integrating Studies of Captive and Free-Ranging Birds." *American Zoologist* 33:117–25.

Cézilly, F. (1993) "Nest Desertion in the Greater Flamingo, *Phoenicopterus ruber roseus*." *Animal Behavior* 45:1038–40.

Cézilly, F., and A. R. Johnson (1995) "Re-Mating Between and Within Breeding Seasons in the Greater Flamingo *Phoenicopterus ruber roseus*." *Ibis* 137:543–46.

*Elbin, S. B., and A. M. Lyles (1994) "Managing Colonial Waterbirds: the Scarlet Ibis *Eudocimus ruber* as a Model Species." *International Zoo Yearbook* 33:85–94.

Kahl, M. P. (1974) "Ritualized Displays." In J. Kear and N. Duplaix-Hall, eds., *Flamingos*, pp. 142–49. Berkhamsted, UK: T. and A. D. Poyser.

*King, C. E. (1996) Personal communication.

*——— (1994) "Management and Research Implications of Selected Behaviors in a Mixed Colony of Flamingos at Rotterdam Zoo." *International Zoo Yearbook* 33:103–13.

*——— (1993a) "Ondergeschoven Kinderen [Supposititious Children]." *Dieren* 10(4):116–19.

*——— (1993b) "Ongelukkige Flamingo Liefdes [Tales of Flamingo-Love Gone Awry]." *Dieren* 10(2):36–39.

Ogilvie, M., and C. Ogilvie (1989) *Flamingos.* Wolfeboro, N.H.: Alan Sutton.

*Shannon, P. (1985) "Flamingo Management at Audubon Park Zoo and the Benefits of Long-Term Research." In *AAZPA Regional Conference Proceedings*, pp. 226–36. Wheeling, W.Va.: AAZPA.

*Stevens, E. F. (1996) Personal communication.

——— (1991) "Flamingo Breeding: The Role of Group Displays." *Zoo Biology* 10:53–63.

*Studer-Thiersch, A. (1975) "Basle Zoo." In Kear and Duplaix-Hall, *Flamingos*, pp. 121–30.

Tourenq, C., A. R. Johnson, and A. Gallo (1995) "Adult Aggressiveness and Crèching Behavior in the Greater Flamingo, *Phoenicopterus ruber roseus*." *Colonial Waterbirds* 18:216–21.

*Wilkinson, R. (1989) "Breeding and Management of Flamingos at Chester Zoo." *Avicultural Magazine* 95:51–61.

Shore Birds

SANDPIPERS AND THEIR RELATIVES

"ARCTIC" SANDPIPERS

RUFF
Philomachus pugnax

BUFF-BREASTED SANDPIPER
Tryngites subruficollis

HOMOSEXUALITY	TRANSGENDER	BEHAVIORS		RANKING	OBSERVED IN
● Female	○ Intersexuality	● Courtship	● Pair-bonding	● Primary	● Wild
● Male	● Transvestism	○ Affectionate	○ Parenting	○ Moderate	○ Semiwild
		● Sexual		○ Incidental	○ Captivity

RUFF
IDENTIFICATION: A large (12 inch) sandpiper with gray or brownish plumage and, in some males, spectacular ruffs and feather tufts on the head that vary widely in color and pattern (see below). DISTRIBUTION: Northern Europe and Asia; winters in Mediterranean, sub-Saharan Africa, Middle East, India. HABITAT: Tundra, lakes, swampy meadows, farms, floodlands. STUDY AREAS: Texel, Schiermonnikoog, Roderwolde, and several other locations in the Netherlands; Oie and Kirr Islands, Germany.

BUFF-BREASTED SANDPIPER
IDENTIFICATION: A medium-sized (7–8 inch) wading bird with a small head and short beak, buff-colored face and underparts, and regular dark brown patterning on the back and crown. DISTRIBUTION: Arctic Canada, Alaska, extreme northeast Siberia; winters in south-central South America. HABITAT: Tundra, grasslands, mudflats. STUDY AREA: Meade River, Alaska.

Social Organization

Ruffs and Buff-breasted Sandpipers are both LEKKING species, which means that males gather together to perform elaborate courtship displays on communal grounds known as LEKS (some Buff-breasts also display solitarily). The mating system is polygamous or promiscuous: males (and sometimes females) mate with multiple partners, and females raise any resulting offspring on their own. Outside of the breeding season these sandpipers tend to associate in flocks, which can number in the thousands among Ruffs.

Description

Behavioral Expression: There are four distinct types or "classes" of male Ruffs, differing in their appearance, social behavior, and sexuality. RESIDENT males generally have dark plumage (with a wide variety of different feather patterns) and defend their own territories on the lek. MARGINAL males look similar to residents but do not have their own territories; they stay on the periphery of the lek and are often attacked by residents. SATELLITE males usually have white or light-colored plumage; they do not own territories, but often visit the lek and associate with particular residents. Finally, NAKED-NAPE males lack the nuptial plumage—ruff and head tufts—of other males, giving them the superficial appearance of females. They are not territorial either, but occasionally visit leks for short periods. Naked-napes may include younger males and/or adults passing through on their migratory journeys prior to developing their breeding plumage. Resident and satellite males also differ genetically from one another.

Homosexual behavior occurs among males of all types and is especially prominent between residents and satellites. While a resident male is displaying on his territory, one or more satellites may approach him and engage in courtship behaviors. Most notable of these is SQUATTING, in which the males lie with their bellies to the ground and expand their ruffs, crouching together while the resident places his bill on top of the satellite's head. This may lead to homosexual copulation, in which either the resident or the satellite mounts the other male and attempts to make genital contact—he lowers himself and spreads his wings while holding the other male's head feathers in his bill. The mounted bird reacts by either remaining crouched or by trying to shake the other male off his back. If more than one satellite male is present on the lek, they sometimes also mount each other. Many satellites have "preferred" resident males with whom they spend most of their time, and residents may also actively entice satellite males onto their display courts.

Females are often drawn to the activities between resident and satellite males, and heterosexual courtship and copulation (involving either residents or satellites) may occur at the same time as homosexual activities (or shortly thereafter). Occasionally, a satellite male will mount a resident male who is in the act of mating with a female; residents and satellites may also try to prevent each other from mating with females. Naked-nape males also engage in homosexual mounting with each other and with residents. When a naked-nape arrives on the lek, the resident male may respond by squatting; the naked-nape approaches him in a horizontal posture or may himself squat. The naked-nape may then try to mount the resident, although he does not usually lower his body to "complete" the copulation; he may also mount in a backwards position with his head facing the resident's tail. Residents also sometimes mount naked-napes, and naked-napes also mount each other. Naked-nape males are sometimes courted by other males during stopovers on the spring migration as well (i.e., outside of the mating season). Although mar-

▲ *A "marginal" male Ruff approaches a crouching "satellite" male* (above) *and then mounts him*

ginal males rarely participate in sexual activity (with either males or females), they have occasionally been seen mounting other males.

Female Ruffs—also known as Reeves—engage in homosexual behavior as well. They often arrive on a lek in groups, and females sometimes mount one another as they begin simultaneously crouching near a resident male during courtship activities. Genital contact may occur, although this is difficult to verify, even for heterosexual (or male homosexual) copulations. Females also occasionally court each other, using some of the same stylized movements such as wing quivering that are seen in heterosexual courtship.

Male Buff-breasted Sandpipers attract other birds to their lek territories with a dramatic WING-UP DISPLAY that can be seen from miles away, in which they raise one wing vertically and flash its brilliant satiny-white underlining. Usually females are attracted to this courtship display, and sometimes up to six of them gather around a displaying male. Often, however, a male from a neighboring territory is drawn to the display as well (or he may "camouflage" himself in a group of females). He may interrupt the courtship when he arrives by mounting the displaying male and trying to copulate with him. He may also aggressively peck the other male on his head and neck while mounted on him, then fly back to his own territory. Sometimes the females follow him, and then the pattern of interruption and mounting is repeated, only this time the other male arrives to disrupt the courtship. This sequence of heterosexual courtship and homosexual mounting may be repeated many times, back and forth for an hour or more. Sometimes, instead of flying back to his own territory with the females, a male simply returns repeatedly to his neighbor's territory, continuously interrupting the male's courtship by mounting him. Homosexual mounting also occurs in other contexts: when not as many females are present on the leks (especially later in the mating season), one or more males may enter a neighbor's territory and simply mount him. As many as four males at a time may participate in such activity.

Frequency: Homosexual mounting occurs regularly in Ruffs, especially at the beginning of the mating season. During one informal three-and-a-half-hour observation period, for example, 3 out of 12 mountings (25 percent) were between males. In Buff-breasted Sandpipers, courtship interruptions by other males are common. Nearly a third of all courtships are disrupted by another male's arrival, although homosexual mounting does not necessarily take place every time—but then neither does heterosexual mounting, since females usually leave without having copulated (even if there has been no interruption).

Orientation: In Ruffs, homosexual behavior is seen primarily in resident males, who constitute about 40 percent of the male population, and satellite males, who make up roughly 15 percent to one-third of all males (on average). Not all of these individuals engage in same-sex mounting—but neither do they all participate in heterosexual mounting. On some leks, just over half the resident males mate with females—and some only copulate once each season—while 40–90 percent of satellites never mate with females (although they may court them). Of those birds that do participate in homosexual behavior, many alternate between same-sex and opposite-sex interactions and are therefore bisexual. This is also true of females, although some Reeves seem to "prefer" homosexual interactions since they ignore males in favor of mounting other females. Naked-nape males—who probably constitute no more than 10 percent of the male population—rarely, if ever, mate with females. Thus, when naked-napes and nonmating residents and satellites are all taken into account, significant portions of the male Ruff population—perhaps more than half—are involved predominantly, if not exclusively, in sexual activity with other males. This homosexuality may be long-term—satellites, for example, almost never become residents during their entire lives (since these two classes differ genetically). Most male Buff-breasted Sandpipers that mount other males are probably functionally bisexual (if not predominantly heterosexual), since they also court and mate with females.

Nonreproductive and Alternative Heterosexualities

Birds who do not mate or breed are a notable feature of both Ruff and Buff-breasted Sandpiper populations (as noted above). More than 60 percent of male Ruffs, on average, do not copulate with females (this includes males of all categories), while more than half of all territorial Buff-breasted males do not mate (and many males in this species are not territorial and hence probably do not reproduce either). In many cases, males are unable to breed because females select which males they want to mate with and often refuse to allow certain males to copulate with them. However, females of both species occasionally choose more than one male to mate with: almost a quarter of all Buff-breasted nests contain eggs fathered by more than one male, while Reeves have been known to copulate with several different males in a row. Sometimes, more than one male will even try to copulate simultaneously with the same female—usually a resident and a satellite together. Cross-species sexual activity has also been observed: male Ruffs occasionally court and try to mount other sandpipers such as red knots (*Calidris canutus*).

Courtship and mating are virtually the only times during the entire breeding season when the two sexes are together: in both species, there is significant separation (both physical and temporal) between males and females. After copulating, female Ruffs often leave the lek and migrate farther north to lay their eggs—sometimes more than 1,800 miles away, and two to three weeks after they last mated. It is thought that females are able to do this because they store sperm in special glands in their reproductive tracts, effectively separating fertilization from insemination. Male Buff-breasts take no part in parenting, and in fact depart from the leks well before the eggs hatch. Male Ruffs also generally leave parenting entirely to the females, who occasionally cooperate amongst themselves in tending and defending their young. In fact, chicks may be killed by males if the two sexes ever interact following the hatching of eggs. Infanticide has not been observed in Buff-breasts, although about 10 percent of nests are abandoned by females if a predator takes some of the eggs. Sex segregation also occurs in Ruffs after the breeding season because males and females have different migratory patterns. Females tend to travel farther south to spend the winter, and at some wintering sites in Africa they may outnumber males 15 to 1.

Sources
asterisked references discuss homosexuality/transgender

Cant, R. G. H. (1961) "Ruff Displaying to Knot." *British Birds* 54:205.

*Cramp, S., and K. E. L. Simmons (eds.) (1983) "Ruff (*Philomachus pugnax*)." In *Handbook of the Birds of Europe, the Middle East, and North Africa*, vol. 3, pp. 385–402. Oxford: Oxford University Press.

*Hogan-Warburg, A.J. (1993) "Female Choice and the Evolution of Mating Strategies in the Ruff *Philomachus pugnax* (L.)." *Ardea* 80:395–403.

*——— (1966) "Social Behavior of the Ruff, *Philomachus pugnax* (L.)." *Ardea* 54:109–229.

Hugie, D. M., and D. B. Lank (1997) "The Resident's Dilemma: A Female Choice Model for the Evolution of Alternative Mating Strategies in Lekking Male Ruffs (*Philomachus pugnax*)." *Behavioral Ecology* 8:218–25.

*Lanctot, R. B. (1995) "A Closer Look: Buff-breasted Sandpiper." *Birding* 27:384–90.

*Lanctot, R. B., and C. D. Laredo (1994) "Buff-breasted Sandpiper (*Tryngites subruficollis*)." In A. Poole and F. Gill, eds., *The Birds of North America: Life Histories for the 21st Century*, no. 91. Philadelphia: Academy of Natural Sciences; Washington, D.C.: American Ornithologists' Union.

Lank, D. B., C. M. Smith, O. Hanotte, T. Burke, and F. Cooke (1995) "Genetic Polymorphism for Alternative Mating Behavior in Lekking Male Ruff *Philomachus pugnax*." *Nature* 378:59–62.

*Myers, J. P. (1989) "Making Sense of Sexual Nonsense." *Audubon* 91:40–45.

——— (1980) "Territoriality and Flocking by Buff-breasted Sandpipers: Variations in Non-breeding Dispersion." *Condor* 82:241–50.

——— (1979) "Leks, Sex, and Buff-breasted Sandpipers." *American Birds* 33:823–25.

Oring, L. W. (1964) "Displays of the Buff-breasted Sandpiper at Norman, Oklahoma." *Auk* 81:83–86.

Pitelka, F. A., R. T. Holmes, and S. F. MacLean, Jr. (1974) "Ecology and Evolution of Social Organization in Arctic Sandpipers." *American Zoologist* 14:185–204.

Prevett, J. P., and J. F. Barr (1976) "Lek Behavior of the Buff-breasted Sandpiper." *Wilson Bulletin* 88:500–503.

Pruett-Jones, S. G. (1988) "Lekking versus Solitary Display: Temporal Variations in Dispersion in the Buff-breasted Sandpiper." *Animal Behavior* 36:1740–52.

*Scheufler, H., and A. Stiefel (1985) *Der Kampfläufer* [The Ruff]. Neue Brehm-Bücherei, 574. Wittenberg Lutherstadt: A. Ziemsen Verlag.

*Selous, E. (1906–7) "Observations Tending to Throw Light on the Question of Sexual Selection in Birds, Including a Day-to-Day Diary on the Breeding Habits of the Ruff (*Machetes pugnax*)." *Zoologist* 10:201–19, 285–94, 419–28; 11:60–65, 161–82, 367–80.

*Stonor, C. R. (1937) "On a Case of a Male Ruff (*Philomachus pugnax*) in the Plumage of an Adult Female." *Proceedings of the Zoological Society of London*, Series A 107:85–88.

*van Rhijn, J. G. (1991) *The Ruff: Individuality in a Gregarious Wading Bird*. London: T. and A. D. Poyser.

——— (1983) "On the Maintenance and Origin of Alternative Strategies in the Ruff *Philomachus pugnax*." *Ibis* 125:482–98.

——— (1973) "Behavioral Dimorphism in Male Ruffs, *Philomachus pugnax* (L.)." *Behavior* 47:153–229.

SHANKS

Shore Bird

(COMMON) GREENSHANK
Tringa nebularia
(COMMON) REDSHANK
Tringa totanus

HOMOSEXUALITY	TRANSGENDER	BEHAVIORS		RANKING	OBSERVED IN
○ Female	○ Intersexuality	● Courtship	○ Pair-bonding	○ Primary	● Wild
● Male	○ Transvestism	○ Affectionate	○ Parenting	● Moderate	○ Semiwild
		● Sexual		○ Incidental	○ Captivity

GREENSHANK
IDENTIFICATION: A large (13–14 inch) sandpiper with streaked and spotted, dark brownish gray plumage; long and slightly upturned bill; greenish yellow legs. DISTRIBUTION: North-central Europe and Asia; winters in western Europe, Africa, Australasia. HABITAT: Marshes, bogs, moors, lakes. STUDY AREAS: Speyside and the northwest highlands of Scotland.

REDSHANK
IDENTIFICATION: Slightly smaller than the Greenshank; plumage grayish brown, with black and dark brown streaks and spots; orange-red legs. DISTRIBUTION: Europe and central Asia; winters in coastal Africa, Middle East, southern Asia. HABITAT: Wet meadows, moors, marshes, lakes, rivers. STUDY AREA: Ribble Marshes National Nature Reserve, Lancashire, England; subspecies *T.t. totanus*.

Social Organization

Outside of the mating season, Greenshanks congregate in flocks of 20–25 birds, while Redshanks are less social and may even be solitary. During the mating season, monogamous pairs are the predominant social unit, although a number of variations occur, including nonbreeding birds (see below).

Description

Behavioral Expression: In both Greenshanks and Redshanks, males sometimes court and copulate with each other. Homosexual courtship in Greenshanks involves spectacular aerial and ground displays, employing patterns also found in heterosexual courtship. One male may pursue another in a NUPTIAL FLIGHT, which starts out as a twisting, careening chase low over the ground, followed by a remarkable ascent of both birds to a great altitude. The two males swerve and turn in unison as they climb higher, sometimes disappearing completely into the clouds; the "sky dance" comes to a dramatic close as the males plummet back to earth in a steep dive. In the ground courtship display, two males bow, fan their tails, flap their wings, and utter deep growling calls or *chip, quip,* and *too-hoo* notes. This may lead

▲ *Homosexual courtship in Redshanks: one male pursuing another in a "ground chase"*

to copulation (often performed on a stump or tree branch), in which one male flutters onto the back of another, lowering his body to make contact with the other while slowly flapping his wings. Homosexual copulation may be briefer than the corresponding heterosexual behavior. Males also sometimes try to copulate with other males that are calling in the treetops, and with males whose female mates are incubating eggs (in the latter cases, the female often makes a threatening or challenging call during the homosexual interaction). Male Greenshanks are also sometimes courted by male Green Sandpipers (*Tringa ochropus*), who approach them from behind with drooping wings and raised, partially fanned tails. The Sandpipers may also try to mount them; in contrast to within-species homosexual matings, male Greenshanks typically resist these sexual advances, shaking the Sandpipers off and violently pecking at them.

Male Redshanks court other males with a GROUND CHASE (also used in heterosexual courtship). In this display, one male pursues another in a series of curves and circles, often running in a distinctive sideways motion—similar to that of a crab—with ruffled feathers and fanned tail. The pursuing male may give a "mate-call" consisting of repeated paired notes: *tyoo-tyoo . . . tyoo-tyoo . . .*, and both birds also sometimes make chipping or trilled calls during the chase. Occasionally one male also mounts the other and tries to copulate with him, although his advances are often rebuked by the other male (as also frequently happens in heterosexual copulations).

Frequency: Homosexual courtship and copulation probably occur only occasionally in Greenshanks and Redshanks, although no systematic long-term studies of their prevalence have yet been undertaken.

Orientation: In many cases, individuals that engage in same-sex activity are bisexual. Male Greenshanks and Redshanks court birds of both sexes, while male Greenshanks are sometimes heterosexually paired fathers when they participate in homosexual copulations with other males. In at least one case, a male Greenshank associated himself with a heterosexual pair and tried to copulate with both the male and the female (although it is not known whether that male himself was single or paired with a female).

Nonreproductive and Alternative Heterosexualities

Sexual behavior in these two Sandpipers often occurs at times when fertilization is not possible: male Greenshanks and Redshanks court and mate with females after eggs are laid, both during incubation and following the hatching of chicks. In addition, REVERSE mounting (in which the female mounts the male) also occurs in Greenshanks. Courtship and copulation with other species of sandpipers have been recorded, including lesser yellowlegs (*Tringa flavipes*) and Green Sandpipers (as noted above). A wide variety of alternative heterosexual mating and parenting arrangements are also found in these birds. Although both Greenshanks and Redshanks are primarily monogamous, males of both species sometimes court and mate with females other than their own mate. In addition, some individuals engage in a number of different polygamous mating arrangements: occasional trios occur in both species, composed of two females and a male. Both females may lay in the same nest, they may have separate nests, or one female may be nonbreeding. Redshanks sometimes also participate in serial polygamy, in which a male mates with a second female, or a female with a second male, after laying a clutch with a first mate. This involves deserting or "divorcing" the first mate, who then typically raises the young as a single parent. (In Greenshanks, another form of "single parenting" occasionally occurs, in which the male partner fails to help the female incubate the eggs, but remains paired with her.) In fact, about 11 percent of Redshanks (and up to a quarter of Greenshanks) change partners between or within breeding seasons (this is more common among males), and only about a third of all males and half of all females mate for life. Some birds may divorce and re-pair with up to four different partners during their lives. In a few cases, female Redshanks have even left their mate to pair with another male, only to return to their "ex" the following season and remain with him for many more years. Adoption or foster-parenting also takes place in Redshanks: females sometimes lay eggs in other females' nests (who then raise all the chicks as their own), and Redshanks have even been seen taking care of chicks of other species of shorebirds, such as the avocet (*Recurvirostra avosetta*).

Some Greenshanks do not participate in breeding at all—about a quarter of all males, on average, do not procreate (and in some years this figure may be higher—nearly half of all males). This includes both single birds and those that are heterosexually paired but do not breed. Such nonbreeding pairs constitute an average of more than 15 percent of pairs, though in some years more than a third do not reproduce. Heterosexual relations may also be marked by unwillingness and aggression between the sexes: female Redshanks sometimes turn on males that are chasing them during courtship, staving off their advances with prolonged fights involving much pecking and scratching. Females of both species may also refuse to allow males to mount them: about a third of all Redshank heterosexual mating attempts, for example, are not completed due to female refusal.

Sources   *asterisked references discuss homosexuality/transgender*

*Cramp, S., and K. E. L. Simmons (eds.) (1983) "Redshank (*Tringa totanus*)." In *Handbook of the Birds of Europe, the Middle East, and North Africa*, vol. 3, pp. 531–35. Oxford: Oxford University Press.

Garner, M. S. (1987) "Lesser Yellowlegs Attempting to Mate with Redshank." *British Birds* 80:283.

Hakansson, G. (1978) "Incubating Redshank, *Tringa totanus*, Warming Young of Avocet, *Avocetta recurvirostra*." *Vår Fågelvårld* 37:137–38.

*Hale, W. G. (1980) *Waders*. London: Collins.

*Hale, W. G., and R. P. Ashcroft (1983) "Studies of the Courtship Behavior of the Redshank *Tringa totanus.*" *Ibis* 125:3–23.

*————— (1982) "Pair Formation and Pair Maintenance in the Redshank *Tringa totanus.*" *Ibis* 124:471–501.

*Nethersole-Thompson, D. (1975) *Pine Crossbills: A Scottish Contribution.* Berkhamsted: T. and A. D. Poyser.

*————— (1951) *The Greenshank.* London: Collins.

*Nethersole-Thompson, D., and M. Nethersole-Thompson (1986) *Waders: Their Breeding, Haunts, and Watchers.* Calton: T. and A. D. Poyser.

*————— (1979) *Greenshanks.* Vermillion, S.D. : Buteo Books.

Thompson, D. B. A., P. S. Thompson, and D. Nethersole-Thompson (1988) "Fidelity and Philopatry in Breeding Redshanks (*Tringa totanus*) and Greenshanks (*T. nebularia*)." In H. Ouellet, ed., *Acta XIX Congressus Internationalis Ornithologici* (Proceedings of the 19th International Ornithological Congress), 1986, Ottawa, vol. I, pp. 563–74. Ottawa: University of Ottawa Press.

————— (1986) "Timing of Breeding and Breeding Performance in a Population of Greenshanks (*Tringa nebularia*)." *Journal of Animal Ecology* 55:181–99.

STILTS   *Shore Bird*

BLACK-WINGED STILT
Himantopus himantopus
BLACK STILT
Himantopus novaezelandiae

HOMOSEXUALITY	TRANSGENDER	BEHAVIORS		RANKING	OBSERVED IN
● Female	○ Intersexuality	● Courtship	● Pair-bonding	○ Primary	● Wild
○ Male	○ Transvestism	○ Affectionate	● Parenting	● Moderate	○ Semiwild
		● Sexual		○ Incidental	○ Captivity

BLACK-WINGED STILT
IDENTIFICATION: A fairly large (12–15 inch) sandpiper-like bird with long pink legs, white plumage with black wings and back, and a slender black bill. DISTRIBUTION: Throughout much of Australasia, Europe, Africa, Central and South America, western and southern United States. HABITAT: Tropical and temperate wetlands. STUDY AREAS: Gyotuku Sanctuary, Ichikawa City, Japan; Morocco and the Belgium/Netherlands border area; subspecies *H.h. himantopus*.

BLACK STILT
IDENTIFICATION: Similar to Black-winged Stilt but with entirely black plumage. DISTRIBUTION: New Zealand; critically endangered. HABITAT: Rivers, lakes, swamps. STUDY AREA: Mackenzie Basin, South Island, New Zealand.

Social Organization

The primary social unit among Stilts is the monogamous mated pair; Black-winged couples often nest in loose colonies containing 2–50 families, while Black

Stilts are less gregarious. Outside of the mating season, the birds gather in flocks of usually up to ten individuals, although assemblies of hundreds of Black-winged Stilts may also occur.

Description

Behavioral Expression: Lesbian pairs occur in both Black-winged and Black Stilts. In these partnerships, two females participate in courtship, copulation, and parenting activities together. Homosexual pairing and courtship in Black-winged Stilts often begins with a ritual NEST DISPLAY activity: each female takes turns symbolically "showing" the other a nest location by squatting on land as if she were incubating eggs, and making pecking motions in the mud as though she were turning the eggs over. Although heterosexual pairings also frequently commence with this activity, in lesbian pairs the two birds may spend considerably more time engaged in nest display. This may lead to full-scale courtship activities, such as DIBBLING— bill dipping and shaking by both partners, involving prominent splashing of water—and ritual preening, in which one female preens the side of her breast nearest to the other female, frequently combined with more splashing activity. Often one female takes up the NECK EXTENDED posture, a stylized pose in which she stands with her legs slightly apart and her neck lowered and extended just above the surface of the water. While one female is standing in this position, the other performs a courtship dance in which she moves back and forth behind her partner, striding in semicircles from one side to the other. The two females may participate in continuous courtship activities for up to three-quarters of an hour at a time. Sexual activity also takes place between members of a lesbian pair, with one female mounting the other as in heterosexual copulation.

Once bonded, the pair vigorously defends their territory against any intruding families and eventually builds a nest together. Because both females sometimes lay eggs, the nests of lesbian pairs often contain SUPERNORMAL CLUTCHES of 7–8 eggs, up to twice as many as those of heterosexual pairs (which usually have only 3–4 eggs). Both females take turns incubating the eggs; in heterosexual pairs the two birds also share incubation duties, but in many cases the female contributes a disproportionately greater amount of time than does the male. If the eggs are eaten by predators, the lesbian pair replaces them by laying a second clutch (as often happens with heterosexual parents as well). Most eggs laid by

A nest with a "supernormal clutch" of eggs belonging to a pair of female Black-winged Stilts in Japan ▶

same-sex pairs are probably infertile. Like heterosexual pairs, some Black-winged Stilt lesbian pairs divorce. This may occur, for example, when one female forms a new pair-bond with another female. Although mate-switching may initially be accompanied by aggression between the separating females, the divorced partner sometimes still remains "friends" with the new pair, being allowed to visit their territory (unlike other birds, which are routinely chased away).

Frequency: In Black-winged Stilts, female pairs may constitute anywhere from 5–17 percent of the total number of pairs (depending on the population), while about 2 percent of Black Stilt pairings are lesbian. Homosexual copulation occurs at fairly high rates in some Black-winged Stilt female pairs: in one case, two females were seen to mate with each other as often as five times in one day.

Orientation: Because the eggs they lay are usually infertile, it is likely that many female Black-winged Stilts in lesbian pairs are exclusively homosexual, i.e., they do not copulate with males (at least for the duration of their bond). In addition, some females show a persistent orientation toward other females, since they remate with another female if their lesbian partnership breaks up.

Nonreproductive and Alternative Heterosexualities

In addition to long-lasting monogamous pairs, a variety of alternative heterosexual family arrangements occur in Stilts. Black Stilts occasionally form trios of two females and one male (with both females laying eggs), while Black-winged Stilt pairs sometimes adopt chicks from other families and foster-parent them along with their own. Divorce and remating may occur in male-female pairs of Black-winged Stilts, and some males engage in courtship and copulation with females other than their mates. In Black Stilts, heterosexual pairs sometimes separate when their young fledge: the male often takes the juveniles with him as a single parent when he migrates, while the female remains behind. On returning, the male may get back together with his previous partner, or the female may find a new mate. In some intact Black Stilt families, fathers may be abusive toward their young, behaving aggressively or rejecting their male offspring (although this has so far only been reported in captivity). In both of these Stilt species, individuals often masturbate by mounting an inanimate object (such as a piece of driftwood) and performing copulatory movements. In Black-winged Stilts this behavior may occur with extraordinary frequency—one bird was recorded making 20–30 such masturbatory mounts in one session, roughly once every 30 seconds. Finally, birds sometimes pair with individuals outside of their species: in some populations, about 30 percent of Black Stilts mate with Black-winged Stilts, and hybrids of the two species are common.

Sources   *asterisked references discuss homosexuality/transgender*

Cramp, S., and K. E. L. Simmons, eds. (1983) "Black-winged Stilt (*Himantopus himantopus*)." In *Handbook of the Birds of Europe, the Middle East, and North Africa*, vol. 3, pp. 36–47. Oxford: Oxford University Press.

Goriup, P. D. (1982) "Behavior of Black-winged Stilts." *British Birds* 75:12–24.

Hamilton, R. B. (1975) *Comparative Behavior of the American Avocet and the Black-necked Stilt (Recurvirostridae).* Ornithological Monographs no. 17. Washington, DC: American Ornithologists' Union.

Kitagawa, T. (1989) "Ethosociological Studies of the Black-winged Stilt *Himantopus himantopus himantopus.* I. Ethogram of the Agonistic Behaviors." *Journal of the Yamashina Institute of Ornithology* 21:52–75.

*——— (1988a) "Ethosociological Studies of the Black-winged Stilt *Himantopus himantopus himantopus.* III. Female-Female Pairing." *Japanese Journal of Ornithology* 37:63–67.

——— (1988b) "Ethosociological Studies of the Black-winged Stilt *Himantopus himantopus himantopus.* II. Social Structure in an Overwintering Population." *Japanese Journal of Ornithology* 37:45–62.

*Pierce, R. J. (1996a) "Recurvirostridae (Stilts and Avocets)." In J. del Hoyo, A. Elliott, and J. Sargatal, eds., *Handbook of the Birds of the World, vol. 3: Hoatzin to Auks,* pp. 332–47. Barcelona: Lynx Ediciòns.

——— (1996b) "Ecology and Management of the Black Stilt *Himantopus novaezelandiae.*" *Bird Conservation International* 6:81–88.

——— (1986) *Black Stilt.* Endangered New Zealand Wildlife Series. Dunedin, New Zealand: John McIndoe and New Zealand Wildlife Service.

*Reed, C. E. M. (1993) "Black Stilt." In S. Marchant and P. J. Higgins, eds., *Handbook of Australian, New Zealand, and Antarctic Birds,* vol. 2, pp. 769–80. Melbourne: Oxford University Press.

OYSTERCATCHERS AND PLOVERS   Shore Bird

(EURASIAN) OYSTERCATCHER
Haematopus ostralegus

(EURASIAN) GOLDEN PLOVER
Pluvialis apricaria

HOMOSEXUALITY	TRANSGENDER	BEHAVIORS		RANKING	OBSERVED IN
● Female	○ Intersexuality	● Courtship	● Pair-bonding	○ Primary	● Wild
● Male	○ Transvestism	● Affectionate	● Parenting	● Moderate	○ Semiwild
		● Sexual		○ Incidental	○ Captivity

OYSTERCATCHER

IDENTIFICATION: A large (17 inch), stocky shore bird with black upperparts, white underparts, and red-orange bill, eyes, and legs. DISTRIBUTION: Throughout Eurasia; winters in Africa, Middle East, southern Asia. HABITAT: Beaches, salt marshes, rocky coasts, mudflats. STUDY AREAS: The islands of Texel, Vlieland, and Schiermonnikoog, the Netherlands; subspecies *H.o. ostralegus.*

GOLDEN PLOVER

IDENTIFICATION: A medium-sized (10 inch) sandpiper-like bird with mottled buff and black plumage; adult males have a black face and underparts bordered with white. DISTRIBUTION: Northern Europe; winters south to Mediterranean and North Africa. HABITAT: Tundra, bogs, moors, heath. STUDY AREA: Dorback Moor, Scotland; subspecies *P.a. apricaria.*

Social Organization

Oystercatchers and Golden Plovers commonly associate in flocks. The mating system typically involves monogamous pair-bonding, although many alternative arrangements also occur (see below). Nonbreeding Oystercatchers tend to aggregate in groups known as CLUBS.

Description

Behavioral Expression: Oystercatchers sometimes participate in same-sex courtship and copulation. This behavior typically occurs within bisexual trios, that is, an association of three birds—two of one sex and one of the other—in which all three members have a bonded sexual relationship. For example, two males and a female sometimes form a trio, and in addition to heterosexual activity between the opposite-sex partners, the two males may court and mount each other. Several different courtship and pair-bonding displays are used in both same-sex and opposite-sex contexts. For example, while walking around each other, two males might perform BALANCING, in which they make seesaw movements with their bodies, or the THICK-SET ATTITUDE, a stylized posture in which the head is drawn down between the shoulders with the tail and back horizontal, all the while bending the legs and making tripping steps. Sometimes two males also perform ritualized nest-building activities as part of their mutual courtship, such as THROWING STRAWS, in which they toss straw and other materials backward, or PRESSING A HOLE, in which they repeatedly sit down, pressing their breasts and wings against the ground as if fashioning a nest. As a prelude to copulation, one male approaches the other in the STEALTHY ATTITUDE, similar to the thick-set attitude except that the head is held to one side and the tail is pressed down and spread. One male may mount and try to copulate with the other, although sometimes his sexual advances are thwarted by an attack from the other male. Interestingly, all three members of such a trio may be nonmonogamous, engaging in heterosexual courtship or copulations with birds other than their primary partners.

Homosexual activities also occur between two female Oystercatchers that form part of a bisexual trio with a male. Most associations of this type start off the way heterosexual trios do, with considerable aggression between the females, but eventually they develop a strong bond with each other. They preen one another while remaining close together and also cooperate (along with their male partner) in mutual defense of their territory. Employing the same behavior patterns seen in heterosexual mating, the two females also regularly copulate with one another: one female approaches the other in a hunched posture, making soft *pip-pip* noises while her partner tosses her tail upward. Then, while mounting, the female flaps her wings to maintain balance and may push her tail under the other female's in order to achieve genital (cloacal) contact, at which point she utters soft *wee-wee* sounds. The two birds may take turns mounting one another, and about 47 percent of lesbian copulations include full genital contact (compared to 67 percent of matings by heterosexual pairs and 74 percent of male-female copulations in heterosexual trios). The females also mate regularly with their male partner, eventually building a joint nest together in which they each lay eggs. This results in a SUPERNORMAL

CLUTCH of up to 7 eggs (compared to a maximum of 4–5 in nests of heterosexual pairs, or in each of the two separate nests of heterosexual trios). All three partners take turns incubating the eggs and they cooperate in raising their chicks. However, because each bird is usually unable to adequately cover all 7 eggs simultaneously, bisexual trios generally hatch and raise fewer offspring than do heterosexual pairs. Bisexual trios can remain together for up to 4–12 years, comparable to Oyster-catcher heterosexual pairs, and are actually more stable and longer-lasting than heterosexual trios (which typically do not extend beyond 4 years).

Male Golden Plovers occasionally court and pair with each other in the early spring. Courtship activities often begin with ground displays, in which one male chases the other with his head lowered, wings half-spread, and back feathers ruffled, all the while raising and lowering his fanned tail. This may develop into a spectacular twisting aerial pursuit flight, in which the two males synchronously dip and climb, careening and skimming over the ground in a dramatic, high-speed chase that may take them far from their home territories.

Frequency: Homosexual behavior occurs occasionally in Oystercatcher and Golden Plover populations. Less than 2 percent of Oystercatchers, for example, live in trios of two females with one male, although 43 percent of such associations involve homosexual bonding and sexual activities. Overall, about 1 in every 185 copulations is between two females; lesbian matings take place roughly once every 6–7 hours within each bisexual trio, compared to roughly once every 3–6 hours for heterosexual matings (in a pair or trio). Likewise, approximately 1 out of every 400 Oystercatcher bonds involves a trio of two males with a female, and only some of these include same-sex activity. As in bisexual trios with two females, however, homosexual behavior may be fairly frequent within the association: in one such trio, for instance, almost two-thirds of all courtship activities were homosexual, and 15–19 percent of all mounting activities were same-sex.

Orientation: Oystercatchers that participate in same-sex activities are usually bisexual, being part of a bonded trio with a member of the opposite sex and sometimes also engaging in promiscuous heterosexual activities. Within the trio, however, one bird may be more homosexually oriented than the other, i.e., it may have a closer bond with a bird of the same sex, while the other may have a stronger heterosexual bond. In one bisexual trio involving two males and a female, for example, 85 percent of one male's courtship activities and more than a third of his mounting activities were homosexual; for the other male, about 70 percent of his courtships and a quarter of his mounting activities were same-sex. Some female Oystercatchers in bisexual trios also end up leaving their trio and pairing with a male, although this occurs less frequently than for females in heterosexual trios.

Nonreproductive and Alternative Heterosexualities

Polygamous heterosexual trios (without same-sex activities) sometimes form in Oystercatchers (as mentioned above), and the same phenomenon also occurs in Golden Plovers. In addition, several other variations on the long-term, monogamous, male-female parenting unit have developed in these species. Although pair-bonds in Oystercatchers and Golden Plovers sometimes last for life, heterosexual partners may divorce and re-pair with new mates. In some Oystercatcher populations 6–10 percent of couples divorce, and the average length of a pair-bond is only two to three years. Some birds (particularly females) divorce repeatedly and may have as many as six or seven different partners during their lives, and only about half of all birds remain with the same partner for life. A female Golden Plover sometimes deserts her mate during the breeding season (often to start a second family with a new male); her former mate must then raise their young on his own. In addition to single parenting, "double-family parenting" sometimes occurs: two Plover families occasionally share the same territory (with one couple breeding earlier than the other) and may help defend each other's brood. Oystercatcher pairs sometimes foster-parent chicks of other related species such as lapwings (*Vanellus vanellus*) and avocets (*Recurvirostra avosetta*), occasionally even "adopting" and hatching foreign eggs.

Infidelity is a prominent feature of Oystercatcher pair-bonds. Up to 7 percent of all copulations are nonmonogamous, often between a paired female and a single male (and usually initiated by the female). Females often have an extended "affair" with a particular male over several years and may eventually leave their mate to pair with him; some females are even unfaithful to their new partner by continuing to copulate with their "ex" after they have remated. However, nonmonogamous mates are not generally more likely to divorce than strictly monogamous pairs, and in fact some evidence suggests that Oystercatchers who engage in outside sexual activity are actually *more likely* to stay together. One study found that 0–5 percent of unfaithful birds divorced, while 11 percent of monogamous ones did. Many nonmonogamous matings are nonreproductive, occurring too early in the breeding season for fertilization to be possible, or between nonbreeders; in fact, only 2–5 percent of all chicks are the result of infidelities. There are several distinct categories of nonbreeders in this species, including nonbreeding pairs with territories (about 5 percent of all pairs) and FLOATERS without any territories. Overall, about 30 percent of the adult population is nonbreeding. Nevertheless, such birds still engage in sexual behavior, both with each other and with paired birds. Nonbreeding pairs and individuals also occur in Golden Plovers, and on average about half of the population is nonreproductive at any time.

Many within-pair copulations are also nonprocreative, with about 40 percent occurring too early or too late in the breeding season (for Oystercatchers), or during incubation. In addition, it has been estimated that each Oystercatcher pair copulates about 700 times during the breeding season—far in excess of the amount required for reproduction. Oystercatchers also sometimes practice nonreproductive REVERSE copulations, in which the female mounts the male. And as mentioned above, a quarter to a third of mounts between heterosexual mates do not involve genital contact; many such copulations are incomplete because the female throws the male off her back or otherwise refuses to participate. Much more rarely, a male will rape or forcibly copulate with a nonconsenting female. Adult-youngster interactions are also sometimes marked by violence and neglect: Oystercatcher chicks have been viciously attacked and even killed when they stray into another bird's territory. In addition, LEAPFROG parents often starve their chicks by failing to bring them enough food. Leapfrog birds are those whose nesting territories are located farther inland, separate from the feeding territories, hence to obtain food they must "leapfrog" over birds that nest directly adjacent to the shore. Studies have shown that the territories of such Oystercatchers do not, however, place undue time or energy constraints on them compared to nonleapfrogs. Thus, the fact that their chicks sometimes starve is due more to inadequate parental care than to their suboptimal territories.

Sources   *asterisked references discuss homosexuality/transgender

Edwards, P. J. (1982) "Plumage Variation, Territoriality, and Breeding Displays of the Golden Plover *Pluvialis apricaria* in Southwest Scotland." *Ibis* 124:88–96.

*Ens, B. J. (1998) "Love Thine Enemy?" *Nature* 391:635–37.

*——— (1996) Personal communication.

——— (1992) "The Social Prisoner: Causes of Natural Variation in Reproductive Success of the Oystercatcher." Ph.D. thesis., University of Groningen.

Ens, B. J., M. Kersten, A. Brenninkmeijer, and J. B. Hulscher (1992) "Territory Quality, Parental Effort, and Reproductive Success of Oystercatchers (*Haematopus ostralegus*)." *Journal of Animal Ecology* 61:703–15.

Ens, B. J., U. N. Safriel, and M. P. Harris (1993) "Divorce in the Long-lived and Monogamous Oystercatcher, *Haematopus ostralegus:* Incompatibility or Choosing the Better Option?" *Animal Behavior* 45:1199–217.

Hampshire, J. S., and F. J. Russell (1993) "Oystercatchers Rearing Northern Lapwing Chick." *British Birds* 86:17–19.

Harris, M. P., U. N. Safriel, M. de L. Brooke, and C. K. Britton (1987) "The Pair Bond and Divorce Among Oystercatchers *Haematopus ostralegus* on Skokholm Island, Wales." *Ibis* 129:45–57.

*Heg, D. (1998) Personal communication.

Heg, D., B. J. Ens, T. Burke, L. Jenkins, and J. P. Kruijt (1993) "Why Does the Typically Monogamous Oystercatcher (*Haematopus ostralegus*) Engage in Extra-Pair Copulations?" *Behavior* 126:247–89.

*Heg, D., and R. van Treuren (1998) "Female-Female Cooperation in Polygynous Oystercatchers." *Nature* 391:687–91.

*Makkink, G. F. (1942) "Contribution to the Knowledge of the Behavior of the Oyster-Catcher (*Haematopus ostralegus* L.)." *Ardea* 31:23–74.

*Nethersole-Thompson, D., and C. Nethersole-Thompson (1961) "The Breeding Behavior of the British Golden Plover." In D. A. Bannerman, ed., *The Birds of the British Isles*, vol.10, pp. 206–14. Edinburgh and London: Oliver and Boyd.

*Nethersole-Thompson, D., and M. Nethersole-Thompson (1986) *Waders: Their Breeding, Haunts, and Watchers.* Calton: T. and A. D. Poyser.

Parr, R. (1992) "Sequential Polyandry by Golden Plovers." *British Birds* 85:309.

——— (1980) "Population Study of Golden Plover *Pluvialis apricaria*, Using Marked Birds." *Ornis Scandinavica* 11:179–89.

——— (1979) "Sequential Breeding by Golden Plovers." *British Birds* 72:499–503.

Tomlinson, D. (1993) "Oystercatcher Chick Probably Killed by Rival Adult." *British Birds* 86:223–25.

GULLS AND TERNS

GULLS

RING-BILLED GULL
Larus delawarensis

COMMON or MEW GULL
Larus canus

Shore Bird

HOMOSEXUALITY	TRANSGENDER	BEHAVIORS		RANKING	OBSERVED IN
● Female	○ Intersexuality	○ Courtship	● Pair-bonding	○ Primary	● Wild
○ Male	○ Transvestism	○ Affectionate	● Parenting	● Moderate	○ Semiwild
		○ Sexual		○ Incidental	○ Captivity

RING-BILLED GULL

IDENTIFICATION: A medium-sized (21 inch) gull with a gray back and wings; spotted black-and-white wing tips; yellow bill, legs, and eyes; and a black band on the bill. DISTRIBUTION: Central Canada, much of United States; winters south to Central America. HABITAT: Coasts, rivers, lakes, prairies. STUDY AREAS: Eastern Oregon and Washington State; Granite Island, Lake Superior; Île de la Couvée, near Montreal; Gull Island, Lake Ontario; other locations on Lakes Erie, Ontario, and Huron.

COMMON GULL

IDENTIFICATION: Similar to Ring-billed Gull, except slightly smaller (up to 18 inches) and with a more slender, plain yellow bill. DISTRIBUTION: Nearly circumpolar in Northern Hemisphere; winters south to North Africa, East Asia, California. HABITAT: Coasts, mudflats, beaches, lakes. STUDY AREA: Fair Isle on the Shetland Islands, Scotland; subspecies *L.c. canus*.

Social Organization

Common Gulls are fairly sociable, often associating in flocks of up to 100 individuals; sometimes tens of thousands of birds congregate outside of the breeding season. Ring-billed Gulls are also gregarious. Birds of both species generally form monogamous pair-bonds (although several variations exist—see below). Nesting colonies in Common Gulls contain a few dozen to several hundred pairs, while Ring-billed colonies can be much larger, in the tens of thousands of pairs.

Description

Behavioral Expression: Both Ring-billed and Common Gull females sometimes develop homosexual bonds, build nests together, lay and incubate eggs, and successfully raise chicks. Same-sex pairs are often long-lasting (as are heterosexual pair-bonds in these species), and females generally return to the same nest site with their female partner each year. Of five homosexual pairs of Ring-billed Gulls tracked over time, for example, all remained together over more than one mating season, while homosexual bonds lasting for at least eight years have been documented in female Common Gulls. Some pairs do divorce between mating seasons, however, as do some heterosexual pairs. In addition, female Ring-billed Gulls on rare occasions switch mates *during* a mating season, first pairing with one female, then with another. Same-sex bonds in this species also show a number of other interesting features that are rarely found in female homosexual bonds in other animals. For example, one or both partners in a homosexual pair are often younger females: couples in which there is an age difference between the two females—one an adult, the other an adolescent or younger adult—are particularly common in some populations. In addition, some female Ring-billed Gulls form homosexual trios consisting of three birds that are all simultaneously bonded to one another. In Common Gulls, homosexual pairs may after several years develop into bisexual trios. This occurs when a male joins them and is accepted into their association, bonding with one or both females (who nevertheless still retain their bond with each other). In Ring-billed Gulls, the opposite scenario may occur: in at least one case, two females in a bisexual trio remained paired to each other after the male left their association. Ring-billed Gulls in homosexual pairs probably do not engage in a great deal of courtship activity (unlike heterosexual or homosexual pairs in other species).

The first breeding season that female Common Gulls begin a pair-bond, they may build a "double nest" consisting of two separate but touching nest cups; in subsequent years, they will build only a single nest in which they both lay eggs (like most Ring-billed female pairs). Since both partners usually lay eggs, nests of homosexual pairs often contain "oversize" or SUPERNORMAL CLUTCHES of 5–8 eggs (Ring-billed Gull) and 6 eggs (Common Gull)—up to twice the number found in heterosexual nests. Some Ring-billed Gulls in female couples may also lay their eggs in the nests of other (heterosexual) pairs. One or both partners in pairs of female Ring-billed Gulls may mate nonmonogamously with a male so that some of their eggs will be fertilized. Female Common Gulls in bisexual trios can also lay fertile eggs by mating with their male partner. Both females share incubation duties and also cooperate in parenting the chicks that they hatch. Homosexual parents in Ring-billed Gulls invest as much time as their heterosexual counterparts do in feeding their chicks, spending time on the nesting territory, and defending their territory. They may actually work harder than male-female pairs in brooding and defending their chicks, with the result that offspring of female pairs often have a faster growth rate than chicks of heterosexual parents.

Nevertheless, chicks belonging to female pairs are often less robust on hatching, and female parents generally raise less than half the number of chicks that male-female pairs do. However, these traits are also characteristic of supernormal clutches belonging to heterosexual trios and are therefore undoubtedly related to the larger clutch size rather than the sex or abilities of the parents. In addition, in some populations female pairs are relegated to smaller, substandard territories on the periphery of the breeding colony or in between other territories (as are less experienced heterosexual pairs). In some cases, homosexual pairs actually appear to form clusters of up to ten nests in close proximity, or else small groups of two or three (sometimes in a straight-line or triangular formation). Many of these nests are located in areas where it is more difficult to parent successfully—places with little or no vegetation, or else away from the beach—and therefore it is remarkable that female pairs are able to successfully raise chicks as often as they do in such less than optimal conditions.

Frequency: There is wide variation in the incidence of female pairing in Ring-billed Gulls. In some populations—especially in growing colonies—as many as 6–12 percent of the pairs are homosexual. In other colonies, they are less common—1–3 percent of all pairs—and in some locations as few as 1 in 700 or 1 in 3,400 nests may belong to a female pair. Overall, pair-bonds between females probably occur only occasionally in Common Gulls, although at one study site, 1 pair out of a total of 12 was homosexual.

Orientation: There is an equally wide variation in the proportion of heterosexual, bisexual, and homosexual orientations among Ring-billed Gulls. In some populations, less than a third of all eggs laid by female pairs are fertile, indicating that the majority of such females are exclusively homosexual (at least for the duration of their pair-bond). In other colonies, egg fertility of female pairs is much higher—two-thirds to nearly 90 percent—indicating a greater prevalence of bisexual activity. Even among such females, however, there is further variation. In some same-sex couples, both females mate with males and lay fertile eggs; in others, only one partner does, or each partner might lay both fertile and infertile eggs at different times, indicating temporal variation in bisexual activity. Similarly, in a Common Gull bisexual trio, one female remained exclusively homosexual even though her female partner mated with the male. In addition, a large number of "heterosexual" Ring-billed females may have a "latent" bisexual potential, since many are able to develop bonds with other females if single males are not available.

Nonreproductive and Alternative Heterosexualities

Heterosexual pairs in Ring-billed and Common Gulls exhibit a variety of bonding and parenting arrangements (like homosexual pairs). Not all males and females couple for life: the heterosexual divorce rate is about 28 percent in both species. Polygamous heterosexual trios—two females bonded to the same male, but not to each other—are also found in both species, as are occasionally even quartets (three females with one male). Common Gull pairs sometimes foster-parent chicks, while another form of "adoption" occurs in these species when females occasionally lay eggs in nests belonging to other pairs or roll eggs from other nests into their own.

Moreover, because of parental ineptitude or inefficiency (such as poor feeding), at least 8 percent of Ring-billed chicks abandon or "run away" from their own families; most of these are adopted and cared for by other families.

About 4 percent of Ring-billed pairs continue to engage in courtship and copulation after the hatching of their eggs—when sexual activity is not directly reproductive—and about 5 percent of adults court and mount chicks. Most of this activity involves females behaving incestuously with their own offspring, including full copulatory REVERSE mounts of young birds. Mounted chicks may be as young as two weeks old, and they usually collapse under the weight of the adult mounting them and cry out in distress. Some individuals appear to be "habitual molesters" in that they repeatedly interact sexually with chicks, including their own. In addition to sexual molestation, Ring-billed chicks are often subjected to vicious attacks from neighboring adults when their parents are away, or if they stray outside of their home territory. About 1 in 300 chicks is killed by such assaults, and infanticide can account for between 5 percent and 80 percent of all chick deaths (depending on the population).

Other Species

Female pairs that lay supernormal clutches also occur in California Gulls (*Larus californicus*), where they constitute about 1 percent of all pairs.

Sources   *asterisked references discuss homosexuality/transgender*

Brown, K. M., M. Woulfe, and R. D. Morris (1995) "Patterns of Adoption in Ring-billed Gulls: Who Is Really Winning the Inter-generational Conflict?" *Animal Behavior* 49:321–31.

*Conover, M. R. (1989) "Parental Care by Male-Female and Female-Female Pairs of Ring-billed Gulls." *Colonial Waterbirds* 12:148–51.

*———— (1984a) "Frequency, Spatial Distribution, and Nest Attendants of Supernormal Clutches in Ring-billed and California Gulls." *Condor* 86:467–71.

*———— (1984b) "Consequences of Mate Loss to Incubating Ring-billed and California Gulls." *Wilson Bulletin* 96:714–16.

*———— (1984c) "Occurrence of Supernormal Clutches in the Laridae." *Wilson Bulletin* 96:249–67.

*Conover, M. R., and D. E. Aylor (1985) "A Mathematical Model to Estimate the Frequency of Female-Female or Other Multi-Female Associations in a Population." *Journal of Field Ornithology* 56:125–30.

*Conover, M. R., and G. L. Hunt, Jr. (1984a) "Female-Female Pairings and Sex Ratios in Gulls: A Historical Perspective." *Wilson Bulletin* 96:619–25.

*———— (1984b) "Experimental Evidence That Female-Female Pairs in Gulls Result From a Shortage of Males." *Condor* 86:472–76.

*Conover, M. R., D.E. Miller, and G. L. Hunt, Jr. (1979) "Female-Female Pairs and Other Unusual Reproductive Associations in Ring-billed and California Gulls." *Auk* 96:6–9.

Emlen, J. R., Jr. (1956) "Juvenile Mortality in a Ring-billed Gull Colony." *Wilson Bulletin* 68:232–38.

Fetterolf, P. M. (1983) "Infanticide and Non-Fatal Attacks on Chicks by Ring-billed Gulls." *Animal Behavior* 31:1018–28.

———— (1984) "Ring-billed Gulls Display Sexually Toward Offspring and Mates During Post-Hatching." *Wilson Bulletin* 96:12–19.

*Fetterolf, P. M., and H. Blokpoel (1984) "An Assessment of Possible Intraspecific Brood Parasitism in Ring-billed Gulls." *Canadian Journal of Zoology* 62:1680–84.

*Fetterolf, P. M., P. Mineau, H. Blokpoel, and G. Tessier (1984) "Incidence, Clustering, and Egg Fertility of Larger Than Normal Clutches in Great Lakes Ring-billed Gulls." *Journal of Field Ornithology* 55:81–88.

*Fox, G. A., and D. Boersma (1983) "Characteristics of Supernormal Ring-billed Gull Clutches and Their Attending Adults." *Wilson Bulletin* 95:552–59.

Kinkel, L. K., and W. E. Southern (1978) "Adult Female Ring-billed Gulls Sexually Molest Juveniles." *Bird-Banding* 49:184–86.

*Kovacs, K. M., and J. P. Ryder (1985) "Morphology and Physiology of Female-Female Pair Members." *Auk* 102:874–78.

*———— (1983) "Reproductive Performance of Female-Female Pairs and Polygynous Trios of Ring-billed Gulls." *Auk* 100:658–69.

*———— (1981) "Nest-site Tenacity and Mate Fidelity in Female-Female Pairs of Ring-billed Gulls." *Auk* 98:625–27.

*Lagrenade, M., and P. Mousseau (1983) "Female-Female Pairs and Polygynous Associations in a Quebec Ring-billed Gull Colony." *Auk* 100:210–12.

Nethersole-Thompson, C., and D. Nethersole-Thompson (1942) "Bigamy in the Common Gull." *British Birds* 36:98–100.

*Riddiford, N. (1995) "Two Common Gulls Sharing a Nest." *British Birds* 88:112–13.

*Ryder, J. P. (1993) "Ring-billed Gull." In A. Poole, P. Stettenheim, and F. Gill, eds., *The Birds of North America: Life Histories for the 21st Century*, no. 33. Philadelphia: Academy of Natural Sciences; Washington, D.C.: American Ornithologists Union.

*Ryder, J. P., and P. L. Somppi (1979) "Female-Female Pairing in Ring-billed Gulls." *Auk* 96:1–5.

Southern, L. K., and W. E. Southern (1982) "Mate Fidelity in Ring-billed Gulls." *Journal of Field Ornithology* 53:170–71.

Trubridge, M. (1980) "Common Gull Rolling Eggs from Adjacent Nest into Own." *British Birds* 73:222–23.

GULLS   Shore Bird

WESTERN GULL
Larus occidentalis
(BLACK-LEGGED) KITTIWAKE
Rissa tridactyla

HOMOSEXUALITY	TRANSGENDER	BEHAVIORS		RANKING	OBSERVED IN
● Female	○ Intersexuality	● Courtship	● Pair-bonding	● Primary	● Wild
○ Male	○ Transvestism	○ Affectionate	● Parenting	○ Moderate	○ Semiwild
		● Sexual		○ Incidental	○ Captivity

WESTERN GULL
IDENTIFICATION: A large gull (up to 27 inches) with a dark gray back and wings; spotted black-and-white wing tips; pink legs; and a yellow bill with a red spot. DISTRIBUTION: Pacific coast of North America. HABITAT: Cliffs, rocky seacoasts, bays. STUDY AREAS: Santa Barbara Island and other Channel Islands, California; subspecies *L.o. wymani*.

KITTIWAKE
IDENTIFICATION: A smaller gull (to 17 inches) with a blue-gray mantle; more pointed black wing tips; relatively short black legs and dark eyes; and a yellowish green bill. DISTRIBUTION: Northern Pacific and Atlantic Oceans; adjacent Arctic Ocean. HABITAT: Oceangoing; breeds on coasts. STUDY AREA: North Shields, Tyne and Wear, England; subspecies *R.t. tridactyla*.

Social Organization

Western Gulls and Kittiwakes form pair-bonds and nest in colonies, some of which contain upwards of 10,000 pairs; Kittiwakes often nest on cliffs. Outside of the breeding season they are less sociable, occasionally gathering in loose aggregations when not solitary.

Description

Behavioral Expression: Female Western Gulls sometimes form homosexual pairs, as do Kittiwakes. In Western Gulls, the females participate in courtship, sexual, and parenting behaviors similar to those of heterosexual pairs in their basic patterns, yet different in many details. Two females court one another by performing HEAD-TOSSING (a stylized bobbing of the head with the bill pointed skyward) and COURTSHIP-FEEDING (in which a small amount of food is regurgitated and offered as a "gift" to the partner). In heterosexual pairs, males usually perform more courtship-feeding and females more head-tossing. In homosexual pairs, both birds perform these behaviors—often with equal frequency—although the overall rate of courtship behaviors for each female is similar to females in heterosexual pairs. Homosexual courtship-feeding differs from the heterosexual pattern in that a female does not offer as large an amount of food to her female partner and may even swallow the "offering" herself rather than give it to her mate. In some female pairs, one partner regularly mounts the other and may even utter the copulation call characteristic of heterosexual matings. Some females adopt unique mounting positions such as sideways or head-to-tail (not seen in heterosexual matings), and genital contact does not usually occur. Like heterosexual pairs, female couples establish territories that they defend against intruders. Both females spend a great deal of time on their territories (typical only of females in heterosexual pairs), while both also exhibit aggressive reactions to intruders (more typical of males in heterosexual pairs). Once a homosexual pair-bond is established, it usually persists for many years, and the two females return to the same territory each season: one study that tracked eight homosexual pairs found that seven of them remained together for more than one breeding season.

Homosexual pairs usually build a nest in which both females lay eggs; the resulting SUPERNORMAL CLUTCH contains 4–6 eggs in Western Gulls, up to twice the number found in nests of heterosexual pairs. Some of these eggs are fertile because females in homosexual pairs occasionally copulate with males (without breaking their same-sex bond). Although eggs laid by female

A homosexual pair of female Western Gulls in California ▶

pairs may be smaller than those of heterosexual females, homosexual parents successfully hatch and raise chicks, sharing all parental duties.

Frequency: As many as 10–15 percent of Western Gull pairs in some populations are homosexual; the percentage is much lower in Kittiwakes, about 2 percent of all pairs.

Orientation: Most female pairs in Kittiwakes are exclusively homosexual, never mating with males and laying only infertile eggs; the same is true for many pairs in Western Gulls. However, up to 15 percent of eggs laid by Western Gull same-sex pairs are fertilized, so at least some females are simultaneously bisexual—copulating with males while retaining their homosexual pair-bond.

Nonreproductive and Alternative Heterosexualities

Not all heterosexual birds in these species form lifelong, monogamous pair-bonds within which they raise their own young. About 30 percent of Kittiwake male-female pairs divorce. Some birds form polygamous trios consisting of one male bonded with two females, each with her own nest (about 3 percent of all bonds). Western Gull pairs (and rarely, single males) sometimes adopt and raise chicks that are not their own, and "stepmothering" occurs when females pair with a male that has lost his mate; foster-parenting also occurs in Kittiwakes (where about 8 percent of all chicks are adopted). In addition, many birds do not reproduce or do so only rarely: 30–40 percent of adult Western Gulls breed only once or twice in their lifetimes, and do not successfully raise any offspring, while 5 percent of all Kittiwakes that attempt to nest do not raise any young during their lives (and nearly two-thirds of all Kittiwakes never produce any offspring, usually because they die before breeding). Female Western Gulls that breed less often (or defer breeding until later in life) actually have higher survival rates than birds that reproduce more frequently. In Kittiwakes, nonbreeding birds form their own flocks or CLUBS on the outskirts of the breeding colonies. Western Gulls in heterosexual pairs also sometimes engage in nonprocreative sexual behaviors, such as REVERSE mounting (where the female mounts the male, typically without genital contact).

Some male Western Gulls are promiscuous, attempting to copulate with females other than their mates (usually birds on neighboring territories), although they are frequently unsuccessful. Occasionally a male will behave aggressively toward a female he has just mated with (nonmonogamously) and may even attack and kill her. Overall, more than 40 percent of aggressive incidents occur between members of the opposite sex. These include females defending themselves against promiscuous males, males attacking neighboring females that are courting them, and territorial disputes. Females may also refuse to copulate with their own mates, either by not allowing them to mount or by walking out from under them during mating. Some pairs, however, begin copulating even before the female's fertile period; this also occurs in Kittiwakes, and many copulations in this species do not involve genital contact (more than 30 percent). In Kittiwakes, 15–27 percent of all heterosexual copulations are harassed and interrupted by other males. Occasionally, adult Western Gulls are violent toward chicks, who may be attacked and even killed if left alone by their parents. Kittiwake parents (especially inexperienced ones) sometimes neglect their chicks (e.g., starving them), and birds may also attack or toss their own or other parents' chicks off cliffs. In fact, many adoptions and chick deaths in both species result from youngsters deserting or "running away" from their biological families as a result of neglect or direct attack. As many as a third of all Kittiwake chicks in some colonies abandon or are driven from their own nests. In addition, adults of both species may eat unattended eggs belonging to other parents.

Sources   *asterisked references discuss homosexuality/transgender

*Baird, P. H. (1994) "Black-legged Kittiwake (*Rissa tridactyla*)." In A. Poole, P. Stettenheim, and F. Gill, eds., *The Birds of North America: Life Histories for the 21st Century*, no. 92. Philadelphia: Academy of Natural Sciences; Washington, D.C.: American Ornithologists' Union.

Carter, L. R., and L. B. Spear (1986) "Costs of Adoption in Western Gulls." *Condor* 88:253–56.

Chardine, J. W. (1987) "The Influence of Pair-Status on the Breeding Behavior of the Kittiwake *Rissa tridactyla* Before Egg-Laying." *Ibis* 129:515–26.

——— (1986) "Interference of Copulation in a Colony of Marked Black-legged Kittiwakes." *Canadian Journal of Zoology* 64:1416–21.

*Conover, M. R. (1984) "Occurrence of Supernormal Clutches in the Laridae." *Wilson Bulletin* 96:249–67.

*Coulson, J. C., and C. S. Thomas (1985) "Changes in the Biology of the Kittiwake *Rissa tridactyla*: A 31 Year Study of a Breeding Colony." *Journal of Animal Ecology* 54:9–26.

——— (1983) "Mate Choice in the Kittiwake Gull." In P. Bateson, ed., *Mate Choice*, pp. 361–76. Cambridge: Cambridge University Press.

Coulson, J. C., and E. White (1958) "The Effect of Age on the Breeding Biology of the Kittiwake *Rissa tridactyla*." *Ibis* 100:40–51.

Cullen, E. (1957) "Adaptations in the Kittiwake to Cliff-Nesting." *Ibis* 99:275–302.

*Fry, D. M., C. K. Toone, S. M. Speich, and R. J. Peard (1987) "Sex Ratio Skew and Breeding Patterns of Gulls: Demographic Toxicological Considerations." *Studies in Avian Biology* 10:26–43.

Hand, J. L. (1986) "Territory Defense and Associated Vocalizations of Western Gulls." *Journal of Field Ornithology* 57:1–15.

*——— (1980) "Nesting Success of Western Gulls on Bird Rock, Santa Catalina Island, California." In D. M. Power, ed., *The California Islands: Proceedings of a Multidisciplinary Symposium*, pp. 467–73. Santa Barbara: Santa Barbara Museum of Natural History.

*Hayward, J. L., and M. Fry (1993) "The Odd Couples/The Rest of the Story." *Living Bird* 12:16–19.

*Hunt, G. L., Jr. (1980) "Mate Selection and Mating Systems in Seabirds." In J. Burger, B. L. Olla, and H. E. Winn, eds., *Behavior of Marine Mammals*, vol.4, pp. 113–51. New York: Plenum Press.

*Hunt, G. L., Jr., and M. W. Hunt (1977) "Female-Female Pairing in Western Gulls (*Larus occidentalis*) in Southern California." *Science* 196:1466–67.

*Hunt, G. L., Jr., A. L. Newman, M. H. Warner, J. C. Wingfield, and J. Kaiwi (1984) "Comparative Behavior of Male-Female and Female-Female Pairs Among Western Gulls Prior to Egg-Laying." *Condor* 86:157–62.

*Hunt, G. L., J. C. Wingfield, A.L. Newman, and D. S. Farner (1980) "Sex Ratio of Western Gulls on Santa Barbara Island, California." *Auk* 97:473–79.

Paludan, K. (1955) "Some Behavior Patterns of *Rissa tridactyla*." *Videnskabelige Meddelelser fra Dansk naturhistorisk Forening* 117:1–21.

Pierotti, R. J. (1991) "Infanticide versus Adoption: An Intergenerational Conflict." *American Naturalist* 138:1140–58.

*——— (1981) "Male and Female Parental Roles in the Western Gull Under Different Environmental Conditions." *Auk* 98:532–49.

——— (1980) "Spite and Altruism in Gulls." *American Naturalist* 115:290–300.

*Pierotti, R. J., and C. A. Annett (1995) "Western Gull (*Larus occidentalis*)." In A. Poole and F. Gill, eds., *The Birds of North America: Life Histories for the 21st Century*, no. 174. Philadelphia: Academy of Natural Sciences; Washington, D.C.: American Ornithologists' Union.

Pierotti, R. J., and E. C. Murphy (1987) "Intergenerational Conflicts in Gulls." *Animal Behavior* 35:435–44.

Pyle, P., N. Nur, W. J. Sydeman, and S.D. Emslie (1997) "Cost of Reproduction and the Evolution of Deferred Breeding in the Western Gull." *Behavioral Ecology* 8:140–47.

Roberts, B. D., and S. A. Hatch (1994) "Chick Movements and Adoption in a Colony of Black-legged Kitti-wakes." *Wilson Bulletin* 106:289–98.

Thomas, C. S., and J. C. Coulson (1988) "Reproductive Success of Kittiwake Gulls, *Rissa tridactyla*." In T. H. Clutton-Brock, ed., *Reproductive Success: Studies of Individual Variation in Contrasting Breeding Systems*, pp. 251–62. Chicago and London: University of Chicago Press.

*Wingfield, J. C., A. L. Newman, G. L. Hunt, Jr., and D. S. Farner (1982) "Endocrine Aspects of Female-Female Pairings in the Western Gull, *Larus occidentalis wymani*." *Animal Behavior* 30:9–22.

*Wingfield, J. C., A. L. Newman, M. W. Hunt, G. L. Hunt, Jr., and D. Farner (1980) "The Origin of Homo-sexual Pairing of Female Western Gulls (*Larus occidentalis wymani*) on Santa Barbara Island." In D. M. Power, ed., *The California Islands: Proceedings of a Multidisciplinary Symposium*, pp. 461–66. Santa Barbara, Calif.: Santa Barbara Museum of Natural History.

GULLS

Shore Bird

SILVER GULL
Larus novaehollandiae

HERRING GULL
Larus argentatus

HOMOSEXUALITY	TRANSGENDER	BEHAVIORS		RANKING	OBSERVED IN
● Female	○ Intersexuality	● Courtship	● Pair-bonding	● Primary	● Wild
● Male	○ Transvestism	○ Affectionate	● Parenting	○ Moderate	○ Semiwild
		● Sexual		○ Incidental	○ Captivity

SILVER GULL
IDENTIFICATION: A medium-sized (16 inch) gull with gray back and wings; spotted black-and-white wing tips; bright red bill and legs; white iris. DISTRIBUTION: Australia, New Zealand, New Caledonia. HABITAT: Coasts, lakes, islands. STUDY AREA: Kaikoura Peninsula, New Zealand; subspecies *L.n. scopulinus*, the Red-billed Gull.

HERRING GULL
IDENTIFICATION: Similar to Silver Gull except larger (2 feet long), legs pinkish, bill yellow with a red spot, and iris yellow. DISTRIBUTION: North America, western Europe, Siberia; winters in Central America, N. Africa, southern Asia. HABITAT: Coasts, bays, lakes, rivers. STUDY AREAS: Gull Island National Wildlife Refuge, Lake Michigan; numerous other island locations in Lake Michigan, Lake Huron, and the Straits of Mackinac; Bird Island, Memmert, Germany; subspecies *L.a. smithsonianus* and *L.a. argentatus*.

Social Organization

Silver and Herring Gulls are usually found in flocks of several hundreds or thousands; they generally form monogamous pair-bonds and nest in colonies containing anywhere from several hundred to tens of thousands of nests.

Description

Behavioral Expression: In both Silver and Herring Gulls, females sometimes form lesbian pairs while males occasionally participate in homosexual mountings. Female pairs may develop between birds who were previously paired to a male, or they may involve birds who have never been paired before. In some cases, single, nonbreeding Herring Gull females visit the territory of a heterosexual pair and court the female, for example by performing HEAD-TOSSING, in which the head is hunched down and then repeatedly flicked upward. The heterosexually paired birds usually respond aggressively, but sometimes this behavior leads to a homosexual pairing the following season. Like heterosexual pairs, homosexual bonds are usually long-lasting and renewed each year: of those Herring Gull females in homosexual pairs that return to the same breeding grounds, 92 percent pair with the same female (compared to 93 percent of birds in heterosexual pairs). Of those that divorce, some remain single while others find a new (female) mate.

Females in same-sex pairs usually build nests and lay eggs. Silver Gull homosexual females generally begin nesting at a younger age than heterosexual females: females paired to other females start on average about a year earlier than females paired to males, and 11 percent of homosexual females begin nesting when they are two years old (heterosexual females never begin this early). Since both females lay eggs, nests belonging to same-sex pairs often have double or more the number of eggs found in nests of heterosexual pairs. These SUPERNORMAL CLUTCHES contain 4 or more eggs in Silver Gulls (compared to 2 eggs for male-female pairs) and 5–7 eggs in Herring Gulls (compared to 3 eggs for heterosexual pairs). Females sometimes mate nonmonogamously with males—or are raped by them (see below)—while still remaining paired to their female partner. Consequently, some of the eggs laid by female pairs are fertile—about a third in Silver Gulls, and 4–30 percent in Herring Gulls. Homosexual parents often successfully hatch these eggs and raise the chicks. Approximately 3–4 percent of all Silver Gull chicks are raised by same-sex pairs, and a further 9 percent of chicks are raised by male-female pairs in which the mother is bisexual. Overall, 7 percent of birds that go on to become breeding adults in this species come from families with two female parents. However, homosexual and bisexual females generally produce fewer offspring during their lifetimes than do heterosexual females.

In both Silver and Herring Gulls, males in heterosexual pairs often try to copulate with birds other than their mates, and in some cases they mount other males. Like females who are mounted by birds other than their mate, male Herring Gulls may respond aggressively to another male's mounting them.

Frequency: About 6 percent of all pair-bonds in Silver Gulls are homosexual, while nesting attempts by female pairs occur in approximately 12 percent of all breeding seasons. In some populations of Herring Gulls, nearly 3 percent of the pairs are homosexual, while in other populations they are much less frequent, about 1 in every 360 pairs. In addition, approximately 2 percent of courtship behavior by unpaired females interacting with heterosexual pairs is directed toward the female partner. Male homosexual mountings account for 10 percent of the nonmonogamous copulations in Silver Gulls, and 2 percent of the total number of copulations; they are probably much less common in Herring Gulls.

Orientation: In Silver Gulls, 21 percent of females pair with another female at least once in their lifetimes; 10 percent are exclusively lesbian, mating only with other females during their lives, while 11 percent are (sequentially) bisexual, pairing with both males and females. In one study of Herring Gull homosexual pairs, six out of eight females had been in heterosexual pairs the previous year and formed same-sex bonds with each other when their male mates did not return. Of the remaining birds, one paired with a female after her male partner re-paired with another female, while the other had been a single nonbreeder prior to developing a same-sex pair-bond. In addition, female Herring Gulls may show a "preference" for homosexual pairings, since they sometimes re-pair with another female following the breakup of a same-sex bond. In both species, some females in homosexual pairs copulate with males in order to fertilize their eggs, still retaining their primary homosexual bond. Males that initiate homosexual mountings, while functionally bisexual, are probably primarily heterosexually oriented, since they are usually paired to females and rarely engage in same-sex behavior.

Nonreproductive and Alternative Heterosexualities

Large numbers of nonreproducing birds are found in both Silver Gulls and Herring Gulls. About half of adult females and 14 percent of males are nonbreeders in some populations of Silver Gulls; more than three-quarters of all birds die before they reproduce, and 85 percent or more never successfully procreate. Many females that do reproduce nevertheless have extended periods of nonbreeding, up to 16 years in some cases. And about a third of females that lose their mates (through divorce or death) never breed again, sometimes living a further decade as single birds. In Herring Gulls, 4–12 percent of males and one-third to two-thirds of females in some populations are nonbreeders; in other populations, more than a third of all birds do not reproduce in any given year. Two distinct types of nonbreeding females occur in this species: FLOATERS, who are truly single and do not consistently associate with any particular gulls, and SECONDARY FEMALES, who maintain a persistent association with a mated heterosexual pair and even help them defend their territory and raise their young (although they themselves do not breed). Herring Gulls and Silver Gulls also sometimes form polygamous trios, in which the two females are both pair-bonded to one male but not to each other. They often build a "double nest" with two cups in which they both lay eggs.

Several other variant family and pairing arrangements are found in these species. About 5–10 percent of Herring Gull pairs adopt chicks, sometimes even from other species such as the lesser black-backed gull (*Larus fuscus*). Herring Gulls occasionally form pair-bonds with adults of this and other gull species as well. Foster parents usually also have young of their own, while some adopted chicks are cared for by more than one foster family simultaneously. Adoption also occasionally occurs in Silver Gulls. In addition, Herring Gulls sometimes form CRÈCHES, in which several adults pool their youngsters and take turns guarding and feeding them. Although most heterosexual pairs are long-lasting, divorce occurs in about 3–7 percent of Herring Gull pairs and 5–10 percent of Silver Gull pairs. Some Silver Gulls have up to seven different mates over their lives, but about a third have only one partner. In addition, many pair-bonds are nonmonogamous: more than 20 percent of all copulations in Silver Gulls are between nonmates, and more than three-quarters of females and one-third of males are "unfaithful."

Most promiscuous copulations in Silver Gulls are nonreproductive: 11 percent involve incubating (nonfertilizable) females, and many are actually "rapes" or forced copulations in which the female is not a willing participant. As a result, only 7 percent of such matings involve genital contact. Many within-pair copulations are also nonprocreative: more than 30 percent occur during times when the female cannot be fertilized (such as too early before egg laying), and more than half do not involve genital contact or sperm transfer. In addition, about 9 percent of copulations between pair members are forced by the male on the female. Several forms of heterosexual family abuse and adult-juvenile violence have also been documented in Herring Gulls. Males sometimes incestuously mount their own chicks and have even been seen breaking and eating their own eggs. In addition, chicks in both species are often pecked at, pummeled, thrown, shaken, and even killed (and occasionally cannibalized) by other adults when they stray away from home. Among Herring Gulls, being eaten by other Gulls (and sometimes even by one's parents) can be a significant mortality factor: in one colony, a quarter of chick deaths—more than 300 youngsters—were the result of cannibalism.

Sources

asterisked references discuss homosexuality/transgender

Burger, J., and M. Gochfeld (1981) "Unequal Sex Ratios and Their Consequences in Herring Gulls." *Behavioral Ecology and Sociobiology* 8:125–28.

Calladine, J., and M. P. Harris (1997) "Intermittent Breeding in the Herring Gull *Larus argentatus* and the Lesser Black-backed Gull *Larus fuscus." Ibis* 139:259–63.

Chardine, J. W., and R. D. Morris (1983) "Herring Gull Males Eat Their Own Eggs." *Wilson Bulletin* 95:477–78.

*Fitch, M. A. (1979) "Monogamy, Polygamy, and Female-Female Pairs in Herring Gulls." *Proceedings of the Colonial Waterbird Group* 3:44–48.

Fitch, M. A., and G. W. Shugart (1984) "Requirements for a Mixed Reproductive Strategy in Avian Species." *American Naturalist* 124:116–26.

——— (1983) "Comparative Biology and Behavior of Monogamous Pairs and One Male–Two Female Trios of Herring Gulls." *Behavioral Ecology and Sociobiology* 14:1–7.

*Goethe, F. (1937) "Beobachtungen und Untersuchungen zur Biologie der Silbermöwe (*Larus a. argentatus* Pontopp.) auf der Vogelinsel Memmerstand [Observations and Investigations on the Biology of the Herring Gull on Bird Island, Memmerstand]." *Journal für Ornithologie* 85:1–119.

Holley, A. J. F. (1981) "Naturally Arising Adoption in the Herring Gull." *Animal Behavior* 29:302–3.

MacRoberts, M. H. (1973) "Extramarital Courting in Lesser Black-backed and Herring Gulls." *Zeitschrift für Tierpsychologie* 32:62–74.

*Mills, J. A. (1994) "Extra-Pair Copulations in the Red-billed Gull: Females with High-Quality, Attentive Males Resist." *Behavior* 128:41–64.

*——— (1991) "Lifetime Production in the Red-billed Gull." *Acta XX Congressus Internationalis Ornithologici, Christchurch, New Zealand* (Proceedings of the 20th International Ornithological Congress), vol. 3, pp. 1522–27. Wellington, N.Z.: New Zealand Ornithological Trust Board.

*———— (1989) "Red-billed Gull." In I. Newton, ed., *Lifetime Reproduction in Birds,* pp. 387–404. London: Academic Press.

———— (1973) "The Influence of Age and Pair-Bond on the Breeding Biology of the Red-billed Gull, *Larus novaehollandiae scopulinus.*" *Journal of Animal Ecology* 42:147–62.

*Mills, J. A., J. W. Yarrall, and D. A. Mills (1996) "Causes and Consequences of Mate Fidelity in Red-billed Gulls." In J. M. Black, ed., *Partnerships in Birds: The Study of Monogamy,* pp. 286–304. Oxford: Oxford University Press.

Nisbet, I. C. T., and W. H. Drury (1984) "Supernormal Clutches in Herring Gulls in New England." *Condor* 86:87–89.

Parsons, J. (1971) "Cannibalism in Herring Gulls." *British Birds* 64:528–37.

Pierotti, R. J. (1980) "Spite and Altruism in Gulls." *American Naturalist* 115:290–300.

Pierotti, R. J., and T. P. Good (1994) "Herring Gull (*Larus argentatus*)." In A. Poole and F. Gill, eds., *The Birds of North America: Life Histories for the 21st Century,* no. 124. Philadelphia: Academy of Natural Sciences; Washington, D.C.: American Ornithologists' Union.

Richards, C. E. (1995) "Attempted Copulation Between Adult and First-Year Herring Gulls." *British Birds* 88:226.

*Shugart, G. W. (1980) "Frequency and Distribution of Polygyny in Great Lakes Herring Gulls in 1978." *Condor* 82:426–29.

*Shugart, G. W., M. A. Fitch, and G. A. Fox (1988) "Female Pairing: A Reproductive Strategy for Herring Gulls?" *Condor* 90:933–35.

*———— (1987) "Female Floaters and Nonbreeding Secondary Females in Herring Gulls." *Condor* 89:902–6.

Tasker, C. R., and J. A. Mills (1981) "A Functional Analysis of Courtship Feeding in the Red-billed Gull, *Larus novaehollandiae scopulinus.*" *Behavior* 77:221–41.

Wheeler, W. R., and I. Watson (1963) "The Silver Gull *Larus novaehollandiae* Stephens." *Emu* 63:99–173.

GULLS

Shore Bird

BLACK-HEADED GULL
Larus ridibundus

HOMOSEXUALITY	TRANSGENDER	BEHAVIORS		RANKING	OBSERVED IN
○ Female	○ Intersexuality	● Courtship	● Pair-bonding	● Primary	● Wild
● Male	○ Transvestism	○ Affectionate	● Parenting	○ Moderate	○ Semiwild
		● Sexual		○ Incidental	● Captivity

IDENTIFICATION: A medium-sized gull (to 17 inches) with a distinctive black or chocolate brown "hood," red legs, and a gray back and wings. DISTRIBUTION: Throughout most of Eurasia; winters south to Africa and southern Asia. HABITAT: Variable, including lakes, swamps, rivers, grassland, coasts, bays. STUDY AREAS: Kiyovo Lake near Moscow, Russia; University of Groningen, the Netherlands.

Social Organization

Black-headed Gulls are gregarious, gathering in large flocks throughout most of the year. They form pair-bonds and nest in dense colonies containing up to several thousand pairs.

Description

Behavioral Expression: Male homosexual pairs are found in Black-headed Gulls, and they are usually initiated with the same courtship behaviors seen in heterosexual pairs. At the beginning of the mating season, one male lands on the territory of another male who is performing an ADVERTISEMENT DISPLAY, consisting of a series of loud, rasping screams sounding like *kreeeee kreeeee . . .* (the LONG CALL). This is made while the bird assumes an OBLIQUE-POSTURE (head extended forward and upward with the bill horizontal). The displaying male may also approach the other male in the FORWARD-POSTURE, in which the head is lowered and held level with the rest of the body, neck kinked, and tail spread. Over the next several weeks, as their pair-bond develops, the two males perform frequent MEETING CERE-MONIES, stylized courtship interactions that include many of the same movements seen in the advertisement display, combined with HEAD-FLAGGING, in which the two birds alternately turn their heads toward and away from their partner. The two birds sometimes also perform COURTSHIP-FEEDING, in which one male ritually begs the other for food, who may respond by regurgitating a ceremonial food "offering" for his mate. Males in homosexual pairs generally perform meeting ceremonies more frequently than birds in heterosexual pairs; their long-calling and head-flagging rates are intermediate between those of males and females in heterosexual pairs, while they generally beg and courtship-feed less than either males or females in opposite-sex pairs.

Once their pair-bond is established, male partners may also engage in sexual behavior. This can involve one male simply mounting the other, but in other cases it consists of full copulation with repeated genital contacts between the two males. In some homosexual couples, both males take turns mounting each other, although in most instances only one partner is the mounter and the other is the mountee. Male couples usually build a nest together, with both birds contributing to its construction. Occasionally the nest is built entirely by one partner (as in heterosexual pairs, where usually only the male builds the nest). Most homosexual pairs are unable to obtain eggs, although it is possible that some may "adopt" eggs abandoned by heterosexual pairs. In addition, females in this species occasionally lay eggs in other birds' nests, so male pairs could acquire eggs in this way as well. When provided with "foster eggs" in captivity, homosexual pairs faithfully incubate and hatch them and in some cases even successfully raise the chicks together.

In addition to monogamous homosexual pairs, several other variations of same-sex bonding occur. About 25 percent of males in homosexual associations are polygamous (comparable to rates for males in heterosexual associations). Such males form trios, bonding either with two males, or with one male and one female (in one case, a same-sex trio later developed into a bisexual "quartet" when a female joined them). Some male couples are nonmonogamous, in that one or both partners court and even copulate with outside partners of either sex. Homosexual pairings are generally not as long-lasting as heterosexual ones: an average of 15 percent of same-sex bonds last for at least two consecutive seasons, compared to 56 percent of heterosexual bonds. Nevertheless, many homosexual bonds are strong, and partners will maintain their attachment to their mate even if they are forcibly separated from one another, resuming their partnership once reunited. Males also frequently engage in serial pairing, forming bonds with three to four different partners over time (both males and females). Sometimes, the permutations of all these different possibilities can be dizzying—in one case, a male who had been in a homosexual couple later paired with a female, during which time he engaged in promiscuous, "incestuous" courtship and mating with a heterosexually paired female that he had previously foster-parented with his male partner!

In addition to a pattern of homosexual pair-bonding, male Black-headed Gulls in heterosexual pairs sometimes copulate with birds other than their mate, including other males. These nonmonogamous copulations are usually "forced" in the sense that the mounted bird does not solicit the sexual activity and usually vigorously pecks at the male and drives him away. Often a promiscuous male will hover over another male, trying unsuccessfully to mount, although in some cases a male may be able to mount and remain on the other male for as long as five minutes (albeit without any genital contact taking place). Sometimes a promiscuous mounting leads to a same-sex pair-bond: one heterosexually paired bird who mounted both partners of a heterosexual couple, for example, formed a homosexual bond the next year with the male partner of that couple.

Frequency: Male couples that court, mate, and nest together have been documented in wild populations, although their overall incidence is not known. In captivity, homosexual pairs can account for 16–18 percent of all bonds, while two out of 9 promiscuous mountings (22 percent) in one study were homosexual. At the height of courtship, individual males in homosexual pairs may perform displays such as ceremonial encounters or long-calling as often as 40–60 or more times an hour.

Orientation: In a study of Black-headed Gulls in captivity, 22 percent of the males formed bonds only with other males, 15 percent formed bonds with both males and females, while 63 percent bonded only with members of the opposite sex. In addition, layered on top of these patterns of pair-bonding are instances of nonmonogamous, bisexual courtship and sexual activity—often same-sex for heterosexually paired birds, and opposite-sex for homosexually paired birds. Furthermore, younger males appear to have a preference for homosexual pairing that declines somewhat as the birds get older: 55–60 percent of bonds are same-sex among one-to-two-year-olds, compared to 30–45 percent in two-to-five-year-olds, and 20 percent in five-to-eight-year-olds. In addition, many males may have a "latent" capacity for bisexuality, being able to form same-sex bonds when females are not available.

Nonreproductive and Alternative Heterosexualities

Most of the variations on long-term, monogamous pair-bonds seen in homosexual contexts also occur in heterosexual associations among Black-headed Gulls. Promiscuous courtships and copulations can be entirely heterosexual: some females court males other than their mate, while males may try to rape females they aren't paired with (within-pair copulations are also sometimes accompanied by aggression or unwillingness between partners). As noted above, divorce (changing of partners between breeding seasons) occurs fairly frequently in opposite-sex pairs. In addition, mate-switching within the breeding season is quite common—more than half of all males form brief liaisons with females that last a few days, mating with up to seven different partners. Polygamous trios also occur in this species: occasionally, a female will join a mated pair, bond with the male, and help them raise their young, although she herself does not breed. Some females lay eggs in the territories of other heterosexual couples, who may then adopt those eggs as their own; adoption of chicks also occurs. A number of nonreproductive sexual behaviors are also characteristic of this species. Heterosexual mounting often does not involve mutual genital contact, females sometimes mount males (REVERSE mounts), and heterosexual pairs occasionally copulate during incubation or after the hatching of their eggs. In addition, males occasionally try to mate with chicks (including their own), who may be only a few days old; juveniles also mount each other. Young birds are also subjected to abuse when they cross territories belonging to other heterosexual pairs, who may attack and even kill them; cannibalism has also been reported in this species.

Sources   *asterisked references discuss homosexuality/transgender*

Axell, H. E. (1969) "Copulatory Behavior of Juvenile Black-headed Gull." *British Birds* 62:445.

Beer, C. G. (1963) "Incubation and Nest-Building Behavior of Black-headed Gulls IV: Nest-Building in the Laying and Incubation Periods." *Behavior* 21:155–76.

*Kharitonov, S. P., and V. A. Zubakin (1984) "Protsess formirovania par u ozyornykh chaek [Pair-bonding in the Black-headed Gull]." *Zoologichesky Zhurnal* 63:95–104.

Kirkman, F. B. (1937) *Bird Behavior*. London: Nelson.

Moynihan, M. (1955) *Some Aspects of Reproductive Behavior in the Black-headed Gull (*Larus ridibundus ridibundus *L.) and Related Species*. Behavior Supplement 4. Leiden: E. J. Brill.

*van Rhijn, J. (1985) "Black-headed Gull or Black-headed Girl? On the Advantage of Concealing Sex by Gulls and Other Colonial Birds." *Netherlands Journal of Zoology* 35:87–102.

*van Rhijn, J., and T. Groothuis (1987) "On the Mechanism of Mate Selection in Black-headed Gulls." *Behavior* 100:134–69.

*——— (1985) "Biparental Care and the Basis for Alternative Bond-Types Among Gulls, with Special Reference to Black-headed Gulls." *Ardea* 73:159–74.

Gulls

Shore Bird

LAUGHING GULL
Larus atricilla

IVORY GULL
Pagophila eburnea

HOMOSEXUALITY	TRANSGENDER	BEHAVIORS		RANKING	OBSERVED IN
○ Female	○ Intersexuality	● Courtship	● Pair-bonding	○ Primary	● Wild
● Male	○ Transvestism	○ Affectionate	● Parenting	● Moderate	○ Semiwild
		● Sexual		○ Incidental	● Captivity

LAUGHING GULL
IDENTIFICATION: A medium-sized (to 18 inches) black-headed gull with white eye-crescents, a dark gray back, and red legs and bill. DISTRIBUTION: Atlantic coast of North America, Caribbean; winters to northern South America. HABITAT: Coastal beaches, islands, salt-marshes. STUDY AREAS: Stone Harbor, New Jersey; National Zoological Park, Washington, D.C.; subspecies *L.a. megalopterus*.

IVORY GULL
IDENTIFICATION: An all-white gull with black legs and a blue-gray bill. DISTRIBUTION: Throughout the high Arctic. HABITAT: Pack ice, cliffs, islands. STUDY AREA: Seymour Island, Northwest Territories, Canada.

Social Organization

Laughing Gulls are highly social and gregarious, forming large flocks at all times of the year, while Ivory Gulls tend to be more solitary or gather in smaller flocks. Their mating system involves pair-bonding, and birds nest in colonies that contain a few dozen pairs in Ivories and several hundred to as many as 25,000 pairs in Laughing Gulls.

Description

Behavioral Expression: Pair-bonds sometimes develop between two male Laughing Gulls, including sexual and parenting activities as well as unique courtship and territorial behaviors. A homosexual bond begins when two males KEEP COMPANY: in this courtship activity, they approach and circle each other while performing elegant FACING-AWAY or HEAD-TOSS displays (in which the birds ritually turn their heads in opposite directions or flick them upward from a hunched posture). This is followed by a period of resting together somewhat apart from the other Gulls. Two males may even begin courting in this way early in the breeding season before heterosexual couples have started their courtships. Male partners also COURTSHIP-FEED each other, in which they present one another with a symbolic "gift" of food. Although this behavior is also found in opposite-sex courtship, in homosexual pairs it has a unique feature. The two males pass the food back and forth between them many times before one or both of them eat it, whereas in heterosexual pairs, the male presents the food to the female, who immediately eats it. Pair-bonded males also mount each other, although genital contact usually does not occur. Mounting is generally not one-sided, although one male may prefer being mounter or mountee more than the other. In one male couple, for example, one partner mounted the other nine times over the season while the other mounted him three times. Homosexual (as well as heterosexual) mounts may be accompanied by distinctive staccato COPULATION CALLS by the mounting male, sounding like *kakakakakaka*. Both males in homosexual pairs build a nest together; in captivity, pairs take turns incubating eggs when these are supplied to the couple. After the eggs hatch, both males share parenting duties such as feeding and protecting the chicks and are able to successfully raise the youngsters.

Male couples can be quite aggressive, repeatedly intruding on neighboring territories of heterosexual pairs, who try, often unsuccessfully, to deter them. During these "raids," the males may engage in sexual behavior with one another, or they may even try to court or mount a neighboring female. Similar territorial invasions are sometimes also made by single males, heterosexually paired males, or "coalitions" of two males that are not necessarily bonded or sexually involved with each other. In these cases, the intruding males often mount the male partner of the heterosexual pair they encounter, sometimes in full view of his incubating female mate (who is usually indifferent). If there are two intruding males, they may take turns in homosexual mounts on the male of the pair. The mounted male usually responds by violently shaking, pecking, and fluttering in order to dislodge the mounter—who may end up getting grabbed by the beak and tossed over his head. Sometimes an intruding male will even mount a male who is copulating with his female partner, creating a three-bird "pile-up."

A similar sort of forced homosexual mounting occurs in Ivory Gulls. Males approach incubating birds whose partners are away from the nest. The intruding male carries a distraction "gift" of nesting material—usually a lump of moss—and circles the nest. He then suddenly jumps on the back of the incubating bird—even if it is a male—and attempts to copulate. As in Black-headed Gulls, the mounted bird usually responds violently.

Frequency: In a zoo population of Laughing Gulls, one pair-bond out of a total of four was between two males; the incidence of homosexual pairs in wild birds is not known, although homosexual mounts by intruding males have been documented in wild birds. Forced mountings on incubating Ivory Gulls are very frequent in some colonies in the wild, although it is not known specifically what proportion are same-sex.

Orientation: Male Laughing Gulls in homosexual pairs may be simultaneously bisexual, since they sometimes court and mount females while maintaining their primary same-sex pair-bond. Intruding males that participate in homosexual mounts may also be bisexual if they are paired to a female, although unpaired males engage in this activity as well. Such males also mount females during their intrusions, but in many cases they are clearly "targeting" the male partner. For example, they may completely ignore the female as she sits incubating, or they may remain mounted on the male after he dismounts from the female during a "pile-up," again ignoring the female once she becomes "available." Most males that are subject to intruding mounts like this probably have a primary heterosexual orientation, since they usually react aggressively to any homosexual activity. However, it must be remembered that females with a primary heterosexual bond also react violently to mounts by intruding males.

Nonreproductive and Alternative Heterosexualities

Female Laughing Gulls occasionally "adopt" and incubate eggs laid in their nests by other females, including females of other species such as clapper rails (*Rallus longirostris*). Heterosexual pairs in this species also have very high copulation rates, mating as often as nine times a day prior to egg laying. As noted above, forced copulation attempts are common in Laughing and Ivory Gulls as well, and many of these are heterosexual. About 30 percent of male Laughing Gull territorial intruders attempt to rape females, although such attacks are usually thwarted by the female and her mate; in addition, most are directed toward incubating females and hence are nonreproductive. Abuse and violence by heterosexual parents against chicks also occur in both species. During raids by predators such as snowy owls (*Nyctea scandiaca*), a pandemonium often develops in Ivory Gull breeding colonies, and groups of adult Gulls mob terrified chicks and sometimes even attack and kill them. The adults also puncture eggs and eat them; eggs containing fully developed embryos about to hatch are smashed and the yolk sac is eaten, but the down-covered embryo is discarded. Laughing Gulls also occasionally attack, kill, and even eat chicks (as well as eggs) on neighboring territories; cannibalism appears to be limited to a few individuals in dense colonies who repeatedly perform this behavior.

Sources   *asterisked references discuss homosexuality/transgender

Bateson, P. P. G., and R. C. Plowright (1959a) "The Breeding Biology of the Ivory Gull in Spitsbergen." *British Birds* 52:105–14.

——— (1959b) "Some Aspects of the Reproductive Behavior of the Ivory Gull." *Ardea* 47:157–76.

Burger, J. (1996) "Laughing Gull (*Larus atricilla*)." In A. Poole and F. Gill, eds., *The Birds of North America: Life Histories for the 21st Century*, no. 225. Philadelphia: Academy of Natural Sciences; Washington, D.C.: American Ornithologists' Union.

——— (1976) "Daily and Seasonal Activity Patterns in Breeding Laughing Gulls." *Auk* 93:308–23.

Burger, J., and C. G. Beer (1975) "Territoriality in the Laughing Gull (*L. atricilla*)." *Behavior* 55:301–20.

Hand, J. L. (1985) "Egalitarian Resolution of Social Conflicts: A Study of Pair-bonded Gulls in Nest Duty and Feeding Contexts." *Zeitschrift für Tierpsychologie* 70:123–47.

*——— (1981) "Sociobiological Implications of Unusual Sexual Behaviors of Gulls: The Genotype/Behavioral Phenotype Problem." *Ethology and Sociobiology* 2:135–45.

*Haney, J. C., and S. D. MacDonald (1995) "Ivory Gull (*Pagophila eburnea*)." In A. Poole and F. Gill, eds., *The Birds of North America: Life Histories for the 21st Century*, no. 175. Philadelphia: Academy of Natural Sciences; Washington, D.C.: American Ornithologists' Union.

MacDonald, S. D. (1976) "Phantoms of the Polar Pack-Ice." *Audubon* 78:2–19.

*Noble, G. K., and M. Wurm (1943) "The Social Behavior of the Laughing Gull." *Annals of the New York Academy of Sciences* 45:179–220.

Segrè, A., J. P. Hailman, and C. G. Beer (1968) "Complex Interactions Between Clapper Rails and Laughing Gulls." *Wilson Bulletin* 80:213–19.

TERNS

CASPIAN TERN
Sterna caspia
ROSEATE TERN
Sterna dougallii

Shore Bird

HOMOSEXUALITY	TRANSGENDER	BEHAVIORS		RANKING	OBSERVED IN
● Female	○ Intersexuality	○ Courtship	● Pair-bonding	○ Primary	● Wild
○ Male	○ Transvestism	○ Affectionate	● Parenting	○ Moderate	○ Semiwild
		○ Sexual		● Incidental	○ Captivity

CASPIAN TERN
IDENTIFICATION: A large (to 22 inches) gull-like bird with a black cap and crest, light gray back and wings, forked tail, and long red bill with a black tip. DISTRIBUTION: Throughout much of North America, Europe, Australasia, Africa. HABITAT: Coasts, lakes, estuaries. STUDY AREA: Columbia River in eastern Washington and Oregon.

ROSEATE TERN
IDENTIFICATION: Similar to Caspian Tern but smaller (to 17 inches) and with a more deeply forked tail and slighter bill. DISTRIBUTION: North Atlantic, Caribbean, Africa, Australasia. HABITAT: Seacoasts, islands. STUDY AREAS: Bird Island, Marion, Massachusetts; Falkner Island, Connecticut; subspecies *S.d. dougallii*.

Social Organization

During the mating season, Caspian and Roseate Terns usually congregate in large colonies which may contain up to 500 pairs in Caspians and several thousand in Roseates. The typical social unit is the monogamous mated pair. Outside of the mating season, Terns are less gregarious and are usually found alone or in small groups.

Description

Behavioral Expression: In both Caspian and Roseate Terns, two females may pair with each other, associating together the way a male-female pair does. Usually such homosexual pairs also build nests and lay eggs. Since both females typically lay, the result is a SUPERNORMAL CLUTCH containing up to twice the number of eggs found in nests of heterosexual pairs—4–6 eggs for Caspian homosexual pairs, and 3–5 eggs for Roseates. Both females take turns incubating the eggs (as do mates in heterosexual pairs). Sometimes the eggs are infertile, but in many cases they do hatch. There are several possible sources for fertile eggs in same-sex pairs: for example, one or both females may copulate with a male while still remaining bonded to her female mate. In addition, it appears that females in some populations occasionally "steal" eggs from others' nests, sometimes transferring eggs from as many as three other nests into their own. Approximately 13 percent of supernormal clutches in some locations have at least one "stolen" egg in them, so it is likely that at least some females in homosexual pairs utilize this strategy. Once the eggs hatch, both females share parenting duties (as do partners in heterosexual pairs), which include feeding the young, protecting the chicks from predators, sheltering them against the sun, and defending the nesting territory.

Frequency: In Caspian Terns, 3–6 percent of pairs are homosexual; in Roseates, about 5 percent of chicks are tended by female pairs in some populations.

Orientation: Females in homosexual pairs that copulate with males in order to fertilize their eggs are functionally bisexual, although they retain their primary bond with the other female. Other female pairs may be exclusively homosexual for the duration of their pair-bond, since they do not lay fertile eggs. Some females are also sequentially bisexual, alternating between male and female partners in different breeding seasons.

Nonreproductive and Alternative Heterosexualities

Terns occasionally "steal" eggs from other clutches, as described above for homosexual pairs; birds in heterosexual pairs probably do so, since some nests with transferred eggs do not have supernormal clutches. In addition, female Roseate Terns sometimes lay eggs in nests other than their own, resulting in "super-supernormal clutches." In one colony, for example, about 1 percent of the nests had 7 eggs—more than twice the number found even in *supernormal* clutches. Most such nests appear to belong to heterosexual pairs. Caspian Terns also have a high divorce rate: more than half of all male-female pairs do not last more than one season. Female Roseate Terns sometimes successfully raise chicks as a single parent when their male partner dies. As in many Gulls, infanticide and aggression toward chicks also occur in some Tern species. Caspian Terns, for example, often violently attack—and may even kill—chicks that wander onto their territories and may break eggs during their squabbles as well. Caspian Terns also commonly form CRÈCHES, dense herds of chicks attended by a few adults who watch over them while their parents are away foraging.

Sources   *asterisked references discuss homosexuality/transgender*

*Conover, M. R. (1983) "Female-Female Pairings in Caspian Terns." *Condor* 85:346–49.

Cramp, S., ed. (1985) "Caspian Tern (*Sterna caspia*)" and "Roseate Tern (*Sterna dougallii*)." In *Handbook of the Birds of Europe, the Middle East, and North Africa*, vol. 4, pp. 17–27, 62–71. Oxford: Oxford University Press.

Cuthbert, F. J. (1985) "Mate Retention in Caspian Terns." *Condor* 87:74–78.

*Gochfeld, M., and J. Burger (1996) "Sternidae (Terns)." In J. del Hoyo, A. Elliott, and J. Sargatal, eds., *Handbook of the Birds of the World, vol. 3: Hoatzin to Auks*, pp. 624–67. Barcelona: Lynx Ediciós.

*Hatch, J. J. (1995) Personal communication.

*———— (1993) "Parental Behavior of Roseate Terns: Comparisons of Male-Female and Multi-Female Groups." *Colonial Waterbird Society Bulletin* 17:43.

Milon, P. (1950) "Quelques observations sur la nidification des sternes dans les eaux de Madagascar [Some Observations on the Nesting of Terns in the Waters of Madagascar]." *Ibis* 92:553.

*Nisbet, I. C. T. (1989) "The Roseate Tern." In W. J. Chandler, ed., *Audubon Wildlife Report 1989/1990*, pp. 478–97. San Diego: Academic Press.

*Nisbet, I. C. T., J. A. Spendelow, J. S. Hatfield, J. M. Zingo, and G. A. Gough (1998) "Variations in Growth of Roseate Tern Chicks: II. Early Growth as an Index of Parental Quality," *Condor* 100:305–15.

Penland, S. T. (1984) "An Alternative Origin of Supernormal Clutches in Caspian Terns." *Condor* 86:496.

*Sabo, T. J., R. Kessell, J. L. Halverson, I. C. T. Nisbet, and J. J. Hatch (1994) "PCR-Based Method for Sexing Roseate Terns (*Sterna dougallii*)." *Auk* 111:1023–27.

Shealer, D. A., and J. G. Zurovchak (1995) "Three Extremely Large Clutches of Roseate Tern Eggs in the Caribbean." *Colonial Waterbirds* 18:105–7.

*Spendelow, J. A., and J. M. Zingo (1997) "Female Roseate Tern Fledges a Chick Following the Death of Her Mate During the Incubation Period." *Colonial Waterbirds* 20:552–55.

Perching Birds and Songbirds

COTINGAS, MANAKINS, AND OTHERS

COTINGAS   *Perching Bird*

GUIANAN COCK-OF-THE-ROCK
Rupicola rupicola

HOMOSEXUALITY	TRANSGENDER	BEHAVIORS		RANKING	OBSERVED IN
○ Female	○ Intersexuality	● Courtship	○ Pair-bonding	● Primary	● Wild
● Male	○ Transvestism	○ Affectionate	○ Parenting	○ Moderate	○ Semiwild
		● Sexual		○ Incidental	○ Captivity

IDENTIFICATION: A small (10 inch) perching bird; adult males are brilliant orange with elaborate fringed wing plumes and an imposing helmetlike feather crest; adolescent males have mottled brown and orange plumage, while females are uniformly dark. DISTRIBUTION: Northern South America, primarily in southern Venezuela, the Guianas, adjacent parts of Brazil. HABITAT: Forests, usually near cliffs, mountains, or rock outcrops (on which females build their nests, giving the bird its name). STUDY AREA: Raleigh Falls–Voltzberg Nature Reserve, Suriname, South America.

Social Organization

The spectacularly plumed Guianan Cock-of-the-Rock has what is known as a LEK social and mating system: males inhabit individual territories, usually clustered in the same area, which are used for display and courtship. Each display "court" consists of a cleared area on the forest floor and surrounding perches. Territories are maintained year-round, but courtship and mating occur only from late December through April. Females (and in this species, young males) visit these territories to choose which males they want to mate with. Other than this, males and females lead virtually separate lives: males do not participate in any aspect of nesting or parental care and rarely encounter females outside of the breeding season.

Description

Behavioral Expression: Homosexual activity between adult and adolescent males, as well as among adolescent males, is a routine occurrence in Guianan Cock-of-the-Rock. Males court and display to each other, as well as engage in homosexual mounting. A typical homosexual encounter begins when the adult males are perched beside their display courts, each glowing like an orange beacon in the jungle gloom. Both females and adolescent (yearling) males are attracted to the lek, which is carefully situated in the forest to take advantage of the ambient light characteristics of the environment (thereby showing off its owner to his best advantage). The courtship sequence commences with a GREETING DISPLAY: the males begin calling raucously, then each drops to the ground with a thump and begins beating his wings violently, flashing patches of black and white. This often produces a whistling sound as air rushes through the specially modified wing feathers.

This attention-grabbing sequence is followed by the GROUND DISPLAY: each adult male crouches down, fanning out the delicate filaments of his wing coverts, puffing out his chest and rump feathers and erecting his crest, resulting in a spectacular visual effect. By this time an adolescent male who is attracted by the display has landed beside the adult and is hopping about the display court, often crouching in a version of the courtship posture himself. The adult keeps his back toward the younger male at all times but is otherwise motionless, showing off his plumage to its best and inviting the adolescent to mount him. During homosexual copulation, the younger male climbs onto the adult's back and perches firmly, moving his tail sideways to try to make genital contact. Often the younger male mounts the older male several times in succession, and courtship and display often alternate with mountings. Sometimes males also mount each other in the trees surrounding a display court. Homosexual interactions differ from heterosexual ones in that both participants perform some version of the ground display; also, unlike females, neither male pecks at or touches the other's rump prior to mounting.

Adolescent males usually visit the display courts of several adult males, although some adults are clearly more "popular" than others because they receive more attention from the younger males. Typically, a yearling has homosexual interactions with anywhere from one to seven different adult males during the mating season. Nor does homosexual activity always involve an adult territory owner and a yearling male: adolescents often mount nonbreeding males who do not have their own display territories and sometimes also mount other adolescent males. Homosexual activity is not separate from heterosexual courtship and copulation,

A younger (adolescent) male Guianan Cock-of-the-Rock mounting a bright orange adult male in the forests of Suriname ▶

but takes place in the same locations and often while male-female interactions are happening in the vicinity. However, homosexual courtship and mounting often take priority over heterosexual interactions. If an adolescent male approaches an adult who is courting a female, the female usually leaves (or is chased away), and the two males turn their attentions to each other. Moreover, if a female encounters a male who is courting or involved sexually with another male, she usually waits until the adolescent male leaves before approaching the adult.

Frequency: Homosexual activity is very common among Guianan Cock-of-the-Rock: in fact, mountings between males are as frequent as male-female mountings, accounting for nearly half of all copulations. About 10 percent of heterosexual courtships are interrupted by adolescent males visiting the courting male; roughly one out of five of these "interruptions" involves courtship or sexual behavior between males. During the breeding season, homosexual activity may occur daily, and an adolescent male will usually have 6–7 homosexual encounters over the season (although some males engage in homosexual mounting more frequently, 15 or more times over a season).

Orientation: Close to 40 percent of the male population participates in some form of homosexual activity. Depending on his age, a bird that has same-sex interactions may or may not also engage in heterosexual activity. Among adult males (three years or older), nearly a quarter (23 percent) are mounted by other males, and 6 percent of these do not mate or court females at all. In fact, those adults who do not mate heterosexually are often the ones most frequently mounted by younger males. Among yearling males (virtually none of whom mount females), almost two-thirds (64 percent) engage in homosexual activity. Two-year-old males, on the other hand, very rarely engage in homosexual activity. Thus, when birds of all ages are taken into account, nearly 20 percent of the male population at any given time is involved exclusively in homosexuality.

Nonreproductive and Alternative Heterosexualities

As described above, male and female Guianan Cock-of-the-Rock have essentially no contact with each other outside of the mating season. Even during the breeding season, their interactions are often unfriendly or overtly aggressive. Males frequently harass females by chasing them around the display court and in some cases attempt to forcibly copulate with them. Females struggle violently during these rape attempts and are usually able to get away, but are visibly stressed following such interactions. In fact, only 20 percent of all lek visits by females result in copulation. There is also a significant proportion of nonbreeding individuals in the population: an average of 20 percent of adult males do not have display territories (and thus rarely, if ever, court or mate with females), while nearly two-thirds of territorial males fail to mate each year. Moreover, males are rarely able to acquire their own territories (and therefore court females) before they are three to four years old. In addition, many females who visit the courtship grounds never actually mate with males during the entire breeding season.

Sources   *asterisked references discuss homosexuality/transgender*

Endler, J. A., and M. Théry (1996) "Interacting Effects of Lek Placement, Display Behavior, Ambient Light, and Color Patterns in Three Neotropical Forest-Dwelling Birds." *American Naturalist* 148:421–52.

Gilliard, E. T. (1962) "On the Breeding Behavior of the Cock-of-the-Rock (Aves, *Rupicola rupicola*)." *Bulletin of the American Museum of Natural History* 124:31–68.

Trail, P. W. (1989) "Active Mate Choice at Cock-of-the-Rock Leks: Tactics of Sampling and Comparison." *Behavioral Ecology and Sociobiology* 25:283–92.

*——— (1985a) "A Lek's Icon: The Courtship Display of a Guianan Cock-of-the-Rock." *American Birds* 39:235–40.

——— (1985b) "Courtship Disruption Modifies Mate Choice in a Lek-Breeding Bird." *Science* 227:778–80.

*——— (1983) "Cock-of-the-Rock: Jungle Dandy." *National Geographic Magazine* 164:831–39.

*Trail, P. W., and D. L. Koutnik (1986) "Courtship Disruption at the Lek in the Guianan Cock-of-the-Rock." *Ethology* 73:197–218.

COTINGAS   Perching Bird

CALFBIRD
Perissocephalus tricolor

HOMOSEXUALITY	TRANSGENDER	BEHAVIORS		RANKING	OBSERVED IN
● Female	○ Intersexuality	● Courtship	● Pair-bonding	○ Primary	● Wild
● Male	○ Transvestism	○ Affectionate	○ Parenting	● Moderate	○ Semiwild
		● Sexual		○ Incidental	○ Captivity

IDENTIFICATION: A crow-sized bird with cinnamon-brown plumage and a bare, blue-gray face. DISTRIBUTION: North-central South America, including Venezuela, the Guianas, Amazonian Brazil. HABITAT: Tropical forest. STUDY AREAS: Brownsberg Nature Park, Suriname; near the Kanuku Mountains, Guyana.

Social Organization

Calfbird males congregate, eight to ten at a time, on a display area or LEK in the understory of the rain forest canopy. Each lek consists of a central perch occupied by only one male, around which the remaining males cluster. Females (and males) visit the central male during courtship, usually at dawn; mating is polygamous or promiscuous, and females nest and raise the young on their own.

Description

Behavioral Expression: Male Calfbirds sometimes court other males who are attracted to the leks. A homosexual courtship begins when one male approaches the display site in a posture that females also adopt when entering the lek: the body is held in a horizontal position and all of the feathers are sleeked down, including the cowl feathers. He is attracted by the dramatic display of the central male, which consists of a loud, droning MOO call (hence the bird's name) sounding something like *grr-aaa-oooo*. While mooing, the displaying male first puffs himself up, fluffing up his cowl feathers, and then sinks back down in a bowing motion while exposing two bright orange feather globes on either side of his tail. The male in horizontal posture gets close to the mooing bird, who may then direct his courtship displays directly at the other male—for example, by hopping closer while mooing intensely, or engaging in ritualized wing-preening. Usually the approaching male is then chased away if he gets too close, and in fact no homosexual copulations have yet been observed in this species. However, heterosexual mating itself is fairly rare (only occurring in 14 percent of dawn courtship visits by females), and displaying males also frequently chase females who approach them. Furthermore, males do sometimes mount or attempt to mount other males outside of the lek, though this may occur in an aggressive context. Calfbirds also typically form "companionships" with each other: the two males perform coordinated courtship displays, almost touching as they perch side by side, while they both moo and bow in precise alternation. In some cases, companions never perform courtship displays without their male partner. Companions also travel together on the lek and may even share a "home" with each other. Calfbirds typically have what is known as a RETREAT, a special location or tree where each male regularly spends time when not on the lek. Display partners sometimes share the same retreat.

Female Calfbirds also develop bonds with one another: the two companions keep each other company while feeding, travel together to and from the lek, and even nest close to each other. This is all the more remarkable considering that there are no heterosexual pair-bonds in this species. Female pairs use a number of distinctive calls to communicate with each other. While feeding together, for example, they maintain contact with soft *wark* calls. When one female sits on her nest, the other may perch nearby uttering a rasping *waaa* call regularly for over an hour, perhaps acting as a lookout for her companion. And two females nesting close to each other sometimes communicate with a unique low, growling call that sounds like *grewer grewer,* which is only heard in this context. Although no same-sex courtship or copulation has yet been observed, females do sometimes perform behaviors that are typically associated with displaying males. For example, one female was seen to call and posture repeatedly in the fashion of a male courting on a lek (though she did this away from the lek, at her nest). Her call was like the first half of the male's mooing, and she displayed the orange tail ornaments usually seen in the male's courtship sequence. Her voice was also raspier than other females', and she built an exceptionally large nest. In addition, females sometimes adopt the male's characteristic upright and fluffed appearance when they visit the lek.

Frequency: Approximately 1–2 percent of all courtship visits involve one male displaying to another, and all males (except for the central one) have male display partners. Mounting attempts between males occur fairly frequently, but it is not known how prevalent female same-sex activities are.

Orientation: Because only the central male ever copulates with females, the remainder of the male population is effectively involved only in same-sex activities for the duration of the breeding season, whether display companionships or courtship/mounting with other males. The central male, however, displays mostly to females and occasionally to males, so his behavior is bisexual—albeit primarily heterosexual. In addition, in the next breeding season another male may become the central one, so at least some of the males exhibit sequential bisexuality, while the remainder may continue to experience longer periods of more exclusive homosexuality.

Nonreproductive and Alternative Heterosexualities

As discussed above, the majority of the male population in any given year does not reproduce: only one male (the central one in the lek) ever breeds with a female. Moreover, only 12 percent of females mate during their dawn courtship visits, and heterosexual relations are frequently fraught with difficulty. Copulations are brief, often incomplete, and accompanied by aggression. Females visiting the lek are constantly chased by the noncentral males, while nearly one-third of male-female copulations and more than half of all courtship visits are disrupted and harassed by other Calfbirds.

Sources   *asterisked references discuss homosexuality/transgender*

*Snow, B. K. (1972) "A Field Study of the Calfbird *Perissocephalus tricolor*." *Ibis* 114:139–62.

——— (1961) "Notes on the Behavior of Three Cotingidae." *Auk* 78:150–61.

Snow, D. (1982) *The Cotingas: Bellbirds, Umbrellabirds, and Other Species.* Ithaca: Cornell University Press.

*——— (1976) *The Web of Adaptation: Bird Studies in the American Tropics.* New York: Quadrangle/New York Times Book Co.

*Trail, P. W. (1990) "Why Should Lek-Breeders Be Monomorphic?" *Evolution* 44:1837–52.

MANAKINS   *Perching Bird*

SWALLOW-TAILED MANAKIN
Chiroxiphia caudata

BLUE-BACKED MANAKIN
Chiroxiphia pareola

HOMOSEXUALITY	TRANSGENDER	BEHAVIORS		RANKING	OBSERVED IN
○ Female	○ Intersexuality	● Courtship	○ Pair-bonding	○ Primary	● Wild
● Male	○ Transvestism	○ Affectionate	○ Parenting	● Moderate	○ Semiwild
		● Sexual		○ Incidental	○ Captivity

SWALLOW-TAILED MANAKIN
IDENTIFICATION: Adult males are bright blue with an orangish red crown and a black head and wings; females and yearling males are all green, while younger adult males are green or blue-green with an orange crown. DISTRIBUTION: Eastern Brazil, Paraguay, northwestern Argentina. HABITAT: Moist forests. STUDY AREA: Southeastern Brazil.

BLUE-BACKED MANAKIN
IDENTIFICATION: Adult males are black with a red crown and a light blue patch on the back; yearling males and females are all green, while younger adult males are green with a reddish crown. DISTRIBUTION: Tobago, the Guianas, the Amazon Basin. HABITAT: Forest undergrowth, woodland. STUDY AREA: Tobago, West Indies; subspecies *C.p. atlantica.*

Social Organization

Male Swallow-tailed Manakins form stable, long-term associations of four to six individuals that spend virtually all of their time together; they remain with each other throughout the mating season and usually from year to year as well. These groups display together on traditional courts or LEKS. The mating system is polygamous, with males (and probably also females) copulating with multiple partners, although large numbers of males are nonbreeders. Blue-backed Manakins have a similar social organization.

Description

Behavioral Expression: Male Swallow-tailed Manakins perform an elaborate group courtship ritual, the JUMP DISPLAY. Sometimes this display is directed toward females, sometimes toward yearling or younger adult males (the latter resembling females in their green plumage, but distinct because of their red-orange caps), and less commonly toward another adult male. To begin a homosexual jump display, a group of two to three adult males gathers on a display perch on their lek, lined up side by side, with the (younger) male they are displaying to perched at the head of this row. While uttering squawking notes that resemble a chorus of frogs, the adult males crouch down, quiver their bodies, and shuffle their feet, forming a collective vibrating mass. The courting males then take turns jumping up and hovering in front of the younger male while giving a sharp *dik-dik-dik* call, landing next to him. As each male jumps, the others slide down the perch toward his former position, repeating the sequence each time to produce a continuous rotation of flying and sliding birds, whose orange-red crowns collectively form a "whirling torch" in front of the courted bird. During this coordinated display, the younger male usually sits motionless in an upright posture. Males also sometimes mount each other and attempt copulation.

A similar revolving courtship display is performed by pairs of male Blue-backed Manakins for a third bird, which is sometimes a younger male. The display begins with a duet between the displaying males in order to attract a potential mate: the two birds perch side by side and utter perfectly synchronized *chup* calls. Once a younger male arrives, they begin courting him with their CATHERINE WHEEL DISPLAY. Each adult male jumps up alternately and flutters backward, "leapfrogging" over his partner, who moves forward to replace him in a precisely timed sequence that resembles two juggling balls. This is repeated up to 60 times to create a "cartwheel" of flying birds, oriented toward the male being courted, that gradually gets faster and faster. All the while the displaying males utter vibrant nasal or buzzing sounds that resemble the twanging of a Jew's harp. At the peak of the dancing frenzy, one of the adult males calls sharply and his partner disappears; the first male then begins courting the younger male one-on-one. He crisscrosses his display perch in a buoyant, butterfly-like flight, alighting in front of the other male every once in a while to lower his head, vibrate his wings and opalescent blue back, and display his brilliant red crest with its two extendable horns. During this performance, the younger male crouches and constantly turns to face the older male courting him, sometimes also responding with a similar bouncing "butterfly" flight across the display perch.

Homosexual courtship in a group of Swallow-tailed Manakins: one male (right) hovers in front of a younger male as the other males prepare to perform their part in the "jump display" ▼

Frequency: The overall proportion of courtships or mounts that occur between males is not yet known for either of these species, although they may be relatively infrequent. However, two of three observed "butterfly" courtship displays in a study of Blue-backed Manakins were directed toward males, while in a two-year study of Swallow-tailed Manakins, only ten heterosexual matings were recorded. It is possible, therefore, that same-sex activity represents a sizable proportion of all courtship and/or sexual activity.

Orientation: Only one out of every four to six male Swallow-tailed Manakins displaying on the lek ever copulates with females; thus, a large proportion of males are not involved directly in male-female sexual (mounting) activity, although they do participate in both heterosexual and homosexual courtships. A similar pattern occurs in Blue-backed Manakins.

Nonreproductive and Alternative Heterosexualities

In addition to significant numbers of nonbreeding males, Swallow-tailed Manakin populations are characterized by several other less-than-optimal heterosexualities. Only about a third of male-female courtships ever result in mating: half of the time, females being courted simply leave, while in the remainder of cases other males disrupt the courtship displays. However, when mating does take place, a female may copulate with the same male up to six times in one visit; the same is also true for Blue-backed Manakins.

Sources   *asterisked references discuss homosexuality/transgender*

Foster, M. S. (1987) "Delayed Maturation, Neoteny, and Social System Differences in Two Manakins of the Genus *Chiroxiphia." Evolution* 41:547–58.

——— (1984) "Jewel Bird Jamboree." *Natural History* 93(7):54–59.

——— (1981) "Cooperative Behavior and Social Organization in the Swallow-tailed Manakin (*Chiroxiphia caudata)." Behavioral Ecology and Sociobiology* 9:167–77.

Gilliard, E. T. (1959) "Notes on the Courtship Behavior of the Blue-backed Manakin (*Chiroxiphia pareola)." American Museum Novitates* 1942:1–19.

*Sick, H. (1967) "Courtship Behavior in the Manakins (Pipridae): A Review." *Living Bird* 6:5–22.

*——— (1959) "Die Balz der Schmuckvögel (Pipridae) [The Mating Ritual of Jewel Birds]." *Journal für Ornithologie* 100:269–302.

*Snow, D. W. (1976) *The Web of Adaptation: Bird Studies in the American Tropics*. New York: Quadrangle/New York Times Book Co.

——— (1971) "Social Organization of the Blue-backed Manakin." *Wilson Bulletin* 83:35–38.

*——— (1963) "The Display of the Blue-backed Manakin, *Chiroxiphia pareola*, in Tobago, W.I." *Zoologica* 48:167–76.

ANTBIRDS

Perching Bird

BICOLORED ANTBIRD
Gymnopithys bicolor
OCELLATED ANTBIRD
Phaenostictus mcleannani

HOMOSEXUALITY	TRANSGENDER	BEHAVIORS		RANKING	OBSERVED IN
○ Female	○ Intersexuality	● Courtship	● Pair-bonding	○ Primary	● Wild
● Male	○ Transvestism	○ Affectionate	○ Parenting	● Moderate	○ Semiwild
		○ Sexual		○ Incidental	○ Captivity

IDENTIFICATION: Small (5—7 inch) birds with brown and rufous plumage and a bluish gray patch around the eyes; Ocellateds have a distinctive scalloped pattern on the back feathers. DISTRIBUTION: Central America and northwestern South America from Honduras to Ecuador. HABITAT: Rain forest undergrowth. STUDY AREA: Barro Colorado Island, Panama.

Social Organization

Antbirds get their name because they follow large swarms of army ants for food, often in flocks containing several different bird species. Both Bicolored and Ocellated Antbirds form monogamous pairs that are generally long-lasting. In addition, Ocellated Antbirds live in a complex extended family or "clan" structure, typically containing up to three generations of males and their mates. Females emigrate from these units to join other clans, while males often return to their extended family once they have found mates.

Description

Behavioral Expression: Male homosexual pairs are a distinctive feature of Bicolored and Ocellated Antbird social life. Usually the pair-bond is initiated and strengthened (as in heterosexual pairs) when one male COURTSHIP-FEEDS the other, that is, ritually presents him with a "gift" of food (usually an insect or spider). In Ocellated Antbirds, males who court each other this way often adopt a characteristic pose (also used in heterosexual courtship)—ruffled throat feathers, stiff neck and upright posture, fluffed-out body feathers, with tail and legs spread. They also CAROL, producing a series of up to 15 whistled notes that decrease in pitch (sounding like *chee chee chew chew*). Unlike heterosexual mates, male partners typically reciprocate courtship feeding by passing the food gift back and forth between them. Homosexual courtship-feeding in Bicolored Antbirds is distinct from the heterosexual pattern in a number of other ways as well: either partner may initiate the activity in a male pair, and courtship-feeding is usually accompanied by CHIRPING—brief, musical *cheup* notes. In heterosexual pairs, only the male feeds the female, and the partners typically utter GROWLS during courtship-feeding (a rapid hissing or grunting noise composed of rough *chauhh* notes). Once paired, two males are constant companions, visiting ant swarms together and foraging side by side much as opposite-sex mated pairs do. Homosexual pair-bonds are sometimes long-lasting associations, persisting for many years. Partners may both be adults, or an older bird may pair with a younger one. In Ocellated Antbirds, father-son courtship-feedings also sometimes occur when a younger male remains within the clan structure.

Frequency: In Bicolored Antbirds, an estimated 2–3 percent of all pairs in some populations are composed of two males, and homosexual pairs may constitute up to 4–6 percent of the total number of pairs in any given year. The incidence of male pairs in Ocellated Antbirds is probably comparable.

Orientation: Approximately 5–14 percent of male Bicolored Antbirds may participate in a homosexual pairing or courtship at some point in their life. In both species, some males exhibit sequential bisexuality. In a few cases, for example, males have been mated to a female prior to their homosexual pairing (they may even have fathered young, and some are widowers), or they may go on to mate with a female when a homosexual pairing breaks up. Other males, however, show no signs of participating in heterosexual mating, and these birds may be involved exclusively in homosexual pairs, at least for a portion of their lives.

Nonreproductive and Alternative Heterosexualities

In both Bicolored and Ocellated Antbirds, significant numbers of birds are nonbreeding. As many as 45 percent of adult Bicolored males may not be heterosexually paired in any given year, and some males fail to acquire a mate for six or more years in a row. Younger males may delay their reproductive careers for up to a year after reaching sexual maturity—by remaining "at home" in their clans (Ocellateds) or wandering solitarily (Bicoloreds). In addition, male Antbirds have been known to live to a relatively old age—more than 11 years in Bicoloreds and Ocellateds. For a few individuals who have lost their female partners (either through death or divorce), this may be a postreproductive period in their lives. In addition to females leaving their mates for younger males, divorce and mate-switching also occasionally happen in adult Bicolored and newly paired Ocellated Antbirds. Often, a divorce is initi-

▲ *A male Ocellated Antbird* (left) *"courtship-feeding" his male pair-mate*

ated by "extramarital" courtship between a mated female and an unpaired male. The extended families of Ocellated Antbirds also sometimes break up: heterosexual pairs wander off from the clan if they have not been able to breed successfully, or grandparents isolate themselves from their relatives. Heterosexual relations within a pair are not always smooth either: in Bicoloreds, for example, males are often distinctly hostile to their female mates. This is especially true early in their pair-bond, when he may aggressively "blast" her off her perch with hissing and snapping. Female Ocellated Antbirds have also been observed steadfastly refusing the courtship and copulatory advances of males. Finally, a number of incestuous activities have been documented in these species, including courtship and attempted copulation between Ocellated males and their mothers, and courtship of Bicolored daughters by their fathers.

Sources
asterisked references discuss homosexuality/transgender

Willis, E.O. (1983) "Longevities of Some Panamanian Forest Birds, with Note of Low Survivorship in Old Spotted Antbirds (*Hylophylax naevioides*)." *Journal of Field Ornithology* 54:413–14.

*———— (1973) *The Behavior of Ocellated Antbirds*. Smithsonian Contributions to Zoology no. 144. Washington, D.C.: Smithsonian Institution Press.

*———— (1972) *The Behavior of Spotted Antbirds*. Ornithological Monographs 10. Washington, D.C.: American Ornithologists' Union.

*———— (1967) *The Behavior of Bicolored Antbirds*. University of California Publications in Zoology 79. Berkeley: University of California Press.

TYRANT FLYCATCHERS
Perching Bird

OCHER-BELLIED FLYCATCHER
Mionectes oleagineus

HOMOSEXUALITY	TRANSGENDER	BEHAVIORS		RANKING	OBSERVED IN
○ Female	○ Intersexuality	● Courtship	○ Pair-bonding	○ Primary	● Wild
● Male	● Transvestism	○ Affectionate	○ Parenting	○ Moderate	○ Semiwild
		○ Sexual		● Incidental	○ Captivity

IDENTIFICATION: A small (5 inch), plain olive-green bird with a long tail and an ocher- or tawny-colored lower breast. DISTRIBUTION: From Mexico south to the Amazon in South America; Trinidad and Tobago. HABITAT: Humid lowland forests, open shrubbery. STUDY AREA: Corcovado National Park, Costa Rica; subspecies *M.o. dyscola*.

Social Organization

Ocher-bellied Flycatchers have a complex social organization, with three distinct categories of males. About 42 percent of males are TERRITORIAL, defending "courts" in the foliage within which they perform courtship displays; sometimes groups of two to six territorial males display close to each other in a LEK formation. Another 10 percent of males are SATELLITES, who associate with territorial males but do not display; they often eventually inherit the territory themselves. Finally, 48 percent of males are FLOATERS, who travel widely and do not hold any territories themselves. The mating system of Ocher-bellies is polygamous or promiscuous. No male-female pair-bonds are formed; instead, males mate with as many females as they can, but the female raises the young on her own.

Description

Behavioral Expression: Female Ocher-bellied Flycatchers are usually attracted to males displaying and singing on their territories, but sometimes another male approaches and is courted by the territorial male. The approaching male behaves much like a female, moving toward the center of the display court flicking his wings while the other male sings more intensely (making whistlelike notes), crouching and wing-flicking. The territorial male then trails the other male, following him closely and sometimes making soft *ipp* calls. The courtship sequence continues as in a heterosexual encounter with a series of three types of flight displays by the territorial male. The HOP DISPLAY involves the male bouncing excitedly back and forth between two perches uttering an *eek* call. In the FLUTTER FLIGHT, the displaying male traverses an arc between two perches with a special, slow wing-fluttering pattern, while in the HOVER FLIGHT, the male slowly rises in a hover above his perch or between two perches, often fairly close to the other male. The courtship sequence typically ends abruptly with the territorial male chasing the other male off while making *chur* calls.

Frequency: Approximately 17 percent of courtship sequences involve a male displaying to another male, and about 5 percent of male visits to territories result in courtship. Although no mountings or attempted copulations between males have been seen, heterosexual matings have rarely been witnessed either. At one study site, for example, only two male-female mountings were observed during more than 560 hours of observation over ten months.

Orientation: It is difficult to determine the relative proportion and "preference" of heterosexual versus homosexual behavior in Ocher-bellied Flycatchers. Some researchers believe that territorial males who court other males do not realize they are displaying to a bird of the same sex, in which case they would be exhibiting superficially heterosexual behavior toward (behaviorally) "transvestite" birds. For the males who approach territorial males, however, the situation is even less clear: many of these are probably floaters who presumably are aware that they are being courted by another male, i.e., they are ostensibly participating in homosexual activity. However, in at least one case the approaching male was a neighboring territorial male who also displayed to females on his own territory, i.e., his courtship interactions were actually bisexual.

Nonreproductive and Alternative Heterosexualities

As noted above, more than half of the male population consists of nonbreeders, since floaters and satellites rarely, if ever, mate heterosexually. Moreover, the absence of breeding activity in these males cannot be attributed to a shortage of available display sites, since more than three-quarters of suitable territories go unused (and nearly a quarter of these are especially prime pieces of "real estate").

Sources   *asterisked references discuss homosexuality/transgender*

Sherry, T. W. (1983) "*Mionectes oleaginea*." In D. H. Janzen, ed., *Costa Rican Natural History*, pp. 586–87. Chicago: University of Chicago Press.

Skutch, A. F. (1960) "*Oleaginous pipromorpha*." In *Life Histories of Central American Birds II*, Pacific Coast Avifauna no. 34, pp. 561–70. Berkeley: Cooper Ornithological Society.

Snow, B. K., and D. W. Snow (1979) "The Ocher-bellied Flycatcher and the Evolution of Lek Behavior." *Condor* 81:286–92.

Westcott, D. A. (1997) "Neighbors, Strangers, and Male-Male Aggression as a Determinant of Lek Size." *Behavioral Ecology and Sociobiology* 40:235–42.

———— (1993) "Habitat Characteristics of Lek Sites and Their Availability for the Ocher-bellied Flycatcher, *Mionectes oleagineus*." *Biotropica* 25:444–51.

———— (1992) "Inter- and Intra-Sexual Selection: The Role of Song in a Lek Mating System." *Animal Behavior* 44:695–703.

*Westcott, D. A., and J. N. M. Smith (1994) "Behavior and Social Organization During the Breeding Season in *Mionectes oleagineus,* a Lekking Flycatcher." *Condor* 96:672–83.

SWALLOWS, WARBLERS, FINCHES, AND OTHERS

Swallows   *Perching Bird*

TREE SWALLOW
Tachycineta bicolor

HOMOSEXUALITY	TRANSGENDER	BEHAVIORS		RANKING	OBSERVED IN
○ Female	○ Intersexuality	○ Courtship	○ Pair-bonding	○ Primary	● Wild
● Male	○ Transvestism	○ Affectionate	○ Parenting	● Moderate	○ Semiwild
		● Sexual		○ Incidental	○ Captivity

IDENTIFICATION: A small to medium-sized swallow with iridescent blue-green upperparts, white underparts, and a tail that is only slightly forked. DISTRIBUTION: Canada and northern United States; winters in southern United States to northwestern South America. HABITAT: Forests, fields, meadows, marshes; usually near water. STUDY AREA: Allendale, Michigan.

Social Organization

Tree Swallows are extremely social birds: outside of the mating season, they gather in large flocks that may contain several hundred thousand individuals, while during the breeding season they form pairs and often nest in aggregations or colonies.

Description

Behavioral Expression: Groups of male Tree Swallows sometimes pursue other males during the mating season in order to copulate with them. When the object of their attentions alights, the males hover in a "cloud" above him, constantly fluttering and making the distinctive *tick-tick-tick* call that is characteristic of males when they are mating with females. If one of them succeeds in mounting the male, a complete homosexual copulation ensues: the male on top holds on to the other male's neck and back feathers with his bill, while the male being mounted lifts his tail so that genital contact can occur. As in heterosexual mating, multiple, repeated genital contacts may occur during a single copulation between two males, which can last for up to a minute (male-female mounts generally last about 30 seconds). The cluster of males may also engage in several consecutive episodes of homosexual mating: when the male who was mounted flies off, the group will continue pursuing him until he lands again, and the whole process is repeated.

Frequency: Homosexual copulations have been observed only occasionally in Tree Swallows. However, heterosexual nonmonogamous matings are also rarely seen in this species, yet they are known to be very common because of the high rate of offspring that result from them (see below). Most such copulations therefore probably occur in locations where (or at times when) they are not readily observed. Homosexual matings (which follow the pattern of nonmonogamous copulations) probably also occur more often than they are observed.

Orientation: Some males that participate in homosexual pursuits and copulations are probably bisexual: for example, one male who was mounted by other males was the father of six-day-old nestlings when he participated in homosexual activity. The males mounting him were not paired with female mates in the same nesting colony, however, and may have been nonbreeders (although they could also have been heterosexually paired birds visiting from another colony).

Nonreproductive and Alternative Heterosexualities

Heterosexual pairs of Tree Swallows sometimes copulate well before the female is fertile, and nonreproductive matings may also occur after the eggs are laid or even following hatching of chicks. Overall, each pair copulates about 50–70 times per clutch of eggs produced. At least 15 percent of matings also occur after the female's fertile period, and more than 20 percent of mounts do not involve genital contact. In addition, a large number of heterosexual copulations that take place during incubation—as well as throughout the breeding season—are nonmonogamous matings between a female and a male other than her mate. Although many pairs are monogamous in this species (about half of all females are strictly faithful), promiscuous copulations are a prominent feature of Tree Swallow heterosexual interactions. Females often solicit such copulations (sometimes from several different males) and are also able to effectively terminate unwanted promiscuous matings. They do this by flying away, refusing to lift their tail for genital contact, or turning their head and snapping or "chattering" at the male. Nonmonogamous matings frequently result in offspring: in some populations, 50–90 percent of all nests contain young that are not genetically related to their mother's mate, and these constitute 40–75 percent of all nestlings. In some families, *all* the offspring are fathered by other males. The opposite situation also sometimes occurs: youngsters may be related to the father but not his female partner. This may result from mate-switching (divorce and remating), or because females occasionally lay eggs in another female's nests (5–9 percent of all nests are PARASITIZED this way).

In some populations, 3–8 percent of males form polygamous trios in which they bond and breed with two females simultaneously. If the two females share a nest, one may help care for the other's nestlings if her own eggs do not hatch. Many populations also have large numbers of nonbreeding birds, sometimes called FLOATERS because they do not occupy their own territories and tend to travel widely. As many as a quarter of all reproductively mature females are floaters. In addition to helping raise unrelated birds of their own species, Tree Swallows sometimes "adopt" nests belonging to other birds such as purple martins (*Progne subis*) or bluebirds (*Sialia* spp.), raising the foster young in addition to their own. More than half of all Tree Swallow heterosexual pairs do not remain together for more than one breeding season. Single parenting also occasionally occurs in this species, for example if one parent is killed or dies during the breeding season. Frequently, however, the widowed parent re-pairs with another mate. If a single parent is laying or incubating eggs, its new mate often adopts the brood, but if a single parent already has nestlings from the previous mate, the new partner often kills them (usually by pecking) in order to begin breeding himself or herself. Infanticide also sometimes occurs when a female kills a paired female's nestlings in order to try to precipitate a divorce and mate with her partner.

Sources   *asterisked references discuss homosexuality/transgender

Barber, C. A., R. J. Robertson, and P. T. Boag (1996) "The High Frequency of Extra-Pair Paternity in Tree Swallows Is Not an Artifact of Nestboxes." *Behavioral Ecology and Sociobiology* 38:425–30.

Chek, A. A., and R. J. Robertson (1991) "Infanticide in Female Tree Swallows: A Role for Sexual Selection." *Condor* 93:454–57.

Dunn, P. O., and R. J. Robertson (1992) "Geographic Variation in the Importance of Male Parental Care and Mating Systems in Tree Swallows." *Behavioral Ecology* 3:291–99.

Dunn, P. O., and R. J. Robertson, D. Michaud-Freeman, and P. T. Boag (1994) "Extra-Pair Paternity in Tree Swallows: Why Do Females Mate with More than One Male?" *Behavioral Ecology and Sociobiology* 35:273–81.

Leffelaar, D., and R. J. Robertson (1985) "Nest Usurpation and Female Competition for Breeding Opportunities by Tree Swallows." *Wilson Bulletin* 97:221–24

——— (1984) "Do Male Tree Swallows Guard Their Mates?" *Behavioral Ecology and Sociobiology* 16:73–79.

Lifjeld, J. T., P. O. Dunn, R. J. Robertson, and P. T. Boag (1993) "Extra-Pair Paternity in Monogamous Tree Swallows." *Animal Behavior* 45:213–29.

Lifjeld, J. T., and R. J. Robertson (1992) "Female Control of Extra-Pair Fertilization in Tree Swallows." *Behavioral Ecology and Sociobiology* 31:89–96.

*Lombardo, M. P. (1996) Personal communication.

——— (1988) "Evidence of Intraspecific Brood Parasitism in the Tree Swallow." *Wilson Bulletin* 100:126–28.

——— (1986) "Extrapair Copulations in the Tree Swallow." *Wilson Bulletin* 98:150–52.

*Lombardo, M. P., R. M. Bosman, C. A. Faro, S. G. Houtteman, and T.S. Kluisza (1994) "Homosexual Copulations by Male Tree Swallows." *Wilson Bulletin* 106:555–57.

Morrill, S. B., and R. J. Robertson (1990) "Occurrence of Extra-Pair Copulation in the Tree Swallow (*Tachycineta bicolor*)." *Behavioral Ecology and Sociobiology* 26:291–96.

Quinney, T. E. (1983) "Tree Swallows Cross a Polygyny Threshold." *Auk* 100:750–54.

Rendell, W. B. (1992) "Peculiar Behavior of a Subadult Female Tree Swallow." *Wilson Bulletin* 104:756–59.

Robertson, R. J. (1990) "Tactics and Counter-Tactics of Sexually Selected Infanticide in Tree Swallows." In J. Blondel, A. Gosler, J.-D. Lebreton, and R. McCleery, eds., *Population Biology of Passerine Birds: An Integrated Approach*, pp. 381–90. Berlin: Springer-Verlag.

Robertson, R. J., B. J. Stutchbury, and R. R. Cohen (1992) "Tree Swallow (*Tachycineta bicolor*)." In A. Poole, P. Stettenheim, and F. Gill, eds., *The Birds of North America: Life Histories for the 21st Century*, no. 11. Philadelphia: Academy of Natural Sciences; Washington, D.C.: American Ornithologists' Union.

Stutchbury, B. J., and R. J. Robertson (1987a) "Signaling Subordinate and Female Status: Two Hypotheses for the Adaptive Significance of Subadult Plumage in Female Tree Swallows." *Auk* 104:717–23.

——— (1987b) "Behavioral Tactics of Subadult Female Floaters in the Tree Swallow." *Behavioral Ecology and Sociobiology* 20:413–19.

——— (1987c) "Two Methods of Sexing Adult Tree Swallows Before They Begin Breeding." *Journal of Field Ornithology* 58:236–42.

——— (1985) "Floating Populations of Female Tree Swallows." *Auk* 102:651–54.

Venier, L. A., P. O. Dunn, J. T. Lifjeld, and R. J. Robertson (1993) "Behavioral Patterns of Extra-Pair Copulation in Tree Swallows." *Animal Behavior* 45:412–15.

Venier, L. A., and R. J. Robertson (1991) "Copulation Behavior of the Tree Swallow, *Tachycineta bicolor*: Paternity Assurance in the Presence of Sperm Competition." *Animal Behavior* 42:939–48.

SWALLOWS   *Perching Bird*

CLIFF SWALLOW
Hirundo pyrrhonota

BANK SWALLOW or SAND MARTIN
Riparia riparia

HOMOSEXUALITY	TRANSGENDER	BEHAVIORS		RANKING	OBSERVED IN
○ Female	○ Intersexuality	○ Courtship	○ Pair-bonding	○ Primary	● Wild
● Male	○ Transvestism	○ Affectionate	○ Parenting	○ Moderate	○ Semiwild
		● Sexual		● Incidental	○ Captivity

CLIFF SWALLOW

IDENTIFICATION: A bluish brown swallow with pale underparts, buff forehead, and a chestnut throat; tail is not forked. DISTRIBUTION: North and Central America; winters in southern South America. HABITAT: Open country, cliffs. STUDY AREAS: Near Jackson Hole (Moran), Wyoming, Lakeview, Kansas, and along the North and South Platte Rivers, Nebraska; subspecies *H.p. hypopolia* and *H.p. pyrrhonota*.

BANK SWALLOW

IDENTIFICATION: A small, sparrow-sized swallow with a slightly forked tail, brown plumage, white underparts, and a brown breast band. DISTRIBUTION: Throughout North America and Eurasia; winters to South America and southern Africa. HABITAT: Open country near water. STUDY AREAS: Near Madison, Wisconsin, and Dunblane, Scotland; subspecies *R.r. riparia*.

Social Organization

Cliff and Bank Swallows are highly gregarious and may flock by the hundreds or even thousands. They generally form mated pairs (although many alternative arrangements also occur, see below) and nest in colonies. Cliff Swallow colonies are the largest of any swallow in the world, often containing a thousand nests (and sometimes up to three times this number).

Description

Behavioral Expression: Male Cliff and Bank Swallows often try to copulate with both males and females that are not their own mates. Unlike in Tree Swallows, these are usually forced copulations or "rapes," since the bird being pursued—whether male or female—does not welcome the sexual advances of the male. Homosexual copulation attempts in Cliff Swallows take place on the ground at social gatherings of birds that are sunning themselves or gathering mud or grass for nests. Anywhere from a handful to several hundred individuals may be present at a time, although such groups usually contain 10–30 birds. At mud-gathering sessions, one male pounces on another male from above, landing on his back and often grabbing his head or neck feathers in his beak. At sunning sites, the male usually lands a few inches away from the other bird and makes threatening HEAD-FORWARD displays prior to jumping on his back. Once mounted, he spreads his tail and moves it sideways, trying to make cloacal (genital) contact, all the while vigorously flapping his wings; the other male usually strongly resists, and sometimes a fight ensues, before both birds fly off. The entire copulation attempt is usually quite brief, though it can last for up to 10 seconds (a relatively long duration for bird mountings). Forcible mountings of females follow this same pattern. When on the ground at mud-gathering sites, birds of both sexes typically flutter their wings above their backs to try to prevent males from mounting them.

Male Bank Swallows also pursue both females and males to try to copulate with them. Males first make many INVESTIGATORY CHASES of unfamiliar birds to determine if they are worth following. If they find a bird they are interested in—which may be another male—a full-fledged SEXUAL CHASE ensues, sometimes drawing several birds into the pursuit as well. Often the targeted bird is able to get away, but sometimes the chase ends with a forced copulation attempt. Homosexual mountings also occur when birds congregate on the ground, for instance to gather nest materials or dust themselves. At times, a veritable "orgy" develops as numerous males frantically try to mount birds of both sexes. Sometimes one or two males will even mount a third male who is already copulating with another bird.

◀ *A male Cliff Swallow mounting another male* (right) *and attempting to copulate with him during a mud-gathering session*

Frequency: In Bank Swallows, 8 percent of sexual chases are homosexual, while 36–40 percent of investigatory chases involve males pursuing other males. Cliff Swallow rape attempts are extremely common, occurring every two to three minutes at some mud-gathering sites; one male may attempt to mount six to eight different birds in a five-to-ten-minute period. These mounts are probably fairly evenly distributed between males and females, although homosexual mounts may in fact occur more often. When presented with stuffed birds of both sexes in identical poses, male Cliff Swallows mounted other males nearly 65 percent of the time.

Orientation: Most, if not all, male Cliff and Bank Swallows that pursue copulations with other males also try to forcibly mount females, and to this extent they are bisexual. However, in Cliff Swallows only a few males appear to engage in such behavior with either males or females. Many of these are unpaired birds, although males who are already heterosexually paired (including fathers) also sometimes participate in promiscuous copulations.

Nonreproductive and Alternative Heterosexualities

Heterosexual social life in Cliff and Bank Swallows is replete with behaviors that deviate from the monogamous-pair/nuclear-family model. As discussed above, a large proportion of heterosexually mated Swallows seek copulations with birds other than their partner, and considerable evidence suggests that these mounting attempts are nonprocreative. Because of the resistance of the bird being mounted, the copulation is often not completed and sperm is rarely transferred. In addition, such attempts may also occur outside the breeding season when there is no possibility of fertilization. Genetic studies have shown that probably only 2–6 percent of Cliff Swallow nests have young that might result from nonmonogamous sexual activity. However, nearly a quarter of all nestlings, in more than half of all nests, are raised by birds other than their biological mother and/or father. This is because Cliff Swallows participate in an extraordinary array of activities that serve to exchange eggs and nestlings between families. For example, as many as 43 percent of all nests contain an egg laid by an outside female—this bird usually has her own nest, but she also PARASITIZES or adds eggs to other nests (and often has eggs added to her own as well). In some cases, this female may even return to the foreign nest to incubate the entire clutch (even though only one egg is hers), but she does not help raise the nestlings once they hatch. Often the intruding female's mate will destroy or toss out eggs in the host nest to make room for their own (up to 20 percent of all nests may suffer egg destruction). In other cases, males appear to destroy eggs in other nests so as to keep the laying female sexually receptive, thereby increasing the opportunities for heterosexual nonmonogamous matings. Birds also occasionally physically carry eggs from their own nest to another—about 6 percent of all nests contain eggs transferred this way. Finally, in a few cases Swallows have even been seen transferring actual baby birds between nests. Infanticide may also occur when birds attack and toss nestlings out of neighbors' nests. In both species, young birds gather into large CRÈCHES or "day-care centers"—sometimes containing up to a thousand youngsters—which give them protection while their parents are away searching for food.

Divorce and single parenting also occur in Bank Swallows: females sometimes desert their mates to start a new family with another male, forcing their first mate to raise the nestlings on his own. In addition to the rape attempts described above, there is considerable aggression between the sexes as well. Ironically, male Bank Swallows often become violent toward their own female partners when trying to protect them from the advances of other males. Sometimes they knock their mate to the ground, fighting her directly, or try to force her back into their burrow. Nonreproductive sexual behaviors are also prevalent in these two species. Besides copulations outside the breeding season and group sexual activity (mentioned above), members of a Cliff Swallow pair often copulate at a rate far in excess of that needed simply to fertilize their eggs. In addition, males of both species have occasionally been seen trying to mate with dead birds, as well as with other species such as barn swallows (*Hirundo rustica*) and Tree Swallows.

Sources   **asterisked references discuss homosexuality/transgender*

*Barlow, J. C., E. E. Klaas, and J. L. Lenz (1963) "Sunning of Bank Swallows and Cliff Swallows." *Condor* 65:438–48.

Beecher, M. D., and I. M. Beecher (1979) "Sociobiology of Bank Swallows: Reproductive Strategy of the Male." *Science* 205:1282–85.

Beecher, M. D., I. M. Beecher, and S. Lumpkin (1981) "Parent-Offspring Recognition in Bank Swallows (*Riparia riparia*): I. Natural History." *Animal Behavior* 29:86–94.

Brewster, W. (1898) "Revival of the Sexual Passion in Birds in Autumn." *Auk* 15:194–95.

*Brown, C. R., and M.B. Brown (1996) *Coloniality in the Cliff Swallow: The Effect of Group Size on Social Behavior.* Chicago: University of Chicago Press.

*——— (1995) "Cliff Swallow (*Hirundo pyrrhonota*)." In A. Poole and F. Gill, eds., *The Birds of North America: Life Histories for the 21st Century,* no. 149. Philadelphia: Academy of Natural Sciences; Washington, D.C.: American Ornithologists' Union.

——— (1989) "Behavioral Dynamics of Intraspecific Brood Parasitism in Colonial Cliff Swallows." *Animal Behavior* 37:777–96.

——— (1988a) "A New Form of Reproductive Parasitism in Cliff Swallows." *Nature* 331:66–68.

——— (1988b) "The Costs and Benefits of Egg Destruction by Conspecifics in Colonial Cliff Swallows." *Auk* 105:737–48.

——— (1988c) "Genetic Evidence of Multiple Parentage in Broods of Cliff Swallows." *Behavioral Ecology and Sociobiology* 23:379–87.

Butler, R. W. (1982) "Wing-fluttering by Mud-gathering Cliff Swallows: Avoidance of 'Rape' Attempts?" *Auk* 99:758–61.

*Carr, D. (1968) "Behavior of Sand Martins on the Ground." *British Birds* 61:416–17.

Cowley, E. (1983) "Multi-Brooding and Mate Infidelity in the Sand Martin." *Bird Study* 30:1–7.

*Emlen, J. T., Jr. (1954) "Territory, Nest Building, and Pair Formation in Cliff Swallows." *Auk* 71:16–35.

——— (1952) "Social Behavior in Nesting Cliff Swallows." *Condor* 54:177–99.

Hoogland, J. L., and P. W. Sherman (1976) "Advantages and Disadvantages of Bank Swallow (*Riparia riparia*) Coloniality." *Ecological Monographs* 46:33–58.

*Jones, G. (1986) "Sexual Chases in Sand Martins (*Riparia riparia*): Cues for Males to Increase Their Reproductive Success." *Behavioral Ecology and Sociobiology* 19:179–85.

*Petersen, A. J. (1955) "The Breeding Cycle in the Bank Swallow." *Wilson Bulletin* 67:235–86.

Thom, A. S. (1947) "Display of Sand-Martin." *British Birds* 40:20–21.

WOOD WARBLERS

Perching Bird

HOODED WARBLER
Wilsonia citrina

HOMOSEXUALITY	TRANSGENDER	BEHAVIORS		RANKING	OBSERVED IN
○ Female	○ Intersexuality	○ Courtship	● Pair-bonding	○ Primary	● Wild
● Male	● Transvestism	○ Affectionate	● Parenting	● Moderate	○ Semiwild
		○ Sexual		○ Incidental	○ Captivity

IDENTIFICATION: A small songbird with bright yellow underparts, olive green upperparts, and a black crown ("hood") and throat in adult males and some females (see below). DISTRIBUTION: Eastern North America; winters in Mexico and Central America. HABITAT: Deciduous forests, cypress swamps. STUDY AREA: Smithsonian Environmental Research Center near Annapolis, Maryland.

Social Organization

During the breeding season, male Hooded Warblers establish and defend territories, attracting mates with whom they usually form pair-bonds. Outside of the mating season, birds migrate south to their winter homes, where the two sexes live largely separate from each other.

Description

Behavioral Expression: Male Hooded Warblers occasionally form homosexual pairs and become joint parents. Same-sex pair-bonds develop early in the mating season when one male is attracted to another male's territory by his singing. In some cases, this is a male he has previously "prospected" on a visit to his territory during the prior mating season. Once a pair-bond is established, the males focus their attentions on parenting duties. Homosexual couples acquire nests and eggs in a variety of ways. Some pairs may build their own nests: although male Hooded Warblers in heterosexual pairs rarely build nests, at least one bird in a homosexual pair was observed carrying grass fibers to a nest and shaping the cup by repeatedly sitting in the nest and shifting his position. It is not known, however, whether he had built the nest or was simply adding material to a nest built by another pair. As for eggs, some pairs incubate eggs laid by another species of bird, the Brown-headed Cowbird. This species is known as a PARASITE because it always lays its eggs in the nests of other birds, "forcing" them to raise its young. Hooded Warblers are particularly susceptible to parasitism by Cowbirds: in some populations, three-quarters of all nests are parasitized, and Cowbirds appear to prefer Hooded Warbler nests over those of other species. Cowbirds occasionally lay eggs in completely empty nests, so some homosexual pairs of Hooded Warblers may build their own

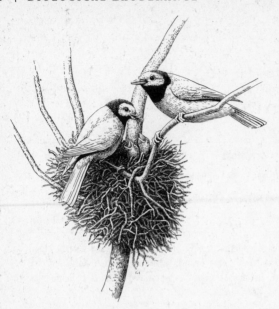

▲ *A pair of male Hooded Warblers tending their chicks*

nest and end up tending only Cowbird eggs. Usually, though, a Cowbird adds its egg(s) to a nest that already has eggs (often removing part of the original clutch). Sometimes a Hooded Warbler mother abandons her nest once it has been disturbed this way, and some homosexual pairs may "adopt" such abandoned nests, or else the father of such a nest may re-pair with a male following the mother's abandonment. At least two male pairs have been observed tending nests that were parasitized, since they each contained both Cowbird and Hooded Warbler chicks. Other male couples probably adopt nests that have been abandoned after attacks by predators. Bluejays and squirrels often prey on Hooded Warbler nestlings, and usually their mother will abandon the entire nest even if only one youngster has been taken. Finally, some homosexual pairs may tend eggs that have been directly laid in their nest by a female Hooded Warbler. In many bird species, females lay eggs in nests belonging to other birds of the same species (this is another form of parasitism); although this rarely occurs in Hooded Warblers, it is a possible source of eggs for homosexual pairs.

Once they have acquired a nestful of eggs, male couples typically divide up the parenting duties: one attends to nest repair, incubation of the eggs, and brooding of the nestlings, while the other feeds his mate and defends the territory. Both birds feed the nestlings insects such as crane flies. Although this division of labor is similar to that in heterosexual pairs—females typically build nests and incubate, males defend territories, and both feed nestlings—there are crucial differences. In homosexual pairs, incubating males are often fed by their mate, which occurs only rarely in heterosexual pairs. In addition, one male who engaged in typically "female" parental duties was later observed performing territorial singing (albeit with a song pattern that differed from that of most other males). Nests belonging to homosexual pairs are often lost entirely to predators, but up to 50 percent or more of all heterosexual nests are lost in the same way. The male couples that have been followed over a longer time do not appear to re-pair with the same mate in subsequent breeding seasons; their divorces may be related to the loss of nests to predators. Heterosexual divorce is also common in Hooded Warblers, with as many as half of all male-female pairs failing to remain together, perhaps also due in large part to loss of nests. It is possible as well that divorce is simply a general feature of pair-bonding in this species (heterosexual or homosexual) independent of nest losses, or that the particular pairs being studied happened to end in divorce without this being indicative of a larger pattern.

Some female Hooded Warblers are transvestite, having the same black hood that males do. In fact there is a continuum of transgendered physical appearance in females: some have no black feathers on their head at all, some have an intermediate amount with a black "bib" around the throat, while others are almost indistinguishable from males. In addition, a few females can sing (typically only males in this species are able to sing). Transgendered females usually mate with males and raise young just like nontransgendered females.

Frequency: The overall incidence of homosexual pairs in Hooded Warblers is not known, since no widespread, systematic study has yet been conducted to determine their prevalence. However, in one population observed over three years, 4 percent of the pair-bonds (3 of 80) were between males. Although overt sexual behavior has not yet been observed between such pair-mates, heterosexual copulations (both within-pair and nonmongamous) are rarely seen in this species either; it is possible, therefore, that homosexual copulations do take place. Among females, plumage transvestism is a regular occurrence, as about 59 percent of females have some degree of malelike black feathers on their head: 40 percent have only a slight amount, 17 percent an intermediate amount, and 2 percent have a nearly complete black hood.

Orientation: Some male Hooded Warblers appear to be exclusively homosexual, pairing only with males; if they divorce a male partner, they re-pair with another male in subsequent breeding seasons. These males often perform parenting duties typically associated with females in heterosexual pairs. Other males, however, are bisexual, alternating between homosexual pairings and heterosexual ones in different breeding seasons.

Nonreproductive and Alternative Heterosexualities

As mentioned above, heterosexual divorce is common in Hooded Warblers, as are a number of other variations on the nuclear family and monogamous pair-bond. About 4 percent of males form trios, mating with two females who both nest simultaneously on the male's territory; 6 percent of males are nonbreeders, and some females remain single as well. Among paired birds, promiscuous copulations also occur very frequently: 30–50 percent of all females copulate with males other than their mates (usually neighboring males), and more than a third of all nestlings in some populations are fathered by a bird other than their mother's mate. In addition, males sometimes adopt young birds from neighboring families whose own parents have finished caring for them; adoptive fathers typically feed these youngsters along with their own nestlings. The adopted birds are usually not genetically related to their foster fathers, i.e., they are not the result of promiscuous matings by the bird who adopts them. Single parenting is also a regular occurrence in Hooded

Warblers: once their young can fly, parents usually separate and each takes care of half the brood (unless the female begins a second family, in which case the male will assume responsibility for all of the youngsters). In fact, single parenting is generally more extensive and longer-lasting than male-female coparenting in this species. Nestlings receive biparental care for only eight or nine days, while single parenting can last for three to six weeks and involves feeding rates that are three to five times higher than that of coparents. As a result of separation of mated pairs, males and females are together for only about one month out of the entire year. During the winter, the two sexes occupy largely segregated habitats, with males preferring forests and females scrub areas.

Sources   *asterisked references discuss homosexuality/transgender*

Evans Ogden, L. J., and B. J. Stutchbury (1997) "Fledgling Care and Male Parental Effort in the Hooded Warbler (*Wilsonia citrina*)." *Canadian Journal of Zoology* 75:576–81.

———— (1994) "Hooded Warbler (*Wilsonia citrina*)." In A. Poole and F. Gill, eds., *The Birds of North America: Life Histories for the 21st Century*, no. 110. Philadelphia: Academy of Natural Sciences; Washington, D.C.: American Ornithologists' Union.

Godard, R. (1993) "Tit for Tat Among Neighboring Hooded Warblers." *Behavioral Ecology and Sociobiology* 33:45–50.

———— (1986) "Long-Term Memory of Individual Neighbors in a Migratory Songbird." *Nature* 350:228–29.

*Lynch, J. F., E. S. Morton, and M. E. Van der Voort (1985) "Habitat Segregation Between the Sexes of Wintering Hooded Warblers (*Wilsonia citrina*)." *Auk* 102:714–21.

*Morton, E. S. (1989) "Female Hooded Warbler Plumage Does Not Become More Male-Like With Age." *Wilson Bulletin* 101:460–62.

*Niven, D. K. (1997) Personal communication.

*———— (1993) "Male-Male Nesting Behavior in Hooded Warblers." *Wilson Bulletin* 105:190–93.

Stutchbury, B. J. M. (1998) "Extra-Pair Mating Effort of Male Hooded Warblers, *Wilsonia citrina*." *Animal Behavior* 55:553–61.

*———— (1994) "Competition for Winter Territories in a Neotropical Migrant: The Role of Age, Sex, and Color." *Auk* 111:63–69.

Stutchbury, B. J., and L. J. Evans Ogden (1996) "Fledgling Adoption in Hooded Warblers (*Wilsonia citrina*): Does Extrapair Paternity Play a Role?" *Auk* 113:218–20.

*Stutchbury, B. J., and J. S. Howlett (1995) "Does Male-Like Coloration of Female Hooded Warblers Increase Nest Predation?" *Condor* 97:559–64.

Stutchbury, B. J., J. M. Rhymer, and E. S. Morton (1994) "Extrapair Paternity in Hooded Warblers." *Behavioral Ecology* 5:384–92.

FINCHES

Perching Bird

CHAFFINCH
Fringilla coelebs
SCOTTISH CROSSBILL
Loxia scotica

HOMOSEXUALITY	TRANSGENDER	BEHAVIORS		RANKING	OBSERVED IN
● Female	○ Intersexuality	● Courtship	● Pair-bonding	○ Primary	● Wild
● Male	○ Transvestism	○ Affectionate	○ Parenting	● Moderate	○ Semiwild
		○ Sexual		○ Incidental	○ Captivity

CHAFFINCH
IDENTIFICATION: A sparrow-sized bird with olive-brown plumage, distinctive white shoulder bars, and (in males) blue-gray crown. DISTRIBUTION: Europe, Siberia, central Asia, North Africa. HABITAT: Forest, farmland. STUDY AREAS: Ylivieska, Finland; near Cambridge, England; subspecies *F.c. coelebs* and *F.c. gengleri*.

SCOTTISH CROSSBILL
IDENTIFICATION: A sparrow-sized bird with olive to orange-red plumage and a distinctive crossed bill. DISTRIBUTION: Northern Scotland. HABITAT: Coniferous forest. STUDY AREAS: Speyside and Sutherland, Scotland.

Social Organization
Chaffinches and Scottish Crossbills commonly associate in flocks; the mating system involves (usually monogamous) pair-bonding.

Description
Behavioral Expression: Female Chaffinches sometimes form homosexual pair-bonds with each other. In this species, usually only males sing; however, in same-sex pairs, one female partner typically sings much in the manner of a male, throwing back her head while perched conspicuously in the trees (this is probably how she attracts her female mate). Her song resembles that of males in duration, loudness, and structure, consisting of a long stream of rapidly trilled notes of descending pitch, finished off with a staccato end phrase or "flourish." It differs from male song, however, in having not quite as ringing a tone. Like male Chaffinches, she may even COUNTERSING in response to a neighboring male's song, the two birds "replying" to each other with alternating or syncopated song phrases. Females in same-sex pairs may also try to sing together, although one partner may only be able to produce an incomplete version of the song. The two females behave like other mated couples, and may even engage in courtship pursuits known as SEXUAL CHASES. In this activity—which is a demonstration of sexual interest—one female zigzags after her partner using a special form of flying known as MOTH FLIGHT (rapid, shallow wing-beats without the pauses or undulating quality typical of regular flight). Occasionally, juvenile males pursue other males in such sexual chases.

Homosexual couples also occur in Scottish Crossbills—but among males. A male attracts potential mates by singing high in a treetop, advertising his presence with a loud stream of notes transcribed as *chip-chip-chip-gee-gee-gee chip-chip-chip*. Most singing males respond aggressively to other males who enter their territory, but occasionally the displaying male does not chase away another male that is attracted to him. The two males may then pair up and associate in much the same way that heterosexual couples do, except that copulation has not been observed between male partners. They synchronize their movements, traveling together between forest clumps, sometimes one leading the other. Occasionally, two neighboring males who are each heterosexually mated become companions. The two forage together and jointly defend clumps of pine trees (their principal food source) while still maintaining their opposite-sex bonds. The two males may even visit their female mates together, attending first to one and then the other, with one male feeding his female partner while his male companion waits for him.

Frequency: In both of these species, homosexual pairs are probably only an occasional occurrence. Although no statistics concerning the incidence of same-sex pairing are available, in Chaffinches about 1 in 150 nests contains a SUPERNORMAL CLUTCH of 7–8 eggs. In many species such larger-than-average clutches are laid by female pairs, and this may also be true in Chaffinches (although their specific association with same-sex pairing has not yet been demonstrated).

Orientation: Individual Chaffinches or Scottish Crossbills who form same-sex associations have not been tracked throughout their entire lives to determine if they only pair with members of the same sex. However, since heterosexual pairs in these species are usually lifelong, it is not unlikely that homosexual pairs would be as well. In addition, some Scottish Crossbill males are bisexual, forming bonds with both males and females simultaneously.

Nonreproductive and Alternative Heterosexualities

A variety of nonprocreative heterosexual behaviors are found in Chaffinches and Scottish Crossbills. Chaffinch couples sometimes copulate during incubation (i.e., after fertilization has occurred), and only 40–50 percent of all matings during the fertilizable period involve full genital contact. Some pairs may attempt to mate more than 200 times for each clutch of eggs they produce, as often as 4 times an hour. Mounts are often not completed for a variety of reasons: the male may slip off the female's back or mount in a reversed head-to-tail position, or (more commonly) either partner may flee from the other out of fear or because of provocation from its mate. In fact, heterosexual courtship and mating in this species often entails a considerable amount of aggressive behavior between the sexes. Many birds do not participate in breeding at all: there are surplus flocks of single birds in many populations (males in Scottish Crossbills, both sexes in Chaffinches), as well as nonbreeding pairs in Scottish Crossbills. Interestingly, outside of the breeding season sexual segregation is also the rule. Males and female tend to migrate separately (wintering flocks are often same-sex), and females generally travel farther than males, often resulting in local populations with more males than females. In fact, this phenomenon has lent the Chaffinch its scientific name: *coelebs* means "bachelor" in Latin.

Scottish Crossbills frequently engage in "symbolic" matings early in the breeding season, in which the male mounts the female but no genital contact occurs. Sometimes these mounts are promiscuous: several males other than the female's mate may participate. Nonmonogamous matings are also fairly common in Chaffinches, often between neighboring birds. About 8 percent of all mating activity involves nonmates, and 17 percent of all offspring (in a quarter of all broods) result from such copulations. Occasionally, a pair will separate or "divorce" during the breeding season, after which the single female may copulate with several already paired males. Eventually, a polygamous trio may develop if one of these males forms a pair-bond with her in addition to his own mate. Some Scottish Crossbills are also polygamous, forming trios of one male with two females. In this species, parents (in monogamous pairs) regularly become single parents: they separate when their young are able to fly, dividing the young between them and raising them on their own. Because the mother and father often travel to widely separated feeding grounds, youngsters belonging to the same family may be raised in very different environments. Occasionally, unrelated birds—both males and females—help feed and "foster-parent" the young birds after their biological parents have separated.

Sources
asterisked references discuss homosexuality/transgender

Adkisson, C. S. (1996) "Red Crossbill (*Loxia curvirostra*)." In A. Poole, P. Stettenheim, and F. Gill, eds., *The Birds of North America: Life Histories for the 21st Century,* no. 256. Philadelphia: Academy of Natural Sciences; Washington, D.C.: American Ornithologists' Union.

Halliday, H. (1948) "Song of Female Chaffinch." *British Birds* 41:343–44.

Hanski, I. K. (1994) "Timing of Copulations and Mate Guarding in the Chaffinch *Fringilla coelebs*." *Ornis Fennica* 71:17–25.

Kling, J. W., and J. Stevenson-Hinde (1977) "Development of Song and Reinforcing Effects of Song in Female Chaffinches." *Animal Behavior* 25:215–20.

Knox, A. G. (1990) "The Sympatric Breeding of Common and Scottish Crossbills *Loxia curvirostra* and *L. scotica* and the Evolution of Crossbills." *Ibis* 132:454–66.

*Marjakangas, A. (1981) "A Singing Chaffinch *Fringilla coelebs* in Female Plumage Paired with Another Female-Plumaged Chaffinch." *Ornis Fennica* 58:90–91.

*Marler, P. (1956) *Behavior of the Chaffinch* Fringilla coelebs. Behavior Supplement V. Leiden: E. J. Brill.

——— (1955) "Studies of Fighting in Chaffinches. 2. The Effect on Dominance Relations of Disguising Females as Males." *British Journal of Animal Behavior* 3:137–46.

*Nethersole-Thompson, D. (1975) *Pine Crossbills: A Scottish Contribution.* Berkhamsted: T. and A. D. Poyser.

Sheldon, B. C. (1994) "Sperm Competition in the Chaffinch: The Role of the Female." *Animal Behavior* 47:163–73.

Sheldon, B. C., and T. Burke (1994) "Copulation Behavior and Paternity in the Chaffinch." *Behavioral Ecology and Sociobiology* 34:149–56.

Svensson, B. V. (1978) "Clutch Dimensions and Aspects of the Breeding Strategy of the Chaffinch *Fringilla coelebs* in Northern Europe: A Study Based on Egg Collections." *Ornis Scandinavica* 9:66–83.

Voous, K. H. (1978) "The Scottish Crossbill *Loxia scotica*." *British Birds* 71:3–10.

SHRIKES, TITS, AND BLUEBIRDS   *Perching Bird*

RED-BACKED SHRIKE
Lanius collurio

BLUE TIT
Parus caeruleus

EASTERN BLUEBIRD
Sialia sialis

HOMOSEXUALITY	TRANSGENDER	BEHAVIORS		RANKING	OBSERVED IN
● Female	○ Intersexuality	● Courtship	● Pair-bonding	○ Primary	● Wild
● Male	○ Transvestism	○ Affectionate	● Parenting	○ Moderate	○ Semiwild
		○ Sexual		● Incidental	○ Captivity

RED-BACKED SHRIKE
IDENTIFICATION: A small (6½ inch) bird with a thick, hooked bill, grayish brown plumage, and a darker facial mask (black in males). DISTRIBUTION: Throughout Eurasia and northeast Africa; winters to southern Africa. HABITAT: Savanna, woodland, scrub, farmland. STUDY AREA: New Forest, Hampshire, England.

BLUE TIT
IDENTIFICATION: A tiny (4½ inch) chickadee-like bird with a bright blue crown, black-and-white face, bluish green plumage, and yellow underparts. DISTRIBUTION: Europe, Middle East, North Africa. HABITAT: Woodland, human habitation. STUDY AREA: Marley Wood, Oxfordshire, England; subspecies *P.c. obscurus*.

EASTERN BLUEBIRD
IDENTIFICATION: A sparrow-sized bird with bright blue plumage, white underparts, and a chestnut throat and breast. DISTRIBUTION: North and Central America. HABITAT: Open woodland, orchards, farmland, pine savanna. STUDY AREA: In the town of Washington, Michigan.

Social Organization

Red-backed Shrikes generally establish monogamous pair-bonds during the mating season and occupy partially overlapping territories. Outside of the breeding season, the birds are more solitary, and males and females typically occupy segregated habitats (males in more open bush country, females in more dense woodland). Blue Tits and Eastern Bluebirds are also territorial and form pair-bonds, but may associate in flocks outside of the breeding season.

Description

Behavioral Expression: In Red-backed Shrikes and Blue Tits, two females sometimes pair with each other, build a nest, and lay eggs. Both partners in Red-backed Shrike homosexual pairs take turns incubating the eggs (in heterosexual pairs, only the female incubates). Sometimes they even sit on the nest together, side by side or one partially covering the other. Both females also lay eggs: as a result, their nests have SUPERNORMAL CLUTCHES of 9–12 eggs, twice the number in most nests belonging to heterosexual pairs. If their nest is robbed by predators, the pair will dutifully build a second one (often in a much higher location, to make it inaccessible to most ground predators) and begin laying eggs again (as do male-female pairs when they lose a nest). However, all of the eggs they lay are typically infertile since neither female has mated with a male, and the

▲ *A nesting pair of female Red-backed Shrikes in England simultaneously incubating their eggs*

pair usually abandons the nest after the eggs have failed to hatch. Blue Tit female pairs are unlike most homosexual pairs in birds because only one female lays eggs. As a result, nests belonging to such pairs contain not supernormal clutches, but exceptionally *small* clutches of 3 or so eggs (nests of male-female pairs typically contain about 11 eggs). Both females incubate the eggs simultaneously, sitting together on the nest facing in the same direction. As in Red-backed Shrike homosexual pairs, the eggs are usually infertile and the pair eventually abandons the nest.

In Eastern Bluebirds, two males sometimes associate with one another in what appears to be a homosexual pair-bond: they travel exclusively in each other's company (perhaps even spending the winter together), jointly inspect nest sites during the early spring, and court one another. The latter activity involves COURTSHIP-FEEDING—a behavior also seen in heterosexual pairs—in which one male offers a symbolic food gift, such as a cutworm, to the other male, often preceded by a distinctive call. Paired males might be related to each other (father-son or brothers). Occasionally, a female who has lost her heterosexual mate is joined by another female, who helps coparent her offspring. The two birds take turns feeding the youngsters and may also be assisted by one or more of their young from a previous brood.

Frequency: Same-sex pairs probably occur only sporadically in these three species: in Red-backed Shrikes, for example, female couples account for perhaps no more than 1 percent of all pairs.

Orientation: Female Red-backed Shrikes and Blue Tits are probably exclusively homosexual, at least for the time that they remain paired with their female partner, since they invariably lay eggs that are infertile (indicating no mating with males). Whether such birds ever subsequently form or have previously formed heterosexual pair-bonds is not known.

Nonreproductive and Alternative Heterosexualities

Approximately 17 percent of all sexual activity in opposite-sex pairs of Blue Tits occurs outside the female's fertile period. Male Eastern Bluebirds also sometimes try to copulate when females are nonfertile, such as during incubation or after the breeding season is over. Other nonprocreative matings involve occasional interspecies encounters: juvenile Eastern Bluebirds have been observed being mounted by great crested flycatchers (*Myiarchus crinitus*). Courtship and mounting between pair-bonded Eastern Bluebirds sometimes involve considerable aggression, with males attacking their partner and violently pecking her head or knocking her over with their feet. Heterosexual copulations are also occasionally interrupted by neighboring birds that attack the mating pair. Within-pair copulation rates are quite high in Red-backed Shrikes, roughly three times a day during the female's fertile period. In addition, nonmonogamous matings are common in these species: approximately 10 percent of Blue Tit copulations are promiscuous, involving birds that are not paired to each other. About a quarter of Bluebird nests and a third to nearly a half of all Blue Tit broods contain youngsters fathered by an outside bird. Promiscuous copulations also occasionally occur in Red-backed Shrikes, including forced matings or rapes in which males attack and mount females or violently push them off their perches. About 5 percent of Red-backed Shrike nestlings are fathered by a bird other than their mother's mate, compared to 8–35 percent in Eastern Bluebirds, and 11–14 percent in Blue Tits.

Several other alternative parenting arrangements occur in these species. Red-backed Shrike youngsters sometimes move to adjacent territories where they are adopted by other families, and a few nonbreeding birds of both sexes help feed chicks belonging to other pairs (sometimes even assuming full responsibility for them later on). "Helper" males such as these may even replace the biological father as a model for song-learning by the youngsters. In addition, late-breeding Red-backed Shrikes sometimes become single parents when their mate deserts them; occasionally two such birds and their offspring join together to form a "blended family." At least 15 percent of Eastern Bluebird mothers in some populations raise unrelated youngsters when outside females lay eggs in their nest. The divorce rate for Blue Tits varies considerably, from 8–85 percent (depending on the population); in addition, about a third of females and up to 20 percent of all males in some areas form polygamous trios (or occasionally quartets). Other more complex family arrangements also occur: one female Blue Tit was part of a polygamous trio with another female and a male; she also copulated promiscuously with a second male, eventually forming a new polygamous bond with both him and yet a third male, all of whom helped care for the young! Overall, however, more than 85 percent of Blue Tits never successfully breed, and about a third of all parents never have grandchildren. Red-backed Shrike parents occasionally cannibalize their own young or eggs.

Other Species

Some male Pied Flycatchers (*Ficedula hypoleuca*), especially younger ones, are transvestite, having the same brownish plumage that females do; these birds are sometimes courted by other males.

Sources   *asterisked references discuss homosexuality/transgender*

Alsop, F. J., III (1971) "Great Crested Flycatcher Observed Copulating with an Immature Eastern Bluebird." *Wilson Bulletin* 83:312.

*Ashby, E. (1958) "Incidents of Bird Life [report on Red-backed Shrikes]." *The Countryman: A Quarterly Review and Miscellany of Rural Life and Progress* 55:272.

Birkhead, T. R., and A. P. Møller (1992) *Sperm Competition in Birds: Evolutionary Causes and Consequences.* London: Academic Press.

*Blakey, J. K. (1996) "Nest-Sharing by Female Blue Tits." *British Birds* 89:279–80.

*Cramp, S., and C. M. Perrins, eds. (1993) "Red-backed Shrike (*Lanius collurio*)." In *Handbook of the Birds of Europe, the Middle East, and North Africa,* vol. 7, pp. 456–78. Oxford: Oxford University Press.

Dhondt, A. A. (1989) "Blue Tit." In I. Newton, ed., *Lifetime Reproduction in Birds,* pp. 15–33. London: Academic Press.

——— (1987) "Reproduction and Survival of Polygynous and Monogamous Blue Tit *Parus caeruleus.*" *Ibis* 129:327–34.

Dhondt, A. A., and F. Adriaensen (1994) "Causes and Effects of Divorce in the Blue Tit *Parus caeruleus.*" *Journal of Animal Ecology* 63:979–987.

Fornasari, L., L. Bottoni, N. Sacchi, and R. Massa (1994) "Home-Range Overlapping and Socio-Sexual Relationships in the Red-backed Shrike *Lanius collurio.*" *Ethology, Ecology, and Evolution* 6:169–177.

Gowaty, P. A., and W. C. Bridges (1991) "Behavioral, Demographic, and Environmental Correlates of Extrapair Fertilizations in Eastern Bluebirds, *Sialia sialis.*" *Behavioral Ecology* 2:339–50.

Gowaty, P. A., and A. A. Karlin (1984) "Multiple Maternity and Paternity in Single Broods of Apparently Monogamous Eastern Bluebirds (*Sialia sialis*)." *Behavioral Ecology and Sociobiology* 15:91–95.

Hartshorne, J. M. (1962) "Behavior of the Eastern Bluebird at the Nest." *Living Bird* 1:131–49.

Herremans, M. (1997) "Habitat Segregation of Male and Female Red-backed Shrikes *Lanius collurio* and Lesser Gray Shrikes *Lanius minor* in the Kalahari Basin, Botswana." *Journal of Avian Biology* 28:240–48.

Jakober, H., and W. Stauber (1994) "Kopulationen und Partnerbewachung beim Neuntöter *Lanius collurio* [Copulation and Mate-Guarding in the Red-backed Shrike]." *Journal für Ornithologie* 135:535–47.

——— (1983) "Zur Phänologie einer Population des Neuntöters (*Lanius collurio*) [On the Phenology of a Population of the Red-backed Shrike]." *Journal für Ornithologie* 124:29–46.

Kempenaers, B. (1994) "Polygyny in the Blue Tit: Unbalanced Sex Ratio and Female Aggression Restrict Mate Choice." *Animal Behavior* 47:943–57.

——— (1993) "A Case of Polyandry in the Blue Tit: Female Extra-Pair Behavior Results in Extra Male Help." *Ornis Scandinavica* 24:246–49.

Kempenaers, B., G. R. Verheyen, and A. A. Dhondt (1997) "Extrapair Paternity in the Blue Tit (*Parus caeruleus*): Female Choice, Male Characteristics, and Offspring Quality." *Behavioral Ecology* 8:481–92.

——— (1995) "Mate Guarding and Copulation Behavior in Monogamous and Polygynous Blue Tits: Do Males Follow a Best-of-a-Bad-Job Strategy?" *Behavioral Ecology and Sociobiology* 36:33–42.

Krieg, D. C. (1971) *The Behavioral Patterns of the Eastern Bluebird (Sialia sialis).* New York State Museum and Science Service Bulletin no. 415. Albany: University of the State of New York.

Massa, R., L. Bottoni, L. Fornasari, and N. Sacchi (1995) "Studies on the Socio-Sexual and Territorial System of the Red-backed Shrike." *Proceedings of the Western Foundation of Vertebrate Zoology* 6:172–75.

Meek, S. B., R. J. Robertson, and P. T. Boag (1994) "Extrapair Paternity and Intraspecific Brood Parasitism in Eastern Bluebirds Revealed by DNA Fingerprinting." *Auk* 111:739–44.

*Owen, J. H. (1946) "The Eggs of the Red-backed Shrike." *Oologists' Record* 20:38–43.

*Pinkowski, B.C. (1977) "'Courtship Feeding' Attempt Between Two Male Eastern Bluebirds." *Jack-Pine Warbler* 55:45–46.

*Pounds, H. E. (1972) "Two Red-backed Shrikes Laying in One Nest." *British Birds* 65:357–58.

*Sætre, G.-P., and T. Slagsvold (1993) "Evidence for Sex Recognition from Plumage Color by the Pied Flycatcher, *Ficedula hypoleuca.*" *Animal Behavior* 44:293–99.

*Slagsvold, T., and G.-P. Sætre (1991) "Evolution of Plumage Color in Male Pied Flycatchers (*Ficedula hypoleuca*): Evidence for Female Mimicry." *Evolution* 45:910–17.

Stanback, M. T., and W. D. Koenig (1992) "Cannibalism in Birds." In M. A. Elgar and B. J. Crespi, eds., *Cannibalism: Ecology and Evolution Among Diverse Taxa,* pp. 277–98. Oxford: Oxford University Press.

*Zeleny, L. (1976) *The Bluebird: How You Can Help Its Fight for Survival.* Bloomington: Indiana University Press.

SPARROWS, BLACKBIRDS, AND CROWS

WEAVERBIRDS

GRAY-CAPPED SOCIAL WEAVER
Pseudonigrita arnaudi
SOCIABLE WEAVER
Philetairus socius
RED BISHOP BIRD
Euplectes orix
ORANGE BISHOP BIRD
Euplectes franciscanus

HOMOSEXUALITY	TRANSGENDER	BEHAVIORS		RANKING	OBSERVED IN
○ Female	○ Intersexuality	● Courtship	● Pair-bonding	○ Primary	● Wild
● Male	○ Transvestism	○ Affectionate	○ Parenting	● Moderate	○ Semiwild
		● Sexual		○ Incidental	● Captivity

GRAY-CAPPED SOCIAL WEAVER
IDENTIFICATION: A grayish buff, sparrow-sized bird with a pale gray-white crown. DISTRIBUTION: Northeast Africa. HABITAT: Bush, acacia savanna. STUDY AREA: Olorgesailie National Prehistoric Site, Kenya; subspecies *P.a. arnaudi*.

SOCIABLE WEAVER
IDENTIFICATION: A drab, sparrowlike bird with brownish gray plumage and a black throat patch. DISTRIBUTION: Southwestern Africa. HABITAT: Scrub, savanna. STUDY AREA: University of California—Los Angeles.

BISHOP BIRDS
IDENTIFICATION: Small, sparrowlike birds with black and brown plumage and various scarlet, red, or reddish orange patches on the chest, nape, crown, and rump. DISTRIBUTION: Sub-Saharan Africa. HABITAT: Moist grassland. STUDY AREAS: Several locations in South Africa, including near Cape Town, Howick, and Bloemfontein; University of Cape Town and University of Bielefeld.

Social Organization

Weaverbirds are named for the intricate—and often colossal—nests they weave. Sociable Weavers build giant condominium-like structures containing many nest chambers; each chamber may house up to 5 birds, while the entire colony can contain as many as 500 birds. Colonies of Gray-capped Social Weavers consist of family groups that each build several hanging nests (one is used as an actual nesting chamber, the others for sleeping). Several groups occupy the same tree, each situated in its own cluster. In both Gray-caps and Sociable Weavers, breeding birds form pair-bonds, and "helper" birds may assist them in their parenting duties. Adult male Bishop Birds establish territories during the breeding season on which they weave elaborate nests and court potential mates. The breeding system of Bishop Birds is polygynous: males mate with numerous females, each of whom lays eggs in one of the nests on his territory, but the male himself does not participate in any parental duties. Outside of the breeding season, Bishop Birds typically socialize in large, often mixed-species flocks.

Description

Behavioral Expression: Male Gray-capped Social Weavers and Sociable Weavers sometimes participate in homosexual copulations. In Gray-caps, birds who participate in same-sex mounting are usually heterosexually paired males, often high-ranking individuals living in the same colony (tree). One male may fly from his home group in the tree to another male's residence, inviting him to mate by holding his body horizontal, raising his head and tail upward, and drooping and vibrating his wings. The other male then mounts him and proceeds to engage in a complete copulation sequence, probably including cloacal (genital) contact. Mounting between males occurs in Sociable Weavers as well, and about 9 percent of such mounts are also full copulations in which the mounted male crouches, quivers his wings, and moves his tail to the side as the mounting male lowers his tail (probably to achieve genital contact). Usually higher-ranking males mount lower-ranking ones, but all males both mount and are mounted by other males (to varying degrees). In addition, some males appear to form "companionships" with each other. Although males do not generally permit other males to roost with them in their nesting chambers, occasionally two males regularly associate with one another and sleep together in the same nest. This can occur even if one of the males is already paired to a female, in which case all three birds occupy the nest together. The two males may remain together for years, even cooperating in attacking other birds together. Some male "companions" participate in mountings and copulations with each other, but in other cases males do not seem to preferentially engage in sexual behavior with their companion.

On their nesting territories, male Red Bishop Birds court both females and males, the latter usually being younger males who have brown plumage (as do females). Courtship consists of two components: a flying display and a perched display. When a young male appears near his territory, an adult male approaches in a distinctive BUMBLE-FLIGHT, in which he fluffs up all of his plumage and flies with slow wingbeats, resembling a bumble-bee. The bird's bright red back or epaulet

▲ *A male Red Bishop Bird displaying fluffed "bumble-bee" plumage, used during courtships with same- and opposite-sex partners*

feathers are prominently displayed, and the bumble-flying may be accompanied by singing or a distinctive noise made by his flapping wings. Sometimes males bumble-fly toward another adult male rather than a younger male. After the display flight, the adult male lands near the other male and begins the SWIVEL DISPLAYS: he hops and twists around his perch, approaching the object of his attentions with ruffled plumage and a rattling call—a continuous stream of notes that sounds like *zik-zik-zik*. No homosexual copulations have been observed in this species in the wild, but heterosexual matings are not commonly seen either. In captivity Orange Bishop Birds—both adults and younger males—sometimes do try to mount younger males (including males of the closely related Golden Bishop Bird, *Euplectes afer*). In a homosexual encounter, one male approaches the other while pumping his body up and down and ruffling his feathers, followed by an attempt to copulate with the other male (who usually rejects his advances).

Frequency: In Sociable Weavers studied in captivity, as much as three-quarters of all mounting activity is between males, and three out of five full copulations are homosexual. The occurrence of same-sex mounting in wild birds of this species is not known, but its prevalence is probably comparable (especially since heterosexual activity is also apparently infrequent). In one study of wild Gray-capped Social Weavers, *all* sexual activity that was observed took place between males. In Red Bishops, approximately 6 percent of courtship bumble-flights are directed by one adult male toward another adult male; courtship of younger males probably occurs more often than this.

Orientation: Male Gray-capped Social Weavers that copulate with other males also mate and pair with females (and in fact may engage in homosexual activity while they are heterosexually paired). The same is true for some Sociable Weavers, although most males in this species mount both males and females, even if they are not paired. However, relatively few males participate in full heterosexual copulations, and those that do appear to have higher rates of homosexual activity as well. To the extent that they court or attempt to mount both males and females, some male Bishops are also bisexual. However, the males they pursue are usually indifferent, at best, to their approaches, indicating perhaps a more heterosexual orientation on their part.

Nonreproductive and Alternative Heterosexualities

In Gray-capped Social Weavers (and occasionally in Sociable Weavers), non-breeding birds often help heterosexual pairs build their nest and feed their young.

Some of these "helpers" are the pairs' young from a previous season who are delaying their own reproductive careers, while others are fully adult birds (who help out in about 18 percent of all feedings). Some nonbreeding youngsters, however, do not help their parents. Sociable Weavers may participate in nonmonogamous heterosexual mountings (in addition to the homosexual matings described above). Although most birds are probably faithful to their partners, some males in captivity have been observed mounting and copulating with females other than their mates. Female Red Bishops occasionally refuse to allow a male to mount them, vigorously pecking and displaying threat postures to repel him. In addition, males often court Bishops of other species, who are not usually attracted by such interspecies displays. Finally, female Red Bishops have been observed cannibalizing both their own and others' nests, eating some or all of the eggs.

Other Species

Adult male Red-shouldered Widowbirds (*Euplectes axillaris*) also sometimes court younger males.

Sources   *asterisked references discuss homosexuality/transgender*

*Collias, E. C., and N. E. Collias (1980) "Individual and Sex Differences in Behavior of the Sociable Weaver *Philetairus socius*." In D. N. Johnson, ed., *Proceedings of the Fourth Pan-African Ornithological Congress (Seychelles, 1976)*, pp. 243–51. Johannesburg: Southern African Ornithological Society.

*———— (1978) "Nest Building and Nesting Behavior of the Sociable Weaver *Philetairus socius*." *Ibis* 120:1–15.

*Collias, N. E., and E. C. Collias (1980) "Behavior of the Gray-capped Social Weaver (*Psuedonigrita arnaudi*) in Kenya." *Auk* 97:213–26.

Craig, A. J. F. K. (1982) "Mate Attraction and Breeding Success in the Red Bishop." *Ostrich* 53:246–48.

*———— (1980) "Behavior and Evolution in the Genus *Euplectes*." *Journal für Ornithologie* 121:144–61.

*———— (1974) "Reproductive Behavior of the Male Red Bishop Bird." *Ostrich* 45:149–60.

Craig, A. J. F. K., and A. J. Manson (1981) "Sexing *Euplectes* Species by Wing-Length." *Ostrich* 52:9–16.

Maclean, G. L. (1973) "The Sociable Weaver." *Ostrich* 44:176–261.

Roberts, C. (1988) "Little Bishop Birds (*Euplectes orix*) in a Lafia Garden—Tom, Dick, Harry, and Fred." *Nigerian Field* 53:11–22.

Skead, C. J. (1959) "A Study of the Redshouldered Widowbird *Coliuspasser axillaris axillaris* (Smith)." *Ostrich* 30:13–21.

———— (1956) "A Study of the Red Bishop." *Ostrich* 27:112–26.

Woodall, P. F. (1971) "Notes on a Rhodesian Colony of the Red Bishop." *Ostrich* 42:205–10.

SPARROWS, BLACKBIRDS,
AND STARLINGS

Perching Bird

HOUSE SPARROW
Passer domesticus

BROWN-HEADED COWBIRD
Molothrus ater

WATTLED STARLING
Creatophora cinerea

HOMOSEXUALITY	TRANSGENDER	BEHAVIORS		RANKING	OBSERVED IN
○ Female	○ Intersexuality	● Courtship	○ Pair-bonding	○ Primary	● Wild
● Male	● Transvestism	○ Affectionate	○ Parenting	○ Moderate	○ Semiwild
		● Sexual		● Incidental	● Captivity

HOUSE SPARROW, BROWN-HEADED COWBIRD
IDENTIFICATION: The familiar black-bibbed sparrow; Cowbird is iridescent black with a dark brown head. DIS-TRIBUTION: Throughout most of North and South America, Eurasia (House Sparrow); North and Central America (Cowbird). HABITAT: Woodland, prairie, farmland, human habitation. STUDY AREAS: Near Stillwater, Oklahoma, and Long Island, New York; subspecies *M.a. ater* and *M.a. artemisiae*.

WATTLED STARLING
IDENTIFICATION: Light gray plumage, black wings and tail, and (in some birds) bare yellow head and fleshy black wattles. DISTRIBUTION: Eastern and southern Africa. HABITAT: Savanna, grassland, woodland. STUDY AREAS: University of Mainz and in Nieder-Olm, Germany.

Social Organization

Wattled Starlings usually associate in small, nomadic flocks, although up to a thousand birds may gather together to pursue locust swarms. Similarly, breeding colonies may contain thousands of nests when locusts are available, but usually birds nest in smaller groups containing a maximum of 400 pairs. Most individuals form monogamous pair-bonds, as do House Sparrows (who also generally nest in colonies). Brown-headed Cowbirds have a highly variable mating system: in many populations birds form (usually monogamous) pair-bonds, while in others they are promiscuous or form polygamous bonds with several individuals.

Description

Behavioral Expression: Male Brown-headed Cowbirds sometimes solicit homosexual copulations from male House Sparrows. Cowbirds commonly invite birds of other species to preen them, but occasionally an interspecies encounter includes homosexual mounting when a House Sparrow is involved. This extraordinary behavior typically begins with a male Cowbird adopting a characteristic

HEAD-DOWN posture next to a Sparrow, in which he bows his head, touching his lower bill to his breast feathers while crouching slightly and raising his wings a bit at the shoulders. The House Sparrow then mounts the Cowbird, grasping his head feathers in his beak while attempting to copulate. If he shows signs of leaving or lack of interest after a single mounting, the Cowbird will immediately resume the invitation posture next to him, insistently nudging the Sparrow with his head and persistently following him until he mounts again. This may continue for an extended time, with repeated homosexual mountings (five or more) occurring in a single session.

Homosexual courtships occasionally occur in Wattled Starlings. Males sometimes select another male as the object of their attentions, displaying to him with a number of stylized postures. Among these are the LATERAL DISPLAY, in which the male turns sideways and lets his wings hang down at his side (exposing their white feathers); the FRONTAL DISPLAY, where the courting male fluffs up his belly and back feathers, raising and quivering his wings while spreading his tail; and the distinctive VULTURE POSTURE, in which he stretches his entire body vertically while puffing out his breast feathers and tightly folding his wings against his sides like a vulture. A form of plumage transvestism also occurs in this species, in which some females develop a male appearance. Most males have a special seasonal nuptial plumage, growing two pendulous wattles from either side of their beaks and losing most of the feathers from their head, thereby exposing the yellow or black skin and two fleshy, comblike growths on the forehead. This feather loss has even been described as a form of "male pattern baldness" akin to the type of hair loss found in humans, and indeed it is regulated by male hormones (as is human baldness). While the majority of females never exhibit these plumage characteristics, a few females do acquire a male appearance with feather loss, wattles, and combs.

Frequency: Brown-headed Cowbirds regularly perform the head-down display toward other species in the wild, and approximately 36 percent of such displays are directed by male Cowbirds toward male House Sparrows; however, Sparrows respond with homosexual mounting probably only sporadically. Similarly, homosexual courtship is in all likelihood only an occasional occurrence in Wattled Starlings. About 2–10 percent of female Wattled Starlings are transgendered, exhibiting fairly complete wattles and/or baldness. Other females appear to fall along a continuum of plumage characteristics, with some individuals showing only partial wattle development or incomplete baldness.

A male Brown-headed Cowbird in the "head-down" posture (right) *inviting a male House Sparrow to mount him* ▶

Orientation: In all three of these species, not enough is yet known about the life histories of individuals participating in homosexual activity to determine their overall sexual orientation profiles. However, at least some male Wattled Starlings appear to preferentially select other males to court.

Nonreproductive and Alternative Heterosexualities

Although most heterosexual Wattled Starlings form monogamous pairs, occasionally males court and mate with females other than their mate. This may result in a female raising her young as a single parent if she is not paired to the male she mates with. The male sexual cycle is especially pronounced in this species, signaled by the seasonal development of wattles and baldness. House Sparrows often engage in multiple copulations during the same mating bout: a male may mount a female and achieve genital contact up to 30 times in rapid succession. In addition, promiscuous matings are quite common in this species: more than a quarter of all nests contain at least one chick sired by a bird other than its mother's mate. Some of these are the result of forced matings that occur during COMMUNAL DISPLAYS, in which "gangs" of up to ten males chase a female, peck at her genitals, and try to mount her. Such displays and the associated sexual activity often take place during nonfertilizable periods as well. Courting male Cowbirds also frequently harass females, and on average only about 12 percent of heterosexual consortships in this species culminate in copulation. In pair-bonding populations of Brown-headed Cowbirds, about 16 percent of courtships are actually between birds not paired to one another, and some promiscuous matings occur as well. In both this species and House Sparrows, a few pairs switch mates during the breeding season, and a subset of birds (about 5–6 percent) are polygamous.

Large numbers of male Cowbirds are nonbreeders: more than half of all males in some populations are unpaired, and only a third of males actually copulate with females in some years. Brown-headed Cowbirds are also BROOD PARASITES, which means that females always lay their eggs in the nests of other bird species and take no part in raising their own young. Infanticide occurs in 9–12 percent of House Sparrow nests, often when a female who has lost her mate pairs with a new male (who pecks her young to death in order to father his own offspring). Females in polygamous trios also occasionally kill one another's nestlings. Sometimes, however, a female whose mate has been replaced by an infanticidal male will stop laying eggs (by interrupting or delaying ovulation) in order not to lose any more young, and some replacement males adopt rather than kill their mate's young.

Other Species

Male Sharp-tailed Sparrows (*Ammodramus caudacutus*), a North American species, sometimes mount other males. Adolescent male Yellow-rumped Caciques (*Cacicus cela*), a South American blackbird, frequently mount fledglings of both sexes. Sexual behavior toward these younger birds is usually part of an overall pattern of harassment, in which the adolescent males (often in groups) chase, peck, attack, and sometimes even knock fledglings from their perches (often resulting in their death by drowning if they fall into water). About 36 percent of such harassments (and the associated sexual behavior) involve same-sex interactions.

Sources   *asterisked references discuss homosexuality/transgender

Craig, A. J. F. K. (1996) "The Annual Cycle of Wing Moult and Breeding in the Wattled Starling *Creatophora cinerea*." *Ibis* 138:448–54.

Darley, J. A. (1978) "Pairing in Captive Brown-headed Cowbirds (*Molothrus ater*)." *Canadian Journal of Zoology* 56:2249–52.

*Dean, W. R. J. (1978) "Plumage, Reproductive Condition, and Moult in Non-Breeding Wattled Starlings." *Ostrich* 49:97–101.

Friedmann, H. (1929) *The Cowbirds: A Study in the Biology of Social Parasitism*. Springfield, Ill.: Charles C. Thomas.

*Greenlaw, J. S., and J. D. Rising (1994) "Sharp-tailed Sparrow (*Ammodramus caudacutus*)." In A. Poole and F. Gill, eds., *The Birds of North America: Life Histories for the 21st Century*, no. 112. Philadelphia: Academy of Natural Sciences; Washington, D.C.: American Ornithologists' Union.

*Griffin, D. N. (1959) "Apparent Homosexual Behavior Between Brown-headed Cowbird and House Sparrow." *Auk* 76:238–39.

*Hamilton, J. B. (1959) "A Male Pattern Baldness in Wattled Starlings Resembling the Condition in Man." *Annals of the New York Academy of Sciences* 83:429–47.

Laskey, A. R. (1950) "Cowbird Behavior." *Wilson Bulletin* 62:157–74.

Liversidge, R. (1961) "The Wattled Starling (*Creatophora cinerea* [Menschen])." *Annals of the Cape Provincial Museums* 1:71–80.

Lowther, P. E. (1993) "Brown-headed Cowbird (*Molothrus ater*)." In A. Poole, P. Stettenheim, and F. Gill, eds., *The Birds of North America: Life Histories for the 21st Century*, no. 47. Philadelphia: Academy of Natural Sciences; Washington, D.C.: American Ornithologists' Union.

Lowther, P. E., and C. L. Cink (1992) "House Sparrow (*Passer domesticus*)." In A. Poole, P. Stettenheim, and F. Gill, eds., *The Birds of North America: Life Histories for the 21st Century*, no. 12. Philadelphia: Academy of Natural Sciences; Washington, D.C.: American Ornithologists' Union.

Møller, A. P. (1987) "House Sparrow, *Passer domesticus*, Communal Displays." *Animal Behavior* 35:203–10.

*Robinson, S. K. (1988) "Anti-Social and Social Behavior of Adolescent Yellow-rumped Caciques (Icterinae: *Cacicus cela*)." *Animal Behavior* 36:1482–95.

Rothstein, S. I. (1980) "The Preening Invitation or Head-Down Display of Parasitic Cowbirds: II. Experimental Analysis and Evidence for Behavioral Mimicry." *Behavior* 75:148–84.

Rothstein, S. I., D. A. Yokel, and R. C. Fleischer (1986) "Social Dominance, Mating and Spacing Systems, Female Fecundity, and Vocal Dialects in Captive and Free-Ranging Brown-headed Cowbirds." *Current Ornithology* 3:127–85.

Scott, T. W., and J. M. Grumstrup-Scott (1983) "Why Do Brown-headed Cowbirds Perform the Head-Down Display?" *Auk* 100:139–48.

*Selander, R. K., and C. J. La Rue, Jr. (1961) "Interspecific Preening Invitation Display of Parasitic Cowbirds." *Auk* 78:473–504.

*Sontag, W. A., Jr. (1991) "Habitusunterschiede, Balzverhalten, Paarbildung, und Paarbindung beim Lappenstar *Creatophora cinerea* [Behavior Differences, Courtship, Pair Formation, and Pair Bonding in the Wattled Starling]." *Acta Biologica Benrodis* 3:99–114.

——— (1978/79) "Remarks Concerning the Social Behavior of Wattled Starlings, *Creatophora cinerea* (Menschen)." *Journal of the Nepal Research Center* 2/3:263–68.

Teather, K. L., and R. J. Robertson (1986) "Pair Bonds and Factors Influencing the Diversity of Mating Systems in Brown-headed Cowbirds." *Condor* 88:63–69.

Uys, C. J. (1977) "Notes on Wattled Starlings in the Western Cape." *Bokmakierie* 28:87–89.

Veiga, J. P. (1993) "Prospective Infanticide and Ovulation Retardation in Free-Living House Sparrows." *Animal Behavior* 45:43–46.

——— (1990) "Infanticide by Male and Female House Sparrows." *Animal Behavior* 39:496–502.

Wetton, J. H., and D. T. Parkin (1991) "An Association Between Fertility and Cuckoldry in the House Sparrow, *Passer domesticus*." *Proceedings of the Royal Society of London*, Series B 245:227–33.

Yokel, D. A. (1986) "Monogamy and Brood Parasitism: An Unlikely Pair." *Animal Behavior* 34:1348–58.

Yokel, D. A., and S. I. Rothstein (1991) "The Basis for Female Choice in an Avian Brood Parasite." *Behavioral Ecology and Sociobiology* 29:39–45.

CROWS   *Perching Bird*

BLACK-BILLED MAGPIE
Pica pica
(EURASIAN) JACKDAW
Corvus monedula
(COMMON) RAVEN
Corvus corax

HOMOSEXUALITY	TRANSGENDER	BEHAVIORS		RANKING	OBSERVED IN
● Female	○ Intersexuality	● Courtship	● Pair-bonding	○ Primary	● Wild
● Male	○ Transvestism	● Affectionate	● Parenting	● Moderate	○ Semiwild
		○ Sexual		○ Incidental	● Captivity

BLACK-BILLED MAGPIE
IDENTIFICATION: A pigeon-sized, crowlike bird with striking iridescent, black-and-white plumage and a long, greenish purple tail. DISTRIBUTION: Eurasia, North Africa, western North America. HABITAT: Woodland, scrub, grassland, savanna. STUDY AREAS: Haren, the Netherlands; University of Groningen, the Netherlands; subspecies *P.p. pica*.

JACKDAW
IDENTIFICATION: A small crow with black plumage and gray on the back of the head. DISTRIBUTION: Eurasia, North Africa. HABITAT: Forest, grassland, farmland. STUDY AREAS: Haren, the Netherlands; Max-Planck Institute; subspecies *C.m. spermologus*.

RAVEN
IDENTIFICATION: A large (2 foot), all-black bird similar to a crow but much bulkier. DISTRIBUTION: Eurasia, North America. HABITAT: Varied, including forest, plains, desert. STUDY AREA: Max-Planck Institute; subspecies *C.c. corax*.

Social Organization

All three of these Crow species are quite gregarious, often associating in flocks and communal roosts. Individuals generally form long-term mated pairs, and Jackdaws usually nest in colonies. Magpies sometimes participate in remarkable group displays known as CEREMONIAL GATHERINGS, noisy aggregations that may be related to territory acquisition.

Description

Behavioral Expression: Black-billed Magpies sometimes court and form pair-bonds with birds of the same sex. Partners include adult males with younger or juvenile males (less than one year old), or else two females or two males of the same age, usually juveniles but also sometimes two adult males. A typical homosexual courtship—for example, between two males—begins with one bird ritually BEGGING the other by crouching in front of him and flapping or quivering his wings while uttering a begging call. The other male responds by hopping in tight circles around him, fluffing up his white feathers and flicking his wings; he may also BAB-BLE-SING, a varied combination of warbling, chattering, and yelping notes. The circling male often adopts a TILTING posture, in which he points his head and tail sideways toward the other male. If the male he is courting flies off, a courtship pursuit known as CHASE-HOPPING may develop, in which one male follows the other while alternately flying and hopping. Sometimes the two birds also HOVER-FLY, one in front of the other, using a rhythmic, undulating flight pattern. The same series of behaviors is seen in courtships between females (as well as in heterosexual interactions). Homosexual courtship sessions can last for up to half an hour. After courting, two birds of the same sex may form a pair-bond. Paired birds stay near each other, follow one another, and often cooperate in evicting intruders from their territory. They also frequently sit close together and preen one another or engage in mutual BILLING, in which they affectionately nibble at each other's beak. Sometimes, homosexual mates also pull or nibble on the same leaf or twig and pass it back and forth between them; this is known as TUGGING. Same-sex pair-bonds are generally of shorter duration than adult heterosexual bonds and last from a few days to several months. However, adult males sometimes form longer-lasting homosexual pairs, and the two birds may even build a nest together (which typically takes five to seven weeks to construct).

Female Jackdaws occasionally develop pair-bonds with other females. In some cases, an older female pairs with a younger one, and the two build a nest together even though the juvenile bird is too young to lay eggs. Later, they might construct a unique "double nest" consisting of two adjacent cups and lay infertile eggs in both cups. Sometimes, a homosexual pair is joined by a male, who may bond with one or both of the females to form a bisexual trio; in this way, the females can lay fertilized eggs. However, they are often unable to successfully care for their offspring, precisely because their bond to each other means that they try to stay together all the time. The two females incubate their eggs and brood their youngsters simultaneously, each sitting on one cup. When the male arrives for his shift, however, they both depart together, leaving the male to try to cover and protect both sets of eggs or nestlings at the same time (which he is usually unable to do). Sometimes, a bi-

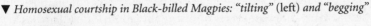

▼ *Homosexual courtship in Black-billed Magpies: "tilting" (left)* and *"begging"*

sexual trio forms when a female joins a heterosexual pair and develops a strong bond with the female partner. The two females engage in courtship and pair-bonding activities such as mutual preening or COURTSHIP-FEEDING, in which one partner begs the other by crouching, fluttering her wings, and quivering her tail. Both females might mate with the male and lay fertile eggs, although the bond between them can end up being stronger than the original heterosexual bond. In fact, in one case the female partners were not able to properly care for their young because the male denied the female "interloper" access to the nestlings. Homosexual bonds also sometimes develop between widowed and nonbreeding females. In these cases, females that lose their male partners during the breeding season may attract unmated females to pair with them; some of these widows are mothers, while others have no offspring. Unlike bisexual trios, which may remain together for years, these female bonds appear to be more transitory, usually lasting only for several weeks until the breeding season is over.

Female homosexual pairs also occasionally occur in Ravens, including incestuous bonds between yearling sisters. Birds in same-sex pairs engage in intense courtship activity similar to heterosexual pairs, such as mutual preening and courtship-feeding.

Frequency: Among Jackdaws in the wild, approximately 5 percent of trios include bonding between the two female partners, while about 10 percent of widowed females form homosexual pairs. Overall, though, same-sex bonds probably represent no more than 1 percent of all pairs/trios. Homosexual activity occurs sporadically in Black-billed Magpies and Ravens as well: about 1 percent of nesting pairs of Magpies, for example, consist of two males. Although homosexual copulations between Magpies have not yet been recorded, heterosexual matings are also infrequently observed (only 9 male-female copulations, for example, were recorded during one 300–hour study period).

Orientation: Homosexual behavior is most prevalent among younger Black-billed Magpies, some of whom also court members of the opposite sex; many of these birds go on to form heterosexual attachments when they become adults, but most are nonbreeders during the time of their same-sex activities. Some adult males, however, continue to court younger males or pair with other adult males. In Jackdaws and Ravens many different forms of bisexuality occur. Some females are simultaneously bonded to both males and females, and such trios may develop out of either an initial heterosexual or homosexual pair-bond. Some females are sequentially bisexual, developing same-sex bonds only after losing their male partners. However, those Jackdaw females who enter into homosexual pairs as nonbreeders may have no prior (and perhaps no subsequent) heterosexual experience.

Nonreproductive and Alternative Heterosexualities

Black-billed Magpies and Jackdaws participate in a number of nonprocreative heterosexual activities. Birds sometimes copulate long after fertilization has taken place—even during incubation or chick-rearing—while courtship and mounting may occur well in advance of the breeding season in Magpies. In both species, birds form pair-bonds as juveniles, long before they are able to breed. Several alternatives to the monogamous pair and nuclear family are also to be found in these species of Crows. In Black-billed Magpies, for example, courtship and copulations with birds outside of the pair-bond can be more frequent than within-pair matings. In addition, some pairs get divorced: about half of all males and just under two-thirds of females change mates, and some may pair with up to three different partners over the course of their lives. Other Magpies, however, do remain in lifelong, faithful pair-bonds. About 6–10 percent of adult Jackdaw pairs, and a third of all juvenile pairs, get divorced. Polygamous trios also sometimes develop in Magpies (1–2 percent of bonds), but are fairly common in Jackdaws, comprising about 14 percent of all bonds. These usually result from an unpaired female joining an existing male-female pair; unlike the bisexual trios described above, she usually bonds only with the male of the pair, not the female. Occasionally, she may oust the former female and start a new family with the male; frequently, however, the newcomer female does not breed even if she remains in the trio for several years.

Nonbreeding couples also occur in Jackdaws and Ravens, as well as a significant population of single, nonreproducing birds. The latter are found in Magpies as well, where they make up 20–60 percent of the population and may be solitary or form their own flocks. In addition, about half of all Magpies that survive to adulthood leave no descendants (even if they try to breed). Some nonbreeding Jackdaw pairs—or couples who have tried unsuccessfully to breed—end up harassing breeding pairs by invading their nests, fighting with them, and even attacking and occasionally killing their young with vicious pecks. Nearly a third of all breeding pairs in some populations are harassed in this way. Cannibalism of young and eggs by neighboring breeding pairs also sometimes occurs in Jackdaws. Magpies and Ravens occasionally adopt unrelated youngsters when they pair with a bird that has lost its mate. Magpie families also sometimes foster-parent chicks as a result of the extraordinary phenomenon of "egg transfer": in some populations, up to 8 percent of nests contain a foreign Magpie egg, but not as a result of being laid there. Rather, another Magpie has physically carried its own egg in its beak to the new nest, placing it there to hatch and be raised by the host parents. Stealing and cannibalism of young or eggs has also been reported among Black-billed Magpies: about 7 percent of nests are robbed by other Magpies. In addition, at least 30 percent of nestlings die of starvation as a result of competition from their siblings, or from direct attacks or cannibalism by their nest-mates.

Other Species

Homosexual courtship occurs in some species of Jays (closely related to Crows), so far observed only in captivity. Male Gray-breasted or Mexican Jays (*Aphelocoma ultramarina*), a species found in the American Southwest and Mexico, have been seen COURTSHIP-FEEDING younger males in mixed-sex groups. This behavior—also part of heterosexual courtships—involves one male offering the other a food item; the latter accepts it while quivering his wings, crouching, and uttering a low *kwa kwa kwa* call. After eating the food or storing it in a cache, he may follow the other male to repeat the episode. Female San Blas Jays (*Cyanocorax sanblasianus*) of Mexico have been known to court other females in same-sex groups in captivity, using the SOTTO VOCE SONG DISPLAY. While making soft, throaty vocalizations, one female sidles up to the other on a perch with her tail raised and belly feathers fluffed.

Sources   *asterisked references discuss homosexuality/transgender*

Antikainen, E. (1981) "The Breeding Success of the Jackdaw *Corvus monedula* in Nesting Cells." *Ornis Fennica* 58:72–77.

*Baeyens, G. (1981a) "Magpie Breeding Success and Carrion Crow Interference." *Ardea* 69:125–39.

——— (1981b) "Functional Aspects of Serial Monogamy: The Magpie Pair-Bond in Relation to Its Territorial System." *Ardea* 69:145–66.

*——— (1979) "Description of the Social Behavior of the Magpie (*Pica pica*)." *Ardea* 67:28–41.

Birkhead, T. (1991) *The Magpies: The Ecology and Behavior of Black-billed and Yellow-billed Magpies*. London: T. and A. D. Poyser.

Birkhead, T., and J. D. Biggins (1987) "Reproductive Synchrony and Extra-Pair Copulation in Birds." *Ethology* 74:320–34.

Birkhead, T., S. F. Eden, K. Clarkson, S. F. Goodburn, and J. Pellatt (1986) "Social Organization of Magpies *Pica pica*." *Ardea* 74:59–68.

Buitron, D. (1988) "Female and Male Specialization in Parental Care and Its Consequences in Black-billed Magpies." *Condor* 90:29–39.

——— (1983) "Extra-Pair Courtship in Black-billed Magpies." *Animal Behavior* 31:211–20.

Coombs, F. (1978) *The Crows: A Study of the Corvids of Europe*. London: B. T. Batsford.

Dhindsa, M. S., and D. A. Boag (1992) "Patterns of Nest Site, Territory, and Mate Switching in Black-billed Magpies (*Pica pica*)." *Canadian Journal of Zoology* 70:633–40.

Dunn, P. O., and S. J. Hannon (1989) "Evidence for Obligate Male Parental Care in Black-billed Magpies." *Auk* 106:635–44.

*Hardy, J. W. (1974) "Behavior and Its Evolution in Neotropical Jays (*Cissilopha*)." *Bird-Banding* 45:253–68.

*——— (1961) "Studies in Behavior and Phylogeny of Certain New World Jays (Garrulinae)." *University of Kansas Science Bulletin* 42:13–149.

Jerzak, L. (1995) "Breeding Ecology of an Urban Magpie *Pica pica* Population in Zielona Góra (SW Poland)." *Acta Ornithologica* 29:123–33.

*Lorenz, K. (1972) "Pair-Formation in Ravens." In H. Friedrich, ed., *Man and Animal: Studies in Behavior*, pp. 17–36. New York: St. Martin's.

*——— (1935) "Der Kumpan in der Umwelt des Vögels." *Journal für Ornithologie* 83:10–213, 289–413. Reprinted as "Companions as Factors in the Bird's Environment." In K. Lorenz (1970) *Studies in Animal and Human Behavior*, vol.1, pp. 101–258. Cambridge, Mass.: Harvard University Press.

Ratcliffe, D. (1997) *The Raven: A Natural History in Britain and Ireland*. London: T. and A. D. Poyser.

Reynolds, P. S. (1996) "Brood Reduction and Siblicide in Black-billed Magpies (*Pica pica*)." *Auk* 113:189–99.

*Röell, A. (1979) "Bigamy in Jackdaws." *Ardea* 67:123–29.

——— (1978) "Social Behavior of the Jackdaw, *Corvus monedula*, in Relation to Its Niche." *Behavior* 64:1–124.

Trost, C. H., and C. L. Webb (1986) "Egg Moving by Two Species of Corvid." *Animal Behavior* 34:294–95.

BIRDS OF PARADISE, BOWERBIRDS, AND OTHERS

RAGGIANA'S BIRD OF PARADISE
Paradisaea raggiana
VICTORIA'S RIFLEBIRD
Ptiloris victoriae

HOMOSEXUALITY	TRANSGENDER	BEHAVIORS		RANKING	OBSERVED IN
● Female	○ Intersexuality	● Courtship	○ Pair-bonding	○ Primary	● Wild
● Male	○ Transvestism	○ Affectionate	○ Parenting	○ Moderate	○ Semiwild
		● Sexual		● Incidental	● Captivity

RAGGIANA'S BIRD OF PARADISE
IDENTIFICATION: A crow-sized bird; male has a bright yellow head, an iridescent green throat, and a long "tail" of orange flank feathers; female has a duller yellow head and brown facial mask and no orange "tail." DISTRIBUTION: Southern and northeastern Papua New Guinea. HABITAT: Lowland and hill forests. STUDY AREA: In captivity at the Baiyer River Sanctuary, Mt. Hagen, Papua New Guinea; subspecies *P.r. augustae-victoriae.*

VICTORIA'S RIFLEBIRD
IDENTIFICATION: Adult males are black with an iridescent sheen and metallic crown, throat, and central tail feathers; females and younger males are drabber, with brown, buff, and off-white plumage. DISTRIBUTION: Northeast Queensland, Australia. HABITAT: Rain forest, eucalyptus forest, swamp woodlands. STUDY AREAS: Southern Atherton Tableland, Queensland, Australia, including near Townsville and in the Ingham and Palmerston National Parks.

Social Organization

Male Raggiana's Birds of Paradise perform courtship displays on communal "courts" or LEKS (located in tree branches) in groups of up to eight individuals at a time; females visit the leks in pairs or small groups. Male Victoria's Riflebirds display singly. Both species have a polygamous or promiscuous mating system, in which males copulate with more than one female but do not participate in any parental duties.

Description

Behavioral Expression: Male Victoria's Riflebirds sometimes court younger males with a series of spectacular displays (also seen in heterosexual courtships). Adult males attract individuals of both sexes to their display perches by repeatedly calling with a loud *yass* note and opening their bills to expose the bright yellow mouth interior. When an interested male (or female) approaches, the courting male begins his CIRCULAR WINGS AND GAPE display, in which he fans his wings to form a perfect circle with the tips meeting above his head while simultaneously raising and lowering his body. The wing circle is held vertically as he gapes his bill and presents his iridescent throat feathers to the other male. If the latter lands on the display perch, the courtship progresses to its final phase, the astounding ALTERNATE WINGS CLAP display. The adult male begins swaying and twisting his body from side to side, alternately raising and lowering each wing so that they "clap" with a dull, thudding sound when they meet above his head. Gradually the wings are brought forward while being clapped so that they encircle the younger male. The tempo of the display is increased until, at its climax, the wing-clapping reaches an astounding speed—each wing alternately encircles the observing male at a rate approaching twice a second. The courted male becomes mesmerized by the brilliant metallic bluish green throat shield of the displaying male as it is jerked violently back and forth, and he may respond by simultaneously performing his own alternate wings clap display. If both males are displaying, they alternate wing movements, gradually

crouching while arching their heads back and quivering all over. They often perform several bouts of mutual display with brief rests between each. Occasionally one male briefly mounts the other; full copulation does not usually occur, though, since the mounted male often simply flies away from underneath the male mounting him.

Courtship displays between female Raggiana's Birds of Paradise involve some of the same postures and movements used in heterosexual courtship in this species. One female arches her wings above her back and

◀ *A male Victoria's Riflebird courting another male with the "circular wings and gape" display*

then claps them against her sides, all the while bobbing up and down and erecting the feathers on her back and flanks as she dances in front of the other female. Every now and then she makes an *ur, ur* call, and she may even hang upside down on a branch while flicking her wings and calling. This dramatic pose is held for a minute or more at a time. Males use this inverted posture to show off their orange "tail" plumes during courtship, but females perform this display even though they do not have such ornate feathers. Sometimes two females engage in mutual display, facing each other while holding their wings stiffly above their backs and bobbing.

Frequency: Courtship displays between male Victoria's Riflebirds occur fairly often at some times of the year, especially during the postmolting period in February and March. Although homosexual mountings are not common in this species, opposite-sex matings have in fact been seen only a few times during more than a hundred courtship displays observed in the wild.

Orientation: At least some male Victoria's Riflebirds are functionally bisexual, courting and attempting to mate with both males and females. In Raggiana's Bird of Paradise, courtship display between females has so far only been observed in captivity in the absence of males, so this behavior may be the expression of a potential or "latent" bisexual capacity. More detailed field observations and life histories are required, though, before any definitive statements can be made regarding the sexual orientation of individuals that engage in this activity.

Nonreproductive and Alternative Heterosexualities

Only about 11 percent of lek visits by female Raggiana's Birds of Paradise actually result in heterosexual mating. More often than not, a courtship interaction is broken off by a female before copulation occurs. In addition, males often chase females and behave aggressively toward them on the lek, which deters them from participating in sexual interactions. When mating does take place, the female often mounts the male prior to copulation. Such REVERSE mounts are then followed by a display in which the male appears to pummel his partner with his wing. The female crouches on her perch absorbing the blows for 20–35 seconds, after which copulation occurs. Nonreproductive matings also take place outside of the breeding season. Many males may delay breeding up to five or more years once becoming sexually mature, since they do not generally acquire their ornate plumage (used in courtship displays) for several years.

Other Species

Male Greater Birds of Paradise (*Paradisaea apoda*), a species closely related to Raggiana's, also sometimes court and mount younger males.

Sources   *asterisked references discuss homosexuality/transgender*
Beehler, B.M. (1989) "The Birds of Paradise." *Scientific American* 261(6):116–23.
———— (1988) "Lek Behavior of the Raggiana Bird of Paradise." *National Geographic Research* 4:343–58.
*Bourke, P. A., and A. F. Austin (1947) "The Atherton Tablelands and Its Avifauna." *Emu* 47:87–116.

Davis, W. E., Jr., and B. M. Beehler (1994) "Nesting Behavior of a Raggiana Bird of Paradise." *Wilson Bulletin* 106:522–30.

*Frith, C. B. (1997) Personal communication.

———— (1981) "Displays of Count Raggi's Bird-of-Paradise *Paradisaea raggiana* and Congeneric Species." *Emu* 81:193–201.

*Frith, C. B., and W. T. Cooper (1996) "Courtship Display and Mating of Victoria's Riflebird *Ptiloris victoriae* with Notes on the Courtship Displays of Congeneric Species." *Emu* 96:102–13.

Frith, C. B., and D. W. Frith (1995) "Notes on the Nesting Biology and Diet of Victoria's Riflebird *Ptiloris victoriae*." *Emu* 95:162–74.

Gilliard, E. T. (1969) "Queen Victoria Rifle Bird" and "Count Raggi's Bird of Paradise." In *Birds of Paradise and Bower Birds*, pp. 112–17, 222–29. Garden City, N.Y.: Natural History Press.

Lecroy, M. (1981) "The Genus *Paradisaea*—Display and Evolution." *American Museum Novitates* 2714:1–52.

*Mackay, M. (1981) "Display Behavior by Female Birds of Paradise in Captivity." *Newsletter of the Papua New Guinea Bird Society* 185/186 (November-December):5.

BOWERBIRDS   *Perching Bird*

REGENT BOWERBIRD
Sericulus chrysocephalus

HOMOSEXUALITY	TRANSGENDER	BEHAVIORS		RANKING	OBSERVED IN
○ Female	○ Intersexuality	● Courtship	○ Pair-bonding	○ Primary	● Wild
● Male	● Transvestism	○ Affectionate	○ Parenting	● Moderate	○ Semiwild
		○ Sexual		○ Incidental	● Captivity

IDENTIFICATION: A thrush-sized bird (12 inches) with velvety black plumage and brilliant golden yellow crown, nape, upper back, and wing feathers. DISTRIBUTION: East-central Australia. HABITAT: Humid forests. STUDY AREA: Sarabah Range, Queensland, Australia.

Social Organization

Regent Bowerbirds have a polygamous or promiscuous mating system: males court and mate with multiple partners on display territories (see below). Following the mating season birds often associate in cosexual flocks of 10–20 individuals.

Description

Behavioral Expression: Male Regent Bowerbirds build elaborate structures called BOWERS, in which they court both males and females. Bowers built by adult males form an "avenue" on the ground consisting of two parallel walls of twigs—10–12 inches high and 7–8 inches long—which are inserted into a platform of twigs on the ground. In some bowers the walls form a sort of arched walkway, while other bowers have triangular-shaped walls of different sizes. The bower is usually decorated with a dozen or so "display objects" or ornaments that are often selected for their color and strewn on the platform. These include brown snail shells, berries, green or pur-

▲ *A bower constructed by a male Regent Bowerbird, used during courtship displays to both males and females*

plish leaves, brown fruits and seeds, cicada husks, occasionally a yellow or pink flower petal, and even scavenged pieces of blue plastic. Remarkably, some birds also "paint" the walls of their bowers by applying, with their bills, macerated plant material mixed with saliva, giving some of the twigs a yellowish green coating. Adult males display in their bowers to females and younger males by flicking their wings and bowing their heads to show off their brilliant orange and yellow neck feathers. Sometimes they also ritually "present" display objects to the bird they are courting by picking up one of the ornaments from the avenue and holding it in their bill while facing their partner.

Juvenile males also build bowers and court both males and females. In some cases, they use the same bower shapes and displays as adults, but in other cases they utilize their own patterns, some of which may be unique to same-sex interactions. Some bowers built by younger males are more in the form of a horseshoe—open only at one end—with the sticks woven horizontally rather than vertically. When courtship-displaying to an adult male, the younger male enters the bower and squats with his tail toward the entrance. The adult male then rushes toward the juvenile, who is behind the closed end of the bower, or sometimes goes around to the entrance and tweaks the younger male's tail. When courting other juvenile birds of both sexes, young males may perform a dance in the center of the bower, picking up an ornament and bobbing up and down with half-opened wings, then tossing the object over the bower wall. The bird(s) watching this display sweep and brush the ground with their wings.

Many younger males are transvestite, having a plumage coloration that more closely resembles that of adult females than adult males. It takes males up to seven years to develop the full yellow, orange, and black feathers typical of adult males, and during that time many actually exhibit plumages that are intermediate between the adult male and female patterns.

Frequency: Among adult male Regent Bowerbirds, 15 percent of their bower courtship time is spent displaying to other males, while 28 percent of juvenile males' display time is same-sex.

Orientation: Regent Bowerbirds that court other males are probably bisexual, since they display to birds of both sexes.

Nonreproductive and Alternative Heterosexualities

A large segment of the male Regent Bowerbird population is nonbreeding: only about a third of all adult males maintain bowers, and of those that do, only a fraction actually mate with females. In addition, heterosexual courtship interactions rarely result in copulation: in only about 7 percent of female visits to bowers does mating actually take place, since the female often leaves while the male is displaying (and in 10 percent of female visits the male does not display at all).

Other Species

Adult male Satin Bowerbirds (*Ptilonorhynchus violaceus*) from Australia have also been observed performing courtship displays toward younger males.

Sources   *asterisked references discuss homosexuality/transgender

Chaffer, N. (1932) "The Regent Bird." *Emu* 32:8–11.
*Gilliard, E. T. (1969) "Australian Regent Bower Bird." In *Birds of Paradise and Bower Birds,* pp. 335–44. Garden City, N.Y.: Natural History Press.
Goddard, M. T. (1947) "Bower-Painting by the Regent Bower-bird." *Emu* 47:73–74.
*Lenz, N. (1994) "Mating Behavior and Sexual Competition in the Regent Bowerbird *Sericulus chrysocephalus.*" *Emu* 94:263–72.
*Marshall, A. J. (1954) "Satin Bower-bird, *Ptilonorhynchus violaceus* (Vieillot)" and "Regent Bower-bird, *Sericulus chrysocephalus* (Lewin)." In *Bower-birds: Their Displays and Breeding Cycles,* pp. 26–71, 109–18. Oxford: Oxford University Press.
*Phillipps, R. (1905) "The Regent Bird (*Sericulus melinus*)." *Avicultural Magazine* (new series) 4:51–68, 88–96, 123–31.
Plomley, K. F. (1935) "Bower of the Regent Bower-bird." *Emu* 34:199.

LYREBIRDS

Perching Bird

SUPERB LYREBIRD
Menura novaehollandiae

HOMOSEXUALITY	TRANSGENDER	BEHAVIORS		RANKING	OBSERVED IN
● Female	○ Intersexuality	● Courtship	● Pair-bonding	● Primary	● Wild
● Male	○ Transvestism	○ Affectionate	○ Parenting	○ Moderate	○ Semiwild
		● Sexual		○ Incidental	● Captivity

IDENTIFICATION: A pheasant-sized bird with brownish gray plumage, powerful legs and claws, and long, ornate tail feathers. DISTRIBUTION: Southeastern Australia. HABITAT: Rain forest, eucalyptus, other forests. STUDY AREAS: Sherbrooke Forest near Melbourne, Australia; Adelaide Zoo, South Australia.

Social Organization

Adult male Superb Lyrebirds establish territories on which they build earth mounds three to five feet wide, used as platforms for courtship displays. The mating system is polygamous or promiscuous: birds mate with multiple partners without developing long-lasting pair-bonds, and males do not contribute to raising their offspring. Adolescent Lyrebirds often associate in small groups, sometimes consisting only of males.

Description

Behavioral Expression: Male Superb Lyrebirds often court younger (adolescent) males on encountering them either in their groups or singly. The adult male closely follows the younger male, sometimes for hours at a time, periodically performing the WING-RAISING DISPLAY, in which he lifts and fans one wing toward his partner. With neck outstretched, he serenades the other male with a variety of extraordinary vocalizations, such as "whisper-song"; chortling; a call that can be rendered phonetically as *clonk clonk clickety clickety click*; another vocalization that sounds like the grinding of a pair of scissors; as well as uncanny imitations of other birds' songs. Sometimes the courtship progresses to spectacular displays of the male's elongated and stunningly beautiful tail feathers. In the dazzling FULL-FACE DISPLAY, for example, the older male arches his tail forward over his head and spreads the silvery filamentary feathers, hiding his body behind a gossamer fan framed by the elegant lyre-shaped, chestnut-striped outer tail feathers (for which the bird is named). Often the entire tail is vibrated to create a shimmering effect directed toward the other male. This may be followed by the INVITATION DISPLAY, in which the male almost "closes up" his quivering tail fan while still holding it forward (the feather tips almost touching the ground in front of him), all the while re-

▲ *An adult male Superb Lyrebird (with molted tail feathers) mounting a younger male in Australia*

peating a *blick blick blick* call. Occasionally, the younger male being courted by an adult is his own offspring. Adult males also sometimes mount adolescent males, even during the nonbreeding season when the adult bird is molting and has shed his elaborate tail plumes. Genital contact probably does not occur, though, since the younger male usually does not facilitate the interaction.

Adolescent male Lyrebirds also sometimes mount one another as well as participate in homosexual courtship displays. Courtship between younger males is usually mutual, with both birds singing and displaying to one another. Occasionally an adult and an adolescent male also engage in mutual display, and adult females (in captivity) have been observed performing a similar courtship display to each other. Typically, the two adolescent males circle around each other on a display mound, tails raised in a fan shape with their feathers intermingled and beaks nearly touching while performing a vocal duet. One male may also perform a full-face or invitation display, and his partner sometimes runs underneath his outstretched tail the way a female does during heterosexual courtship. Sometimes the two males engage in what appears to be a form of COURTSHIP-FEEDING: in response to begging from his partner, one male regurgitates a worm or other food item as an offering to the other, who promptly eats it. This behavior appears to be unique to homosexual courtships. Two younger males often form a "companionship": in addition to courting each other, they follow one another, feed together (even

Two younger male Superb Lyrebirds performing a mutual courtship display ▶

digging in the same hole for food items), roost next to each other, and share bathing pools. These male pair-bonds usually last for only a few days, and adolescent males often form multiple serial attachments of this sort.

Frequency: Homosexual courtship occurs fairly often in Superb Lyrebirds. Adult males approach groups of adolescent males approximately once every three days during the breeding season and roughly once every day and a half outside the breeding season, and 93 percent of these encounters include courtship. In comparison, heterosexual encounters occur about four times as often as homosexual encounters during the breeding season, and about twice as often outside the breeding season. Adult males associate with adolescent males (in groups or singly) more than half of the time they are away from their display mounds. Same-sex mounting occurs less frequently than courtship interactions between males.

Orientation: Most adult males are functionally bisexual, courting and mounting both females and younger males. Adolescent males appear to be more exclusively homosexual: most individuals form same-sex companionships and engage in homosexual courtship (and sometimes mounting) for several years prior to mating heterosexually. Some females probably have a bisexual potential that manifests itself in the absence of males (for example, in captivity).

Nonreproductive and Alternative Heterosexualities

Male and female Superb Lyrebirds lead largely separate lives. Other than brief encounters during the breeding season for courtship and copulation, the two sexes rarely interact: only about 8–10 percent of all Lyrebird sightings are of males and females together. Because males do not contribute at all to parenting, incubation and chick-raising often become burdensome for females—and by extension, potentially harmful for the eggs and/or chicks. During the early stages of incubation, females regularly leave their eggs unattended during the day for up to seven hours at a time to feed, causing the egg temperature to drop dramatically. Overall, females are relatively "inattentive" parents, incubating their eggs for only about 27–45 percent of the available daylight hours; this is significantly less than in other perching birds, who generally spend 60–80 percent. Because the eggs are laid and incubated during the coldest months of winter, they are consequently exposed to dangerously low (sometimes even sub-freezing) temperatures that generally slow embryonic development. After hatching, nestlings occasionally die from overexposure when their mother has been away from the nest for too long.

Sources   *asterisked references discuss homosexuality/transgender*

Kenyon, R. F. (1972) "Polygyny Among Superb Lyrebirds in Sherbrooke Forest Park, Kallista, Victoria." *Emu* 72:70–76.

Lill, A. (1986) "Time-Energy Budgets During Reproduction and the Evolution of Single Parenting in the Superb Lyrebird." *Australian Journal of Zoology* 34:351–71.

*———— (1979a) "An Assessment of Male Parental Investment and Pair Bonding in the Polygamous Superb Lyrebird." *Auk* 96:489–98.

——— (1979b) "Nest Inattentiveness and Its Influence on Development of the Young in the Superb Lyrebird." *Condor* 81:225–31.

Reilly, P. (1988) *The Lyrebird: A Natural History.* Kensington, Australia: New South Wales University Press.

*Smith, L. H. (1996–97) Personal communication.

*——— (1988) *The Life of the Lyrebird.* Richmond, Australia: William Heinemann.

——— (1982) "Molting Sequences in the Development of the Tail Plumage of the Superb Lyrebird, *Menura novae-hollandiae.*" *Australian Wildlife Research* 9:311–30.

*——— (1968) *The Lyrebird.* Melbourne: Lansdowne Press.

Watson, I. M. (1965) "Mating of the Superb Lyrebird, *Menura novae-hollandiae.*" *Emu* 65:129–32.

Other Birds

FLIGHTLESS BIRDS

RATITES

OSTRICH
Struthio camelus

EMU
Dromaius novaehollandiae

GREATER RHEA
Rhea americana

HOMOSEXUALITY	TRANSGENDER	BEHAVIORS		RANKING	OBSERVED IN
○ Female	○ Intersexuality	● Courtship	● Pair-bonding	○ Primary	● Wild
● Male	● Transvestism	○ Affectionate	● Parenting	● Moderate	○ Semiwild
		● Sexual		○ Incidental	● Captivity

OSTRICH
IDENTIFICATION: The largest living bird (over 6 feet tall), with striking black-and-white plumage in the male and powerful legs and claws. DISTRIBUTION: Southern, eastern, and west-central Africa. HABITAT: Open savanna, dry veld, steppe, semidesert. STUDY AREA: Namib Game Reserve, Namibia; subspecies *S.c. australis*, the South African Ostrich.

EMU
IDENTIFICATION: The second-largest living bird (5–6 feet tall), with shaggy, brown plumage and bare patches of blue skin on the face and neck. DISTRIBUTION: Australia. HABITAT: Arid plains, semidesert, scrub, open woodland. STUDY AREAS: Barcoo River and Alice Downs areas, Central Queensland, Australia; Division of Wildlife Research, Helena Valley, Western Australia; Berlin Zoo and Melbourne Zoological Gardens.

GREATER RHEA
IDENTIFICATION: Similar to Ostrich but smaller (up to 4½ feet tall) and with overall grayish brown plumage in both sexes. DISTRIBUTION: Southeastern South America. HABITAT: Open brush, grassland, plains. STUDY AREA: Near General Lavalle, Buenos Aires Province, Argentina; subspecies *R.a. albescens*.

Social Organization

Ostriches associate in flocks and frequently form sex-segregated groups. All-male flocks may contain up to 40 individuals, many of them juveniles, who travel with each other for long periods. Emus generally associate in pairs or groups of 3–10 birds, while Greater Rheas gather in flocks of 15–40 birds outside of the mating season. All three species have a wide variety of mating systems (discussed below). These are notable for their various forms of POLYANDRY (females mating with several males) and the fact that—in Emus and Greater Rheas—all incubation and chick-rearing is performed by males without any help from females.

Description

Behavioral Expression: Male Ostriches perform a homosexual courtship dance to each other that is distinct from heterosexual interactions. Same-sex courtships consist of a sequence of three activities, performed by sexually mature adult males in full nuptial plumage (black-and-white feathers, with a red flush on the face and legs). First, there is a dramatic APPROACH in which one male runs rapidly toward his chosen partner—often achieving speeds of 25–30 mph—and stops abruptly just short of the other male. Then he launches into frenzied PIROUETTE DANCING, a high-speed, energetic circling in place beside his partner. This whirling may occur in a series of bursts, each lasting for several minutes. Finally, in KANTLING, the male drops to the ground next to his partner and rocks steadily from side to side, fluffing out his tail and sweeping the ground with his wings in an exaggerated fashion. All the while, he twists his head and neck in a continuous corkscrew action and repeatedly inflates and deflates his throat. The male being displayed to may acknowledge the dance with his posture, or he may simply maintain a calm stance devoid of alarm or aggression. Homosexual courtships are distinct from heterosexual ones in a number of respects: neither the running approach nor pirouette dancing occur in male-female interactions. Kantling is performed in heterosexual contexts but differs because it is usually accompanied by singing (when males display to females, they frequently produce a booming call), and it is significantly shorter. Same-sex displays last for 10–20 minutes, while opposite-sex ones rarely exceed three minutes. Also, symbolic feeding and nest-site displays are com-

◀ *A male Ostrich* (right, on the ground) *courting another male with the "kantling" display*

ponents of heterosexual but not homosexual courtships.

Although no copulation takes place between courting male Ostriches, homosexual mating has been observed in pairs of male Emus. A sexual interaction begins with one male approaching the other, stretching his neck upward and erecting his neck feathers so that they stand out horizontally, while grunting deeply. The two birds begin following and chasing each other; if the male who initiated the activity is behind the other, he may make treading movements with his feet, indicating his intention to mount the other. Often, however, it is the initiating male who lies flat on the ground as an

▲ *Joint parenting in male Greater Rheas: two males in Argentina* (above) *sitting on their double nest , which contains two sets of eggs* (below) *that are frequently rolled between the coparents*

invitation for the other male to mount. The males may also take turns mounting each other. The mounting male lies down behind his partner, resting his breast on the other's rump, and uses his heels to slide forward until he covers most of the other male. While copulation is taking place, the mountee makes soft grunting noises (not usually heard during heterosexual matings), and the mounter gently toys with the feathers on his partner's upper back. After mating, his erect penis is often visible: the male Emu, along with other ratites, is one of the few birds in the world that has a penis (most male birds simply have a cloacal, or genital, opening).

Male Emus also sometimes coparent with each other: two (and occasionally three) males may attend one nest at the same time, incubating all the eggs together. Such nests often contain SUPERNORMAL CLUTCHES of 14–16, and sometimes more than 20, eggs. This is over twice the number found in nests attended by single males, probably because more than one female has laid in them. Unlike single fathers, male coparents are able to take a break from incubating while their partner sits on the nest; they also sometimes roll the eggs between them while on the nest together. Although they are probably not sexually involved with one another, the two fathers cooperate in raising their chicks together, calling to them with "purr-growls" and jointly defending them from predators. A similar phenomenon is found in Greater Rheas: pairs of males occasionally sit on "double nests" that are close to or touching one another; they incubate the eggs together and jointly parent the chicks when they hatch. Most such nests begin as standard nests with only one male incubating, after which another male joins him and begins transferring eggs to his half of the nest; later, eggs may be transferred back and forth between the twin nests. Unlike Emu nests belonging to male coparents, Rhea double nests usually have a combined number of eggs that is the same as for single nests. Male coparents are different from male nest helpers, which are also found in Rheas. About a quarter of breeding males are assisted by an adolescent male, who incubates and raises (on his own) a clutch of eggs fathered by the adult while the latter goes off to start a new family. This differs from male coparenting in that the two nests are widely separated from one another, each contains the full clutch size of a single nest, the two males never share parenting duties, and the helper is always an adolescent male.

Female Ostriches are occasionally transvestite, having full black-and-white male plumage (along with underdeveloped ovaries).

Frequency: Homosexual courtship in Ostriches is quite common in some populations, occurring two to four times a day (usually in the morning). Sexual behavior between male Emus has so far only been observed in captivity, but it does occur repeatedly between partners. Among Greater Rheas, joint parenting between males occurs in about 3 percent of all nests; Emu coparenting probably occurs at a similar rate.

Orientation: In Ostriches, 1–2 percent of all adult males engage in homosexual courtship in some populations. Male Ostriches who court other males typically ignore any females that may be present; they are probably solitary birds that participate in little, if any, heterosexual interactions. Most Emus and Rheas that participate in male coparenting have probably mated and/or paired with females earlier in the season prior to parenting with another male. Male Emus may also have a latent capacity for bisexuality, as evidenced by the occurrence of sexual behavior between captive males (at least one of whom had previously mated heterosexually). However, individual life histories and the full patterns of sexual orientation have not yet been systematically studied in this species.

Nonreproductive and Alternative Heterosexualities

Heterosexual mating in Ratites occurs in the context of an extraordinary variety of complex social arrangements that deviate significantly from the nuclear-family model. Ostriches have a mating system that has been described as SEMIPROMISCUOUS MONOGAMY. Male and female Ostriches form a type of pair-bond with each other that one biologist describes as a sort of "open marriage," since both partners also copulate with a number of other birds besides their primary partner (birds often mate with a different primary partner each year as well). In addition, females often lay eggs in nests other than their own, especially if they are not the primary partner of a male. As a result, many of the eggs that a pair incubates (and the young they raise) are not necessarily their own. Adoption also occurs when broods are combined to form nursery groups or CRÈCHES—sometimes containing hundreds of chicks—that are looked after by one or more adults. Emus utilize SERIAL POLYANDRY in their mating system: a male pairs with one female who remains with him until incubation begins, at which point the female leaves her partner and pairs with a new male to begin a second clutch. Many females also seek nonmonogamous matings, copulating with males other than their pair-mate. One study found that the majority of copulations—nearly three-quarters—are promiscuous. In addition, copulation between pair members may be nonprocreative, occurring several months before egg laying. Greater Rheas have a variable mating system that can be characterized as SERIAL POLYGYNANDRY: a male associates with a "harem" of three to ten females, all of whom he mates with. The females lay their eggs communally in one nest; after the male begins to incubate, the females then move on to another male, repeating the process with up to seven different males. As noted above, most Emu and Rhea males are single parents, which can be an arduous task. While tending the eggs, male Rheas rarely leave the nest for more than a few minutes during the six-week incubation period. Male Emus often become severely emaciated and weakened from not eating, drinking, defecating, or leaving the nest during their entire eight-week incubation period. Nonbreeding and failed breeding attempts also occur at high rates among Greater Rheas: less than 20 percent of males even try to reproduce each year, and overall only 5–6 percent of males are successful at breeding each year.

As discussed above, sex-segregated flocks are common among Ostriches, many of whom are not involved in heterosexual pursuits. Also, heterosexual courtship is often not synchronized: females typically begin approaching males several weeks before the latter become sexually interested, and during this time the males often appear to ignore or be indifferent to the females' advances. The onset of the males' sexual cycle is marked by a red flush on the legs and face, as well as enlargement and erection of the penis, which is often displayed in a special "penis-swinging" ceremony. However, once males begin courting, nearly a third of their advances are, in turn, refused by females. Among Emus and Rheas, outright hostility often develops between the sexes once the male starts incubating. Fathers typically threaten, chase, or attack females that try to approach them, while female Emus have been seen responding with vicious double-footed kicks that can tumble males head over heels. Infanticide also sometimes occurs: females that are able to get close to a male tending his chicks may end up killing the youngsters. Egg abandonment or destruction takes place among Ostriches, often of eggs laid by another female. Abandonment also occurs in Greater Rheas, where nearly two-thirds of nests are deserted by males during incubation. In addition, female Rheas who are unable to find a nest and male caretaker for their eggs often lay them in the open and then abandon them; these are known as ORPHAN EGGS. Once (nonorphan) eggs hatch, fathers often adopt youngsters from other broods, raising them alongside their own. Nearly a quarter of male Rheas are adoptive parents, and up to 37 percent of each of their broods may be composed of foster chicks. Researchers have found that adopted young actually have a better chance of surviving than do their stepsiblings.

Sources   **asterisked references discuss homosexuality/transgender*

Bertram, B. C. (1992) *The Ostrich Communal Nesting System.* Princeton: Princeton University Press.

Brown, J. L. (1987) *Helping and Communal Breeding in Birds: Ecology and Evolution.* Princeton: Princeton University Press.

Bruning, D. F. (1974) "Social Structure and Reproductive Behavior in the Greater Rhea." *Living Bird* 13:251–94.

Coddington, C. L., and A. Cockburn (1995) "The Mating System of Free-Living Emus." *Australian Journal of Zoology* 43:365–72.

Codenotti, T. L., and F. Alvarez (1998) "Adoption of Unrelated Young by Greater Rheas." *Journal of Field Ornithology* 69:58–65.

——— (1997) "Cooperative Breeding Between Males in the Greater Rhea *Rhea americana*." *Ibis* 139:568–71.

*Curry, P. J. (1979) "The Young Emu and Its Family Life in Captivity." Master's thesis, University of Melbourne.

Fernández, G. J., and J. C. Reboreda (1998) "Effects of Clutch Size and Timing of Breeding on Reproductive Success of Greater Rheas." *Auk* 115:340–48.

*——— (1995) "Adjacent Nesting and Egg Stealing Between Males of the Greater Rhea *Rhea americana*." *Journal of Avian Biology* 26:321–24.

Fleay, D. (1936) "Nesting of the Emu." *Emu* 35:202–10.

Folch, A. (1992) "Order Struthioniformes." In J. del Hoyo, A. Elliott, and J. Sargatal, eds., *Handbook of the Birds of the World, vol. 1: Ostrich to Ducks,* pp. 76–110. Barcelona: Lynx Edicións.

*Gaukrodger, D. W. (1925) "The Emu at Home." *Emu* 25:53–57.

*Heinroth, O. (1927) "Berichtigung zu 'Die Begattung des Emus (*Dromaeus novae-hollandiae*)' [Correction to 'Mating Behavior of Emus']." *Ornithologische Monatsberichte* 35:117–18.

*——— (1924) "Die Begattung des Emus, *Dromaeus novae-hollandiae* [Mating Behavior of Emus]." *Ornithologische Monatsberichte* 32:29–30.

*Hiramatsu, H., K. Tasaka, S. Shichiri, and F. Hashizaki (1991) "A Case of Masculinization in a Female Ostrich." *Journal of Japanese Association of Zoological Gardens and Aquariums* 33:81–84.

Navarro, J. L., M. B. Martella, and M. B. Cabrera (1998) "Fertility of Greater Rhea Orphan Eggs: Conservation and Management Implications." *Journal of Field Ornithology* 69:117–20.

O'Brien, R. M. (1990) "Emu, *Dromaius novaehollandiae*." In S. Marchant and P. J. Higgins, eds., *Handbook of Australian, New Zealand, and Antarctic Birds,* vol. 1, part A, pp. 47–58. Melbourne: Oxford University Press.

Raikow, R. J. (1968) "Sexual and Agonistic Behavior of the Common Rhea." *Wilson Bulletin* 81:196–206.

*Sauer, E. G. F. (1972) "Aberrant Sexual Behavior in the South African Ostrich." *Auk* 89:717–37.

Sauer, E. G. F., and E. M. Sauer (1966) "The Behavior and Ecology of the South African Ostrich." *Living Bird* 5:45–75.

Penguins   *Other Birds*

HUMBOLDT PENGUIN
Spheniscus humboldti
KING PENGUIN
Aptenodytes patagonicus
GENTOO PENGUIN
Pygoscelis papua

HOMOSEXUALITY	TRANSGENDER	BEHAVIORS		RANKING	OBSERVED IN
● Female	○ Intersexuality	● Courtship	● Pair-bonding	○ Primary	● Wild
● Male	○ Transvestism	● Affectionate	● Parenting	● Moderate	○ Semiwild
		● Sexual		○ Incidental	● Captivity

HUMBOLDT PENGUIN
IDENTIFICATION: A small penguin (approximately 2 feet tall) with a black band on its chest and patches of red skin at the base of its bill. DISTRIBUTION: Coastal Peru to central Chile. HABITAT: Marine areas; nests on islands or rocky coasts. STUDY AREAS: Emmen Zoo, the Netherlands; Washington Park Zoo, Portland, Oregon.

KING PENGUIN
IDENTIFICATION: A large (3 foot tall) penguin with orange ear patches and a yellow-orange wash on the breast. DISTRIBUTION: Sub-Antarctic seas. HABITAT: Oceans; nests on islands and beaches. STUDY AREA: Edinburgh Zoo, Scotland; subspecies *A.p. patagonicus*.

GENTOO PENGUIN
IDENTIFICATION: A medium-sized penguin (up to 2½ feet) with a white patch above the eye. DISTRIBUTION: Circumpolar in Southern Hemisphere. HABITAT: Oceans; nests on islands and coasts. STUDY AREAS: South Georgia, Falkland Islands; Edinburgh Zoo, Scotland; subspecies *P.p. papua*.

Social Organization

Humboldt Penguins form mated pairs during the breeding season and nest in small colonies; they travel and feed at sea in social groups of 10–60 birds. King Penguins are highly gregarious, breeding in enormous colonies—some numbering 300,000 pairs—and generally form monogamous pair-bonds. Gentoo Penguins have a similar social system, although their nesting colonies are not as large.

Description

Behavioral Expression: Lifelong homosexual pair-bonds sometimes develop between male Humboldt Penguins. Like heterosexual pairs, same-sex partners remain together for many years: some male couples have stayed together for up to six years, until the death of one of the partners. Same-sex pairs (like opposite-sex pairs) spend much of their time close together, often touching. They also usually live together in a nest that they have built—either an underground burrow, a shallow bowl dug in the ground, or a rock niche lined with twigs. Unlike male pairs in other birds, though, homosexual pairs of Humboldt Penguins never acquire any eggs. Courtship and pair-bonding activities are also a prominent aspect of homosexual partnerships. This includes the ECSTATIC DISPLAY, in which a male stretches his head and neck upward, spreading his flippers wide and flapping them while emitting several long, very loud donkeylike brays. Sometimes this is performed mutually by both males standing side by side. Homosexual partners also ALLO-PREEN each other, affectionately running their bills through one another's feathers. Occasional same-sex BOWING also occurs, in which one male points his beak down toward his partner and vibrates his head from side to side. As a prelude to copulation, one male approaches the other from behind, pressing against his body and vibrating his flippers against his partner; this distinctive display is known as the ARMS ACT. Homosexual copulation occurs when the bird in front lies down on his chest, allowing the other male to climb onto his back; genital contact may occur when the male being mounted holds his tail up or to the side and exposes his cloaca. Homosexual mountings are sometimes briefer than heterosexual ones, but often the two males take turns mounting each other. Not all same-sex courtship and sexual activity occurs between birds in homosexual pairs. Males who are paired to females also sometimes court and copulate with other heterosexually paired males (as well as with females other than their own mate).

In King Penguins, same-sex pairs also occur, in both males and females. These bonds are probably not as long-lasting as homosexual pairs in Humboldts, since same-sex partners sometimes divorce each other after being together for only one season (which also occurs commonly in heterosexual pairs in this species). Courtship activities are a part of King Penguin homosexual pair-bonds, especially between males. One such display is BOWING, in which one bird approaches the other while making courtly bows, often leading to mutual bowing. Another display is DABBLING, in which the birds face each other while rapidly clapping their bills and gently nibbling or preening one another's feathers, sometimes accompanied by quivering of the flippers and tail. This may lead to homosexual copulation, in which one bird urges the other to lie down by pressing on its back, then mounts; this occurs among both males and females. In addition, female pairs sometimes lay an (infertile) egg, which they take turns incubating.

Homosexual courtship also occurs early in the breeding season among Gentoo Penguins. A male or a female brings an "offering" of pebbles or grass and lays it at the feet of another bird of the same sex,

◀ A female King Penguin urging another female to lie down prior to mounting her

bowing and making slight hissing noises. The other bird, if interested, may respond with bowing or arranging the material into a nest. Females that pair with each other usually lay eggs in the nest that they tend together; because these birds do not typically mate with males, their eggs are infertile. However, female pairs can become successful foster parents in captivity, incubating and hatching fertile eggs when provided and successfully raising the resulting chicks.

Frequency: In some zoo populations of Humboldt Penguins, at least 5 percent of all pairs are homosexual, and 12 percent of all copulations are between males. Among paired birds, 10 percent of mountings take place in male couples, while 15 percent of promiscuous matings (between nonmates) are homosexual. Of courtship displays performed by males to birds other than their partner, about a quarter of all arms acts are homosexual, and about 2 percent of courtship bows are same-sex. In one zoo colony consisting of five King Penguins, 2 out of 10 bonds that formed among the birds over a period of nine years were homosexual. Although same-sex matings have not yet been observed in these species in the wild, homosexual courtship has been seen in wild Gentoo Penguins: in one informal survey, 3 out of 13 courtships (23 percent) by Gentoos were same-sex.

Orientation: Some male Humboldt Penguins are exclusively homosexual, remaining with their male partners for their entire lives, or else re-pairing with another male should they lose their original partner. Other males are sequentially bisexual, pairing with a male after having lost one or more previous female mates. Still other males are simultaneously bisexual, engaging in both same-sex and opposite-sex courtship and copulation. Of these, some have a primary heterosexual bond but occasionally engage in homosexual activity with another breeding male: about 47 percent of all same-sex copulations are of this type (as opposed to occurring between bonded partners). In a few cases, the opposite occurs: a male with a primary homosexual pair-bond occasionally participates in a heterosexual copulation. Among King Penguins, birds in same-sex pairs are probably exclusively homosexual for the duration of their pair-bonds (since any eggs that are laid are infertile), and birds exhibit a "preference" for same-sex mates even when unpaired birds of the opposite sex are available. Over the course of their lives, however, most such birds are sequentially bisexual, since following the breakup of a homosexual pair they may go on to form heterosexual pair-bonds and even raise a family. Most Gentoo Penguins that participate in homosexual courtship are probably bisexual, since they court both males and females, albeit with a primary heterosexual orientation (since most go on to breed with birds of the opposite sex). Females that pair with each other are exclusively homosexual for the duration of their bond (which may last for one or both birds' lives); some females pair with a heterosexual mate after the death of their female partner.

Nonreproductive and Alternative Heterosexualities

As noted above, promiscuous matings by heterosexually paired birds are common in Humboldt Penguins: one-third to one-half of all heterosexual copulations are between nonpair members, and courtship of birds other than one's mate is also frequent. Promiscuous courtship and copulation occur in King and Gentoo Penguins as well. Even within pairs, sexual behavior may be nonprocreative. In Humboldts, copulation occurs both early and late in the breeding season, when the chances of fertilization are low or nonexistent, while heterosexual mounts (like homosexual ones) sometimes do not include genital contact or sperm transfer (in both Humboldts and Gentoos). Female Gentoo Penguins sometimes mount their mates (REVERSE mounting), while male Gentoos occasionally masturbate by mounting and copulating with clumps of grass. Males have also been observed trying to copulate with dead Penguins.

Several other variations on the lifetime monogamous pair-bond and nuclear family also occur. About a quarter of all Humboldt male-female pairs divorce, often when the female leaves her mate for another male. Divorce also occurs in 10–50 percent of Gentoo pairs, and in some years no pairs remain together. It is especially common in King Penguins, where only about 30 percent of birds retain the same mate from one season to the next. In addition, some King Penguins abandon their mates during the breeding season, and about 6 percent of chicks are reared by single parents (either abandoned or widowed). Humboldts occasionally form trios consisting of either one male and two females or two females and one male; these make up about 5 percent of all heterosexual bonds. In King Penguins, nonbreeding females may associate with a heterosexual pair and help them raise their chick, who recognizes all three birds as its parents; single parenting is also common. Nonbreeders that aren't part of trios also occasionally feed chicks belonging to other birds, particularly when the chicks are in CRÈCHES. These large nursery groups, sometimes containing thousands of chicks, form while the parents are away. Crèches also occur in Gentoos, where they are often attended by several adult "guardians." During the winter, King parents are often gone for long periods on fishing trips, and chicks may not be fed for weeks or months at a time. As many as 10 percent of them perish from this prolonged fasting and starvation. Some parents abandon their chicks or eggs (especially in severe weather), and chicks may also be killed in squabbles between their parents and nonbreeding birds that are trying to "kidnap" them. King Penguins also occasionally "steal" other pairs' eggs.

Breeding can take its toll on adults as well: male King Penguins fast for more than fifty days during courtship and incubation, losing 10–12 percent of their body weight. In addition, heterosexual copulations are sometimes harassed, with throngs of neighboring birds converging on mating pairs, attacking them and trying to interrupt the sexual activity. Many birds forgo breeding altogether: more than 40 percent of the population each year consists of nonbreeders, and birds generally do not breed every year (primarily because of the unusually long 16–month breeding cycle). Extensive nonbreeding is also a feature of Gentoo populations: up to a quarter of the adults may skip breeding each year, and more than 15 percent of birds breeding late in the season lay infertile clutches. In addition, breeding is delayed for one to two years in younger King and Gentoo Penguins, due to both physiological and social factors. Some Humboldt Penguins remain single and nonbreeding as well, although they may still engage in sexual behavior with other birds.

Other Species

Reciprocal homosexual copulations—involving full genital (cloacal) contact—also occur among male Adélie Penguins (*Pygoscelis adeliae*) in Antarctica, accompanied by courtship displays such as DEEP BOWING and the ARMS ACT. Following ejaculation by the mounter, the mountee contracts his cloaca, perhaps facilitating movement of his partner's semen in his genital tract and/or indicating orgasm. Some males who participate in homosexual activity also mate heterosexually.

Sources   *asterisked references discuss homosexuality/transgender*

Bagshawe, T. W. (1938) "Notes on the Habits of the Gentoo and Ringed or Antarctic Penguins." *Transactions of the Zoological Society of London* 24:185–306.

Bost, C. A., and P. Jouventin (1991) "The Breeding Performance of the Gentoo Penguin *Pygoscelis papua* at the Northern Edge of Its Range." *Ibis* 133:14–25.

*Davis, L. S., F. M. Hunter, R. G. Harcourt, and S. M. Heath (1998) "Reciprocal Homosexual Mounting in Adélie Penguins *Pygoscelis adeliae*." *Emu* 98:136–37.

*Gillespie, T. H. (1932) *A Book of King Penguins*. London: Herbert Jenkins Ltd.

Kojima, I. (1978) "Breeding Humboldt's Penguins *Spheniscus humboldti* at Kyoto Zoo." *International Zoo Yearbook* 18:53–59.

*Merritt, K., and N. E. King (1987) "Behavioral Sex Differences and Activity Patterns of Captive Humboldt Penguins." *Zoo Biology* 6:129–38.

*Murphy, R. C. (1936) *Oceanic Birds of South America*, vol. 1, p.340. New York: American Museum of Natural History.

Olsson, O. (1996) "Seasonal Effects of Timing and Reproduction in the King Penguin: A Unique Breeding Cycle." *Journal of Avian Biology* 27:7–14.

*Roberts, B. (1934) "The Breeding Behavior of Penguins, with Special Reference to *Pygoscelis papua* (Forster)." *British Graham Land Expedition* Science Report 1:195–254.

Schmidt, C. R. (1978) "Humboldt's Penguins *Spheniscus humboldti* at Zurich Zoo." *International Zoo Yearbook* 18:47–52.

*Scholten, C. J. (1996) Personal communication.

*——— (1992) "Choice of Nest-site and Mate in Humboldt Penguins (*Spheniscus humboldti*)." *SPN (Spheniscus Penguin Newsletter)* 5:3–13.

*——— (1987) "Breeding Biology of the Humboldt Penguin *Spheniscus humboldti* at Emmen Zoo." *International Zoo Yearbook* 26:198–204.

*Stevenson, M. F. (1983) "Penguins in Captivity." *Avicultural Magazine* 89:189–203 (reprinted in *International Zoo News* 189 [1985]:17–28).

Stonehouse, B. (1960) "The King Penguin *Aptenodytes patagonica* of South Georgia. I. Breeding Behavior and Development." *Falkland Islands Dependencies Survey Scientific Reports* 23:1–81.

van Zinderen Bakker, E. M., Jr. (1971) "A Behavior Analysis of the Gentoo Penguin *Pygoscelis papua*." In E. M. van Zinderen Bakker Sr., J. M. Winterbottom, and R. A. Dyer, eds., *Marion and Prince Edward Islands: Report on the South African Biological and Geological Expedition*, pp. 251–72. Cape Town: A. A. Balkema.

Weimerskirch, H., J. C. Stahl, and P. Jouventin (1992) "The Breeding Biology and Population Dynamics of King Penguins *Aptenodytes patagonica* on the Crozet Islands." *Ibis* 134:107–17.

*Wheater, R. J. (1976) "The Breeding of Gentoo Penguins *Pygoscelis papua* in Edinburgh Zoo." *International Zoo Yearbook* 16:89–91.

Williams, T. D. (1996) "Mate Fidelity in Penguins." In J. M. Black, ed., *Partnerships in Birds: The Study of Monogamy*, pp. 268–85. Oxford: Oxford University Press.

——— (1995) *The Penguins: Spheniscidae*. Oxford: Oxford University Press.

Williams, T. D., and S. Rodwell (1992) "Annual Variation in Return Rate, Mate and Nest-Site Fidelity in Breeding Gentoo and Macaroni Penguins." *Condor* 94:636–45.

Wilson, R. P., and M.-P. T. Wilson (1990) "Foraging Ecology of Breeding *Spheniscus* Penguins." In L. S. Davis and J. T. Darby, eds., *Penguin Biology*, pp. 181–206. San Diego: Academic Press.

BIRDS OF PREY
AND GAME BIRDS

RAPTORS

(COMMON) **KESTREL**
Falco tinnunculus

(EURASIAN) **GRIFFON VULTURE**
Gyps fulvus

Other Birds 🐦

HOMOSEXUALITY	TRANSGENDER	BEHAVIORS		RANKING	OBSERVED IN
● Female	○ Intersexuality	● Courtship	● Pair-bonding	○ Primary	● Wild
● Male	○ Transvestism	○ Affectionate	○ Parenting	● Moderate	○ Semiwild
		● Sexual		○ Incidental	● Captivity

KESTREL
IDENTIFICATION: A small falcon (12–15 inches) having chestnut plumage spotted with black, and a gray head and tail in males. DISTRIBUTION: Throughout Eurasia and Africa. HABITAT: Variable, including plains, steppe, woodland, wetlands. STUDY AREA: Nivå, Denmark; subspecies *F.t. tinnunculus*.

GRIFFON VULTURE
IDENTIFICATION: A large vulture (wingspan up to 9 feet) with a white head and neck and brown plumage. DISTRIBUTION: Southern Europe, North Africa, Middle East to Himalayas. HABITAT: Mountains, steppe, forest. STUDY AREAS: Berlin Zoo; Jonte Gorge and other regions of the Massif Central Mountains, France; Lumbier, Spain; subspecies *G.f. fulvus*.

Social Organization

During early spring through summer, Kestrels associate as mated pairs that each have their own territory; there is also a significant subpopulation of non-breeding birds. Outside of the mating season, males and females are often segregated from each other and largely solitary: sometimes only one member of a pair—typically the female—migrates, though males that migrate often travel farther than females. During the winter, males and females also tend to occupy separate habitats, with males generally in more wooded areas. Griffon Vultures are much more social and tend to nest in colonies containing 15–20 pairs, sometimes as many as 50–100. As in Kestrels, mated pairs often last for many years.

Description

Behavioral Expression: In Kestrels and Griffon Vultures, two birds of the same sex—usually males—occasionally bond with each other and become a mated pair. Male Kestrels in a homosexual pair often soar together in the early spring, performing dramatic courtship display flights that reinforce their pair-bond (these displays are also found in heterosexual pairs). One such display is the ROCKING FLIGHT, in which the two partners fly at an immense height and rock from side to side, using flicking wingbeats. Another display is the slower WINNOWING FLIGHT, in which the wings beat with shallow, almost vibrating strokes, giving the impression that only the tips are moving or "shivering." Both displays are accompanied by distinctive calls, such as the TSIK CALL—a series of clipped notes sounding like *tsick* or *kit*—and the LAHN CALL, a sequence of high-pitched trills transcribed as *quirrr-rr quirrr-rr*. The two males sometimes display together, or one male might soar while the other sits on his perch. Same-sex partners also copulate with each other, making the distinctive copulation call sounding like *kee-kee-kee* or *kik-kik-kik;* homosexual mounts last for 10–15 seconds (comparable to heterosexual matings).

Male Griffon Vultures in homosexual pairs also mate with each other repeatedly beginning in December (the onset of the mating season), and such pairs may remain together for years. The two males sometimes build a nest together each year—typically a flat assemblage of sticks on a crag, two to three feet across. Like Kestrels, pairs of Griffon Vultures perform a spectacular aerial pair-bonding display called TANDEM FLYING. The two birds spiral upward to a great height on a thermal, then glide downward in a path that will bring them extremely close to each other, "riding" for a few seconds one above the other, until they separate again. Although most tandem flights are by heterosexually paired birds, Vultures of the same sex also engage in this activity.

Frequency: Homosexual pairs probably occur only occasionally in Kestrels, although no systematic study of their frequency has been undertaken. Male pairs of Griffon Vultures have not yet been fully verified in the wild; however, tandem flights between same-sex partners (both males and females) account for about 20 percent of all display flights in the wild, and some of these probably represent homosexual pairings.

Orientation: No detailed studies of the life histories of birds of prey in homosexual pairs have yet been conducted. However, at least some male Griffon Vultures in same-sex tandem flights have female mates, suggesting a possible form of bisexuality, while at least some younger females in same-sex tandems likely have no prior heterosexual experience.

Nonreproductive and Alternative Heterosexualities

In any given year, many Kestrels do not breed: about a third of all birds in some populations are unpaired, while 6–13 percent of heterosexually mated birds do not lay eggs. Male-female pairs of Griffon Vultures, too, may abstain from procreating—some pairs go for as long as eight or nine consecutive years without reproducing. Nonlaying pairs, as well as younger Griffon Vultures that have not yet begun breeding, nevertheless still engage in sexual activity, often mating with each other near the breeding colonies. Several other types of nonprocreative copulations are also prominent in these species. Both Kestrels and Griffon Vultures sometimes mate outside of the breeding season (in the autumn and winter) and during the breeding season when there is no chance of fertilization. This includes during incubation, chick-raising, or very early in the season. Outside of the breeding season, though, Griffon females may refuse to participate in copulations, attacking their male partner when he tries to mount. Kestrel males and females very often live separately during the winter (as discussed above). In addition, heterosexual mating in both species occurs at astonishingly high rates, indicating that it is not simply procreative activity: Griffon heterosexual pairs sometimes mate every half hour, while Kestrels average a copulation once every 45 minutes, or seven to eight times per day during the breeding season. Even higher rates have been recorded for some Kestrels—up to three times per hour—and it is estimated that each Kestrel pair mates as many as 230 times during the breeding season alone. Male Kestrels also sometimes court and attempt to mate with females other than their mate; they usually do not succeed, though, owing to resistance by the female and defense by her mate. Nevertheless, 5–7 percent of all broods contain chicks fathered by a bird other than its mother's mate, and in a few cases none of the nestlings are genetically related to their caretaking father. Nonmonogamous copulations probably also occur in Griffon Vultures.

Alternative heterosexual family arrangements are widespread in Kestrels: up to 10 percent of males in some years have two female mates (they usually each have families in separate nests), while a female sometimes forms a trio with two males. Divorce is fairly common in Kestrels: about 17 percent of females and 6 percent of males change partners between breeding seasons, and sometimes a pair splits during the breeding season as well. In Griffon Vultures the divorce rate is about 5 percent. Some Kestrel males are unable to provide their mates with enough food during incubation, resulting in desertion and loss of the entire clutch (accounting for more than half of all nesting failures). Finally, cannibalism has been documented in these species: Kestrel nestlings sometimes kill and eat their siblings, while parents of both species cannibalize their own chicks on rare occasions (usually if the chick has already died).

Other Species

Same-sex pairing and coparenting have been observed in other birds of prey in captivity. Female Barn Owls (*Tyto alba*) that are raised together occasionally bond with one another, ignoring any available males. They may even nest together, each laying a clutch of infertile eggs that they incubate side by side. Female coparents share parenting duties and can successfully raise foster young. Courtship, pairbonding, nesting, and coparenting of foster chicks have also been documented in a pair of female Powerful Owls (*Ninox strenua*) from Australia. In addition, a pair of male Steller's Sea Eagles (*Haliaeetus pelagicus*)—a species native to Siberia and East Asia—courted one another and built a nest together. They even incubated and hatched another eagle's egg and successfully raised the chick together.

Sources   *asterisked references discuss homosexuality/transgender

Blanco, G., and F. Martinez (1996) "Sex Difference in Breeding Age of Griffon Vultures (*Gyps fulvus*)." *Auk* 113:247–48.

Bonin, B., and L. Strenna (1986) "The Biology of the Kestrel *Falco tinnunculus* in Auxois, France." *Alauda* 54:241–62.

Brown, L., and D. Amadon (1968) "*Gyps fulvus*, Griffon Vulture." In *Eagles, Hawks, and Falcons of the World*, pp. 325–28. New York: McGraw-Hill.

Cramp, S., and K. E. L. Simmons, eds. (1980) "Griffon Vulture (*Gyps fulvus*)" and "Kestrel (*Falco tinnunculus*)." In *Handbook of the Birds of Europe, the Middle East, and North Africa*, vol. 2, pp. 73–81, 289–300. Oxford: Oxford University Press.

*Fleay, D. (1968) *Nightwatchmen of Bush and Plain: Australian Owls and Owl-like Birds*. Brisbane: Jacaranda Press.

*Heinroth, O., and M. Heinroth (1926) "Der Gänsegeier (*Gyps fulvus* Habl.) [The Griffon Vulture]." In *Die Vögel Mitteleuropas*, vol. 2, pp. 66–69. Berlin and Lichterfeld: Bermühler.

*Jones, C. G. (1981) "Abnormal and Maladaptive Behavior in Captive Raptors." In J. E. Cooper and A. G. Greenwood, eds., *Recent Advances in the Study of Raptor Diseases (Proceedings of the International Symposium on Diseases of Birds of Prey, London, 1980)*, pp. 53–59. West Yorkshire: Chiron Publications.

Korpimäki, E. (1988) "Factors Promoting Polygyny in European Birds of Prey—A Hypothesis." *Oecologia* 77:278–85.

Korpimäki, E., K. Lahti, C. A. May, D. T. Parkin, G. B. Powell, P. Tolonen, and J. H. Wetton (1996) "Copulatory Behavior and Paternity Determined by DNA Fingerprinting in Kestrels: Effects of Cyclic Food Abundance." *Animal Behavior* 51:945–55.

Mendelssohn, H., and Y. Leshem (1983) "Observations on Reproduction and Growth of Old World Vultures." In S.R. Wilbur and J. A. Jackson, eds., *Vulture Biology and Management*, pp. 214–41. Berkeley: University of California Press.

*Mouze, M., and C. Bagnolini (1995) "Le vol en tandem chez le vautour fauve (*Gyps fulvus*) [Tandem Flying in the Griffon Vulture]." *Canadian Journal of Zoology* 73:2144–53.

*Olsen, K. M. (1985) "Pair of Apparently Adult Male Kestrels." *British Birds* 78:452–53.

Packham, C. (1985) "Bigamy by the Kestrel." *British Birds* 78:194.

*Pringle, A. (1987) "Birds of Prey at Tierpark Berlin, DDR." *Avicultural Magazine* 93:102–6.

Sarrazin, F., C. Bagnolini, J. L. Pinna, and E. Danchin (1996) "Breeding Biology During Establishment of a Reintroduced Griffon Vulture *Gyps fulvus* Population." *Ibis* 138:315–25.

Stanback, M. T., and W. D. Koenig (1992) "Cannibalism in Birds." In M. A. Elgar and B. J. Crespi, eds., *Cannibalism: Ecology and Evolution Among Diverse Taxa*, pp. 277–98. Oxford: Oxford University Press.

Terrasse, J. F., M. Terrasse, and Y. Boudoint (1960) "Observations sur la reproduction du vautour fauve, du percnoptère, et du gypaète barbu dans les Basses-Pyrénées [Observations on the Reproduction of the Griffon Vulture, the Egyptian Vulture, and the Bearded Vulture in the Lower Pyrenees]." *Alauda* 28:241–57.

Village, A. (1990) *The Kestrel*. London: T. and A. D. Poyser.

GROUSE

Other Birds 🐦

SAGE GROUSE
Centrocercus urophasianus

RUFFED GROUSE
Bonasa umbellus

HOMOSEXUALITY	TRANSGENDER	BEHAVIORS		RANKING	OBSERVED IN
● Female	○ Intersexuality	● Courtship	○ Pair-bonding	○ Primary	● Wild
● Male	○ Transvestism	○ Affectionate	● Parenting	● Moderate	○ Semiwild
		● Sexual		○ Incidental	● Captivity

SAGE GROUSE
IDENTIFICATION: A gray-brown grouse with speckled plumage, pointed tail feathers, and inflatable air sacs in the breast. DISTRIBUTION: Western North America. HABITAT: Sage grassland, semidesert. STUDY AREAS: Green River Basin and Laramie Plains, Wyoming; Long Valley, California.

RUFFED GROUSE
IDENTIFICATION: A large grouse with a banded, fan-shaped tail and distinctive black ruffs on the side of the neck. DISTRIBUTION: Northern and central North America. HABITAT: Forest. STUDY AREA: In captivity in Ithaca, New York.

Social Organization

During the breeding season, male Ruffed and Sage Grouse display on territories—large, communal "strutting grounds" or LEKS in Sage Grouse, and individual "drumming logs" in Ruffed Grouse. Both species have a promiscuous mating system, in which birds mate with multiple partners, do not form pair-bonds, and females care for the young with no male assistance. Outside of the mating season, birds sometimes congregate in mixed-sex flocks.

Description

Behavioral Expression: At dawn on their prairie display grounds, female Sage Grouse gather in groups of eight to ten (or sometimes more) known as CLUSTERS. Although many are there to mate with males, some females also court and mate with each other. Homosexual courtship display is similar to that used by males and is called STRUTTING. The female takes short steps forward and turns while presenting herself in a spectacular posture—long tail feathers fanned in a circle, air sacs in the breast expanded, and neck feathers erected and rustling. Unlike males, however, strutting females do not make the characteristic "plopping" sound during courtship, owing to the smaller size of their air sacs. When the female has finished her strut, another female may solicit a copulation from her by crouching down, arching her wings and fanning the wing feathers on the ground. The other female mounts her and often performs a complete mating sequence, spreading her wings on either side of the mounted female for balance, treading on her back with her feet, and lowering and rotating the tail for cloacal (genital) contact. Some females also chase others in the group and try to mount them, and "pile-ups" of three or four females all mounted on each other sometimes develop. During most lesbian matings,

▲ *Homosexual courtship in Sage Grouse: a female performing the "strutting" display on the plains of Wyoming*

males pay no attention to the females. Sometimes, however, they try to disrupt a lesbian mounting, or they may even try to join in by mounting a female who is herself mounted on another female. In one case, a male even mounted another male who was himself mounted on two females that were mounting each other! Later in the breeding season, two females also sometimes jointly parent their offspring, combining all their chicks into a single brood that they both look after.

Male Ruffed Grouse also court and mount each other. Deep in the forest, each male advertises his presence by DRUMMING at dawn or dusk, producing a throbbing, drumroll-like sound by rapidly "beating" the air with his wings. If another Grouse lands on the drumming log, he begins his strutting display. Fanning his tail like a turkey, he lowers his wings, erects his neck ruffs, and rotates his head vigorously, all the while emitting a hissing sound as he approaches the other bird. If the other bird is a female, mating takes place, whereas if it is another male, typically a fight will ensue. However, in some cases the other male does not challenge the displaying male. The courting male then makes a "gentle" approach, sleeking down his feathers, dragging his tail on the ground, and occasionally shaking his head. He sometimes pecks softly at the base of the other male's bill or places his foot on the other's back and may mount him as in heterosexual copulation (although the mounted bird does not usually adopt the female's typical copulatory posture).

◀ *Homosexual group copulation in Sage Grouse: a female crouching with outstretched wings* (left) *is being mounted by another female* (center, upright bird), *who is simultaneously being mounted by a third female* (right, with its neck extended over the back of the second female). *A fourth female* (foreground, facing front), *is not part of this "pile-up."*

Frequency: Approximately 2–3 percent of Sage Grouse females participate in homosexual courtships and copulations, which occur in perhaps one out of five female visits to the lek (on average); the prevalence of homosexual behavior in wild Ruffed Grouse is not known.

Orientation: Some female Sage Grouse that mate with females do not apparently engage in heterosexual copulation; others alternate between homosexual and heterosexual activity (sometimes within a few hours or minutes). As described above, bisexual "pile-ups" involving birds of the same and opposite sexes mounting each other simultaneously also occur. In Ruffed Grouse, little is known about homosexual activity in the wild, but it is likely that there is a similar combination of bisexuality with occasional exclusive homosexuality. Displaying males that mount other males probably also court and mate with females, while those males who approach displaying males are most likely nondrumming "alternate" males (see below) that do not mate heterosexually.

Nonreproductive and Alternative Heterosexualities

In both of these species, significant portions of the population are nonbreeding. As many as 30 percent of male Ruffed Grouse are nondrummers who do not mate heterosexually, and some birds never breed during their entire lives. In fact, one researcher found that nonbreeders live longer and have a better survival rate than breeders. Many nonbreeders are younger males who have yet to acquire a drumming log; others are ALTERNATE males that tend to associate with another male on his display site without themselves drumming. Still others give up or "abdicate" their display territories and become nonbreeders. Up to 25 percent of female Ruffed Grouse may not nest in any given year, as is true for 20–32 percent of female Sage Grouse in some populations. Moreover, 14–16 percent of female Sage Grouse abandon their nests (especially if they have been disturbed), which means that any eggs or chicks they have will not survive; this also occasionally occurs in Ruffed Grouse. The majority of Sage Grouse copulations are performed by only a small fraction of the male population, and one-half to two-thirds of males never mate at all; during each breeding season, 3–6 percent of females do not ovulate either.

Even among birds that do mate, heterosexual copulation is often complicated by a host of factors: female Sage Grouse may refuse to be mounted, males often ignore females' solicitations to mate (especially later in the breeding season), and 10–18 percent of copulations are disrupted by neighboring males who attack mating birds. In addition, males and females are often physically separated from each other: in both species, typically the only contact the two sexes have with each other during the breeding season is mating. Since each female usually copulates only once, hers is a largely male-free existence. Even on the display grounds, Sage Grouse are typically sex-segregated when not actually mating. Several types of alternative sexual behavior also occur in these species. Male Sage Grouse often "masturbate" by mounting a pile of dirt or a dunghill and performing all the motions of a full copulation. Both male Ruffed and Sage Grouse occasionally court and mate with females of other grouse species. And male Sage Grouse sometimes mount females without attempting to inseminate them (no genital contact). Moreover, even though most females mate only once (that is, the minimum required to fertilize their eggs), multiple copulations also occasionally occur: one female, for example, was mounted more than 22 times in one hour. Female Sage Grouse sometimes combine their youngsters into what is known as a GANG BROOD, a communal "nursery flock" of sorts.

Other Species

Homosexual activity occurs in several species of pigeons. Feral Rock Doves (*Columba livia*), for example, form both male and female same-sex pairs that engage in a full suite of courtship, pair-bonding, sexual, and nesting activities. Homosexual pairs of female Ring Doves (*Streptopelia risoria*) in captivity are generally more devoted incubators than heterosexual pairs, being less likely to abandon their eggs.

Sources   *asterisked references discuss homosexuality/transgender*
*Allen, A. A. (1934) "Sex Rhythm in the Ruffed Grouse (*Bonasa umbellus* Linn.) and Other Birds." *Auk* 51:180–99.
*Allen, T. O., and C. J. Erickson (1982) "Social Aspects of the Termination of Incubation Behavior in the Ring Dove (*Streptopelia risoria*)." *Animal Behavior* 30:345–51.
Bergerud, A. T., and M. W. Gratson (1988) "Survival and Breeding Strategies of Grouse." In A. T. Bergerud and M. W. Gratson, eds., *Adaptive Strategies and Population Ecology of Northern Grouse*, pp. 473–577. Minneapolis: University of Minnesota Press.
*Brackbill, H. (1941) "Possible Homosexual Mating of the Rock Dove." *Auk* 58:581.
*Gibson, R. M., and J. W. Bradbury (1986) "Male and Female Mating Strategies on Sage Grouse Leks." In D. I. Rubenstein and R. W. Wrangham, eds., *Ecological Aspects of Social Evolution*, pp. 379–98. Princeton: Princeton University Press.
Gullion, G. W. (1981) "Non-Drumming Males in a Ruffed Grouse Population." *Wilson Bulletin* 93:372–82.
——— (1967) "Selection and Use of Drumming Sites by Male Ruffed Grouse." *Auk* 84:87–112.
Hartzler, J. E. (1972) "An Analysis of Sage Grouse Lek Behavior." Ph.D. thesis, University of Montana.
Hartzler, J. E., and D. A. Jenni (1988) "Mate Choice by Female Sage Grouse." In A. T. Bergerud and M. W. Gratson, eds., *Adaptive Strategies and Population Ecology of Northern Grouse*, pp. 240–69. Minneapolis: University of Minnesota Press.
Johnsgard, P. A. (1989) "Courtship and Mating." In S. Atwater and J. Schnell, eds., *Ruffed Grouse*, pp. 112–17. Harrisburg, Pa.: Stackpole Books.
*Johnston, R. F., and M. Janiga (1995) *Feral Pigeons*. New York: Oxford University Press.
Lumsden, H. G. (1968) "The Displays of the Sage Grouse." *Ontario Department of Lands and Forests Research Report* (Wildlife) 83:1–94.
*Patterson, R. L. (1952) *The Sage Grouse in Wyoming*. Denver: Sage Books.
Schroeder, M. A. (1997) "Unusually High Reproductive Effort by Sage Grouse in a Fragmented Habitat in North-Central Washington." *Condor* 99:933–41.
*Scott, J. W. (1942) "Mating Behavior of the Sage Grouse." *Auk* 59:477–98.
Simon, J. R. (1940) "Mating Performance of the Sage Grouse." *Auk* 57:467–71.
Wallestad, R. (1975) *Life History and Habitat Requirements of Sage Grouse in Central Montana*. Helena: Montana Department of Fish and Game.
*Wiley, R. H. (1973) "Territoriality and Non-Random Mating in Sage Grouse, *Centrocercus urophasianus*." *Animal Behavior Monographs* 6:87–169.

HUMMINGBIRDS, WOODPECKERS, AND OTHERS

HUMMINGBIRDS

Other Birds

LONG-TAILED HERMIT HUMMINGBIRD
Phaethornis superciliosus
ANNA'S HUMMINGBIRD
Calypte anna

HOMOSEXUALITY	TRANSGENDER	BEHAVIORS		RANKING	OBSERVED IN
○ Female	○ Intersexuality	● Courtship	○ Pair-bonding	○ Primary	● Wild
● Male	○ Transvestism	○ Affectionate	○ Parenting	● Moderate	○ Semiwild
		● Sexual		○ Incidental	○ Captivity

LONG-TAILED HERMIT HUMMINGBIRD
IDENTIFICATION: A medium-sized hummingbird with purplish or greenish bronze upperparts, a striped face, a long, downward-curving bill, and elongated tail feathers. DISTRIBUTION: Southwestern Mexico, Central America, northwestern South America. HABITAT: Tropical forest undergrowth. STUDY AREA: La Selva Biological Reserve, Sarapiquí, Costa Rica.

ANNA'S HUMMINGBIRD
IDENTIFICATION: A medium-sized hummingbird (up to 4 inches long) with an iridescent, rose-colored throat and crown (in males), and a bronze-green back. DISTRIBUTION: Western United States to northwestern Mexico. HABITAT: Woodland, chaparral, scrub, meadows. STUDY AREA: Franklin Canyon, Santa Monica Mountains, California.

Social Organization

Long-tailed Hermit Hummingbirds form singing assemblies or LEKS composed of about a dozen males and have a polygamous or promiscuous mating system (in which birds mate with multiple partners). Anna's Hummingbirds are not particularly social: each bird defends its own territory and does not generally associate with others. No pair formation occurs as part of the mating system; instead, males and probably also females mate with several different partners.

Description

Behavioral Expression: Male Long-tailed Hermit Hummingbirds gather on their leks or courtship display territories in dense, stream-side thickets, singing to advertise their presence and attract birds to mate with. Their monotonous songs consist of single notes of various types—sometimes transliterated as *kaching, churk, shree,* or *chrrik*—repeated for up to 30 minutes at a time. Females and males visit the leks, and both sexes may be courted and mounted by the territorial males. In a typical homosexual encounter, a male approaches another male that has landed on his territory and performs an aerial maneuver known as the FLOAT. In this display, he slowly flies back and forth in front of the perched male, pivoting his body from side to side. Often he holds his bill wide open, exposing his bright orange mouth lining and striking facial stripes, which combine to produce an arresting visual pattern. The perched bird may respond by gaping his own bill and "tracking" the movements of the swiveling and hovering male in front of him, always keeping his bill pointed toward him. The courting male then circles behind the other male and copulates with him: he alights on the other male's back, quivering his wings while twisting and vibrating his tail to achieve cloacal (genital) contact. Homosexual copulations are generally somewhat briefer than the three-to-five-second duration of heterosexual matings, and the mountee may fail to cooperate (for example by not twisting his own tail to facilitate genital contact).

Male Anna's Hummingbirds also court and mount both females and males (including juvenile males). These birds usually visit the male's territory to feed on his supply of nectar-rich currant and gooseberry blossoms. If a visiting male lands on a perch, the territorial male usually performs a spectacular DIVE DISPLAY toward him. He first hovers above the other male and utters a few *bzz* notes, then climbs nearly vertically in a wavering path of 150 feet or more, peering down at the other male. At the top of his climb, he suddenly dives downward at immense speed, making a shrill, metallic popping or squeaking sound just as he swoops over the other male. He then repeats the entire performance several more times. The startlingly loud sound at the end of his dives is produced by air rushing through his tail feathers and is often preceded by vocalizations such as various trilled or buzzing notes. A dive-bombing male actually orients his acrobatic display precisely to face the sun, dazzling the object of his attentions with the shimmering, iridescent, rose-colored feathers of his crown and

A male Long-tailed Hermit Hummingbird (right) *courting another male with the "float" display* ▶

throat. On cloudy days, he rarely performs such dives since the mesmerizing visual effect cannot be achieved. After a dive display the other male usually flies off—with the territorial male in close pursuit—and seeks refuge by perching in a low clump of vegetation away from the territory. The pursuing male sings intensely at him, uttering a loud and complex sequence of notes that sounds like *bzz-bzz-bzz chur-ZWEE dzi! dzi! bzz-bzz-bzz.* He may also perform a SHUTTLE DISPLAY (similar to the Long-tailed Hermit's float), flying back and forth above the other male, tracing a series of arcs with his body. A homosexual copulation attempt may then follow, with the male landing on the other's back as in a heterosexual mount. If the mounted male tries to get away, the pursuing male may knock him down, grappling and tumbling with him while emitting low-pitched, gurgling *brrrt* notes (similar aggressive interactions are also characteristic of heterosexual mating attempts; see below).

Frequency: Although homosexual copulations are not frequent in these species, neither are heterosexual ones, and a relatively high proportion of sexual activity—up to 25 percent—actually occurs between males. During several extensive studies, two out of eight observed copulations in Long-tailed Hermits were between males, while one out of four sexual encounters in Anna's Hummingbirds (where the sexes of the birds could reliably be determined) was homosexual. Moreover, when male Anna's Hummingbirds are presented with stuffed birds of both sexes, they court and mount the males as frequently as they do the females.

Orientation: In Long-tailed Hermit Hummingbirds, approximately 7 percent of territorial males and 11 percent of all males participate in homosexual activity. Territorial males in both of these hummingbird species are probably bisexual, pursuing, courting, and mounting both females and males. Some of the male Long-tailed Hermits who visit other males' territories are nonbreeders (they do not have their own territories), which means they probably do not participate in any heterosexual activity (at least for the duration of that breeding season). Male Anna's Hummingbirds usually strongly resist being mounted by other males, perhaps indicating a more heterosexual orientation on their part (although females also sometimes resist heterosexual mating attempts).

Nonreproductive and Alternative Heterosexualities

Heterosexual mating in Anna's Hummingbirds can have all of the aggressive and even violent characteristics described above for homosexual matings—males pursue females in high-speed chases and sometimes even strike them in midair, forcing them down in order to copulate. Some matings are also nonreproductive since they take place outside of the breeding season. Males in this species have their own distinct seasonal sexual cycle, with their sperm production and hormone levels greatly reduced from July through November. Male Anna's Hummingbirds also frequently court females of other species such as the Allen hummingbird (*Selasphorus sasin*) and Costa's hummingbird (*Calypte costae*). Among Long-tailed Hermit Hummingbirds (as well as other species of hermit hummingbirds), males often "masturbate" by mounting and copulating with small, inanimate objects (including leaves suspended in spiderwebs).

Other than when mating, however, males and females in both of these species rarely meet. In Anna's Hummingbirds, the two sexes occupy distinct habitats during the breeding season—males frequent open areas such as hill slopes or the sides of canyons, females occupy more covered, forested areas. Each female Long-tailed Hermit usually encounters males only once every two to four weeks when she visits the lekking areas prior to nesting. Males of both species take no part in nesting or raising of young. In addition, a significant number of birds are nonbreeders: nearly a quarter of all Long-tailed Hermit males are nonterritorial and therefore do not participate in heterosexual courtship or copulation, while of those who hold territories, only some get to mate with females.

Sources   *asterisked references discuss homosexuality/transgender

Gohier, F., and N. Simmons-Christie (1986) "Portrait of Anna's Hummingbird." *Animal Kingdom* 89:30–33.

Hamilton, W. J., III (1965) "Sun-Oriented Display of the Anna's Hummingbird." *Wilson Bulletin* 77:38–44.

*Johnsgard, P. A. (1997) "Long-tailed Hermit" and "Anna Hummingbird." In *The Hummingbirds of North America,* 2nd ed., pp. 65–69, 195–99. Washington, D.C.: Smithsonian Institution Press.

Ortiz-Crespo, F. I. (1972) "A New Method to Separate Immature and Adult Hummingbirds." *Auk* 89:851–57.

Russell, S. M. (1996) "Anna's Hummingbird (*Calypte anna*)." In A. Poole and F. Gill, eds., *The Birds of North America: Life Histories for the 21st Century,* no. 226. Philadelphia: Academy of Natural Sciences; Washington, D.C.: American Ornithologists' Union.

Snow, B. K. (1974) "Lek Behavior and Breeding of Guy's Hermit Hummingbird *Phaethornis guy*." *Ibis* 116:278–97.

——— (1973) "The Behavior and Ecology of Hermit Hummingbirds in the Kanaku Mountains, Guyana." *Wilson Bulletin* 85:163–77.

Stiles, F. G. (1983) "*Phaethornis superciliosus*." In D. H. Janzen, ed., *Costa Rican Natural History,* pp. 597–599. Chicago: University of Chicago Press.

*——— (1982) "Aggressive and Courtship Displays of the Male Anna's Hummingbird." *Condor* 84:208–25.

*Stiles, F. G., and L. L. Wolf (1979) *Ecology and Evolution of Lek Mating Behavior in the Long-tailed Hermit Hummingbird.* Ornithological Monographs no. 27. Washington, D.C.: American Ornithologists' Union.

Tyrell, E. Q., and R. A. Tyrell (1985) *Hummingbirds: Their Life and Behavior.* New York: Crown Publishers.

Wells, S., and L. F. Baptista (1979) "Displays and Morphology of an Anna × Allen Hummingbird Hybrid." *Wilson Bulletin* 91:524–32.

Wells, S., L. F. Baptista, S. F. Bailey, and H. M. Horblit (1996) "Age and Sex Determination in Anna's Hummingbird by Means of Tail Pattern." *Western Birds* 27:204–6.

Wells, S., R. A. Bradley, and L. F. Baptista (1978) "Hybridization in *Calypte* Hummingbirds." *Auk* 95:537–49.

Williamson, F. S. L. (1956) "The Molt and Testis Cycle of the Anna Hummingbird." *Condor* 58:342–66.

BLACK-RUMPED FLAMEBACK
Dinopium benghalense

ACORN WOODPECKER
Melanerpes formicivorus

HOMOSEXUALITY	TRANSGENDER	BEHAVIORS		RANKING	OBSERVED IN
● Female	○ Intersexuality	● Courtship	○ Pair-bonding	○ Primary	● Wild
● Male	○ Transvestism	○ Affectionate	● Parenting	● Moderate	○ Semiwild
		● Sexual		○ Incidental	○ Captivity

BLACK-RUMPED FLAMEBACK
IDENTIFICATION: A medium-sized, red-crested woodpecker with a golden back, black rump, and black-and-white patterning on the face and neck. DISTRIBUTION: India, Pakistan, Sri Lanka. HABITAT: Woodland, scrub, gardens. STUDY AREA: Near Chittur, India; subspecies *D.b. puncticolle.*

ACORN WOODPECKER
IDENTIFICATION: A red-capped woodpecker with a striking black-and-white face, black upperparts, white underparts, and a black breast band. DISTRIBUTION: Pacific and southwest United States, Mexico through Colombia. HABITAT: Oak and pine woodland. STUDY AREAS: Hastings Natural History Reservation (Monterey) and near Los Altos, California.

Social Organization

Acorn Woodpeckers have an extraordinarily varied and complex social organization. In many populations, birds live in communal family groups containing up to 15 individuals—typically there are as many as 4 breeding males and 3 breeding females in such groups (though nonbreeding groups also occur—see below). The remaining birds in a group are nonbreeding "helpers" that may share in the parenting duties. Within groups, the mating system is known as POLYGYNANDRY, that is, each male mates and bonds with several females and vice versa. In other populations, monogamous pairs as well as other variations on polygamy occur. Little is known about the social organization of Black-rumped Flamebacks, although it is thought that they form monogamous mated pairs.

Description

Behavioral Expression: Male Black-rumped Flamebacks sometimes copulate with each other. One male mounts the back of the other (as in heterosexual mating), bending his tail down and thrusting it under the belly of the other male to make cloacal (genital) contact. Reciprocal mounting may occur, in which a male copulates with a male that has just mounted him. A bird involved in same-sex mounting activity may adopt a distinctive posture, in which his body is perpendicular to the branch he is perching on and his wing tips are arched toward the ground and hanging below his feet. Males also sometimes drum against a tree trunk prior to homosexual mounting.

Acorn Woodpeckers participate in a fascinating group display that involves ritualized sexual and courtship behavior, including homosexual mounting. At dusk, the members of a group gather together prior to roosting in their tree holes. As more and more birds arrive, they begin mounting each other in all combinations—males mount females and other males, females mount males and other females, young Woodpeckers mount older ones and vice versa. The mounting behavior resembles heterosexual mating, except it is usually briefer and cloacal contact is generally not involved (although genital contact does sometimes occur). Reciprocal mountings are common, and sometimes two Woodpeckers will try to mount the same bird simultaneously. Following the display, group members fly off to their roost holes to sleep. Ritualized mounting may also occur at dawn when the birds emerge from their roost holes. Because many group members are related to each other, at least some of this mounting is incestuous. Female Acorn Woodpeckers often coparent together, both laying eggs in the same nest cavity. Such "joint nesters" are often related (mother and daughter, or sisters), but sometimes two unrelated females nest and parent together as well. Joint-nesting females may continue to associate even if they happen not to breed in a particular year.

Frequency: Homosexual behavior in Black-rumped Flamebacks probably occurs only occasionally; however, heterosexual mating has never been observed in this species in the wild, so much remains to be learned about the behavior of this Woodpecker. The mounting display of Acorn Woodpeckers—including homosexual mounting—is a regular feature of the social life of this species at all times of the year, occurring daily in most groups. More than a third of all female Acorn Woodpeckers nest jointly, and about a quarter of all groups have joint-nesting females; 14 percent of these joint nests involve unrelated females as coparents.

Orientation: To the extent that they mount both males and females in the group display, Acorn Woodpeckers are bisexual (although it must be remembered that such mounting is often ritualized, i.e., it may not always involve genital contact). Not enough is known about the life histories of individual Black-rumped Flamebacks to make any generalizations about their sexual orientation.

Nonreproductive and Alternative Heterosexualities

As described above, Acorn Woodpeckers have an unusual communal family organization that can involve different forms of polygamy. In addition, many birds are nonbreeding: more than a third of all groups may not reproduce in a given year, and one-quarter to one-half of all adult birds do not procreate. In some populations the proportion of nonbreeders may be as high as 85 percent. Many of these are birds who remain with their family group for several years after they become sexually mature, helping their parents raise young; some delay reproducing for three or four years. Other nonbreeders (as many as one-quarter) do not in any way help to raise young. Some groups are nonreproductive because all their adult members are of the same sex: nearly 15 percent of nonbreeding groups have no adult females and nearly 4 percent have no adult males. In addition to the nonprocreative heterosexual behaviors mentioned above (REVERSE mounting, group sexual activity, mounting without genital contact), female Acorn Woodpeckers also sometimes copulate with more than one male in quick succession. About 3 percent of families contain offspring that result from promiscuous matings with males outside the group. Incestuous heterosexual matings occasionally occur as well, although they seem to be avoided—in fact, incest avoidance may lead to a group's forgoing breeding for an extended time. Parenting in this species is notable for a variety of counterreproductive and violent behaviors. Egg destruction is common—particularly among joint-nesting females, who often break (and eat) each other's and their own eggs until they begin laying synchronously. Males also sometimes destroy eggs of their own group. In addition, infanticide and cannibalism occur regularly in Acorn Woodpeckers. A common pattern seems to be for a new bird in a group—often a female—to peck the nestlings to death and eat some of them in order to breed with the other adults in the group. Parents also regularly starve any chicks that hatch later than a day after the others do.

Sources   *asterisked references discuss homosexuality/transgender

*Koenig, W. D. (1995–96) Personal communication.

Koenig, W. D., and R. L. Mumme (1987) *Population Ecology of the Cooperatively Breeding Acorn Woodpecker*. Princeton, N.J. : Princeton University Press.

Koenig, W. D., R. L. Mumme, M. T. Stanback, and F. A. Pitelka (1995) "Patterns and Consequences of Egg Destruction Among Joint-Nesting Acorn Woodpeckers." *Animal Behavior* 50:607–21.

Koenig, W. D., and P. B. Stacey (1990) "Acorn Woodpeckers: Group-Living and Food Storage Under Contrasting Ecological Conditions." In P. B. Stacey and W. D. Koenig, eds., *Cooperative Breeding in Birds: Long-Term Studies of Ecology and Behavior*, pp. 415–53. Cambridge: Cambridge University Press.

Koenig, W. D., and F. A. Pitelka (1979) "Relatedness and Inbreeding Avoidance: Counterploys in the Communally Nesting Acorn Woodpecker." *Science* 206:1103–5.

*MacRoberts, M. H., and B. R. MacRoberts (1976) *Social Organization and Behavior of the Acorn Woodpecker in Central Coastal California*. Ornithological Monographs no. 21. Washington, D.C.: American Ornithologists' Union.

Mumme, R. L., W. D. Koenig, and F. A. Pitelka (1988) "Costs and Benefits of Joint Nesting in the Acorn Woodpecker." *American Naturalist* 131:654–77.

——— (1983) "Reproductive Competition in the Communal Acorn Woodpecker: Sisters Destroy Each Other's Eggs." *Nature* 306:583–84.

Mumme, R. L., W. D. Koenig, R. M. Zink, and J.A. Marten (1985) "Genetic Variation and Parentage in a California Population of Acorn Woodpeckers." *Auk* 102:305–12.

*Neelakantan, K. K. (1962) "Drumming by, and an Instance of Homo-sexual Behavior in, the Lesser Goldenbacked Woodpecker (*Dinopium benghalense*)." *Journal of the Bombay Natural History Society* 59:288–90.

Short, L.L. (1982) *Woodpeckers of the World*. Delaware Museum of Natural History Monograph Series no. 4. Greenville, Del.: Delaware Museum of Natural History.

——— (1973) "Habits of Some Asian Woodpeckers (Aves, Pisidae)." *Bulletin of the American Museum of Natural History* 152:253–364.

Stacey, P. B. (1979) "Kinship, Promiscuity, and Communal Breeding in the Acorn Woodpecker." *Behavioral Ecology and Sociobiology* 6:53–66.

Stacey, P. B., and T. C. Edwards, Jr. (1983) "Possible Cases of Infanticide by Immigrant Females in a Group-breeding Bird." *Auk* 100:731–33.

Stacey, P. B., and W. D. Koenig (1984) "Cooperative Breeding in the Acorn Woodpecker." *Scientific American* 251:114–21.

Stanback, M. T. (1994) "Dominance Within Broods of the Cooperatively Breeding Acorn Woodpecker." *Animal Behavior* 47:1121–26.

*Troetschler, R. G. (1976) "Acorn Woodpecker Breeding Strategy as Affected by Starling Nest-Hole Competition." *Condor* 78:151–65.

Winkler, H., D. A. Christie, and D. Nurney (1995) "Black-rumped Flameback (*Dinopium benghalense*)." In *Woodpeckers: A Guide to the Woodpeckers of the World,* pp. 375–77. Boston: Houghton Mifflin.

KINGFISHERS AND ROLLERS   *Other Birds*

PIED KINGFISHER
Ceryle rudis

BLUE-BELLIED ROLLER
Coracias cyanogaster

HOMOSEXUALITY	TRANSGENDER	BEHAVIORS		RANKING	OBSERVED IN
● Female	○ Intersexuality	○ Courtship	● Pair-bonding	○ Primary	● Wild
● Male	○ Transvestism	○ Affectionate	○ Parenting	○ Moderate	○ Semiwild
		● Sexual		● Incidental	○ Captivity

PIED KINGFISHER
IDENTIFICATION: A robin-sized, crested bird with speckled black-and-white plumage and a long bill. DISTRIBUTION: Sub-Saharan Africa, the Middle East, India, Southeast Asia. HABITAT: Lakes and rivers. STUDY AREA: Basse Casamance region of Senegal; subspecies *C.r. rudis.*

BLUE-BELLIED ROLLER
IDENTIFICATION: A stocky, 14-inch bird with dark blue plumage, a long, turquoise, forked tail, and a creamy white head and breast. DISTRIBUTION: West Africa. HABITAT: Savanna woodland. STUDY AREA: Basse Casamance region of Senegal.

Social Organization

Pied Kingfishers sometimes gather in flocks of 80 or more birds, and outside of the mating season they associate in small groups. Breeding birds form monogamous pairs, but there is a large population of nonbreeding males as well, many of whom help heterosexual pairs raise their young. Blue-bellied Rollers live in pairs or small groups of 3–13 birds, which are probably extended families or clans; mating may occur promiscuously among several group members.

Description

Behavioral Expression: In Pied Kingfishers, two males sometimes develop a pair-bond and may engage in homosexual mounting and copulation attempts. Homosexual mounting can also occur among males that are not bonded to each other. In all cases, homosexual activity is found among nonbreeding males, of which there are several distinct categories. Some males are HELPERS, who assist heterosexual pairs in raising their young. There are two types of such helpers: PRIMARY helpers, adult birds who help their parents; and SECONDARY helpers, who are unrelated to the pairs they help. In addition, some nonbreeding birds are nonhelpers, who do not assist heterosexual pairs at all. Homosexual pairing probably occurs mostly in the latter group, since primary helpers are devoted to assisting their parents and are also often hostile toward secondary helpers, openly attacking and fighting with them. Some homosexual behavior may also take place among secondary helpers, although this is less likely, since such males are usually preoccupied with feeding females in the pairs they assist (though their parenting duties are usually less extensive than those of primary helpers).

A remarkable form of ritualized sexual behavior occurs among Blue-bellied Rollers, and in some cases the participating birds are of the same sex. One bird mounts the other as in regular copulation, beating its wings and sometimes grabbing in its bill the neck or head feathers of its partner. The mounter lowers its tail while the mountee droops its wings and raises its tail, in some cases achieving cloacal (genital) contact. In almost three-quarters of the cases, mounting is reciprocal (the mountee becoming the mounter and vice versa); reciprocal mounting may be more common between birds of the opposite sex, however. Sometimes, mounting with exchange of positions is performed repeatedly, with as many as 28 mounts alternating between the partners in succession. This mounting behavior is often a ritualized display performed for other birds, and sometimes the tail movements and other gestures characteristic of full sexual behavior are more stylized or attenuated. Mounting may be accompanied by a number of dramatic aerial displays (often considered signs of aggression), including acrobatic chases, SOARS (rapid ascents with wings angled in a V-shape, just prior to being "caught" by a pursuing bird), and swoops (breathtaking plummets with folded wings). Birds may also utter loud, mechanical-sounding RATTLES as well as screaming RASP notes during mounting or the associated aerial displays.

Frequency: Homosexual bonding and mounting probably occur only occasionally among Pied Kingfishers. Ritual mounting behavior is common among Blue-bellied Rollers, occurring throughout the year; the exact proportion of mounting that is same-sex, however, is not known.

Orientation: In some populations of Pied Kingfishers, about 30 percent of the birds are neither breeders nor helpers, while about 18 percent are secondary helpers—these are the segments in which male homosexual activity is found, although probably only a fraction of these birds are involved. Although secondary helpers often go on to mate heterosexually, it is not known whether the same is true of nonhelpers or birds that participate in homosexual activity. However, because of the relatively short life span (one to three years) and high mortality rate of this species, it is likely that at least some males are involved in homosexual activity for most of their lives without ever mating heterosexually.

Nonreproductive and Alternative Heterosexualities

As discussed above, there is a large segment of nonbreeders in the Pied King-fisher population: as many as 45–60 percent of males do not mate heterosexually. Remarkably, studies have shown that the reproductive systems of primary helpers are actually physiologically suppressed, since they have reduced male hormone lev-els, small testes, and no sperm production. Only one in three primary helpers goes on to mate after being a helper, and it is likely that some never breed for their entire lives. In contrast, secondary helpers do not have dormant reproductive systems, but are in most cases simply unable to find female mates due to the greater proportion of males in most populations. Because secondary helpers are not genetically related to the birds they assist, a large number of Pied Kingfishers are involved in "foster-parenting."

Sexual activity between male and female Blue-bellied Rollers is notable for its nonreproductive components: it occurs at all times of the year (not just during the breeding season), and it often involves nonprocreative REVERSE mounts or mount-ing without genital contact. In addition, multiple copulations—far in excess of what is required for fertilization—are common. Not only do birds mount each other repeatedly in a single session (dozens of times, as mentioned above), but both males and females may copulate with many partners, sometimes several times each with up to three birds in a row.

Sources   *asterisked references discuss homosexuality/transgender*

Douthwaite, R. J. (1978) "Breeding Biology of the Pied Kingfisher *Ceryle rudis* in Lake Victoria." *Journal of the East African Natural History Society* 166:1–12.

Dumbacher, J. (1991) Review of Moynihan (1990). *Auk* 108:457–58.

Fry, C. H., and K. Fry (1992) *Kingfishers, Bee-eaters, and Rollers*. London: Christopher Helm.

*Moynihan, M. (1990) *Social, Sexual, and Pseudosexual Behavior of the Blue-bellied Roller*, Coracias cyanogaster: The Consequences of Crowding or Concentration. Smithsonian Contributions to Zoology 491. Washington, D.C.: Smithsonian Institution Press.

Reyer, H.-U. (1986) "Breeder-Helper-Interactions in the Pied Kingfisher Reflect the Costs and Benefits of Cooperative Breeding." *Behavior* 82:277–303.

——— (1984) "Investment and Relatedness: A Cost/Benefit Analysis of Breeding and Helping in the Pied Kingfisher (*Ceryle rudis*)." *Animal Behavior* 32:1163–78.

——— (1980) "Flexible Helper Structure as an Ecological Adaptation in the Pied Kingfisher (*Ceryle rudis rudis* L.)." *Behavioral Ecology and Sociobiology* 6:219–27.

Reyer, H.-U., J. P. Dittami, and M. R. Hall (1986) "Avian Helpers at the Nest: Are They Psychologically Cas-trated?" *Ethology* 71:216–28.

Thiollay, J.-M. (1985) "Stratégies adaptatives comparées des Rolliers (*Coracias* sp.) sédentaires et migra-teurs dans une Savane Guinéenne [Comparative Adaptive Strategies of Sedentary and Migratory Rollers in a Guinean Savanna]." *Revue d'Écologie* 40:355–78.

PARROTS

Other Birds

GALAH or ROSEATE COCKATOO
Eolophus roseicapillus
PEACH-FACED LOVEBIRD
Agapornis roseicollis
ORANGE-FRONTED PARAKEET
Aratinga canicularis

HOMOSEXUALITY	TRANSGENDER	BEHAVIORS		RANKING	OBSERVED IN
● Female	○ Intersexuality	● Courtship	● Pair-bonding	● Primary	● Wild
● Male	○ Transvestism	● Affectionate	● Parenting	○ Moderate	○ Semiwild
		● Sexual		○ Incidental	● Captivity

GALAH
IDENTIFICATION: A medium-sized parrot (about 14 inches) with a pale pink forehead and crest, rose-pink face and underparts, and gray upperparts. DISTRIBUTION: Interior Australia. HABITAT: Savanna woodland, grassland, scrub. STUDY AREAS: Healesville Sanctuary and Monash University, Victoria, Australia; Helena Valley, Western Australia.

PEACH-FACED LOVEBIRD
IDENTIFICATION: A small parrot (6 inches) with a short tail, green plumage, blue rump, and a red or pink breast and face. DISTRIBUTION: Southwestern Africa. HABITAT: Savanna. STUDY AREAS: Cornell University, New York; University of Bielefeld, Germany.

ORANGE-FRONTED PARAKEET
IDENTIFICATION: A small parrot with green plumage, a long tail, and an orange forehead. DISTRIBUTION: Western Central America from Mexico through Costa Rica. HABITAT: Tropical and scrub forests. STUDY AREAS: Near Managua, Nicaragua; University of Kansas and University of California—Los Angeles; subspecies *A.c. canicularis* and *A.c. eburnirostrum*.

Social Organization

Galahs and Peach-faced Lovebirds are gregarious birds, gathering in large flocks that can number up to several hundred in Lovebirds and up to a thousand in Galahs. They typically form mated pairs, and Peach-faced Lovebirds usually nest in colonies. In addition, there are nomadic flocks of juveniles and younger nonbreeding adult Galahs. Orange-fronted Parakeets are also highly social, traveling in groups of 12–15 birds (often composed of mated pairs) and sometimes forming flocks of 50–200. During the breeding season, pairs generally separate from the flock to nest, although they periodically recongregate in small groups.

Description

Behavioral Expression: Galahs of both sexes form stable, long-lasting homosexual pairs that participate in courtship, sexual, and pair-bonding activities.

Same-sex bonds are strong, often developing in juvenile birds and then continuing for the rest of their lives (as do most heterosexual bonds). Homosexual pairs of at least six years' duration have been documented in captivity. If one partner dies before the other, the remaining Galah may stay single or may eventually form a homosexual (or heterosexual) partnership with another bird. Pair-bonded Galahs almost always stay close to each other (rarely more than a few inches apart), feeding and roosting together both day and night. When one bird flies to a new location, it calls after its mate to join it, using a special two-syllable warbled call that sounds like *sip-sip* or *lik-lik*. If another bird intrudes between them, both partners threaten the intruder and may force it to leave by edging it out or simultaneously stabbing their beaks at it. Pair-mates spend considerable time preening each other; this intimate behavior, sometimes known as ALLOPREENING, involves one bird lowering its head in front of the other, allowing its mate to gently nibble and run its bill through the feathers. After a short time the birds switch positions, and often this develops into a playful fencing bout, in which the birds gently clash beaks and dodge each other.

Homosexual (and heterosexual) Galah pairs also perform a number of synchronized, highly stylized displays while perching side by side or facing toward or away from each other. One of the most elegant of these is WING-STRETCHING, in which each bird simultaneously fans open one of its wings. Often, one bird fans its left wing while the other opens its right to give a strikingly symmetrical, "mirror-image" effect, while in other cases each pair member fans the same wing in a parallel, but nonsymmetrical, pattern. Other synchronized displays include HEAD-BOBBING (in which the birds dip their heads down and to the side) as well as crest-raising and feather ruffling. In addition, such activities as self-preening, feeding, and leaf- and bark-stripping can also be performed in unison by pair members—in fact, homosexual pairs synchronize their behaviors about 65 percent of the time. Galahs in same-sex pairs may also court and copulate with each other. Courtship includes a sideways shuffling movement toward the partner with crest raised and facial feathers fanned forward, followed by head-bobbing and BREAST POINTING (in which mates touch their own or their partner's breast feathers with their beaks). Sexual activity involves one bird mounting the other and making pelvic thrusts against its mate; this may occur even when the birds are still juveniles.

Peach-faced Lovebirds also sometimes form stable homosexual pairs; as in Galahs, these are probably lifelong bonds that usually originate

A homosexual pair of female Galahs performing a mirror-image "wing-stretching" display ▶

▲ *A male Orange-fronted Para-keet nuzzling his male pair-mate*

while the birds are still youngsters. Same-sex pairs of Lovebirds also engage in frequent mutual preening. In their courtship and sexual activity, some homosexual pairs combine elements of male and female behavior. In female pairs, for example, each partner may feed the other (a typically male activity in heterosexual pairs) or invite the other to mount (a typically female activity). Other pairs are more role-differentiated, with one bird performing the behaviors most typically associated with males while the other exhibits the patterns of a female. However, in their parenting behaviors both members of female homosexual pairs adopt typically "female" duties. After having investigated several potential nest sites, they jointly select a suitable cavity that they occupy together, and each female contributes to building the nest. In Peach-faced Lovebirds, this involves a unique method of collecting nesting material: long strips of bark, grass, or leaves are tucked directly into the birds' back and rump feathers to be carried back to the nest. Both partners lay eggs (usually infertile) and simultaneously incubate them. In contrast, male homosexual pairs never build nests.

Both male and female homosexual pairs also occur in Orange-fronted Parakeets. Same-sex couples often sit side by side—sometimes for half an hour or more at a time—preening and nuzzling each other while fluffing their plumage. Males also sometimes engage in sexual behavior with their pair-mate: one bird mounts the other, usually preceded by a display known as FAWNING or CLAWING, in which he lifts one foot in the air and places it gently on the other male's back or wing. Female partners often COURTSHIP-FEED each other: one bird regurgitates some food and feeds it to her mate, and the two partners interlock their bills while jerking their heads back and forth. Unlike in heterosexual pairs, either bird may feed the other. This is usually accompanied by several stylized visual and vocal displays such as head-bobbing, rubbing or kneading of the bill on the branch (often producing a distinctive popping sound), fluffing of the cheek feathers, and flashing the iris of the eyes while whistling. The two mates sometimes play-fight with one another by BILL-SPARRING, in which they grasp and tug at each other's beak. Female pairs may also jointly prepare a nest, usually constructed in arboreal termite nests: one partner excavates the entrance tunnel (as do males in heterosexual pairs) while the other hollows out the nest chamber (as do females in heterosexual pairs). Some female couples successfully compete against heterosexual pairs for nesting sites; pair-bonded females often become powerful allies that support one another and may even come to dominate opposite-sex pairs through attacks and threat behavior.

Frequency: Homosexual bonds are common in these Parrot species. In some captive populations of Galahs, for instance, one-half to two-thirds of pair-bonds are between birds of the same sex. A study of Peach-faced Lovebirds in captivity revealed that 4 of 12 pair-bonds (33 percent) were between females; a similar study of Orange-fronted Parakeets found 5 of 9 pair-bonds (56 percent) between females. Although male homosexual pairs of Orange-fronted Parakeets have been documented in wild birds, the overall incidence of same-sex bonds in the wild is not known for any of these species. However, about 1 out of every 180 nests of wild Galahs contains a SUPERNORMAL CLUTCH of 10–11 eggs, double the number of most other nests. These are undoubtedly laid by two females, perhaps members of a homosexual pair (although they could also result from alternative heterosexual arrangements—see below).

Orientation: Among captive Galahs that form some sort of pair-bond, about 44 percent of birds only form homosexual bonds, another 44 percent only develop heterosexual pairings, while about 11 percent bond with birds of both sexes. Of the latter group, a variety of bisexualities occur, usually in the context of trios or quartets of birds. One female, for example, formed a polygamous bond with two males, then went on to develop a bond with another female while maintaining her trio bond. Another female was bonded to a male who also had another female partner; she then "divorced" him and paired with another female. One male in a homosexual pair later also developed simultaneous bonds with two females. Similar bisexual trios and quartets also occur in Orange-fronted Parakeets, and females in this species sometimes compete successfully against both males and other females for the attentions of another female. In wild Galahs (especially males), it is possible that many birds form same-sex pairings in the juvenile flocks, with some later developing opposite-sex associations as adults. In Lovebirds (as well as in Galahs and Parakeets that only form same-sex pairs), at least some birds in homosexual bonds appear to be more exclusively same-sex oriented. For instance, bonds sometimes form between females even if there are available unpaired males, and one widowed female Lovebird who lost her female partner went on to form another homosexual pairing.

Nonreproductive and Alternative Heterosexualities

Although most heterosexual bonds in Galahs are lifelong and between only two birds, several other variations occur: some birds form polygamous trios (as discussed above), while 6–10 percent of Galahs divorce their partners and seek new mates. Infidelity in the form of copulations by paired Galahs with birds other than their mate is common, especially during the incubation period. Sometimes a mated pair changes partners after their eggs have been laid, usually resulting in loss of the eggs (accounting for about 2 percent of all clutches that fail to hatch). In addition, parenting in this species is not confined to a strictly nuclear-family structure: youngsters from several different families are pooled together into a CRÈCHE or "day-care" flock as soon as they are old enough to leave the nest. This allows their parents to forage on their own (although a few tending parents are always on hand in the crèche). Youngsters spend nearly as much time in these crèches as they do being cared for solely by their parents in the nest. Sometimes two pairs share the same nest hollow, with both females laying eggs in a combined clutch.

Although divorce and alternative bonding arrangements are not as prominent among Lovebirds, there is nevertheless often considerable antagonism between the sexes in heterosexual pairs. Males tend to become sexually ready each season before females; as a result, their courtship advances are frequently ignored or rejected, and females may even respond with overt aggression. Heterosexual copulations often do not culminate in ejaculation because the female refuses to allow the male to remain mounted, walking or flying out from under him while making threatening displays at him. A similar asynchrony in male and female sexual cycles may also occur in Galahs: in some years, males become ready to mate before females, but by the time the latter are ready, the males often lose interest and many pairs end up not breeding at all. Besides pairs that do not reproduce in a particular year, there is always a sizable proportion of single birds. As many as 60 percent of the adult Galahs in foraging flocks are nonbreeding birds from the nomadic population. Sometimes a nonbreeding female associates regularly with a breeding pair, "tagging along" with the male when he leaves the nest (perhaps in the hope of pairing with him). Such birds are known as "aunts" even though they are probably not related to the birds they associate with.

When a female Lovebird does consent to mate, nonreproductive REVERSE mounts may occur: as a part of courtship, males sometimes solicit females, who briefly mount them the way a male would. Other forms of nonprocreative sexual behavior also take place. Both Lovebirds and Galahs form pair-bonds as juveniles, long before they begin to reproduce—Galahs, in fact, may commence pairing and copulation up to three years before they can breed. In addition, some of the mating activity among younger Galahs involves birds mounting and thrusting against their partner's head rather than the genital region. Among adults, at least 12 percent of all copulations occur well before fertilization is possible (four to five weeks prior to egg laying).

Other Species

Homosexual pairs occur in several Parrot species related to these, including Masked Lovebirds (*Agapornis personata*) and Red-faced Lovebirds (*Agapornis pullaria*), from Africa, and female Aztec Parakeets (*Aratinga astec*), from central America. Same-sex pairs (both male and female) have also been documented in numerous other species of Parrots (usually in captivity). Canary-winged Parakeets (*Brotogeris versicolorus*) and Rose-ringed Parakeets (*Psittacula krameri*) in homosexual pairs engage in frequent mutual preening and soft grasping and nibbling of each other's beak. Female pairs of Elegant Parrots (*Neophema elegans*) from Australia courtship-feed and mount one another, as do male pairs of Senegal Parrots (*Poicephalus senegalus*) and White-fronted Amazon Parrots (*Amazona albifrons*). Male pairs have also been reported in Mealy Amazon Parrots (*Amazona farinosa*). Homosexual pairing is found as well in Ornate Lorikeets (*Trichoglossus ornatus*) and several other species of Lories from the islands of Southeast Asia, in which courtship-feeding and copulation are regular features of same-sex pairing. One male pair of Yellow-backed (Chattering) Lorikeets (*Lorius garrulus flavopalliatus*) remained together for more than 14 years.

Sources
asterisked references discuss homosexuality/transgender

Arrowood, P. C. (1991) "Male-Male, Female-Female, and Male-Female Interactions Within Captive Canary-winged Parakeet *Brotogeris v. versicolurus* Flocks." *Acta XX Congressus Internationalis Ornithologici, Christchurch, New Zealand* (Proceedings of the 20th International Ornithological Congress), vol. 2, pp. 666–72. Wellington, N.Z.: New Zealand Ornithological Trust Board.

*——— (1988) "Duetting, Pair Bonding, and Agonistic Display in Parakeet Pairs." *Behavior* 106:129–57.

*Buchanan, O. M. (1966) "Homosexual Behavior in Wild Orange-fronted Parakeets." *Condor* 68:399–400.

*Callaghan, E. (1982) "Breeding the Senegal Parrot *Poicephalus senegalus*." *Avicultural Magazine* 88:130–34.

*Clarke, P. (1982) "Breeding the Spectacled (White-fronted) Amazon Parrot *Amazona albifrons nana*." *Avicultural Magazine* 88:71–74.

*Dilger, W. C. (1960) "The Comparative Ethology of the African Parrot Genus *Agapornis*." *Zeitschrift für Tierpsychologie* 17:649–85.

*Fischdick, G., V. Hahn, and K. Immelmann (1984) "Die Sozialisation beim Rosenköpfchen *Agapornis roseicollis* [Socialization in the Peach-faced Lovebird]." *Journal für Ornithologie* 125:307–19.

Forshaw, J. M. (1989) *Parrots of the World*. 3rd ed. London: Blandford Press.

*Goodwin, D. (1983) "Notes on Feral Rose-ringed Parakeets." *Avicultural Magazine* 89:84–93.

*Hampe, H. (1940) "Beobachtungen bei Schmuck- und Feinsittichen, *Neophema elegans* und *chrysostomus* [Observations on Blue-winged and Elegant Parrots]." *Journal für Ornithologie* 88:587–99.

*Hardy, J. W. (1966) "Physical and Behavioral Factors in Sociality and Evolution of Certain Parrots (*Aratinga*)." *Auk* 83:66–83.

*——— (1965) "Flock Social Behavior of the Orange-fronted Parakeet." *Condor* 67:140–56.

*——— (1963) "Epigamic and Reproductive Behavior of the Orange-fronted Parakeet." *Condor* 65:169–99.

*Kavanau, J. L. (1987) *Lovebirds, Cockatiels, Budgerigars: Behavior and Evolution*. Los Angeles: Science Software Systems.

*Lack, D. (1940) "Courtship Feeding in Birds." *Auk* 57:169–78.

*Lantermann, W. (1990) "Breeding the Mealy Amazon Parrot *Amazona farinosa farinosa* (Boddaert) at Oberhausen Ornithological Institute, West Germany." *Avicultural Magazine* 96:126–29.

*Low, R. (1977) *Lories and Lorikeets: The Brush-Tongued Parrots*. London: Paul Elek.

Pidgeon, R. (1981) "Calls of the Galah *Cacatua roseicapilla* and Some Comparisons with Four Other Species of Australian Parrots." *Emu* 81:158–68.

*Rogers, L. J., and H. McCulloch (1981) "Pair-bonding in the Galah *Cacatua roseicapilla*." *Bird Behavior* 3:80–92.

*Rowley, I. (1990) *Behavioral Ecology of the Galah* Eolophus roseicapillus *in the Wheatbelt of Western Australia*. Chipping Norton, NSW: Surrey Beatty.

Appendix: Other Species

The following is a partial list of other species in which homosexual behavior has been documented, including reptiles/amphibians, fishes, insects and other invertebrates, and domesticated animals.

Abbreviations are used to indicate the general type of homosexuality (and in some cases, transgender) for each species:

F = female homosexuality   FTvM = female-to-male transvestism   P = parthenogenesis
M = male homosexuality   MTvF = male-to-female transvestism

Common Name	Scientific Name	Type	Sources
Reptiles and Amphibians			
Common Ameiva	*Ameiva chrysolaema*	M	Noble & Bradley 1933
Jamaican Giant Anole	*Anolis garmani*	MTvF,M	Trivers 1976
Green Anole	*Anolis carolinensis*	M	Jenssen & Hovde 1993; Noble & Bradley 1933
Brown Anole	*Anolis sagrei*	M	Noble & Bradley 1933
Cuban Green Anole	*Anolis porcatus*	M	Noble & Bradley 1933
Largehead Anole	*Anolis cybotes*	M	Noble & Bradley 1933
Anole sp.	*Anolis inaguae*	M	Noble & Bradley 1933
Checkered Whiptail Lizard	*Cnemidophorus tesselatus*	P,F	Cole & Townsend 1993; Crews & Fitzgerald 1980; Crews & Young 1991; Young 1991; Crews et al. 1983; Eifler 1983
Desert Grassland Whiptail Lizard	*Cnemidophorus uniparens*	P,F	
Plateau Striped Whiptail Lizard	*Cnemidophorus velox*	P,F	
Chihuahuan Spotted Whiptail Lizard	*Cnemidophorus exsanguis*	P,F	
Laredo Striped Whiptail Lizard	*Cnemidophorus laredoensis*	P,F	Paulissen & Walker 1989

Common Name	Scientific Name	Type	Sources
Broad-headed Skink	*Eumeces laticeps*	F,M	Moehn 1986; Noble & Bradley 1933
Five-lined Skink	*Eumeces fasciatus*	M	Noble & Bradley 1933
Red-tailed Skink	*Eumeces egregius*	F	Mount 1963
Inagua Curlytail Lizard	*Leiocephalus inaguae*	M	Noble & Bradley 1933
Mourning Gecko	*Lepidodactylus lugubris*	P,F	McCoid & Hensley 1991; Werner 1980
Western Banded Gecko	*Coleonyx variegatus*	M	Greenberg 1943
Fence Lizard	*Sceloporus undulatus*	M	Noble & Bradley 1933
Side-blotched Lizard	*Uta stansburiana*	M,F	Tinkle 1967
Wood Turtle	*Clemmys insculpta*	M	Kaufmann 1992
Desert Tortoise	*Gopherus agassizii*	M	Bulova 1994; Niblick et al. 1994
Tengger Desert Toad	*Bufo raddei*	M	Liu 1931
Black-spotted Frog	*Rana nigromaculata*	M	Liu 1931
Appalachian Woodland Salamander	*Plethodon (metcalfi) jordani*	M,MTvF	Arnold 1976; Organ 1958
Mountain Dusky Salamander	*Desmognathus ochrophaeus*	M,MTvF	Verrell & Donovan 1991
Common Garter Snake	*Thamnophis sirtalis*	MTvF,M	Mason 1993; Mason & Crews 1985; Mason et al. 1989; Noble 1937
Pine (Gopher) Snake	*Pituophis melanoleucus*	M	Hansen 1950; Shaw 1951
Water Moccasin	*Agkistrodon piscivorus*	M	Shaw 1951
Red Diamond Rattlesnake	*Crotalus ruber*	M	Shaw 1948
Speckled Rattlesnake	*Crotalus mitchelli*	M	Klauber 1972
Western Rattlesnake	*Crotalus viridis*	M	Klauber 1972

Fishes

Blackstripe Topminnow	*Fundulus notatus*	F	Carranza & Winn 1954
Green Swordtail	*Xiphophorous helleri*	M	Schlosberg et al. 1949
Southern Platyfish	*Platypoecilus maculatus*	M	Schlosberg et al. 1949
Amazon Molly	*Poecilia formosa*	P,F	Döbler et al. 1997; Schlupp et al. 1992
European Bitterling	*Rhodeus amarus*	F	Duyvené de Wit 1955
Ten-spined Stickleback	*Pygosteus pungitius*	M	Morris 1952
Three-spined Stickleback	*Gasterosteus aculeatus*	M	Morris 1952
Bluegill Sunfish	*Lepomis macrochirus*	MTvF,M	Dominey 1980
Guiana Leaffish	*Polycentrus schomburgkii*	MTvF,M	Barlow 1967
Least Darter	*Microperca punctulata*	M	Petravicz 1936
Jewel Fish	*Hemichromus bimaculatus*	F	Greenberg 1961
Mouthbreeding Fish sp.	*Tilapia macrocephala*	F	Aronson 1948

Common Name	Scientific Name	Type	Sources
Char	*Salmo alpinus*	M,F	Fabricius 1953; Fabricius & Gustafson 1954
Houting Whitefish	*Coregonus lavaretus*	M,F	Fabricius & Lindroth 1954
Grayling	*Thymallus thymallus*	M,F	Fabricius & Gustafson 1955

Insects

Clubtail Dragonfly spp.	*Trigomphus melampus*	M	
	Hagenius brevistylus	M	
	Gomphus adelphus	M	
	G. apomyius	M	
	G. geminatus	M	
	G. parvidens	M	Utzeri & Belfiore 1990,
	G. viridifrons	M	Dunkle 1991
	G. dilatatus	M	
	G. lineatifrons	M	
	G. modestus	M	
	G. ozarkensis	M	
	G. vastus	M	
Broadwinged Damselfly sp.	*Calopteryx haemorrhoidalis*	M	Utzeri & Belfiore 1990
Spreadwinged Damselfly spp.	*Lestes disjunctus*	M	Utzeri & Belfiore 1990
	L. sponsa	M	Utzeri & Belfiore 1990
	L. barbarus	M	Utzeri & Belfiore 1990
	L. viridis	M	Utzeri & Belfiore 1990
	Sympecma paedisca	M	Utzeri & Belfiore 1990
	Enallagma cyathigerum	M	Utzeri & Belfiore 1990
Narrow-winged Damselfly spp.	*Ischnura graellsii*	M	Utzeri & Belfiore 1990
	I. elegans	M	Utzeri & Belfiore 1990
	I. senegalensis	M	Utzeri & Belfiore 1990
	Ceriagrion tenellum	M	Utzeri & Belfiore 1990
	C. nipponicum	M	Utzeri & Belfiore 1990
	Nehalennia gracilis	M	Utzeri & Belfiore 1990
Common Skimmer Dragonfly spp.	*Sympetrum striolatum*	M	Utzeri & Belfiore 1990
	Leucorrhinia hudsonica	M	Utzeri & Belfiore 1990
	L. dubia	M	Utzeri & Belfiore 1990
	L. rubicunda	M	Utzeri & Belfiore 1990
	L. caudalis	M	Utzeri & Belfiore 1990
	Cercion hieroglyphicum	M	Utzeri & Belfiore 1990
Field Cricket sp.	*Acheta firmus*	M	Alexander 1961
Migratory Locust	*Locusta migratoria*	M	Mika 1959
Cockroach spp.	*Byrsotria fumigata*	M	Barth 1964
	Periplaneta americana	M	Simon & Barth 1977
	P. fuliginosa	M	Simon & Barth 1977
	Petasodes dominica	M	Matthiesen 1990
	Nauphoeta cinerea	M	Rocha 1991
	Henchoustedenia flexivitta	M	Rocha 1991

Common Name	Scientific Name	Type	Sources
Large Milkweed Bug	*Oncopeltus fasciatus*	M	Loher & Gordon 1968
Southern Green Stink Bug	*Nezara viridula*	M	Harris & Todd 1980
Water Boatman Bug	*Palmacorixa nana*	M	Aiken 1981
Creeping Water Bug sp.	*Ambrysus occidentalis*	M	Constanz 1973
Water Strider spp.	*Limnoporus dissortis*	M	Spence & Wilcox 1986
	L. notabilis	M	Spence & Wilcox 1986
Bedbug and other Bug spp.	*Xylocoris maculipennis*	M	Carayon 1974
	Alloeorhynchus furens	M	Carayon 1966
	Latrocimex spectans	M	Carayon 1966
	Cimex lectularius	M	Carayon 1966
	Afrocimex sp.	M	Carayon 1966
	Embiophila sp.	M	Carayon 1966
	Hesperoctenes sp.	M	Carayon 1966
Green Lacewing	*Chrysopa carnea*	M	Henry 1979
Queen Butterfly	*Danaus gilippus*	M	Brower et al. 1965
Monarch Butterfly	*Danaus plexippus*	M	Leong 1995; Leong et al. 1995; Rothschild 1978; Tilden 1981; Urquhart 1960, 1987
Checkerspot Butterfly	*Euphydryas editha*	M	Shah et al. 1986
Mexican White	*Eucheira socialis*	M	Shapiro 1989
Cabbage (Small) White	*Pieris rapae*	M	Obara 1970; Obara & Hidaka 1964
Large White	*Pieris brassicae*	M	David & Gardiner 1961
Glasswing Butterfly	*Acraea andromacha*	M	Alcock 1993
Mazarine Blue	*Cyaniris semiargus*	M	Tennent 1987
Spruce Budworm Moth	*Choristoneura fumiferana*	M	Palaniswamy et al. 1979; Sanders 1975
Larch Bud Moth	*Zeiraphera diniana*	M	Benz 1973
Grape Berry Moth	*Eupoecilia ambiguella*	M	Schmieder-Wenzel & Schruft 1990
Silkworm Moth	*Bombyx mori*	M	Berlese 1909
Blister Beetle spp.	*Meloe angusticollis*	F	Pinto & Selander 1970
	M. dianella	F	Pinto & Selander 1970
	M. proscarabaeus	M, F	Bologna & Marangoni 1986
	M. violaceus	M,F	Bologna & Marangoni 1986
Rove Beetle spp.	*Aleochara curtula*	M, MTvF	Peschke 1987
	Leistotrophus versicolor	M, MTvF	Forsyth & Alcock 1990
Stag Beetle spp.	*Lucanus* sp., *Dorcus* sp.,	M	Mathieu 1967
	Platycerus sp., *Ceruchus* sp.	M	Mathieu 1967
Red Flour Beetle	*Tribolium castaneum*	M	Castro et al. 1994; Rich 1989; Serrano et al. 1991; Spratt 1980

Common Name	Scientific Name	Type	Sources
Eucalyptus Longhorned Borer	*Phoracantha semipunctata*	M	Hanks et al. 1996
Grape Borer	*Xylotrechus pyrrhoderus*	F	Iwabuchi 1987
Alfalfa Weevil	*Hypera postica*	F	LeCato & Pienkowski 1970
Bean Weevil sp.	*Callosobruchus chinensis*	F	Nakamura 1969
Other Weevils	*Otiorrhynchus pupillatus*	F,P	Pardi 1987
	Diaprepes abbreviatus	F	Harari 1997
Southern One-Year Canegrub	*Antitrogus consanguineus*	M	Allsopp & Morgan 1991
Japanese Scarab Beetle	*Popillia japonica*	M,F	Barrows & Gordh 1978; Fleming 1972
Rosechafer	*Macrodactylus subspinosa*	M	Bennett 1974
Southern Masked Chafer	*Cyclocephala immaculata*	M	Hayes 1927
Other Melolonthine Scarab Beetles	*Polyphylla hammondi*	M	Hayes 1927
	Cotalpa lanigera	M	Hayes 1927
	Pelidnota punctata	M	Hayes 1927
	Melolontha vulgaris	M	Gadeau de Kerville 1896; Laboulmène 1859; Maze 1884; Noel 1895
Parsnip Leaf Miner	*Euleia fratria*	M	Tauber & Toschi 1965
House Fly	*Musca domestica*	M	Schlein et al. 1981
Tsetse Fly	*Glossina morsitans*	M	Schlein et al. 1981
Stable Fly sp.	*Fannia femoralis*	M	Tauber 1968
Blowfly	*Protophormia terrae-novae*	M	Parker 1968
Midge sp.	*Stictochironomus crassiforceps*	M	Syrajämäki 1964
Pomace Fly	*Euarestoides acutangulus*	M	Piper 1976
Long-legged Fly spp.	*Dolichopus popularis*	M	Qvick 1984
	Medetera spp.	M	Dyte 1989
Reindeer Warble Fly	*Hypoderma tarandi*	M	Anderson et al. 1994
Screwworm Fly	*Cochliomyia hominivorax*	M,F	Fletcher et al. 1966
Mediterranean Fruit Fly	*Ceratitis capitata*	M,F	Arita & Kaneshiro 1983; Prokopy & Hendrichs 1979
Other Fruit Fly spp.	*Drosophila melanogaster*	M,F	Cook 1975; Finley et al. 1997; Gill 1963; McRobert & Tompkins 1983, 1988, Napolitano & Tompkins 1989; Schaner et al. 1989; Tompkins 1989, 1988; Vaias et al. 1993
	D. affinis	M	
Other Fly spp.	*Fucomyia frigida*	M	Bristowe 1929
	Fucellia maritima	M	Bristowe 1929
	Scatella sp.	M	Bristowe 1929

Common Name	Scientific Name	Type	Sources
Hen Flea	*Ceratophyllus gallinae*	M	Humphries 1967
Digger Bee	*Centris pallida*	M	Alcock & Buchmann 1985
Southeastern Blueberry Bee	*Habropoda laboriosa*	M	Cane 1994
Eastern Giant Ichneumon	*Megarhyssa macrurus*	M	Heatwole et al. 1962
Ichneumon Wasp sp.	*Megarhyssa atrata*	M	Heatwole et al. 1962
Australian Parasitic Wasp sp.	*Cotesia rubecula*	MTvF,M	Field & Keller 1993
Red Ant sp.	*Formica subpolita*	M	O'Neill 1994

Spiders and Other Invertebrates

Common Name	Scientific Name	Type	Sources
Hawaiian Orb-Weaver	*Doryonychus raptor*	M	Gillespie 1991
Jumping Spider sp.	*Portia fimbriata*	M,F	Jackson 1982
Harvest Spider sp.	(not identified)	M	Bristowe 1929, 1939
Box Crab	*Calappa lophos*	M	Kazmi & Tirmizi 1987
Mite sp.	*Promesomachilis hispanica*	M	Sturm 1992
Acanthocephalan Worms	*Moniliformis dubius*	M	Abele & Gilchrist 1977
	Acanthocephalus parksidei	M	Abele & Gilchrist 1977
	Echinorhynchus truttae	M	Abele & Gilchrist 1977
	Polymorphus minutus	M	Abele & Gilchrist 1977
Incirrate Octopus spp.	(2 unidentified spp., family Incirrata)	M	Lutz & Voight 1994; Mirsky 1995

Domesticated Animals

Common Name	Scientific Name	Type	Sources
Sheep	*Ovis aries*	M,F	Banks 1964; Grubb 1974; Perkins et al. 1992, 1995; Resko et al. 1996
Goat	*Capra hircus*	M,F	Collias 1956; Schaller & Laurie 1974; Shank 1972
Cattle	*Bos taurus*	M,F	Blackshaw et al. 1997; Hurnik et al. 1975; Klemm et al. 1983, Mylrea & Beilharz 1964; Reinhardt 1983; van Vliet & van Eerdenburg 1996
Horse, Donkey	*Equus caballus,*	M,F	Feist & McCullough 1976; Ford & Beach 1951; McDonnell & Haviland 1995
	E. asinus	M,F	
Pig	*Sus scrofa*	M,F	Craig 1981; Signoret et al. 1975

Common Name	Scientific Name	Type	Sources
Cat	*Felis catus*	M,F	Aronson 1949; Green et al. 1957; Leyhausen 1979; Michael 1961; Prescott 1970; Rosenblatt Schneirla 1962
Dog	*Canis familiaris*	M,F	Beach et al. 1968; Fuller & DuBuis 1962; King 1954; LeBoeuf 1967; Vasey 1996
Guinea Pig	*Cavia porcellus*	M,F	Rood 1972; Young et al. 1938
Rabbit	*Oryctolagus cuniculus*	M,F	Kawai 1955; Mykytowycz & Hesterman 1975; Verberne & Blom 1981
Rat	*Rattus norvegicus*	M,F	Beach & Rasquin 1942; Grant & Chance 1958
Mouse	*Mus musculus*	M,F	van Oortmerssen 1971
Hamster	*Mesocricetus auratus*	F	Beach 1971; Tiefer 1970
Bengalese Finch	*Lonchura striata*	F	Masatomi 1957, 1959; Jefferies 1967
Zebra Finch	*Taeniopygia guttata*	M,F	Burley 1981; Immelmann et al. 1982; Morris 1954
Turkey	*Meleagris gallopavo*	F	Hale 1955; Hale & Schein 1962
Chicken	*Gallus gallus*	F	Guhl 1948
Pigeon	*Columba livia*	M,F	Craig 1909; Whitman 1919
Budgerigar	*Melopsittacus undulatus*	M,F	Brockway 1967; Kavanau 1987

References

Reptiles and Amphibians

Arnold, S. J. (1976) "Sexual Behavior, Sexual Interference, and Sexual Defense in the Salamanders *Ambystoma maculatum, Ambystoma trigrinum,* and *Plethodon jordani." Zeitschrift für Tierpsychologie* 42:247–300.

Bulova, S. J. (1994) "Patterns of Burrow Use by Desert Tortoises: Gender Differences and Seasonal Trends." *Herpetological Monographs* 8:133–43.

Cole, C. J., and C. R. Townsend (1983) "Sexual Behavior in Unisexual Lizards." *Animal Behavior* 31:724–28.

Crews, D., and K. T. Fitzgerald (1980) "'Sexual' Behavior in Parthenogenetic Lizards (Cnemidophorous)." *Proceedings of the National Academy of Sciences* 77:499–502.

Crews, D., and L. J. Young (1991) "Pseudocopulation in Nature in a Unisexual Whiptail Lizard." *Animal Behavior* 42:512–14.

Crews, D., J. E. Gustafson, and R. R. Tokarz (1983) "Psychobiology of Parthenogenesis." In R. B. Huey, E. R. Pianka, and T. W. Schoener, eds., *Lizard Ecology: Studies of a Model Organism,* pp. 205–31. Cambridge: Harvard University Press.

Eifler, D. A. (1993) "*Cnemidophorus uniparens* (Desert Grassland Whiptail). Behavior." *Herpetological Review* 24:150.

Greenberg, B. (1943) "Social Behavior of the Western Banded Gecko, *Coleonyx variegatus* Baird." *Physiological Zoology* 16:110–22.

Hansen, R. M. (1950) "Sexual Behavior in Two Male Gopher Snakes." *Herpetologica* 6:120.

Jenssen, T. A., and E. A. Hovde (1993) "*Anolis carolinensis* (Green Anole). Social Pathology." *Herpetological Review* 24:58–59.

Kaufmann, J. H. (1992) "The Social Behavior of Wood Turtles, *Clemmys insculpta*, in Central Pennsylvania." *Herpetological Monographs* 6:1–25.

Klauber, L. M. (1972) *Rattlesnakes: Their Habits, Life Histories, and Influence on Mankind.* Vol. 1. Berkeley: University of California Press.

Liu, Ch'eng-Chao (1931) "Sexual Behavior in the Siberian Toad, *Bufo raddei* and the Pond Frog, *Rana nigromaculata*." *Peking Natural History Bulletin* 6:43–60.

Mason, R. T. (1993) "Chemical Ecology of the Red-Sided Garter Snake, *Thamnophis sirtalis parietalis*." *Brain Behavior and Evolution* 41:261–68.

Mason, R. T., and D. Crews (1985) "Female Mimicry in Garter Snakes." *Nature* 316:59–60.

Mason, R. T., H. M Fales, T. H. Jones, L. K. Pannell, J. W. Chinn, and D. Crews (1989) "Sex Pheromones in Snakes." *Science* 245:290–93.

McCoid, M. J., and R. A. Hensley (1991) "Pseudocopulation in *Lepidodactylus lugubris*." *Herpetological Review* 22:8–9.

Moehn, L. D. (1986) "Pseudocopulation in *Eumeces laticeps*." *Herpetological Review* 17:40–41.

Moore, M. C., J. M. Whittier, A. J. Billy, and D. Crews (1985) "Male-like Behavior in an All-Female Lizard: Relationship to Ovarian Cycle." *Animal Behavior* 33:284–89.

Mount, R. H. (1963) "The Natural History of the Red-Tailed Skink, *Eumeces egregius* Baird." *American Midland Naturalist* 70:356–85.

Niblick, H. A., D. C. Rostal, and T. Classen (1994) "Role of Male-Male Interactions and Female Choice in the Mating System of the Desert Tortoise, *Gopherus agassizii*." *Herpetological Monographs* 8:124–32.

Noble, G. K. (1937) "The Sense Organs Involved in the Courtship of *Storeria, Thamnophis,* and Other Snakes." *Bulletin of the American Museum of Natural History* 73:673–725.

Noble, G. K., and H. T. Bradley (1933) "The Mating Behavior of Lizards; Its Bearing on the Theory of Sexual Selection." *Annals of the New York Academy of Sciences* 35:25–100.

Organ, J. A. (1958) "Courtship and Spermatophore of *Plethodon jordani metcalfi*." *Copeia* 1958:251–59.

Paulissen, M. A., and J. M. Walker (1989) "Pseudocopulation in the Parthenogenetic Whiptail Lizard *Cneimidophorous laredoensis* (Teiidae)." *Southwestern Naturalist* 34:296–98.

Shaw, C. E. (1951) "Male Combat in American Colubrid Snakes With Remarks on Combat in Other Colubrid and Elapid Snakes." *Herpetologica* 7:149–68.

——— (1948) "The Male Combat 'Dance' of Some Crotalid Snakes." *Herpetologica* 4:137–45.

Tinkle, D. W. (1967) *The Life and Demography of the Side-blotched Lizard,* Uta stansburiana. Miscellaneous Publications, Museum of Zoology, no. 132. Ann Arbor: University of Michigan.

Trivers, R. L. (1976) "Sexual Selection and Resource-Accruing Abilities in *Anolis garmani*." *Evolution* 30:253–69.

Verrell, P., and A. Donovan (1991) "Male-Male Aggression in the Plethodontid Salamander *Desmognathus ochrophaeus*." *Journal of Zoology* 223:203–12.

Werner, Y. L. (1980) "Apparent Homosexual Behavior in an All-Female Population of a Lizard, *Lepidodactylus lugubris* and Its Probable Interpretation." *Zeitschrift für Tierpsychologie* 54:144–50.

Fishes

Aronson, L. R. (1948) "Problems in the Behavior and Physiology of a Species of African Mouthbreeding Fish." *Transactions of the New York Academy of Sciences* 2:33–42.

Barlow, G. W. (1967) "Social Behavior of a South American Leaf Fish, *Polycentrus schomburgkii*, with an Account of Recurring Pseudofemale Behavior." *American Midland Naturalist* 78:215–34.

Carranza, J., and H. E. Winn (1954) "Reproductive Behavior of the Blackstripe Topminnow, *Fundulus notatus*." *Copeia* 4:273–78.

Döbler, M., I. Schlupp, and J. Parzefall (1997) "Changes in Mate Choice with Spontaneous Masculinization in *Poecilia formosa*." In M. Taborsky and B. Taborsky, eds., *Contributions to the XXV International Ethological Conference,* p. 204. Advances in Ethology no. 32. Berlin: Blackwell Wissenschafts-Verlag.

Dominey, W. J. (1980) "Female Mimicry in Male Bluegill Sunfish—a Genetic Polymorphism?" *Nature* 284:546–48.

Duyvené de Wit, J. J. (1955) "Some Observations on the European Bitterling (*Rhodeus amarus*)." *South African Journal of Science* 51:249–51.

Fabricius, E. (1953) "Aquarium Observations on the Spawning Behavior of the Char, *Salmo alpinus*." *Institute of Freshwater Research, Drottningholm,* report 35:14–48.

Fabricius, E., and K.-J. Gustafson (1955) Observations on the Spawning Behavior of the Grayling, *Thymallus thymallus* (L.)." *Institute of Freshwater Research, Drottningholm,* report 36:75–103.

——— (1954) "Further Aquarium Observations on the Spawning Behavior of the Char, *Salmo alpinus* L." *Institute of Freshwater Research, Drottningholm,* report 35:58–104.

Fabricius, E., and A. Lindroth (1954) "Experimental Observations on the Spawning of the Whitefish, *Coregonus lavaretus* L., in the Stream Aquarium of the Hölle Laboratory at River Indalsälven." *Institute of Freshwater Research, Drottningholm,* report 35:105–12.

Greenberg, B. (1961) "Spawning and Parental Behavior in Female Pairs of the Jewel Fish, *Hemichromis bimaculatus* Gill." *Behavior* 18:44–61.

Morris, D. (1952) "Homosexuality in the Ten-Spined Stickleback." *Behavior* 4:233–61.

Petravicz, J. J. (1936) "The Breeding Habits of the Least Darter, *Microperca punctulata* Putnam." *Copeia* 1936:77–82.

Schlosberg, H., M. C. Duncan, and B. Daitch (1949) "Mating Behavior of Two Live-bearing Fish *Xiphophorous helleri* and *Platypoecilus maculatus*." *Physiological Zoology* 22:148–61.

Schlupp, I., J. Parzefall, J. T. Epplen, I. Nanda, M. Schmid, and M. Schartl (1992) "Pseudomale Behavior and Spontaneous Masculinization in the All-Female Teleost *Poecilia formosa* (Teleostei: Poeciliidae)." *Behavior* 122:88–104.

Insects

Aiken, R. B. (1981) "The Relationship Between Body Weight and Homosexual Mounting in *Palmacorixa nana* Walley (Heteroptera: Corixidae)." *Florida Entomologist* 64:267–71.

Alcock, J. (1993) "Male Mate-Locating Behavior in Two Australian Butterflies, *Anaphaeis java teutonia* (Fabricius) (Pieridae) and *Acraea andromacha andromacha* (Fabricius) (Nymphalidae)." *Journal of Research on the Lepidoptera* 32:1–7.

Alcock, J., and S. L. Buchmann (1985) "The Significance of Post-Insemination Display by Males of *Centris pallida* (Hymenoptera: Anthophoridae)." *Zeitschrift für Tierpsychologie* 68:231–43.

Alexander, R. D. (1961) "Aggressiveness, Territoriality, and Sexual Behavior in Field Crickets (Orthoptera: Gryllidae)." *Behavior* 17:130–223.

Allsopp, P. G., and T. A. Morgan (1991) "Male-Male Copulation in *Antitrogus consanguineus* (Blackburn) (Coleoptera: Scarabaeidae)." *Australian Entomological Magazine* 18(4):147–48.

Anderson, J. R., A. C. Nilssen, and I. Folstad (1994) "Mating Behavior and Thermoregulation of the Reindeer Warble Fly, *Hypoderma tarandi* L. (Diptera: Oestridae)." *Journal of Insect Behavior* 7:679–706.

Arita, L. H., and K. Y. Kaneshiro (1983) "Pseudomale Courtship Behavior of the Female Mediterranean Fruit Fly, *Ceratitis capitata* (Wiedemann)." *Proceedings of the Hawaiian Entomological Society* 24:205–10.

Barrows, E. M., and G. Gordh (1978) "Sexual Behavior in the Japanese Beetle, *Popillia japonica*, and Comparative Notes on Sexual Behavior of Other Scarabs (Coleoptera: Scarabaeidae)." *Behavioral Biology* 23:341–54.

Barth, R. H., Jr. (1964) "Mating Behavior of *Byrsotria fumigata* (Guerin) (Blattidae, Blaberinea)." *Behavior* 23:1–30.

Bennett, G. (1974) "Mating Behavior of the Rosechafer in Northern Michigan (Coleoptera: Scarabaeidae)." *Coleopterists Bulletin* 28:167–68.

Benz, G. (1973) "Role of Sex Pheromone and Its Insignificance for Heterosexual and Homosexual Behavior of Larch Bud Moth." *Experientia* 29:553–54.

Berlese, A. (1909) *Gli insetti: loro organizzazione, sviluppo, abitudini, e rapporti coll'uomo.* Vol. 2 (*Vita e costumi*). Milano: Società Editrice Libraria.

Bologna, M. A., and C. Marangoni (1986) "Sexual Behavior in Some Palaearctic Species of *Meloe* (Coleoptera, Meloidae)." *Bollettino della Società Entomologica Italiana* 118(4–7):65–82.

Bristowe, W. S. (1929) "The Mating Habits of Spiders with Special Reference to the Problems Surrounding Sex Dimorphism." *Proceedings of the Zoological Society of London* 1929:309–58.

Brower, L. P., J. V. Z. Brower, and F. P. Cranston (1965) "Courtship Behavior of the Queen Butterfly, *Danaus gilippus berenice* (Cramer)." *Zoologica* 50:1–39.

Cane, J. H. (1994) "Nesting Biology and Mating Behavior of the Southeastern Blueberry Bee, *Habropoda laboriosa* (Hymenoptera: Apoidea)." *Journal of the Kansas Entomological Society* 67:236–41.

Carayon, J. (1974) "Insémination traumatique hétérosexuelle et homosexuelle chez *Xylocoris maculipennis* (Hem. Anthocoridae)." *Comptes rendus hebdomadaires des séances de l'Académie des Sciences, Série D—Sciences naturelles* 278:2803–6.

——— (1966) "Les inséminations traumatiques accidentelles chez certains Hémiptères *Cimicoidea.*" *Comptes rendus hebdomadaires des séances de l'Académie des Sciences, Série D—Sciences naturelles* 262:2176–79.

Castro, L., M. A. Toro, and C. Fanjul-Lopez (1994) "The Genetic Properties of Homosexual Copulation Behavior in *Tribolium castaneum:* Artificial Selection." *Genetics Selection Evolution* (Paris) 26:361–67.

Constanz, G. (1973) "The Mating Behavior of a Creeping Water Bug, *Ambrysus occidentalis* (Hemiptera: Naucoridae)." *American Midland Naturalist* 92:230–39.

Cook, R. (1975) "'Lesbian' Phenotype of *Drosophila melanogaster?*" *Nature* 254:241–42.

David, W. A. L., and B. O. C. Gardiner (1961) "The Mating Behavior of *Pieris brassicae* (L.) in a Laboratory Culture." *Bulletin of Entomological Research* 52:263–80.

Dunkle, S. W. (1991) "Head Damage from Mating Attempts in Dragonflies (Odonata: Anisoptera)." *Entomological News* 102:37–41.

Dyte, C. E. (1989) "Gay Courtship in Medetera." *Empid and Dolichopodid Study Group Newsheet* 6:2–3.

Field, S. A., and M. A. Keller (1993) "Alternative Mating Tactics and Female Mimicry as Post-Copulatory Mate-Guarding Behavior in the Parasitic Wasp *Cotesia rubecula.*" *Animal Behavior* 46:1183–89.

Finley, K. D., B. J. Taylor, M. Milstein, and M. McKeown (1997) "*dissatisfaction,* a Gene Involved in Sex-Specific Behavior and Neural Development of *Drosophila melanogaster.*" *Proceedings of the National Academy of Sciences* 94:913–18.

Fleming, W. E. (1972) *Biology of the Japanese Beetle.* U.S. Department of Agriculture Technical Bulletin no. 1449. Washington, D.C.: Agricultural Research Service.

Fletcher, L. W., J. J. O'Grady Jr., H. V. Claborn, and O. H. Graham (1966) "A Pheromone from Male Screwworm Flies." *Journal of Economic Entomology* 59:142–43.

Forsyth, A., and J. Alcock (1990) "Female Mimicry and Resource Defense Polygyny by Males of a Tropical Rove Beetle, *Leistotrophus versicolor* (Coleoptera: Staphylinidae)." *Behavioral Ecology and Sociobiology* 26:325–30.

Gadeau de Kerville, H. (1896) "Perversion sexuelle chez des Coléoptères mâles." *Bulletin de la Société Entomologique de France* 1896:85–87.

Gill, K. S. (1963) "A Mutation Causing Abnormal Courtship and Mating Behavior in Males of *Drosophila melanogaster.*" *American Zoologist* 3:507.

Hanks, L. M., J. G. Millar, and T. D. Paine (1996) "Mating Behavior of the Eucalyptus Longhorned Borer (Coleoptera: Cerambycidae) and the Adaptive Significance of Long 'Horns.' " *Journal of Insect Behavior* 9:383–93.

Harari, A. R. (1997) "Mating Strategies of Female *Diaprepes abbreviatus* (L.)." In M. Taborsky and B. Taborsky, eds., *Contributions to the XXV International Ethological Conference,* p. 222. Advances in Ethology no. 32. Berlin: Blackwell Wissenschafts-Verlag.

Harris, V. E., and J. W. Todd (1980) "Temporal and Numerical Patterns of Reproductive Behavior in the Southern Green Stinkbug, *Nezara viridula* (Hemiptera: Pentatomidae)." *Entomologia Experimentalis et Applicata* 27:105–16.

Hayes, W. P. (1927) "Congeneric and Intergeneric Pederasty in the Scarabaeidae (Coleop.)." *Entomological News* 38:216–18.

Heatwole, H., D. M. Davis, and A. M. Wenner (1962) "The Behavior of *Megarhyssa,* a Genus of Parasitic Hymenopterans (Ichneumonidae: Ephilatinae)." *Zeitschrift für Tierpsychologie* 19:652–64.

Henry, C. (1979) "Acoustical Communication During Courtship and Mating in the Green Lacewing, *Chrysopa carnea* (Neuroptera; Chrysopidae)." *Annals of the Entomological Society of America* 72:68–79.

Humphries, D. A. (1967) "The Mating Behavior of the Hen Flea *Ceratophyllus gallinae* (Schrank) (Siphonaptera: Insecta)." *Animal Behavior* 15:82–90.

Iwabuchi, K. (1987) "Mating Behavior of *Xylotrechus pyrrhoderus* Bates (Coleoptera: Cerambycidae). 5. Female Mounting Behavior." *Journal of Ethology* 5:131–36.

Laboulmène, A. (1859) "Examen anatomique de deux *Melolontha vulgaris* trouvés accouplés et paraissant du sexe mâle." *Annales de la Société Entomologique de France* 1859:567–70.

LeCato, G. L., III, and R. L. Pienkowski (1970) "Laboratory Mating Behavior of the Alfalfa Weevil, *Hypera postica.*" *Annals of the Entomological Society of America* 63:1000–7.

Leong, K. L. H. (1995) "Initiation of Mating Activity at the Tree Canopy Level Among Overwintering Monarch Butterflies in California." *Pan-Pacific Entomologist* 71:66–68.

Leong, K. L. H., E. O'Brien, K. Lowerisen, and M. Colleran (1995) "Mating Activity and Status of Over-wintering Monarch Butterflies (Lepidoptera: Danaidae) in Central California." *Annals of the Entomological Society of America* 88:45–50.

Loher, W., and H. T. Gordon (1968) "The Maturation of Sexual Behavior in a New Strain of the Large Milk-weed Bug *Oncopeltus fasciatus.*" *Annals of the Entomological Society of America* 61:1566–72.

Mathieu, J. (1967) "Mating Behavior of Five Species of Lucanidae (Coleoptera: Insecta)." *American Zoologist* 7:206.

Matthiesen, F. A. (1990) "Comportamento sexuale outros aspectos biologicos da barata selvagem, *Petasodes dominicana* Burmeister, 1839 (Dictyoptera, Blaberidae, Blaberinae)." *Revista Brasileira de Entomologia* 34(2):261–66.

Maze, A. (1884) "Communication." *Journal officiel de la République française* 2:2103.

McRobert, S., and L. Tompkins (1988) "Two Consequences of Homosexual Courtship Performed by *Drosophila melanogaster* and *Drosophila affinis* Males." *Evolution* 42:1093–97.

——— (1983) "Courtship of Young Males Is Ubiquitous in *Drosophila melanogaster.*" *Behavior Genetics* 13:517–23.

Mika, G. (1959) "Über das Paarungsverhalten der Wanderheuschrecke *Locusta migratoria* R. und F. und deren Abhängigkeit vom Zustand der inneren Geschlechtsorgane." *Zoologische Beiträge* 4:153–203.

Nakamura, H. (1969) "Comparative Studies on the Mating Behavior of Two Species of *Callosobruchus* (Coleoptera: Bruchidae)." *Japanese Journal of Ecology* 19:20–26.

Napolitano, L. M., and L. Tompkins (1989) "Neural Control of Homosexual Courtship in *Drosophila melanogaster.*" *Journal of Neurogenetics* 6:87–94.

Noel, P. (1895) "Accouplements anormaux chez les insectes." *Miscellanea entomologica* 1:114.

Obara, Y. (1970) "Studies on the Mating Behavior of the White Cabbage Butterfly, *Pieris rapae crucivora* Boisduval. III. Near-Ultra-Violet Reflection as the Signal of Intraspecific Communication." *Zeitschrift für vergleichende Physiologie* 69:99–116.

Obara, Y., and T. Hidaka (1964) "Mating Behavior of the Cabbage White, *Pieris rapae crucivora*. I. The 'Flutter Response' of the Resting Male to Flying Males." *Zoological Magazine* (*Dobutsugaku Zasshi*) 73:131–35.

O'Neill, K. M. (1994) "The Male Mating Strategy of the Ant *Formica subpolita* Mayr (Hymenoptera: Formicidae): Swarming, Mating, and Predation Risk." *Psyche* 101:93–108.

Palaniswamy, P., W. D. Seabrook, and R. Ross (1979) "Precopulatory Behavior of Males and Perception of Potential Male Pheromone in Spruce Budworm, *Choristoneura fumiferana.*" *Annals of the Entomological Society of America* 72:544–51.

Pardi, L. (1987) "La 'pseudocopula' delle femmine di *Otiorrhynchus pupillatus* cyclophtalmus (Sol.) (Coleoptera Curculionidae)." *Bollettino dell'Istituto di Entomologia "Guido Grandi" della Università degli Studi di Bologna* 41:355–63.

Parker, G. A. (1968) "The Sexual Behavior of the Blowfly, *Protophormia terraenovae* R.-D." *Behavior* 32:291–308.

Peschke, K. (1987) "Male Aggression, Female Mimicry and Female Choice in the Rove Beetle, *Aleochara curtula* (Coleoptera, Staphylinidae)." *Ethology* 75:265–84.

Pinto, J. D., and R. B. Selander (1970) *The Bionomics of Blister Beetles of the Genus* Meloe *and a Classification of the New World Species*. University of Illinois Biological Monographs no. 42. Urbana: University of Illinois Press.

Piper, G. L. (1976) "Bionomics of *Euarestoides acutangulus* (Diptera: Tephritidae)." *Annals of the Entomological Society of America* 69:381–86.

Prokopy, R. J., and J. Hendrichs (1979) "Mating Behavior of *Ceratitis capitata* on a Field-Caged Host Tree." *Annals of the Entomological Society of America* 72:642–48.

Qvick, U. (1984) "A Case of Abnormal Mating Behavior of *Dolichopus popularis* Wied. (Diptera, Dolichopodidae)." *Notulae Entomologicae* 64:93.

Rich, E. (1989) "Homosexual Behavior in Three Melanic Mutants of *Tribolium castaneum.*" *Tribolium Information Bulletin* 29:99–101.

Rocha, I. R. D. (1991) "Relationship Between Homosexuality and Dominance in the Cockroaches, *Nauphoeta cinerea* and *Henchoustedenia flexivitta* (Dictyoptera, Blaberidae)." *Revista Brasileira de Entomologia* 35(1):1–8.

Rothschild, M. (1978) "Hell's Angels." *Antenna* 2:38–39.

Sanders, C. J. (1975) "Factors Affecting Adult Emergence and Mating Behavior of the Eastern Spruce Budworm, *Choristoneura fumiferana* (Lepidoptera: Totricidae)." *Canadian Entomologist* 107:967–77.

Schaner, A. M., P. D. Dixon, K. J. Graham, and L. L. Jackson (1989) "Components of the Courtship-Stimulating Pheromone Blend of Young Male *Drosophila melanogaster:* (Z)-13-tritriacontene and (Z)-11-tritriacontene." *Journal of Insect Physiology* 35:341–45.

Schlein, Y., R. Galun, and M. N. Ben-Eliahu (1981) "Abstinons: Male-Produced Deterrents of Mating in Flies." *Journal of Chemical Ecology* 7:285–90.

Schmieder-Wenzel, C., and G. Schruft (1990) "Courtship Behavior of the European Grape Berry Moth, *Eupoecilia ambiguella* Hb. (Lepidoptera, Tortricidae) in Regard to Pheromonal and Tactile Stimuli." *Journal of Applied Entomology* 109:341–46.

Serrano, J. M., L. Castro, M. A. Torro, and C. López-Fanjul (1991) "The Genetic Properties of Homosexual Copulation Behavior in *Tribolium castaneum*: Diallel Analysis." *Behavior Genetics* 21:547–58.

Shah, N. K., M. C. Singer, and D. R. Syna (1986) "Occurrence of Homosexual Mating Pairs in a Checkerspot Butterfly." *Journal of Research on the Lepidoptera* 24:393.

Shapiro, A. M. (1989) "Homosexual Pseudocopulation in *Eucheira socialis* (Pieridae)." *Journal of Research on the Lepidoptera* 27:262.

Simon, D., and R. H. Barth (1977) "Sexual Behavior in the Cockroach Genera *Periplaneta* and *Blatta*. I. Descriptive Aspects. II. Sex Pheromones and Behavioral Responses." *Zeitschrift für Tierpsychologie* 44:80–107, 162–77.

Spence, J. R., and R. S. Wilcox (1986) "The Mating System of Two Hybridizing Species of Water Striders (Gerridae)." *Behavioral Ecology and Sociobiology* 19:87–95.

Spratt, E. C. (1980) "Male Homosexual Behavior and Other Factors Influencing Adult Longevity in *Tribolium castaneum* (Herbst) and *T. confusum* Duval." *Journal of Stored Products Research* 16:109–14.

Syrajämäki, J. (1964) "Swarming and Mating Behavior of *Allochironomus crassiforceps* Kieff. (Dipt., Chironomidae)." *Annales Zoologici Fennici* 1:125–45.

Tauber, M. J. (1968) "Biology, Behavior, and Emergence Rhythm of Two Species of *Fannia* (Diptera: Muscidae)." *University of California Publications in Entomology* 50:1–86.

Tauber, M. J., and C. Toschi (1965) "Bionomics of *Euleia fratria* (Loew) (Diptera: Tephritidae). I. Life History and Mating Behavior." *Canadian Journal of Zoology* 43:369–79.

Tennent, W. J. (1987) "A Note on the Apparent Lowering of Moral Standards in the Lepidoptera." *Entomologist's Record and Journal of Variation* 99:81–83.

Tilden, J. W. (1981) "Attempted Mating Between Male Monarchs." *Journal of Research on the Lepidoptera* 18:2.

Tompkins, L. (1989) "Homosexual Courtship in *Drosophila*." *MBL (Marine Biology Laboratory) Lectures in Biology* (Woods Hole) 10:229–48.

Urquhart, F. (1987) *The Monarch Butterfly: International Traveler*. Chicago: Nelson-Hall.

——— (1960) *The Monarch Butterfly*. Toronto: University of Toronto Press.

Utzeri, C., and C. Belfiore (1990) "Tandem anomali fra Odonati (Odonata)." *Fragmenta Entomologica* 22:271–87.

Vaias, L. J., L. M. Napolitano, and L. Tomkins (1993) "Identification of Stimuli that Mediate Experience-Dependent Modification of Homosexual Courtship in *Drosphila melanogaster*." *Behavior Genetics* 23:91–97.

Spiders and Other Invertebrates

Abele, L. G., and S. Gilchrist (1977) "Homosexual Rape and Sexual Selection in Acanthocephalan Worms." *Science* 197:81–83.

Bristowe, W. S. (1939) *The Comity of Spiders*. London: Ray Society.

——— (1929) "The Mating Habits of Spiders with Special Reference to the Problems Surrounding Sex Dimorphism." *Proceedings of the Zoological Society of London* 1929:309–58.

Gillespie, R. G. (1991) "Homosexual Mating Behavior in Male *Doryonychus raptor* (Araneae, Tetragnathidae)." *Journal of Arachnology* 19:229–30.

Jackson, R. R. (1982) "The Biology of *Portia fimbriata*, a Web-building Jumping Spider (Araneae, Salticidae) from Queensland: Intraspecific Interactions." *Journal of Zoology, London* 196:295–305.

Kazmi, Q. B., and N. M. Tirmizi (1987) "An Unusual Behavior in Box Crabs (Decapoda, Brachyura, Calappidae)." *Crustaceana* 53:313–14.

Lutz, R. A., and J. R. Voight (1994) "Close Encounter in the Deep." *Nature* 371:563.

Mirsky, S. (1995) "Armed and Amorous." *Wildlife Conservation* 98(6):72.

Sturm, H. (1992) "Mating Behavior and Sexual Dimorphism in *Promesomachilis hispanica* Silvestri, 1923 (Machilidae, Archaeognatha, Insecta)." *Zoologischer Anzeiger* 228:60–73.

Domesticated Animals

Aronson, L. R. (1949) "Behavior Resembling Spontaneous Emissions in the Domestic Cat." *Journal of Comparative and Physiological Psychology* 42:226–27.

Banks, E. M. (1964) "Some Aspects of Sexual Behavior in Domestic Sheep, *Ovis aries*." *Behavior* 23:249–79.

Beach, F. A. (1971) "Hormonal Factors Controlling the Differentiation, Development, and Display of Copulatory Behavior in the Hamster and Related Species." In E. Tobach, L. R. Aronson, and E. Shaw, eds., *The Biopsychology of Development*, pp. 249–96. New York: Academic Press.

Beach, F. A., and P. Rasquin (1942) "Masculine Copulatory Behavior in Intact and Castrated Female Rats." *Endocrinology* 31:393–409.

Beach, F. A., C. M. Rogers, and B. J. LeBoeuf (1968) "Coital Behavior in Dogs: Effects of Estrogen on Mounting by Females." *Journal of Comparative and Physiological Psychology* 66:296–307.

Blackshaw, J. K., A. W. Blackshaw, and J. J. McGlone (1997) "Buller Steer Syndrome Review." *Applied Animal Behavior Science* 54:97–108.

Brockway, B. F. (1967) "Social and Experimental Influences of Nestbox-Oriented Behavior and Gonadal Activity of Female Budgerigars (*Melopsittacus undulatus* Shaw)." *Behavior* 29:63–82.

Burley, N. (1981) "Sex Ratio Manipulation and Selection for Attractiveness." *Science* 211:721–22.

Collias, N. E. (1956) "The Analysis of Socialization in Sheep and Goats." *Ecology* 37:228–39.

Craig, J. V. (1981) *Domestic Animal Behavior: Causes and Implications for Animal Care and Management.* Englewood Cliffs, N.J.: Prentice-Hall.

Craig, W. (1909) "The Voices of Pigeons Regarded as a Means of Social Control." *American Journal of Sociology* 14:86–100.

Feist, J. D., and D. R. McCullough (1976) "Behavior Patterns and Communication in Feral Horses." *Zeitschrift für Tierpsychologie* 41:337–71.

Ford, C. S., and F. A. Beach (1951) *Patterns of Sexual Behavior.* New York: Harper and Row.

Fuller, J. L., and E. M. DuBuis (1962) "The Behavior of Dogs." In E. S. E. Hafez, ed., *The Behavior of Domestic Animals*, pp. 415–52. Baltimore: Williams and Wilkins.

Grant, E. C., and M. R. A. Chance (1958) "Rank Order in Caged Rats." *Animal Behavior* 6:183–94.

Green, J. D., C. D. Clemente, and J. de Groot (1957) "Rhinencephalic Lesions and Behavior in Cats: An Analysis of the Klüver-Bucy Syndrome with Particular Reference to Normal and Abnormal Sexual Behavior." *Journal of Comparative Neurology* 108:505–36.

Grubb, P. (1974) "Mating Activity and the Social Significance of Rams in a Feral Sheep Community." In V. Geist and F. Walther, eds., *Behavior in Ungulates and Its Relation to Management*, vol. 1, pp. 457–76. Morges, Switzerland: International Union for Conservation of Nature and Natural Resources.

Guhl, A. M. (1948) "Unisexual Mating in a Flock of White Leghorn Hens." *Transactions of the Kansas Academy of Science* 5:107–11.

Hale, E. B. (1955) "Defects in Sexual Behavior as Factors Affecting Fertility in Turkeys." *Poultry Science* 34:1059–67.

Hale, E. B., and M. W. Schein (1962) "The Behavior of Turkeys." In E. S. E. Hafez, ed., *The Behavior of Domestic Animals*, pp. 531–64. Baltimore: Williams and Wilkins.

Hulet, C. V., G. Alexander, and E. S. E. Hafez (1975) "The Behavior of Sheep." In E. S. E. Hafez, ed., *The Behavior of Domestic Animals*, 3rd ed., pp. 246–94. London: Baillière Tindall.

Hurnik, J. F., G. J. King, and H. A. Robertson (1975) "Estrous and Related Behavior in Postpartum Holstein Cows." *Applied Animal Ethology* 2:55–68.

Immelmann, K., J. P. Hailman, and J. R. Baylis (1982) "Reputed Band Attractiveness and Sex Manipulation in Zebra Finches." *Science* 215:422.

Jefferies, D. J. (1967) "The Delay in Ovulation Produced by pp'-DDT and Its Possible Significance in the Field." *Ibis* 109:266–72.

Kavanau, J. L. (1987) *Lovebirds, Cockatiels, Budgerigars: Behavior and Evolution.* Los Angeles: Science Software Systems.

Kawai, M. (1955) "The Dominance Hierarchy and Homosexual Behavior Observed in a Male Rabbit Group." *Dobutsu shinrigaku nenpo (Annual of Animal Psychology)* 5:13–24.

King, J. A. (1954) "Closed Social Groups Among Domestic Dogs." *Proceedings of the American Philosophical Society* 98:327–36.

Klemm, W. R., C. J. Sherry, L. M. Schake, and R. F. Sis (1983) "Homosexual Behavior in Feedlot Steers: An Aggression Hypothesis." *Applied Animal Ethology* 11:187–95.

LeBoeuf, B. J. (1967) "Interindividual Associations in Dogs." *Behavior* 29:268–95.

Leyhausen, P. (1979) *Cat Behavior: The Predatory and Social Behavior of Domestic and Wild Cats.* New York and London: Garland STPM Press.

Masatomi, H. (1959) "Attacking Behavior in Homosexual Groups of the Bengalee, *Uroloncha striata* var. *domestica* Flower." *Journal of the Faculty of Science, Hokkaido University* (Series 6) 14:234–51.

——— (1957) "Pseudomale Behavior in a Female Bengalee." *Journal of the Faculty of Science, Hokkaido University* (Series 6) 13:187–91.

McDonnell, S. M., and J. C. S. Haviland (1995) "Agonistic Ethogram of the Equid Bachelor Band." *Applied Animal Behavior Science* 43:147–88.

Michael, R. P. (1961) "Observations Upon the Sexual Behavior of the Domestic Cat (*Felis catus* L.) Under Laboratory Conditions." *Behavior* 18:1–24.

Morris, D. (1954) "The Reproductive Behavior of the Zebra Finch (*Poephila guttata*), with Special Reference to Pseudofemale Behavior and Displacement Activities." *Behavior* 6:271–322.

Mykytowycz, R., and E. R. Hesterman (1975) "An Experimental Study of Aggression in Captive European Rabbits, *Oryctolagus cuniculus* (L.)." *Behavior* 52:104–23.

Mylrea, P. J., and R. G. Beilharz (1964) "The Manifestation and Detection of Oestrus in Heifers." *Animal Behavior* 12:25–30.

Perkins, A., J. A. Fitzgerald, and G. E. Moss (1995) "A Comparison of LH Secretion and Brain Estradiol Receptors in Heterosexual and Homosexual Rams and Female Sheep." *Hormones and Behavior* 29:31–41.

Perkins, A., J. A. Fitzgerald, and E.O Price (1992) "Luteinizing Hormone and Testosterone Response of Sexually Active and Inactive Rams." *Journal of Animal Science* 70:2086–93.

Prescott, R. G. W. (1970) "Mounting Behavior in the Female Cat." *Nature* 228:1106–7.

Reinhardt, V. (1983) "Flehmen, Mounting, and Copulation Among Members of a Semi-Wild Cattle Herd." *Animal Behavior* 31:641–50.

Resko, J. A., A. Perkins, C. E. Roselli, J. A. Fitzgerald, J. V .A. Choate, and F. Stormshak (1996) "Endocrine Correlates of Partner Preference in Rams." *Biology of Reproduction* 55:120–26.

Rood, J. P. (1972) "Ecological and Behavioral Comparisons of Three Genera of Argentine Cavies." *Animal Behavior Monographs* 5:1–83.

Rosenblatt, J. S., and T. C. Schneirla (1962) "The Behavior of Cats." In E. S. E. Hafez, ed., *The Behavior of Domestic Animals*, pp. 453–88. Baltimore: Williams and Wilkins.

Schaller, G. B., and A. Laurie (1974) "Courtship Behavior of the Wild Goat." *Zeitschrift für Säugetierkunde* 39:115–27.

Shank, C. C. (1972) "Some Aspects of Social Behavior in a Population of Feral Goats (*Capra hircus* L.)." *Zeitschrift für Tierpsychologie* 30:488–528.

Signoret, J. P., B. A. Baldwin, D. Fraser, and E. S. E. Hafez (1975) "The Behavior of Swine." In E. S. E. Hafez, ed., *The Behavior of Domestic Animals*, 3rd ed., pp. 295–329. London: Baillière Tindall.

Tiefer, L. (1970) "Gonadal Hormones and Mating Behavior in the Adult Golden Hamster." *Hormones and Behavior* 1:189–202.

van Oortmerssen, G. A. (1971) "Biological Significance, Genetics, and Evolutionary Origin of Variability in Behavior Within and Between Inbred Strains of Mice (*Mus musculus*)." *Behavior* 38:1–92.

van Vliet, J. H., and F. J. C. M. van Eerdenburg (1996) "Sexual Activities and Oestrus Detection in Lactating Holstein Cows." *Applied Animal Behavior Science* 50:57–69.

Vasey, P. L. (1996) Personal communication.

Verberne, G., and F. Blom (1981) "Scentmarking, Dominance, and Territorial Behavior in Male Domestic Rabbits." In K. Myers and C. D. MacInnes, eds., *Proceedings of the World Lagomorph Conference*, pp. 280–90. Guelph: University of Guelph.

Whitman, C. O. (1919) *The Behavior of Pigeons*. Posthumous Works of C. O. Whitman, vol. 3. Washington, D.C.: Carnegie Institution of Washington.

Young, W. C., E. W. Dempsey, and H. I. Myers (1935) "Cyclic Reproductive Behavior in the Female Guinea Pig." *Journal of Comparative Psychology* 19:313–35.

Notes to Part I

When no specific references are noted for a particular species in part 1, the information and sources will be found in the profile itself in part 2. When a note is included for a profiled species (e.g., to provide more detailed information), the citation format includes the species name, author, year, and (in most cases) page numbers of the source, referring to the full references in the profile. References for species that are not profiled in part 2 are included directly in notes.

Introduction

1. Einstein, A. (1930) "What I Believe," *Forum and Century* 84(4):193–94.

Chapter 1. The Birds and the Bees

1. Haldane, J. B. S. (1928) *Possible Worlds and Other Papers,* p. 298 (New York: Harper & Brothers).
2. Animal names that are capitalized refer to a species or group of closely related species that is profiled in part 2, or whose references are included in the appendix.
3. Homosexuality among primates, for example, has been traced back to at least the Oligocene epoch, 24–37 million years ago (based on its distribution among contemporary primates; Vasey 1995:195). Some scientists place its original appearance even earlier in the evolutionary line leading to mammals, at around 200 million years ago (Baker and Bellis 1995:5), and it has probably existed for much longer among other animal groups. Vasey, P. L. (1995) "Homosexual Behavior in Primates: A Review of Evidence and Theory," *International Journal of Primatology* 16:173–204; Baker, R., and M. A. Bellis (1995) *Human Sperm Competition: Copulation, Masturbation, and Infidelity* (London: Chapman and Hall).
4. See note 29, as well as part 2 and the appendix, for more detailed tabulations (including discussion of species *not* included in this tally).
5. For further discussion of sexual orientation in animals, as well as comparisons between animal and human homosexuality, see chapter 2. Following Vasey ("Homosexual Behavior in Primates," p. 175), the term *homosexual* is used to designate primarily the *form* of behaviors without necessarily implying anything about their "function or context or the actors' ages and motivation." For further consideration of the terminology used to describe same-sex activity in animals, including discussion of alternative definitions of the term *homosexual(ity)* as it is applied to animals (and some of the controversies that have surrounded its use in the zoological literature), see chapter 3. For more on the "functions" and contexts of homosexual behavior, see chapters 4–5.
6. Guianan Cock-of-the-Rock (Endler and Théry 1996); Anna's Hummingbird (Hamilton 1965); Buff-breasted Sandpiper (Myers 1989).
7. For a general survey of play-fighting, see Aldis, O. (1975) *Play Fighting.* (New York: Academic Press).
8. Spinner Dolphin (Norris et al. 1994:250).

9. Ingestion of semen by both males and females during masturbation in heterosexual contexts also occurs among Golden Monkeys (Clarke 1991:371).

10. Supernormal clutches have also been reported for pairs of male Emus, probably because more than one female has laid in their nest. What might be termed "subnormal" clutches—i.e., nests containing fewer eggs than are usually found for heterosexual pairs—are reported for female pairs of Blue Tits. And "super-supernormal" clutches occasionally occur in heterosexual pairs of Roseate Terns: as a result of within-species parasitism and possibly also egg transfer (see chapter 5 for more on these phenomena), some nests contain more than double the number of eggs found even in supernormal clutches (as is also true for "dump" nests in many Ducks and Geese).

11. For discussion, and refutation, of the idea that same-sex pairs form in species such as these solely for the purpose of raising offspring, see chapter 5. In some birds such as grouse (e.g., sharp-tailed grouse, prairie chickens, white-tailed ptarmigan) and ducks (e.g., eiders, buffleheads) broods from more than one female are combined or "amalgamated" but no same-sex coparenting occurs (one female, or a heterosexual pair, look after all the offspring; cf. Bergerud and Gratson 1988:545 (Grouse); Afton 1993 (Ducks); Eadie, J. McA., F. P. Kehoe, and T. D. Nudds (1988) "Pre-Hatch and Post-Hatch Brood Amalgamation in North American Anatidae: A Review of Hypotheses," *Canadian Journal of Zoology* 66:1709–21.

12. Ring-billed Gull (Conover 1989:148).

13. In some bird species in which same-sex pairs are unable to obtain fertile eggs on their own (or in which homosexual parenting has yet to be observed in the wild), parenting skills have been demonstrated by supplying homosexual pairs with "foster" eggs or young in captivity. Same-sex pairs of Flamingos, White Storks, Black-headed Gulls, Steller's Sea Eagles, Barn Owls, and Gentoo Penguins, for example, have all successfully hatched such eggs and/or raised foster chicks.

14. Black Swan (Braithwaite 1981:140–42); for more details, see chapter 5 and part 2.

15. Ring-billed Gull (Conover 1989:148); Western Gull (Hayward and Fry 1993:17–18); see chapter 2 for further discussion of same-sex pairs being limited to nonoptimal territories. Several other studies point to the possibility of more "attentive" parenting by female homosexual pairs. Researchers have found that female Ring-billed Gulls in same-sex pairs, for example, may have higher levels of progesterone—a female hormone associated with nest-building and incubation behavior—than females in heterosexual pairs (Kovacs and Ryder 1985); see chapter 4 for more on the hormonal profiles of animals involved in same-sex activity. In a related set of observations, some investigators have documented more "intense" nesting behavior in female homosexual pairs than heterosexual pairs in some captive studies. Allen and Erickson (1982:346, 350), for instance, found that female pairs of Ring Doves are more persistent incubators than heterosexual pairs, being less likely to abandon their nests and terminate incubation when they have infertile eggs than are heterosexual pairs. Brockway (1967:76) found that female Budgerigars in homosexual pairs begin continuous occupation of their nests significantly sooner than females in heterosexual pairs. However, because female pairs begin noncontinuous occupation of their nests significantly later than heterosexual pairs in this species, the overall amount of their nesting activity and the timing of their egg-laying essentially evens out.

16. See chapter 5 for further discussion of homosexual activity in communal groups and the often complex relationship between "helpers" and same-sex activity.

17. In many species, young may also be raised in heterosexual trios, i.e., family units with three parents in which only opposite-sex bonding is present between the adults. See chapter 5 for some examples.

18. For discussion of single parenting in animals where two (heterosexual) parents usually raise the young, as well as examples of other heterosexual parenting arrangements that deviate from the species-typical pattern, see chapter 5.

19. For additional discussion of male bias in biological studies, see chapters 3 and 5.

20. Rhesus Macaque (Altmann 1962:383; Lindburg 1971:69); Hamadryas Baboon (Abegglen 1984:63); Gelada Baboon (Bernstein 1970:94); Tasmanian Native Hen (Ridpath 1972:30); Gray-headed Flying Fox (Nelson 1965:546).

21. Pukeko (Jamieson and Craig 1987a:1251); Bonobo (Thompson-Handler et al. 1984:349; Kano 1992:187; Kitamura 1989:55–57); Stumptail Macaque (Chevalier-Skolnikoff 1974:101, 110); Red Deer (Hall 1983:278); Red-necked Wallaby (LaFollette 1971:96); Northern Quoll (Dempster 1995:29).

22. Pig-tailed Macaque (Oi 1990a:350–51); Galah (Rogers and McCulloch 1981); Pronghorn (Kitchen 1974:44).

23. Gorilla (Fischer and Nadler 1978:660–61; Yamagiwa 1987a:12, 1987b:37).

24. Pukeko (Jamieson and Craig 1987a:1251–52); Flamingo (C. E. King, personal communication). In Lesser Flamingos, however, the reverse appears to be true: males but not females achieve cloacal contact during their homosexual mounts (Alraun and Hewston 1997:176).

25. Japanese Macaque (Hanby 1974:838–40; Wolfe 1984:149; Fedigan 1982:143).

26. For further discussion of comparisons between animal and human homosexuality, see chapter 2.

27. These formulas are also used to estimate the number of bisexual/heterosexual trios in a population; see Conover and Aylor 1985:127 (Ring-billed Gull).

28. Kob (Buechner and Schloeth 1965); Long-tailed Hermit (Stiles and Wolf 1979). Likewise, up to 30 homosexual pairs of Herring Gulls have been counted in some locations—a relatively high number of same-sex associations to be present at one time—but in colonies that number more than 10,000 pairs, this amounts to less than 1 percent of the total number of pairs (Shugart 1980:426–27).

29. Same-sex courtship, sexual, pair-bonding, and /or parenting behaviors have been documented in the scientific literature in at least 167 species of mammals, 132 birds, 32 reptiles and amphibians, 15 fishes, and 125 insects and other invertebrates, for a total of 471 species (see part 2 and appendix for a complete list). These figures do not include domesticated animals (at least another 19 species; see the appendix), nor species in which only sexually immature animals/juveniles engage in homosexual activities (for a survey of the latter in mammals, see Dagg 1984). For a number of reasons, this tally is likely to be an underestimate (especially for species other than mammals and birds, which are not as thoroughly covered); see chapter 3 for further discussion. It should also be pointed out that species totals may differ depending on the classificatory schema used; in some taxonomies, for example, animals that in this book are lumped together as subspecies are considered separate species (e.g., the various subspecies of Savanna Baboons, Flamingos, or Wapiti/Red Deer). This roster also excludes a wide variety of other cases in which the evidence for homosexual activities is not definitive, such as:

(1) species in which homosexual activity is suspected (and sometimes included in comprehensive surveys of homosexual behavior, such as Dagg 1984) but in which the sex of participants has not yet been confirmed (e.g., one-horned rhinoceros [Laurie 1982:323], yellow-bellied marmot [Armitage 1962:327], South African cliff swallow [Earlé 1985:46], band-tailed barbthroat hummingbird [Harms and Ahumada 1992], calliope hummingbird [Armstrong 1988], ringed parakeet [Hardy 1964]).

(2) bird species in which supernormal clutches have been documented without any direct evidence of same-sex pairs (e.g., numerous gulls and other bird species; see note 70, chapter 4).

(3) same-sex trios or joint parenting arrangements with little or no conclusive evidence of courtship, sex, or pair-bonding between the like-sexed coparents (e.g., bobolink [Bollinger et al. 1986]; various ducks, grouse [cf. note 11, this chapter]).

(4) bird species in which males associate in "pairs" or form "partnerships" with other males for joint displays during heterosexual courtships, but in which no overt courtship or sexual behavior occurs between such partners or other same-sex individuals (e.g., several manakins of the genera *Chiroxiphia, Pipra, Machaeropterus,* and *Masius*—note however that males in these species often court "female-plumaged" birds, the sex of most of which has not been determined, while in two other species, some of these individuals have been determined to be males; wild turkey; king bird of paradise and possibly other birds of paradise. For further references, see McDonald 1989:1007 and Trainer and McDonald 1993:779).

(5) species in which the only form of documented "same-sex" activity involves individuals mounting heterosexual copulating pairs, such that the mounting activity is not necessarily limited to like-sexed individuals or the same-sex motivation/orientation is not clear (e.g., camel [Gauthier-Pilters and Dagg 1981:92], Buller's albatross [Warham 1967:129]).

(6) species in which the only same-sex activity is mounting that appears to be exclusively aggressive in character with no sexual component (e.g., collared lemming, degu, ground squirrel; cf. Dagg 1984 and sources cited therein; see also chapter 3 for further discussion of aggressive or "dominance" mounting and the difficulty of distinguishing this from sexual mounting); and species in which the only same-sex activities are "affectionate" behaviors or "platonic" companionships unaccompanied by either signs of sexual arousal or overt courtship or sexual behaviors.

(7) other inconclusive cases, such as species reported in secondary sources as exhibiting homosexuality but whose original sources do not definitively document same-sex activity (e.g., avocets, reported in Terres [1980:813], with no mention of source, as engaging in homosexual mounting; Makkink [1936] and Hamilton [1975]—the most comprehensive primary field studies of this species and the most likely sources for this information—describe ritual mountings and masturbation ["eruptive copulations"] but no homosexual mounting).

Armitage, K. B. (1962) "Social Behavior of a Colony of the Yellow-bellied Marmot (*Marmota flaviventris*)." *Animal Behavior* 10:319–31; Armstrong, D. P. (1988) "Persistent Attempts by a Male Calliope Hummingbird, *Stellula calliope,* to Copulate with Newly Fledged Conspecifics," *Canadian Field-Naturalist* 102:259–60; Bollinger, E. K., T. A. Gavin, C. J. Hibbard, and J. T. Wootton (1986) "Two Male Bobolinks Feed Young at the Same Nest," *Wilson Bulletin* 98:154–56; Dagg, A. I. (1984) "Homosexual Behavior and Female-Male Mounting in Mammals—a First Survey," *Mammal Review* 14:155–85; Earlé, R. A. (1985) "A Description of the Social, Aggressive, and Maintenance Behavior of the South African Cliff Swallow *Hirundo spilodera* (Aves: Hirundinidae)," *Navorsinge van die nasionale Museum, Bloemfontein* 5:37–50; Gauthier-Pilters, H., and A. I. Dagg (1981) *The Camel: Its Evolution, Ecology, Behavior, and Relationship to Man* (Chicago: University of Chicago Press); Hardy, J. W. (1964) "Ringed Parakeets Nesting in Los Angeles, California," *Condor* 66:445–47; Harms, K. E., and J. A. Ahumada (1992) "Observations of an Adult Hummingbird Provisioning an Incubating Adult," *Wilson Bulletin* 104:369–70; Laurie, A. (1982) "Behavioral Ecology of the Greater One-horned Rhinoceros (*Rhinoceros unicornis*)," *Journal of Zoology, London* 196:307–41; Makkink, G. F. (1936) "An Attempt at an Ethogram of the European Avocet (*Recurvirostra avosetta* L.), With Ethological and Psychological Remarks," *Ardea* 25:1–63; McDonald, D. B. (1989) "Correlates of Male Mating Success in a Lekking Bird with Male-Male Cooperation," *Animal Behavior* 37:1007–22; Terres, J. K. (1980) *The Audubon Society Encyclopedia of North American Birds* (New York: Alfred A. Knopf); Trainer, J. M., and D. B. McDonald (1993) "Vocal Repertoire of the Long-tailed Manakin and Its Relation to Male-Male Cooperation," *Condor* 95:769–81; Warham, J. (1967) "Snares Island Birds," *Notornis* 14:122–39.

30. According to Wilson (1992), approximately 1,032,000 species of animals are currently known to science, although the number of actually occurring species is undoubtedly much higher—on the order of 10–100 million—and there are many complexities in estimating the total number of species. For further discussion, see Wilson, E. O. (1992) *The Diversity of Life*, pp. 131ff (Cambridge, Mass.: Belknap Press); Wilson, E. O. (1988) "The Current State of Biological Diversity," in E. O. Wilson, ed., *BioDiversity*, pp. 3–18. (Washington, D.C.: National Academy Press); May, R. M. (1988) "How Many Species Are There on Earth?" *Science* 241:1441–49.

31. Le Boeuf and Mesnick 1991:155 (Elephant Seal); see also Wilson, E. O. (1975) *Sociobiology: The New Synthesis* (Cambridge and London: Belknap Press). This figure is borne out by the data on homosexuality: the average number of observation hours for scientific studies in which homosexual behavior has been observed is approximately 1,050 hours (based on data from 47 species in which the number of observation hours has been recorded).

32. Marten, M., J. May, and R. Taylor (1982) *Weird and Wonderful Wildlife*, p. 7. (San Francisco: Chronicle Books). A somewhat more precise estimate of the number of species that have been adequately studied can be obtained for a subset of animals by using the *Zoological Record* (a comprehensive electronic database that indexes more than a million zoological source documents, including articles from over 6,000 journals worldwide, over the past 20 years). The *Zoological Record* for the period 1978–97 lists 825 mammal species in which at least some aspects of courtship, sexual, pair-bonding, mating-system, and/or parenting behaviors have been studied (the behavior categories in which homosexuality, if present, is likely to be found). Homosexual behavior has been documented in 133 of these species, or approximately 16 percent—comparable to the lower range obtained using the estimate of Marten et al. (The following subject headings/behavior categories indexed by *Zoological Record* were used in compiling this estimate: Courtship, Lek, Sexual Display, Precopulatory Behavior, Copulation, Mating, Pair Formation, Monogamy, Polygamy, Cooperative Breeding, Breeding Habits, Parental Care, Care of Young, Homosexuality).

33. See chapter 4 for further discussion of these factors.

34. For species that do not engage in "heterosexual" mating at all, e.g., parthenogenetic or hermaphrodite animals, see the next section.

35. Clapham, P. J. (1996) "The Social and Reproductive Biology of Humpback Whales: An Ecological Perspective," p. 37, *Mammal Review* 26:27–49.

36. Scott, P. E. (1994) "Lucifer Hummingbird (*Calothorax lucifer*)," in A. Poole and F. Gill, eds., *The Birds of North America: Life Histories for the 21st Century*, no. 134, p. 9. (Philadelphia: Academy of Natural Sciences; Washington, D.C.: American Ornithologists' Union); DeJong, M. J. (1996) "Northern Rough-winged Swallow (*Stelgidopteryx serripennis*)," in Poole and Gill, *The Birds of North America*, no. 234, p. 9; Kricher, J. C. (1995) "Black-and-white Warbler (*Phaeucticus melanocephalus*)," in Poole and Gill, *The Birds of North America*, no. 158, p. 9; O'Brien, R. M. (1990) "Red-tailed Tropicbird (*Phaethon rubricauda*)", in S. Marchant and P. J. Higgins, eds., *Handbook of Australian, New Zealand, and Antarctic Birds*, vol. 1, part B, p. 940 (Melbourne: Oxford University Press); Johnsgard, P. A. (1983) *Cranes of the World* (Bloomington: Indiana University Press); Powers, D. R. (1996) "Magnificent Hummingbird (*Eugenes fulgens*)," in Poole and Gill, *The Birds of North America*, no. 221, p. 10; Hill, G. E. (1994) "Black-headed Grosbeak (*Pheucticus melanocephalus*)," in Poole and Gill, *The Birds of North America*, no. 143, p. 8; Victoria's Riflebird (Frith and Cooper 1996:43); Gilliard 1969:13); Cheetah (Caro 1994:42); Lepson, J. K., and L. A. Freed (1995) "Variation in Male Plumage and Behavior of the Hawaii Akepa," *Auk* 112:402–14; Spotted Hyena (Frank 1996:117); Agile Wallaby (Stirrat and Fuller 1997:75); Birds of Paradise (Davis and Beehler 1994:522); Nelson, S. K., and S. G. Sealy (1995) "Biology of the Marbled Murrelet: Inland and at Sea (Symposium Introduction)," *Northwestern Naturalist* 76:1–3; Orang-utan (Schürmann 1982; Schürmann and van Hoof 1986; Maple 1980); Rowe, S., and R. Empson (1996) "Observations on the Breeding Behavior of the Tanga'eo or Mangaia Kingfisher (*Halcyon tuta ruficollaris*)," *Notornis* 43:43–48; Common Chimpanzee (Gagneux et al. 1997; Wrangham 1997); Harbor Seal (Perry and Amos 1998).

37. Emu (Coddington and Cockburn 1995; Heinroth 1924); Black-rumped Flameback (Neelakantan 1962); Nilgiri Langur (Poirier 1970a,b); Harbor Seal (Johnson 1976:45); Northern Quoll (Dempster 1995); Gray-capped Social Weaver (Collias and Collias 1980); Walrus (Miller 1975; Fay et al. 1984); Acorn Woodpecker (Koenig and Stacey 1990:427); Australian Shelduck (Riggert 1977:20); Killer Whale (Jacobsen 1990:78; Rose 1992:1–2). See also Lutz and Voight 1994 for the first documentation of copulatory behavior—between two males—in two previously unknown species of deep-sea octopuses (a group in which heterosexual mating has yet to be observed in any species). Other animals in which same-sex activity has been documented and in which heterosexual activity has rarely been observed include Musk-oxen (Smith 1976:62), Red-necked Wallabies (Johnson 1989a:275), Vicuñas (Koford 1957:182–84), Musk Ducks (Lowe 1966:285), and Ruffed Grouse (Johnsgard 1983:295). See chapter 4 for further discussion of the often insurmountable difficulties in attempting to observe and study sexual activity under field conditions.

38. Ring-billed Gull (Conover and Aylor 1985). See chapter 4 for discussion of some of the pitfalls of equating homosexual pairings with supernormal clutches.

39. Dragonflies (Dunkle 1991).

40. Pukeko (Craig 1980); Pronghorn (Kitchen 1974). The behavior was probably classified as rare in Pronghorn because the amount of same-sex activity was not being compared to the amount of opposite-sex activity, but rather to the total amount of "dominance" behavior (which it was classified as), consisting primarily of nonsexual activities. For further discussion and critique of homosexuality interpreted as a "dominance" activity, see chapter 3.

41. Killer Whale (Rose 1992:116); Regent Bowerbird (Lenz 1994:266 [table 2]); White-handed Gibbon (Edwards and Todd 1991:233 [table 1]); Crab-eating Macaque (Thompson 1969:465).

42. Giraffe (Pratt and Anderson 1985, 1982, 1979). In a study of another population of Giraffes, only three mounts between males were recorded, but only 400 hours of observation were involved (Dagg and Foster 1976:124).

43. Mountain Sheep (Geist 1968:210–11 [tables 4, 6]); 1971:152 [table 30]).

44. Gray Heron (Ramo 1993:116–17). In some species (e.g., Little Egrets, Little Blue Herons) quantitative information is only available for the proportion of promiscuous copulations that are homosexual (the higher figure). Where both proportions are available (e.g., Cattle Egrets, Gray Herons), an average of the two is taken when calculating cross-species comparisons of frequency (see below). Different frequencies can also be obtained depending on whether a distinction between copulatory and noncopulatory (or "ritualized") mounting is taken into account. (See chapter 3 for discussion of the—sometimes arbitrary—distinction between these two types of mounting.) In some cases, such as Pig-tailed and Crested Black Macaques, a sizable difference obtains. In Pigtails, 82 percent of mounting is same-sex if only noncopulatory mounts are considered (Oi 1990a:350–51 [table 4]), whereas 7–23 percent of all mounts are same-sex if "full" heterosexual copulations are included in the calculation (Bernstein 1967:226–7; Oi 1996:345). In Crested Blacks, roughly a third of noncopulatory mounts are between males, but overall this constitutes only about 8 percent of mounting activity when combined with copulatory (heterosexual) mounts (C. Reed, personal communication). In cases such as these, the latter (smaller) percentage is taken to be the overall rate of same-sex activity. For other species, however, the two rates are comparable. In Common Chimpanzees, for example, Nishida and Hosaka (1996:122, 129 [tables 9.7, 9.17a]) recorded 61 ritualized mounts between males compared to 123 male-female copulations (noncopulatory), yielding a rate of 33 percent same-sex mountings. This is comparable to the exclusively noncopulatory figures of Bygott (1974; cited in Hanby 1974:845 [Japanese Macaque]), who found that 4 of 14 ritualized mounts (29 percent) occurred between males.

45. Of course, these cross-species comparisons refer only to those animals in which homosexual behavior has been observed and in which the appropriate quantitative information is available. In many species same-sex behavior is more or less common than the maximum and minimum figures obtained from these measures, but it has not been quantified and therefore cannot be compared to these examples. For these calculations, if multiple frequency proportions were available for the same species—either because of population, seasonal, or behavioral differences (as discussed above)—these were averaged prior to being combined with the figures for other species. Proportions for courtship behavior are based on quantitative information from 21 species (avg = 23 percent same-sex activity), for sexual behavior from 77 species (avg = 26 percent), for pairing behavior from 45 species (avg = 14 percent), and for population percentages from 56 species (avg = 27 percent). For the purposes of comparison, tallies of observed homosexual and heterosexual behaviors in each species are assumed to be representative of actually occurring frequencies. The statistics for pairing and sexual behavior do not include the many species in which the *only* form of pair-bonding, coparenting, or observed sexual behavior is between same-sex individuals, i.e., in which 100 percent of pairs, coparents, or observed sexual interactions are homosexual (were these to be included, the proportions would be considerably, and perhaps unrepresentatively, higher). In five of these cases, however—Northern Elephant Seals, Cheetahs, Grizzly Bears, Lesser Scaup Ducks, and Greater Rheas—the proportion of all families or nests that are tended by same-sex pairs or trios (as opposed to single individuals) is substituted for the proportion of same-sex pairs. For population calculations, figures represent only the sex or activity that has been quantified (e.g., for species in which only females form same-sex pairs, or in which only female pairs have been tallied, only the proportion of females in such pairs is included). Moreover, for many bird species—especially those in which only a small fraction of individuals participate in same-sex activity—population percentages are not available. However, in species with same-sex pairing, the proportion of homosexual pairs is roughly comparable to the proportion of individuals engaging in same-sex activity (the two differ, of course, if there are sizable numbers of nonbreeding birds that do not form pairs with either sex). In order not to bias the sample toward species with relatively high population percentages, therefore, pairing proportions for wild-bird populations have been substituted. In species where no other population data are available, these figures are taken as a rough estimate of the proportion of individuals involved in same-sex activity.

46. The term *transgender,* when it is applied to people, has two uses: as a cover term that refers to a wide variety of gender-crossing or gender-mixing phenomena, including transsexuality (with various degrees of hormonal and/or surgical transformation), transvestism (including cross-dressers, drag queens and kings, and female and male impersonators), intersexuality (including various forms of hermaphroditism), and even extremes of butch-femme presentation. It is also used as a designation for a specific form of gender mixing in which an individual lives full-time in the gender opposite to his or her anatomy (e.g., a man who passes for a woman without undergoing the full physical transitioning of a transsexual). For more discussion and exemplification see Feinberg, L. (1996) *Transgender Warriors: Making History from Joan of Arc to RuPaul* (Boston: Beacon Press).

47. Foltz, D. W., H. Ochman, J. S. Jones, S. M. Evangelisti, and R. K. Selander (1982) "Genetic Population Structure and Breeding Systems in Arionid Slugs (Mollusca: Pulmonata)," *Biological Journal of the Linnean Society* 17:225–41.

48. Of course, the term *transvestism,* when applied to people, refers primarily the wearing clothing of the opposite sex (and all of the attendant social and cultural repercussions). In its zoological usage, however, it simply refers to physical or behavioral attributes that are typical of the opposite sex in that species. For scientific

use of this term, see the references in the notes for this section, as well as Weinrich, J. D. (1980) "Homosexual Behavior in Animals: (A New Review of Observations from the Wild, and Their Relationship to Human Sexuality," in R. Forleo and W. Pasini, eds., *Medical Sexology: The Third International Congress*, pp. 288–95. (Littleton, Mass.: PSG Publishing).

49. Owen, D. F. (1988) "Mimicry and Transvestism in *Papilio phorcas* (Lepidoptera: Papilionidae)," *Journal of the Entomological Society of Southern Africa* 51:294–96; Weldon, P. J., and G. M. Burghardt (1984) "Deception Divergence and Sexual Selection," *Zeitschrift für Tierpsychologie* 65:89–102.

50. Rohwer, S., S. D. Fretwell, and D. M. Niles (1980) "Delayed Maturation in Passerine Plumages and the Deceptive Acquistion of Resources," *American Naturalist* 115:400–437.

51. Estes, R. D. (1991) "The Significance of Horns and Other Male Secondary Sexual Characters in Female Bovids," *Applied Animal Behavior Science* 29:403–51; Guthrie, R. D., and R. G. Petocz (1970) "Weapon Automimicry Among Mammals," *American Naturalist* 104:585–88.

52. Kirwan, G. M. (1996) "Rostratulidae (Painted-Snipes)," p. 297, in J. del Hoyo, A. Elliott, and J. Sargatal, eds., *Handbook of the Birds of the World*, vol. 3, Hoatzin to Auks, pp. 292–301 (Barcelona: Lynx Ediciòns). Examples such as these are often termed sex-role reversal by biologists.

53. See the discussion of homosexual gender roles and interpretations of homosexuality as "pseudoheterosexuality" in chapter 4.

54. Bighorn Sheep (Berger 1985). See also chapter 2 for further discussion of human and animal comparisons.

55. Policansky, D. (1982) "Sex Change in Plants and Animals," *Annual Review of Ecology and Systematics* 13:471–95; Forsyth, A. (1986) *A Natural History of Sex: The Ecology and Evolution of Mating Behavior*, chapter 13. (New York: Scribner's).

56. For surveys of transsexuality in fishes, see Potts, G. W., and R. J. Wootton, eds., (1984) *Fish Reproduction: Strategies and Tactics* (London: Academic Press); Warner, R. R. (1978) "The Evolution of Hermaphroditism and Unisexuality in Aquatic and Terrestrial Vertebrates," in E. S. Reese and F. J. Lighter, eds., *Contrasts in Behavior: Adaptations in the Aquatic and Terrestrial Environments*, pp. 77–101 (New York: Wiley); Warner, R. R. (1975) "The Adaptive Significance of Sequential Hermaphroditism in Animals," *American Naturalist* 109:61–82; Warner, R. R. (1984) "Mating Behavior and Hermaphroditism in Coral Reef Fishes," *American Scientist* 72:128–36; Policansky, "Sex Change"; Armstrong, C. N. (1964) *Intersexuality in Vertebrates Including Man* (London: Academic); Smith, C. L. (1975) "The Evolution of Hermaphroditism in Fishes," in R. Reinboth, ed., *Intersexuality in the Animal Kingdom*, pp. 295–310 (New York: Springer-Verlag); Smith, C. L. (1967) "Contribution to a Theory of Hermaphroditism," *Journal of Theoretical Biology* 17:76–90.

57. Robertson, D. R., and R. R. Warner (1978) "Sexual Patterns in the Labroid Fishes of the Western Caribbean, II: The Parrotfishes (Scaridae)," *Smithsonian Contributions to Zoology* 255:1–26; Warner, R.R., and I. F. Downs (1977) "Comparative Life Histories: Growth versus Reproduction in Normal Males and Sex-changing Hermaphrodites in the Striped Parrotfish, *Scarus croicensis*," *Proceedings of the Third International Symposium on Coral Reefs* 1(Biology):275–82; Thresher, R. E. (1984) *Reproduction in Reef Fishes* (Neptune City, N.J.: T.F.H. Publications).

58. Paketi: Jones, G. P. (1980) "Growth and Reproduction in the Protogynous Hermaphrodite *Pseudolabrus celidotus* (Pisces: Labridae) in New Zealand," *Copeia* 1980:660–75; Ayling, T. (1982) *Sea Fishes of New Zealand*, p. 255 (Auckland: Collins). Humbug damselfish: Coates, D. (1982) "Some Observations on the Sexuality of Humbug Damselfish, *Dascyllus aruanus* (Pisces, Pomacentridae) in the Field," *Zeitschrift für Tierpsychologie* 59:7–18. Red Sea anemonefish: Fricke, H. W. (1979) "Mating System, Resource Defence, and Sex Change in the Anemonefish *Amphiprion akallopisos*," *Zeitschrift für Tierpsychologie* 50:313–26. Lantern bass and others: Petersen, C. W., and E. A. Fischer (1986) "Mating System of the Hermaphroditic Coral-reef Fish, *Serranus baldwini*," *Behavioral Ecology and Sociobiology* 19:171–78; Nakashima, Y., K. Karino, T. Kuwamura, and Y. Sakai (1997) "A Protogynous Wrasse May Have a Functionally Simultaneous Hermaphrodite Phase," in M. Taborsky and B. Taborsky, eds., *Contributions to the XXV International Ethological Conference*, p. 214, Advances in Ethology no. 32 (Berlin: Blackwell Wissenschafts-Verlag). Coral goby: Kuwamura, T., Y. Nakashima, and Y. Yogo (1994) "Sex Change in Either Direction by Growth-Rate Advantage in the Monogamous Coral Goby, *Paragobiodon echinocephalus*," *Behavioral Ecology* 5:434–38; Nakashima, Y., T. Kuwamura, and Y. Yogo (1995) "Why Be a Both-ways Sex Changer?" *Ethology* 101:301–7.

Chapter 2. Humanistic Animals, Animalistic Humans

1. Names for individual animals in each species, and the activities they engaged in, are from the following sources: Gorillas (Yamagiwa 1987a, Stewart 1977); Bottlenose Dolphins (Tavolga 1966); West Indian Manatees (Hartman 1971); Siamangs (Fox 1977); Bonobos (Idani 1991); Crested Black Macaques (Poirier 1964); Rhesus Macaques (Reinhardt et al. 1986); Japanese Macaques (Sugiyama 1960); Crab-eating Macaques (Hamilton 1914); Asiatic Mouflons (Pfeffer 1967); Grizzly Bears (Craighead 1979); Long-eared Hedgehogs (Poduschka 1981); Greylag Geese (Lorenz 1991); White-handed Gibbons (Edwards and Todd 1991); Orangutans (Rijksen 1978).

2. For cross-cultural and other surveys of the wide variety of human homosexualities, see Ford, C. S., and F. A. Beach (1951) *Patterns of Sexual Behavior* (New York: Harper and Row); Bell, A. P., and M. S. Weinberg (1978) *Homosexualities: A Study of Diversity Among Men and Women* (New York: Simon and Schuster); Blackwood,

E., ed., (1986) *The Many Faces of Homosexuality: Anthropological Approaches to Homosexual Behavior* (New York: Harrington Park Press); Greenberg, D. F. (1988) *The Construction of Homosexuality* (Chicago and London: University of Chicago Press); Murray, S. O., ed., (1992) *Oceanic Homosexualitites* (New York: Garland); Plummer, K., ed., (1992) *Modern Homosexualities: Fragments of Lesbian and Gay Experience* (London: Routledge); Murray, S. O. (1995) *Latin American Homosexualities* (Albuquerque: University of New Mexico Press); Murray, S., and W. Roscoe (1997) *Islamic Homosexualities: Culture, History, and Literature* (New York: New York University Press).

3. Kangaroos: Dagg, A. I. (1984) "Homosexual Behavior and Female-Male Mounting in Mammals—a First Survey," p. 179, *Mammal Review* 14:155–85. Bighorn Sheep: Weinrich, J. D. (1987) *Sexual Landscapes*, p. 294 (New York: Charles Scribner's Sons). Bottlenose Dolphins (Caldwell and Caldwell 1977:804).

4. For some discussion of the parameters, complexities, and variations found in prison homosexuality (including male "pairing" as opposed to "rape"), see Donaldson, S. (1993) "A Million Jockers, Punks, and Queens: Sex Among American Male Prisoners and Its Implications for Concepts of Sexual Orientation," lecture delivered at the Columbia University Seminar on Homosexualities; Wooden, W. S., and J. Parker (1982) *Men Behind Bars: Sexual Exploitation in Prison* (New York: Plenum). For discussion of similar factors in other types of "situational" homosexuality (i.e., evidence for the nonmonolithic nature of sexual activity in all-male groups), see Williams, W. L. (1986) "Seafarers, Cowboys, and Indians: Male Marriage in Fringe Societies on the Anglo-American Frontier," chapter 8 in *The Spirit and the Flesh: Sexual Diversity in American Indian Culture*, pp. 152–74 (Boston: Beacon Press).

5. Donaldson, S., and W. R Dynes (1990) "Typology of Homosexuality," in W. R. Dynes, ed., *Encyclopedia of Homosexuality*, vol. 2, pp. 1332–37 (New York and London: Garland). Donaldson and Dynes's typology uses three main axes, representing sexual orientation, gender expression, and temporal or chronological patterning. This triaxial schema has been expanded here to include a number of other axes.

6. For discussion of "axes" not specifically considered here, such as gendered homosexual interactions and the complex manifestation of gender roles in same-sex contexts, see chapter 4.

7. Weinrich, J. D. (1982) "Is Homosexuality Biologically Natural?" in W. Paul, J. D. Weinrich, J. C. Gonsiorek, and M. E. Hotveldt, eds., *Homosexuality: Social, Psychological, and Biological Issues*, pp. 197–211 (Beverly Hills, Calif.: SAGE Publications).

8. Gadpaille, W. J. (1980) "Cross-Species and Cross-Cultural Contributions to Understanding Homosexual Activity," *Archives of General Psychiatry* 37:349–56; Dagg, "Homosexual Behavior and Female-Male Mounting in Mammals"; Vasey, P. L. (1995) "Homosexual Behavior in Primates: A Review of Evidence and Theory," *International Journal of Primatology* 16:173–204.

9. This does not include domesticated species, in which the evidence for exclusive homosexuality is sometimes even more conclusive, as in the recent behavioral and physiological studies of domesticated sheep; see Adler, T. (1996) "Animals' Fancies (Why Members of Some Species Prefer Their Own Sex," *Science News* 151:8–9; Resko et al. 1996; Perkins et al. 1992, 1995. The question of homosexual orientation or "preference" also ties in to the common misconception that animal homosexuality is largely a matter of "necessity" or "last resort," i.e., a response to the absence or unavailability of the opposite sex. This issue will be addressed more fully in chapter 4.

10. Silver Gull (Mills 1991); Greylag Goose (Huber and Martys 1993); Humboldt Penguin (Scholten 1992 and personal communication). In addition, a pair-bond between (captive) male Yellow-backed Lories has been documented as lasting more than 14 years (Low 1977:134), although (unlike the other species) this occurred in the absence of birds of the opposite sex.

11. Galah (Rogers and McCulloch 1981); Common Gull (Riddiford 1995); Black-headed Gull (van Rhijn and Groothuis 1985, 1987); Great Cormorant (Fukuda 1992); Bicolored Antbird (Willis 1967, 1972); Black Swan (Braithwaite 1981); Ring-billed Gull (Kovacs and Ryder 1981); Western Gull (Hunt and Hunt 1977); Hooded Warbler (Niven 1993). See also Clarke (1982:71) for documentation of a pair-bond between captive male White-fronted Amazon Parrots that lasted for at least two years.

12. Bottlenose Dolphin (Wells 1991, 1995; Wells et al. 1987). Another possible example is male Cheetahs, who form life-long pair-bonds or "coalitions" with each other. Although many such individuals mate with females, and sexual activity specifically between male pair-members has only recently been verified in captivity (Ruiz-Miranda et al. 1998), it is likely that at least some paired males have few if any heterosexual contacts for significant portions of their lives (especially in view of the fact that opposite-sex mating opportunities for such males are often reduced [Caro 1994:252, 304]).

13. Similarities and differences between homosexual and heterosexual patterns are discussed in more detail in the following section "Sexual Virtuosos."

14. Kleiman, D. G. (1977) "Monogamy in Mammals," *Quarterly Review of Biology* 52:39–69; Clutton-Brock, T. G. (1989) "Mammalian Mating Systems," *Proceedings of the Royal Society of London*, Series B 235:339–72. In most of these pair-forming mammals homosexuality has not been reported; however, same-sex activity does occur among Gibbons, Rufous-naped Tamarins, and Wolves, but not between animals that are bonded to each other as mates. In Gibbons, homosexual interactions are incestuous—between father and son(s). In an oblique sense, then, this pattern of homosexuality involves a sort of "monogamy," in that sexual activity is not sought outside of the primary relationship or family unit. Interestingly, Gibbons do sometimes seek promiscuous copulations with partners other than their mate—but all such cases reported so far involve heterosexual, rather than homosexual, infidelities (see, for example, Palombit 1994a,b).

15. Gorilla (Robbins 1995:29, 30, 38); Hanuman Langur (Rajpurohit et al. 1995:292).

16. Of course, it has often been claimed that homosexuality in such contexts is strictly "situational" or "by de-fault," i.e., due to the absence of females. This is a great oversimplification, assuming as it does that males are necessarily "forced" into living in same-sex groups and engaging in homosexual activities; for further dis-cussion and evidence against this interpretation, see chapter 5. Moreover, homosexual behavior "by default" is still homosexual behavior. If the "motivation" or desire of participants is to be factored in so as to distin-guish "genuine" homosexuality, then the same must be done for opposite-sex interactions. The fact is that heterosexual behavior in many contexts is also "situational," not actively sought out by one or both partners, or even overtly resisted, yet it still falls under the umbrella of "heterosexuality" (see chapters 4 and 5 for fur-ther discussion and exemplification).

17. Mountain Zebra (Rasa and Lloyd 1994:172); American Bison (Komers et al. 1992:197, 201).

18. Nilgiri Langur (Hohmann 1989:445–47); Ruff (van Rhijn 1991; Hogan-Warburg 1966); White-handed Gib-bon (Edwards and Todd 1991); Red Fox (Macdonald 1980:137; Schantz 1984:200; Storm and Montgomery 1975:239).

19. For further examples of these various forms of sequential bisexuality, consult the index. Most, if not all, of these patterns are also attested in human bisexualities; for an example of seasonal bisexuality (one of the less well known patterns), see the description of a medieval Persian practice in which men had female partners in winter and male ones in the summer (Murray and Roscoe, *Islamic Homosexualities,* p. 139).

20. Kinsey, A. C., W. B. Pomeroy, and C. E. Martin (1948) *Sexual Behavior in the Human Male,* pp. 638–41 (Philadelphia: W. B. Saunders).

21. Calculations are based on data in the following sources: Bonobo (Idani 1991:90–91 [tables 5–6]); Red Deer (Hall 1983:278 [table 2]); Bonnet Macaque (Sugiyama 1971:259–60 [tables 8–9]); Pig-tailed Macaque (Tokuda et al. 1968:288, 290 [tables 3, 5]); Kob (Buechner and Schloeth 1965:219 [table 2]).

22. R. Wrangham, quoted in Bull, C. (1997) "Monkey Love," *The Advocate,* June 10, 735:58. This is but one ex-ample of the often misleading statements about animal homosexuality that are perpetuated by both scien-tists and the popular press. See chapter 3 for further discussion.

23. Females had an average of 5.2 different female partners and 4.1 male partners and ranged 4–9 female part-ners (out of a pool of 10) and 1–9 male partners (out of a pool of 10) (Idani 1991:91). Of course, all of these figures represent only a relatively short "snapshot" of Bonobo behavior (covering three months), but it is likely that longer-term or lifetime patterns exhibit a comparable spectrum of variation. Because female Bonobos are neither exclusively heterosexual nor exclusively homosexual, de Waal (1997:192) advocates use of the term *pansexual* to descibe their sexual orientation. This characterization is as appropriate as *bisexual* as long as it is understood that individuals exhibit a range of same-sex versus opposite-sex interactions (i.e., there are many gradations of "pansexuality" or "bisexuality" in this species).

24. Of course, many other factors besides sexual "preference" are involved in the choice of mates, especially with regard to the availability and specific characteristics of partners. The fact that only some animals ever engage in homosexual or heterosexual activity, however, is an important indication that differences in sexual orien-tation probably also exist at an individual level. For further discussion of the role that partner availability may play in the occurrence of homosexual (and heterosexual) activity, see chapter 4.

25. Silver Gull: data for 131 females tracked in the wild over their entire lives, from Mills 1991:1525 (table 1); Black-headed Gull: data for 27 males in a captive population studied for seven years, based on van Rhijn and Groothuis 1985:161 (table 3); Japanese Macaque: averages for 46–58 females over two consecutive years in a semiwild population, from Wolfe 1979:526; Galah: based on pair-bonding data over six years pooled from two captive populations comprising 27 birds, from Rogers and McCulloch 1981.

26. Vasey, "Homosexual Behavior in Primates," p 197.

27. For explicit observations of the nonchalant responses of surrounding animals, including the heterosexual mates or parents of individuals involved in same-sex activity, see Common Chimpanzee (Kortlandt 1962:132); Gorilla (Harcourt et al. 1981:276); White-handed Gibbon (Edwards and Todd 1991:232–33); Japanese Macaque (Wolfe 1984; Vasey 1995:190); Killer Whale (Jacobsen 1986:152); Gray Whale (Darling 1978:55); Northern Fur Seal (Bartholomew 1959:168); African Buffalo (Mloszewski 1983:186); Lion (Cha-van 1981:364); Rufous Rat Kangaroo (Johnson 1980:356); Dwarf Cavy (Rood 1970:442); Laughing Gull (Noble and Wurm 1943:205); Sage Grouse (Scott 1942:495). In the majority of cases where homosexual ac-tivities draw no response from surrounding animals, scientific observers simply make no comment about the behavior of the other animals. In one species, the Blue-bellied Roller, same-sex (and opposite-sex) "dis-play" mounting is *only* performed when other birds are present to watch (but not intervene).

28. African Buffalo (Mloszewski 1983:186); Musk-ox (Tener 1965:75).

29. Other species in which harassment of heterosexual but not homosexual activity has been reported include Proboscis Monkeys (Yeager 1990a:224), Squirrel Monkeys (DuMond 1968:121–22; Baldwin and Baldwin 1981:304), Lechwe (Nefdt 1995), Wolves (Zimen 1976, 1981; Derix et al. 1993), Red-necked Wallabies (John-son 1989:275), Gray Squirrels (Koprowski 1992a:393; 1993:167–68), Kittiwakes (Chardine 1986), and King Penguins (Stonehouse 1960:32). In Hanuman Langurs, more than 83 percent of heterosexual copulations are harassed while harassment of homosexual ones only occurs occasionally (Sommer 1989a:208; Srivastava et al. 1991:497); for greater interruption of heterosexual mounts in Japanese Macaques, see Hanby 1974:840; in Moor Macaques, see Matsumura and Okamoto 1998:227–28. See also chapter 5 for further discussion of harassment of heterosexual matings.

30. Bonobo (de Waal 1995:48, 1997:117, 120; Hashimoto 1997:12); Jackdaw (Röell 1978:29); Guianan Cock-of-the-Rock (Trail and Koutnik 1986:210–11). In a number of species (e.g., Rhesus and Crab-eating Macaques,

Spotted Hyenas) a phenomenon known as scapegoating sometimes occurs, in which several individuals combine forces to attack another individual not directly involved in a dispute. Notably, individuals engaging in same-sex activity are not specifically targeted as scapegoats, and this behavior is not in fact generally related to sexual activity at all (Harcourt, A. H., and F. B. M. de Waal, eds., (1992) *Coalitions and Alliances in Humans and Other Animals*, pp. 87, 91, 129, 240 [Oxford: Oxford University Press]).

31. Savanna Baboon (Marais 1922/1969); Red Deer (Darling 1937); Common Garter Snake (Mason and Crews 1985).

32. Greylag Goose (Lorenz 1979, 1991; Huber and Martys 1993; Schönfeld 1985); Black Swan (Braithwaite 1981).

33. For example, Japanese Macaques, Savanna Baboons, Kob, Mute Swans, Black-winged Stilts, Caspian Terns, Black-billed Magpies.

34. Greylag Goose (Huber and Martys 1993); Black Swan (Braithwaite 1981); Flamingo (King 1994, 1993a,b; E. Stevens, personal communication); Orange-fronted Parakeet (Hardy 1963:187, 1965:150); Laughing Gull (Noble and Wurm 1943:205; Hand 1981:138–39); Rose-ringed Parakeet (Goodwin 1983:87); Nilgiri Langur (Hohmann 1989:452); Lion (Cooper 1942:27–28); Rhesus Macaque (Fairbanks et al. 1977:247); Japanese Macaque (Vasey 1998); Common Chimpanzee (de Waal 1982:64–66); Livingstone's Fruit Bat (Courts 1996:27); Savanna Baboon (Marais 1922/1969:205–6). Homosexual pairs in the domesticated Bengalese Finch also attack other birds (Masatomi 1959). Additionally, in a number of mammals (e.g., Common Chimpanzees, Bonnet Macaques, Savanna Baboons, Bottlenose and Atlantic Spotted Dolphins, Cheetahs), paired "coalitions" or "alliances" of males that "solidify" their partnership through overt or ritualized sexual, affectionate, and bonding behaviors often cooperate in challenging and attacking other individuals; a similar phenomenon ocurs in homosexually bonded female Oystercatchers that are part of bisexual trios.

35. Brown Capuchin (Linn et al. 1995:50); Rufous Rat Kangaroo (Ganslosser and Fuchs 1988:311); Sage Grouse (Patterson 1952:155–56); Gorilla (Harcourt et al. 1981:276; Fisher and Nadler 1978:660–61); Bonobo (de Waal 1997:114, 130); Canada Goose (Allen 1934:197–98); Wapiti (Franklin and Lieb 1979:188–89); Japanese Macaque (Vasey 1998); Rhesus Macaque (Akers and Conaway 1979:76); Jackdaw (Röell 1979:124–25). In Greenshanks, the female partner of a male involved in homosexual copulation made a threatening call during a same-sex interaction but did not interfere (Nethersole-Thompson 1951:109–10).

36. White-tailed Deer (Thomas et al. 1965). Another possible example of transgendered animals being harassed is found in the paketi (a New Zealand fish), in which Ayling (1982:255) claims that transvestite fish are attacked ("beat up") when their true sex is discovered; however, Jones (1980), the original source on which this account is based, does not in fact mention such behavior (Ayling, T. [1982] *Sea Fishes of New Zealand* [Auckland: Collins]; Jones, G. P. [1980] "Growth and Reproduction in the Protogynous Hermaphrodite *Pseudolabrus celidotus* [Pisces: Labridae] in New Zealand," *Copeia* 1980:660–75).

37. Manakadan, R. (1991) "A Flock of One-Legged Greenshanks *Tringa nebularia*," *Journal of the Bombay Natural History Society* 88:452.

38. Ring-billed Gull (Kovacs and Ryder 1983; Fetterolf et al. 1984); Japanese Macaque (Wolfe 1986:272; Gouzoules and Goy 1983:41); Greylag Goose (Huber and Martys 1993:161–62); Mallard Duck (Schutz 1965). Heg and van Treuren (1998:689) also found that bisexual trios of Oystercatchers are just as common on optimal as suboptimal territories.

39. Weinrich, *Sexual Landscapes*, p. 308.

40. Bonobo (de Waal 1997:107); Bottlenose Dolphin (Wells et al. 1987:294); Orang-utan (Galdikas 1981:285, 297, 1995:172; Kaplan and Rogers 1994:82).

41. Undoubtedly other species will be discovered that also exhibit this full range of characteristics. Many of these features are already known to characterize Stumptail Macaque sexuality, for example, including hidden estrous cycles (cf. de Waal 1989:150), anal and oral intercourse, and pairlike "sexual friendships" or "preferred partners" (and much remains to be learned about this species in the wild). Similarly, Japanese Macaques have pair-bonded consortships, face-to-face sexual encounters, and "social class" differences in sexual/pairing activity (cf. Corradino 1990:360), while Gorillas have face-to-face copulation, bonding or "preferred partners," hidden estrous cycles (cf. Wolfe 1991:125), and oral sexual activities. Certain of these characteristics also occur individually in animal groups other than primates and cetaceans: a face-to-face mating position, for instance, is occasionally used by snow leopards, while Ruffs have a highly structured "class" system among males involving (among other features) differing sexual behaviors (Freeman, H. [1983] "Behavior in Adult Pairs of Captive Snow Leopards [*Panthera uncia*]," *Zoo Biology* 2:1–22). Another erroneous claim about human uniqueness is that no animal exhibits a type of homosexuality sometimes known as "(mutual) androphilia," an interaction involving two adult males neither of whom adopts a stereotypically "feminine" gender presentation or sexual behavior (on the supposed absence of this in animals, see Houser, W. [1990] "Animal Homosexuality," in W. R. Dynes, ed., *Encyclopedia of Homosexuality*, vol. 1, pp. 60–63 [New York and London: Garland]). In fact, exactly this sort of homosexuality occurs in Greylag Goose and Mallard Duck male pairs, as well as in a number of other species; see chapter 4 for further discussion of gender roles (or their absence) in homosexual interactions.

42. Weinrich, *Sexual Landscapes*, p. 305 (where this idea is formulated as the "technique puzzle" and characterized as "a disturbing generalization"); Masters, W. H., and V. E. Johnson (1979) *Homosexuality in Perspective* (Boston: Little, Brown).

43. Likewise, the durations of homosexual as opposed to heterosexual acts (such as mounting) are usually comparable. In some species, however, homosexual interactions generally last longer (e.g., Gorillas, White-handed Gibbons, American Bison, West Indian Manatees), while in others heterosexual encounters typically

last longer (e.g., Harbor Seals, Red Foxes, Humboldt Penguins, Long-tailed Hermit Hummingbirds). In many species homosexual interactions do exhibit greater variability or flexibility in terms of the role differentiation of partners (see chapter 4 for further discussion).

44. Bonobo (Kitamura 1989:53–57, 61; Kano 1992:187); Gorilla (Fischer and Nadler 1978:660–61; Yamagiwa 1987a:12–14, 1987b:37; Harcourt and Stewart 1978:611–12); Hanuman Langur (Weber and Vogel 1970:76; Srivastava et al. 1991:496–97).

45. Japanese Macaque (Hanby and Brown 1974:164; Hanby 1974:838–40).

46. Flamingo (C. E. King, personal communication).

47. The head-to-tail position does occur in interspecies homosexual interactions with Tucuxi Dolphins. Same-species versus cross-species differences in mounting position (independent of the sex of the partner) are also found in other cetaceans. Among Bottlenose Dolphins, for example, a belly-to-belly mating position is typical of same-species contacts, both homosexual and heterosexual (cf. McBride and Hebb 1948:115, among others), while a sideways, dorsoventral position occurs in interspecies encounters with Atlantic Spotted Dolphins of both sexes (Herzing and Johnson 1997:92, 96).

48. Anderson, S. (1993) "Stitchbirds Copulate Front to Front," *Notornis* 40:14; Tyrrell, E. Q. (1990) *Hummingbirds of the Caribbean*, pp. 114, 155 (New York: Crown Publishers); "Red-capped Plover, *Charadrius ruficapillus*," in S. Marchant and P. J. Higgins, eds. (1993) *Handbook of Australian, New Zealand, and Antarctic Birds*, vol. 2, pp. 836–47 (Melbourne: Oxford University Press); Wilkinson, R., and T. R. Birkhead (1995) "Copulation Behavior in the Vasa Parrots *Coracopsis vasa* and *C. nigra*," *Ibis* 137:117–19; Kilham, L. (1983) *Life History Studies of Woodpeckers of Eastern North America*, pp. 49–50, 143, 160 (Cambridge, Mass.: Nuttall Ornithological Club); Southern, W. E. (1960) "Copulatory Behavior of the Red-headed Woodpecker," *Auk* 77:218–19.

49. Vasey, "Homosexual Behavior in Primates," p. 195.

50. For further details see the primate profiles in part 2, as well as the discussion of nonreproductive heterosexualities in chapter 5.

51. For discussion of cultural traditions in animals, including references to many specific cases, see Bonner, J. T. (1980) *The Evolution of Culture in Animals* (Princeton, N.J.: Princeton University Press); Galef, B. G., Jr. (1995) "Why Behavior Patterns That Animals Learn Socially Are Locally Adaptive," *Animal Behavior* 49:1325–34; Lefebvre, L. (1995) "The Opening of Milk Bottles by Birds: Evidence for Accelerating Learning Rates, but Against the Wave-of-Advance Model of Cultural Transmission," *Behavioral Processes* 34:43–54; Menzel, E.W., Jr., ed. (1973) *Precultural Primate Behavior* (Symposia of the Fourth International Congress of Primatology, vol. 1) (Basel: S. Karger); Gould, J. L., and C. G. Gould (1994) *The Animal Mind* (New York: Scientific American Library). For an excellent survey of animal cultural traditions, see Mundinger, P. C. (1980) "Animal Cultures and a General Theory of Cultural Evolution," *Ethology and Sociobiology* 1:183–223.

52. Japanese Macaque (Itani 1959; Gouzoules and Goy 1983:47; Eaton 1978; Wolfe 1984:152); Stumptail Macaque (Chevalier-Skolnikoff 1976:512; Bertrand 1969:193–94); Savanna Baboon (Ransom 1981:139). In Hanuman Langurs, mounting between females may also have a cultural component, since it exhibits wide variability not only between individuals but also between geographic areas. It occurs frequently in some regions (e.g., Jodhpur, India), less frequently in others (e.g., Abu and Sariska in India), rarely in others (e.g., Sri Lanka), and not at all in still others (e.g., some parts of Nepal) (Srivastava et al.1991:504–5 [table V]). Heterosexual courtship patterns in Common Chimpanzees also exhibit cultural variations (cf. Nishida 1997:394, among others).

53. Bonobo (Savage-Rumbaugh et al. 1977; Savage-Rumbaugh and Wilkerson 1978; Savage and Bakeman 1978; Roth 1995; S. Savage-Rumbaugh, personal communication). Drawings, verbal descriptions, and "glosses" of the hand signals and their meanings in the accompanying illustration are based on these sources. For alternate descriptions of some of these gestures, as well as gestures used in nonsexual situations, see de Waal 1988:214–21.

54. Bonobo (Savage-Rumbaugh and Wilkerson 1978:334; Roth 1995:75, 88).

55. Bonobo (Savage-Rumbaugh et al. 1977:108).

56. Linguists studying the structure of American Sign Language, for example, have identified a continuum of iconicity in signs, ranging from *transparent* signs (quasi-mimetic gestures whose meaning is readily identifiable from their form, even to nonsigners) to *translucent* signs (gestures in which a connection between meaning and form can be discerned but not automatically identified without knowing the meaning of the sign) to *opaque* signs (gestures in which all form-meaning correspondences have been lost). According to these criteria, the Bonobo gestures would fall primarily in the transparent-translucent range. For further discussion see Klima, E.S., and U. Bellugi (1979) *The Signs of Language*, especially chapter 1, "Iconicity in Signs and Signing" (Cambridge: Harvard University Press).

57. This gestural system has only been observed in captivity, albeit in "untrained" Bonobos. Studies of wild Bonobos have so far revealed a less elaborate communicative repertoire associated with sexual interactions, although researchers have identified similar types of communicative exchanges prior to some episodes of sexual activity (e.g., Kitamura 1989:54–55; Enomoto 1990:473–75). It must also be remembered that many behaviors are easily missed in the field (especially given the particular difficulties of observing wild Bonobos; cf. de Waal 1997:12, 63–64, 70, 76–77); hence it is possible that more elaborate gestural repertoires do occur in wild Bonobos but have yet to be observed. For more on the issue of behaviors that are only observed in captivity as opposed to the wild, see chapter 4.

58. Hewes, G. W. (1973) "Primate Communication and the Gestural Origin of Language," *Current Anthropology* 14:5–24; Hewes, G. W. (1976) "The Current Status of the Gestural Theory of Language Origin," in S. Harnad, H. Steklis, and J. Lancaster, eds., *Origins and Evolution of Language and Speech.* Annals of the New York Academy of Science, vol. 280, pp. 482–504 (New York: New York Academy of Sciences).

59. Bonobo (Roth 1995:4–45).

60. Beck, B. B. (1980) *Animal Tool Behavior: The Use and Manufacture of Tools by Animals* (New York: Garland); Goodall 1986:545–48, 559 (Common Chimpanzee); van Lawick-Goodall, J., H. van Lawick, and C. Packer (1973) "Tool-Use in Free-living Baboons in the Gombe National Park, Tanzania," *Nature* 241:212–13; McGrew, W. C. (1992) *Chimpanzee Material Culture: Implications for Human Evolution* (Cambridge: Cambridge University Press); Berthelet, A., and J. Chavaillon, eds., (1993) *The Use of Tools by Human and Nonhuman Primates* (Oxford: Clarendon Press); Weinberg, S.M., and D.K. Candland (1981) "'Stone-Grooming' in *Macaca fuscata,*" *American Journal of Primatology* 1:465–68.

61. Orang-utan (Rijksen 1978:262–63; Nadler 1982:241; Harrison 1961:61).

62. Common Chimpanzee (Bingham 1928:148–50; Kollar et al. 1968:456–57; Goodall 1986:559–60; McGrew, *Chimpanzee Material Culture,* p. 183); Bonobo (Takeshita and Walraven 1996:428; Walraven et al. 1993:28, 30; Becker, C. [1984] *Orang-Utans und Bonobos im Spiel: Untersuchungen zum Spielverhalten von Menschenaffen* [Orang-utans and Bonobos at Play: Investigations on the Play Behavior of Apes], pp. 149, 152, 193–94 [Munich: Profil-Verlag]). Female Japanese Macaques are also reported to use inanimate objects for masturbation (Rendall and Taylor 1991:321), although it is not clear whether this involves use of "tools" or simply rubbing of genitals against a surface (as is found in many other species). Masturbatory tool use is also occasionally reported for animals other than primates; see, for example, Shadle's description of male and female Porcupines holding sticks in their forepaws while straddling the object in order to stimulate their genitals (Shadle, A. R. [1946] "Copulation in the Porcupine," *Journal of Wildlife Management* 10:159–62; Shadle, A. R., M. Smelzer, and M. Metz [1946] "The Sex Reactions of Porcupines (*Erethizon d. dorsatum*) Before and After Copulation," *Journal of Mammalogy* 27:116–21). Objects or "tools" are also sometimes employed by Common Chimpanzees and Bonobos during heterosexual courtship and solicitations (cf. McGrew, *Chimpanzee Material Culture,* pp. 82, 188; Nishida 1997:385, 394 [Common Chimpanzee]; de Waal 1997:120 [Bonobo]).

63. Bonnet Macaque (Sinha 1997). Sinha (1997:23) believes that this female was using the tools to "scratch" her vagina, possibly because of "some irritation," whose presence, however, was never confirmed. Sexual stimulation is also compatible with the observed behaviors (instead of, or along with, "scratching"), especially considering that masturbation without the use of tools occurs regularly in Bonnet Macaques of both sexes (cf. Makwana 1980:11; Kaufman and Rosenblum 1966:221; Rahaman and Parthasarathy 1969:155).

64. See, for example, (Rawson, P. (1973) *Primitive Erotic Art,* especially pp. 20, 71 (New York: G. P. Putnam's Sons); Kinsey, A. C., W. B. Pomeroy, C. E. Martin, and P. H. Gebhard (1953) *Sexual Behavior in the Human Female,* p. 136 (Philadelphia: W. B. Saunders). Examples of tools utilized for sexual stimulation of partners (rather than self-stimulation) have yet to be reported for any nonhuman species. For a recent discussion of the role of sexual pleasure in the evolution of tool use among both nonhuman primates and early humans, see Vasey, P. L. (1998) "Intimate Sexual Relations in Prehistory: Lessons from Japanese Macaques," *World Archaeology* 29:407–25.

65. For further discussion of these (and other) examples as well as cultural variation in the occurrence of incest and its taboos, see Leavitt, G. C. (1990) "Sociobiological Explanations of Incest Avoidance: A Critical Review of Evidential Claims," *American Anthropologist* 92:971–93; Arens, W. (1986) *The Original Sin: Incest and Its Meaning* (New York and Oxford: Oxford University Press); Livingstone, F. B. (1980) "Cultural Causes of Genetic Change," in G. W. Barlow and J. Silverberg, eds., *Sociobiology: Beyond Nature/Nurture?* pp. 307–29, AAAS Selected Symposium, no. 35 (Boulder: Westview Press); Schneider, D. M. (1976) "The Meaning of Incest," *Journal of the Polynesian Society* 85:149–69.

66. For an overview of a variety of kinship restrictions on Melanesian homosexual relations, see Schwimmer, E. (1984) "Male Couples in New Guinea," pp. 276–77, in G. H. Herdt, ed., *Ritualized Homosexuality in Melanesia,* pp. 248–91 (Berkeley: University of California Press); Murray, S. O. (1992) "Age-Stratified Homosexuality: Introduction," pp. 10–12, in Murray, *Oceanic Homosexualities,* pp. 293–327. For more on New Guinean homosexualities, see chapter 6.

67. Leavitt, "Sociobiological Explanations," pp. 974–75; Livingstone, "Cultural Causes," p. 318. For arguments that in animals some forms of inbreeding (such as between cousins) may actually be *beneficial* genetic and social effects and are preferred in some species (e.g., Vervet monkeys, Japanese quail), see Moore and Ali 1984 (Bonnet Macaque); Bateson, P. (1982) "Preferences for Cousins in Japanese Quail," *Nature* 295:236–37; Shields, W. M. (1982) *Philopatry, Inbreeding, and the Evolution of Sex* (Albany: State University of New York Press); Cheney, D. M., and R. M. Seyfarth (1982) "Recognition of Individuals Within and Between Groups of Free-Ranging Vervet Monkeys," *American Zoologist* 22:519–30.

68. Japanese Macaque (Wolfe 1979; Chapais and Mignault 1991; Vasey 1996:543; Chapais et al. 1997); Hanuman Langur (Srivastava et al. 1991:509 [table II]; Sommer and Rajpurohit 1989:304, 310); Bonobo (Hashimoto et al. 1996:315–16). Bonobo mother-daughter homosexual relations occasionally occur (Thompson-Handler et al. 1984:355).

69. Savanna Baboon (Smuts and Watanabe 1990:167–70).

70. Berndt, R., and C. Berndt (1945) "A Preliminary Report of Field Work in the Ooldea Region, Western South Australia," pp. 245, 260–66, *Oceania* 15:239–75; Meggitt, M. J. (1962) *Desert People: A Study of the Walbiri*

Aborigines of Central Australia, pp. 262–63. (Sydney: Angus and Robertson); Eibl-Eibesfeldt, I. (1977) "Patterns of Greeting in New Guinea," pp. 221, 226, in S. A. Wurm, ed., *New Guinea Area Languages and Language Study,* Vol. 3: *Language, Culture, Society, and the Modern World,* pp. 209–47, Pacific Linguistics Series C, no. 40 (Canberra: Australian National University Press). For more on ritualized homosexuality among the Bedamini and other New Guinean peoples, see chapter 6.

71. Interestingly, homosexual activity in another group of highly intelligent creatures—whales and dolphins—also has many of the hallmarks of cultural activity identified here. Same-sex activity varies considerably between individuals, populations, and time periods in a number of cetaceans. For example, sexual interactions between male Killer Whales appear to differ in frequency and occurrence depending on the geographic area (Rose 1992:7), while male pairs of Bottlenose Dolphins exhibit different characteristics in various populations (see chapter 5). In addition, "incest taboos" appear to be operative in most male Killer Whale homosexual interactions (Rose 1992:112), sexual activity in Bottlenose Dolphins sometimes has a ritualistic or "greeting" component (Östman 1991:313), while Bottlenose Dolphins of both sexes have also been observed "masturbating" or stimulating their genitals using inanimate objects (Caldwell and Caldwell 1972:430).

72. Hamer, D., and P. Copeland (1994) *The Science of Desire: The Search for the Gay Gene and the Biology of Behavior,* p. 213 (New York: Simon and Schuster).

73. Ward, J. (1987) "The Nature of Heterosexuality," in G. E. Hanscombe and M. Humphries, eds., *Heterosexuality,* pp. 145–69. (London: GMP Publishers).

74. Weinrich, J. D. (1982) "Is Homosexuality Biologically Natural?" in W. Paul, J. D. Weinrich, J. C. Gonsiorek, and M. E. Hotvedt, eds., *Homosexuality: Social, Psychological, and Biological Issues,* pp. 197–208 (Beverly Hills, Calif.: SAGE Publications). For an early discussion of animal homosexuality in relation to the question of "naturalness," see Gide, A. (1911/1950) *Corydon* (New York: Farrar, Straus, and Co.).

75. Weinrich, ibid.; Plant, R. (1986) *The Pink Triangle: The Nazi War Against Homosexuals,* pp. 27, 185 (New York: Henry Holt); Grau, G., ed., (1995) *Hidden Holocaust? Gay and Lesbian Persecution in Germany 1933–45,* p. 284 (London: Cassell); Mann, M. (1797/1866) *The Female Review: Life of Deborah Sampson, the Female Soldier in the War of the Revolution,* p. 225 (Boston: J. K. Wiggin & W. P. Lunt) [excerpts reprinted in Katz, J. (1976) *Gay American History,* pp. 212–214 (New York: Thomas Y. Crowell)]. Boswell, J. (1980) *Christianity, Social Tolerance, and Homosexuality: Gay People in Western Europe from the Beginning of the Christian Era to the Fourteenth Century,* p. 309 (Chicago: University of Chicago Press).

76. For a summary and overview of such experimental studies (e.g., involving hormones), see Mondimore, F. M. (1996) *A Natural History of Homosexuality,* pp. 111–13, 129–30 (Baltimore: Johns Hopkins University Press). These studies, typically involving laboratory rats, also invariably overlook the fact that the homosexual behaviors "induced" by hormones and other experimental treatments occur spontaneously in the wild ancestors of the laboratory animals involved, e.g., (European) Brown Rats (cf. Barnett 1958). Concerning further pitfalls in extrapolating from laboratory animals, as well as a general discussion of the "nature versus nurture" controversy, see Byne, W. (1994) "The Biological Evidence Challenged," *Scientific American* 270(5):50–55; LeVay, S., and D. H. Hamer (1994) "Evidence for a Biological Influence in Male Homosexuality," *Scientific American* 270(5):44–49.

77. Weinrich, "Is Homosexuality Biologically Natural?" p. 207.

78. See chapter 5, as well as the animal profiles in part 2, for specific examples.

79. For explicit statements by gay-bashers to the effect that homosexuality is "not natural," see Comstock, G. D. (1991) *Violence Against Lesbians and Gay Men,* p. 74 (New York: Columbia University Press).

80. Middleton, S., and D. Liittschwager (1996) "Parting Shots?" *Sierra* 81(1):40–45.

81. For documentation of these activities, see the following sources: Ligon, J. D. (1970) "Behavior and Breeding Biology of the Red-cockaded Woodpecker," *Auk* 87:255–78; Lennartz, M. R., R.G. Hooper, and R. F. Harlow (1987) "Sociality and Cooperative Breeding of Red-cockaded Woodpeckers, *Picoides borealis,*" *Behavioral Ecology and Sociobiology* 20:77–88; Walters, J. R., P. D. Doerr, and J. H. Carter III (1988) "The Cooperative Breeding System of the Red-cockaded Woodpecker," *Ethology* 78:275–305; Walters, J. R. (1990) "Red-cockaded Woodpeckers: A 'Primitive' Cooperative Breeder," in P. B. Stacey and W. D. Koenig, eds., *Cooperative Breeding in Birds: Long-Term Studies of Ecology and Behavior,* pp. 69–101 (Cambridge: Cambridge University Press); Haig, S. M., J. R. Walters, and J. H. Plissner (1994) "Genetic Evidence for Monogamy in the Cooperatively Breeding Red-cockaded Woodpecker," *Behavioral Ecology and Sociobiology* 34:295–303; Rossell, C. R., Jr., and J. J. Britcher (1994) "Evidence of Plural Breeding by Red-cockaded Woodpeckers," *Wilson Bulletin* 106:557–59.

82. For a complete list of references, see the appendix. For an example of an anecdotal, nonscientific account, see O'Donoghue, B. P. (1996) *My Lead Dog Was a Lesbian: Mushing Across Alaska in the Iditarod—the World's Most Grueling Race,* p. 42 (New York: Vintage). Interestingly, a female dog that showed interest in both females and males is described in this book as "sexually confused" and willing to mount "any dog within reach"—some of the same subjective characterizations that appear in the scientific descriptions of bisexuality and homosexuality among wild animals (see chapters 3 and 4).

83. Ford and Beach, *Patterns of Sexual Behavior,* p. 142; Denniston, R. H. (1980) "Ambisexuality in Animals," p. 34, in J. Marmor, ed., *Homosexual Behavior: A Modern Reappraisal,* pp. 25–40 (New York: Basic Books).

84. Kelley, K. (1978) *Playboy* interview: Anita Bryant, *Playboy,* May 1978, p. 82. Quoted in Weinrich, "Is Homosexuality Biologically Natural?" p. 198.

85. Lillian Faderman, interviewed in the *Seattle Gay News*, October 21, 1994, p. 26. See also Faderman's mention of same-sex activities in her pet terriers in Faderman, L. (1998) "Setting Love Straight," *The Advocate*, February 17, 753:72.

Chapter 3. Two Hundred Years of Looking at Homosexual Wildlife

1. Edwards, G. (1758–64) *Gleanings of Natural History. Exhibiting figures of quadrupeds, birds, insects, plants, etc., many of which have not, till now, been either figured or described*, vol. 3, p. xxi (London: Royal College of Physicians); Orang-utan (Morris 1964:502); Tree Swallow (Lombardo et al. 1994:555).

2. For discussion of observations of animal homosexuality in nonwestern scientific traditions, particularly those of indigenous cultures, see chapter 6.

3. Horapollo (1835) *Hieroglyphica*, Greek text edited by Conradus Leemans (Amsterdam: J. Muller; English translation by George Boas [New York: Pantheon, 1950]); Cory, A. T., ed. and trans., (1840) *The Hieroglyphics of Horapollo Nilous* (London: Pickering); Buffon, G. L. L. Count de (1749–67) *Histoire naturelle générale et particulière* (Natural History, General and Particular), 15 vols. (Paris: De l'Imprimerie royale); Edwards, *Gleanings of Natural History.* For an annotated bibliography of these and other early references on animal homosexuality, see Dynes, W. R. (1987) "Animal Homosexuality," in *Homosexuality: A Research Guide*, pp. 743–49 (New York and London: Garland Publishing). For further discussions of Aristotle, Horapollo, and others, see Boswell, J. (1980) *Christianity, Social Tolerance, and Homosexuality: Gay People in Western Europe from the Beginning of the Christian Era to the Fourteenth Century*, especially chapters 6 and 11 (Chicago and London: University of Chicago Press).

4. Laboulmène 1859, Gadeau de Kerville 1896 (insects); Rollinat and Trouessart 1895, 1896 (Bats); Whitaker 1885 (Mute Swan); Selous 1906–7 (Ruff); Karsch, F. (1900) "Päderastie und Tribadie bei den Tieren auf Grund der Literatur" (Pederasty and Tribadism Among Animals on the Basis of the Literature), *Jahrbuch für sexuelle Zwischenstufen* 2:126–60.

5. Morris 1964 (Orang-utan), Morris 1954 (Zebra Finches), Morris 1952 (Ten-spined Stickleback); Fossey 1983, 1990, Harcourt, Stewart, and Fossey 1981, Harcourt, Fossey, Stewart, and Watts 1980 (Gorilla); Lorenz 1979, 1991 (Greylag Goose), Lorenz 1935, 1972 (Jackdaw, Raven).

6. Mute Swan (Low and M. of Tavistock 1935:147).

7. Snow Goose (Quinn et al. 1989); Oystercatcher (Heg and van Treueren 1998); Bonobo (Hashimoto et al. 1996; Roth 1995; Savage-Rumbaugh et al. 1977); Roseate Tern (Sabo et al. 1994); Ruff (Lank et al. 1995); Silver Gull (Mills 1989, 1991); Bottlenose Dolphin (Wells 1991, 1995; Wells et al. 1987); Red Fox (Macdonald 1980; Storm and Montgomery 1975); Spotted Hyena (Mills 1990); Grizzly Bear (Craighead and Craighead 1972; Craighead et al. 1995); Griffon Vulture (Mouze and Bagnolini 1995); Victoria's Riflebird (Frith and Cooper 1996); Black-winged Stilt (Kitagawa 1988a).

8. Separation—Rhesus Macaque (Erwin and Maple 1976); Bottlenose Dolphin (McBride and Hebb 1948); Cheetah (Ruiz-Miranda et al. 1998); Long-eared Hedgehog (Poduschka 1981); Black-headed Gull (van Rhijn 1985; van Rhijn and Groothuis 1987); see also Clarke 1982:71 (White-fronted Amazon Parrot); removal—Orange-fronted Parakeet (Hardy 1963:187); electrodes—Stumptail Macaque (Goldfoot et al. 1980); deafening—Squirrel Monkey (Talmage-Riggs and Anschel 1973); castration—Crab-eating, Rhesus Macaques (Hamilton 1914); White-tailed Deer (Taylor et al. 1964); lobotomy—Domestic Cats (Green et al. 1957); killing, tissue collection—Common Garter Snake (Noble 1937); Hooded Warbler (Niven 1993); Gentoo Penguin (Roberts 1934). For primate hormonal treatment studies relating to homosexuality, see the literature survey in Vasey, P. L. (1995) "Homosexual Behavior in Primates: A Review of Evidence and Theory," *International Journal of Primatology* 16:173–204. For examples of hormonal treatments administered to transgendered animals, see Savanna Baboon (Bielert 1984b, 1985); White-tailed Deer (Thomas et al. 1970).

9. Wolfe, L. D. (1991) "Human Evolution and the Sexual Behavior of Female Primates," p. 130, in J. D. Loy and C. B. Peters, eds., *Understanding Behavior: What Primate Studies Tell Us About Human Behavior*, pp. 121–51 (New York: Oxford University Press). For another example of the extent to which scientific information about animal homosexuality remains unpublished (thereby perpetuating inaccuracies), see Weinrich's account of how he had to obtain much of his information from personal conversations and letters with zoologists—a procedure that was still necessary, a decade later, in the preparation of this book (Weinrich, J. D. [1987] *Sexual Landscapes*, p. 308 [New York: Charles Scribner's Sons]).

10. See, for example, Hubbard, R., M. Henifin, and B. Fried, eds., (1979) *Women Look at Biology Looking at Women: A Collection of Feminist Critiques* (Cambridge: Schenkman); Hrdy, S. B., and G. C. Williams (1983) "Behavioral Biology and the Double Standard," in S. K. Wasser, ed., *Social Behavior of Female Vertebrates*, pp. 3–17 (New York: Academic Press); Shaw, E., and J. Darling (1985) *Female Strategies* (New York: Walker and Company); Kevles, B. (1986) *Females of the Species: Sex and Survival in the Animal Kingdom* (Cambridge, Mass.: Harvard University Press); Haraway, D. (1989) *Primate Visions: Gender, Race, and Nature in the World of Modern Science* (New York: Routledge); Gowaty, P. A., ed. (1996) *Feminism and Evolutionary Biology: Boundaries, Intersections, and Frontiers* (New York: Chapman Hall); Cunningham, E., and T. Birkhead (1997) "Female Roles in Perspective," *Trends in Ecology and Evolution* 12:337–38. On the general male-centeredness of most biological theorizing, see Eberhard, W. G. (1996) *Female Control: Sexual Selection by Cryptic Female Choice*, pp. 34–36. (Princeton: Princeton University Press); Batten, M. (1992) *Sexual Strategies* (New York:

Putnam's); Gowaty, P. A. (1997) "Principles of Females' Perspectives in Avian Behavioral Ecology," *Journal of Avian Biology* 28:95–102.

11. This is not to suggest, of course, that only scientists who are themselves homosexual can deal with the subject in an unbiased way. Certainly many contemporary heterosexual biologists do not harbor negative views about homosexuality, while some gay and lesbian zoologists have undoubtedly perpetuated the silences and prejudices of their field. (There are also those who believe that being homosexual actually invalidates a gay or lesbian scientist's objectivity on the subject. However, if sexual orientation resulted in such bias, then heterosexual zoologists should confine themselves only to research topics that have nothing to do with breeding or male-female relations.) Nevertheless, sexism and male bias in biology have been exposed most directly through the work of women and feminist scientists, and it is likely that similar insights regarding heterosexism and homophobia will be forthcoming from openly gay, lesbian, or bisexual zoologists—that is, once such people no longer have to fear losing tenure, research grants, or jobs because of their outspokenness. Regardless of their own sexual orientation, however, many zoologists have avoided studying homosexuality or speaking widely about their results because the topic is still far from being considered a "legitimate" area of inquiry (see, for example, Wolfe's commentary above; also, Anne Perkins's decision not to discuss her findings on homosexuality in domestic sheep until after she had secured tenure, reported in "Counting Sheep," *Advocate*, July 8, 1997, 737:21). A parallel situation exists in the fields of anthropology and history, where denial, omission, suppression, and condemnation of information about human homosexuality have long been carried out by researchers studying other cultures or historical periods. For a particularly good discussion of this phenomenon, see Read, K. E. (1984) "The *Nama* Cult Recalled," in G. H. Herdt, ed., *Ritualized Homosexuality in Melanesia*, pp. 211–47 (Berkeley: University of California Press). On the myth of observer "objectivity" where discussion of homosexuality by anthropologists is concerned, see Lewin, E., and W. L. Leap, eds. (1996) *Out in the Field: Reflections of Lesbian and Gay Anthropologists* (Urbana and Chicago: University of Illinois Press). For further discussion of indigenous human homosexualities, see chapter 6.

12. Dagg, A. I. (1984) "Homosexual Behavior and Female-Male Mounting in Mammals—a First Survey," *Mammal Review* 14:155–85; Vasey, "Homosexual Behavior in Primates"; Vasey 1996, 1998 (Japanese Macaque); Vasey, P. L. (in press) "Homosexual Behavior in Male Birds," "Homosexual Behavior in Male Primates," in W. R. Dynes, ed., *Encyclopedia of Homosexuality*, 2nd ed., vol. 1: Male Homosexuality (New York: Garland Press). See also the recent bibliography: Williams, J. B. (1992) *Homosexuality in Nonhuman Primates: A Bibliography: 1940–1992* (Seattle: Primate Information Center). For descriptions of animal homosexuality that are relatively value neutral (i.e., that do not view homosexual behavior as inherently problematic), or for accounts that are not overly concerned with finding a "cause" or "explanation" for the behavior, see, for example, Yeager 1990a (Proboscis Monkey); Marlow 1975 (Australian, New Zealand Sea Lions); Sowls 1974, 1984 (Collared Peccary); Schaller 1967 (Blackbuck, Barasingha); Braithwaite 1981 (Black Swan); King 1994 (Flamingo); Riddiford 1995 (Common Gull); Smith 1988 (Lyrebird); Neelakantan 1962 (Black-rumped Flameback); and Rogers and McCulloch 1981, Rowley 1990 (Galah). For descriptions of homosexual activity that recognize it as a routine or "normal" behavioral phenomenon, see Porton and White 1996 (Gorilla); Akers and Conaway 1979 (Rhesus Macaque); Eaton 1978, Fedigan 1982, Wolfe 1984, 1986, Chapais and Mignault 1991, Vasey 1996 (Japanese Macaque); Chevalier-Skolnikoff 1976 (Stumptail Macaque); Wells et al. 1987, Wells 1991, Wells et al. 1998 (Bottlenose Dolphin); Rose 1992 (Killer Whale); Hartman 1971, 1979 (West Indian Manatee); Lott 1983 (American Bison); and Coe 1967 (Giraffe). In addition, a number of zoologists in their personal communications with me have been refreshingly free of the negative judgments or interpretations that unfortunately characterize most of the field; among them, B. J. Ens (Oystercatchers), C. B. Frith (Birds of Paradise), M. Fujioka (Egrets), M. Fukuda (Great Cormorants), D. Heg (Oystercatchers), D. L. Herzing (Dolphins), C. E. King (Flamingos), W. D. Koenig (Acorn Woodpeckers), D. F. Lott (American Bison), M. Martys (Greylag Geese), M. G. L. Mills (Spotted Hyenas), C. Reed (Crested Black Macaques), S. Savage-Rumbaugh (Bonobos), C. J. Scholten (Humboldt Penguins), L. H. Smith (Superb Lyrebirds), Y. Sugiyama (primates), and P. L. Vasey (Japanese Macaques, other species).

13. While the word *homophobia* means, literally, an irrational fear of homosexuality, the term is also applied to instances of disgust, revulsion, hatred, or open hostility, as well as more subtle prejudicial feelings of discomfort, distaste, or dislike toward homosexuality or homosexual individuals (not necessarily accompanied by fear). For more discussion and further references on the nature and consequences of homophobia, see Blumenfeld, W. J., ed. (1992) *Homophobia: How We All Pay the Price* (Boston: Beacon Press).

14. Ruff (Selous 1906–7:420, 423); American Bison (McHugh 1958:25); Waterbuck (Spinage 1982:118).

15. The appellations *abnormal, aberrant, unnatural,* or *perverted,* for example, have been applied by scientists to homosexual behavior or transgender in at least 30 different species of mammals and birds (often in multiple sources for each species), and as recently as the mid-1980s in some published accounts (Kittiwake [Coulson and Thomas 1985:20]; Bighorn Sheep [Berger 1985]). Even more recently (Finley et al. 1997:914–15, 917), same-sex courtship and sexual activity in Fruit Flies (as well as refusal of heterosexual advances) have been characterized as "abnormal," "aberrant," and a "defect," and similar terms have also been used by some zoologists in their personal communications with me. Somewhat less derogatory designations such as *odd* (including *odd couples*), *peculiar, irregular,* or *bizarre* have been used to describe homosexuality or transgender in at least 15 other species of mammals and birds. Many other examples can of course be found in descriptions of reptiles, amphibians, fishes, insects, and other creatures. Heterosexual behaviors or individuals are characterized as "normal" in opposition to homosexual behaviors or individuals in the following published

scientific sources, among others: Common Chimpanzee (Adang et al. 1987:242); Gorilla (Harcourt 1988:59); Kob (Buechner and Schloeth 1965:2219); Canada Goose (Collias and Jahn 1959:484); Black-winged Stilt (Kitagawa 1988a:64); Black-headed Gull (van Rhijn and Groothuis 1985:161); Lovebirds (Dilger 1960:667); Hooded Warbler (Niven 1994:192); Ostrich (Sauer 1972:729).

16. Gadeau de Kerville (1896); Grollet and L. Lepinay (1908) "L'inversion sexuelle chez les animaux" (Sexual Inversion in Animals), *Revue de l'hypnotisme* 23:34–37; Savanna Baboon (Marais 1922/1969); Bengalese Finch (Masatomi 1957); Ostrich (Sauer 1972); Long-eared Hedgehog (Poduschka 1981); Whiptail Lizard (Crews and Young 1991).

17. Mazarine Blue (Tennent 1987:81–82).

18. Domestic Cattle (Klemm et al. 1983:187); Elephants (Rosse 1892:799); Lion (Cooper 1942:26–28); Buff-breasted Sandpiper (Myers 1989); Domestic Turkey (Hale 1955:1059); Spinner Dolphin (Wells 1984:470); Killer Whale (Rose 1992:112); Caribou (Bergerud 1974:420); Adélie Penguin (Davis et al. 1998:137); Black-billed Magpie (Baeyens 1979:39–40); Guianan Cock-of-the-Rock (Trail 1985a:238–39); Sage Grouse (Scott 1942:494). Other terms, while not necessarily derogatory, reflect scientists' particular interpretations of such behavior as substitute or counterproductive activities: same-sex mounting in Gorillas is called "vicarious" sexual activity (Fossey 1983:74, 188–89), and homosexual mounting in African Buffalo is categorized as "barren sexual behavior" (Mloszewski 1983:186). See also the subsequent section "Mock Courtships and Sham Matings" for discussion of the widespread use of terms such as *false* or *mock* sexual behavior to characterize homosexual activity, and chapters 4 and 5 for other interpretations of homosexuality.

19. Long-eared Hedgehog (Poduschka 1981:84, 87); Eastern Gray Kangaroo (Grant 1974:74); Black-crowned Night Heron (Noble et al. 1938:29); King Penguin (Gillespie 1932:95, 98); Gorilla (Harcourt 1988:59); Lorikeets (Low 1977:24); Red Fox (Macdonald 1987:101); Greenshank (Nethersole-Thompson and Nethersole-Thompson 1979:112–13; Nethersole-Thompson 1951:109).

20. This is not to say, of course, that homosexual "advances" are never unwanted. Various forms of nonconsensual courtship or sexual approaches between animals of the same sex have been reported in about a quarter of the mammal and bird species exhibiting homosexuality. However, in many cases they co-occur with "consensual" homosexual interactions in the same species, from which they are clearly distinguished by behavioral indications of unwillingness on the part of one partner. As in nonconsensual *heterosexual* interactions (which are reported in more than a third of the species in which homosexual behavior has been documented and in general are equally, if not more, prevalent in animals—see chapter 5), there is actually a continuum of disinterest and "refusal" behavior. An animal may signal its unwillingness by not permitting any sexual approaches or contact at all, by permitting sexual contact but not facilitating the interaction, or by actively interrupting contact (either by trying to get away or by attacking the other animal). Assertions by scientists of "unwanted" homosexual attentions are usually anthropomorphic projections made regardless of whether such behavioral evidence is present (or what degree of nonconsensuality is involved).

21. Mountain Sheep (Geist 1975:100); Rhesus Macaque (Carpenter 1942:137, 151–52); Laughing Gull (Noble and Wurm 1943:205–6); Cattle Egret (Fujioka and Yamagishi 1981:139); Sage Grouse (Gibson and Bradbury 1986:396); Orang-utan (Rijksen 1978:264–65); Kob (Buechner and Schloeth 1965:211–12, 217, 219); Ostrich (Sauer 1972:729, 733); Guianan Cock-of-the-Rock (Trail and Koutnik 1986:210–11, 215); Mallard Duck (Schutz 1965:458); Rhesus Macaque (Kempf 1917:136). One zoologist also reveals something of his own misconceptions concerning both homosexual and heterosexual intercourse when he expresses surprise that a female Bonobo "on the bottom" during a lesbian interaction does not appear to mind—in fact, visibly enjoys—being in that position: "If we were on the bottom being held down, we would probably feel submissive and inferior, but female pygmy chimpanzees seem not to take it that way . . . the female on the bottom . . . looks proud and affectionate" (Kano 1992:193).

22. Greylag Goose (Huber and Martys 1993:161); see Lorenz (1991:241–42) on gander pairs being more closely bonded than heterosexual pairs.

23. Ocellated Antbird (Willis 1973:31); on heterosexual divorce in Antbirds, see Willis (1983:414).

24. Gorilla (Fischer and Nadler 1978:660–61); Western Gull (Hunt et al. 1984:160); Guianan Cock-of-the-Rock (Trail 1985a:238, 240); Red Fox (Macdonald 1987:101); de Waal 1989a:25 (Bonobo). For descriptions of nonstandard mounting positions (lateral, head-to-tail) in heterosexual contexts, see (for example) Japanese Macaque (Hanby and Brown 1974:156, 164); Boto (Best and da Silva 1989:15); Bottlenose/Spotted Dolphins (Herzing and Johnson 1997:92, 96); Waterbuck (Spinage 1969:41–42); Mountain Sheep (Geist 1971:139–40); Mountain Goat (Hutchins 1984:268); Takhi (Boyd and Houpt 1994:202); Collared Peccary (Byers and Bekoff 1981:771); Warthog (Cumming 1975:118–19); Koala (Smith 1980c:48); Ruff (Hogan-Warburg 1966:176); Hammerhead (Brown 1982:171; Campbell 1983:11); Flamingo (Shannon 1985:229); Chaffinch (Marler 1956:114); red-winged blackbird (Monnett, C., L. M. Rotterman, C. Worlein, and K. Halupka [1984] "Copulation Patterns of Red-winged Blackbirds [*Agelaius phoeniceus*]," p. 759, *American Naturalist* 124:757–64). Of these, subjective or derogatory terms are only used in Monnett et al. 1984 ("abnormal," "aberrant") and Hutchins 1984 ("clumsy," "awkward"). Nonstandard homosexual mounting positions such as sideways or head-to-tail mounts have usually been classified as "mistakes" or "incomplete" mounting attempts by zoologists who insist on viewing homosexual interactions only in terms of how closely they resemble "standard" heterosexual intercourse. In other words, anything that deviates from genital penetration (or cloacal contact in birds) in the front-to-back position used by males with females is an "error." Because these mounting positions are often used by female animals (when they mount individuals of either sex), a further,

sexist, interpretation is also frequently overlaid on these behaviors: it is claimed that they represent an "imperfect" attempt on the part of females to imitate male mounting behavior. An equally valid perspective, however, is that these represent alternative or more "fluid" sexual interchanges—not bound by the "requirement" of genital penetration—rather than flawed imitations of heterosexual postures. A parallel example can be found in the behavior of "sideways presenting" in female Crab-eating Macaques: previously classified as "disoriented" or "inadequate," this posture was later found to be a systematic behavioral variant (Emory and Harris 1978). For further discussion and a critique of the widespread view that homosexuality is an imperfect approximation of heterosexuality, see chapter 4. For evidence that heterosexual sex is not focused exclusively on vaginal penetration and ejaculation, see chapter 5.

25. Laughing Gull (Hand 1981:138–39); Black-headed Gull (van Rhijn and Groothuis 1985:161); Herring Gull (Shugart et al. 1988:934); inclusion of infertile eggs in hatching rates of female pairs: Kovacs and Ryder 1983:661–62, Ryder and Somppi 1979:3 (Ring-billed Gull); Burger, J., and M. Gochfeld (1996) "Laridae (Gulls)," p. 584, in J. del Hoyo, A. Elliott, and J. Sargatal, eds., *Handbook of the Birds of the World*, vol. 3, Hoatzin to Auks, pp. 572–623 (Barcelona: Lynx Edicións); shared characteristics of heterosexual and homosexual supernormal clutches: Kovacs and Ryder 1983:660–62, Lagrenade and Mousseau 1983, Ryder and Somppi 1979:3 (Ring-billed Gull and other species) (on the lower productivity of supernormal clutches attended by heterosexual pairs in species other than Gulls, see Sordahl, T. A. [1997] "Breeding Biology of the American Avocet and Black-necked Stilt in Northern Utah," pp. 350, 352, *Southwestern Naturalist* 41:348–54); equivalent parenting abilities of homosexual and heterosexual pairs: Hunt and Hunt 1977:1467, Hayward and Fry 1993:17–18 (Western Gull); Conover 1989:148 (Ring-billed Gull); Nisbet et al. 1998:314 (Roseate Tern); "runaways" from heterosexual parents: Pierotti and Murphy 1987 (Western Gull and other species); Brown et al. 1995 (Ring-billed Gull); Roberts and Hatch 1994 (Kittiwake).
26. Gray Whale (Darling 1977:10–11).
27. In fact, it can safely be said that no scientific study of wild animals has yet been undertaken with the *expectation* that homosexual activity would be observed—same-sex behavior is invariably a "surprise." In contrast, many a field study has been initiated for the express purpose of studying heterosexual mating—and has quite often been treated to the unexpected occurrence of same-sex activity and/or the absence (or rarity) of opposite-sex interactions.
28. Laughing Gull (Burger and Beer 1975:312); Common Murre (Hatchwell 1988:167); Kittiwake (Chardine 1986:1416, 1987:516); Griffon Vulture (Blanco and Martinez 1996:247).
29. Grebes (Nuechterlein and Storer 1989:344–45).
30. For a recent example concerning a little-known species, see Dyrcz, A. (1994) "Breeding Biology and Behavior of the Willie Wagtail *Rhipidura leucophrys* in the Mdang Region, Papua New Guinea," *Emu* 94:17–26.
31. Emu (Heinroth 1924, 1927); Regent Bowerbird (Gilliard 1969:341); Dugong (Jones 1967; Nair et al. 1975:14). The visual resemblance between younger male and adult female Superb Lyrebirds has also resulted in some misidentifications and revised interpretations of this species' behavior in the wild. Although Smith (1968:88–89, 1988:30–32, 75–78) and Lill (1979a:496) state clearly (and offer photographic documentation) that adult males court (and even mount) younger males, the identification of some individuals has not been so straightforward. One bird photographed as it was being courted by an adult male (including full courtship displays) was first identified as "possibly" a male (Smith 1968:60), then as a female (Smith 1988:30). However, after a careful review of the plumage characteristics of adult females and younger males, L. H. Smith has confirmed (personal communication) that the younger bird in this case was indeed a male and in fact was most likely the adult male's own son. Unfortunately, the earlier published reports in which the sex of the younger bird was unclear may have led Reilly (1988:32) to state erroneously that males never perform full courtship displays toward other males. For additional photographs of males performing full displays to other males, see Smith (1988:77) and p. 13 (this book).
32. King Penguin (Gillespie 1932:96–120).
33. Snow Goose (Quinn et al. 1989); Ring-billed Gull (Kovacs and Ryder 1981); Red-backed Shrike (Pounds 1972); Blue Tit (Blakey 1996); Guianan Cock-of-the-Rock (Trail and Koutnik 1986); Stumptail Macaque (Chevalier-Skolnikoff 1976:522 [table III]); Jackdaw (Röell 1979:126–27); Bonobo (Parish 1996:65, 86; de Waal 1997:112–15). Similarly, in citing Hartman's (1971) original descriptions of homosexuality in West Indian Manatees, Ronald et al. (1978:37) focus on examples of same-sex activity that occur in conjunction with heterosexual behaviors and downplay those that are independent of opposite-sex encounters (even though such independent encounters are equally, if not more, prevalent). On a related point, genes that are thought to control homosexual activity in Fruit Flies have been given names by scientists that refer solely to their (negative) effect on heterosexuality and breeding. One gene has been labeled *dissatisfaction* (alluding to the fact that carriers of this gene, in addition to being interested in homosexual activity, typically refuse or are "dissatisfied" with heterosexual advances), while another has been called *fruitless* (alluding to the fact that carriers, in addition to courting individuals of both sexes, are infertile) (Finley et al. 1997:917).
34. Savanna (Olive) Baboon (Owens 1976:254); Right Whale (Clark 1983:169); Moose (Van Ballenberghe and Miquelle 1993:1688); Cattle Egret (Fujioka and Yamagishi 1981:136).
35. Squirrel Monkey (Talmage-Riggs and Anschel 1973:70–71); Bonobo (Savage-Rumbaugh and Wilkerson 1978:338; Savage and Bakeman 1978:614); Spotted Hyena (Burr 1996:118–19). For conflicting information on the occurrence of clitoral penetration in Spotted Hyenas, see Glickman (1995). See also Morris (1956:261), who defines courtship as "the heterosexual reproductive communication system leading up to

the consummatory sexual act" (Morris, D. [1956] "The Function and Causation of Courtship Ceremonies," in M. Autuori and Fondation Singer-Polignac, *L'instinct dans le comportement des animaux et de l'homme* [Paris: Masson et Cie.])

36. Savanna (Chacma) Baboon (Marais 1922/1969:215).

37. Ruff (van Rhijn 1991:21); Bonnet Macaque (Nolte 1955:179).

38. Walrus (Miller and Boness 1983:305); African Elephant (Shelton 1965:163–64); Gorilla (Maple, T. [1977] "Unusual Sexual Behavior of Nonhuman Primates," in J. Money and H. Musaph, eds., *Handbook of Sexology*, pp. 1169–70 [Amsterdam: Excerpta Medica]); Sage Grouse (Scott 1942:495); Hanuman Langur (Mohnot 1984:349); Common Chimpanzee (Kortlandt 1962:132); Musk-ox (Reinhardt 1985:297–98); Mallard Duck (Lebret 1961:111–12); Blue-bellied Roller (Moynihan 1990:17); Lion (Cooper 1942:26–28); Orang-utan (Rijksen 1978:257); Savanna Baboon (Noë 1992:295, 311); Mule Deer (Halford et al. 1987:107); Hammerhead (Brown 1982:171; Campbell 1983:11); Bonobo (Thompson-Handler et al. 1984:358; de Waal 1987:319, 1997:102); Japanese Macaque (Green 1975:14); Rhesus Macaque (Reinhardt et al. 1986:56); Red Fox (Macdonald 1980:137); Squirrels (Ferron 1980; Horwich 1972; Reilly 1972). A few of these terms are also applied to nonreproductive heterosexual activities, in which case the attribution of "falseness" refers to the fact that the behavior does not result in procreation rather than to a same-sex context per se. See chapter 5 for further discussion of the parallel treatment of nonreproductive heterosexual behaviors as "abnormal" in the history of zoology.

39. The categorization of homosexual activity as less than "genuine" sexual activity is an important issue, and the various ways that same-sex activity is desexualized will be examined in greater detail in the next section.

40. Bonobo (Kano 1992); Common Chimpanzee (de Waal 1982); Snow Goose (Diamond 1989); Lesser Flamingo (Alraun and Hewston 1997); Oystercatcher (Heg and van Treuren 1998); Black-billed Magpie (Baeyens 1979); Black Stilt (Reed 1993); Fruit Flies (Cook 1975); Long-legged Fly sp. (Dyte 1989).

41. Gowaty, P. A. (1982) "Sexual Terms in Sociobiology: Emotionally Evocative and Paradoxically, Jargon," *Animal Behavior* 30:630–31. The title of the article in question (Abele and Gilchrist 1977, on Acanthocephalan Worms) also contained the word *rape*, so it is possible that scientists were "snickering" at this as well. Gowaty suggests replacing, along with *unisexual* for *homosexual*, all "loaded" terminology with more "neutral" words, e.g., *forced copulation* for *rape*, *kleptogamy* for *cuckoldry*, *one-male social unit* for *harem*. Many of her arguments for such alternate terminology are valid, e.g., that the "loaded" terms are often scientifically inaccurate. Notably, however, her principal argument against the word *homosexual* is not that it is inaccurate, but that use of this term is "sensationalistic" and triggers the prejudices of other scientists, thereby preventing them from seeing past the word to what it describes. It should also be pointed out that many formerly controversial terms for heterosexual behaviors are now acceptable in scientific circles. The word *divorce*, for example, was first greeted with an "uproar" when used to describe the break-up of pair-bonds in birds, and numerous scientists suggested replacing it with more "neutral" words; yet the term is now widely used in the ornithological literature (Milius, S. [1998] "When Birds Divorce: Who Splits, Who Benefits, and Who Gets the Nest," p. 153, *Science News* 153:153–55).

42. Giraffe (Coe 1967:320; Leuthold, W. [1977] *African Ungulates: A Comparative Review of Their Ethology and Behavioral Ecology*, p. 130 [Berlin: Springer-Verlag]).

43. Connor, J. (1997) "Courtship Testing," *Living Bird* 16(3)31–32; Depraz, V., G. Leboucher, L. Nagle, and M. Kreutzer (1997) "'Sexy' Songs of Male Canaries: Are They Necessary for Female Nest-Building?" in M. Taborsky and B. Taborsky, eds., *Contributions to the XXV International Ethological Conference*, p. 122, Advances in Ethology no. 32 (Berlin: Blackwell Wissenschafts-Verlag); Emlen, S. T., and N. J. Demong (1996) "All in the Family," *Living Bird* 15(3):30–34; Savanna Baboon (Smuts 1985:223, 1987:39, 43); Tasmanian Native Hen (Goldizen et al. 1998); Mirande, C. M., and G. Archibald (1990) "Sexual Maturity and Pair Formation in Captive Cranes at the International Crane Foundation," in *AAZPA Annual Conference Proceedings*, pp. 216–25 (Wheeling, W.Va.: American Association of Zoological Parks and Aquariums); Bonobo (de Waal 1997:117); Eisner, T., M. A. Goetz, D. E. Hill, S. R. Smedley, and J. Meinwald (1997) "Firefly 'Femmes Fatales' Acquire Defensive Steroids (Lucibufagins) from Their Firefly Prey," *Proceedings of the National Academy of Sciences* 94:9723–28; Domestic Goat (Shank 1972:500). See also the discussion of red-cockaded woodpecker "family values" in chapter 2.

44. Greylag Goose (Lorenz 1991:241–43) (see also Lorenz's own assertion, in this same passage, that such male pairs are not simply platonic "friendships" between males but are equivalent to male-female mated pairs). Analogously, Kortlandt (1949) (Great Cormorant) labels same-sex pairs "pseudohomosexual" rather than "homosexual" if their members later form heterosexual bonds, once again equating "true" homosexuality with lifetime, exclusive same-sex pairing. See chapter 2 for more on the dubious notion of "true" homosexuality and its relation to more sophisticated characterizations of sexual orientation. Lorenz's unwillingness to apply the term *homosexual* to gander pairs and thereby invite human-animal comparisons (or imply full heterosexual-homosexual equivalence) is especially problematic in light of his activities during the Third Reich. As a member of the Nazi party in Austria and an official lecturer for its Office of Race Policy, Lorenz did not hesitate to draw analogies between animals and people to support and develop the doctrines of "biological degeneracy," "racial purity," and the "elimination" of "inferior" or "asocial" elements (Deichmann, U. [1996] *Biologists under Hitler*, especially "Konrad Lorenz, Ethology, and National Socialist Racial Doctrine," pp. 179–205. Cambridge, Mass.: Harvard University Press). Among his most blatant assertions in this regard are statements (in 1943) that physical and moral "decay" in people is "identical" to the effects of domestication on animals and (in 1940) that the "defective type" among humans is like "the domesticated animal that

can be bred in the dirtiest stable and with any sexual partner" (ibid., pp. 186, 188; cf. Lorenz's [1935/1970:203] surmise that same-sex pairing in Jackdaws only occurs in captive animals and is not a feature of "natural" populations). He also asserted (in 1941) that "Precisely in the large field of instinctive behavior, humans and animals can be directly compared.... We confidently venture to predict that these studies will be fruitful for both theoretical as well as practical concerns of race policy" (ibid., p. 186). The subject of how the antihomosexual climate of Nazi Germany and the Nazi sympathies of some biologists helped shape the scientific discourse on animal homosexuality deserves further investigation. Many zoological studies of this phenomenon, after all, were written in Germany and Austria during this period or were heavily influenced by work that had its genesis during this time. Moreover, one of the earliest scientific surveys of animal homosexuality (Karsh, "Päderastie und Tribadie bei den Tieren" [1990]), appeared in the periodical *Jahrbuch für sexuelle Zwischenstufen* (Yearbook for Sexual Intermediate Types), published by the noted Jewish homosexual Magnus Hirschfield, whose mammoth archives and library of sexology were later destroyed by the Nazis.

45. Western Gull (Hayward and Fry 1993). See also Diamond and Burns, who suggest that same-sex pairing in Gulls should be referred to as "joint brooding" or "coparenting" rather than as homosexuality, thereby emphasizing its supposed reproductive functions (Diamond, M., and J. A. Burns [1995] "Human-Nonhuman Comparisons in Sex: Valid and Invalid," paper presented at the 24th International Ethological Conference, Honolulu, Hawaii). For arguments that same-sex pairing is *not* primarily a reproductive behavior, see chapters 4 and 5.

46. For examples of scientists who use the term *homosexual* (or *lesbian* or *gay*) even when no overt sexual activity is involved (i.e., to refer to related behaviors such as courtship, pairing, or parenting), see Sauer 1972 (Ostrich), Nethersole-Thompson 1975 (Scottish Crossbill), Wingfield et al. 1980 (Western Gull), Braithwaite 1981 (Black Swan), Smith 1988 (Lyrebird), Diamond 1989 (Snow Goose), Reed 1993 (Black Stilt).

47. And in fact just such a "broad" definition of heterosexuality is *required* in many cases. "Heterosexual pairs" in which little or no sexual activity occurs between partners have been reported for Greylag Geese (as mentioned above) and Lesser Scaup Ducks (Afton 1985:150), among others; see also Loy (1971:26) for "sexual" bonds between male and female Rhesus Macaques that do not involve mounting or copulation, and Smuts (1985:18, 163–66, 199, 213) on the platonic "pair-bonds" or "friendships" between male and female Savanna Baboons. In addition, in some "heterosexual pairs" of splendid fairy-wrens all offspring are fathered by males other than the female's pair-bonded mate (i.e., she does not copulate—or at least is not fertilized by— her partner); see Russell, E., and I. Rowley (1996) "Partnerships in Promiscuous Splendid Fairy-wrens," in J. M. Black, ed., *Partnerships in Birds: The Study of Monogamy*, pp. 162–73 (Oxford: Oxford University Press). For an example of a "broad" definition of (hetero)sexuality that encompasses courtship activities in addition to overt copulatory behavior, see Tinbergen, N. (1965) "Some Recent Studies of the Evolution of Sexual Behavior," in F. A. Beach, ed., *Sex and Behavior*, pp. 1–33 (New York: John Wiley and Sons).

48. In discussing the possible dangers of anthropomorphism in terminology, the comments of biologist John Bonner are instructive: "An anthropologist might find the use of words such as *slaves* or *castes* for ant colonies most undesirable.... For instance, it implies that the most repugnant human morals are ascribed to the members of some species of ant.... Much worse, it could imply that if ants have slavery, it is a natural thing to do and therefore quite justified in a human society. These arguments are not quite rational and can only be advanced under extreme fervor of one sort or another. A more reasoned objection would be that the motivations of ants and men might differ radically, but by using the same words this distinction is lost. A biologist, on the other hand, feels that the points made above are too obvious to interfere with the dual use of the words. He does not see any problem: in both ant and human slavery individuals forcibly capture members of their own species or related species and cause their captives to do work for the benefit of the captors. It is unnecessary to drag in all the possible political, psychological, or strictly human nuances; a very simple definition of the word is sufficient. There is no need to be tyrannized by words. If a biologist may not use the common words, he will be forced to invent a whole new set of jargon terms for nonhuman societies, an unfortunate direction since there are too many jargon words in any science as it is. I hope it will be sufficient if I make it clear in the beginning that words either invented or frequently used for human societies will also be used for animal societies with the understanding that I am not implying anything human in their meaning; they are to be considered simple descriptions of conditions." (Bonner, J. T. [1980] *The Evolution of Culture in Animals*, pp. 9–10. [Princeton, N.J.: Princeton University Press].) Unfortunately, this eminently reasonable position has not been adopted by most biologists where homosexuality is concerned; for a counterview, see Gowaty, "Sexual Terms in Sociobiology."

49. Examples of species in which homosexual activity is given only cursory treatment compared to heterosexual activity are too numerous to list, but include White-tailed Deer (Hirth 1977), Wapiti (Harper et al. 1967), Fat-tailed Dunnart (Ewer 1968), Matschie's Tree Kangaroo (Hutchins et al. 1991), Wattled Starling (Sontag 1991), Sage Grouse (Wiley 1973, Gibson and Bradbury 1986), and Canary-winged Parakeet (Arrowood 1988). In a few studies, however, detailed quantitative and descriptive information is provided on homosexual behavior; see, for example, Kitamura 1989, Kano 1992, de Waal 1987, 1995, 1997 (Bonobo); Edwards and Todd 1991 (White-handed Gibbon); Hanby 1974, Eaton 1978, Chapais and Mignault 1991, Vasey 1996 (Japanese Macaque); Pratt and Anderson 1985 (Giraffe); Jamieson and Craig 1987a (Pukeko); van Rhijn and Groothuis 1985, 1987 (Black-headed Gull); Rogers and McCulloch 1981 (Galah). For further discussion of how same-sex activity has frequently not been considered "genuine" sexual behavior, see the next section.

50. Spinner Dolphin (Wells 1984:468; Bateson 1974); Kob (Buechner and Schloeth 1965:219 [table 2]); Crested Black Macaque (Dixson 1977); Brown Capuchin (Linn et al. 1995); Giraffe (Dagg and Foster 1976:75–77).

51. Western Gull (Hunt et al. 1984:160) (see Hayward and Fry [1993:16, 18] for a recent reiteration of the findings of this study, in which sexual activity is once again downplayed); Black-crowned Night Heron (Noble et al. 1938:28–29); on comparable levels of crowding in wild colonies, see Gross 1923:13–15; Davis 1993:6; Kazantzidis et al. 1997:512); Laughing Gull (Hand 1985:128); Canary-winged Parakeet (Arrowood 1988, 1991); Greater Rhea (Fernández and Reboreda 1998:341); Zebra Finch (Burley 1981:722).

52. Gorilla (Harcourt 1979a:255). Harcourt et al. (1981:266) also characterize heterosexual copulation as "rare." In addition, they report directly observing only 69 episodes of heterosexual mating (other copulations were heard but not seen) compared to 10 episodes between females, which yields an even higher proportion of nearly 13 percent homosexual activity.

53. Western Gull (Hunt et al 1980:474); Spotted Hyena (Glickman 1995; Burr 1996:118–19); for further discussion of comparisons between wild and captive animals, see the next chapter.

54. Tree Swallow (M. P. Lombardo, personal communication; Venier et al. 1993:413; Lombardo 1986; Leffelaar and Robertson 1984:78). Similarly, homosexual mounting is claimed to be very rare in Northern Fur Seals, yet most *heterosexual* matings in this species are missed by observers because they occur at night (Gentry 1998:75–77, 107, 145).

55. For specific examples, see Nilgiri Langur (Poirier 1970; Hohmann 1989), White-tailed Deer (Hirth 1977; Rue 1989), Mule Deer (Geist 1981; Halford et al. 1987; Wong and Parker 1988), Red Deer (Lincoln et al. 1970; Guiness et al. 1971; Hall 1983), American Bison (McHugh 1958; Lott 1983 and personal communication), Red Squirrel (Layne 1954; Smith 1968; Ferron 1980), Mallard Duck (Ramsay 1956; Lebret 1961; Schutz 1965; Bossema and Roemers 1985; Geh 1987), Ruff (Selous 1906–7; Hogan-Warburg 1966; Scheufler and Stiefel 1985; van Rhijn 1991), Oystercatcher (Makkink 1942; Heg and van Treuren 1998), Hooded Warbler (Niven 1994 and personal communication), Cliff Swallow (Emlen 1954; Barlow et al. 1963; Brown and Brown 1996), Red-backed Shrike (Owen 1946; Ashby 1958; Pounds 1972), Victoria's Riflebird (Bourke and Austin 1947; Frith and Cooper 1996; C. B. Frith, personal communication), Sage Grouse (Scott 1942; Patterson 1952; Wiley 1973; Gibson and Bradbury 1986), Acorn Woodpecker (MacRoberts and MacRoberts 1976; Troetschler 1976; W. D. Koenig, personal communication), Gentoo Penguin (Roberts 1934; Wheater 1976; Stevenson 1983), and the examples of wild versus captive observations in notes 99–100, chapter 4.

56. Pukeko (Craig 1980:594; Jamieson and Craig 1987a:1252); Black-headed Gull (van Rhijn and Groothuis 1985:161, 165).

57. For example, Vasey ("Homosexual Behavior in Primates," p. 181) sets up a general frequency scale in which homosexual behavior is classified as "rare" if it occurs "5 percent or less frequently as heterosexual behavior" and "occasional" if it occurs "6–24 percent as frequently as heterosexual behavior"; it is regarded as "frequent" only if it occurs "25 percent or more frequently." Certainly this scale is to be commended for its standardization and multipoint assessment criteria (which also include nonquantitative measures); yet (like most scales) it is not without arbitrariness, and it is at odds with the heterosexual "5 percent" criterion. The "polygyny threshold" model, which recognizes a frequency of ≥5 percent as significant for "minority" heterosexual mating systems (i.e., polygamy in otherwise monogamous species) was originally proposed by Verner, J., and M. F. Willson (1966) "The Influence of Habitats on Mating Systems of North American Passerine Birds," *Ecology* 47:143–47. The 5 percent threshold continues to be widely used as a criterion for "regular" polygyny—for more recent examples, see Quinney 1983 (Tree Swallow); Møller, A. P. (1986) "Mating Systems Among European Passerines: A Review," *Ibis* 128:234–50; Petit, L. J. (1991) "Experimentally Induced Polygyny in a Monogamous Bird Species: Prothonotary Warblers and the Polygyny Threshold," *Behavioral Ecology and Sociobiology* 29:177–87.

58. House Sparrow/Cowbird (Griffin 1959); Savanna Baboon (Marais 1922/1969:214–18); Kestrel (Olsen 1985). Regarding the House Sparrow/Cowbird case, a number of subsequent researchers (e.g., Selander and LaRue 1961; Rothstein 1980) have also interpreted this behavior as "aggression" or "appeasement." Aside from the fact that the activity involving homosexual mounting is not identical to strictly "aggressive" or "preening invitation" displays in Cowbirds (cf. Laskey 1950), a "nonsexual" interpretation cannot explain why Cowbirds "tolerate" homosexual mountings from Sparrows and even actively solicit them. Moreover, the function(s) of these "head-down" displays remain controversial and speculative independent of any homosexual activity (cf. Scott and Grumstrup-Scott 1983). Specific arguments against an "aggressive" or "appeasement" interpretation of these types of behaviors (regardless of whether any same-sex mounting is involved) are presented in Verbeek, N. A. M., R. W. Butler, and H. Richardson (1981) "Interspecific Allopreening Solicitation in Female Brewer's Blackbirds," *Condor* 83:179–80. A parallel example involves Stonor (1937:88), who "reinterprets" Selous's (1906–7) early descriptions of homosexual mountings by female Ruffs as involving heterosexual activity by "female-plumaged" males. More recent observers (e.g., Hogan-Warburg 1966; van Rhijn 1991) have corroborated Selous's original observations, confirming not only the existence of both female and male homosexual activity, but also "female-plumaged" males (i.e., the so-called naked-nape males) that participate in *homosexuality*.

59. Chaffinch (Marjakangas 1981); Regent Bowerbird (Phillipps 1905; Marshall 1954). Similarly, early reports of courtship activity between male Swallow-tailed Manakins by Sick (1959, 1967) were discounted by Foster (1981:174), who tried to claim that the younger male birds being courted by adult males were actually females that had malelike plumage or were male observers or participants in nonsexual aggressive displays.

However, Sick (1959:286) verified the male sex of these birds by dissecting them, and he stated explicitly (Sick 1967:17) that no aggression was involved in the displays. Moreover, it is clear from his descriptions (Sick 1959:286) that the display type that Foster (1981) claimed was aggressive occurs in the absence of younger males, not in their presence. Foster's categorization of such displays as aggressive also appears to be based primarily on the fact that they occur between males, rather than on any inherent differences in the behaviors: as Foster (1981:172; 1984:58) admits, such displays are "extremely similar to" and "strongly reminiscent" of courtship behaviors. That Foster was unable to directly observe courtship displays of the type that Sick reported between males may also be due to geographic or subspecies differences in behaviors: Sick studied a population in Brazil while Foster observed birds in Paraguay. Other elements of the courtship displays between the two populations do appear to differ significantly, such as the vocalizations used and the direction in which males fly during the display (in Brazil, the male farthest from the courted bird begins the courtship "wheel," while in Paraguay the bird closest to the courted bird begins). It should also be pointed out that Snow (1963) independently observed courtship between males in the closely related Blue-backed Manakin.

60. Vasey "Homosexual Behavior in Primates," p. 197; for a similar observation, see Wolfe, "Human Evolution and the Sexual Behavior of Female Primates," p. 130.

61. Hyde, H. M. (1970) *The Love That Dared Not Speak Its Name: A Candid History of Homosexuality in Britain,* p. 1 (Boston: Little, Brown and Company).

62. Killer Whale (Balcomb et al. 1979:23); published version: Balcomb, K. C., III, J. R. Boran, R. W. Osborne, and N. J. Haenel (1980) "Observations of Killer Whales (*Orcinus orca*) in Greater Puget Sound, State of Washington," report no. MMC-78/13 to U.S. Marine Mammal Commission, NTIS# PB80–224728. (Washington, D.C.: U.S. Department of Commerce).

63. Musk-ox (Smith 1976; Tener 1965; Reinhardt 1985); Walrus (Miller 1976); Harbor Seal (Johnson 1974, 1976; Johnson and Johnson 1977).

64. Halls, L. K., ed., (1984) *White-tailed Deer: Ecology and Management* (Harrisburg, Pa.: Stackpole Books); Gerlach, D., S. Atwater, and J. Schnell, eds., (1994) *Deer* (Mechanicsburg, Pa.: Stackpole Books); Jones, M. L., S. L. Swartz, and S. Leatherwood, eds., (1984) *The Gray Whale,* Eschrictius robustus (Orlando: Academic Press). In contrast, a similarly comprehensive book on Mule Deer *does* mention homosexual activity (Geist 1981), as does another volume on White-tailed Deer (Rue 1989).

65. Woodpeckers (Short 1982; Winkler et al. 1995); Skutch, A. F. (1985) *Life of the Woodpecker,* p. 44 (Santa Monica: Ibis Publishing). For a similar omission of all information on homosexuality in Parrots by the standard "comprehensive" guide to this bird family, see Forshaw (1989).

66. See, for example, Fay (1982) on Walruses, Birkhead (1991) on Magpies, Lowther and Cink (1992) on House Sparrows, Davis (1993) on Black-crowned Night Herons, Lowther (1993) on Brown-headed Cowbirds, Telfair (1994) on Cattle Egrets, Burger (1996) on Laughing Gulls, Russell (1996) on Anna's Hummingbirds, and Ciaranca et al. (1997) on Mute Swans.

67. Hooded Warbler (Niven 1993:190); Antbirds (Willis 1967, 1972, 1973); Orange-fronted Parakeet (Buchanan 1966); Golden Plover (Nethersole-Thompson and Nethersole-Thompson 1961, 1986); Mallard Duck (Lebret 1961); Black Swan (Braithwaite 1970, 1981); Scottish Crossbill (Nethersole-Thompson 1975); Black-billed Magpie (Baeyens 1981a); Pied Kingfisher (Moynihan 1990). Similar statements have been made by Konrad Lorenz (1991:241 [Greylag Goose]), who claimed that long-term pair-bonding between males only occurs in Geese and Ducks; and Hunt and Hunt (1977:1467 [Western Gull]), who were unaware of any previous reports of homosexual pairing in wild birds.

68. Black-headed Gull (van Rhijn and Groothuis 1985:165; Kharitonov and Zubakin 1984); Adélie Penguin (Davis et al. 1998:136); Humboldt Penguin (Scholten 1992:8); Kestrel (Olsen 452). Similar statements have been issued by scientists studying other species—Sylvestre (1985:64), for example, reported not being aware of any previous records of homosexual activity in Botos, even though fairly extensive descriptions were available in Layne and Caldwell (1964), Caldwell et al. (1966), Spotte (1967), and Pilleri et al. (1980). Walther (1990:308) claimed that courtship betweeen male hoofed mammals had not been observed in the wild, when in fact such behavior had been reported in numerous prior studies, including in Pronghorn, Blackbuck, Mountain Sheep (Bighorn, Thinhorn, Asiatic Mouflon), Mountain Goats, Musk-oxen, Bharal, and Markhor (Walther, F. R. [1990] "Bovids: Introduction," in *Grzimek's Encyclopedia of Mammals,* vol. 5, pp. 290–324 [New York: McGraw-Hill]).

69. See, for example, Takahata et al. (1996:149), who ask, "Is GG-rubbing a sexual behavior?" and conclude that its "nonsexual" aspects are more prominent, because of its association with tension reduction, feeding, reassurance, participation by nonestrous females, and the fact that Bonobos (unlike Japanese Macaques) do not form "exclusive homosexual female-female pairs." None of these characteristics, in fact, negate a fully "sexual" interpretation. In particular, the fact that Bonobos do not form same-sex pairs or consortships hardly argues against the sexual nature of their genital rubbing—it simply indicates that homosexual interactions in this species do not involve extensive pair-bonding. By these criteria, Bonobo heterosexual interactions would have to be considered nonsexual as well, since they are often associated with the same "social" or "nonsexual" situations, nor do individuals form "exclusive heterosexual male-female pairs." See also Kuroda (1980:190), who considers genital rubbing between females to be "uninterpretable" when it occurs in contexts that are not related to tension reduction or food exchange; and Kano (1980:253–54, 1992:139,1990:66–67, 69), who classifies same-sex activities in Bonobos as primarily "social" rather than "sexual" and ascribes to them the primary "functions" of greeting, reassurance, reconciliation, and food-sharing (while neverthe-

less recognizing that sexual aspects may be secondarily involved in some cases). As recently as 1997, researchers were still speculating about, and emphasizing, the nonsexual "functions" of Bonobo homosexual activity (Hohmann and Fruth 1997).

70. Mountain Sheep (Geist 1975:97–98).

71. Vasey, P. L. (1997, August 19) "Summary: Homosexual or Dominance Behavior? (Discussion)," message posted to *Primate Talk* (on-line discussion list).

72. Rhesus Macaque (Hamilton 1914). A standard and widely cited exposition of the dominance interpretation is Wickler, W. (1967) "Socio-sexual Signals and Their Intra-specific Imitation Among Primates," in D. Morris, ed., *Primate Ethology*, pp. 69–147 (London: Weidenfield and Nicolson).

73. On the occurrence of dominance hierarchies in various mammals and birds without homosexuality, and further references, see Wilson, E. O. (1975) *Sociobiology: The New Synthesis*, p. 283 (Cambridge and London: Belknap Press); Welty, J. C., and L. Baptista (1988) *The Life of Birds*, 4th ed., pp. 206–210 (New York: W. B. Saunders).

74. For explicit statements on the absence, unimportance, or irrelevance of dominance hierarchies in these species or populations, see Gorilla (Yamagiwa 1987a:25; Robbins 1996:957); Savanna (Olive) Baboon (Rowell 1967b:507–8); Bottlenose Dolphin (Shane et al. 1986:42); Zebras (Penzhorn 1984:113; Schilder 1988:300); Musk-ox (Smith 1976:92–93); Koala (Smith 1980:187); Buff-breasted Sandpiper (Lanctot and Laredo 1992:7); Tree Swallow (Lombardo et al. 1994:556). In Gorilla all-male groups, dominance is not a central organizing feature of social interactions (including homosexual interactions) even though some semblance of a dominance "hierarchy" may exist and males clearly have different ranks. The same may also be true for Hanuman Langur all-male groups (Weber and Vogel 1970:75) and Collared Peccary mixed-sex groups (Sowls 1997:151–53) in which same-sex interactions occur. In Buff-breasted Sandpipers, although mounting between males may be accompanied by aggression and therefore superficially appears related to "dominance," there is in fact no evidence that a dominance hierarchy exists in this species or constitutes an important aspect of its social organization. In some of these species (e.g., Zebras, Musk-oxen, Bottlenose Dolphins) dominance hierarchies are more prominent in captivity, although homosexual activity occurs in both wild and captive contexts. Finally, J. Steenberg (personal communication) suggests that mounting between female Matschie's Tree Kangaroos is a dominance display, yet Hutchins et al. (1991:154–56, 161) found no clear-cut dominance hierarchy in the study population where this behavior was observed.

75. Tasmanian Native Hen (Ridpath 1972:81) (in this species, hierarchies *can* be induced in wild birds by provisioning them with food, but dominance plays no role in their unprovisioned activities—including homosexual mounting, which is not associated in any way with induced dominance); Little Blue Heron (Werschkul 1982:383–84); white-browed sparrow weaver and other weavers (Collias, N. E., and E. C. Collias [1978] "Co-operative Breeding Behavior in the White-browed Sparrow Weaver," *Auk* 95:472–84; Collias, N. E., and E. C. Collias [1978] "Group Territory, Dominance Hierarchy, Co-operative Breeding in Birds, and a New Factor," *Animal Behavior* 26:308–9). Likewise, dominance systems occur in most Macaques, yet homosexual behavior is apparently absent in some species, e.g., the Barbary Macaque (*Macaca sylvanus*)—see Vasey, "Homosexual Behavior in Primates," pp. 178–79; for an extensive summary of research on this species with no mention of same-sex mounting, see Fa, J. E., ed. (1984) *The Barbary Macaque: A Case Study in Conservation* (New York: Plenum Press). However, recent work seems to suggest that same-sex mounting may in fact occur: Di Trani, C. M. P. (1998) "Conflict Causes and Resolution in Semi-Free-Ranging Barbary Macaques (*Macaca sylvanus*)," *Folia Primatologica* 69:47–48. Therefore, this example must be interpreted with caution, like many other instances involving an apparent "absence" of homosexual behavior (see chapter 4 for further discussion).

76. Wolf (Zimen 1976, 1981); Spotted Hyena (Frank 1986); Squirrel Monkey (Baldwin and Baldwin 1981:294–95; Castell and Heinrich 1971:187–88); Bottlenose Dolphin (Samuels and Gifford 1997:82, 88–90). In Red Squirrels, both sexes have dominance systems yet same-sex mounting is much more prominent among males (Ferron 1980:135–36); in Bonobos, a dominance system is much more developed or important among males (de Waal 1997:72–74), yet homosexual activities occur in both sexes. A related observation is that in Bighorn Sheep, both sexes have well-defined dominance systems and exhibit same-sex mounting, yet only among males does it have some correlation with homosexual activity.

77. For examples of animals that participate in interspecies homosexual mounting, see the profiles for Crab-eating Macaque, Bottlenose Dolphin, Walrus, Greenshank, Orange Bishop Bird, and House Sparrow. On the occurrence of interspecies dominance hierarchies, see, for example, Fisler, G. F. (1977) "Interspecific Hierarchy at an Artificial Food Source," *Animal Behavior* 25:240–44; Morse, D. H. (1974) "Niche Breadth as a Function of Social Dominance," *American Naturalist* 108:818–30.

78. Rhesus Macaque (Reinhardt et al. 1986:56); Japanese Macaque (Chapais and Mignault 1991:175–76; Vasey et al. 1998); Common Chimpanzee (Nishida and Hosaka 1996:122 [table 9.7]). See also Bygott 1974—cited in Hanby 1974:845 [Japanese Macaque]—who found that 59 percent of mounts between male Chimps were by subordinates on dominants or by equally ranked participants.

79. Musk-ox (Reinhardt 1985:298). In Cattle Egrets, Fujioka and Yamagishi (1981:139) stated that males attempting homosexual copulations always rank higher than or equal to the males they mount. Yet two males in their study population who mounted other males were apparently not part of the dominance hierarchy (cf. their table 3), while the highest-ranking male did not participate in any same-sex mounts. M. Fujioka (personal communication) concedes that the rank of the males may not actually be an important factor in their homosexual mounting.

80. Crested Black Macaque (Dixson 1977:77; Poirier 1964:96); American Bison (Reinhardt 1985:218, 222, 1987:8); Pig-tailed Macaque (Oi 1990a:350); Red Deer (Hall 1983:278); Pukeko (Jamieson and Craig 1987b:319–22); Japanese Macaque (Chapais and Mignault 1991:175–76); Bighorn Sheep (Shackleton 1991:179–80).

81. A further argument is provided by "pile-up mounts," i.e., when three individuals are all mounted (stacked) on each other. In this case, the mounter-mountee relations rarely if ever follow dominance lines: either they occur in species without dominance hierarchies (e.g., Sage Grouse, Common Murre), or else it is not the case that the middle animal is both higher-ranking than the animal it is mounting but lower-ranking than the animal who is mounting it (e.g., Wolf, Bonobo). For more on pile-up mounts, see chapter 4.

82. Common Chimpanzee (Nishida and Hosaka 1996:122; Bygott 1974 [cited in Hanby 1974:845 (Japanese Macaque)]); White-faced Capuchin (Manson et al. 1997:771, 780); Blackbuck (Dubost and Feer 1981:89–90); Cavies (Rood 1972:36); Gray-capped Social Weaver (Collias and Collias 1980:218, 220). Although mounting between male Musk-oxen in captivity seems to follow dominance lines (Reinhardt 1985), in wild herds Smith (1976) found no dominance hierarchy within (as opposed to between) sex/age classes. Same-sex mounting in the wild occurs among age-mates (who are therefore essentially equal in rank, e.g., two–year-old males mount each other).

83. See, for example, Bertrand 1969:191 (Stumptail Macaque); Simonds 1965:183, Sugiyama 1971:259 (Bonnet Macaque); Bernstein 1972:406 (Pig-tailed Macaque); Dixson et al. 1975:195–96 (Talapoin Monkey); Kaufmann 1974:309 (Whiptail Wallaby).

84. A distinction between consensual and nonconsensual homosexual mounts is found in more than 30 different species of mammals and birds. Direct evidence of sexual arousal and stimulation on the part of animals being mounted is also available in many species, including orgasmic (and other) responses in female Japanese, Rhesus, Stumptail, and Pig-tailed Macaques being mounted; erection and masturbation by male mountee Rhesus, Pig-tailed, and Crested Black Macaques; thrusting by male Bonobos being mounted; and stimulation of the mountee's clitoris by her partner's thrusting in Hanuman Langurs and Japanese Macaques. In addition to direct and indirect genital stimulation during mounting, it is quite likely that male animals being penetrated during anal intercourse also experience stimulation of the prostate gland (which presses against the wall of the rectum). In human males, direct stimulation of the prostate—for instance, during anal intercourse—can be highly arousing and may precipitate or enhance orgasm. A similar capacity is probably present in all male mammals. Although direct evidence (in the form of firsthand accounts) of the pleasurable or arousing nature of this activity is, of course, lacking in nonhuman animals, there is some indirect evidence. A standard technique of inducing erection and ejaculation (for purposes of artificial insemination) in male mammals is through anal and/or prostate stimulation. Known as electroejaculation, this technique involves insertion of an anal probe and stimulation of the rectum—especially in the area of the prostate gland—with a mild electrical current as well as back-and-forth (thrusting) movements of the probe. This technique has proven effective in numerous species of mammals, including virtually all of those in which male homosexual mounting and /or anal penetration occur. For further information on electroejaculation, see Watson, P. F., ed. (1978) *Artificial Breeding of Non-Domestic Animals,* Symposia of the Zoological Society of London no. 43, especially pp. 109, 129, 208–10, 221, 295 (London: Academic Press).

85. Hanuman Langur (Srivastava et al. 1991:506–7); for a similar assessment with regard to homosexual activity between males in this species, see Weber and Vogel (1970:77–78). See also Rowell (1967a:23), who states that "sexual" and "dominance" mounts in Savanna (Yellow) Baboons are virtually indistinguishable, and Enomoto (1990:473), who remarks on the difficulty of discriminating between sexual and ritualized dominance mounting in Bonobos because of the gradation between the two. Weinrich (*Sexual Landscapes,* p. 294), in discussing mounting between male Mountain Sheep, also points out how sexuality and dominance can both be part of the same behavior and suggests an analogy with human sexuality. Indeed, elements of consensual "dominance" or power-play, although rarely acknowledged, are often a part of human lovemaking and sexual pleasure, ranging along a continuum from gentle "love bites" to full sadomasochism (and nonconsensual dominance also figures prominently in many human sexual interactions, especially heterosexual ones).

86. Japanese Macaque (Wolfe 1986:268); Rhesus Macaque (Akers and Conaway 1979:78); Greylag Goose (Lorenz 1991:206); Black-winged Stilt (Kitagawa 1989:65, 69) (see also the distinction between same-sex courtship and aggressive/appeasing kantling in Ostriches [Sauer 1972:731; Bertram 1992:15, 50–51]). For species such as these that have a clear distinction between mounts in sexual and nonsexual contexts, only the former are considered (in this book and in most sources) to be homosexual behavior. As noted in chapter 1, some species classified by Dagg (1984) as exhibiting homosexuality (e.g., bush squirrels and degus) are excluded from our roster on the basis of this criterion, because all same-sex mounting in these species appears to fall into this genuinely nonsexual category; see Viljoen, S. (1977) "Behavior of the Bush Squirrel, *Paraxerus cepapi cepapi,*" *Mammalia* 41:119–66; Fulk, G. W. (1976) "Notes on the Activity, Reproduction, and Social Behavior of *Octodon degus,*" *Journal of Mammology* 57:495–505.

87. Walrus (Miller 1975:607); Gray Seal (Anderson and Fedak 1985); Oystercatcher (Ens, B. J., and J. D. Goss-Custard [1986] "Piping as a Display of Dominance by Wintering Oystercatchers *Haematopus ostralegus,*" *Ibis* 128:382–91). Early observers of this species (e.g., Makkink 1942) misinterpreted the piping display as a courtship activity because it often occurs between males and females.

88. For details of the way that dominance is expressed in these species, see Savanna (Yellow) Baboon (Maxim and Buettner-Janusch 1963:169); Hamadryas Baboon (Stammbach, E. [1978] "On Social Differentiation in

Groups of Captive Female Hamadryas Baboons," *Behavior* 67:322–38); Bottlenose Dolphin (Samuels and Gifford 1997); Killer Whale (Rose 1992:108–9); Caribou (Espmark, J. [1964] "Studies in Dominance-Subordination Relationship in a Group of Semi-Domestic Reindeer (*Rangifer tarandus* L.)," *Animal Behavior* 12:420–26); Blackbuck (Dubost and Feer 1981:97–100); Wolf (Zimen 1976, 1981); Bush Dog (Macdonald 1996); Spotted Hyena (Frank 1986:1511); Grizzly Bear (Craighead et al. 1995:109ff); Black Bear (Stonorov and Stokes 1972:235, 242); Red-necked Wallaby (Johnson 1989:267); Canada Goose (Collias and Jahn 1959:500–501); Scottish Crossbill (Nethersole-Thompson 1975:53); Black-billed Magpie (Birkhead 1991); Jackdaw (Röell 1978); Acorn Woodpecker (Stanback 1994); Galah (Rowley 1990:57). In Pronghorns, mounting between males was originally claimed to represent a dominance activity (Kitchen 1974), yet more recent studies of dominance in this species have not included same-sex mounting (Bromley 1991).

89. In some cases, sexual behaviors other than mounting can be correlated with dominance. For example, grooming between males in Nilgiri Langurs and Crested Black Macaques is often performed by a subordinate animal on a more dominant one. Nevertheless, it is apparent that this activity has a clearly sexual component as well: one or both males may become intensely aroused, developing an erection and even ejaculating during the grooming (see Poirier 1970a:334 for Nilgiri Langurs and Poirier 1964:146–47 for Crested Black Macaques). Similarly, adult (dominant) Bonobos often masturbate or massage the genitals of adolescent (subordinate) males, but again, the activity involves clear sexual stimulation (cf. de Waal 1987, 1995, 1997). Also, Squirrel Monkey genital displays are sometimes correlated with dominance, but there are also cases where the association is less than definitive, or where they occur in clearly sexual contexts between animals of the same sex (cf. Talmage-Riggs and Anschel 1973:70; Travis and Holmes 1974:55; Baldwin and Baldwin 1981:295–97; Castell and Heinrich 1971:187–88).

90. One cannot help but surmise that it is the heterosexism of many biologists that has led them to focus on mounting behavior to the exclusion of other activities in their appeal to dominance factors—for only in mounting can the positions of the participants be clearly analogized to those of a male and female in a heterosexual interaction. As Fedigan (1982:101 [Japanese Macaque]) points out, underlying the entire discussion of dominance in same-sex interactions is the assumption that homosexual mounting is essentially a transposition from heterosexual copulation—and that males "dominate" females in such interactions. For further evidence against this view, see the discussion of homosexuality as a form of "pseudoheterosexuality" in chapter 4.

91. Possible exceptions are same-sex courtship interactions in Mountain Sheep (Geist 1968, 1971), Musk-oxen (Reinhardt 1985), and Cavies (Rood 1972), which have been interpreted as reflecting dominance. Additionally, mounting or other sexual behaviors within a same-sex pair-bond—common in many bird species—does not fit easily into a dominance interpretation, since this usually involves ongoing interaction with only one other animal (rather than the establishment of hierarchical positions within a network of individuals).

92. Giraffe (Pratt and Anderson 1985:774–75, 780–81); Crested Black Macaque (Dixson 1977:77–78; Reed et al. 1997:255); Stumptail Macaque (Bernstein 1980:40); Pig-tailed Macaque (Giacoma and Messeri 1992:187); Savanna (Olive) Baboon (Owens 1976:250–51); Squirrel Monkey (Baldwin and Baldwin 1981:295–97; Baldwin 1968:296, 311); Red Squirrel (Ferron 1980:136); Spinifex Hopping Mouse (Happold 1976:147); American Bison (Reinhardt 1985:222–23); Pukeko (Lambert et al. 1994); Sociable Weaver (Collias and Collias 1980:246, 248; in the latter instance, the inconsistency in dominance status was not one of the cases of temporary reversals of dominance that were occasionally seen in this species). In female Squirrel Monkeys, dominance hierarchies are not considered to be a salient feature of social organization in the wild (Baldwin and Baldwin 1981:294–95). However, even when dominance systems appear to develop (e.g., in some captive situations), investigators have found that the rank of females based on their homosexual activities does not agree with other measures of rank (Anschel and Talmage-Riggs 1978:602 [table 1]).

93. For some reevaluation and/or critiques of the concept of dominance, see Gartlan, J, S. (1968) "Structure and Function in Primate Society," *Folia Primatologica* 8:89–120; Bernstein 1970 (Crab-eating Macaque); Richards, S. M. (1974) "The Concept of Dominance and Methods of Assessment," *Animal Behavior* 22:914–30; Ralls, K. (1976) "Mammals in Which Females Are Larger Than Males," *Quarterly Review of Biology* 51:245–76; Lockwood, R. (1979) "Dominance in Wolves: Useful Construct or Bad Habit?" in E. Klinghammer, ed., *Behavior and Ecology of Wolves*, pp. 225–44 (New York: Garland); Baldwin and Baldwin 1981 (Squirrel Monkey); Bernstein, I. S. (1981) "Dominance: The Baby and the Bathwater," *Behavioral and Brain Sciences* 4:419–57; Hand, J. L. (1986) "Resolution of Social Conflicts: Dominance, Egalitarianism, Spheres of Dominance, and Game Theory," *Quarterly Review of Biology* 61:201–20; Walters, J. R., and R. M. Seyfarth (1987) "Conflict and Cooperation," in B. B. Smuts, D. L. Cheney, R. M. Seyfarth, R. W. Wrangham, and T. T. Struhsaker, eds., *Primate Societies*, pp. 306–17 (Chicago and London: University of Chicago Press); Drews, C. (1993) "The Concept and Definition of Dominance in Animal Behavior," *Behavior* 125:283–313; Lambert et al. 1994 (Pukeko).

94. Fedigan 1982:92–93 (Japanese Macaque).

95. Bonobo (Kano 1992:253–54; Kitamura 1989:57, 63); Gorilla (Harcourt et al. 1981:276; Yamagiwa 1987a:25; Harcourt 1988:59); Hanuman Langur (J. J. Moore, in Weinrich 1980:292); Japanese Macaque (Vasey 1996:549; Chapais and Mignault 1991:175–76; Tartabini 1978:433, 435; Hanby 1974:841); Rhesus Macaque (Akers and Conaway 1979:78; Reinhardt et al. 1986:55; Gordon and Bernstein 1973:224); Pig-tailed Macaque (Tokuda et al. 1968:293); Crested Black Macaque (Dixson 1977:77–78; Poirier 1964:20, 49; Reed et al. 1997:255); Savanna Baboon (Owens 1976:256); Gelada Baboon (Mori 1979:134–35; R.Wrangham, in

Weinrich 1980:291); Squirrel Monkey (Talmage-Riggs and Anschel 1973:70); Bottlenose Dolphin (Caldwell and Caldwell 1972:427); Blackbuck (Dubost and Feer 1981:89–90); Giraffe (Pratt and Anderson 1985:774–75, 780); American Bison (Reinhardt 1985:222, 1987:8); Red Squirrel (Ferron 1980:136); Little Blue Heron (Werschkul 1982:383–84); Tree Swallow (Lombardo et al. 1994:556).

96. For examples of earlier claims of a dominance connection being refuted by later studies, see Common Chimpanzee (Yerkes 1939:126–27; Nishida 1970:57—Bygott 1974 [cited in Hanby 1974:845 (Japanese Macaque)]; Nishida and Hosaka 1996:122 [table 9.7]); Hanuman Langur (Weber 1973:484—Srivastava et al. 1991:506–7; J. J. Moore, in Weinrich 1980:292); Rhesus Macaque (Carpenter 1942—Akers and Conaway 1979:78; Reinhardt et al. 1986; Gordon and Bernstein 1973:224); Japanese Macaque (Sugiyama 1960:136—Hanby 1974:841; Chapais and Mignault 1991:175–76); Bonnet Macaque (Rahaman and Parthasarathy 1968:68, 263—Makwana 1980:10); Pig-tailed Macaque (Tokuda et al. 1968—Oi 1990a:353–54); Killer Whale (Balcomb et al. 1979:23—Rose 1992:108–9); Giraffe (Dagg and Foster 1976, Leuthold 1979:27–29—Pratt and Anderson 1985:774–75); Blackbuck (Schaller 1967—Dubost and Feer 1981:89–90); American Bison (Lott 1974:391—Reinhardt 1986:222–23); Wolf (Schenkel 1947—van Hooff and Wensing 1987:232). In addition, a parallel example in Laughing Gulls involves an indirect refutation of the relevance of dominance. Noble and Wurm (1943:205–6) linked homosexual mounting in Laughing Gulls to the supposedly lower rank of the male being mounted, citing as evidence of his lower status the fact that the mounted male did not "dominate" his female mate. In a more recent detailed study of interactions between partners in heterosexual pairs, however, Hand (1985) concluded that males do not in general dominate their female mates in this species—thus invalidating the earlier claim that being mounted homosexually was correlated with "lower status." Studies that attribute homosexual activity to dominance with little or no supporting evidence include Orang-utan (Rijksen 1978:257); Squirrel Monkey (DuMond 1968:124); West Indian Manatee (Rathbun et al. 1995:150); Pied Kingfisher (Moynihan 1990:19).

97. Whiptail Wallaby (Kaufmann 1974:307, 309); Rhesus Macaque (Gordon and Bernstein 1973:224). Kaufmann concluded that Whiptail Wallaby homosexual mountings are themselves probably not dominance-related, however, because dominant animals generally invite subordinate ones to mount (the opposite of the "usual" dominance pattern).

98. Bighorn Sheep (Hogg 1987:120; Hass and Jenni 1991:471); Crested Black Macaque (Poirier 1964:54).

99. Vasey, "Homosexual Behavior in Primates," p. 191.

100. Orang-utan (Maple 1980:118).

101. West Indian Manatee (Rathbun et al. 1995:150). See also the suggestion in Buss (1990:19–21) that sexual arousal in male African Elephants during same-sex play-fighting serves to dull pain. While this is possible, it is rather far-fetched, considering that such ritual fights (described by Buss as "erotic") are rarely violent.

102. Vasey, P. L. (in press) "Homosexual Behavior in Male Birds," in W. R. Dynes, ed., *Encyclopedia of Homosexuality,* 2nd ed., vol. 1: Male Homosexuality (New York: Garland Press).

103. American Bison (Reinhardt 1985:222) (cf. also Kaufmann [1974:107] on Whiptail Wallabies, who asserts, "Though tail-lashing seems clearly a sign of sexual arousal, it was occasionally performed by males when they were approached by subordinate males in nonsexual situations"); Asiatic Mouflon (McClelland 1991:80); Stumptail Macaque (O'Keefe and Lifshitz 1985:149); Dugong (Nair et al. 1975:14); Laysan Albatross (Frings and Frings 1961:311); Dwarf Mongoose (Rasa 1979a:365); Bonnet Macaque (Nolte 1955:179). Similarly, Frank et al. (1990:308) state that genital erections in Spotted Hyenas have no "sexual significance" unless displayed by a male toward a female during courtship. The "desexing" of this behavior stems, in large part, from the fact that erections are frequently displayed between animals of the same sex (especially females) and in situations that do not involve (heterosexual) mounting (e.g., during the "meeting ceremony"). While erections undoubtedly have "nonsexual" connotations outside of a mounting context (see, for example, East et al. 1993), it seems overly restrictive to eliminate all "sexual significance" from situations that do not fall into the category of heterosexual courtship and mating.

104. Redshank (Hale and Ashcroft 1983:21). For a summary of the historical interpretation of this behavior, see also Cramp and Simmons 1983:533.

105. Crested Black Macaque (Dixson 1977:71, 76; Poirier 1964:147). Dixson (1977:77) does concede that the distinction between sexual and nonsexual mounts and solicitations is a subjective one, but only in *heterosexual* contexts—homosexual interactions are assumed to be self-evidently nonsexual.

106. Vicuña (Koford 1957:183, 184); Musk-ox (Smith 1976:51); Giraffe (Dagg and Foster 1976:127; Pratt and Anderson 1985:777–78; Leuthold 1979:27, 29); Bank Swallow (Beecher and Beecher 1979:1284); Savanna Baboon (Smuts 1985:18, 148–49, 163–66, 199, 213); Rhesus Macaque (Loy 1971:26); Oystercatcher (Makkink 1942; Ens and Goss-Custard, "Piping as a Display of Dominance").

107. Crested Black Macaque (Dixson 1977:70–71); Bottlenose Dolphin (Östman 1991:313; Dudok van Heel and Mettivier 1974:12; Saayman and Tayler 1973); Spinner Dolphin (Norris and Dohl 1980a:845; Norris et al. 1994:199); Common Murre (Birkhead 1978a:326); Blue-bellied Roller (Moynihan 1990).

108. Rhesus Macaque (see, for example, Sade 1968:32–33); Japanese Macaque (Hanby 1974:843, 845; Wolfe, "Human Evolution and the Sexual Behavior of Female Primates," p. 129).

109. For further discussion see chapter 5. On a related point, aggressive behaviors may accompany homosexual interactions in some species and are therefore used to argue that such behavior is not "really" sexual. However, aggression is also characteristic of heterosexual relations in many species, where such male-female interactions are still classified as "sexual."

110. Kob (Buechner and Schloeth 1965:218); Giraffe (Pratt and Anderson 1985:774–75); northern jacana (del Hoyo, J., A. Elliott, and J. Sargatal, eds. [1996] *Handbook of the Birds of the World*, vol. 3: Hoatzin to Auks, p. 282. [Barcelona: Lynx Edición]); Orang-utan (Galdikas 1981:286).

111. Walrus (Dittrich 1987:168); Musk-ox (Smith 1976:62); Bighorn Sheep (Hogg 1984:527; Geist 1971:139); Asiatic Mouflon (McClelland 1991:81); Grizzly Bear (Craighead et al. 1995:161); Olympic Marmot (Barash 1973:212); White-tailed Deer (Hirth 1977:43); Orang-utan (Galdikas 1981:286); White-faced Capuchin (Manson et al. 1997:775); Northern Fur Seal (Gentry 1998:172); Ruff (Hogan-Warburg 1966:167–68). Additionally, in one study of Matschie's Tree Kangaroos—a species in which researchers deny that mounting between females is (homo)sexual (J. Steenberg, personal communication)—all mounts observed between animals of the opposite sex were "incomplete" in that they did not involve penetration or thrusting (Hutchins et al. 1991:158). Another study of the same population found both that "full" copulations between males and females were infrequent, and that in heterosexual contexts females showed few overt signs of sexual interest, since the behavioral cues for female sexual arousal are extremely subtle (Dabek 1994:84, 93–94, 116).

112. Morrill and Robertson 1990 (Tree Swallow); Scott, M. P., and T. N. Tan (1985) "A Radiotracer Technique for the Determination of Male Mating Success in Natural Populations," *Behavioral Ecology and Sociobiology* 17:29–33. More recently, a copulation-verification technique using fluorescent powder has been tested for rodents. Dusted on males, the powder is transferred to females during mating and can be checked using ultraviolet light. Ironically, during the testing of this procedure, pairs of females were used as "controls" since it was assumed that they would not engage in mounting behavior with one another. Nevertheless, 12 percent of female pairs showed transfer of powder—but of course this was interpreted by researchers as evidence of nonsexual contact between such females (Ebensperger, L. A., and R. H. Tamarin [1997] "Use of Fluorescent Powder to Infer Mating Activity of Male Rodents," *Journal of Mammalogy* 78:888–93).

113. Rhesus Macàque (Erwin and Maple 1976); field report of penetration and ejaculation (Sade 1968:27); see also Kempf (1917:134) for an even earlier documentation of anal penetration between (captive) male Rhesus Macaques. Walther (1990:308) makes a parallel claim that mounting activity between male hoofed mammals does not constitute (homo)sexual behavior because erection and anal penetration are not always observed (Walther, "Bovids: Introduction"). On a related point, Tuttle (1986:289) takes great pains to point out that rump-rubbing and mounting between male Bonobos do not "qualify" as genital contact because, "pace certain sodomites, the anus is not a genital organ (International Anatomical Nomenclature Commitee, 1977, p. A49)." Tuttle does, however, accept that sexual activity between females—which he calls "bizarre homosexual hunching" (ibid., p. 282)—qualifies as genital contact (Tuttle, R. H. [1986] *Apes of the World: Their Social Behavior, Communication, Mentality, and Ecology* [Park Ridge, N.J.: Noyes Publications]). For a recent survey of homosexual behavior in primates that (wisely) drops the occurrence of penetration, arousal, and/or orgasm as a defining criterion of the behavior, see Vasey, "Homosexual Behavior in Primates," p. 175.

114. On a similar gradation of mounting behavior in male birds, see Moynihan 1955:105 (Black-headed Gull).

115. For specific arguments against homosexual activity as a form of tension reduction in various species, see Yamagiwa 1987a:23, 1987b:37 (Gorilla); Edwards and Todd 1991:234–35 (White-handed Gibbon); Vasey 1996:549–50 (Japanese Macaque); R. Wrangham, in Weinrich 1980:291 (Gelada Baboon). Against homosexuality as a form of play, see Talmage-Riggs and Anschel 1973:71 (Squirrel Monkey); Lombardo et al. 1994:556 (Tree Swallow). Against homosexuality as reconciliation or reassurance behavior, see Vasey 1996:550 (Japanese Macaque); Akers and Conaway 1979:78 (Rhesus Macaque); Lombardo et al. 1994:556 (Tree Swallow). Against homosexual activities as a means of forging coalitions or alliances, see Silk 1994:285–87 (Bonnet Macaque) (and also Silk 1993:187 for arguments that coalition-bonding between males in this species is not "functional" in terms of enhancing the males' status, access to resources, or inclusive fitness). Against homosexuality as a gesture of appeasement or placation, see Manson et al. 1997:783 (White-faced Capuchin); Ferron 1980:136 (Red Squirrel); Lombardo et al. 1994:556 (Tree Swallow). Against homosexual relations as "kinship alliances" between individuals who associate with each other primarily because they are related (so-called kin selection), see Fernández and Reboreda 1995:323 (Greater Rhea); Heg and van Treuren 1998:688–89, Ens 1998:635 (Oystercatcher); Afton 1993:232 (Lesser Scaup Duck); Rose 1992:104, 112 (Killer Whale); Hashimoto et al. 1996:316 (Bonobo). See also Vasey, "Homosexual Behavior in Primates," for a summary and review of the evidence against many of these nonsexual "explanations."

116. Japanese Macaque (Vasey 1996).

117. Bonobo (de Waal 1987, 1995 [among others]); Savage-Rumbaugh and Lewin 1994:110); Gorilla (Yamagiwa 1987a:23, 1987b:37).

118. See Silk 1994:285–87 (Bonnet Macaque) for more detailed discussion.

119. Signs of sexual arousal such as these have been documented for homosexual interactions in more than 90 species of mammals and birds. In addition, a number of scientists have themselves asserted the clearly sexual character of same-sex interactions (in addition to, or instead of, nonsexual aspects); see, for example, de Waal 1995:45–46 (Bonobo); Yamagiwa 1987a, Harcourt 1988:59, Porton and White 1996:724 (Gorilla); Edwards and Todd 1991 (White-handed Gibbon); Weber and Vogel 1970:76–77 (Hanuman Langur); Vasey 1996:550, Rendall and Taylor 1991:324, Wolfe 1984:147 (Japanese Macaque); Akers and Conaway 1979:78–79 (Rhesus Macaque); Chevalier-Skolnikoff 1976:525 (Stumptail Macaque); Srivastava et al. 1991 (Hanuman Langur); R. Wrangham, in Weinrich 1980:291 (Gelada Baboon); Manson et al. 1997:775–76 (White-faced Capuchin);

Herzing and Johnson 1997:85, 90 (Bottlenose/Atlantic Spotted Dolphins); Saulitis 1993:58 (Killer Whale); Darling 1978:60, 1977:10–11 (Gray Whale); Coe 1967:320 (Giraffe); Rue 1989:313 (White-tailed Deer); Buss 1990:20 (African Elephant); Heg and van Treuren 1998:688 (Oystercatcher); Davis et al. 1998 (Adélie Penguin); Stiles 1982:216 (Anna's Hummingbird). For use of the word *erotic* to characterize same-sex interactions, see, for example, de Waal 1987:323, 1997:103–4, Kano 1992:192, 1990:66 (Bonobo); Darling 1977:10–11 (Gray Whale); Mathews 1983:72 (Walrus); Buss 1990:19 (African Elephant).

120. Occasionally, however, multiple "functions" are granted to heterosexual behavior; see, for example, Lindburg 1971 (Rhesus Macaque); de Waal 1987, 1995, 1997, Kano 1990:67 (Bonobo); Manson et al. 1997 (White-faced Capuchin); Hanby, J. (1976) "Sociosexual Development in Primates," in P. P. G. Bateson and P. H. Klopfer, eds., *Perspectives in Ethology*, vol. 2, pp. 1–67 (New York: Plenum Press).

Chapter 4. Explaining (Away) Animal Homosexuality

1. M. Grober, opening remarks to the plenary session on Sexual Orientation, 24th International Ethological Conference, Honolulu, Hawaii, August 12, 1995.

2. Among the attendees who had previously documented or written extensively on animal homosexual behavior, but who were not speaking on this topic, were B. Le Boeuf (Northern Elephant Seals), C. Clark (Right Whales), W. D. Koenig (Acorn Woodpeckers), M. Moynihan (Rufous-naped Tamarins, Pied Kingfishers, Blue-bellied Rollers), A. Srivastava (Hanuman Langurs), F. B. M. de Waal (Bonobos, other primates), and J. C. Wingfield (Gulls). A number of other disconcerting trends were also in evidence among the papers presented during this symposium: for example, many were based on studies of laboratory or captive animals to the exclusion of information on homosexuality/transgender in wild animals. One presenter (Ulibarri) actually went so far as to state that no information was available in English on any behavior of wild Mongolian gerbils, when in fact at least one such study had been published several years earlier in a prominent zoology journal (Ulibarri, C. [1995] "Gonadal Steroid Regulation of Differentiation of Neuroanatomical Structures Underlying Sexual Dimorphic Behavior in Gerbils," paper presented at the 24th International Ethological Conference, Honolulu, Hawaii; Ågren, G., Q. Zhou, and W. Zhong [1989] "Ecology and Social Behavior of Mongolian Gerbils, *Meriones unguiculatus*, at Xilinhot, Inner Mongolia, China," *Animal Behavior* 37:11–27).

3. Caprio, F. S. (1954) *Female Homosexuality*, pp. 19, 76 (New York: Grove); Northern Fur Seal (Bartholomew 1959:168).

4. This idea appears in the descriptions of homosexuality in more than 40 different species of mammals and birds.

5. Homosexuality and related phenomena in animals have even been labeled "heterotypical behavior" (cf. Haug, M., P. F. Brain, and C. Aron, eds., [1991] *Heterotypical Behavior in Man and Animals* [New York: Chapman and Hall]). The intended meaning of this term is that the behavior of at least one of the partners during same-sex interactions is supposedly "typical" of participants in heterosexual activity, but transposed onto a same-sex context—in other words homosexuality is simply recast as a modified version of heterosexuality.

6. Orange-fronted/Aztec Parakeets (Hardy 1966:77, 1963:171). In a related vein, the vocal and sexual responses of female Stumptail Macaques during orgasm were studied primarily in homosexual, rather than heterosexual, interactions; this information was then generalized or extrapolated to opposite-sex contexts (cf. Goldfoot et al. 1980; Leinonen et al. 1991:245). Likewise, the synchronization of pair-bonding activities in Galahs was typified in one study with quantitative information from same-sex rather than opposite-sex pairs (Rogers and McCulloch 1981:87).

7. Freud, S. (1905/1961) *Drei Abhandlungen zur Sexualtheorie* (Frankfurt: Fischer); see also Ellis, H. (1936) *Sexual Inversion: Studies in the Psychology of Sex* (New York: Random House).

8. Morris 1954 (Zebra Finch); Morris 1952 (Ten-spined Stickleback). For a more recent article, see Schlupp et al. 1992 (Amazon Molly). See also Lorenz 1972:21 (Raven) for an early (errroneous) statement to the effect that during same-sex interactions animals only exhibit "purely" masculine or feminine behaviors (as defined by a heterosexual context) rather than any intermediate forms.

9. Takhi (Boyd 1986:661); Mallard Duck (Ramsay 1956:277); Snow Goose (Starkey 1972). Another notable example of the conflation of "inverted" gender traits (and other "deviant" characteristics) with playing the "opposite-sex" role in homosexual interactions involves the Common Chimpanzee. A female Chimp that was apparently exclusively lesbian for many years (and consorted with otherwise "heterosexual" females) was described by a scientist—in addition to being sexually "aberrant"—as having a "burly manner," being "masculine-looking," "two-faced and mean," "malevolent," and "deceitful." Comments from untrained observers that compared her to a witch were also repeated without qualification (de Waal 1982:64–65). While some of these traits may have reflected genuine aspects of her physical appearance, behavior, and personality, it is striking how loaded and anthropomorphic these descriptions are, and how many of the characteristics singled out for mention correspond precisely to the negative and distorted stereotypes of "butch" lesbians among humans. Moreover, in many animals, (heterosexual) females may display greater levels of aggression when they are in "heat"—one scientist even described female Chimpanzees as being "masculinized" by the onset of their estrus (Nishida 1979:103). Aside from being inappropriate in specific cases, then, it is inaccurate to ascribe greater aggression solely to "malelike" females in homosexual contexts when this may in fact be an independent feature of female sexual arousal. In addition, a recent comprehensive survey of over 700 mammal species found no correlation between the occurrence of "masculinized" female genitalia

and female aggression or dominance (Teltscher, C., H. Hofer, and M. L. East [1997] "Virilized Genitalia are Not Required for the Evolution of Female Dominance," in M. Taborsky and B. Taborsky, eds., *Contributions to the XXV International Ethological Conference,* p. 281, Advances in Ethology no. 32. [Berlin: Blackwell Wissenschafts-Verlag]). Incidentally, the female Chimpanzee referred to above was also nicknamed "the Madam" because of her apparent regulation of the sexual activity of other females, echoing an earlier nicknaming of an intersexual Savanna Baboon as "the Prostitute" (Marais 1922/1969:205–6). These examples offer striking parallels to the association, among humans, of female homosexuality/gender variance with prostitution. Both are seen as "deviant" activities and are linked not only in the mythic and popular imagination, but also sometimes in actual historical and social realities (cf. Nestle, J. [1987] "Lesbians and Prostitutes: A Historical Sisterhood," in *A Restricted Country,* pp. 157–77 [Ithaca: Firebrand Books]; Salessi, J. [1997] "Medics, Crooks, and Tango Queens: The National Appropriation of a Gay Tango," pp. 151, 161–62, in C. F. Delgado and J. E. Muñoz, eds., *Everynight Life: Culture and Dance in Latin/o America,* pp. 141–74 [Durham: Duke University Press]).

10. Although many zoologists have uncritically advocated such an "explanation" or interpretation of homosexuality, a few scientists have presented explicit arguments against such an analysis: Wolfe 1979:532, Lunardini 1989:183 (Japanese Macaque); Srivastava et al. 1991:506–7 (Hanuman Langur); Huber and Martys 1993:160 (Greylag Goose); Hunt et al. 1984 (Western Gull); Rogers and McCulloch 1981:90 (Galah).

11. This is especially true for "penis fencing" between male Bonobos, less so for mutual genital rubbing between females in this species. The latter usually involves one female "mounting" or embracing the other in a face-to-face position, hence it could be analogized with positions used in heterosexual interactions.

12. Even in some of these cases, however, a "pseudoheterosexual" framework has been imposed on the behavior. Mutual rump rubbing, in which two animals back up toward each other and rub their anal and genital regions together, has been interpreted as both animals adopting a "female" heterosexual invitation-to-mate posture in some species (e.g., Bonobos [Kitamura 1989:56–57]; Stumptail Macaques [Chevalier-Skolnikoff 1976:518]). This ignores the fact that both participants often actively rub their rumps together and make pelvic thrusts rather than simply passively presenting their hindquarters, and the two animals may also simultaneously fondle and stimulate each other's genitals with their hands—clearly making this a distinct sexual activity rather than simply a version of a heterosexual practice or posture.

13. Bottlenose Dolphin (Östman 1991). For more on reverse heterosexual mounting, see chapter 5 and the species profiles in part 2.

14. See, for example, Huber and Martys (1993:160) for explicit refutation of the idea that one member of a Greylag gander pair adopts a "pseudofemale" role.

15. This is true, for example, in Mallard Ducks, Black-crowned Night Herons, Black-headed Gulls, Emus, and Jackdaws.

16. Red Deer (based on table 2, Hall 1983:278).

17. Byne, W. (1994) "The Biological Evidence Challenged," p. 53, *Scientific American* 270(5):50–55.

18. Northern jacana (del Hoyo, J., A. Elliott, and J. Sargatal, eds. [1996] *Handbook of the Birds of the World,* vol. 3, Hoatzin to Auks, p. 282 [Barcelona: Lynx Edicións]); arctic tern and other species (Weldon, P. J., and G. M. Burghardt [1984] "Deception Divergence and Sexual Selection," *Zeitschrift für Tierpsychologie* 65:89–102, especially table 1).

19. Mountain Zebra (Penzhorn 1984:119); Chaffinch (Marler 1956:69, 96–97, 119) (Marler misleadingly labels some cases of opposite-sex mimicry as "homosexual behavior" while noting explicitly that no same-sex mounting occurs in these contexts); Rufous-naped Tamarin (Moynihan 1970:48, 50); Black-crowned Night Heron (Noble and Wurm 1942:216); Kittiwake (Paludan 1955:16–17); Koala (Smith 1980:49). Two species in which opposite-sex mimicry does appear to be a component of at least some homosexual interactions are Buff-breasted Sandpipers and Ocher-bellied Flycatchers.

20. Northern Elephant Seal (Le Boeuf 1974:173); Black-headed Gull (van Rhijn 1985:87, 100); Red Deer (Darling 1937:170); Common Garter Snake (Mason and Crews 1985:59). Researchers have also found that transvestite paketi (a fish species) have huge testes that are about five times larger than that of nontransvestite males and are thus able to fertilize more eggs (Ayling, T. [1982] *Sea Fishes of New Zealand,* p. 255 [Auckland: Collins]; Jones, G. P. [1980] "Growth and Reproduction in the Protogynous Hermaphrodite *Pseudolabrus celidotus* [Pisces: Labridae] in New Zealand," *Copeia* 1980:660–75).

21. Tasmanian Native Hen (Ridpath 1972:30); Rhesus Macaque (Akers and Conaway 1979:76). On a related point, male Laysan Albatrosses may be stimulated to mount birds of either sex when the latter happen to assume a posture that resembles a female's invitation to mate (typically involving drooping and spread wings)—to the extent that if only a bird's right wing is drooping, for example, males on the bird's right side will attempt to mount while those on the left will not. However, this "triggering" effect can only be a partial explanation, since males do not generally try to mount females who are sitting on a nest, even though the posture and drooping wings of such birds greatly resemble the mating invitation. Researchers studying this species (e.g., Fisher 1971:45–46) have expressed puzzlement over the apparent failure of the triggering effect in this context, suggesting that perhaps the height of the incubating females (nests in this species are six to eight inches high) is an inhibiting factor. This is not consistent, however, with the fact that males sometimes mount even taller "stacks" of up to three other males that are simultaneously mounting one another. Similarly, scientists once observed a Red Deer stag mount another male whose posture, as it was beginning to undergo the effects of a tranquilizer, supposedly "resembled" a female's (Lincoln et al. 1970:101; cf. Klingel [1990:578] for a similar observation concerning anesthetized Plains Zebra stallions). Consequently, they

attributed the homosexual behavior to the "triggering" effect of the supposedly femalelike visual cues presented by the other animal. Aside from the fact that the resemblance between a female Red Deer ready to mate and a drugged male is questionable, same-sex mounting in this species occurs commonly in contexts that have nothing to do with opposite-sex "resemblance" (cf. Hall 1983, Guiness et al. 1971; the same holds for Zebras).

22. Bighorn Sheep (Berger 1985:334; Geist 1971:161–63, 185, 219). Another possible case of heterosexual interactions being modeled on homosexual ones occurs in Atlantic Spotted Dolphins: during heterosexual copulations some individuals have been observed apparently "mimicking" the sideways mounting position used during interspecies homosexual copulations with Bottlenose Dolphins (Herzing and Johnson 1997:96). Interestingly, the patterning of heterosexual relations after homosexual ones is also found in some human cultures. In medieval Baghdad and Andalusia, for example, the preeminence of (largely intergenerational) homosexual relations was such that heterosexual women often cross-dressed as male youths—sometimes even with painted mustaches—in order to compete with boys for the attentions of men (Murray and Roscoe, *Islamic Homosexualities*, pp. 99, 151). In contemporary North America, some men cross-dress as women when having sex with their wives/girlfriends because they enjoy imagining themselves as a lesbian couple, or they become transsexual/transgendered women and live with their female partners in a lesbian relationship (see, for example, Money, J. [1988] *Gay, Straight, and In-Between: The Sexology of Erotic Orientation*, pp. 105–6 [New York: Oxford University Press]; Bolin, A. [1994] "Transcending and Transgendering: Male-to-Female Transsexuals, Dichotomy and Diversity," p. 484, in G. Herdt, ed., *Third Sex, Third Gender: Beyond Sexual Dimorphism in Culture and History*, pp. 447–85 [New York: Zone Books]; Rothblatt, M. [1995] *The Apartheid of Sex: A Manifesto on the Freedom of Gender*, pp. 159–60 [New York: Crown]).

23. This is to some extent an arbitrary classification, since these three "types" may overlap with each other or even co-occur to varying degrees within the same species or individual. Nevertheless, they represent broad patterns that are a useful point of departure for discussion.

24. In the words of the scientists studing this species, "Female sexual displays formed a continuum from male-behaving females to normal females" (Buechner and Schloeth 1965:219).

25. Gorilla (Yamagiwa 1987a:13 [table 7], 1987b:36–37 [table 4]); Hanuman Langur (Srivastava et al. 1991:492–93 [table II]); Bonnet Macaque (Sugiyama 1971:260 [table 9]); Pig-tailed Macaque (Tokuda et al. 1968:291 [table 7]).

26. Western Gull (Hunt et al. 1984).

27. On the rarity of incubation feeding in male-female pairs, see Evans Ogden and Stutchbury 1994:8.

28. Some cases of apparently role-differentiated behavior are not so clearly gendered when examined in more detail. Kitagawa (1988a:65–66) suggests that females in homosexual pairs of Black-winged Stilts can be divided into "malelike" and "femalelike" partners. However, many of the courtship and pair-bonding behaviors that are used to make this distinction, such as "splashing water" or "irrelevant preening," are described by other sources (e.g., Goriup 1982; Hamilton 1975) as being performed by *both* sexes in heterosexual pairs. Even if we accept Kitagawa's classification of some behaviors as more typical of males or females, though, in at least one of the homosexual pairs described, it is difficult to see how this translates into gendered behavior. Both partners in this case performed putatively female activities such as "extending neck" and egg laying, putatively male behaviors such as "half-circling round," and putatively nongendered activities such as "showing nest spot" and incubation.

29. This behavior is exhibited by females when initiating pair-directed courtships, and by males when pursuing promiscuous matings (cf. Coddington and Cockburn 1995).

30. Swallow-tailed Manakin (Foster 1987:555; Sick 1967:17, 1959:286).

31. On the role differentiation of these parental duties in heterosexual pairs, see Martin et al. 1985:258.

32. Black-headed Gull (based on figs. 3–6, van Rhijn 1985:92–94). These comparisons are drawn from studies of captive birds; however, the behavior of wild Gulls appears to be similar—in a homosexual pair observed in the wild by Kharitonov and Zubakin (1984:103), for example, at least one partner exhibited a combination of both "male" and "female" behaviors.

33. For more extensive discussion of the full complexity and diversity of lesbian butch-femme, see Nestle, J. (1981) "Butch-Fem Relationships: Sexual Courage in the 1950's," *Heresies No. 12*, 3(4):21–24; Nestle, J., ed. (1992) *The Persistent Desire: A Femme-Butch Reader* (Boston: Alyson); Burana, L., Roxxie, and L. Due, eds. (1994) *Dagger: On Butch Women* (Pittsburgh and San Francisco: Cleis Press); Newman, L. (1995) *The Femme Mystique* (Boston: Alyson); Pratt, M. B. (1995) *S/HE* (Ithaca: Firebrand Books); Harris, L., and E. Crocker, eds. (1997) *Femme: Feminists, Lesbians, and Bad Girls* (New York: Routledge).

34. Australian Shelduck (Riggert 1977:60–61); Ring-billed Gull (Conover and Hunt 1984a); Mute Swan (Kear 1972:85–86); Mountain Sheep (Geist 1971:162); Bottlenose Dolphin (Tavolga 1966:729–30); Killer Whale (Rose 1992:112); White-handed Gibbon (Edwards and Todd 1991:234); West Indian Manatee (Hartman 1979:107–8); Hanuman Langur (Srivastava et al. 1991:508–9); Asiatic Elephant (Ramachandran 1984); Lion (Chavan 1981); Sage Grouse (Scott 1942:488).

35. For explicit rejection of (and evidence against) the shortage hypothesis by various zoologists studying animal homosexuality, see Gorilla (Harcourt et al. 1981:276); Japanese Macaque (Fedigan and Gouzoules 1978:494; Vasey 1996:550, 1998:17); Rhesus Macaque (Akers and Conaway 1979:77); Flamingo (King 1994:107); Common Gull (Riddiford 1995:112); Jackdaw (Röell 1978:103); Galah (Rogers and McCulloch 1981:90; Rowley 1990:59–60).

36. Orang-utan (Rijksen 1978:259); Japanese Macaque (Vasey 1996 and personal communication; Corradino 1990:360; Wolfe 1984); Stumptail Macaque (Chevalier-Skolikoff 1976:520); Rhesus Macaque (Akers and Conaway 1979:76–77); Common Gull (Riddiford 1995:112); Black-headed Gull (van Rhijn 1985:91–93); King Penguin (Murphy 1936:340–41); Galah (Rogers and McCulloch 1981:90; Rowley 1990:59–60).

37. Bottlenose Dolphin (Östman 1991:310); Squirrel Monkey (Mendoza and Mason 1991:476–77; Travis and Holmes 1974:55, 63); Bonobo (Kano 1992:149; Savage-Rumbaugh and Wilkerson 1979:338); Stumptail Macaque (Chevalier-Skolikoff 1976:524); Savanna (Yellow) Baboon (Maxim and Buettner-Janusch 1963:176); West Indian Manatee (Hartman 1979:101, 106); Pukeko (Jamieson and Craig 1987a:1251); Common Murre (Birkhead et al. 1985:614); Sociable Weaver (Collias and Collias 1980b:248); Bonnet Macaque (Sugiyama 1971:252, 259–60); Japanese Macaque (Vasey 1996:543 and personal communication). Homosexual mounting rates can also be independent of the presence of animals of the opposite sex: in an all-female group of Pig-tailed Macaques, for example, the rate of same-sex mounting was virtually identical both before and after introduction of a male into the group (Giacoma and Messeri 1992:183 [table 1]). The finding of a positive correlation between homosexual and heterosexual rates is paralleled by some data on humans in sex-segregated environments. Researchers found that married men in prisons who receive conjugal visits with their wives are actually *more* likely to have sex with other male prisoners than men without conjugal visits (Wooden, W. S., and J. Parker [1982] *Men Behind Bars: Sexual Exploitation in Prison,* pp. 55–56 [New York: Plenum]).

38. For species with skewed sex ratios but no homosexuality, see Welty, J. C., and L. Baptista (1988) *The Life of Birds,* 4th ed., p. 154 (New York: W. B. Saunders); Newton, I. (1986) *The Sparrowhawk,* pp. 37, 151 (Calton, England: T. and A. D. Poyser); Taborsky, B., and M. Taborsky (1991) "Social Organization of North Island Brown Kiwi: Long-Term Pairs and Three Types of Male Spacing Behavior," *Ethology* 89:47–62. For verification of balanced sex ratios in populations exhibiting homosexuality, see Bonobo (Thompson-Handler et al. 1984:349); Bonnet Macaque (Simonds 1965); West Indian Manatee (Hartman 1979:139); Snow Goose (Quinn et al. 1989:184); California Gull (Conover et al. 1979); Pukeko (Craig 1980:594).

39. For sex ratios of various Seals and Sea Lions, see Fay 1982:256 (Walrus); for lunulated and salvin's antbirds, see Willis, E. O. (1968) "Studies of the Behavior of Lunulated and Salvin's Antbirds," *Condor* 70:128–48; for other antbird species with a "surplus" of unmated males but no homosexual pairs, see Willis, E. O. (1969) "On the Behavior of Five Species of *Rhegmatorhina,* Ant-Following Antbirds of the Amazon Basin," *Wilson Bulletin* 81:363–95.

40. Crab-eating Macaque (Poirier and Smith 1974); Pukeko (Craig 1980:594); Rhesus Macaque (Lindburg 1971:14, 69); Tree Swallow (Stutchbury and Robertson 1985, 1987b); Galah (Rogers and McCulloch 1981:90); Scarlet Ibis (Elbin and Lyles 1994:90–91); Flamingo (King 1994:104–5); Nilgiri Langur (Hohmann 1989:449); Little Egret (M. Fujioka, personal communication); Little Blue Heron (Werschkul 1982:382).

41. Black Stilt (Reed 1993:772); Humboldt Penguin (Scholten 1992:6 and personal communication); Savanna (Yellow) Baboon (Rowell 1967a:16, 22–23 [tables 2, 3]); Mallard Duck (Lebret 1961:108 [table I]).

42. Pig-tailed Macaque (Oi 1990a:340); Bottlenose Dolphin (Wells 1991:222); Cheetah (Eaton and Craig 1973:252); Koala (Smith 1980:184); Canada Goose (Collias and Jahn 1959:484); Flamingo (C. E. King, personal communication); Lesser Flamingo (Alraun and Hewston 1997:175–76).

43. Japanese Macaque (Chapais and Mignault 1991:172; Wolfe 1984:155); Giraffe (Dagg and Foster 1976:28, 124, 144); Greylag Goose (Huber and Martys 1993:160). Likewise, in Northern Fur Seal populations with up to 40 or more females for every male, a number of behavioral and other factors insure that nearly every female is still able to mate heterosexually (Gentry 1998:167, 192–93). For Macaques, some researchers have suggested that females resort to homosexuality when deprived of "novel" male partners rather than of males per se (i.e., when they "run out" of new partners or become overly familiar with them) (Wolfe 1984:155, 1986:274 [Japanese Macaque]; Huynen 1997 [Rhesus Macaque]). As Vasey (1996:550) points out, however, this explanation is flawed because the females they turn to are no more "novel" than the males are (and probably even less so, owing to the high levels of female bonding and familiarity in these species). In addition, some females continue to choose other females as partners even in populations that have novel males.

44. Gorilla (Robbins 1996; Fossey 1983, 1984; Harcourt et al. 1981); Hanuman Langur (Weber and Vogel 1970); Crested Black Macaque (Reed et al. 1997; Dixson 1977); Squirrel Monkey (DuMond 1968; Travis and Holmes 1974; Akers and Conaway 1979; Denniston 1980; Mendoza and Mason 1991); Walrus (Miller and Boness 1983; Sjare and Stirling 1996); Lion (Schaller 1972; Chavan 1981); Mallard Duck (Bossema and Roemers 1985; Schutz 1965:457–59); Black-headed Gull (Kharitonov and Zubakin 1984); West Indian Manatee (Hartman 1971, 1979); Cheetah (Eaton and Craig 1973; Eaton 1974a). In some of these cases (e.g., Gorillas, Hanuman Langurs) homosexual activity among males is much more common in same-sex groups although it still occurs sporadically or "residually" in mixed groups; in other cases (e.g., Squirrel Monkeys, Crested Black Macaques) homosexual activity is equally if not more common in at least some mixed-sex groups.

45. Squirrel Monkey (Talmage-Riggs and Anschel 1973:68, 71); Long-eared Hedgehog (Poduschka 1981:81).

46. Silver Gull (Mills 1991:1523, 1526); Mallard Duck (Schutz 1965:442); Canada Goose (Collias and Jahn 1959:500); Jackdaw (Röell 1979:126, table 1); Lesser Scaup Duck (Bellrose 1976:344); Caribou (Bergerud 1974:432).

47. Flamingo (Wilkinson 1989:53–54; King 1994:105; C. E. King, personal communication); Laughing Gull (Hand 1981:138–39); Humboldt Penguin (Scholten 1992:5); Gentoo Penguin (Stevenson 1983:192); Pied Kingfisher (Moynihan 1990:19; Reyer 80:220); Peach-faced Lovebird (Fischdick et al 1984:314); Galah (Rogers and McCulloch 1981:90); Bicolored Antbird (Willis 1967:112).

48. Cattle Egret (Fujioka 1986b:421–22); emperor and other penguins (Williams, T. D. [1995] *The Penguins: Spheniscidae,* pp. 80, 160 [Oxford: Oxford University Press]); dipper (Wilson, J. D. [1996] "The Breeding Biology and Population History of the Dipper *Cinclus cinclus* on a Scottish River System," *Bird Study* 43:108–18); Oystercatcher (Heg and van Treuren 1998); Australian noisy miner (Dow, D. D., and M. J. Whitmore [1990] "Noisy Miners: Variations on the Theme of Communality," in P. B. Stacey and W. D. Koenig, eds., *Cooperative Breeding in Birds: Long-Term Studies in Behavior,* pp. 559–92 [Cambridge: Cambridge University Press]); spotted sandpiper (Oring, L. W., J. M. Reed, and S. J. Maxson [1994] "Copulation Patterns and Mate Guarding in the Sex-Role Reversed, Polyandrous Spotted Sandpiper, *Actitis macularia*," *Animal Behavior* 47:1065–72).

49. Redshank (Nethersole-Thompson and Nethersole-Thompson 1986:228); mustached warbler (Fessl, B., S. Kleindorfer, and H. Hoi [1996] "Extra Male Parental Behavior: Evidence for an Alternative Mating Strategy in the Moustached Warbler *Acrocephalus melanopogon*," *Journal of Avian Biology* 27:88–91); Ostrich (Bertram 1992:125–26, 178); Greater Rhea (Navarro et al. 1998:117–18); Tree Swallow (Leffelaar and Robertson 1985); tropical house wren (Freed, L. A. [1986] "Territory Takeover and Sexually Selected Infanticide in Tropical House Wrens," *Behavioral Ecology and Sociobiology* 19:197–206); barn swallow (Crook, J. R., and W. M. Shields [1985] "Sexually Selected Infanticide by Adult Male Barn Swallows," *Animal Behavior* 33:754–61); Black Stilt (Pierce 1996:85); Silver Gull (Mills 1989:388); Herring Gull (Burger and Gochfeld 1981:128); African Elephant (Buss and Smith 1966:385–86; Kühme 1963:117).

50. White-handed Gibbon (Edwards and Todd 1991:234; Reichard 1995 a,b; Mootnick and Baker 1994); Ostrich (Sauer 1972:737); Buff-breasted Sandpiper (Lanctot and Laredo 1994:8; Pruett-Jones 1988:1748).

51. American Bison (Komers et al. 1994:324; D. F. Lott, personal communication); Bonobo (Hashimoto 1997:12–13).

52. Musk-ox (Smith 1976:37, 56, 75–77; Gray 1979; Reinhardt 1985); Asiatic Elephant (Poole et al. 1997:304, 306–7 [fig. 5]); New Zealand Sea Lion (Marlow 1975:186, 203); Wolf (Zimen 1981:140); Killer Whale (Rose 1992:73, 83–84, 112, 116).

53. Ruff (Hogan-Warburg 1966:178–79, 199–200; van Rhijn 1991:69); Pukeko (Jamieson et al. 1994:271; Jamieson and Craig 1987a); Ocher-bellied Flycatcher (Westcott 1993:450); Ruffed Grouse (Gullion 1981:377, 379–80); Oystercatcher (Heg and van Treueren 1998: 689–90); Brown-headed Cowbird (Rothstein et al. 1986:150, 154–55, 167; Darley 1978); Guianan Cock-of-the-Rock (Trail and Koutnik 1986:209).

54. Giraffe (Dagg and Foster 1976:123; Innis 1958:258–60); Japanese Macaque (Vasey 1996 and personal communication; Corradino 1990:360; Wolfe 1984); Hanuman Langur (Srivastava et al. 1991); Gray Seal (Backhouse 1960:310); Killer Whale (Jacobsen 1990:75–78); Zebras (Rasa and Lloyd 1994:186); Great Cormorant (Kortlandt 1949); Orange-fronted Parakeet (Hardy 1965:152–53); Wapiti (Lieb 1973:61; Graf 1955:73; Harper et al. 1967:37); Ducks (McKinney et al. 1983). Most of these cases are also examples of a "preference" for homosexual activity in the participating individuals.

55. White-fronted Amazon Parrot (Clarke 1982:71); Long-eared Hedgehog (Poduschka 1981:81); Steller's Sea Eagle (Pringle 1987:104); Barn Owl (Jones 1981:54); Rhesus Macaque (Erwin and Maple 1976:12–13); Crab-eating Macaque (Hamilton 1914:307–8); Bottlenose Dolphin (McBride and Hebb 1948:121); Cheetah (Ruiz-Miranda et al. 1998:7, 12); Black-headed Gull (van Rhijn and Groothuis 1987:142–43; van Rhijn 1985:91–93); Mallard Duck (Schutz 1965:442, 449–50, 460).

56. Ring-billed Gull (Conover and Hunt 1984a); Greylag Goose (Huber and Martys 1993:157[fig.1]).

57. Willson, M. F., and E. R. Pianka (1963) "Sexual Selection, Sex Ratio, and Mating Systems," *American Naturalist* 97:405–7; Verner, J. (1964) "Evolution of Polygamy in the Long-billed Marsh Wren," *Evolution* 18:252–61; Verner, J., and M. F. Willson (1966) "The Influence of Habitats on Mating Systems of North American Passerine Birds," *Ecology* 47:143–47; Wittenberger, J. F. (1976) "The Ecological Factors Selecting for Polygyny in Altricial Birds," *American Naturalist* 109:779–99; Wittenberger, J. F. (1979) "The Evolution of Vertebrate Mating Systems," in P. Marler and J. Vandenbergh, eds., *Handbook of Neurobiology: Social Behavior and Communication,* pp. 271–349 (New York: Plenum Press); Goldizen et al 1998 (Tasmanian Native Hen). For examples of (heterosexual) mating systems actually determining the sex ratio rather than vice versa, see Hamilton, W. D. (1967) "Extraordinary Sex Ratios," *Science* 156:477–88; Wilson, D. S., and R. K. Colwell (1981) "Evolution of Sex Ratio in Structured Demes," *Evolution* 35:882–97.

58. In Roseate Terns, for example, homosexual pairs were initially taken as evidence of skewed sex ratios, even though the sex ratio in this species had not yet been reliably determined (owing to the difficulty, until recently, of accurately determining the sex of individuals) (Sabo et al. 1994:1023, 1026).

59. Western Gull (Hunt and Hunt 1977; Hunt et al. 1980; Wingfield et al. 1980; Fry and Toone 1981; Fry et al. 1987; Hayward and Fry 1993); Herring Gull (Fitch 1979; Shugart et al. 1987, 1988; Pierotti and Good 1994).

60. For explicit refutation of an association between female homosexual pairs and environmental toxins, see Hunt 1980 (Western Gull); Lagrenade and Mousseau 1983; and Conover 1984c (Ring-billed Gull).

61. Fry et al. 1987; Fry, D. M., and C. K. Toone (1981) "DDT-induced Feminization of Gull Embryos," *Science* 213:922–24.

62. Fry et al. 1987:37, 39; Fry and Toone 1981:923. Behavioral changes that could potentially be relevant have only been observed in other bird species, and only as a result of direct injection with estrogen, a female hormone, and not as a result of exposure to toxins (which mimic some of the effects of estrogen).

63. Indeed, if toxin-induced "feminization" resulted in behavioral changes, one might even expect this to be manifested directly as *male* homosexuality (especially under a "pseudoheterosexual" interpretation, or one in which homosexuality is equated with intersexuality), yet this has not been reported for these populations

either. Even if such homosexuality were to occur, however, it would not necessarily argue for reduced numbers of breeding males: homosexually paired males in several bird species (including Black-headed and Laughing Gulls) sometimes continue to copulate with females (i.e., they are functionally bisexual and their same-sex pair bonds are nonmonogamous).

64. Herring Gull and other species (Fitch and Shugart 1983:6).

65. Western Gull (Fry et al. 1987); Herring Gull (Burger and Gochfeld 1981; Nisbet and Drury 1984:88). In these populations scientists have suggested that perhaps a cofactor is involved: availability of nest sites (Fry et al. 1987:40). The hypothesis is that homosexual pairs will only form in sex-skewed populations if there are vacant nest sites, since female pairs presumably are less able to compete for territories in dense colonies. However, Hand (1980:471) argues that homosexual pairs can effectively obtain (and defend) territories even in dense colonies. In addition, Fetterolf et al. (1984) show that female pairs of Ring-billed Gulls in crowded colonies are simply relegated to less optimal nest sites, rather than failing to form in the first place (or disbanding) because of competition or crowding. This "cofactor" is also of limited applicability to other bird species. In Orange-fronted Parakeets, for example, female pairs compete successfully against heterosexual pairs for possession of nest sites (Hardy 1963:187), while in many species female pairs form regardless of whether they acquire nesting sites (i.e., homosexual pair-formation is independent of nesting).

66. Herring Gull (Shugart et al. 1987, 1988); Ring-billed Gull (Conover and Hunt 1984a,b).

67. Watson, A., and D. Jenkins (1968) "Experiments on Population Control by Territorial Behavior in Red Grouse," *Journal of Animal Ecology* 37:595–614; Weatherhead, P. J. (1979) "Ecological Correlates of Monogamy in Tundra-Breeding Savannah Sparrows," *Auk* 96:391–401; Smith, J. N. M., Y. Yom-Tov, and R. Moses (1982) "Polygyny, Male Parental Care, and Sex Ratio in Song Sparrows: An Experimental Study," *Auk* 99:555–64; Hannon, S. J. (1984) "Factors Limiting Polygyny in the Willow Ptarmigan," *Animal Behavior* 32:153–61; Greenlaw, J. S., and W. Post (1985) "Evolution of Monogamy in Seaside Sparrows, *Ammodramus maritimus:* Tests of Hypotheses," *Animal Behavior* 33:373–83; Gauthier, G. (1986) "Experimentally-Induced Polygyny in Buffleheads: Evidence for a Mixed Reproductive Strategy?" *Animal Behavior* 34:300–302; Björklund, M., and B. Westman (1986) "Adaptive Advantages of Monogamy in the Great Tit (*Parus major*): An Experimental Test of the Polygyny Threshold Model," *Animal Behavior* 34:1436–40; Stenmark, G., T. Slagsvold, and J. T. Lifjeld (1988) "Polygyny in the Pied Flycatcher, *Ficedula hypoleuca:* A Test of the Deception Hypothesis," *Animal Behavior* 36:1646–57; Brown-headed Cowbird (Yokel and Rothstein 1991).

68. Western Gull (Hunt and Hunt 1977); Herring Gull (Shugart et al. 1988). Fertility rates for homosexual pairs in other Gull species (not associated with environmental toxins) vary considerably, from 0 percent fertile eggs in Kittiwake female pairs (Coulson and Thomas 1985), 33 percent for Silver Gulls (Mills 1991), and 8–94 percent for Ring-billed Gulls (Ryder and Somppi 1979; Kovacs and Ryder 1983). Incidentally, only some of the males that copulate with female Western Gulls in homosexual pairs are known to be already paired; the remainder may in fact be single males that females are bypassing for pair-bonding, while utilizing them to fertilize their eggs (see Pierotti 1981:538–39). Also, some Silver Gulls in homosexual pairs may be raped by males, i.e., their participation in breeding may be "forced" rather than "consensual" (Mills 1989:397).

69. Herring Gull (Fitch and Shugart 1984:123); Ring-billed Gull (Conover 1984b:714–16; Fetterolf and Blokpoel 1984:1682); Western Gull (Pierotti 1980:292); Roseate Tern (Spendelow and Zingo 1997:553). In Roseate Terns, females with proven single-parenting abilities nevertheless sometimes still form homosexual pairs, indicating that their same-sex partnership is not due solely to the "necessity" of finding a coparent (e.g., one female formed a homosexual pair even though she had successfully raised a chick on her own when her male partner died the previous year).

70. For an extensive list of species in which supernormal clutches have been found—only a fraction of which involve verified female pairs—see Conover 1984c (Ring-billed Gull). For other sources of supernormal clutches (and the occurrence of female pairs with regular-sized clutches), see Western Gull and other species (Conover 1984); Ring-billed Gull and other species (Conover and Aylor 1985; Conover and Hunt 1984; Ryder and Somppi 1979); Common Gull (Trubridge 1980); Terns (Penland 1984; Shealer and Zurovchak 1995; Gochfeld and Burger 1996:631); loons (McNicholl, M. K. [1993] "Supernumerary Clutches of Common Loons, *Gavia immer,* in Ontario," *Canadian Field-Naturalist* 107:356–58); sandpipers and related species (Mundahl, J. T., O. L. Johnson, and M. L. Johnson [1981] "Observations at a Twenty-Egg Killdeer Nest," *Condor* 83:180–82; Sordahl, T. A. [1997] "Breeding Biology of the American Avocet and Black-necked Stilt in Northern Utah," pp. 350, 352, *Southwestern Naturalist* 41:348–54); Laysan Albatross (Fisher 1968). On the nonoccurrence of female pairs in certain species with supernormal clutches, see Narita, A. (1994) "Occurrence of Super Normal Clutches in the Black-tailed Gull *Larus crassirostris,*" *Journal of the Yamashina Institute of Ornithology* 26:132–34; Chardine, J. W., and R. D. Morris (1996) "Brown Noddy (*Anous stolidus*)," in A. Poole and F. Gill, eds., *The Birds of North America: Life Histories for the 21st Century* no. 220, pp. 10, 18 (Philadelphia: Academy of Natural Sciences; Washington, D.C.: American Ornithologists' Union).

71. And sometimes the correlation between the two end points is itself suspect. For example, it has been claimed that supernormal clutches are more common in Great Lakes populations of Herring Gulls than they are in New England, a fact that is attributed to greater levels of DDT poisoning in the Great Lakes (Conover 1984c:254) and/or the absence of available nesting sites in New England (Fry et al. 1987:40; see note 65 above). However, Nisbet and Drury (1984:88) show that the "higher rate" of supernormal clutches in the Great Lakes can be traced to only one particular colony site; in three other Great Lakes areas censused, the prevalence of supernormal clutches was no greater than in New England. Moreover, even if such clutches are

more common in the Great Lakes area, the fact that they still occur in New England indicates that their presence cannot be due entirely to pollutant-related (or nest-site availability) factors.

72. In Western Gulls, the correlation between toxins and supernormal clutches is claimed to be supported by chronological evidence: larger clutches were supposedly not common prior to the widespread use of pesticides in the 1950s–1970s in southern California, while female pairs are claimed to occur at a "much lower" rate (Hayward and Fry 1993:19) or to have all but disappeared (Pierotti and Annett 1995:11) now that pesticide use has stopped. However, no comprehensive survey of the affected areas has in fact been conducted to assess the actual incidence of female pairs today (even if such a study were to find consistently low levels, this would still be significant, since it would demonstrate a "residual" component of same-sex activity that is independent of toxin effects and of a "shortage" of the opposite sex, as is true for many other species). Nor have detailed longitudinal or geographic studies been conducted to track the putative correlations during this entire five-to-six-decade period. In fact, records of supernormal clutches in Ring-billed Gulls go back much earlier, to the early 1900s (and in other species back to the late 1800s), while in some Terns their frequency has actually *decreased* since the 1950s (Conover 1984c), so the chronological question is far from resolved. At least one researcher who has addressed the temporal issue rejects the DDT (or other pollutant) connection for the majority of cases: Conover (1984c:254) conducted an extensive survey of the occurrence of supernormal clutches in 34 species, including comparing pre- and post-1950 rates, and concluded that their frequency is not higher since the 1940s for most Gull and Tern species. Finally, no studies have yet determined the incidence of homosexual pairing/supernormal clutches in other regions of the world that have the highest levels of contamination from DDT and related pollutants, such as the Baltic Sea, the Waddensee, the Irish Sea, the Gulf of St. Lawrence, and the northern Gulf of Mexico (Nisbet, I. C. T. [1994] "Effects of Pollution on Marine Birds," p. 13, in D. N. Nettleship, J. Burger, and M. Gochfeld, eds. *Seabirds on Islands: Threats, Case Studies, and Action Plans*, pp. 8–25. [Cambridge: BirdLife International]).

73. Hayward and Fry 1993:19; Luoma, J. R. (1995) "Havoc in the Hormones," *Audubon* 97(4):60–67; Robson, B. (1997) "A Chemical Imbalance," *Nature Canada* 26(1):29–33; see also Coulson 1983 (Caspian Tern). The equating of homosexuality with environmental and physiological "havoc" has also entered the more popular discourse, as in a recent public radio broadcast that referred to lesbian pairs in Gulls as evidence of hormonal imbalances caused by environmental contamination ("Gator Envy," *All Things Considered*, National Public Radio, February 1, 1995). Some things *not* considered in this report were the broader context of same-sex pairing in other species and the intricacies of the specific cases. For more on the pathologizing of homosexuality, see the following section "Gross Abnormalities of Behavior."

74. See, for example, Aiken (1981) on Water Boatman Bugs. Even this case is somewhat less than definitive, however, since *more* than half of all mating attempts in this species are by males on other males.

75. Guianan Cock-of-the-Rock (Trail 1985a:238, 240); Giraffe (Spinage 1968:130); Black-billed Magpie (Baeyens 1979:39–40); Mountain Sheep (Geist 1968:208). For examples of homosexual interactions that are explicitly labeled "mistakes" or "errors" (including, but not limited to, cases of sex misrecognition), see Asiatic Mouflon (Schaller and Mirza 1974:318–20); Common Murre (Birkhead et al. 1985:610–11); Oystercatcher (Makkink 1942:60); Laughing Gull (Hand 1981:139–40); Greater Rhea (Fernández and Reboreda 1995:323).

76. Redshank (Hale and Ashcroft 1982:471). Other species in which the occurrence of homosexuality is taken as the sole evidence of faulty sex recognition or "indiscriminate" mating or courtship include Cavies (Rood 1970:449), Little Brown Bats (Thomas et al. 1979:134), Shags (Snow 1963:93–94), Little Egrets (Fujioka 1988), Oystercatchers (Makkink 1942:67–68), Black-headed Gulls (van Rhijn 1985:87, 93), Superb Lyrebirds (Lill 1979a:496), and King Penguins (Murphy 1936:340). It should also be pointed out that the claim of "indiscriminate" sexual activity is often quite exaggerated: it is not uncommon for the mere existence of same-sex activity to be interpreted as evidence that the sex of the partner is immaterial, even when the animals show clear partner preferences, sometimes even favoring homosexual activity. For example, Trail and Koutnik (1986:210–11) claim that yearling Guianan Cock-of-the-Rock will mount *any* bird that sits still long enough; in fact, only one attempted heterosexual mount by a yearling was recorded during their study, compared to hundreds of homosexual mounts, and certain adult males were clearly mounted more often than others (ibid., 211–12, 215).

77. Yellow-eyed penguin (Richdale, L. E. [1951] *Sexual Behavior in Penguins*, p. 73 [Lawrence, Kans.: University of Kansas Press]); Silvery Grebe (Nuechterlein and Storer 1989:344); Red-faced Lovebird (Dilger 1960:667).

78. For species with adult-female/younger-male resemblances, see Rohwer, S., S. D. Fretwell, and D. M. Niles (1980) "Delayed Maturation in Passerine Plumages and the Deceptive Acquistion of Resources," *American Naturalist* 115:400–437; for species with adult-female/adult-male resemblances, see Burley, N. (1981) "The Evolution of Sexual Indistinguishability," in R. D. Alexander and D. W. Tinkle, eds., *Natural Selection and Social Behavior*, pp. 121–37 (New York: Chiron Press). A caveat about these cases is that the *absence* of homosexuality in a species is not necessarily a reliable form of evidence, since (as discussed in chapters 1–3) homosexual behavior is often hard to observe, easy to overlook, or deliberately ignored in the field.

79. Blackbuck (Dubost and Feer 1981:74–75); Guianan Cock-of-the-Rock (Trail and Koutnik 1986:199; Trail 1983); Swallow-tailed Manakin (Foster 1987:549; Sick 1967:17; 1959:286); Blue-backed Manakin (Snow 1963:172); Raggiana's Bird of Paradise, Victoria's Riflebird (Gilliard 1969:113, 223); Regent Bowerbird (Gilliard 1969:337); Superb Lyrebird (Smith 1982 and personal communication).

80. Mountain Goat (Chadwick 1983:14, 189–91); Bishop Birds (Craig and Manson 1981:13); Galah (Rogers and McCulloch 1981:81; Rowley 1990:4); Humboldt Penguin (Scholten 1987:200); King Penguin (Stonehouse

1960:11); Superb Lyrebird (Smith 1982 and personal communication); Ocher-bellied Flycatcher (Westcott and Smith 1994:678, 681; Snow and Snow 1979:286); Tree Swallow (Stutchbury and Robertson 1987c); Anna's Hummingbird (Ortiz-Crespo 1972; Wells et al. 1996).

81. Andersson, S., J. Örnborg, and M. Andersson (1998) "Ultraviolet Sexual Dimorphism and Assortative Mating in Blue Tits," *Proceedings of the Royal Society of London,* Series B 265:445–50; Hunt, S., A. T. D. Bennett, I. C. Cuthill, and R. Griffiths (1998) "Blue Tits Are Ultraviolet Tits," *Proceedings of the Royal Society of London,* Series B 265:451–55; Witte, K., and M. J. Ryan (1997) "Ultraviolet Ornamentation and Mate Choice in Bluethroats," in M. Taborsky and B. Taborsky, eds., *Contributions to the XXV International Ethological Conference,* p. 201, Advances in Ethology no. 32 (Berlin: Blackwell Wissenschafts-Verlag); Roper, T. J. (1997) "How Birds Use Sight and Smell," *Journal of Zoology, London* 243:211–13; Bennett, A. T. D., I. C. Cuthill, J. C. Partridge, and E. J. Maier (1996) "Ultraviolet Vision and Mate Choice in Zebra Finches," *Nature* 380:433–35; Waldvogel, J. A. (1990) "The Bird's Eye View," *American Scientist* 78:342–53; Cabbage White Butterfly (Obara 1970 and personal communication; Obara, Y. [1995] "The Mating Behavior of the Cabbage White Butterfly," paper presented at the 24th International Ethological Conference, Honolulu, Hawaii); Superb Lyrebird (Reilly 1988:45).

82. Mountain Goat (Geist 1964:565); Musk-ox (Smith 1976:56); Cavies (Rood 1972:27, 54, 1970:443); Bighorn Sheep (Geist 1968:208); Common Murre (Birkhead et al. 1985:610–11); Flamingo (C. E. King, personal communication); Pronghorn (Kitchen 1974:44 [table 22]). In addition, some homosexual activity in Mountain Goats and Pronghorns also involves age-mates interacting with each other (adult males in Mountain Goats, younger males in Pronghorns). See also Wagner (1996) on Razorbills.

83. Swallow-tailed Manakin (Foster 1987:555); Laughing Gull (Noble and Wurm 1943:205); Black-headed Gull (van Rhijn and Groothuis 1985:163). Conversely, homosexuality has sometimes been attributed to behavioral *identity* between males and females. In Ruffed Grouse, for example, the nonaggressive or "submissive" posture of a male is similar to the behavior of a female during courtship, and that males court both sexes is attributed to their inability to distinguish "female-acting" males from actual females (Allen 1934:185; see also the discussion of "pseudoheterosexuality" earlier in this chapter). Aside from the fact that males and females are very different *visually* from each other in this species and therefore "there is no excuse for a male not recognizing a female" (as Allen [1934:180–81] observes), in the related red grouse there is a parallel identity between male "submissive" and female courtship behavior, yet males do not court other males in this species (Watson, A., and D. Jenkins [1964] "Notes on the Behavior of the Red Grouse," *British Birds* 57:157).

84. Tree Swallow (Stutchbury and Robertson 1987a:719–20, 1987b:418). It is also unlikely that homosexual activity between adult males results from their mistaking one another for (adult) females. As Lombardo et al. (1994) point out, although the two sexes in this species look similar, the sex of at least one male involved in homosexual activity was nevertheless identifiable from his cloacal (genital) protuberance, lack of brood patch, and wing length. Most adult females are also visually distinct from males owing to the presence of a brown patch on the forehead (shorter wings also distinguish subadult females from subadult males) (Stutchbury and Robertson 1987c). In addition, same-sex copulations appear to be fairly uncommon in this species (Lombardo, personal communication)—certainly they are not nearly as frequent as one would expect if "mistakes" in sex recognition were prevalent.

85. Black-headed Gull (van Rhijn 1985:87, 100).

86. Hooded Warbler (Niven 1993:191) (cf. Lynch et al. [1985:718] for mean dimensions of other males). Niven (1993 and personal communication) suggests that it was also the female behavior patterns of this male that "triggered" the homosexual pairing, yet this bird's behavior was actually a mixture of male and female patterns, involving, for example, incubation (female duties) as well as singing (male). Moreover, male Hooded Warblers are particularly attuned to differences in song pattern, using this information to recognize individual birds and then storing it in long-term memory for future use (Godard 1991). Because this male's singing was highly distinctive, it is improbable that other males simply "disregarded" this aspect of his behavior or were "unaware" of his male status (especially given his physical characteristics). Furthermore, all "female" behaviors recorded in this individual occurred after the formation of the pair-bond—since pairs were not observed early in the breeding season, we do not in fact know whether this individual exhibited any (or only) "femalelike" patterns during courtship and pair-formation.

87. Hooded Warbler: differential attacking of males (Stutchbury 1994:65–67); mating success of malelike females (as evidenced by the fact that nests are fairly equally distributed between dark and light females) (Stutchbury et al. 1994:389[fig.6]; Stutchbury and Howlett 1995:95); promiscuous mating attempts on hooded females (Stutchbury et al. 1994:388).

88. Common Garter Snake (Mason 1993:261, 264; Mason et al. 1989:292; Mason and Crews 1985; Noble 1937:710–11); other species (Muma, K., and P. J. Weatherhead [1989] "Male Traits Expressed in Females: Direct or Indirect Sexual Selection?" *Behavioral Ecology and Sociobiology* 25:23–31; Potti, J. [1993] "A Male Trait Expressed in Female Pied Flycatchers *Ficedula hypoleuca*: The White Forehead Patch," *Animal Behavior* 45:1245–47 [cf. also Sætre and Slagsvold 1992:295–96]; Tella, J. L., M. G. Forero, J. A. Donázar, and F. Hiraldo [1997] "Is the Expression of Male Traits in Female Lesser Kestrels Related to Sexual Selection?" *Ethology* 103:72–81; McDonald, D. B. [1993] "Delayed Plumage Maturation and Orderly Queues for Status: A Manakin Mannequin Experiment," p. 38, *Ethology* 94:31–45). Experimental "disguising" of individuals to look like the opposite sex does not automatically induce "homosexual" behavior either. Female Chaffinches whose plumage has been painted to resemble male patterns, for example, are not courted by (nor do they form pair-bonds with) other females that "mistake" them for males (Marler 1955). Homosexual pairing *does*

occur in this species, but between females that do not look like males. Likewise, yellowthroats (a bird species) are able to recognize the "true" sex of both males and females whose facial coloration has been manipulated to make them resemble the opposite sex. Similar results have been found for damselflies (Lewis, D. M. [1972] "Importance of Face-Mask in Sexual Recognition and Territorial Behavior in the Yellowthroat," *Jack-Pine Warbler* 50:98–109; Gorb, S. N. [1997] "Directionality of Tandem Response by Males of a Damselfly, *Coenagrion puella*," in M. Taborsky and B. Taborsky, eds., *Contributions to the XXV International Ethological Conference*, p. 138, Advances in Ethology no. 32 [Berlin: Blackwell Wissenschafts-Verlag]). In addition, in species such as lazuli buntings where juvenile males resemble adult females, experimental studies have demonstrated that adult males are in fact consistently able to distinguish the two sexes (Muehter, V. R., E. Greene, and L. Ratcliffe [1997] "Delayed Plumage Maturation in Lazuli Buntings: Tests of the Female Mimicry and Status Signalling Hypotheses," *Behavioral Ecology and Sociobiology* 41:281–90).

89. Tree Swallow (Lombardo et al. 1994:555–56; Venier et al. 1993; Venier and Robertson 1991); Black-crowned Night Heron (Noble et al. 1938:29); Regent Bowerbird (Marshall 1954:114–16); Greenshank (Nethersole-Thompson and Nethersole-Thompson 1979:114; Nethersole-Thompson 1951:104). In Tree Swallows, it is also unlikely that males cooperate during homosexual copulations in order to "appease" the birds mounting them and thereby avoid attack or injury (as suggested by Lombardo et al. 1994:556). Aggressive attacks in this species are characterized by a number of distinctive behavioral elements on the part of both the attacker (e.g., threat displays, grappling, pecking) and the bird being attacked (e.g., appeasement displays, submissive and distress calling) (cf. Robertson et al. 1992:6, 8)—and homosexual mountings exhibit none of these hallmarks.

90. Such cases contrast markedly with ones in which the pursued animal is clearly not a willing participant, such as Mountain Goats, Common Murres, or Anna's Hummingbirds. In these instances, however, there are other arguments against a sex misrecognition analysis (as mentioned previously).

91. Swans (Kear 1972:85–86).

92. Wattled Starling (Sontag 1991:6); Common Chimpanzee (Kollar et al. 1968:444, 458); Gorilla (Coffin 1978:67); Stumptail Macaque (Bernstein 1980:32); Musk-ox (Reinhardt 1985:298–99); Koala (Smith 1980:186); Long-eared Hedgehog (Poduschka 1981:81; Reeve 1994:189); Vampire Bat (Greenhall 1965:442); Black-crowned Night Heron (Noble et al. 1938:14, 28–29). Factors such as stress or crowding have also been invoked for wild animals, such as Blue-bellied Rollers (Moynihan 1990).

93. Dolphins (Pilleri, G. [1983] "Cetaceans in Captivity," *Investigations on Cetacea* 15:221–49); Barn Owl (Jones 1981); Rhesus Macaque (Strobel, D. [1979] "Behavior and Malnutrition in Primates," in D. A. Levitsky, ed., *Malnutrition, Environment, and Behavior: New Perspectives*, pp. 193–218 [Ithaca: Cornell University Press]). Many reports of animal homosexuality and transgender have appeared in medical journals and other publications dealing with pathology. See, for example, the descriptions of same-sex activity among Common Chimpanzees in Kollar et al. 1968 (characterized as "perverse sexual acts"), which appeared in the *Journal of Nervous and Mental Disease*.

94. Cheetah (Eaton 1974a:116); Zebra Finch (Immelmann et al. 1982:422). The assessment for Cheetahs is particularly inappropriate in light of the fact that heterosexual activity is extremely difficult to observe in this species in the wild. As mentioned in chapter 1, during one ten-year study of Cheetahs, no heterosexual matings were seen over 5,000 hours of observation, and copulation has only been observed a total of five times in the wild during the entire scientific study of this species (Caro 1994:42). It is hardly surprising, therefore, that homosexual courtship and mating activity has so far only been seen in captivity. It should also be pointed out that male "coalitions" (bonded pairs or trios) *have* been observed in both wild and captive Cheetahs (wild [Caro and Collins 1986, 1987; Caro 1994]; captive [Eaton and Craig 1973:223; Ruiz-Miranda et al. 1998]). The assumption that sex segregation is completely "artificial" for male Cheetahs living in captivity is also false (see discussion below).

95. Fedigan 1982:143 (Japanese Macaque). See also Crews et al. (1983:228–30) and Crews and Young (1991:514) for similar statements challenging the supposed "abnormalcy" of same-sex copulation among Whiptail Lizards in captivity versus the wild.

96. In a few cases, specific homosexual activities, rather than the occurrence of homosexuality itself, have been observed in only wild or captive conditions. In Bonobos, for example, penis-fencing (a form of genital rubbing) has only been seen in the wild, while fellatio has only been observed in captivity (de Waal 1997:103–4). In addition, the duration of sexual acts can vary contextually: for example, de Waal (1987:326) found that episodes of genital rubbing between female Bonobos were considerably shorter in captivity (averaging around 9 seconds) than in the wild (averaging around 15 seconds).

97. Orang-utan (Nadler 1988:107); Hamadryas Baboon (Kummer and Kurt 1965:74); Mule Deer (Halford et al. 1987:107); Musk-ox (Reinhardt 1985:298).

98. Bonobo, wild (Kano 1992:187 [table 24], 140; Kitamura 1989:53, 55–57, 61); Bonobo, captive (de Waal 1995:41[table 3.1]); Black Swan (Braithwaite 1981:141–42). Five other species for which the relevant quantitative information is available are Pig-tailed, Crested Black, and Stumptail Macaques, Common Chimpanzees, and Vervets. Although the wild (or semi-wild) and captive figures in these cases are more difficult to compare (due to differences in group size and composition, observed behaviors, length of study periods, etc.), they also generally show fairly comparable rates. For Pig-tailed Macaques in the wild, 7–23 percent of mounting is same-sex, compared to about 25 percent in captivity (rates in the wild based on information in Oi 1990a:350–1 [including table IV], Oi 1996:345, and Bernstein 1967:226–27; captivity—Tokuda et al. 1968:287, 291 [table 7]). Among captive Crested Black Macaques, about 5 percent of mounting is between

males (Dixson 1977:74, 77), compared to an estimated 8 percent in the wild (C. Reed, personal communication; figures for both of these species combine "copulatory" with "noncopulatory" mounts). However, another study (Bernstein 1970:94 [table IV]) yielded a much higher rate of same-sex mountings in captivity for these species—49 percent for Pigtails, 22 percent for Crested Blacks—demonstrating that there can be considerable differences between individual studies and/or populations (see also Bernstein [1967:228] for more on wild/captive comparisons in Pigtails). In Stumptail Macaques, 25 percent of sexual interactions (of all types) in captivity are homosexual (Chevalier-Skolnikoff 1974:100–101, 110), compared to 30–40 percent for mounting in a semi-wild troop (formerly captive animals that were transplanted and released) (Estrada et al. 1977:667 [fig.14]; Estrada and Estrada 1978:672 [table 4]). In Common Chimpanzees, same-sex mounting actually occurs more frequently in the wild: de Waal and van Hooff (1981:182 [table 2]) found that mounting between males in captivity constitutes only 1–2 percent of the behaviors involved in reassurance, enlistment of support, and other activities during conflicts, while Nishida and Hosaka (1996:120–21 [tables 9.5–9.6]) found that mounting accounts for one-third to one-half of such behaviors in wild Chimps. Likewise, Bernstein (1970:94 [table IV]) found that 9 percent of mounting activity in captive Vervets is same-sex, while Gartlan (1969:144, 146) and Struhsaker (1967:21, 27 [tables 8, 10]) both recorded 11 percent same-sex mounting in the wild. Rowell (1967b) also conducted a detailed quantitative comparison of behavioral frequency rates in the wild and captivity among Savanna (Olive) Baboons; unfortunately, mounting (and other sexual-behavior) rates between males in the wild could not be compared to rates in captivity because males were too aggressive to be kept together in captivity. On a related point, Rasa (1979b:321) found no substantial differences in Dwarf Mongoose same-sex (and opposite-sex) mounting rates when their behavior in crowded versus noncrowded captive conditions was compared (based on controlled observational regimes). Likewise, Heg and van Treuren (1998:689–90) did not find significantly higher rates of homosexual bonding (in the form of bisexual trios) when population densities increased among wild Oystercatchers.

99. Bottlenose Dolphin (McBride and Hebb 1948:114, 122; Wells et al. 1987; Wells 1991; Wells et al. 1998:65–67); Gorilla (Schaller 1963:278; Stewart 1977; Yamagiwa 1987a,b; Harcourt 1988; Porton and White 1996:723–24).

100. Jackdaw (Lorenz 1935/1970; Röell 1979); Elephants (Rosse 1892; Shelton 1965); Crested Black Macaque (Poirier 1964:147; Dixson 1977; Reed et al. 1997); Orange-fronted Parakeet (Buchanan 1966); Lion (Cooper 1942; Chavan 1981); Great Cormorant (Kortlandt 1949; Fukuda 1992); Regent Bowerbird (Phillipps 1905; Lenz 1994); Dolphins (Brown et al. 1966; Herzing and Johnson 1997). Similar erroneous assertions are sometimes made regarding transgender. Payne (1984:14), for example, claims that female-plumaged or transvestite male Ruffs occur only in captivity (citing Stonor 1937). In fact, female-plumaged males—generally referred to in this species as naked-nape males—are now known to be a regular feature of wild Ruff populations (cf. van Rhijn 1991) and have been discussed as such in the scientific literature since at least Hogan-Warburg (1966). Payne, R. B. (1984) *Sexual Selection, Lek and Arena Behavior, and Sexual Size Dimorphism in Birds*, Ornithological Mongraphs no. 33 (Washington, D.C.: American Ornithologists' Union).

101. In this regard, homosexual activity in some species is also claimed to be "caused" by unusual or abnormal environmental or climatic conditions, such as severe winter snowstorms that disrupt "normal" pairing in Golden Plovers (Nethersole-Thompson and Nethersole-Thompson 1961:207–8), or exceptionally rainy seasons that somehow "overstimulate" Ostriches (Sauer 1972:717). Assuming that ecological factors of this sort could be involved (which is debatable), an equally valid interpretation is that such species possess an inherent flexibility in their social and sexual systems that manifests itself during times of ecological flux or stress. Rather than being the "product" of "abnormal" conditions, then, such behavioral plasticity allows the species to respond "creatively"—in ways that, obviously, are not yet fully understood—to the vagaries of an ever-changing environment. See chapter 6 for further discussion.

102. Cheetah (Herdman 1972:112, 123; Caro 1993:27–28, 1994:362; Ruiz-Miranda et al. 1998:1, 13). For more on the false dichotomy of "wild" versus "captive" studies of animals, and the general compatibility and continuity between the two, see de Waal 1989a:27–33, 1997:11 (Bonobo).

103. Boto (Best and da Silva 1989:12–13); Orang-utan (van Schaik, C. P., E.A. Fox, and A. F. Sitompul [1996] "Manufacture and Use of Tools in Wild Sumatran Orangutans: Implications for Human Evolution," *Naturwissenschaften* 83:186–88); Savanna (Olive) Baboon (DeVore, I. [1965] "Male Dominance and Mating Behavior in Baboons," p. 286, in F. A. Beach, ed., *Sex and Behavior*, pp. 266–89 [New York: John Wiley and Sons]); Thomson's Gazelle (Walther 1995:30–31); King Penguin (Gillespie 1932; Stonehouse 1960); Blackheaded Gull (Kharitonov and Zubakin 1984:103; van Rhijn and Groothuis 1987:144); Flamingo (Cézilly and Johnson 1995).

104. Griffon Vulture (Blanco and Martinez 1996:247; Sarrazin et al. 1996:316); King Penguin (Weimerskirch et al. 1992:108); Gentoo Penguin (Williams and Rodwell 1992:637; Bost and Jouventin 1991:14); Flamingo (A. R. Johnson, personal communication; Dugong (Anderson 1997:440, 458; Preen 1989:384). See also chapter 3 for further discussion of heterosexual bias in the methods of sex determination employed during field studies of these and other species.

105. Canids (Macdonald 1980, 1996); Macaques (Oi 1990a; Reed et al. 1997); Gibbons (Fox 1977; Edwards and Todd 1991); Rose 1992:1–2 (Killer Whale); Aperea (Rood 1972:42); Rufous Bettong (Johnson 1980:347).

106. Orang-utan (Schürmann 1982:270–71, 282); Oystercatcher (Angier, N. [1998] "Birds' Design for Living Offers Clues to Polygamy," *New York Times* March 3, pp. B11–12).

107. van Lawick-Goodall, J. (1970) "Tool-Using in Primates and Other Vertebrates," p. 208, *Advances in the Study of Behavior* 3:195–249.

108. Sage Grouse (Scott 1942:495); Rhesus Macaque (Carpenter 1942:150); Fat-tailed Dunnart (Ewer 1968:351); Long-eared Hedgehog (Poduschka 1981:84); Takhi (Boyd 1986:660).

109. Common Garter Snake (Noble 1937:710–11); Hooded Warbler (Niven 1993:192).

110. African Elephant (Sikes 1971:265–66); Snow/Canada Goose (Starkey 1972:456–57).

111. Western Gull (Wingfield et al. 1982); Ring-billed Gull (Kovacs and Ryder 1985). See also the examples of more "intense" nesting behavior in female pairs of Ring Doves and Budgerigars discussed in note 15, chapter 1, which might also be correlated with hormonal effects.

112. For a summary of these results, see Vasey, P. L. (1995) "Homosexual Behavior in Primates: A Review of Evidence and Theory," *International Journal of Primatology* 16:173–204. Some of the species in which hormone levels have been studied in association with homosexual behavior are Rhesus Macaques (Akers and Conaway 1979; Turner et al. 1989) and Hanuman Langurs (Srivastava et al. 1991). (Turner, J. J., J. G. Herndon, M.-C. Ruiz de Elvira, and D. C. Collins [1989] "A Ten-Month Study of Endogenous Testosterone Levels and Behavior in Outdoor-Living Female Rhesus Monkeys [*Macaca mulatta*]," *Primates* 30:523–30.) For a discussion of the problematic nature of studies on laboratory rats that purport to show an association between homosexual behavior and hormones, see Mondimore, F. M. (1996) *A Natural History of Homosexuality*, pp. 111–13, 129–30 (Baltimore: Johns Hopkins University Press); Byne, W. (1994) "The Biological Evidence Challenged," *Scientific American* 270(5):50–55.

113. Pied Kingfisher (Reyer et al. 1986:216); Orang-utan (Kingsley 1982:227); Spotted Hyena (Frank 1996; Frank et al. 1985, 1995; Glickman et al. 1993); Western Gull (Wingfield et al. 1982). See also Mloszewski (1983:186), who indicates that masculinized female African Buffalo—i.e., those with "pronounced male secondary sexual characteristics," likely due in part to a differing hormonal profile—do not participate in homosexual activity any more often than do nontransgendered females (and perhaps do so even less often). For other species in which a subset of individuals have different hormone profiles (not associated with homosexual activity), see Solomon, N. G., and J. A. French, eds. (1997) *Cooperative Breeding in Mammals*, pp. 241, 304–5, 370 (Cambridge: Cambridge University Press).

114. Takhi (Boyd 1986:660). Although detailed hormonal studies of Takhi during pregnancy have been conducted, they did not involve sampling of androgens or other male hormones; see Monfort et al. 1994; Monfort, S. L., N. P. Arthur, and D. E. Wildt (1991) "Monitoring Ovarian Function and Pregnancy by Evaluating Excretion of Urinary Oestrogen Conjugates in Semi-Free-Ranging Przewalski's Horses (*Equus przewalskii*)," *Journal of Reproduction and Fertility* 91:155–64.

115. Domestic Horses (McDonnell, S. [1986] "Reproductive Behavior of the Stallion," especially p. 550, in S. L. Crowell-Davis and K. A. Houpt, eds., *Behavior*, pp. 535–55. Veterinary Clinics of North America: Equine Practice 2[3] [Philadelphia: W. B. Saunders]).

116. Recent work on the sexual orientation of Domestic Sheep has begun to move away from this paradigm, to the extent that hormonal profiles are assessed for males who prefer mounting other males, rather than simply for the ("gender-atypical") males who are themselves mounted by other males. In this case, there do appear to be some differences between homosexual and heterosexual sheep (cf. Adler, T. [1996] "Animals' Fancies: Why Members of Some Species Prefer Their Own Sex," *Science News* 151:8–9; Resko et al. 1996; Perkins et al. 1992, 1995). However, rarely (if ever) is the two-way influence of biology and behavior discussed in these studies, i.e., biology (hormones, brain structure) is invariably assumed to determine sexual behavior, when in fact it is also possible for behavior (and other social factors) to alter or affect an animal's hormonal profile or brain structure. Moreover, the search for hormonal differences is little more than a continuation of the need to find a physiological "cause" for homosexuality. Within an overall framework in which any nonreproductive behavior is still seen as anomalous, this is only a few steps removed from the overt pathologizing of homosexuality so characteristic of earlier studies.

117. Savanna Baboon (Marais 1922/1969:205); Baker, J. R. (1929) *Man and Animals in the New Hebrides*, pp. 22, 117 (London: George Routledge and Sons).

118. Bighorn Sheep (Berger 1985:334–35); White-tailed Deer (Thomas et al. 1964:236; see also Taylor et al. 1964; Thomas et al. 1965, 1970); Savanna Baboon (Marais 1922/1969; Bielert 1984b, 1985).

119. For early descriptions of intersexual Savanna Baboons, see Marais 1922/1969, 1926. For a summary of early observations of velvet-horns and other gender-mixing Deer, see Thomas et al. 1970:3 (White-tailed Deer) and Anderson 1981:94–95 (Mule Deer).

120. Northern Elephant Seal (Le Boeuf 1974:173); Red Deer (Darling 1937:170); Black-headed Gull (van Rhijn 1985:87, 100); Common Garter Snake (Mason and Crews 1985:59).

Chapter 5. Not for Breeding Only: Reproduction on the Periphery of Life

1. Hutchinson, G. E. (1959) "A Speculative Consideration of Certain Possible Forms of Sexual Selection in Man," *American Naturalist* 93:81–91.

2. According to sociobiologist James Weinrich, biological "mistakes" such as genetically transmitted diseases occur at very low rates, roughly 1 in 10,000 or less (Weinrich, J. D. [1987] *Sexual Landscapes*, p. 334 [New York: Charles Scribner's Sons]). Moreover, such genetic "defects," rather than being uniformly detrimental, sometimes confer unique abilities on their carriers. People with the genetic "disorder" of William's syndrome, for example—which occurs in about 1 in 20,000 people—often display extraordinary musical abilities, remarkable verbal skills, and exceptionally empathetic personalities, although they typically also have

low IQs and some medical complications (Lenhoff, H. M., P. P. Wang, F. Greenberg, and U. Bellugi [1997] "William's Syndrome and the Brain," *Scientific American* 277[6]:68–73).

3. As in most other "explanations" of homosexuality, these include both "proximate" and "ultimate" factors (a distinction widely employed in evolutionary biology). "Proximate" explanations focus on the immediate behavioral, social, physiological, demographic, environmental, and other factors that supposedly "trigger" or lead to homosexual activity, while "ultimate" explanations focus on the wider reproductive and evolutionary benefits that supposedly accrue from such activity.

4. Weinrich, *Sexual Landscapes;* Ruse, M. (1982) "Are There Gay Genes? Sociobiology and Homosexuality," *Journal of Homosexuality* 6:5–34; Kirsch, J. A. W., and J. E. Rodman (1982) "Selection and Sexuality: The Darwinian View of Homosexuality," in W. Paul, J. D. Weinrich, J. C. Gonsiorek, and M. E. Hotveldt, eds., *Homosexuality: Social, Psychological, and Biological Issues,* pp. 183–95 (Beverly Hills, Calif.: SAGE Publications); Wilson, E. O. (1978) *On Human Nature,* pp. 142–47 (Cambridge, Mass.: Harvard University Press); Trivers, R. L. (1974) "Parent-Offspring Conflict," pp. 260–62, *American Zoologist* 14:249–64. For a critique of these theories as applied to humans, see Futuyama, D. J., and S. J. Risch (1984) "Sexual Orientation, Sociobiology, and Evolution," *Journal of Homosexuality* 9:157–68. For specific examples of homosexuality cited as a possible population-regulation mechanism—including nonreproductive sexuality as a stress-induced response to overpopulation in some species, and homosexuality as a form of "birth control" in humans—see Calhoun, J. B. (1962) "Population Density and Social Pathology," *Scientific American* 206(2):139–48; von Holst, D. (1974) "Social Stress in the Tree-Shrew: Its Causes and Physiological and Ethological Consequences," in R. D. Martin, G. A. Doyle, and A. C. Walker, eds., *Prosimian Biology,* pp. 389–411 (Pittsburgh: University of Pittsburgh Press); Denniston 1980:38 (Squirrel Monkey); Harris, M. (1980) *Culture, People, and Nature,* p. 208 (New York: Harper and Row). For more on the special "role" of homosexual and transgendered humans in some indigenous cultures, see chapter 6.

5. See the discussion of same-sex parenting in chapter 1.

6. See pp. 206–7 for further discussion of these and other alternate parenting arrangements.

7. For a complete list of bird species with helpers, see Brown, J. L. (1987) *Helping and Communal Breeding in Birds,* pp. 18–24 (table 2.2) (Princeton: Princeton University Press). Three other species in which homosexual behavior occurs (Ostriches, House Sparrows, and Sociable Weavers) are classified by Brown as having helpers, but it is not clear that these represent genuine cases of helping. Even if they did, however, they would still not support the "helper" theory of homosexuality because homosexuality is either not limited to helpers in these species, or else not all helpers engage in homosexual behavior. In Ostriches "helping" behavior actually consists of foster-parenting by breeding pairs of males and females (ibid., p. 161); homosexuality only occurs in males in this species, and probably nonbreeders at that. In House Sparrows helping occurs occasionally among juveniles, probably of both sexes, and in only some populations (p. 31), while homosexual behavior only occurs in (a few) adult males. And in Sociable Weavers, breeding pairs are assisted in building communal nests, probably by birds of both sexes, but such birds do not help feed their young (see Maclean 1973); homosexuality occurs in both breeders and nonbreeders, but only males. Recently, helping behavior by adolescent males has also been discovered in Greater Rheas; however, this phenomenon is distinct from same-sex coparenting (and sexual activity) in this species, which involves adult males (Codenotti and Alvarez 1997:570). For other surveys of the phenomena of communal breeding and helpers in birds, see Skutch, A. F. (1987) *Helpers at Birds' Nests: A Worldwide Survey of Cooperative Breeding and Related Behavior* (Iowa City: University of Iowa Press); Stacey, P. B., and W. D. Koenig, eds. (1990) *Cooperative Breeding in Birds* (Cambridge: Cambridge University Press).

8. See chapters 1 and 4 for discussion of the fact that many cases of homosexuality in animals have probably been missed, overlooked, or remain to be discovered.

9. Moynihan (1990:19) states that homosexual pairing and/or mounting is found among nonbreeding Pied Kingfisher males, but does not further specify which categories of nonbreeders known to exist in this species (primary helpers, secondary helpers, or nonhelpers) are involved. However, the likelihood that they are nonhelpers can be deduced from independent descriptions of the behavior of each of these categories. Homosexuality probably does not take place between breeding males and secondary helpers, since the former are antagonistic to the latter, engaging in "intense and prolonged fights" with them (Reyer 1986:288). Likewise for primary and secondary helpers: the former often attack and fight the latter (Reyer 1986:291). Thus, homosexuality probably occurs largely among nonhelping nonbreeders, or among secondary helpers—the latter is less likely, though, since their attentions are usually focused on feeding females, often as potential mates for the next season (Reyer 1984:1170; Reyer 1980:222). Patterns of helping, breeding, and homosexual participation analogous to the bird examples also occur among mammals. In Red Foxes, for example, same-sex mounting occurs both among younger females (nonbreeders and/or helpers) and between them and older breeding females, but only a subset of each; in Bush Dogs, nonbreeders of both sexes act as helpers (Macdonald 1996:535), yet only males occasionally participate in same-sex mounting.

10. In fact, the only possible cases of adoption by homosexual pairs are in Hooded Warblers (where some male pairs may take over nests abandoned by females after they have been parasitized or robbed by predators), Black-headed Gulls (in which adoption of eggs by male pairs has been suggested [van Rhijn and Groothuis 1985:165–66] but not yet documented), and Cheetahs (in which paired males have occasionally been observed temporarily looking after lost cubs [Caro 1994:45, 91]). Coparenting of adopted pups by two females also occurs in Northern Elephant Seals, Gray Seals, and Spotted Seals, although the two females do not appear to have a "pair-bonded" or sexual relationship with each other.

11. For additional examples, see Squirrel Monkey, Common Murre, and Herring Gull. Another type of "helper" arrangement involves hierarchical societies in which only a small fraction of animals breed and the remainder assist them, often in a complex "caste" system in which each class of nonbreeders has its own specialized duties. This is typical of many social insects such as ants or honeybees, but is also found in some mammals such as naked mole-rats. Again, there is no particular association of homosexuality with these systems: homosexual behavior has been reported for perhaps only a handful of insect species with this type of social organization and is not specifically associated with helpers in these species. In fact, in most social insects helpers are asexual (and genetically sterile), and homosexual behavior is actually found among breeders, for example among fertile males participating in mating swarms (cf. O'Neill 1994 on Red Ants).

12. Hanuman Langur (Srivastava et al. 1991:506). For specific evidence or argumentation against the idea that homosexual relations are a form of "kin selection" (i.e., an association between indivduals who interact with or help one another primarily because they are related and will therefore potentially be "benefiting" their own genes, albeit indirectly), see Fernández and Reboreda 1995:323 (Greater Rhea), Afton 1993:232 (Lesser Scaup Duck), Rose 1992:104, 112 (Killer Whale), Hashimoto et al. 1996:316 (Bonobo), Ens 1998:635h (Oystercatcher), as well as the numerous species with nonincestuous homosexual relations and/or incest taboos.

13. The general concept of a "population control" mechanism in animals would also be rejected by most biologists on theoretical grounds because it relies on the generally discredited notion of "group selection," which maintains that an animal's behavior sometimes benefits the population as a whole rather than the individual. This contradicts one of the most fundamental principles of evolutionary biology, that organisms act only in their self-interest. Some scientists, however, have strongly advocated the concept of group selection, and it remains an intriguing and controversial proposal. See, for example, Wynne-Edwards, V. C. (1986) *Evolution through Group Selection* (Oxford: Blackwell Scientific). For an overall critique of the notion of population regulation in humans, see Bates, D. G., and S. H. Lees (1979) "The Myth of Population Regulation," in N. A. Chagnon and W. Irons, eds., *Evolutionary Biology and Human Social Behavior: An Anthropological Perspective,* pp. 273–89 (North Scituate, Mass.: Duxbury Press).

14. Damaraland mole-rat (Bennett, N.C. [1994] "Reproductive Suppression in Social *Cryptomys damarensis* Colonies—a Lifetime of Socially-Induced Sterility in Males and Females," *Journal of Zoology, London* 234:25–39); Killer Whale (Olesiuk et al. 1990:209). Long-term study of a stable Silver Gull population revealed that 93 percent of all eggs fail to produce birds that survive to breed, only 3 percent of the birds produce half of all surviving offspring, and 84–86 percent of the birds never produce any offspring who go on to breed themselves. In a number of other bird species, the proportion of "noncontributing" individuals is similarly high, ranging from 62–87 percent (Mills 1991:1525–26). Species with more than 50 percent nonbreeders in at least one sex, at any given time, include Bison (54 percent; based on figures in Lott 1981:98), Regent Bowerbirds (67 percent; based on figures in Lenz 1994:264, 267), Pronghorns (75 percent; based on figures in Kitchen 1974:11, 48, 50), and Grant's Gazelles (92 percent; based on figures in table 2, Walther 1972:358). See pp. 196–99 for further examples.

15. Mammals (Macdonald, D. W., ed. [1993] *The Encyclopedia of Mammals,* pp. 633, 646, 654, 656–57, 722–23 [New York: Facts on File]); Birds (Piersma, T. [1996] "Scolopacidae [Snipes, Sandpipers, and Phalaropes]," p. 476, in J. del Hoyo, A. Elliott, and J. Sargatal, eds., *Handbook of the Birds of the World,* vol. 3: Hoatzin to Auks, pp. 444–533 [Barcelona: Lynx Edicións]); Grouse (Bergerud, A. T. [1988] "Population Ecology of North American Grouse," in A. T. Bergerud and M. W. Gratson, eds., *Adaptive Strategies and Population Ecology of Northern Grouse,* pp. 578–685. [Minneapolis: University of Minnesota Press]).

16. Nor does the occurrence of homosexual bonding in Oystercatchers fluctuate along with the environmentally induced population fluctuations that occur in this species (Heg and van Treuren 1998:689–90). On the other hand, the incidence of velvet-horn (transgendered) White-tailed Deer might be associated with overpopulation or drought cycles. Anecdotal reports from ranchers and longtime residents of some regions suggest that the occurrence of such Deer (who are infertile) is cyclic and related to the ending of drought periods (Thomas et al. 1970:3). Scientists studying one population found that, overall, the reproductive rate was not reduced by the presence of so many nonbreeding bucks (ibid., p. 19)—in fact, their data show that such populations actually had *elevated* reproductive rates. However, this skew might accord with a population regulation/fluctuation hypothesis. In populations with significant numbers of velvet-horns, there were *higher* ovulation rates, pregnancy rates, and numbers of does with fawns among both adult and yearling females (at least one of which—the ovulation rate for adult females in 1960—was statistically significant). Scientists were in fact puzzled over this apparently "opposite" finding: "the results are contrary to what is expected if reproduction . . . was adversely affected" by the presence of velvet-horns in the herd (ibid., p. 17). In fact, we might expect a slightly delayed, rather than immediate, effect on the number of velvet-horns if their prevalence is a response to population pressure. In 1959–61 the population in this region was significantly elevated, and velvet-horn numbers actually peaked several years later in 1962 (at 9.4 percent). Taylor et al. do report periods of drought and overpopulation in the Deer herds of this region during this time (ibid., p. 25). In addition, if the *overall* reproductive rate is the same between populations with and without velvet-horns, the effect of the velvet-horns could still be to reduce population growth during times when the population is in fact increasing at a faster rate. Of course, much more systematic long-term investigation is required before any conclusions can be drawn about these possible connections.

17. Tallies and designations of threatened species are based on the official roster of the World Conservation Union. The three categories (critically endangered, endangered, and vulnerable) represent points along a

continuum, based on a set of five quantitative criteria that encompass the species' rate of population decline, restricted geographic distribution, extent of population fluctuation, age distribution, effects of human disturbances (pollutants, introduced species, exploitation), and so on. See Baillie, J., and B. Groombridge, eds. (1996) *1996 IUCN Red List of Threatened Animals* (Gland, Switzerland, and Cambridge, UK: IUCN–World Conservation Union).

18. Needless to say, the near extinction of this New Zealand bird is not a result of homosexuality in this species, but rather is due to the destructive effects of human activities—habitat loss because of drainage and hydroelectric development, as well as severe depletion by nonnative species introduced to the islands (Reed 1993:771).

19. For a review of some of these strategies, and information on other possible mechanisms, see the discussion in the following section "Nonreproductive and Alternative Heterosexualities in Animals," as well as the following references: Cohen, M. N., R. S. Malpass, and H. G. Klein, eds. (1980) *Biosocial Mechanisms of Population Regulation* (New Haven: Yale University Press); Wilson, E. O. (1975) *Sociobiology: The New Synthesis,* pp. 82–90 (Cambridge, Mass.: Belknap Press); Wynne-Edwards, V. C. (1965) "Social Organization as a Population Regulator," in P. Ellis, ed., *Social Organization of Animal Communities,* pp. 173–80, Symposia of the Zoological Society of London no. 14 (London: Academic Press); Wynne-Edwards, V. C. (1959) "The Control of Population-Density Through Social Behavior: A Hypothesis," *Ibis* 101:436–41.

20. For various statements of this hypothesis, see Hutchinson, "A Speculative Consideration of Certain Possible Forms of Sexual Selection in Man"; Kirsch and Rodman, "Selection and Sexuality: The Darwinian View of Homosexuality"; for a refutation, see Futuyama and Risch, "Sexual Orientation, Sociobiology, and Evolution." It is also possible that the homosexual gene would be recessive, i.e., not expressed when combined with the heterosexual gene—such individuals would therefore not be bisexual, but could still have a reproductive advantage. However, in the absence of any actual genetic information, there is no way to evaluate this version of the hypothesis, since individuals with a recessive homosexual gene would presumably be (superficially) indistinguishable from those with two heterosexual genes (for an alternate view and several other versions of this hypothesis, see McKnight, J. [1997] *Straight Science? Homosexuality, Evolution, and Adaptation* [London: Routledge]). Therefore, the following discussion is confined to assessing the version in which such individuals are actually behaviorally bisexual (e.g., Weinrich's 1987 version). In the spirit of the "bisexual superiority" hypothesis, see also Caldwell and Caldwell's (1967:15) suggestion that bisexuality in Bottlenose Dolphins represents a "more evolved" state because their sexuality is neither limited to reproductive activity nor confined to partners of only one sex. These scientists suggest that Dolphins may be more advanced than humans in this regard, based on the (erroneous) belief that Dolphins do not exhibit exclusive homosexuality, or (in their words) are not "fixated on a biologically inappropriate stimulus to the exclusion of the biologically appropriate one." For more on the myth of human uniqueness with regard to exclusive homosexuality, see chapter 2.

21. Based on data in fig. 2, Braithwaite 1981:140; on heterosexual partitioning of incubation duties, and the possible advantages of greater male participation, see O'Brien 1990:1186 and Brugger and Taborsky 1994. Another possible case of bisexual pairs being more successful at reproduction concerns the Snow Goose. Diamond (1989:101) had speculated that female pairs (in this and other species) that fertilize their eggs by mating with males might be able to produce more offspring than heterosexual pairs. However, this does not appear to be a genuine case: the initial suggestion was entirely conjectural and not based on actual long-term studies of the reproductive output of same-sex versus opposite-sex pairs. Furthermore, this idea was later shown to be based on faulty reasoning, since the critical factor for comparing reproductive advantage is the number of goslings produced by each female in the pair, not by the pair as a whole (as the females are usually not related to one another). See Conover (1989) and Grether and Weaver (1990) for further discussion.

22. Ruff (Hogan-Warburg 1966:179; van Rhijn 1973:197, 1991:76; Hugie and Lank 1997:220); Greylag Goose (Lorenz 1979:59–60); Pukeko (Jamieson and Craig 1987a:1251); Guianan Cock-of-the-Rock (Trail and Koutnik 1986:211, 215); Oystercatcher (Heg and van Treuren 1998:690; Ens 1998:635).

23. Sociable Weaver (based on data in Collias and Collias 1980:248[table 5]); Bonnet Macaque (based on data in Sugiyama 1971:252, 259–60 [tables 2, 8, 9]); Asiatic Elephant (based on data in Poole et al. 1997:306–7 [fig.5]); Japanese Macaque (Hanby 1974:838; Vasey 1996 and personal communication).

24. Some "exclusively lesbian" females copulate with males to fertilize their eggs and thus are technically bisexual in their sexual behavior. However, in terms of pair-bonding these females only choose other females as partners, and therefore I follow Mills in not classifying these individuals as bisexual for the purpose of assessing their reproductive output. However, since exclusively homosexual females produce even fewer offspring than bisexual ones, including them in the bisexual category would not alter the overall conclusion that bisexual females are less prolific breeders.

25. Silver Gull (Mills 1991:1525).

26. Silver Gull (Mills 1989:397–98 [table 23.5]).

27. Kirsch and Rodman (1982:189) state that "it would be difficult to construct a crucial experiment" to test this hypothesis, while Futuyama and Risch (1984:158) note that "it is hard to see how some of these theories could ever be subjected to proper scientific testing." They are primarily considering investigations on human homosexuality and bisexuality, yet (as we have seen) studies of homosexuality in wild animals can often provide exactly the type of information needed to evaluate these ideas.

28. Kirsch and Rodman, "Selection and Sexuality," p. 189.

29. The few studies that have been conducted on bisexuality and reproductive output in humans also tend to agree with the Silver Gull (and other animal) findings. Two surveys of bisexual women (in Los Angeles and the UK) found that they had either less or statistically equivalent numbers of children over their lifetime than did exclusively heterosexual women (one study did find that before the age of 25, bisexual women generally have more children than heterosexual women, but this difference evens out once lifetime reproductive rates are considered). (Baker, R. R., and M. A. Bellis [1995] *Human Sperm Competition: Copulation, Masturbation, and Infidelity,* pp. 117–18 [London: Chapman and Hall]; Essock-Vitale, S. M., and M. T. McGuire [1985] "Women's Lives Viewed from an Evolutionary Perspective: I. Sexual Histories, Reproductive Success, and Demographic Characteristics of a Random Subsample of American Women," *Ethology and Sociobiology* 6:137–54.) This is one of the few examples of the relevant quantitative data in humans being available for testing the "bisexual superiority" hypothesis. Although Baker and Bellis (1995) address the question of how homosexuality affects reproductive output, their primary concern is in evaluating the hypothesis that bisexuality *reduces* rather than improves reproductive output, i.e., they are not specifically addressing the "bisexual superiority" hypothesis.

30. Jackdaw (Lorenz 1970:202–3); Canada Goose (Allen 1934:187–88); Oystercatcher (Heg and van Treuren 1998: 688–89; Ens 1998:635); Calfbird (Snow 1972:156; Snow 1976:108); Buff-breasted Sandpiper (Myers 1989:44–45); Cheetah (Caro and Collins 1987:59, 62; Caro 1993:25, 1994:252, 304).

31. Silver Gull (Mills 1991:1525 [table 1]); Black-headed Gull (based on table 3, van Rhijn and Groothuis 1985:161); Galah (based on figures in Rogers and McCulloch 1981:83–85). See also the discussion of sexual orientation profiles in chapter 2.

32. Kob (Buechner and Schloeth 1965:219 [based on table 2]); Bonobo (Idani 1991:90–91 [based on tables 5–6]); Japanese Macaque (Chapais and Mignault 1991:175 [based on table II]); Pig-tailed Macaque (Tokuda et al. 1968:288, 290 [based on tables 3 and 5]).

33. For example, an animal could participate in a large number of heterosexual copulations, only a few of which would actually lead to fertilization (not to mention successful birth or rearing of offspring), while an animal with fewer heterosexual encounters could have a higher proportion of fertilizations or successful pregnancies or could be a better parent. Moreover, females who mate repeatedly during one breeding season can only get pregnant or be fertilized once, effectively equalizing the difference between greater and lesser participation in heterosexual mating (unless promiscuity is positively correlated with parenting success). For further discussion of how copulation frequency does not necessarily reflect reproductive output, see Eberhard, W. G. (1996) *Female Control: Sexual Selection by Cryptic Female Choice,* especially pp. 418ff (Princeton: Princeton University Press).

34. It should also be reiterated that detailed longitudinal studies of breeding success and sexual orientation (comparble to that done on Silver Gulls) have not been conducted on any of these species to verify possible connections between bisexuality and reproduction. Moreover, all of these cases involve homosexuality among members of only one gender, which again is inconsistent with a "bisexual superiority" hypothesis.

35. Bonobo (Hashimoto 1997:12–13); Gorilla (Fossey 1990:460, 1983:74, 188–89); Squirrel Monkey (Mendoza and Mason 1991:476–77); Wolf (Zimen 1976:311, 1981:140); Common Tree Shrew (Kaufmann 1965:72); Bottlenose Dolphin (Östman 1991:310). For arguments that this is not merely "displaced," "redirected," or "vicarious" heterosexuality (as Fossey [1990:460] and others have labeled it), see the discussion of the "shortage" hypothesis in chapter 4.

36. Guianan Cock-of-the-Rock (Trail and Koutnik 1986:215); Oystercatcher (Heg and van Treuren 1998:690; D. Heg, personal communication).

37. Kob (Buechner and Schloeth 1965:219 [table 2]).

38. Cattle Egret (Fujioka and Yamagishi 1981:136, 139 [including tables 1, 3, 4]).

39. See the profile of the Ruff for further details and illustrations.

40. Ruff (Van Rhijn 1991:87; Hogan-Warburg 1966:176).

41. Ruff (Lank et al. 1995). Nearly 30 years previously, Hogan-Warburg (1966) and van Rhijn (1973) had suggested that there might be genetic differences between various categories of males, based on the indirect evidence of their plumage and behavioral distinctions as well as the constancy of their category status. This hypothesis was subsequently confirmed by DNA and heredity studies.

42. Red Flour Beetle (Castro et al. 1994; Serrano et al. 1991); Fruit Flies (numerous references, summarized in Finley et al. 1997). See also Hamer and Copeland (1994) on the role of genetics in human homosexuality.

43. This has been suggested for species such as Stumptail Macaques, White-faced Capuchins, Killer Whales, Northern Elephant Seals, West Indian Manatees, Giraffes, Gray-headed Flying Foxes, Ring-billed Gulls, Black-headed Gulls, Ocher-bellied Flycatchers, Guianan Cock-of-the-Rock, Calfbirds, Superb Lyrebirds, and Adélie Penguins. In addition, in some species where homosexual behavior is classified as "play," the implication is also that it functions as practice for "real" (i.e., heterosexual) activity.

44. Rhesus Macaque (Akers and Conaway 1979:76–77); Tree Swallow (Lombardo et al. 1994:556).

45. In a number of species, though, homosexual activity is restricted to juvenile or younger animals (see Dagg, "Homosexual Behavior and Female-Male Mounting in Mammals," for a survey of such cases).

46. See Baker and Bellis, *Human Sperm Competition,* pp. 118–19, where this "explanation" is proposed for both humans and nonhumans.

47. Guianan Cock-of-the-Rock (Trail and Koutnik 1986:209, 215). These scientists admit that they have no

specific data to support the conjecture that courtship interactions between males actually improve the subsequent heterosexual performance of younger males.

48. See below for discussion and refutation of the (related) idea that homosexuality is a form of courtship "disruption" in this species.

49. See Rose et al. (1991:188) for a statement to this effect concerning Northern Elephant Seals. The Ocherbellied Flycatcher is another species where the relative infrequency of this behavior is curiously at odds with its putative "practice" function (see Westcott and Smith [1994:681] for the suggestion that courtship interactions between males in this species may allow younger birds to gain courtship experience).

50. Sage Grouse (Patterson 1952:153–54); Pandolfi, M. (1996) "Play Activity in Young Montagu's Harriers (*Circus pygargus*)," *Auk* 113:935–38.

51. For a similar conclusion regarding this "explanation" for primates, see Vasey, "Homosexual Behavior in Primates," p. 192. See also Wagner (1996:212) on Razorbills.

52. The only example specifically involving females is the Ring-billed Gull, and notably in this case it is experience with parenting or pair-bonding, not sexual behavior, that females are claimed to acquire through homosexual partnerships (Fox and Boersma 1983:555).

53. On the myth of female passivity in sexual interactions, as well as the generally sexist interpretations of female behavior and physiology in this regard, see Eberhard, *Female Control*, pp. 34–41, 238, 420–21; Batten, M. (1992) *Sexual Strategies* (New York: Putnam's); Gowaty, P. A. (1997) "Principles of Females' Perspectives in Avian Behavioral Ecology," *Journal of Avian Biology* 28:95–102; and the numerous other references in note 10, chapter 3.

54. For examples of scientists who have argued (or suggested) that homosexuality promotes or strengthens social bonds or general social cohesion and stability, see Kano 1992:192 (Bonobo); Yamagiwa 1987a:1, 23, 1987b:37, Robbins 1996:944 (Gorilla); Weber and Vogel 1970:79 (Hanuman Langur); Reinhardt et al. 1986:55 (Rhesus Macaque); Rose 1992:97–98, 116–17 (Killer Whale); Coe 1967:319 (Giraffe); Nelson 1965:552 (Gray-headed Flying Fox); Heg and van Treuren 1998:688, Ens 1998:635 (Oystercatcher); Sauer 1972:735 (Ostrich); Rogers and McCulloch 1981:90 (Galah). One scientist suggests that homosexual activity in Bonobos, although promoting bonding between same-sex individuals, actually serves a more important role in *heterosexual* relations: "Homosexual activity became a way of tying males and females together in larger aggregations" (de Waal 1997:138); for a refutation of this type of heterocentric interpretation, see Parish (1996:65). For examples of homosexuality claimed to be a strategy of alliance or coalition building, including for the purpose of acquiring heterosexual mates, see Kano 1992, Idani 1991 (Bonobo); Vasey 1996 (Japanese Macaque, other species); Bernstein 1980:40 (Stumptail Macaque); Smuts and Watanabe 1990 (Savanna Baboon); Colmenares 1991 (Hamadryas Baboon); R. Wrangham and S. B. Hrdy, in Weinrich 1980:291 (Gelada Baboon, Hanuman Langur); Wells 1991:218–20 (Bottlenose Dolphin).

55. See chapter 4 for further discussion and refutation of the idea that this is the motivation (or adaptive function) for homosexual associations.

56. Rhesus Macaque (Carpenter 1942:149); Bottlenose Dolphin (Wells 1991:220); both of these cases are highly speculative. The Rhesus example is based on an isolated observation of a single consortship and is a questionable interpretation, while the Bottlenose case is considerably more complex than it initially appears (see discussion below).

57. Parker, G. A., and R. G. Pearson (1976) "A Possible Origin and Adaptive Significance of the Mounting Behavior Shown by Some Female Mammals in Oestrus," *Journal of Natural History* 10:241–45; Thompson-Handler et al. 1984:355–57 (Bonobo); African Elephant (Buss 1990); Greenshank, Golden Plover (Nethersole-Thompson 1975:55).

58. Species in which homosexual activity among females has *only* been reported outside the breeding season (or when females are not in heat) include Wapiti, Barasingha, Waterbuck, and Gray-headed Flying Foxes.

59. Japanese Macaque (Gouzoules and Goy 1983:47); Hanuman Langur (Srivastava et al. 1991:508).

60. Hanuman Langur (Srivastava et al. 1991:508); Calfbird (Trail 1990:1849–50); Cheetah (Caro and Collins 1987:59, 62; Caro 1993:25, 1994:252, 304); Savanna Baboon (Noë 1992:295). For specific descriptions of animals of the opposite sex being "disinterested" or not attracted by homosexual activity, see Gorilla (Harcourt et al. 1981:276); White-handed Gibbon (Edwards and Todd 1991:232–33); Japanese Macaque (Wolfe 1984, Vasey 1995:190; Corradino 1990:360); Killer Whale (Jacobsen 1986:152); Gray Whale (Darling 1978:51–52); Northern Fur Seal (Bartholomew 1959:168); African Buffalo (Mloszewski 1983:186); Rufous Rat Kangaroo (Johnson 1980:356); Dwarf Cavy (Rood 1970:442); Laughing Gull (Noble and Wurm 1943:205); Sage Grouse (Scott 1942:495).

61. Bottlenose Dolphin (Florida—Wells 1991:219–20, Wells 1995; Ecuador—Félix 1997:14; Australia—Connor et al. 1992:419, 426; Bahamas—Herzing and Johnson 1997).

62. Squirrel Monkey (Travis and Holmes 1974:55); Stumptail Macaque (Chevalier-Skolnikoff 1976:524); Wolf (Zimen 1981:140); Savanna (Yellow) Baboon (Maxim and Buettner-Janusch 1963:176); Mountain Sheep (Geist 1971:162). For arguments against this being simple "displacement" or "redirected" (hetero)sexual activity, see chapter 4. On a related point, scientists have observed that male Oystercatchers in trios are unable to influence or "promote" homosexual activity among their female partners, yet males may suffer reproductive losses without the cooperation between females entailed by such same-sex activity (Heg and van Treuren 1998:690). Thus, males are essentially powerless to cultivate homosexual activity in females even when this activity may benefit them.

63. Dagg, "Homosexual Behavior and Female-Male Mounting in Mammals," p. 179.
64. The one exception is R. Wrangham (quoted in Weinrich 1980:291), who suggests that male Gelada Baboons may essentially "perform" homosexual mounts in front of, and for the benefit of, females to demonstrate their mating "prowess." This activity is not, however, claimed to be sexually stimulating for females in the same way that female homosexuality is claimed to be for males.
65. See discussion in chapter 4.
66. Kittiwake (Coulson and Thomas 1985); Western Gull (Hunt and Hunt 1977); Herring Gull (Shugart et al. 1988); Silver Gull (Mills 1991); Ring-billed Gull (Ryder and Somppi 1979; Kovacs and Ryder 1983). See also chapter 4 for further evidence against the claim that female pairs in Gulls form primarily as a breeding strategy.
67. Western Gull (Hunt et al. 1984); Black-winged Stilt (Kitagawa 1988); Lesser Scaup Duck (Afton 1993; Munro 1941); Acorn Woodpecker (W. D. Koenig, personal communication); Squirrel Monkey (Ploog 1967:159–60); Greylag Goose (Lorenz 1979, 1991); Oystercatcher (Heg and van Treuren 1998). Although female coparents in Acorn Woodpeckers are "platonic" in that they do not specifically engage in courtship or sexual behavior with one another, they still participate in the group mounting displays characteristic of this species (which usually include homosexual mounting and may actually involve mounting of their coparent).
68. Greater Rhea (Fernández and Reboreda 1995:323, 1998:340–46); Lesser Scaup Duck (Afton 1993). The supposed "benefits" of having male nest helpers in Greater Rheas are also not readily apparent: researchers were able to demonstrate few, if any, statistically significant differences in breeding success between solitary males and those with helpers (these results are still preliminary, though, as the phenomenon has only recently been discovered; cf. Codenotti and Alvarez 1997).
69. Snow Goose (Martin et al. 1985:262–63); Black-billed Magpie (Dunn and Hannon 1989; Buitron 1988).
70. Superb Lyrebird (Lill 1979b, 1986). For a survey of mate and offspring desertion by one parent in a wide variety of bird species, see Székely, T., J. N. Webb, A. I. Houston, and J. M. McNamara (1996) "An Evolutionary Approach to Offspring Desertion in Birds," especially pp. 275–76, 310, in V. Nolan Jr., and E.D. Ketterson, eds., *Current Ornithology*, vol. 13, pp. 271–330 (New York: Plenum Press). For a summary of the effects of mate removal in more than 15 bird species, see Bart, J., and A. Tornes (1989) "Importance of Monogamous Male Birds in Determining Reproductive Success: Evidence for House Wrens and a Review of Male-Removal Studies," *Behavioral Ecology and Sociobiology* 24:109–16.
71. Calfbird (Snow 1972:156, 1976:108); Japanese Macaque (Vasey 1998:13–14, 16); Oystercatcher (Heg and van Treuren 1998:688–89; Ens 1998:635); Jackdaw (Lorenz 1970:202–3); Lesser Scaup Duck (Munro 1941:130–31); Canada Goose (Allen 1934:187–88).
72. See, for example, Srivastava et al. (1991:508–9) on female Hanuman Langurs, Huynen (1997:211) on female Rhesus Macaques, Gibson and Bradbury (1986:396) on female Sage Grouse, Jamieson and Craig (1987a:1252) on male Pukeko, and Wagner (1996:213) on male Razorbills.
73. See Gouzoules and Goy (1983:47) for an explicit refutation of this hypothesis in Japanese Macaques, and Vasey, "Homosexual Behavior in Primates," for a more general refutation for primates. See also the discussion of mountee facilitation or initiation of homosexual interactions in chapters 3 and 4.
74. Pukeko (Jamieson and Craig 1987a:1252, 1987b:321–23; Jamieson et al. 1994:275–76); Ocher-bellied Flycatcher: only 4 out of 12 courtship interactions between males occurred when a female was present (Westcott and Smith 1994:680); Guianan Cock-of-the-Rock (Trail and Koutnik 1986).
75. Buff-breasted Sandpiper (Myers 1989:44–45; Pruett-Jones 1988:1745–47; Lanctot and Laredo 1994:9).
76. A. P. Møller, in Lombardo et al. 1994:556 (Tree Swallow).
77. As pointed out by Lombardo et al. (1994:556). In Tree Swallows, there are further arguments that reproductively oriented "sperm-swapping" is probably not involved. In one observation of homosexual mating in this species, the bird that other males were copulating with was already tending chicks, i.e., his mate could no longer be fertilized (the homosexual copulations occurred fairly late in the breeding season). Although some females may still have been fertile at that point because they had not yet laid eggs (M. P. Lombardo, personal communication), and reproductive copulations can occur fairly late in the season in this species (Robertson et al. 1992:11), it seems unlikely that homosexual matings are generally timed to take advantage of reproductive opportunities. In particular, they do not appear to be more prevalent earlier in the breeding season when putative "sperm-swapping" would be more likely to result in fertilizations (Lombardo, personal communication).
78. Pukeko: See Craig (1980:593, 601–2) for speculation on the "possible swapping of sperm during female homosexual cloacal contacts" as well as synchronization of egg laying. On the mechanisms that independently insure obscured paternity and shared parenting, see Jamieson et al. 1994:274–76; Jamieson and Craig 1987b:323–25.
79. Best, R. I., and M. A. O'Brien (1967) *The Book of Leinster*, vol. 5, lines 35670–35710. (Dublin: Dublin Institute for Advanced Studies); Greene, D. (1976) "The 'Act of Truth' in a Middle-Irish Story," *Saga och Sed* (Kungliga Gustav Adolfs Akademiens Årsbok) 1976:30–37.
80. Boswell, J. (1994) *Same-Sex Unions in Premodern Europe*, pp. xxviii–xxix (New York: Villard Books). In discussing this story, Greene (1976:33–34) cites several "extremely rare" examples from the late 1800s, of questionable validity, in which women supposedly became pregnant from homosexual activity in this way. Regardless of whether conception by this means has been "documented" or is even biologically possible, what stands out in these descriptions of both human and animal homosexuality is their concern with *heterosexuality*. Namely, the putative role of same-sex activity in facilitating insemination is emphasized, and there is an insistence on ascribing a reproductive function to homosexual activity.

81. For a good summary and survey of some current strains of thought in this area, see Abramson, P. A., and S. D. Pinkerton, eds. (1995) *Sexual Nature, Sexual Culture* (Chicago: University of Chicago Press).

82. African Elephant (Sikes 1971:266).

83. Cordero, A. (1995) "Correlates of Male Mating Success in Two Natural Populations of the Damselfly *Ischnura graellsii* (Odonata: Coenagrionidae)," *Ecological Entomology* 20:213–22.

84. See the profiles for more information and references. Among those 48 species in which homosexuality has been documented *and* quantitative information on nonbreeders is available, an average of half of the population (or of one sex) does not participate in reproduction (independent of homosexuality).

85. Squirrel Monkey (Baldwin and Baldwin 1981:295; Baldwin 1968:296, 311); Grizzly Bear (Craighead et al. 1995:139).

86. Chalmers, N. R. (1968) "Group Composition, Ecology, and Daily Activities of Free-Living Mangabeys in Uganda," *Folia Primatologica* 8:247–62; Musk-ox (Gray 1973:170–71).

87. Searcy, W. A., and K. Yasukawa (1995) *Polygyny and Sexual Selection in Red-winged Blackbirds*, pp. 6, 169 (Princeton: Princeton University Press). For the other species, see the profiles and the discussion of sexual orientation in chapter 2.

88. Bennett, N. C. (1994) "Reproductive Suppression in Social *Cryptomys damarensis* Colonies—a Lifetime of Socially-Induced Sterility in Males and Females," *Journal of Zoology, London* 234:25–39; Northern Elephant Seal (Le Boeuf and Reiter 1988:351). In mole-rats, large numbers of adults are "permanently" nonbreeding, while in Northern Elephant Seals, many males simply do not survive to the relatively advanced age when breeding typically begins, and of those that do, less than half actually breed.

89. Waser, P. M. (1978) "Postreproductive Survival and Behavior in a Free-Ranging Female Mangabey," *Folia Primatologica* 29:142–60; Ratnayeke, S. (1994) "The Behavior of Postreproductive Females in a Wild Population of Toque Macaques (*Macaca sinica*) in Sri Lanka," *International Journal of Primatology* 15:445–69; Bester, M. N. (1995) "Reproduction in the Female Subantarctic Fur Seal, *Arctocephalus tropicalis*," *Marine Mammal Science* 11:362–75. For further examples, see profiles of species indexed under "postreproductive individuals."

90. Marsh, H., and T. Kasuya (1991) "An Overview of the Changes in the Role of a Female Pilot Whale With Age," in K. Pryor and K. S. Norris, eds., *Dolphin Societies: Discoveries and Puzzles*, pp. 281–85 (Berkeley: University of California Press).

91. Canada Goose (Collias and Jahn 1959:505). It is not the case that these birds were simply "trying harder" to reproduce, since some of these pairs produced eggs but failed to incubate them. Rather, it appears that as nonparents, they were able to "indulge" in more sexual behavior.

92. Birkhead, T. R., and A. P. Møller (1993) "Why Do Male Birds Stop Copulating While Their Partners Are Still Fertile?" *Animal Behavior* 45:105–18; Eberhard, *Female Control*, p. 395.

93. Wasser, S. K., and D. P. Barash (1983) "Reproductive Suppression Among Female Mammals: Implications for Biomedecine and Sexual Selection Theory," *Quarterly Review of Biology* 58:513–38; Abbott, D. H. (1987) "Behaviorally Mediated Suppression of Reproduction in Female Primates," *Journal of Zoology, London* 213:455–70; Reyer et al. 1986 (Pied Kingfisher); Macdonald and Moehlman 1982 (Wild Dogs); Jennions, M. D., and D. W. Macdonald (1994) "Cooperative Breeding in Mammals," *Trends in Ecology and Evolution* 9:89–93; Creel and Macdonald 1995 (Wild Dogs); Solomon, N. G., and J. A. French, eds. (1997) *Cooperative Breeding in Mammals*, pp. 304–5 (Cambridge: Cambridge University Press).

94. American Bison (Komers et al. 1994:324 [see also discussion in chapter 4]); Pied Kingfisher (Reyer et al. 1986:216); tamarins and marmosets (Snowdon, C. T. [1996] "Infant Care in Cooperatively Breeding Species," *Advances in the Study of Behavior* 25:643–89, especially pp. 677–80); other species (Solomon and French, *Cooperative Breeding in Mammals*, p. 5).

95. Rohrbach, C. (1982) "Investigation of the Bruce Effect in the Mongolian Gerbil (*Meriones unguiculatus*)," *Journal of Reproduction and Fertility* 65:411–17.

96. Bighorn Sheep (Geist 1971:181, 295); Red Deer (Clutton-Brock et al. 1983:371–72); Northern Quoll and other carnivorous marsupials (Dickman and Braithwaite 1992); Ruffed Grouse (Gullion 1981:379–80); Western Gull (Pyle et al. 1997:140, 145); Spotted Hyena (Frank and Glickman 1994). For further discussion of the avoidance of reproduction because of its stressful and potentially injurious effects on the individual, see Hand 1981:140–42 (Laughing Gull).

97. Wagner, R. H. (1991) "The Use of Extrapair Copulations for Mate Appraisal by Razorbills, *Alca torda*," *Behavioral Ecology* 2:198–203; Sheldon, B. C. (1993) "Sexually Transmitted Disease in Birds: Occurrence and Evolutionary Significance," *Philosophical Transactions of the Royal Society of London*, Series B 339:491–97; Hamilton, W. D. (1990) "Mate Choice Near or Far," *American Zoologist* 30:341–52; Freeland, W. J. (1976) "Pathogens and the Evolution of Primate Sociality," *Biotropica* 8:12–24. See also Birkhead, T. R., and A. P. Møller (1992) *Sperm Competition in Birds: Evolutionary Causes and Consequences*, p. 194 (London: Academic Press); Eberhard, *Female Control*, p. 111.

98. Watson, L. (1981) *Sea Guide to Whales of the World*, p. 174 (New York: E.P. Dutton).

99. For further discussion, see Peterson 1968, Gentry 1981 (Northern Fur Seal); Smith 1976:71 (Musk-ox).

100. Lee and Cockburn 1985:87–90, 163–70 (Carnivorous Marsupials).

101. Birkhead, T. R., and A. P. Møller (1993) "Sexual Selection and the Temporal Separation of Reproductive Events: Sperm Storage Data from Reptiles, Birds, and Mammals," *Biological Journal of the Linnaean Society* 50:295–311; Birkhead and Møller, *Sperm Competition in Birds*; Shugart, G. W. (1988) "Uterovaginal Sperm-storage Glands in Sixteen Species with Comments on Morphological Differences," *Auk* 105:379–84; Stewart,

G. R. (1972) "An Unusual Record of Sperm Storage in a Female Garter Snake (Genus *Thamnophis*)," *Herpetologica* 28:346–47; Racey, P. A. (1979) "The Prolonged Storage and Survival of Spermatozoa in Chiroptera," *Journal of Reproduction and Fertility* 56:391–96; Baker and Bellis, *Human Sperm Competition*, pp. 42–43; Eberhard, *Female Control*, pp. 50–61, 167–69.

102. Sandell, M. (1990) "The Evolution of Seasonal Delayed Implantation," *Quarterly Review of Biology* 65:23–42; York and Scheffer 1997:680 (Northern Fur Seal); Renfree, M. B., and J. H. Calaby (1981) "Background to Delayed Implantation and Embryonic Diapause," in A. P. F. Flint, M. B. Renfree, and B. J. Weir, eds., *Embryonic Diapause in Mammals, Journal of Reproduction and Fertility,* supplement no. 29:1–9; Riedman, M. (1990) *The Pinnipeds: Seals, Sea Lions, and Walruses*, pp. 224–25 (Berkeley: University of California Press).

103. Greylag Goose (Lorenz 1979:74).

104. Francis, C. M., E. L. P. Anthony, J. A. Brunton, and T. H. Kunz (1994) "Lactation in Male Fruit Bats," *Nature* 367:691–92.

105. McVean, G., and L. D. Hurst (1996) "Genetic Conflicts and the Paradox of Sex Determination: Three Paths to the Evolution of Female Intersexuality in a Mammal," *Journal of Theoretical Biology* 179:199–211; King, A. S. (1981) "Phallus," in A. S. King and J. McLelland, eds., *Form and Function in Birds*, vol. 2, pp. 107–47 (London: Academic Press).

106. Walrus (Fay 1982:39–40); Layne, J.N. (1954) "The Os Clitoridis of Some North American Sciuridae," *Journal of Mammalogy* 35:357–66; Bray, K. (1996) "Size Is Nothing at All: Female Fish Has Novel Way to Adapt to Mate's Lack of Penis," *BBC Wildlife* 14(11):15.

107. Chaffinch (Marler 1956:113–14, 163 [table XI]); African jacana (Jenni, D. A. [1996] "Jacanidae [Jacanas]," p. 282, in J. del Hoyo, A. Elliott, and J. Sargatal, eds., *Handbook of the Birds of the World*, vol. 3: Hoatzin to Auks, pp. 276–91 [Barcelona: Lynx Edicións]). For further examples and statistics on the widespread occurrence of matings that "fail" to result in insemination, see Eberhard, *Female Control*, pp. 399–403.

108. For a general survey of mating harassment in primates, see Niemeyer, C. L., and J. R. Anderson (1983) "Primate Harassment of Matings," *Ethology and Sociobiology* 4:205–20.

109. Asiatic Elephant (Eisenberg et al. 1971:205). For specific examples of male and female genitalia that do not "fit," see Eberhard, W. G. (1985) *Sexual Selection and Animal Genitalia* (Cambridge, Mass.: Harvard University Press). On the hostility of the female's reproductive tract to sperm, see Birkhead, T. R., A. P. Møller, and W. J. Sutherland (1993) "Why Do Females Make It So Difficult for Males to Fertilize Their Eggs?" *Journal of Theoretical Biology* 161:51–60; Birkhead, T., and A. Møller (1993) "Female Control of Paternity," *Trends in Ecology and Evolution* 8:100–104; Eberhard, *Female Control*, pp. 331–49.

110. Musk-ox (Smith 1976:54–55).

111. Clutton-Brock, T. H., and G. A. Parker (1995) "Sexual Coercion in Animal Societies," *Animal Behavior* 49:1345–65; Smuts, B. B., and R. W. Smuts (1993) "Male Aggression and Sexual Coercion of Females in Nonhuman Primates and Other Mammals: Evidence and Theoretical Implications," *Advances in the Study of Behavior* 22:1–63; Palmer, C. T. (1989) "Rape in Nonhuman Animal Species: Definitions, Evidence, and Implications," *Journal of Sex Research* 26:355–74; McKinney et al. 1983 (Ducks).

112. For further examples and references, see Le Boeuf and Mesnick 1991 (Northern Elephant Seal); Miller et al. 1996 (Northern Fur Seal).

113. Pronghorn (Geist 1990:283).

114. Besides mating during the nonbreeding season or during menstruation or pregnancy, many female mammals also copulate during anovulatory cycles, that is, menstrual cycles during which ovulation has not taken place (Baker and Bellis, *Human Sperm Competition*, pp. 69–70; Eberhard, *Female Control*, pp. 133–39).

115. Eberhard, *Female Control*, pp. 3–5, 202.

116. Birkhead et al., "Why Do Females Make It So Difficult for Males to Fertilize Their Eggs?" p. 52; Birkhead and Møller, "Female Control of Paternity," p. 101; Ginsberg, J. R., and U. W. Huck (1989) "Sperm Competition in Mammals," *Trends in Ecology and Evolution* 4:74–79; Eberhard, *Female Control*, pp. 81–94.

117. Rodents (Voss, R. S. [1979] "Male Accessory Glands and the Evolution of Copulatory Plugs in Rodents," *Occasional Papers of the Museum of Zoology, University of Michigan* 689:1–27; Baumgardner, D. J., T. G. Hartung, D. K. Sawrey, D. G. Webster, and D. A. Dewsbury [1982] "Muroid Copulatory Plugs and Female Reproductive Tracts: A Comparative Investigation," *Journal of Mammalogy* 63:110–17); Squirrel Monkey (Srivastava et al. 1970:129–30); Hedgehogs (Reeve 1994:178; Deansley, R. [1934] "The Reproductive Processes of Certain Mammals. VI. The Reproductive Cycle of the Female Hedgehog," especially p. 267, *Philosophical Transactions of the Royal Society of London*, Series B 223:239–76); lemurs and other prosimians (Dixson, A. F. [1995] "Sexual Selection and the Evolution of Copulatory Behavior in Nocturnal Prosimians," in L. Alterman, G. A. Doyle, and M. K. Izard, eds., *Creatures of the Dark: The Nocturnal Prosimians*, pp. 93–118 [New York: Plenum Press]); Dolphins (Harrison, R. J. [1969] "Reproduction and Reproductive Organs," p. 272, in H. T. Andersen, ed., *The Biology of Marine Mammals*, pp. 253–348 [New York and London: Academic Press]); on "chastity plugs" in Bats, see Fenton, M. B. (1984) "Sperm Competition? The Case of Vespertilionid and Rhinolophid Bats," in Smith, R. L. (1984) *Sperm Competition and the Evolution of Animal Mating Systems*, pp. 573–87 (Orlando: Academic Press); Squirrels (Koprowski 1992). For additional species, as well as other examples of females removing plugs, see Eberhard, *Female Control*, pp. 146–55.

118. Common Chimpanzee (Dahl et al. 1996).

119. Bruce, H. M. (1960) "A Block to Pregnancy in the Mouse Caused by Proximity of Strange Males," *Journal of Reproduction and Fertility* 1:96–103; Schwagmeyer, P. L. (1979) "The Bruce Effect: An Evaluation of Male/

Female Advantages," *American Naturalist* 114:932–38; Labov, J. B. (1981) "Pregnancy Blocking in Rodents: Adaptive Advantages for Females," *American Naturalist* 118:361–71. See also Eberhard, *Female Control,* pp. 162–66.

120. Springer, S. (1948) "Oviphagous Embryos of the Sand Shark, *Carcharias taurus,*" *Copeia* 1948:153–57; Gilmore, R. G., J. W. Dodrill, and P. A. Linley (1983) "Reproduction and Embryonic Development of the Sand Tiger Shark, *Odontaspis taurus* (Rafinesque)," *Fishery Bulletin U.S.* 81:201–25; Gilmore, R. G. (1991) "The Reproductive Biology of Lamnoid Sharks," *Underwater Naturalist* 19:64–67; Kuzmin, S. L. (1994) "Feeding Ecology of *Salamandra* and *Mertensiella:* A Review of Data and Ontogenetic Evolutionary Trends," *Mertensiella* 4:271–86.

121. Geist, V. (1971) "A Behavioral Approach to the Management of Wild Ungulates," in E. Duffey and A. S. Watt, eds., *The Scientific Management of Animal and Plant Communities for Conservation,* pp. 413–24 (London: Blackwell).

122. California sea lion (Le Boeuf, B. J., R. J. Whiting, and R. F. Gantt [1972] "Perinatal Behavior of Northern Elephant Seal Females and Their Young," p. 129, *Behavior* 43:121–56; Odell, D. K. [1970] "Premature Pupping in the California Sea Lion," in *Proceedings of the Seventh Annual Conference on Biological Sonar and Diving Mammals,* pp. 185–90 [Menlo Park, Calif.: Stanford Research Institute]). On selective abortion as a mechanism females use to control paternity, see Birkhead and Møller, "Female Control of Paternity," p. 102. On possible deliberate ingestion of abortifacient plants by primates, see Bewley, D. (1997) "Healing Meals?" *BBC Wildlife* 15(9):63; Garey, J. D. (1997) "The Consumption of Human Medicinal Plants, Including Abortifacients, by Wild Primates," *American Journal of Primatology* 42:111. On abortion in other species not profiled in part 2, see Stehn, R. A., and F. J. Jannett, Jr. (1981) "Male-induced Abortion in Various Microtine Rodents," *Journal of Mammalogy* 62:369–72; Gosling, L. M. (1986) "Selective Abortion of Entire Litters in the Coypu: Adaptive Control of Offspring Production in Relation to Quality and Sex," *American Naturalist* 127:772–95; Berger, J. (1983) "Induced Abortion and Social Factors in Wild Horses," *Nature* 303:59–61; Kozlowski, J., and S. C. Stearns (1989) "Hypotheses for the Production of Excess Zygotes: Models of Bet-Hedging and Selective Abortion," *Evolution* 43:1369–77; Schadker, M. H. (1981) "Postimplantation Abortion in Pine Voles (*Microtus pinetorum*) Induced by Strange Males and Pheromones of Strange Males," *Biology of Reproduction* 25:295–97.

123. On ovicide, see Heinsohn, R. G. (1988) "Inter-group Ovicide and Nest Destruction in Cooperatively Breeding White-winged Choughs," *Animal Behavior* 36:1856–58. On egg ejection, see St. Clair, C. C., J. R. Waas, R. C. St. Clair, and P. T. Boag (1995) "Unfit Mothers? Maternal Infanticide in Royal Penguins," *Animal Behavior* 50:1177–85.

124. Hausfater, G., and S. B. Hrdy, eds. (1984) *Infanticide: Comparative and Evolutionary Perspective* (New York: Aldine Press).

125. This strategy is also sometimes employed by females: see Acorn Woodpecker, Little Egret; and Ichikawa, N. (1995) "Male Counterstrategy Against Infanticide of the Female Giant Water Bug *Lethocerus deyrollei* (Hemiptera: Belostomatidae)," *Journal of Insect Behavior* 8:181–88; Stephens, M. L. (1982) "Mate Takeover and Possible Infanticide by a Female Northern Jacana (*Jacana spinosa*)," *Animal Behavior* 30:1253–54.

126. Hoagland, J. L. (1995) *The Black-tailed Prairie Dog: Social Life of a Burrowing Mammal* (Chicago: University of Chicago Press). For further discussion of the often neglected topic of female infanticide, see Digby, L. (1995) "Infant Care, Infanticide, and Female Reproductive Strategies in Polygynous Groups of Common Marmosets (*Callithrix jacchus*)," *Behavioral Ecology and Sociobiology* 37:51–61; Digby, L., M. Y. Merrill, and E. T. Davis (1997) "Infanticide by Female Mammals. Part I: Primates," *American Journal of Primatology* 42:105.

127. For general surveys of cannibalism among animals, see Elgar, M. A., and B. J. Crespi, eds. (1992) *Cannibalism: Ecology and Evolution Among Diverse Taxa* (Oxford: Oxford University Press); Jones, J. S. (1982) "Of Cannibals and Kin," *Nature* 299:202–3; Polis, G. (1981) "The Evolution and Dynamics of Intraspecific Predation," *Annual Review of Ecology and Systematics* 12:225–51; Fox, L. R. (1975) Cannibalism in Natural Populations," *Annual Review of Ecology and Systematics* 6:87–106.

128. Daly, M., and M. I. Wilson (1981) "Abuse and Neglect of Children in Evolutionary Perspective," in R. D. Alexander and D. W. Tinkle, eds., *Natural Selection and Social Behavior: Recent Research and New Theory,* pp. 405–16 (New York: Chiron Press); Reite, M., and N.G. Caine, eds., (1983) *Child Abuse: The Nonhuman Primate Data.* Monographs in Primatology, vol.1 (New York: Alan R. Liss); Székely et al., "An Evolutionary Approach to Offspring Desertion in Birds."

129. Stoleson, S. H., and S. R. Beissinger (1995) "Hatching Asynchrony and the Onset of Incubation in Birds, Revisited: When Is the Critical Period?" in D. M. Power, ed., *Current Ornithology,* vol. 12, pp. 191–270 (New York: Plenum Press); Evans, R. M., and S. C. Lee (1991) "Terminal-Egg Neglect: Brood Reduction Strategy or Cost of Asynchronous Hatching?" *Acta XX Congressus Internationalis Ornithologici* (Proceedings of the 20th International Ornithological Congress, Christchurch, New Zealand), vol. 3, pp. 1734–40 (Wellington, NZ: New Zealand Ornithological Trust Board); Mock, D. W. (1984) "Siblicidal Aggression and Resource Monopolization in Birds," *Science* 225:731–32; O'Connor, R. J. (1978) "Brood Reduction in Birds: Selection for Fratricide, Infanticide, or Suicide?" *Animal Behavior* 26:79–96.

130. Skeel and Mallory (1996) "Whimbrel (*Numenius phaerops*)," in A. Poole and F. Gill, eds., *The Birds of North America: Life Histories for the 21st Century,* no. 219, p. 17 (Philadelphia: Academy of Natural Sciences; Washington, D.C.: American Ornithologists' Union); Skutch, A. F. (1976) *Parent Birds and Their Young,* pp. 349–50 (Austin: University of Texas Press); Anthonisen, K., C. Krokene, and J. T. Lifjeld (1997) "Brood Division Is

Associated with Fledgling Dispersion in the Bluethroat (*Luscinia s. svecica*)," *Auk* 114:553–61; Székely et al., "An Evolutionary Approach to Offspring Desertion in Birds," pp. 275–76.

131. See discussion on p. 171 and the references in note 7 (this chapter).

132. Pierotti and Murphy 1987 (Western Gull/Kittiwake); Redondo, T., F. S. Tortosa, and L. A. de Reyna (1995) "Nest Switching and Alloparental Care in Colonial White Storks," *Animal Behavior* 49:1097–110; Tella, J. L., M. G. Forero, J. A. Donázar, J. J. Negro, and F. Hiraldo (1997) "Non-Adaptive Adoptions of Nestlings in the Colonial Lesser Kestrel: Proximate Causes and Fitness Consequences," *Behavioral Ecology and Sociobiology* 40:253–60. For egg transfer with adoption, see Black-billed Magpie, Caspian Tern, Cliff Swallow; for egg transfer through swallowing and regurgitation, see Vermeer, K. (1967) "Foreign Eggs in Nests of California Gulls," *Wilson Bulletin* 79:341; for a case of egg transfer that does not necessarily involve adoption, see Truslow, F. K. (1967) "Egg-Carrying by the Pileated Woodpecker," *Living Bird* 6:227–36.

133. For further examples of animals caring for offspring other than their own, see the index and the following articles: Riedman, M. L. (1982) "The Evolution of Alloparental Care and Adoption in Mammals and Birds," *Quarterly Review of Biology* 57:405–35; Lank, D. B., M. A. Bousfield, F. Cooke, and R. F. Rockwell (1991) "Why Do Snow Geese Adopt Eggs?" *Behavioral Ecology and Sociobiology* 2:181–87; Andersson, M. (1984) "Brood Parasitism Within Species," in C. J. Barnard, ed., *Producers and Scroungers: Strategies of Exploitation and Parasitism,* pp. 195–228 (London: Croom Helm); Yom-Tov, Y. (1980) "Intraspecific Nest Parasitism in Birds," *Biological Reviews* 55:93–108; Quiatt, D. (1979) "Aunts and Mothers: Adaptive Implications of Allo-maternal Behavior of Nonhuman Primates," *American Anthropologist* 81:310–19; Packer, C., S. Lewis, and A. Pusey (1992) "A Comparative Analysis of Non-Offspring Nursing," *Animal Behavior* 43:265–81; Solomon and French, *Cooperative Breeding in Mammals,* especially pp. 335–63.

134. For surveys of various types of mating systems, see Rowland, R. (1966) *Comparative Biology of Reproduction in Mammals* (Orlando: Academic Press); Slater, P. J. B., and T. R. Halliday, eds. (1994) *Behavior and Evolution* (Cambridge: Cambridge University Press); Clutton-Brock, T. G. (1989) "Mammalian Mating Systems," *Proceedings of the Royal Society of London,* Series B 235:339–72.

135. See, for example, Palombit (1994a,b, 1996), especially with regard to reevaluating the nature and diversity of pair-bonding, fidelity, and monogamy in Gibbons. It should also be pointed out that because the occurrence of infidelity between pair-bonded partners has only been appreciated relatively recently, the term *monogamy* is often used in the zoological literature simply as a synonym for *pair-bonding.*

136. Monogamy (absolute, or near absolute): Gyllensten, U. B., S. Jakobsson, and H. Temrin (1990) "No Evidence for Illegitimate Young in Monogamous and Polygynous Warblers," *Nature* 343:168–70; Holthuijzen, A. M. A. (1992) "Frequency and Timing of Copulations in the Prairie Falcon," *Wilson Bulletin* 104:333–38; Decker, M. D., P. G. Parker, D. J. Minchella, and K. N. Rabenold (1993) "Monogamy in Black Vultures: Genetic Evidence from DNA Fingerprinting," *Behavioral Ecology* 4:29–35; Vincent, A. C. J., and L. M. Sadler (1995) "Faithful Pair Bonds in Wild Seahorses, *Hippocampus whitei*," *Animal Behavior* 50:1557–69; Mauck, R. A., T. A. Waite, and P. G. Parker (1995) "Monogamy in Leach's Storm-Petrel: DNA-Fingerprinting Evidence," *Auk* 112:473–82; Haydock, J., P. G. Parker, and K. N. Rabenold (1996) "Extra-Pair Paternity Uncommon in the Cooperatively Breeding Bicolored Wren," *Behavioral Ecology and Sociobiology* 38:1–16; Fleischer, R. C., C. L. Tarr, E. S. Morton, A. Sangmeister, and K. C. Derrickson (1997) "Mating System of the Dusky Antbird, a Tropical Passerine, as Assessed by DNA Fingerprinting," *Condor* 99:512–14; Piper, W. H., D. C. Evers, M. W. Meyer, K. B. Tischler, J. D. Kaplan, and R. C. Fleischer (1997) "Genetic Monogamy in the Common Loon (*Gavia immer*)," *Behavioral Ecology and Sociobiology* 41:25–31; Kleiman, D. G. (1977) "Monogamy in Mammals," *Quarterly Review of Biology* 52:39–69; Foltz, D. W. (1981) "Genetic Evidence for Long-Term Monogamy in a Small Rodent, *Peromyscus polionotus*," *American Naturalist* 117:665–75; Ribble, D. O. (1991) "The Monogamous Mating System of *Peromyscus californicus* As Revealed by DNA Fingerprinting," *Behavioral Ecology and Sociobiology* 29:161–66; Brotherton, P. N. M., J. M. Pemberton, P. E. Komers, and G. Malarky (1997) "Genetic and Behavioral Evidence of Monogamy in a Mammal, Kirk's Dik-dik (*Madoqua kirkii*)," *Proceedings of the Royal Society of London,* Series B 264:675–81. Infidelity or nonmonogamy: Gladstone, D. E. (1979) "Promiscuity in Monogamous Colonial Birds," *American Naturalist* 114:545–57; Gowaty, P. A., and D. W. Mock, eds., (1985) *Avian Monogamy* (Washington, D.C.: American Ornithologists' Union); Birkhead, T. R., L. Atkin, and A. P. Møller (1986) "Copulation Behavior of Birds," *Behavior* 101:101–38; Westneat, D. F., P. W. Sherman, and M. L. Morton (1990) "The Ecology and Evolution of Extra-pair Copulations in Birds," *Current Ornithology* 7:331–69; Black, J. M., ed. (1996) *Partnerships in Birds: The Study of Monogamy* (Oxford: Oxford University Press); Richardson, P. R. K. (1987) "Aardwolf Mating System: Overt Cuckoldry in an Apparently Monogamous Mammal," *South African Journal of Science* 83:405–10; Palombit 1994a,b (Gibbons); Sillero-Zubiri, C., D. Gottelli, and D. W. Macdonald (1996) "Male Philopatry, Extra-Pack Copulations, and Inbreeding Avoidance in Ethiopian Wolves (*Canis simensis*)," *Behavioral Ecology and Sociobiology* 38:331–40.

137. As noted earlier, females also avoid STDs by refraining from genital contact during such mountings. For both Razorbills and spotted sandpipers, see Wagner, R. H. (1991) "The Use of Extrapair Copulations for Mate Appraisal by Razorbills, *Alca torda*," *Behavioral Ecology* 2:198–203. See also Koala for an example of a species with high rates of STDs in wild populations (Brown et al. 1987; Weigler et al. 1988). For other species in which significant portions of nonmonogamous matings are nonprocreative, see the profiles of Snow Goose, Lesser Scaup Duck, Common Murre, Oystercatcher, Silver Gull, and Swallows.

138. In addition to pair-bonding species in which nonmonogamous or alternate parenting arrangements are adopted by some individuals, the opposite situation also occurs. In some species in which the mating

arrangement is typically polygamous or in which males do not usually participate in parenting, some individuals deviate from this pattern. Monogamous pair-bonding occurs in some Gray Seals (Amos et al. 1995) and Ruffs (Cramp and Simmons 1983:391), for example, even though most individuals are polygamous in these species, while some male Mallards (Losito and Baldassarre 1996:692) and Lyrebirds (Smith 1988:37–38) occasionally parent their offspring even though males of these species generally do not contribute to parental duties.

139. Based on data from 140 populations of 76 different bird species, the average divorce rate is about 20 percent; only about 11 percent of these populations have no heterosexual divorce at all or rates of less than 1 percent. See appendix 19.1 in Ens, B. J., S. Choudhury, and J. M. Black (1996) "Mate Fidelity and Divorce in Monogamous Birds," in J. M. Black, ed., *Partnerships in Birds: The Study of Monogamy*, pp. 344–401 (Oxford: Oxford University Press). For further discussion of divorce, see Choudhury, S. (1995) "Divorce in Birds: A Review of the Hypotheses," *Animal Behavior* 50:413–29; Rowley, I. (1983) "Re-Mating in Birds," in P. Bateson, ed., *Mate Choice*, pp. 331–60 (Cambridge: Cambridge University Press).

140. Oystercatcher (Harris et al. 1987:47, 55); Ocellated Antbird (Willis 1973:35–36); Warthog (Cumming 1975:89–90); White-tailed Deer (Gerlach, D., S. Atwater, and J. Schnell, eds. [1994] *Deer*, pp. 145, 150 [Mechanicsburg, Pa.: Stackpole Books]); Snow Goose (Prevett and MacInnes 1980:25, 43).

141. Siamang (Fox 1977:409, 413–14).

142. Common Murre (based on figures in Hatchwell 1988:161, 164, 168); Kleiman, D. G., and D. S. Mack (1977) "A Peak in Sexual Activity During Mid-Pregnancy in the Golden Lion Tamarin, *Leontopithecus rosalia* (Primates: Callitrichidae)," *Journal of Mammalogy* 58:657–60; Proboscis Monkey (Gorzitze 1996:77).

143. Rhesus Macaque (Rowell et al. 1964:219); Mountain Goat (Hutchins 1984:45); addax antelope (Manski, D.A. [1982] "Herding of and Sexual Advances Toward Females in Late Stages of Pregnancy in Addax Antelope, *Addax nasomaculatus*," *Zoologische Garten* 52:106–12; wildebeest (Watson, R. M. (1969) "Reproduction of Wildebeest, *Connochaetes taurinus albojubatus* Thomas, in the Serengeti Region, and Its Significance to Conservation," p. 292, *Journal of Reproduction and Fertility*, supp. 6:287–310. One scientist (Loy 1970:294) goes so far as to suggest that the term *estrus* (meaning, roughly, the period when the female is "in heat") should be redefined for Rhesus Macaques so as to make no reference to ovulation, since nonreproductive heterosexual behaviors are so prevalent in this species (traditionally, *estrus* is defined strictly in relation to the "reproductive" event of ovulation).

144. See the index for examples of profiled species that engage in these activities. For cross-species surveys and additional examples, see also Rose et al. 1991 (Northern Elephant Seal); Robinson, S. K. (1988) "Anti-Social and Social Behavior of Adolescent Yellow-rumped Caciques (Icterinae: *Cacicus cela*)," *Animal Behavior* 36:1482–95; Thornhill, N. W. (1992) *The Natural History of Inbreeding and Outbreeding: Theoretical and Empirical Perspectives* (Chicago: University of Chicago Press); Krizek, G. O. (1992) "Unusual Interaction Between a Butterfly and a Beetle: 'Sexual Paraphilia' in Insects?" *Tropical Lepidoptera* 3(2):118; Ishikawa, H. (1985) "An Abnormal Connection Between *Indolestes peregrinus* and *Cercion hieroglyphicum*," *Tombo* (Tokyo) 28(1–4):39; Matsui, M., and T. Satow (1975) "Abnormal Amplexus Found in the Breeding Japanese Toad," *Niigata Herpetological Journal* 2:4–5; Riedman, M. (1990) *The Pinnipeds: Seals, Sea Lions, and Walruses*, pp. 216–17 (Berkeley: University of California Press).

145. Lion (Eaton 1978; Bertram 1975:479); Raptors (Korpimäki et al. 1996).

146. Oystercatcher (Heg et al. 1993:256); Kob (Buechner and Schloeth 1965:218–19).

147. Such mounts are often described as "incomplete" or are viewed as nothing more than a component or prelude to "full" copulations. This implies that the "goal" of all sexual mounting is penetration, ejaculation, and ultimately, fertilization—certainly true for a great deal of mounting behavior, but by no means a uniform characterization of all sexual activity. For further discussion of what one biologist has aptly termed "fertilization myopia"—i.e., the narrowness and bias of most scientific descriptions of animal copulation, which focus only on "successful" matings (those that lead to fertilization)—see Eberhard, *Female Control*, pp. 28–34. For an example of "display" copulations in a bird species not profiled in part 2, as well as examples from other species, see Eberhard, *Female Control*, pp. 94–102; Strahl, S. D., and A. Schmitz (1990) "Hoatzins: Cooperative Breeding in a Folivorous Neotropical Bird," p. 145, in P. B. Stacey and W. D. Koenig, eds., *Cooperative Breeding in Birds: Long-term Studies of Ecology and Behavior*, pp. 131–56 (Cambridge: Cambridge University Press).

148. For a survey of mammal species where reverse mounting occurs, see Dagg (1984). Reverse mounting usually involves the female climbing on top of the male (and rarely includes penetration [in mammals] or cloacal contact [in birds]). Because heterosexual mating in Dolphins typically occurs with the male in an upside-down position underneath the female, however, "reverse" mounting in these species involves the female assuming a position underneath the male.

149. In addition to the references for species profiled in part 2, descriptions and discussion of masturbation in a wide variety of other animals can be found in the following articles: Shadle, A. R. (1946) "Copulation in the Porcupine," *Journal of Wildlife Management* 10:159–62; Ficken, M. S., and W. C. Dilger (1960) "Comments on Redirection with Examples of Avian Copulations with Substitute Objects," *Animal Behavior* 8:219–22; Snow, B. K. (1977) "Comparison of the Leks of Guy's Hermit Hummingbird *Phaethornis guy* in Costa Rica and Trinidad," *Ibis* 119:211–14; Buechner, H. K., and S. F. Mackler (1978) "Breeding Behavior in Captive Indian Rhinoceros," *Zoologische Garten* 48:305–22; Harger, M., and D. Lyon (1980) "Further Observations of Lek Behavior of the Green Hermit Hummingbird *Phaethornis guy* at Monteverde, Costa Rica," *Ibis* 122:525–

30; Wallis, S. J. (1983) "Sexual Behavior and Reproduction of *Cercocebus albigena johnstonii* in Kibale Forest, Western Uganda," *International Journal of Primatology* 4:153–66; Poglayen-Neuwall, I., and I. Poglayen-Neuwall (1985) "Observations of Masturbation in Two Carnivora," *Zoologische Garten* 1985 55:347–348; Frith, C. B., and D. W. Frith (1993) "Courtship Display of the Tooth-billed Bowerbird *Scenopoeetes dentirostris* and Its Behavioral and Systematic Significance," *Emu* 93:129–36; Post, W. (1994) "Redirected Copulation by Male Boat-tailed Grackles," *Wilson Bulletin* 106:770–71; Frith, C. B., and D. W. Frith (1997) "Courtship and Mating of the King of Saxony Bird of Paradise *Pteridophora alberti* in New Guinea with Comment on their Taxonomic Significance," *Emu* 97:185–93.

150. As a rough measure of the overwhelming attention devoted to male as opposed to female genitalia, for example, the *Zoological Record* for 1978–97 lists 539 articles that deal with the penis, compared to only 7 for the clitoris (the *Zoological Record* is a comprehensive electronic database that indexes more than a million zoological source documents, including articles from over 6,000 journals worldwide; the following keywords/search terms were used in compiling this estimate: *penis/penile/penial/penes, phallus/phallic, baculum, hemipenes, clitoris/clitoral/clitorides, (os) clitoridis*).

151. Stumptail Macaque (Goldfoot et al. 1980); Rhesus Macaque (Zumpe, D., and R. P. Michael [1968] "The Clutching Reaction and Orgasm in the Female Rhesus Monkey [*Macaca mulatta*]," *Journal of Endocrinology* 40:117–23). In what is perhaps the most extreme "experiment" of this type, female Rhesus Macaques were strapped to an apparatus made of iron and wood and forced to undergo stimulation with a dildo or "penis substitute" while their responses were monitored with electrodes (Burton, F. D. [1971] "Sexual Climax in Female *Macaca mulatta*," in J. Biegert and W. Leutenegger, eds., *Proceedings of the 3rd International Congress of Primatology*, vol. 3, pp. 180–91 [Basel: S. Karger]).

152. For a sample of some of this debate, see Allen, M. L., and W. B. Lemmon (1981) "Orgasm in Female Primates," *American Journal of Primatology* 1:15–34; Rancour-Laferrière, D. (1983) "Four Adaptive Aspects of the Female Orgasm," *Journal of Social and Biological Structures* 6:319–33; Baker, R., and M. A. Bellis (1995) *Human Sperm Competition: Copulation, Masturbation, and Infidelity*, pp. 234–49 (London: Chapman and Hall); Hrdy, S. B. (1996) "The Evolution of Female Orgasms: Logic Please but No Atavism," *Animal Behavior* 52:851–52; Thornhill, R., and S. W. Gangstead (1996) "Human Female Copulatory Orgasm: A Human Adaptation or Phylogenetic Holdover," *Animal Behavior* 52:853–55. For recent discussions that sidestep the question of sexual pleasure with regard to the "function" of the clitoris, as well as in relation to a variety of specific sexual behaviors (e.g., stimulatory movements such as thrusting during intercourse, multiple ejaculations, lengthy copulations, etc.), see Baker and Bellis, *Human Sperm Competition*, pp. 126–31; Eberhard, *Female Control*, pp. 142–46, 204–45, 248–54.

153. A similar conundrum pertains to the "function" of the male copulatory organ in birds. Most male birds do not have a penis—insemination is achieved through simple contact of male and female genital apertures—and therefore its occurrence in some birds would appear to be, from a functional standpoint, "superfluous" (which could perhaps also be said about its occurrence in all other species). Moreover, in those species that do have a phallus (about 3 percent of all birds), its precise role in ejaculation and transporting semen remains unclear (see King, A. S. [1981] "Phallus," in A. S. King and J. McLelland, eds., *Form and Function in Birds*, vol. 2, pp. 107–47 [London: Academic Press]; Briskie, J. V., and R. Montgomerie [1997] "Sexual Selection and the Intromittent Organ of Birds," *Journal of Avian Biology* 28:73–86). In ratites such as Ostriches, Rheas, and Emus, as well as in Ducks and Geese, for example, the penis does not have an orifice connected to the male's internal reproductive organs, and he simply ejaculates through his cloaca (at the base of the penis) as do all other male birds without a phallus. Although it carries a groove on its outside surface that may help direct semen during penetration, the penis does not transport semen internally. Moreover, in some birds such as buffalo weavers, the phallus has no such groove whatsoever (nor any internal ducts) and its role in sperm transport is even less clear. Consequently, the phallus's reproductive "function" in these species is nearly as puzzling to biologists as that of the clitoris—the possibility that it could give sexual pleasure (to male and/or female) is rarely, if ever, even considered. Indeed, it is perhaps just as appropriate to speak of a male "clitoris" as it is of an actual "penis" in these cases, since the anatomy and function(s) of this organ may not be directly related to insemination (i.e., sperm transport). In addition, display of the phallus may also be an important element of courtship (as opposed to copulatory) activity in some species, as in the male Ostrich's "penis-swinging" ceremony (Sauer and Sauer 1966:56–57) and possible penile displays in the white-billed buffalo weaver (Birkhead, T. R., M. T. Stanback, and R. E. Simmons [1993] "The Phalloid Organ of Buffalo Weavers *Bubalornis*," p. 330, *Ibis* 135:326–31).

154. Scientists who have recognized that sexual pleasure (or related aspects such as sexual arousal, gratification, or libido, and/or sexual, affectionate, or "erotic" attraction) may play a significant role in homosexual and/or heterosexual interactions include: primates (Wolfe, "Human Evolution and the Sexual Behavior of Female Primates," p. 144; Vasey, "Homosexual Behavior in Primates," p. 196); Bonobo (Kano 1992:195–96, 1990:66; Thompson-Handler et al. 1984; de Waal 1995:45–46, 1997:1,4,104, 111, 158); Orang-utan (Maple 1980:158–59); Rhesus Macaque (Hamilton 1914:317–18; Akers and Conaway 1979:78–79; Erwin and Maple 1976:13); Japanese Macaque (Vasey 1996); Stumptail Macaque (Chevalier-Skolnikoff 1976:525); Killer Whale (Rose 1992:116–17); Gray Whale (Darling 1978:60; 1977:10); Northern Elephant Seal (Rose et al. 1991:186); African Elephant (Buss 1990:20); Silver Gull (Mills 1994:57–58); Laughing Gull (Hand 1981:139–40); Sage Grouse (Scott 1942:495). See also M. O'Neil's and J. D. Paterson's replies to Small (Small, M. F. (1988) "Female Primate Sexual Behavior and Conception: Are There Really Sperm to Spare?" pp. 91–92, *Current An-*

thropology 29:81–100), and P. Vasey's recent comments in Adler, T. (1996) "Animals' Fancies: Why Members of Some Species Prefer Their Own Sex," *Science News* 151:8–9.

155. Birkhead, T. (1995) "The Birds in the Trees Do It," *BBC Wildlife* 13(2):46–50; Brown-headed Cowbird (Rothstein et al. 1986:127–28).

156. For some specific examples, see Marais 1922/1969:196–97 (Savanna Baboon); Frädrich 1965:379 (Warthog); Greenhall 1965:450 (Vampire Bat); Kear 1972:85–86 (Swans); Kharitonov and Zubakin 1984:103 (Blackheaded Gull), Coulson and Thomas 1985:20 (Kittiwake); Nuechterlein and Storer 1989:341 (Grebes); Székely et al., "An Evolutionary Approach to Offspring Desertion in Birds," pp. 272–73.

157. Common Murre (Birkhead and Nettleship 1984:2123–25).

158. Virtually any of the references provided in the preceding notes will offer a sense of the ongoing debate and confusion about the "function" of each of these phenomena. For further examples, see:

Adoption—Hansen, T. F. (1995) "Does Adoption Make Evolutionary Sense?" *Animal Behavior* 51: 474–75.

Nonreproductive copulations—Hatchwell 1988 (Common Murre); Small, "Female Primate Sexual Behavior and Conception."

Multiple copulations—Gowaty, P. A. (1996) "Battles of the Sexes and Origins of Monogamy," in J. M. Black, ed., *Partnerships in Birds: The Study of Monogamy*, pp. 21–52 (Oxford: Oxford University Press); Hunter, F. M., M. Petrie, M. Otronen, T. Birkhead, and A. P. Møller (1993) "Why Do Females Copulate Repeatedly With One Male?" *Trends in Ecology and Evolution* 8:21–26; Petrie, M. (1992) "Copulation Behavior in Birds: Why Do Females Copulate More Than Once with the Same Male?" *Animal Behavior* 44:790–92.

Infanticide—Hrdy, S. B., C. Janson, and C. van Schaik (1994/1995) "Infanticide: Let's Not Throw Out the Baby with the Bath Water," *Evolutionary Anthropology* 3:151–54; Sussman, R. W., J. M. Cheverud, and T. Q. Bartlett (1984/1985) "Infant Killing as an Evolutionary Strategy: Reality or Myth?" *Evolutionary Anthropology* 3:149–51; Small, "Female Primate Sexual Behavior and Conception."

Sex segregation (including migratory)— Miquelle et al. 1992 (Moose); Myers, J. P. (1981) "A Test of Three Hypotheses for Latitudinal Segregation of the Sexes in Wintering Birds," *Canadian Journal of Zoology* 59:1527–34; Stewart and DeLong 1995 (Northern Elephant Seal).

Masturbation—Baker, R. R., and M. A. Bellis (1993) "Human Sperm Competition (Ejaculate Adjustment by Males and the Function of Masturbation," *Animal Behavior* 46:861–85; Wikelski, M., and S. Bäurle (1996) "Pre-Copulatory Ejaculation Solves Time Constraints During Copulations in Marine Iguanas," *Proceedings of the Royal Society of London*, Series B 263:439–44.

On a related point, a number of insightful analyses of otherwise puzzling aspects of sexual and reproductive behavior are now being offered by two relatively recent (and complementary) strains in biological thinking. One of these is the theory of "sperm competition," which contends that reproductive anatomy, physiology, and behavior are fundamentally shaped by the phenomenon of sperm from different males competing for fertilization by being present simultaneously in the female's reproductive tract. The other is the theory of "cryptic female choice," which argues that females themselves exert considerable influence on paternity after mating takes place by controlling whether and how sperm is utilized for fertilization. However, the complete absence of any discussion of sexual pleasure in these analyses (even where human beings are concerned) is notable. Not only is sexual pleasure as a "motivating force" compatible with many "sperm competition" and "cryptic female choice" analyses (and should therefore be considered as an important cofactor), it also offers significant insights into phenomena that continue to elude even these approaches (such as the extraordinarily high copulation rates of monogamous raptors, or mating far in advance of sperm storage periods in birds, or extrapair copulations with nonfertilizable females). For some discussion of these theories, see Baker and Bellis, *Human Sperm Competition;* Birkhead and Møller, *Sperm Competition in Birds;* Ginsberg and Huck, "Sperm Competition in Mammals"; Smith, ed., *Sperm Competition and the Evolution of Animal Mating Systems;* Eberhard, *Female Control;* Birkhead and Møller, "Female Control of Paternity." For a critique of the general male-centeredness of most sperm-competition studies, see Gowaty, P. A. (1997) "Principles of Females' Perspectives in Avian Behavioral Ecology," pp. 97–98, *Journal of Avian Biology* 28:95–102. For additional observations on the limitations of sperm competition (and sexual selection) theory as applied to species such as Oystercatchers, see Ens (1998:637).

159. On the "function" of kissing in various species, see Common Chimpanzee (Nishida 1970:51–52); Orangutan (Rijksen 1978:204–6); Squirrel Monkey (Peters 1970); West Indian Manatee (Moore 1956; Hartman 1979:110). For similar analyses applied to human kissing in various cultures, see Eibl-Eibesfeldt, I. (1972) *Love and Hate: The Natural History of Behavior Patterns,* pp. 134–39 (New York: Holt, Rinehart, and Winston).

160. cummings, e. e. (1963) *Complete Poems 1913–1962,* p. 556 (New York and London: Harcourt Brace Jovanovich).

161. Dawson, W. L. (1923) *The Birds of California,* pp. 1090–91 (San Diego: South Moulton Co.); Jehl, J. R., Jr. (1987) "A Historical Explanation for Polyandry in Wilson's Phalarope," *Auk* 104:555–56. Likewise, an even more "innocuous" phenomenon—the existence of female choice in mating among a wide variety of organisms—was considered "controversial" less than 20 years ago (Eberhard, *Female Control,* pp. 420–21), owing to the widespread belief among biologists that females are merely passive participants or "receptacles" in mating activities. Unfortunately, this idea still persists among many biologists today (cf. Gowaty, "Principles of Females' Perspectives in Avian Behavioral Ecology"). Similarly, de Waal (1997:76) suggests that cultural

biases and sexism may have contributed to scientists' denial, until 1992, of the occurrence of female dominance in Bonobos. Indeed, he points out that if any scientists had proposed this thirty years ago—along with
the full set of traits now known to be a part of Bonobo life (including a richly elaborated nonreproductive
sexuality)—they would simply have been "laughed out of the halls of academe" (ibid., p. 160).

Chapter 6. A New Paradigm: Biological Exuberance

1. Boswell, J. (1980) *Christianity, Social Tolerance, and Homosexuality: Gay People in Western Europe from the
Beginning of the Christian Era to the Fourteenth Century,* pp. 48–49 (Chicago: University of Chicago Press);
Carse, J. P. (1986) *Finite and Infinite Games,* pp. 75, 159 (New York: Ballantine Books).

2. Homosexuality and transgender of various types have also been reported from numerous indigenous cultures of South America, Asia, Africa, the Pacific islands, and Australia, and many of these cultures deserve
further investigation in terms of how they perceive systems of gender and sexuality in animals. Two potentially rich sources of knowledge about animal homosexuality/transgender are the many aboriginal cultures
of Africa and South America. The Mongandu people of Congo (Zaire), for example, have long known of the
sexual activity (genito-genital rubbing) between female Bonobos, which they call *hoka-hoka*. Among the
Hausa of Nigeria, transgendered men known as *'yan daudu* (who are effeminate, usually married to women,
and also sometimes have homosexual relations) are culturally linked to Cattle Egrets, a species in which heterosexually paired males do sometimes copulate with other males (Wrangham, R., and D. Peterson [1996]
Demonic Males: Apes and the Origins of Human Violence, p. 209 [New York: Houghton Mifflin]; Gaudio, R.
P. [1997] "Not Talking Straight in Hausa," p. 420–22, in A. Livia and K. Hall, eds., *Queerly Phrased: Language,
Gender, and Sexuality,* pp. 416–29 [New York: Oxford University Press]). In South America, the U'wa people
of Columbia have a myth involving copulation between a male fox and a male opossum, as well as various
forms of gender mixing such as pregnancy in the male fox and transformation into a woman by the male
opossum (Osborn, A. [1990] "Eat and Be Eaten: Animals in U'wa [Tunebo] Oral Tradition," pp. 152–53, in
R. Wills, ed., *Signifying Animals: Human Meaning in the Natural World,* pp. 140–58 [London: Unwin Hy-
man]). The creation myth cycle of the Mundurucú people of the Amazon includes images of birds as symbols of anal birth and a male homosexual reproductive capacity, and the male tapir as a creature with
symbolically female sexual organs, undergoing anal penetration and being sexually attracted to a man dis-
guised as a woman (Nadelson, L. [1981] "Pigs, Women, and the Men's House in Amazonia: An Analysis of
Six Mundurucú Myths," pp. 250, 254, 260–61, 270, in S. B. Ortner and H. Whitehead, eds., *Sexual Meanings:
The Cultural Construction of Gender and Sexuality,* pp. 240–72 [Cambridge: Cambridge University Press]).
And among the Waiwai and other cultures, the scent gland on the backs of both male and female peccaries
is considered to have androgynous sexual functions (Morton, J. [1984] "The Domestication of the Savage
Pig: The Role of Peccaries in Tropical South and Central America and Their Relevance for the Understanding of Pig Domestication in Melanesia," pp. 43–44, 63, *Canberra Anthropology* 7:20–70). Undoubtedly many
other similar examples remain to be discovered and studied, even within the culture areas surveyed here
(New Guinea, Siberia/Arctic, and indigenous North America), since this topic has yet to be systematically investigated in the anthropological literature.

3. Of course these four themes are not discrete or mutually exclusive, since they often overlap or interconnect
in a particular culture, nor are they uniform either between or within cultures. They are used here simply as
a way of organizing and discussing a wide range of beliefs and practices, thereby highlighting a number of
their salient features. Throughout this section the "ethnographic present tense" is used, i.e., indigenous beliefs and practices are described as ongoing, contemporary occurrences even though some have been (or are
being) actively suppressed and/or eradicated by colonizer and majority cultures and their legacy of homophobic attitudes (particularly in North America and Siberia). In spite of severe declines and disappearances
in the face of nearly insurmountable obstacles, however, many of these traditions continue in altered form
or are undergoing wholesale cultural revival; they should be considered neither "dead" nor "lost."

4. For more information on Native American two-spirit, see, for example, Callender, C., and L. M. Kochems
(1983) "The North American Berdache," *Current Anthropology* 24:443–70; Williams, W. L. (1986) *The Spirit
and the Flesh: Sexual Diversity in American Indian Culture* (Boston: Beacon Press); Allen, P. G. (1986)
"Hwame, Koshkalaka, and the Rest: Lesbians in American Indian Cultures," in *The Sacred Hoop: Recovering
the Feminine in American Indian Traditions,* pp. 245–61 (Boston: Beacon Press); Gay American Indians
(GAI) and W. Roscoe, coordinating ed., (1988) *Living the Spirit: A Gay American Indian Anthology* (New
York: St. Martin's Press); Jacobs, S.-E., W. Thomas, and S. Lang, eds., (1997) *Two-Spirit People: Native Amer-
ican Gender Identity, Sexuality, and Spirituality* (Urbana: University of Illinois Press); Roscoe, W. (1998)
Changing Ones: Third and Fourth Genders in Native North America (New York: St. Martin's Press).

5. Whitman, W. (1937) *The Oto,* pp. 22, 29, 30, 50 (New York: Columbia University Press); Callender and
Kochems, "The North American Berdache," p. 452.

6. Cushing, F. H. (1896) "Outlines of Zuñi Creation Myths," pp. 401–2, *Bureau of American Ethnology Annual
Report* 13:321–447; Parsons, E. C. (1916) "The Zuñi La'mana," p. 524, *American Anthropologist* 18:521–28.

7. Boas, F. (1898) "The Mythology of the Bella Coola Indians," *Memoirs of the American Museum of Natural
History* 2(2):38–40 (reprinted in GAI and Roscoe, *Living the Spirit,* pp. 81–84); McIlwraith, T. F. (1948) *The
Bella Coola Indians* (Toronto: University of Toronto Press); Gifford, E. W. (1931) "The Kamia of Imperial
Valley," pp. 79–80, *Bureau of American Ethnology Bulletin* 97:1–94. The names of two other birds encoun-

tered by the Kamia two-spirit are also mentioned in this story *(tokwil* and *kusaul),* but Gifford does not identify which species these are.

8. Haile, B., I. W. Goossen, and K. W. Luckert (1978) *Love-Magic and Butterfly People: The Slim Curly Version of
    the* Ajiłee *and Mothway Myths,* pp. 82–90, 161. American Tribal Religions, vol. 2 (Flagstaff: Museum of
    Northern Arizona Press); Luckert, K. W. (1975) *The Navajo Hunter Tradition,* pp. 176–77 (Tucson: Univer-
    sity of Arizona Press); Levy, J. E., R. Neutra, and D. Parker (1987) *Hand Trembling, Frenzy Witchcraft, and
    Moth Madness: A Study of Navajo Seizure Disorders,* p. 46 (Tucson: University of Arizona Press).

9. Wissler, C. (1916) "Societies and Ceremonial Associations in the Oglala Division of the Teton-Dakota," pp.
    92–94, *Anthropological Papers of the American Museum of Natural History* 11:1–99; Howard, J. H. (1965)
    "The Ponca Tribe," pp. 142–43, *Bureau of American Ethnology Bulletin* 195:572–97; Powers, W. (1977) *Oglala
    Religion,* pp. 58–59 (Lincoln: University of Nebraska Press); Thayer, J. S. (1980) "The Berdache of the North-
    ern Plains: A Socioreligious Perspective," p. 289, *Journal of Anthropological Research* 36:287–93; Williams,
    Spirit and the Flesh, pp. 28–29; Allen, "Hwame, Koshkalaka, and the Rest"; GAI and Roscoe, *Living the Spirit,*
    pp. 87–89; Fletcher, A. C., and F. La Flesche (1911) "The Omaha Tribe," p. 133, *Bureau of American Ethnology
    Annual Report* 27:16–672.

10. Kenny, M. (1975–76) "Tinselled Bucks: A Historical Study in Indian Homosexuality," *Gay Sunshine* 26–
    27:15–17 (reprinted in GAI and Roscoe, *Living the Spirit,* pp. 15–31); Grinnell, G. B. (1923) *The Cheyenne In-
    dians: Their History and Ways of Life,* vol. 2, pp. 79–86 (New Haven: Yale University Press); Moore, J. H.
    (1986) "The Ornithology of Cheyenne Religionists," pp. 181–82, *Plains Anthropologist* 31:177–92; Tafoya, T.
    (1997) "M. Dragonfly: Two-Spirit and the Tafoya Principle of Uncertainty," p. 194, in Jacobs et al., *Two-Spirit
    People,* pp. 192–200.

11. Kroeber, A. (1902–7) "The Arapaho," pp. 19–20, *Bulletin of the American Museum of Natural History* 18:1–
    229; Bowers, A. W. (1992) *Hidatsa Social and Ceremonial Organization,* pp. 325, 427 (reprint of the Bureau
    of American Ethnology Bulletin no. 194, 1965) (Lincoln: University of Nebraska Press).

12. Pilling mentions the "wolf power" attributed to the well-known cross-dressing Tolowa shaman, also known
    as Doctor Medicine (Pilling, A. R. [1997] "Cross-Dressing and Shamanism among Selected Western North
    American Tribes," p. 84, in Jacobs et al., *Two-Spirit People,* pp. 69–99). Turner reports the well-known Sno-
    qualmie shaman who, though biologically male, was "like a woman" and had Grizzly Bear and Rainbow
    powers (Turner, H. [1976] "Ethnozoology of the Snoqualmie," p. 84 [unpublished manuscript, available in
    the Special Collections Division, University of Washington Library, Seattle, Wash.]). Another possible asso-
    ciation of Bears with sexual and gender variance has been reported (and widely cited) for the Kaska Indians:
    Honigmann mentions that cross-dressing women who were raised as boys, perform male tasks, and may
    have homosexual relationships with other women wear an amulet made of the dried ovaries of a Bear, tied
    to their inner belt and worn for life, to prevent conception (Honigmann, J. J. [1954] *The Kaska Indians: An
    Ethnographic Reconstruction,* p. 130, Yale University Publications in Anthropology no. 51 [New Haven: Yale
    University Press]). However, Goulet has challenged and reinterpreted this example, specifically with regard
    to the claims of cross-dressing, homosexual involvements, and the uniqueness of the Bear amulet to these
    supposedly gender-mixing females (Goulet, J.-G. A. [1997] "The Northern Athapaskan 'Berdache' Reconsid-
    ered: On Reading More Than There Is in the Ethnographic Record," in Jacobs et al., *Two-Spirit People,* pp.
    45–68).

13. Miller, J. (1982) "People, Berdaches, and Left-Handed Bears: Human Variation in Native America," *Journal of
    Anthropological Research* 38:274–87.

14. Among the Hopi people, a parallel view exists regarding hawks and eagles: these creatures are all thought of
    as mothers, and individual raptors are sometimes even given names such as Female Bear for this reason
    (Tyler, H. A. [1979] *Pueblo Birds and Myths,* p. 54 [Norman: University of Oklahoma Press]).

15. For indigenous views on bears and menstruation, as well as further information on the Bear Mother figure,
    see Rockwell, D. (1991) *Giving Voice to Bear: North American Indian Rituals, Myths, and Images of the Bear,*
    pp. 14–17, 123–25, 133 (Niwot, Colo.: Roberts Rinehart Publishers); Buckley, T., and A. Gottlieb (1988)
    Blood Magic: The Anthropology of Menstruation, p. 22 (Berkeley: University of California Press); Shepard, P.,
    and B. Sanders (1985) *The Sacred Paw: The Bear in Nature, Myth, and Literature,* pp. 55–59 (New York:
    Viking); Hallowell, A. I. (1926) "Bear Ceremonialism in the Northern Hemisphere," *American Anthropologist*
    28:1–175; Rennicke, J. (1987) *Bears of Alaska in Life and Legend* (Boulder, Colo.: Roberts Rinehart).

16. Miller, "People, Berdaches, and Left-Handed Bears," pp. 277–78; Drucker, P. (1951) *The Northern and Cen-
    tral Nootkan Tribes,* p. 130, Bureau of American Ethnology Bulletin no. 144 (Washington, D.C.: Smithsonian
    Institution); Clutesi, G. (1967) "Ko-ishin-mit Invites Chims-meet to Dinner," in *Son of Raven, Son of Deer:
    Fables of the Tse-shaht People,* pp. 62–69 (Sidney, B.C.: Gray's Publishing); Sapir, E. (1915) *Abnormal Types of
    Speech in Nootka,* Geological Survey, Memoir 62, Anthropological Series no. 5 (Ottawa: Government Print-
    ing Bureau).

17. Teit, J. A. (1917) "Okanagon Tales," *Memoirs of the American Folk-Lore Society* 11:75–76 (reprinted in GAI
    and Roscoe, *Living the Spirit,* pp. 89–91); Mandelbaum, M. (1938) "The Individual Life Cycle," p. 119, in L.
    Spier, ed., *The Sinkaietk or Southern Okanagon of Washington,* pp. 101–29, General Series in Anthropology
    no. 6 (Menasha, Wis.: George Banta); Brooks, C., and M. Mandelbaum (1938) "Coyote Tricks Cougar into
    Providing Food," in Spier, *The Sinkaietk,* pp. 232–33, 257; Kroeber, "The Arapaho," p. 19; Kenny, "Tinselled
    Bucks," p. 22; Jones, W. (1907) "The Turtle Brings Ruin Upon Himself," in *Fox Texts,* pp. 314–31, Publications
    of the American Ethnological Society no. 1 (Leyden: E. J. Brill); Radin, P. (1956) *The Trickster: A Study in
    American Indian Mythology,* pp. 20–24, 137–39 (New York: Greenwood Press). Other, more tangential, asso-

ciations between homosexuality and turtles occur among the Fox people. In a cautionary tale of two women who had an affair with each other, for example, the erect clitoris of one woman during lesbian sex is described as being like a turtle's penis, while the child that resulted from their union is compared to a soft-shell turtle ("Two Maidens Who Played the Harlot with Each Other," Jones, *Fox Texts*, pp. 151–53).

18. Brant, B. (Degonwadonti) (1985) "Coyote Learns a New Trick," in *Mohawk Trail*, pp. 31–35 (Ithaca: Firebrand Books) (reprinted in GAI and Roscoe, *Living the Spirit*, pp. 163–66); Steward, D.-H. (1988) "Coyote and Tehoma," in GAI and Roscoe, *Living the Spirit*, pp. 157–62; Cameron, A. (1981) "Song of Bear," in *Daughters of Copper Woman*, pp. 115–19 (Vancouver: Press Gang); Tafoya, "M. Dragonfly"; Robertson, D. V. (1997) "I Ask You to Listen to Who I Am," p. 231, in Jacobs et al., *Two-Spirit People*, pp. 228–35; Brant, B. (1994) *Writing as Witness: Essay and Talk*, pp. 61, 69–70, 75, 108 (Toronto: Women's Press); Chrystos (1988) *Not Vanishing* (Vancouver: Press Gang); Chrystos (1991) *Dream On* (Vancouver: Press Gang); Chrystos (1995) *Fire Power* (Vancouver: Press Gang).

19. George Catlin's original 1867 description of the ritual homosexuality and other sexual imagery in this ceremony was considered so scandalous at the time that it was eliminated from most published versions of his monograph. Only a few copies of the first edition of the book that were delivered to scholars included this material, and even then it was set aside in a special appendix. Catlin, G. (1867/1967) *O-kee-pa: A Religious Ceremony and Other Customs of the Mandans*, pp. 83–85, centennial edition, edited and with an introduction by J. C. Ewers (New Haven and London: Yale University Press); Bowers, A. W. (1950/1991) *Mandan Social and Ceremonial Organization*, pp. 131, 145–46 (reprint of the 1950 University of Chicago Press edition) (Moscow, Idaho: University of Idaho Press); Campbell, J. (1988) *Historical Atlas of World Mythology, Vol. 1: The Way of the Animal Powers, Part 2: Mythologies of the Great Hunt*, pp. 226–31 (New York: Harper & Row).

20. Extraordinary as it may seem, rites like this may be far more ancient and widespread than previously imagined. Among the Paleolithic cave paintings of Lascaux in France, for example, imagery combining anal penetration of bison bulls, shamanic and sexual ecstasy, hunting motifs, and hermaphroditic animal figures can be found—a striking echo of certain elements in the Okipa ceremony and other Native American belief systems. One picture, regarded as among the most important in the entire Lascaux complex, is of a shaman lying in rapture, with erect penis, in front of a bison bull. Penetrating the bull from behind is a spear that, according to Joseph Campbell, has "transfixed its anus and emerged through its sexual organ." The phallic imagery of the bison is also combined with vulvar symbolism in the shape of the spilled entrails or wound of the beast. Elsewhere in the Lascaux caves, a startling and enigmatic figure of an apparently gender-mixing hoofed mammal appears prominently in one fresco. On the wall of a grotto known as the Rotunda is the image of a pregnant bull whose "two long, straight horns point directly forward from its head . . . and [whose] gravid belly hangs nearly to the ground." Dating from around 12,000 B.C., these are probably the earliest known depictions of gender-mixing animals, and they are testimony to an ancient and profound association between variant forms of gender and sexual expression in animals and humans (see Campbell, *Historical Atlas of World Mythology*, pp. 58–66, for further discussion of these images). Campbell also draws a parallel between some of these figures and the contemporary shamanic practices of the Aranda people of Australia, which involve uncanny correspondences in terms of their mixture of phallic, anal, and male-female imagery. Perhaps not uncoincidentally, the Aranda also participate in a variety of homosexual practices, both overt and "ritualized" (see chapter 2 for discussion of Aranda penis-handling as a ritualized "greeting" gesture between men; for overt homosexual activities, see Ford, C. S., and F. A. Beach [1951] *Patterns of Sexual Behavior*, p. 132 [New York: Harper and Brothers]; Berndt, R., and C. Berndt [1943] "A Preliminary Report of Field Work in the Ooldea Region, Western South Australia," pp. 276–77, *Oceania* 13:239–75; Murray, S. O. [1992] "Age-Stratified Homosexuality: Introduction," pp. 5–6, in S. O. Murray, ed., *Oceanic Homosexualities*, pp. 293–327 [New York: Garland]).

21. Schlesier, K. H. (1987) *The Wolves of Heaven: Cheyenne Shamanism, Ceremonies, and Prehistoric Origins*, pp. 7, 14–15, 66–73, 78–111 (Norman: University of Oklahoma Press); Grinnell, *The Cheyenne Indians*, vol. 2, pp. 285–336; Hoebel, E. A. (1960) *The Cheyennes: Indians of the Great Plains*, pp. 16–17 (New York: Holt, Rinehart, and Winston).

22. Powers, M. N. (1980) "Menstruation and Reproduction: An Oglala Case," p. 61, *Signs* 6:54–65; Parsons, E. C. (1939) *Pueblo Indian Religion*, pp. 831–32 (Chicago: University of Chicago Press); Tyler, H. A. (1975) *Pueblo Animals and Myths*, pp. 98, 131, 148–50 (Norman: University of Oklahoma Press); Duberman, M. B., F. Eggan, and R. O. Clemmer (1979) "Documents in Hopi Indian Sexuality: Imperialism, Culture, and Resistance," pp. 119–20, *Radical History Review* 20:99–130; Du Bois, C.A. (1935) "Wintu Ethnography," p. 50, *University of California Publications in American Archaeology and Ethnology* 36:1–148.

23. Hill, W. W. (1935) "The Status of the Hermaphrodite and Transvestite in Navaho Culture," p. 274, *American Anthropologist* 37:273–79; Haile et al., *Love-Magic and Butterfly People*, p. 163; Luckert, *The Navajo Hunter Tradition*, pp. 176–77; Hill, W. W. (1938) *The Agricultural and Hunting Methods of the Navaho Indians*, pp. 99, 110, 119, 126–27, Yale University Publications in Anthropology no. 18 (New Haven: Yale University Press).

24. For overviews of ritual homosexuality and alternate gender systems in New Guinea and Melanesia, see Herdt, G. H. (1981) *Guardians of the Flutes: Idioms of Masculinity* (New York: McGraw-Hill); Herdt, G. H., ed., (1984) *Ritualized Homosexuality in Melanesia* (Berkeley: University of California Press). On the "third sex" category, see Herdt, G. (1994) "Mistaken Sex: Culture, Biology, and the Third Sex in New Guinea," in G. Herdt, ed., *Third Sex, Third Gender: Beyond Sexual Dimorphism in Culture and History*, pp. 419–45 (New

York: Zone Books); Poole, F. J. P. (1996) "The Procreative and Ritual Constitution of Female, Male, and Other: Androgynous Beings in the Cultural Imagination of the Bimin-Kuskusmin of Papua New Gunea," in S. P. Ramet, ed., *Gender Reversals and Gender Cultures: Anthropological and Historical Perspectives*, pp. 197–218 (London: Routledge). For ceremonial transvestism and "male menstruation," see, for example, Schwimmer, E. (1984) "Male Couples in New Guinea," in Herdt, *Ritualized Homosexuality in Melanesia*, pp. 248–91; Lutkehaus, N. C., and P. B. Roscoe, eds., (1995) *Gender Rituals: Female Initiation in Melanesia*, pp. 16–17, 36, 49, 69, 107, 120, 198–200, 229 (New York: Routledge); A. Strathern, in Callender and Kochems, "The North American Berdache," p. 464.

25. Herdt, *Guardians of the Flutes*, p. 94; Schwimmer, "Male Couples in New Guinea," p. 271; Van Baal, J. (1984) "The Dialectics of Sex in Marind-anim Culture," in Herdt, *Ritualized Homosexuality in Melanesia*, pp. 128–66.

26. Herdt, *Guardians of the Flutes*, pp. 87–94; Poole, "The Procreative and Ritual Constitution of Female, Male, and Other," pp. 205, 217; Sørum, A. (1984) "Growth and Decay: Bedamini Notions of Sexuality" in Herdt, *Ritualized Homosexuality in Melanesia*, pp. 318–36; Lindenbaum, S. (1984) "Variations on a Sociosexual Theme in Melanesia," in Herdt, *Ritualized Homosexuality in Melanesia*, pp. 83–126.

27. An echo of these beliefs can also be found in native North America: the Cherokee maintain that female opossums (a North American marsupial) are essentially parthenogenetic, i.e., they reproduce without males (Fradkin, A. [1990] *Cherokee Folk Zoology: The Animal World of a Native American People, 1700–1838*, pp. 377–78 [New York: Garland]).

28. Herdt (*Guardians of the Flutes*, p. 91) tentatively identifies this as the "crested bird of paradise"; however, the description of its round display platforms (constructed of twigs and straw, with a central pole) strongly suggests that this is actually a species of bowerbird. Most likely it is MacGregor's bowerbird (*Amblyornis macgregoriae*), whose "maypole" bower type matches this description, and whose orange crest also fits the description of this species provided by Herdt. For further details, see Gilliard, E. T. (1969) "MacGregor's Gardener Bower Bird," in *Birds of Paradise and Bower Birds*, pp. 300–311 (Garden City, N.Y.: Natural History Press); Johnsgard, P. A. (1994) *Arena Birds: Sexual Selection and Behavior*, pp. 206, 211–12 (Washington, D.C., and London: Smithsonian Institution Press). Among the Kaluli people, the (male) Raggiana's Bird of Paradise and other brightly colored birds are also considered female; men adorn themselves with their plumes to acquire the beauty of these supposedly feminine creatures (Feld, S. [1982] *Sound and Sentiment: Birds, Weeping, Poetics, and Song in Kaluli Expression*, pp. 55, 65–66 [Philadelphia: University of Pennsylvania Press]).

29. Although Poole (1996:205) identifies this only as the "night bird," it is most likely a species of nightjar (family Caprimulgidae), frogmouth (family Podargidae), or owlet-nightjar (family Aegothelidae).

30. Herdt, *Guardians of the Flutes*, pp. 131–57; Gardner, D. S. (1984) "A Note on the Androgynous Qualities of the Cassowary: Or Why the Mianmin Say It Is Not a Bird," *Oceania* 55:137–45; Bulmer, R. N. H. (1967) "Why Is the Cassowary Not a Bird? A Problem of Zoological Taxonomy Among the Karam of the New Guinea Highlands," *Man* 2:5–25; Juillerat, B., ed., (1992) *Shooting the Sun: Ritual and Meaning in West Sepik*, pp. 65, 282 (Washington, D.C.: Smithsonian Institution Press); Feld, *Sound and Sentiment*, pp. 68–71; Tuzin, D. (1997) *The Cassowary's Revenge: The Life and Death of Masculinity in a New Guinea Society*, pp. 80–82, 94, 209–10 (Chicago: University of Chicago Press). Some Australian Aboriginal peoples hold parallel beliefs about a related bird, the Emu, being all-female or having ambiguous or simultaneous genders (Maddock, K. [1975] "The Emu Anomaly," pp. 112–13, 118, 121, in L. R. Hiatt, ed., *Australian Aboriginal Mythology*, pp. 102–22 [Canberra: Australian Institute of Aboriginal Studies]).

31. Gell, A. (1975) *Metamorphosis of the Cassowaries: Umeda Society, Language, and Ritual*, pp. 180, 182, 184, 225–26, 233–34, 239–40, 250, L.S.E. Monographs on Social Anthropology no. 51 (London: Athlone Press); Gell, A. (1971) "Penis Sheathing and Ritual Status in a West Sepik Village," pp. 174–75, *Man* 6:165–81.

32. These individuals are "born with labial folds, reared as girls, and then recognized as being the descendants of Yomnok when distinctive but diminutive male genitalia descend into view on the eve of puberty." This type of intersexuality (known medically as 5–alpha reductase male pseudo-hermaphroditism) also occurs fairly frequently among the Sambia, where it is recognized as a "third sex" (Poole, "The Procreative and Ritual Constitution of Female, Male, and Other," pp. 209, 218; Herdt, "Mistaken Sex"). The species of echidna referred to is probably the long-beaked echidna, *Zaglossus bruijni;* for more on indigenous views of echidnas in New Guinea, see Jorgensen, D. (1991) "Echidna and *Kuyaam:* Classification and Anomalous Animals in Telefolmin," *Journal of the Polynesian Society* 100:365–80.

33. Poole, "The Procreative and Ritual Constitution of Female, Male, and Other," pp. 197, 203–5, 209–10, 216–17; Poole, F. J. P. (1981) "Transforming 'Natural' Woman: Female Ritual Leaders and Gender Ideology Among Bimin-Kuskusmin," pp. 117, 120, 153–60, in S. B. Ortner and H. Whitehead, eds., *Sexual Meanings: The Cultural Construction of Gender and Sexuality*, pp. 116–65 (Cambridge: Cambridge University Press); Poole, F. J. P. (1982) "The Ritual Forging of Identity: Aspects of Person and Self in Bimin-Kuskusmin Male Initiation, pp. 125–31, in G. H. Herdt, ed., *Rituals of Manhood: Male Initiation in Papua New Guinea*, pp. 99–154 (Berkeley: University of California Press).

34. Layard, J. (1942) *Stone Men of Malekula*, especially pp. 482–94 (London: Chatto and Windus); Allen, M. (1981) "Innovation, Inversion, and Revolution as Political Tactics in West Aoba," in M. Allen, ed., *Vanuatu: Politics, Economics, and Ritual in Island Melanesia*, pp. 105–34 (Sydney: Academic Press); Allen, M. R. (1984) "Ritualized Homosexuality, Male Power, and Political Organization in North Vanuatu: A Comparative Analysis," in Herdt, *Ritualized Homosexuality in Melanesia*, pp. 83–126; Battaglia, D. (1991) "Punishing the

Yams: Leadership and Gender Ambivalence on Sabarl Island," p. 94, in M. Godelier and M. Strathern, eds., *Big Men and Great Men: Personifications of Power in Melanesia*, pp. 83–96 (Cambridge: Cambridge University Press).

35. Baker, J. R. (1925) "On Sex-Intergrade Pigs: Their Anatomy, Genetics, and Developmental Physiology," *British Journal of Experimental Biology* 2:247–63; Baker, J. R. (1928) "Notes on New Hebridean Customs, with Special Reference to the Intersex Pig," *Man* 28:113–18; Baker, J. R. (1928) "A New Type of Mammalian Intersexuality," *British Journal of Experimental Biology* 6:56–64; Baker, J. R. (1929) *Man and Animals in the New Hebrides*, pp. 22, 30–31, 115–30 (London: George Routledge & Sons); Jolly, M. (1984) "The Anatomy of Pig Love: Substance, Spirit, and Gender in South Pentecost, Vanuatu," pp. 84–85, 101, 104–5, *Canberra Anthropology* 7:78–108; Jolly, M. (1991) "Soaring Hawks and Grounded Persons: The Politics of Rank and Gender in North Vanuatu," pp. 54, 59, 67, 71, in Godelier and Strathern, *Big Men and Great Men*, pp. 48–80; Rodman, W. (1996) "The Boars of Bali Ha'i: Pigs in Paradise," in J. Bonnemaison, C. Kaufmann, K. Huffman, and D. Tryon, eds., *Arts of Vanuatu*, pp. 158–67 (Honolulu: University of Hawaii Press); Huffman, K. W. (1996) "Trading, Cultural Exchange, and Copyright: Important Aspects of Vanuatu Arts" and "Plates and Bowls from Northern and Central Vanuatu," pp. 183, 192, 228, in Bonnemaison et al., *Arts of Vanuatu*, pp. 182–94, 226–31.

36. In accordance with many anthropological treatments, North American Inuit cultures are here included with the Siberian culture complex, with which they share many features. They also, of course, show a number of similarities to non-Inuit Native American peoples (as do many Siberian cultures), as well as a large number of unique features, and this arrangement is largely a matter of exposition rather than a reflection of actual or perceived cultural relationships.

37. Balzer, M. M. (1996) "Sacred Genders in Siberia: Shamans, Bear Festivals, and Androgyny," in Ramet, *Gender Reversals and Gender Cultures*, pp. 164–82; Bogoras, W. (1904–9) *The Chukchee*, pp. 448–57, Memoirs of the American Museum of Natural History, vol. 11, Publications of the Jesup North Pacific Expedition, vol. 7 (Leiden: E. J. Brill; New York: G. E. Stechert [reprinted in 1975, New York: AMS Press]); Jochelson, W. (1908) *The Koryak*, pp. 47, 65, 469, 502, 525, 733, Memoirs of the American Museum of Natural History, vol. 10, Publications of the Jesup North Pacific Expedition, vol. 6 (Leiden: E. J. Brill; New York: G. E. Stechert [reprinted in 1975, New York: AMS Press]); Murray, S. O. (1992) "Vladimir Bogoraz's Account of Chukchi Transformed Shamans" and "Vladimir Iokalson's Reports of Northeastern Siberian Transformed Shamans," in S. O. Murray, ed., *Oceanic Homosexualities*, pp. 293–327 (New York: Garland).

38. Serov, S. I. (1988) "Guardians and Spirit-Masters of Siberia," pp. 241, 247–49, in W. W. Fitzhugh and A. Crowell, eds., *Crossroads of Continents: Cultures of Siberia and Alaska*, pp. 241–55 (Washington, D.C.: Smithsonian Institution Press); Pavlinskaya, L. R. (1994) "The Shaman Costume: Image and Myth," in G. Seaman and J. S. Day, eds., *Ancient Traditions: Shamanism in Central Asia and the Americas*, pp. 257–64 (Niwot, Colo.: University Press of Colorado); Zornickaja, M. J. (1978) "Dances of Yakut Shamans," in V. Diószegi and M. Hoppál, eds., *Shamanism in Siberia*, pp. 299–307 (Budapest: Akadémiai Kiadó); Hamayon, R. N. (1992) "Game and Games, Fortune and Dualism in Siberian Shamanism," in M. Hoppál and J. Pentikäinen, eds., *Northern Religions and Shamanism*, pp. 134–37 (Budapest: Akadémiai Kiadó); Bogoras, *The Chukchee*, pp. 268–9.

39. Saladin d'Anglure, B. (1986) "Du fœtus au chamane: la construction d'un 'troisième sexe' inuit" (From Fetus to Shaman: The Construction of an Inuit "Third Sex"), especially pp. 72, 84, 86, *Études/Inuit/Studies* 10:25–113 (selections translated into English and reprinted in A. Mills and R. Slobodin, eds., [1994] *Amerindian Rebirth: Reincarnation Belief among North American Indians and Inuit*, pp. 82–106 [Toronto: University of Toronto Press]); Saladin d'Anglure, B. (1983) "Ijiqqat: voyage au pays de l'invisible inuit (Ijiqqat: Travel to the Land of the Inuit Invisible), pp. 72, 81, *Études/Inuit/Studies* 7:67–83; Saladin d'Anglure, B. (1990) "Frère-lune (Taqqiq), sœur-soleil (Siqiniq), et l'intelligence du monde (Sila): Cosmologie inuit, cosmographie arctique, et espace-temps chamanique" (Brother-Moon [Taqqiq], Sister-Sun [Siqiniq], and the Intelligence of the World [Sila]: Inuit Cosmology, Arctic Cosmography, and Shamanistic Space-Time), pp. 96–98, *Études/Inuit/Studies* 14:75–139; Boas, F. (1901–7) "The Eskimo of Baffin Land and Hudson Bay," p. 509, *Bulletin of the American Museum of Natural History* 15:1–570.

40. Saladin d'Anglure, B. (1990) "Nanook, Super-Male: The Polar Bear in the Imaginary Space and Social Time of the Inuit of the Canadian Arctic," especially pp. 190, 193, in R. Wills, ed., *Signifying Animals: Human Meaning in the Natural World*, pp. 178–95 (London: Unwin Hyman).

41. Balzer, "Sacred Genders in Siberia," pp. 169–74.

42. Fienup-Riordan, A. (1994) *Boundaries and Passages: Rule and Ritual in Yup'ik Eskimo Oral Tradition*, pp. 114, 139, 274–79, 293, 297–98, 307–12, 320, 345–50 (Norman: University of Oklahoma Press); Kaplan, S. A. (1984) "Note," in E. S. Burch Jr., ed., *The Central Yup'ik Eskimos*, supplementary issue of *Études/Inuit/Studies* 8:2; Morrow, P. (1984) "It Is Time for Drumming: A Summary of Recent Research on Yup'ik Ceremonialism," pp. 119, 138, in E. S. Burch Jr., ed., *The Central Yup'ik Eskimos*, supplementary issue of *Études/Inuit/Studies* 8:113–40; Fienup-Riordan, A. (1996) *The Living Tradition of Yup'ik Masks: Agayuliyararput (Our Way of Making Prayer)*, pp. 39, 63, 87–88, 92, 98, 100, 176 (Seattle: University of Washington Press); Chaussonnet, V. (1988) "Needles and Animals: Women's Magic," p. 216, in Fitzhugh and Crowell, *Crossroads of Continents*, pp. 209–26. Among the Cumberland Sound Inuit of eastern Canada, the spirit-guardian and mother of sea mammals, *Sedna*, has an attendant named *Qailertetang* who is also represented during ceremonies by a man dressed in a woman's costume (Boas, "The Eskimo of Baffin Land and Hudson Bay," pp. 139–40).

43. Bogoras, *The Chukchee*, pp. 79, 84; Diachenko, V. (1994) "The Horse in Yakut Shamanism," pp. 268–69, in Seaman and Day, *Ancient Traditions*, pp. 265–71.

44. On handedness/laterality in various animals, see Marino, L., and J. Stowe (1997) "Lateralized Behavior in Two Captive Bottlenose Dolphins (*Tursiops truncatus*)," *Zoo Biology* 16:173–77; Marino, L., and J. Stowe (1997) "Lateralized Behavior in a Captive Beluga Whale (*Delphinapterus leucas*)," *Aquatic Mammals* 23:101–3; McGrew, W. C., and L. F. Marchant (1996) "On Which Side of the Apes? Ethological Study of Laterality of Hand Use," in W. C. McGrew, L. F. Marchant, and T. Nishida, eds., *Great Ape Societies*, pp. 255–72 (Cambridge: Cambridge University Press); Clapham, P. J., E. Leimkuhler, B. K. Gray, and D. K. Mattila (1995) "Do Humpback Whales Exhibit Lateralized Behavior?" *Animal Behavior* 50:73–82; Morgan, M. J. (1992) "On the Evolutionary Origin of Right-Handedness," *Current Biology* 2:15–17; MacNeilage, P. F., M. G. Studdert-Kennedy, and B. Lindblom (1987) "Primate Handedness Reconsidered," *Behavioral and Brain Sciences* 10:247–303; Rogers, L. J. (1980) "Lateralization in the Avian Brain," *Bird Behavior* 2:1–12; Cole, J. (1955) "Paw Preference in Cats Related to Hand Preference in Animals and Man," *Journal of Comparative and Physiological Psychology* 48:337–45; Friedman, H., and M. Davis (1938) "'Left Handedness' in Parrots," *Auk* 55:478–80.

45. Beck, B. B. (1980) *Animal Tool Behavior: The Use and Manufacture of Tools by Animals*, p. 39 (New York: Garland); Koch, T. J. (1975) *The Year of the Polar Bear*, p. 32 (Indianapolis and New York: Bobbs-Merrill); Bruemmer, F. (1972) *Experiences with Arctic Animals*, p. 92 (Toronto: McGraw-Hill Ryerson); Perry, R. (1966) *The World of the Polar Bear*, pp. 11, 76 (Seattle: University of Washington Press); Haig-Thomas, D. (1939) *Tracks in the Snow*, p. 230 (New York: Oxford University Press).

46. Lindesay, J. (1987) "Laterality Shift in Homosexual Men," *Neuropsychologia* 25:965–69; McCormick, C. M., S. F. Witelson, and E. Kinstone (1990) "Left-handedness in Homosexual Men and Women: Neuroendocrine Implications," *Psychoneuroendocrinology* 1:69–76; Watson, D. B., and S. Coren (1992) "Left-handedness in Male-to-Female Transsexuals," *JAMA (Journal of the American Medical Association)* 267:1342; Coren, S. (1992) *The Left-Hander Syndrome: The Causes and Consequences of Left-Handedness*, pp. 199–202 (New York: Free Press).

47. For scientific experiments, see Cushing, B. S. (1983) "Responses of Polar Bears to Human Menstrual Odors," in E. C. Meslow, ed., *Proceedings of the Fifth International Conference on Bear Research and Management (1980)*, pp. 270–274 (West Glacier, Mont.: International Association for Bear Research and Management); Cushing, B. S. (1980) *The Effects of Human Menstrual Odors, Other Scents, and Ringed Seal Vocalizations on the Polar Bear* (master's thesis, University of Montana). For additional discussion of the phenomenon, see March, K. S. (1980) "Deer, Bears, and Blood: A Note on Nonhuman Animal Response to Menstrual Odor," *American Anthropologist* 82:125–27. For an alternative evaluation of the scientific evidence and discussion of the way these findings have been misinterpreted to mean that bears are more likely to attack women—and therefore used to justify policies excluding women from certain forestry jobs—see Byrd, C. P. (1988) *Of Bears and Women: Investigating the Hypothesis That Menstruation Attracts Bears* (master's thesis, University of Montana).

48. Bears (Cattet 1988).

49. Common Chimpanzee (Egozcue 1972); Rhesus Macaque (Sullivan and Drobeck 1966; Weiss et al. 1973); Savanna Baboon (Bielert 1984; Bielert et al. 1980; Wadsworth et al. 1978); Bowhead Whale and other whales and dolphins (Tarpley et al. 1995); Eastern Gray Kangaroo and other marsupials (Sharman et al. 1990).

50. Another set of terms used by biologists to describe certain types of gender mixing are specific to Deer, where they often refer to the unusual antler configurations of these individuals. Such animals are called *velvet-horns* in White-tailed Deer, *cactus bucks* in Mule Deer, *perukes* in Moose and various European deer, and *hummels* in Red Deer. See the animal profiles in part 2 for further information.

51. Benirschke, K. (1981) "Hermaphrodites, Freemartins, Mosaics, and Chimaeras in Animals," in C. R. Austin and R. G. Edwards, eds., *Mechanisms of Sex Differentiation in Animals and Man*, pp. 421–63 (London: Academic Press); Reinboth, R., ed., (1975) *Intersexuality in the Animal Kingdom* (New York: Springer-Verlag); Perry, J. S. (1969) *Intersexuality* (Proceedings of the Third Symposium of the Society for the Study of Fertility), *Journal of Reproduction and Fertility* supplement no. 7 (Oxford: Blackwell Scientific Publications); Armstrong, C. N., and A. J. Marshall, eds., (1964) *Intersexuality in Vertebrates Including Man* (London and New York: Academic Press). For an overview of intersexuality in humans, see Fausto-Stirling, A. (1993) "The Five Sexes: Why Male and Female Are Not Enough," *The Sciences* 33(2):20–24.

52. Graves, G. R. (1996) "Comments on a Probable Gynandromorphic Black-throated Blue Warbler," *Wilson Bulletin* 108:178–80; Stratton, G. E. (1995) "A Gynandromorphic *Schizocosa* (Araneae, Lycosidae)," *Journal of Arachnology* 23:130–33; Patten, M. A. (1993) "A Probable Gynandromorphic Black-throated Blue Warbler," *Wilson Bulletin* 105:695–98; Kumerloeve, H. (1987) "Le gynandromorphisme chez les oiseaux—récapitulation des données connues," *Alauda* 55:1–9; Dexter, R. W. (1985) "Nesting History of a Banded Hermaphroditic Chimney Swift," *North American Bird Bander* 10:39; Hannah-Alava, A. (1960) "Genetic Mosaics," *Scientific American* 202(5):118–30; Kumerloeve, H. (1954) "On Gynandromorphism in Birds," *Emu* 54:71–72.

53. Fredga, K. (1994) "Bizarre Mammalian Sex-Determining Mechanisms," in R. V. Short and E. Balaban, eds., *The Differences Between the Sexes*, pp. 419–31 (Cambridge: Cambridge University Press); Ishihara, M. (1994) "Persistence of Abnormal Females That Produce Only Female Progeny with Occasional Recovery to Normal Females in Lepidoptera," *Researches on Population Ecology* 36:261–69.

54. Moles (Jimenez, R., M. Burgos, L. Caballero, and R. Diaz de la Guardia [1988] "Sex Reversal in a Wild Population of *Talpa occidentalis* [Insectivora, Mammalia]," *Genetical Research* 52[2]:135–40; McVean, G., and L.

D. Hurst [1996] "Genetic Conflicts and the Paradox of Sex Determination: Three Paths to the Evolution of Female Intersexuality in a Mammal," *Journal of Theoretical Biology* 179:199–211); mole voles (Fredga, "Bizarre Mammalian Sex-Determining Mechanisms"); Orang-utan (Dutrillaux et al. 1975; Turleau et al. 1975); Hanuman Langur (Egozcue 1972).

55. Johnsgard, *Arena Birds*, p. 242.

56. On the cassowary mating system, see Crome, F. H. J. (1976) "Some Observations on the Biology of the Cassowary in Northern Queensland," *Emu* 76:8–14.

57. There are actually three distinct, but closely related, species of cassowaries; this genital configuration is based on descriptions of the moruk or Bennett's cassowary (*Casuarius bennettii*) in King, A. S. (1981) "Phallus," in A. S. King and J. McLelland, eds., *Form and Function in Birds*, vol. 2, pp. 107–47 (London: Academic Press). Males and females of a number of other birds, including related flightless species such as Ostriches and Rheas, as well as ducks and geese, also possess a similar genital/anal configuration. Incidentally, the phallus/clitoris of the cassowary (as well as of these other birds) consistently bends to the left when erect (owing to the asymmetrical arrangement of its internal tissues), and males are said to mount females from the left side because of the curvature of their organs. These anatomical and behavioral facts suggest an interesting parallel to Native American beliefs about the left-handedness of (gender-mixing) Bears. Although there are no reports of indigenous New Guinean beliefs about "left-sidedness" in cassowaries, the Arapesh people do represent the cassowary mother figure as the left foot of an ancestral spirit (Tuzin, *The Cassowary's Revenge*, p. 115); the existence of other such connections is worth investigating.

58. Callender and Kochems, "The North American Berdache," pp. 448–49; Roscoe, *Changing Ones*, p. 9; Allen, "Ritualized Homosexuality, Male Power, and Political Organization in North Vanuatu," p. 117; American Bison (Roe 1970:63–64); Savanna (Chacma) Baboon (Marais 1922/1969:205–6); Bielert et al. 1980:4–5); Hooded Warbler (Niven 1993:191 [cf. Lynch et al. 1985:718]); Northern Elephant Seal (Le Boeuf 1974:173); Red Deer (Darling 1937:170); Black-headed Gull (van Rhijn 1985:87, 100); Common Garter Snake (Mason and Crews 1985:59; Mason 1993:264); Bighorn Sheep (Berger 1985:334). "Hypermasculinity" also characterizes (some forms of) male homosexuality in other cultures, most notably contemporary North America. As one recent observer of the gay scene comments, "It's like a very intense male bonding thing . . . it's the ultimate in masculinity. People think faggots are queers; they're fairies. No way. They're more men than anybody, 'cause they're totally homoerotic. How much more masculine can you get?" ("Walter," quoted in Devor, H. [1997] *FTM: Female-to-Male Transsexuals in Society*, p. 504 [Bloomington: Indiana University Press]).

59. Wilson, E. O. (1992) *The Diversity of Life* (Cambridge, Mass.: Belknap/Harvard University Press). For another example of a New Guinean (Foré) indigenous bird taxonomy that nearly matches that of western ornithologists, see Diamond, J. (1966) "Zoological Classification System of a Primitive People," *Science* 151:1102–4.

60. Milton M. R. Freeman, quoted in Mander, J. (1991) *In the Absence of the Sacred: The Failure of Technology and the Survival of the Indian Nations*, p. 259 (San Francisco: Sierra Club Books).

61. Walrus: throat pouches (Fay 1960; Schevill et al. 1966); adoption (Fay 1982; Eley 1978); all-male herds (Miller 1975; 1976); stampedes (Fay and Kelly 1980).

62. Musk-ox (Smith 1976:126–27; Tener 1965:89–90). See also discussion in Freeman, M. M. R. (1984) "New/Old Approaches to Renewable Resource Management in the North," in *Northern Frontier Development—Alaska/Canada Perspectives* (Twenty-Third Annual Meeting of the Western Regional Science Association, Monterey, Calif., February 1984); Freeman, M. M. R. (1986) "Renewable Resources, Economics, and Native Communities," in *Native People and Renewable Resource Management*, 1986 Symposium of the Alberta Society of Professional Biologists (Edmonton: Alberta Society of Professional Biologists); Mander, *In the Absence of the Sacred*, pp. 257–60.

63. Norris, K. S., and K. Pryor (1991) "Some Thoughts on Grandmothers," in K. Pryor and K. S. Norris, eds., *Dolphin Societies: Discoveries and Puzzles*, pp. 287–89 (Berkeley: University of California Press).

64. Feit, H. A. (1986) "James Bay Cree Indian Management and Moral Consideration of Fur-Bearers," in *Native People and Renewable Resource Management*, pp. 49–62; Mander, *In the Absence of the Sacred*, pp. 59–61.

65. Miller, "People, Berdaches, and Left-handed Bears," p. 286. Whether direct knowledge of animal homosexuality (rather than transgender) has contributed to indigenous belief systems remains an open question, although it seems quite likely that observation of a species' same-sex activity may also have been a factor in its status as a shamanic "power animal." Although there are no specific reports of this in the ethnographic literature (which is, however, notoriously incomplete with regard to matters of sexuality, particularly homosexuality), there are several suggestive cases. In a number of Native American cultures, animals are selected as symbolically important for shamanistic practices because their biology and behavior exhibit particularly salient or "unusual" features. In the Pacific Northwest culture region, for example, "animals that shamans relied on as spirit helpers [including shore birds, sea mammals, otters, and Mountain Goats] were those that inhabit border areas of the environment such as the shoreline, the water's surface, or the tops of trees. Their behavior was thought to represent their supernatural ability to move through the different zones of the cosmos"—echoing the shaman's ability to traverse different worlds. (This also corresponds to the well-established ecological principle in Western science that the greatest diversity, flexibility, and environmental richness is to be found in the border zones between major ecosystems, such as the region where forest meets grassland.) This is especially true for the (American) oystercatcher, whose preeminent status as a spirit animal in Tlingit shamanism is based not only on its inhabiting border zones, but

also its furtive behavior and habit of being among the first creatures to sound alarm at the approach of danger (likened to the shaman's function as "guardian" for his or her people) (Wardwell, A. [1996] *Tangible Visions: Northwest Coast Indian Shamanism and Its Art,* pp. 40–43, 96, 239 [New York: Monacelli Press]; for similar observations concerning totemic or shamanic animals in Yup'ik and New Guinean cultures, see Fienup-Riordan, *Boundaries and Passages,* pp. 124, 130–31, and Jorgensen, "Echidna and *Kuyaam,*" pp. 374, 378). Homosexuality is also part of the biological repertoires of many of these species (e.g., various shore birds, sea mammals, and Mountain Goats) or else of their close relatives (e.g., the [Eurasian] Oystercatcher). It is possible, therefore, that observed sexual variance in animals—paralleling the shaman's straddling of sexual boundaries—might also have contributed to the spiritual importance of such creatures. Another interesting example concerns red ants, which feature prominently as shamanic helpers in a number of indigenous cultures of south-central California (all of which, incidentally, recognize two-spirit people). The religious and cultural importance of ants is tied to their powerful medicinal and hallucinogenic properties as well as their use in ritual activities. This includes the extraordinary practice of swallowing large quantities of live ants to induce visions and the acquistion of spirit-animal helpers. Although no homosexual activity has yet been reported for these species (identified as belonging to the genus *Pogonomyrmex*), nor is human gender or sexual variance directly associated with these ant-related beliefs or practices, there are some intriguing clues. Recently, for example, homosexual activity was discovered in a different species of Red Ant (*Formica subpolita*) endemic to the semidesert regions of the western United States (O'Neill 1994:96). Moreover, among the Kawaiisu people (where shamanic ant practices are especially prominent), unusual habits of animals are singled out as a potent spiritual sign, and two-spirit people (who may occupy positions of power, e.g., as chiefs) are reported to be particularly attuned to such animal behaviors (Groark, K. P. [1996] "Ritual and Therapeutic Use of 'Hallucinogenic' Harvester Ants [*Pogonomyrmex*] in Native South-Central California," *Journal of Ethnobiology* 16:1–29; Zigmond, M. [1977] "The Supernatural World of the Kawaiisu," pp. 60–61, 74, in T. C. Blackburn, ed., *Flowers of the Wind: Papers on Ritual, Myth, and Symbolism in California and the Southwest,* pp. 59–95 [Socorro, N.Mex.: Ballena Press]). Once again, it is not unreasonable to suppose that indigenous knowledge or observations of homosexuality (or other sexual variance) in red ants might have been an additional factor in their elevation to religious prominence. Certainly these examples are highly speculative, but they suggest some fascinating connections between animal biology, shamanic practices, and two-spiritedness that deserve further investigation.

66. Roe 1970:63–64 (American Bison); Powers, *Oglala Religion,* p. 58; Wissler, "Societies and Ceremonial Associations in the Oglala Division of the Teton Dakota," p. 92; Dorsey, J. O. (1890) "A Study of Siouan Cults," p. 379, *Bureau of American Ethnology Annual Report* 11:361–544.

67. Haile et al., *Love-Magic and Butterfly People,* p. 163. The term *nádleeh* is also applied to intersexual goats, horses, cattle, and (presumably) other wild game animals. There is also evidence in the Tsistsistas language of possible recognition of transgender in animals: the proper name *Šemoz* is translated as "effeminate bull" (Petter, R. C. [1915] *English-Cheyenne Dictionary,* p. 196 [Kettle Falls, Wash.: Valdo Petter]). This is not, however, related to the Tsistsistas term for human two-spiritness, *hemaneh,* although it is possible that this is the name of a two-spirited person.

68. Reid, B. (1979) "History of Domestication of the Cassowary in Mendi Valley, Southern Highlands Papua New Guinea," *Ethnomedizin/Ethnomedicine* 5:407–32; Reid, B. (1981/82) "The Cassowary and the Highlanders: Present Day Contribution and Value to Village Life of a Traditionally Important Wildlife Resource in Papua New Guinea," *Ethnomedizin/Ethnomedicine* 7:149–240.

69. Gardner, "A Note on the Androgynous Qualities of the Cassowary," p. 143. The Sambia and Arapesh, however, are apparently not aware of the bird's penis (Herdt, *Guardians of the Flutes,* p. 145; Tuzin, *The Cassowary's Revenge,* pp. 80–82). There is no mention of the male cassowary's phallus in the standard Western scientific reference for sexual behavior in this species (Crome 1976), nor mention of the female's phallus/clitoris in the species profiles found in comprehensive ornithological handbooks such as Folch, A. (1992) "Casuariidae (Cassowaries)," in J. del Hoyo, A. Elliott, and J. Sargatal, eds., *Handbook of the Birds of the World,* vol. 1: Ostrich to Ducks, pp. 90–97 (Barcelona: Lynx Edicións); Marchant, S., and P. J. Higgins, eds., (1990) *Handbook of Australian, New Zealand, and Antarctic Birds,* vol. 1, pp. 60–67 (Melbourne: Oxford University Press).

70. Koch, *Year of the Polar Bear,* p. 32; Harington, C. R. (1962) "A Bear Fable?" *The Beaver: A Magazine of the North* 293:4–7; Perry, *World of the Polar Bear,* p. 91; Miller, "People, Berdaches, and Left-Handed Bears," p. 286.

71. Roe 1970 (especially appendix D: "Albinism in Buffalo," pp. 715–28); McHugh 1972: 123–29; Banfield, A. W. F. (1974) *The Mammals of Canada,* p. 405 (Toronto: University of Toronto Press); Berger, J., and M. C. Pearl (1994) *Bison: Mating and Conservation in Small Populations,* p. 34 (New York: Columbia University Press); Pickering, R. B. (1997) *Seeing the White Buffalo* (Denver: Denver Museum of Natural History Press; Boulder: Johnson Books).

72. The poorwill—along with a number of other birds such as the related common nighthawk and other goatsuckers, as well as some hummingbirds—also sometimes enters daily or nocturnal periods of torpor that typically last less than 24 hours. The poorwill, however, is the only species that exhibits extended periods of torpor. See Jaeger, E. C. (1949) "Further Observations on the Hibernation of the Poor-will," *Condor* 51:105–9; Jaeger, E. C. (1948) "Does the Poor-will 'Hibernate'?" *Condor* 50:45–46; Brigham, R. M. (1992) "Daily Torpor in a Free-Ranging Goatsucker, the Common Poorwill (*Phalaenoptilus nuttallii*)," *Physiological Zoology* 65:457–72; Kissner, K. J., and R. M. Brigham (1993) "Evidence for the Use of Torpor by Incubating and Brooding Common Poorwills *Phalaenoptilus nuttallii,*" *Ornis Scandinivica* 24:333–34; Csada, R. D., and

R. M. Brigham (1994) "Reproduction Constrains the Use of Daily Torpor by Free-ranging Common Poor-wills (*Phalaenoptilus nuttallii*)," *Journal of Zoology, London* 234:209–16; Brigham, R. M., K. H. Morgan, and P. C. James (1995) "Evidence That Free-Ranging Common Nighthawks May Enter Torpor," *Northwestern Naturalist* 76:149–50.

73. Russell, F. (1975) *The Pima Indians*, p. x (Tucson: University of Arizona Press); Grant, V., and K. A. Grant (1983) "Behavior of Hawkmoths on Flowers of *Datura meteloides*," *Botanical Gazette* 144:280–84; Nabham, G. P., and S. St. Antoine (1993) "The Loss of Floral and Faunal Story: The Extinction of Experience," in S. R. Kellert and E. O. Wilson, eds., *The Biophilia Hypothesis*, pp. 229–50 (Washington, D.C.: Island Press).

74. Bulmer, R. (1968) "Worms That Croak and Other Mysteries of Karam [sic] Natural History," *Mankind* 6:621–39. Among the worm species identified as particularly "vocal" is *Pheretima musica* of Indonesia. Bul-mer points out, however, that frogs rather than earthworms are the more likely source of the actual sounds associated by the Kalam with worms.

75. Bauer, A. M., and A. P. Russell (1987) "*Hoplodactylus delcourti* (Reptilia: Gekkonidae) and the *Kawekaweau* of Maori Folklore," *Journal of Ethnobiology* 7:83–91.

76. The plant, identified as *Ligusticum porteri*, is widely used as an indigenous herbal medicine throughout the southwestern United States and Mexcio, where it is known by various names including *oshá, chuchupa(s)te*, and *smelly root*. Sigstedt, S. (1990) "Bear Medicine: 'Self-Medication' by Animals," *Journal of Ethnobiology* 10:257; Clayton, D. H., and N. D. Wolfe (1993) "The Adaptive Significance of Self-Medication," *Trends in Ecology and Evolution* 8:60–63; Rodriguez, E., and R. Wrangham (1993) "Zoopharmacognosy: The Use of Medicinal Plants by Animals," in K. R. Downum, J. T. Romeo, and H. A. Stafford, eds, *Phytochemical Poten-tial of Tropical Plants*, pp. 89–105, Recent Advances in Phytochemistry vol. 27 (New York: Plenum Press); Beck, J. J., and F. R. Stermitz (1995) "Addition of Methyl Thioglycolate and Benzylamine to (Z)-Ligustilide, a Bioactive Unsaturated Lactone Constituent of Several Herbal Medicines," *Journal of Natural Products* 58:1047–55; Linares, E., and R. A. Bye Jr. (1987) "A Study of Four Medicinal Plant Complexes of Mexico and Adjacent United States," *Journal of Ethnopharmacology* 19:153–83.

77. Arima, E. Y. (1983) *The West Coast People: The Nootka of Vancouver Island and Cape Flattery*, British Colum-bia Provincial Museum Special Publication no. 6, pp. 2, 102 (Victoria: British Columbia Provincial Mu-seum). This culture (like most other indigenous cultures) was "interrupted" relatively recently, of course, by the disease, genocide, and cultural suppression brought on by European immigrants—forces that have nev-ertheless failed to obliterate these people or their traditions.

78. As some researchers have pointed out, this is largely because most Western scientists consider traditional aboriginal knowledge to be "unscientific" and difficult to separate from its cultural context (which often in-cludes "fantastic" or "mythological" elements that are seemingly at odds with orthodox Western scientific principles). For further discussion of this view as well as the potential for collaboration between indigenous and Western scientists, see Pearson, D., and the Ngaanyatjarra Council (1997) "Aboriginal Involvement in the Survey and Management of Rock-Wallabies," *Australian Mammalogy* 19:249–56.

79. Dumbacher, J. P., B. M. Beeler, T. F. Spande, H. M. Garrafo, and J. W. Daly (1992) "Homobatrachotoxin in the Genus *Pitohui*: Chemical Defense in Birds?" *Science* 258:799–801; Dumbacher, J. P. (1994) "Chemical De-fense in New Guinean Birds," *Journal für Ornithologie* 135:407; Majnep, I. S., and R. Bulmer (1977) *Birds of My Kalam Country* (Mnmon Yad Kalam Yakt), p. 103 (Aukland: Aukland University Press); Dumbacher, J. P., and S. Pruett-Jones (1996) "Avian Chemical Defense," in V. Nolan Jr., and E. D. Ketterson, eds., *Current Or-nithology*, vol. 13, pp. 137–74 (New York: Plenum Press). See also the inclusion of indigenous New Guinean observations on the courtship behaviors of Birds of Paradise in Frith, C. B., and D. W. Frith (1997) "Courtship and Mating of the King of Saxony Bird of Paradise *Pteridophora alberti* in New Guinea with Comment on Their Taxonomic Significance," pp. 186, 190–91, *Emu* 97:185–93.

80. Stephenson, R. O., and R. T. Ahgook (1975) "The Eskimo Hunter's View of Wolf Ecology and Behavior," in M. W. Fox, ed., *The Wild Canids: Their Systematics, Behavioral Ecology, and Evolution*, pp. 286–91 (New York: Van Nostrand Reinhold). See also the inclusion of Inuit observations on the behavior and distribution of Orcas in Reeves and Mitchell (1988).

81. From a letter written to Dean Hamer and excerpted (anonymously) in his book *The Science of Desire: The Search for the Gay Gene and the Biology of Behavior*, p. 213 (New York: Simon and Schuster, 1994).

82. Steward, "Coyote and Tehoma," p. 160.

83. Beston, H. (1928) *The Outermost House: A Year of Life on the Great Beach of Cape Cod*, p. 25 (New York: Rine-hart); Bey, H. (1994) *Immediatism*, p. 1 (Edinburgh and San Francisco: AK Press).

84. R. Pirsig, quoted in Carse, *Finite and Infinite Games*.

85. Ibid., p. 127.

86. Worster, D. (1990) "The Ecology of Chaos and Harmony," *Environmental History Review* 14:1–18.

87. Bunyard P., and E. Goldsmith, eds., (1989) "Towards a Post-Darwinian Concept of Evolution," in P. Bunyard and E. Goldsmith, eds., *Gaia and Evolution*, Proceedings of the Second Annual Camelford Conference on the Implications of the Gaia Thesis, pp. 146–51 (Camelford: Wadebridge Ecological Centre). This school of thought is also sometimes called "post-neo-Darwinian" evolution, to emphasize its divergence from other, less recent, evolutionary theorizing that has occurred subsequent to Darwin (since the latter is generally characterized as "neo-Darwinian").

88. Ho, M.-W., and P. T. Saunders (1984) "Pluralism and Convergence in Evolutionary Theory" and preface, in M.-W. Ho and P. T. Saunders, eds., *Beyond Neo-Darwinism: An Introduction to the New Evolutionary Para-digm*, pp. ix-x, 3–12 (London: Academic Press).

89. For further discussion and exemplification, see Ho, M.-W., P. Saunders, and S. Fox (1986) "A New Paradigm for Evolution," *New Scientist* 109(1497):41–43; and the numerous articles in Ho and Saunders, *Beyond Neo-Darwinism*. For a more recent summary of some new ideas emerging in post-neo-Darwinian thought, see Wieser, W. (1997) "A Major Transition in Darwinism," *Trends in Ecology and Evolution* 12:367–70.

90. See, for example, the numerous contributors to Barlow, C. (1994) *Evolution Extended: Biological Debates on the Meaning of Life* (Cambridge, Mass.: MIT Press).

91. Wilson, E. O. (1978) *On Human Nature*, p. 201 (Cambridge, Mass.: Harvard University Press).

92. von Bertalanffy, L. (1969) "Chance or Law," in A. Koestler and R. M. Smithies, eds., *Beyond Reductionism* (London: Hutchinson); Lewontin, R., and S. J. Gould (1979) "The Spandrels of San Marco and the Pan-glossian Paradigm: A Critique of the Adaptationist Programme," *Proceedings of the Royal Society of London* Series B 205:581–98; Hamilton, M. (1984) "Revising Evolutionary Narratives: A Consideration of Evolutionary Assumptions About Sexual Selection and Competition for Mates," *American Anthropologist* 86:651–62; Levins, R., and R. C. Lewontin (1985) *The Dialectical Biologist* (Cambridge, Mass.: Harvard University Press); Rowell, T. (1979) "How Would We Know If Social Organization Were Not Adaptive?" in I. Bernstein and E. Smith, eds., *Primate Ecology and Human Origins*, pp. 1–22 (New York: Garland). See also the discussion in Ho et al., "A New Paradigm for Evolution," and in Ho and Saunders, *Beyond Neo-Darwinism*.

93. May, R. (1989) "The Chaotic Rhythms of Life," *New Scientist* 124(1691):37–41; Ford quote in Gleick, J. (1987) *Chaos: Making a New Science*, p. 314 (New York: Viking); Ferrière, R., and G. A. Fox (1995) "Chaos and Evolution," *Trends in Ecology and Evolution* 10:480–85; Robertson, R., and A. Combs, eds., (1995) *Chaos Theory in Psychology and the Life Sciences* (Mahwah, N.J.: Lawrence Erlbaum Associates); Degn, H., A. V. Holden, and L. F. Olsen, eds., (1987) *Chaos in Biological Systems* (New York: Plenum Press); see also Abraham, R. (1994) *Chaos, Gaia, Eros: A Chaos Pioneer Uncovers the Three Great Streams of History* (San Francisco: HarperSanFrancisco).

94. Alados, C. L., J. M. Escos, and J. M. Emlen (1996) "Fractal Structure of Sequential Behavior Patterns: An Indicator of Stress," *Animal Behavior* 51:437–43; Cole, B. J. (1995) "Fractal Time in Animal Behavior: The Movement Activity of *Drosophila*," *Animal Behavior* 50:1317–24; Erlandsson, J., and V. Kostylev (1995) "Trail Following, Speed, and Fractal Dimension of Movement in a Marine Prosobranch, *Littorina littorea*, During a Mating and a Non-Mating Season," *Marine Biology* 122:87–94; Solé, R. V., O. Miramontes, and B. C. Goodwin (1993) "Oscillations and Chaos in Ant Societies," *Journal of Theoretical Biology* 161:343–57; Fourcassie, V., D. Coughlin, and J. F. A. Traniello (1992) "Fractal Analysis of Search Behavior in Ants," *Naturwissenschaften* 79:87–89; Camazine, S. (1991) "Self-Organizing Pattern Formation on the Combs of Honey Bee Colonies," *Behavioral Ecology and Sociobiology* 28:61–76; Cole, B. J. (1991) "Is Animal Behavior Chaotic? Evidence from the Activity of Ants," *Proceedings of the Royal Society of London*, Series B 244:253–59.

95. Gleick, *Chaos: Making a New Science*; Botkin, D. B. (1990) *Discordant Harmonies: A New Ecology for the Twenty-first Century* (New York: Oxford University Press).

96. Savalli, U. M. (1995) "The Evolution of Bird Coloration and Plumage Elaboration: A Review of Hypotheses," in D. M. Power, ed., *Current Ornithology*, vol. 12, pp. 141–90 (New York: Plenum Press).

97. For a promising direction of research in this regard, see the proposal that a wide range of animal coat patterns may be generable from a single, simple mathematical equation (based on the work of Alan Turing): Murray, J. D. (1988) "How the Leopard Gets Its Spots," *Scientific American* 258(3):80–87.

98. Goerner, S. (1995) "Chaos, Evolution, and Deep Ecology," in Robertson and Combs, *Chaos Theory in Psychology and the Life Sciences*, pp. 17–38; Worster, "The Ecology of Chaos and Harmony," p. 14; Haldane, J. B. S. (1928) *Possible Worlds and Other Papers*, p. 298 (New York: Harper & Brothers).

99. Goerner, "Chaos, Evolution, and Deep Ecology," p. 24.

100. Lovelock, J. E. (1988) "The Earth as a Living Organism," in E. O. Wilson, ed., *BioDiversity*, pp. 486–489 (Washington, D.C.: National Academy Press).

101. Lovelock, J. E. (1979) *Gaia: A New Look at Life on Earth* (Oxford: Oxford University Press); Margulis, L., and D. Sagan (1986) *Microcosmos: Four Billion Years of Microbial Evolution* (New York: Summit Books); Bunyard, P., and E. Goldsmith, eds., (1988) *Gaia: The Thesis, the Mechanisms, and the Implications*, Proceedings of the First Annual Camelford Conference on the Implications of the Gaia Hypothesis (Camelford: Wadebridge Ecological Centre); Lovelock, J. E. (1988) *The Ages of Gaia: A Biography of Our Living Earth* (New York: W. W. Norton and Company); Bunyard and Goldsmith, *Gaia and Evolution*; Schneider, S. H., and P. J. Boston, eds., (1991) *Scientists on Gaia*, Proceedings of the American Geophysical Union's Annual Chapman Conference (Cambridge, Mass.: MIT Press); Williams, G. R. (1996) *The Molecular Biology of Gaia* (New York: Columbia University Press).

102. Lambert, D., and R. Newcomb (1989) "Gaia, Organisms, and a Structuralist View of Nature," in Bunyard and Goldsmith, *Gaia and Evolution*, pp. 75–76.

103. Lovelock, "The Earth as a Living Organism," p. 488.

104. Tilman, D., and J. A. Downing (1994) "Biodiversity and Stability in Grasslands," *Nature* 367:363–65.

105. Technically, this group comprises 13 distinct families of birds, combined into a higher-level grouping (or "suborder") known as the Charadrii. For information on the heterosexual mating systems in these families, see del Hoyo, J., A. Elliott, and J. Sargatal, eds., (1996) *Handbook of the Birds of the World*, vol. 3: Hoatzin to Auks, pp. 276–555 (Barcelona: Lynx Edicións); Paton, P. W. C. (1995) "Breeding Biology of Snowy Plovers at Great Salt Lake, Utah," *Wilson Bulletin* 107:275–88; Nethersole-Thompson, D., and M. Nethersole-Thompson (1986) *Waders: Their Breeding, Haunts, and Watchers* (Calton: T. and A. D. Poyser); Pitelka, F. A., R. T. Holmes, and S. F. MacLean Jr. (1974) "Ecology and Evolution of Social Organization in Arctic Sandpipers," *American*

Zoologist 14:185–204. For details of species involving homosexual activity, see the profiles and references in part 2.

106. Carranza, J., S. J. Hidalgo de Trucios, and V. Ena (1989) "Mating System Flexibility in the Great Bustard: A Comparative Study," *Bird Study* 36:192–98. For further discussion of the possible benefits provided by behavioral plasticity, and variable sexual behaviors as a response to environmental or social variability, see Komers, P. E. (1997) "Behavioral Plasticity in Variable Environments," *Canadian Journal of Zoology* 75:161–69; Carroll, S. P., and P. S. Corneli (1995) "Divergence in Male Mating Tactics Between Two Populations of the Soapberry Bug: II. Genetic Change and the Evolution of a Plastic Reaction Norm in a Variable Social Environment," *Behavioral Ecology* 6:46–56; Rodd, F. H., and M. B. Sokolowski (1995) "Complex Origins of Variation in the Sexual Behavior of Male Trinidadian Guppies, *Poecilia reticulata*: Interactions Between Social Environment, Heredity, Body Size, and Age," *Animal Behavior* 49:1139–59. For an analysis of nonbreeding as an adaptive response to environmental variability, see, for example, Aebischer and Wanless 1992 (Shag).

107. Golden Plover (Nethersole-Thompson and Nethersole-Thompson 1961:207–8 [on the possibility that "disruption" of heterosexual pairing in related species of plovers is due to late snow-melts, see Johnson, O. W., P. M. Johnson, P. L. Bruner, A. E. Bruner, R. J. Kienholz, and P. A. Brusseau (1997) "Male-Biased Breeding Ground Fidelity and Longevity in American Golden-Plovers," *Wilson Bulletin* 109:348–351]); Grizzly Bear (Craighead et al. 1995:216–17; J. W. Craighead, personal communication); Ostrich (Sauer 1972:717); Ring-billed and California Gulls (Conover et al. 1979); Rhesus Macaque (Fairbanks et al. 1977:247–48); Stumptail Macaque and other primates (Bernstein 1980:32; Vasey, "Homosexual Behavior in Primates," pp. 193–94). See also Hand (1985) for the suggestion that environmental "stresses" may call forth "plastic" social and sexual responses (such as homosexual pairing) in Laughing Gulls and other species. As noted in chapter 4, the occasional association of homosexuality with "unusual" ecological (or other) conditions is typically interpreted by scientists in a negative way, as evidence of a "disturbed" biological or social order rather than of a flexible response to (or synergy with) ongoing environmental flux. Moreover, the evidence for many of these cases—while intriguing—is anecdotal at best, and more systematic investigation will be necessary before any conclusions or even further speculations can be put forward in this regard.

108. Japanese Macaque (Eaton 1978:55–56). See also Vasey's ("Homosexual Behavior in Primates," p. 196) suggestion that homosexuality may not be adaptive itself, but may represent a neutral behavioral "by-product" of some other trait that is adaptive, such as behavioral plasticity. For more on cultural and protocultural phenomena in animals, see chapter 2.

109. Bataille, G. (1991) *The Accursed Share*, vol. 1, p. 33 (New York: Zone Books).

110. Gleick, *Chaos: Making a New Science*, pp. 4, 221, 306.

111. Wilson, *Diversity of Life*, pp. 201, 210.

112. Catchpole, C. K., and P. J. B. Slater (1995) *Bird Song: Themes and Variations*, pp. 187, 189 (Cambridge: Cambridge University Press).

113. Eberhard, W. G. (1996) *Female Control: Sexual Selection by Cryptic Female Choice*, pp. 55, 81 (Princeton: Princeton University Press); Eberhard, W. G. (1985) *Sexual Selection and Animal Genitalia*, p. 17 (Cambridge, Mass.: Harvard University Press).

114. Weldon, P. J., and G. M. Burghardt (1984) "Deception Divergence and Sexual Selection," *Zeitschrift für Tierpsychologie* 65:89–102.

115. Bataille, *Accursed Share*.

116. For example, it is often erroneously thought that indigenous "subsistence" cultures are characterized by a scarcity of resources and an arduous, even desperate, struggle for survival, in contrast to modern industrial societies that have an abundance of resources and ample leisure time—when in fact the actual circumstances are usually reversed. Industrial society is essentially a system of enforced scarcity, in which basic necessities such as housing, food, and shelter are denied to the vast majority of people except in exchange for labor that occupies 40–60 hours a week of an adult's time. In contrast, detailed studies of the economies of a number of hunter-gatherer societies (including those living in the most "arduous" of environments such as the deserts of southern Africa) have revealed a "workweek" of only 15–25 hours for *all* adults (not just a privileged few). So abundant are the basic resources, minimal the material needs, and equitable the forms of social organization (which make resources freely available to all) that the remainder of people's time in such societies is occupied by "leisure activities." For further discussion, see Sahlins, M. (1972) *Stone Age Economics* (Chicago: Aldine Publishing); Lee, R. B. (1979) *The !Kung San: Men, Women, and Work in a Foraging Society* (Cambridge: Cambridge University Press); Mander, "Lessons in Stone-Age Economics," chapter 14 in *In the Absence of the Sacred*.

117. cummings, e. e. (1963) *Complete Poems 1913–1962*, p. 749 (New York and London: Harcourt Brace Jovanovich).

118. For more on the "problem" of sexual reproduction, see Dunbrack, R. L., C. Coffin, and R. Howe (1995) "The Cost of Males and the Paradox of Sex: An Experimental Investigation of the Short-Term Competitive Advantages of Evolution in Sexual Populations," *Proceedings of the Royal Society of London*, Series B 262:45–49; Collins, R. J. (1994) "Artificial Evolution and the Paradox of Sex," in R. Parton, ed., *Computing With Biological Metaphors*, pp. 244–63 (London: Chapman & Hall); Slater, P. J. B., and T. R. Halliday, eds., (1994) *Behavior and Evolution* (Cambridge: Cambridge University Press); Michod, R. E., and B. R. Levin, eds., (1987) *The Evolution of Sex: An Examination of Current Ideas* (Sunderland, Mass.: Sinauer Associates); Alexander, R. D., and D. W. Tinkle (1981) *Natural Selection and Social Behavior: Recent Research and New Theory* (New York: Chiron Press); Daly, M. (1978) "The Cost of Mating," *American Naturalist* 112:771–74.

119. In fact, a number of zoologists have independently characterized homosexual (and alternate heterosexual) activities as "energetically expensive," "wasteful," "inefficient," or "excessive." See, for example, Fry et al. (1987:40) on same-sex pairing in Western Gulls; Schlein et al. (1981:285) on homosexual courtship in Tsetse and House Flies; Moynihan (1990:17) on noncopulatory mounting in Blue-bellied Rollers; Thomas et al. (1979:135) on the "wasting" of sperm during male homosexual interactions in Little Brown Bats; Møller (1987:207–8) on the "communal displays" (group courtship and promiscuous sexual activity) of House Sparrows; Ens (1992:72) on the "spectacular ceremonies" among nonbreeding Oystercatchers and Black-billed Magpies that involve the expenditure of "vast amounts of energy"; J. D. Paterson in Small (p. 92), on the "excessive" nonreproductive heterosexual activity of female primates that entails considerable "inefficiency" and "energy wastage" (Small, M. F. [1988] "Female Primate Sexual Behavior and Conception: Are There Really Sperm to Spare?" *Current Anthropology* 29:81–100); and Miller et al. (1996:468) on the "excess" sexual selection involved in the violent, often nonreproductive heterosexual matings between different species of fur seals. For an early characterization of some animal behaviors being motivated by an "excess" of sexual (and other) drives, see Tinbergen, N. (1952) "'Derived' Activities: Their Causation, Biological Significance, Origin, and Emancipation During Evolution," especially pp. 15, 24, *Quarterly Review of Biology* 27:1–32. For an early, nonscientific theory of (male) homosexuality as the expression of natural "superabundance," "excess," and "prodigality," see Gide, A. (1925/1983) *Corydon*, especially pp. 41, 48, 68 (New York: Farrar Straus Giroux).
120. von Hildebrand, M. (1988) "An Amazonian Tribe's View of Cosmology," in Bunyard and Goldsmith, *Gaia: The Thesis, the Mechanisms, and the Implications*, pp. 186–195.
121. Bataille, *Accursed Share*, vol. 1, p. 28.
122. Wilson, E. O., *Diversity of Life*, pp. 43, 350ff.
123. Abraham, *Chaos, Gaia, Eros*, p. 63. For discussion of the possibility that fractal or chaotic patterns may underlie some Native American and New Guinean cultures, see Bütz, M. R., E. Duran, and B. R. Tong (1995) "Cross-Cultural Chaos," in Robertson and Combs, *Chaos Theory in Psychology and the Life Sciences*, pp. 319–30; Wagner, R. (1991) "The Fractal Person," in Godelier and Strathern, *Big Men and Great Men*, pp. 159–73.
124. See, for example, Ehrlich, P. R. (1988) "The Loss of Diversity: Causes and Consequences," in Wilson, *BioDiversity*, pp. 21–27; Takacs, D. (1996) *The Idea of Biodiversity: Philosophies of Paradise*, pp. 254–70 (Baltimore and London: Johns Hopkins University Press); Wilson, *On Human Nature*. For a recent overview of the "spiritualization" of science, and the controversy it has engendered, see Easterbrook, G. (1997) "Science and God: A Warming Trend?" *Science* 277:890–93.
125. Nelson, R. (1993) "Searching for the Lost Arrow: Physical and Spiritual Ecology in the Hunter's World," in Kellert and Wilson, *The Biophilia Hypothesis*, pp. 202–28; Nabham and St. Antoine, "The Loss of Floral and Faunal Story"; Diamond, J. (1993) "New Guineans and Their Natural World," in Kellert and Wilson, *The Biophilia Hypothesis*, pp. 251–71.
126. Chadwick 1983:15 (Mountain Goat); Grumbie, R. E. (1992) *Ghost Bears: Exploring the Biodiversity Crisis*, pp. 69–71 (Washington, D.C.: Island Press); Soulé, M. E. (1988) "Mind in the Biosphere; Mind of the Biosphere," in Wilson, *BioDiversity*, pp. 465–69.
127. Goldsmith, E. (1989) "Gaia and Evolution," in Bunyard and Goldsmith, *Gaia and Evolution*, p. 8; Bunyard, P. (1988) "Gaia: Its Implications for Industrialized Society," in Bunyard and Goldsmith, *Gaia: The Thesis, the Mechanisms, and the Implications*, pp. 218–20.
128. LaPena, F. (1987) *The World Is a Gift* (San Francisco: Limestone Press); see also Theodoratus, D. J., and F. LaPena (1992) "Wintu Sacred Geography," in L. J. Bean, ed., *California Indian Shamanism*, pp. 211–25 (Menlo Park, Calif.: Ballena Press).
129. Littlebird, L. (1988) "Cold Water Spirit," in Wilson, *BioDiversity*, pp. 476–80.
130. Miller, "People, Berdaches, and Left-Handed Bears," pp. 278–80; Lange, C. H. (1959) *Cochiti: A New Mexico Pueblo, Past and Present*, pp. 135, 256 (Austin: University of Texas Press). On the *kokwimu* or two-spirit, see Gutiérrez, R. A. (1991) *When Jesus Came, the Corn Mothers Went Away: Marriage, Sexuality, and Power in New Mexico, 1500–1846*, pp. 33–35 (Stanford: Stanford University Press); Parsons, E. C. (1923) "Laguna Genealogies," p. 166, *Anthropological Papers of the American Museum of Natural History* 19:133–292; Parsons, E. C. (1918) "Notes on Acoma and Laguna," pp. 181–82, *American Anthropologist* 20:162–86.
131. Although the exact species is not named in Littlebird's story, it is possible to identify it with a fair degree of certainty based on a number of characteristics mentioned in the story, including its appearance (it has dark gray lines running down a green back); habits (it lifts its chest up and down rhythmically while moving its throat, is a swift runner, frequents dry and dusty areas, and seeks shelter under branches of tumbleweed); and location (west-central New Mexico). Herpetologist Donald Miles has confirmed (personal communication) that this is most likely a species of Whiptail Lizard, probably the Desert Grassland Whiptail (*Cnemidophorus uniparens*). For parthenogenesis and homosexual copulation in this and other Whiptail Lizards, see the references for these species in the appendix.
132. Anguksuar (Richard LaFortune) (1997) "A Postcolonial Colonial Perspective on Western (Mis)Conceptions of the Cosmos and the Restoration of Indigenous Taxonomies," p. 219, in Jacobs et al., *Two-Spirit People*, pp. 217–22.
133. Barlow, *Evolution Extended*, pp. 292–93, 298, 300.
134. Harjo, J. (1988) "The Woman Hanging from the Thirteenth Floor Window," in C. Morse and J. Larkin, eds., *Gay & Lesbian Poetry in Our Time*, pp. 179–81 (New York: St. Martin's Press); Harjo, J. (1990) *In Mad Love and War* (Middletown, Conn.: Wesleyan University Press); Harjo, J. (1994) *The Woman Who Fell from the Sky*

(New York: W. W. Norton and Company); Harjo, J. (1996) *The Spiral of Memory: Interviews* (Ann Arbor: University of Michigan Press), pp. 28, 57, 68, 108, 115–17, 126, 129; Randall, M. (1990) "Nothing to Lose," *Women's Review of Books* 7:17–18.

135. Geist, V. (1996) *Buffalo Nation: History and Legend of the North American Bison,* p. 55 (Stillwater, Minn.: Voyageur Press). The picture shows a three-year-old male mounting another three-year-old male; the sex and age of the mountee can be discerned from the shape and size of its horns and head, and the presence of a prominent preputial (penis) tuft (D. F. Lott, personal communication).

136. Brant, B. (1994) "Anodynes and Amulets," in Brant, *Writing as Witness,* pp. 25–34; Shaw, C. (1995) "A Theft of Spirit?" *New Age Journal* (July/August 1995):84–92.

137. Sørum, A. (1984) "Growth and Decay: Bedamini Notions of Sexuality," in Herdt, *Ritualized Homosexuality in Melanesia,* pp. 318–36.

138. Schlesier, *The Wolves of Heaven,* pp. 13–14, 66–67, 190.

139. Nataf, Z. I. (1996) *Lesbians Talk Transgender,* p.55 (London: Scarlet Press); with quotations from Smith, S. A. (1993) "Morphing, Materialism, and the Marketing of *Xenogenesis,*" *Genders* 18:67–86.

140. cummings, e. e. *Complete Poems,* p. 809.

141. Monarch Butterfly (Leong et al. 1995; Leong 1995; Urquhart 1987; Tilden 1981; Rothschild 1978; Malcolm, S. B., and M. P. Zalucki, eds., [1993] *Biology and Conservation of the Monarch Butterfly,* Science Series no. 38 [Los Angeles: Natural History Museum of Los Angeles County]).

142. Bey, H. (1991) *T.A.Z.: The Temporary Autonomous Zone, Ontological Anarchy, Poetic Terrorism,* p. 137 (New York: Autonomedia).

143. The image of locating a conceptual position on the trajectory between distinct but related "points" is borrowed from Hakim Bey (*Immediatism,* p. 32).

144. MacNeice, L. (1966) "Snow," in *Collected Poems* (Oxford: Oxford University Press).

145. Bey, *T.A.Z.,* pp. 23, 55.

Credits and Permissions

Grateful acknowledgment is made to the publishers, authors, and photographers listed below for permission to reproduce material copyrighted or controlled by them. Full citations for sources listed below are to be found in the bibliographies of each of the relevant species profiled in part 2.

"Snow" by Louis MacNeice, from *Collected Poems* (Faber & Faber). Reprinted with the permission of the Estate of Louis MacNeice. © 1966 by the Estate of Louis MacNeice.

The lines from "now that, more nearest even than your fate," copyright © 1962, 1990, 1991 by the Trustees for the E. E. Cummings Trust, from *Complete Poems: 1904–1962* by e.e. cummings, edited by George J. Firmage. Reprinted by permission of Liveright Publishing Corporation.

"Pied Beauty," by Gerard Manley Hopkins, from *Poems of Gerard Manley Hopkins* (4th ed., 1967; Oxford University Press).

Page 13 (Superb Lyrebird): Photo by L. H. Smith. Printed with the permission of L. H. Smith. © 1999 by L. H. Smith.

Pages 16, 106, 272 (Bonobo): Photos by Frans de Waal. Reprinted from de Waal (1995: Fig. 3.2, p. 43; Fig. 3.4, p. 45; Fig. 3.5, p. 46) with the permission of the University of Chicago Press and Frans de Waal. © 1995 by The University of Chicago. All rights reserved.

Page 19 (Giraffe): Photo by Bristol Foster. Printed with the permission of J. Bristol Foster. © 1999 by J. Bristol Foster.

Page 21 (Canada Goose): Photo by Arthur A. Allen. Reprinted from Allen (1934: Plate XI-A, p. 196) with the permission of David G. Allen and *The Auk*/American Ornithologists' Union. © 1934 by *The Auk*/American Ornithologists' Union, Arthur A. Allen, and David G. Allen.

Pages 24, 525 (Flamingo): Photos by Ron Entius. Printed with the permission of Ron Entius. © 1999 by Ron Entius.

Page 39 (Bighorn Sheep): Photo by Len Rue Jr. Printed with the permission of Leonard Rue Enterprises. © 1999 by Leonard Lee Rue III and Len Rue Jr.

Pages 50, 210, 371 (Walrus): Photos by Edward H. Miller. Reprinted from Miller (1975: Fig. 11, p. 595; Fig. 12, p. 596; Fig. 14, p. 596) with the permission of NRC Research Press. © 1975 by the National Research Council of Canada.

Pages 58, 492, 493 (Mallard Duck): Photos by Friedrich Schutz. Reprinted from Schutz (1965: Fig. 3, p. 443; Figs. 4–5, p. 446) with the permission of Springer-Verlag. © 1965 by Springer-Verlag.

Pages 70, 286 (Orang-utan): Photos by H. D. Rijksen. Reprinted from Rijksen (1978: Fig. 128b, p. 263; Fig. 129a, p. 264) with the permission of H. D. Rijksen. © 1978 by H. D. Rijksen.

Page 325 (Savanna Baboon): Photo by Timothy W. Ransom. Reprinted from Ransom (1981: Plate 5-A[3], p. 84) with the permission of Associated University Presses. © 1981 by Associated University Presses, Inc.

Page 326 (Savanna Baboon): Photo by K. R. L. Hall. Reprinted from Hall (1962: Plate 1, Fig. 2, p. 326) with the permission of Cambridge University Press. © 1962 by the Zoological Society of London.

Page 331 (Squirrel Monkey): Photo by Frank DuMond. Reprinted from DuMond (1968: Fig. 12, p. 125) with the permission of Academic Press, Inc. © 1968 by Academic Press, Inc.

Page 341 (Boto): Photo by James N. Layne. Reprinted from Layne and Caldwell (1964: Plate 1, Fig. 2, p. 108) with the permission of James N. Layne. © 1964 by James N. Layne.

Page 345 (Bottlenose/Atlantic Spotted Dolphins): Photo by Wild Dolphin Project. Reprinted from Herzing and Johnson (1997: Fig. 8, p. 92) with the permission of Denise Herzing and *Aquatic Mammals*. © 1997 by Wild Dolphin Project, Inc.

Page 379 (White-tailed Deer): Photo by Len Rue Jr. Printed with the permission of Leonard Rue Enterprises. © 1999 by Leonard Rue III and Len Rue Jr.

Page 384 (Wapiti/Red Deer): Reprinted from Guiness et al. (1971: Plate 2, Fig. 4, p. 435) with the permission of the *Journal of Reproduction and Fertility*. © 1971 by the *Journal of Reproduction and Fertility*.

Page 392 (Giraffe): Photo by Stephen Maka. Printed with the permission of Stephen G. Maka. © 1999 by Stephen G. Maka.

Page 406 (Bighorn Sheep): Photo by Valerius Geist. Reprinted from Geist (1975:p.106) with the permission of Valerius Geist. © 1975 by Valerius Geist.

Page 428 (African Elephant): Photo by Mitch Reardon. Printed with the permission of Photo Researchers, Inc. © 1999 by Photo Researchers, Inc.

Page 442 (Grizzly Bear): Photo by John J. Craighead. Printed with the permission of Craighead Wildlife-Wildlands Institute. © 1999 by John Craighead.

Page 446 (Spotted Hyena): Photos by Gus Mills. Reprinted from Mills (1990: Fig. 5.4, p. 173; Fig. 7.6, p. 243) with the permission of M. G. L. Mills. © 1990 by Gus Mills.

Page 480 (above) (Greylag Goose): Photo by Robert Huber. Reprinted from Huber and Martys (1993: Fig. 2, p. 158) with the permission of Michael Martys. © 1993 by Michael Martys.

Page 480 (below) (Greylag Goose): Photo by Michael Martys. Printed with the permission of Michael Martys. © 1999 by Michael Martys.

Page 530 (Ruff): Photos by Horst Scheufler. Reprinted from Scheufler and Stiefel (1985: Fig. 23a–b, p. 57) with the permission of Westarp-Wissenschaften. © 1985 by Westarp-Wissenschaften/A. Ziemsen Verlag.

Page 549 (Western Gull): Photo by George L. Hunt Jr. Printed with the permission of George L. Hunt Jr. © 1999 by George L. Hunt Jr.

Page 567 (Guianan Cock-of-the-Rock): Photo by Pepper W. Trail. Reprinted from Trail (1985: Fig. 16, p. 240) with the permission of Pepper W. Trail. © 1985 by Pepper W. Trail.

Page 618 (above) (Superb Lyrebird): Photo by L. H. Smith. Reprinted from Smith (1988) with the permission of L. H. Smith. © 1988 by L. H. Smith.

Page 618 (below) (Superb Lyrebird): Photo by L. H. Smith. Reprinted from Smith (1968: p. 92) with the permission of L. H. Smith. © 1968 by L. H . Smith.

Page 623 (Greater Rhea): Photos by Juan C. Reboreda. Reprinted from Fernández and Reboreda (1995: Fig. 1B–C, p. 323) with the permission of Munksgaard International Publishers Ltd., Gustavo J. Fernández, and Juan C. Reboreda. © 1995 by the *Journal of Avian Biology*.

Page 637 (Sage Grouse): Photos by John W. Scott. Reprinted from Scott (1942: Plate 17, Figs. 17–18, p. 498) with the permission of *The Auk*/American Ornithologists' Union and John P. Scott. © 1942 by *The Auk*/American Ornithologists' Union and John P. Scott.

Cartography by Ortelius Design; icon design by Phyllis Wood.

All others drawings © 1999 by John Megahan.

Animal Index

Names of species that exhibit homosexuality/transgender (i.e., animals that are included in part 2 or the appendix) are capitalized. Page numbers where animals are profiled are indicated with **boldface**. Pages with illustrations are indicated with *italics*. Names for groups of closely related species (e.g., Antbirds) are indexed under the names of the individual species (i.e., Bicolored Antbird, Ocellated Antbird).

Subject Index

Pages with illustrations are indicated with *italics*. Subjects discussed in notes are indicated with an n and the note number following the page number.

About the Author

BRUCE BAGEMIHL is a biologist, researcher, and author. He earned his bachelor's degree in biogeography from the University of Wisconsin–Milwaukee and his doctorate in linguistics and cognitive science from the University of British Columbia, where he also served on the faculty. He has published diverse essays and scientific articles on issues pertaining to language, biology, gender, and sexuality. He lives in Seattle, Washington.